## Formula Weights

| | | | |
|---|---|---|---|
| $AgBr$ | 187.78 | $K_2Cr_2O_7$ | 294.19 |
| $AgCl$ | 143.32 | $K_3Fe(CN)_6$ | 329.26 |
| $Ag_2CrO_4$ | 331.73 | $K_4Fe(CN)_6$ | 368.38 |
| $AgI$ | 234.77 | $KHC_8H_4O_4$ (phthalate) | 204.23 |
| $AgNO_3$ | 169.87 | $KH(IO_3)_2$ | 389.92 |
| $AgSCN$ | 165.95 | $K_2HPO_4$ | 174.18 |
| $Al_2O_3$ | 101.96 | $KH_2PO_4$ | 136.09 |
| $Al_2(SO_4)_3$ | 342.14 | $KHSO_4$ | 136.17 |
| $As_2O_3$ | 197.85 | $KI$ | 166.01 |
| $B_2O_3$ | 69.62 | $KIO_3$ | 214.00 |
| $BaCO_3$ | 197.35 | $KIO_4$ | 230.00 |
| $BaCl_2 \cdot 2H_2O$ | 244.28 | $KMnO_4$ | 158.04 |
| $BaCrO_4$ | 253.33 | $KNO_3$ | 101.11 |
| $Ba(IO_3)_2$ | 487.14 | $KOH$ | 56.11 |
| $Ba(OH)_2$ | 171.36 | $KSCN$ | 97.18 |
| $BaSO_4$ | 233.40 | $K_2SO_4$ | 174.27 |
| $Bi_2O_3$ | 466.0 | $La(IO_3)_3$ | 663.62 |
| $CO_2$ | 44.01 | $Mg(C_9H_6ON)_2$ | 312.59 |
| $CaCO_3$ | 100.09 | $MgCO_3$ | 84.32 |
| $CaC_2O_4$ | 128.10 | $MgNH_4PO_4$ | 137.35 |
| $CaF_2$ | 78.08 | $MgO$ | 40.31 |
| $CaO$ | 56.08 | $Mg_2P_2O_7$ | 222.57 |
| $CaSO_4$ | 136.14 | $MgSO_4$ | 120.37 |
| $Ce(HSO_4)_4$ | 528.4 | $MnO_2$ | 86.94 |
| $CeO_2$ | 172.12 | $Mn_2O_3$ | 157.88 |
| $Ce(SO_4)_2$ | 332.25 | $Mn_3O_4$ | 228.81 |
| $(NH_4)_2Ce(NO_3)_6$ | 548.23 | $Na_2B_4O_7 \cdot 10H_2O$ | 381.37 |
| $(NH_4)_4Ce(SO_4)_4 \cdot 2H_2O$ | 632.6 | $NaBr$ | 102.90 |
| $Cr_2O_3$ | 151.99 | $NaC_2H_3O_2$ | 82.03 |
| $CuO$ | 79.54 | $Na_2C_2O_4$ | 134.00 |
| $Cu_2O$ | 143.08 | $NaCl$ | 58.44 |
| $CuSO_4$ | 159.60 | $NaCN$ | 49.01 |
| $Fe(NH_4)_2(SO_4)_2 \cdot 6H_2O$ | 392.14 | $Na_2CO_3$ | 105.99 |
| $FeO$ | 71.85 | $NaHCO_3$ | 84.01 |
| $Fe_2O_3$ | 159.69 | $Na_2H_2EDTA \cdot 2H_2O$ | 372.2 |
| $Fe_3O_4$ | 231.54 | $Na_2O_2$ | 77.98 |
| $HBr$ | 80.92 | $NaOH$ | 40.00 |
| $HC_2H_3O_2$ (acetic acid) | 60.05 | $NaSCN$ | 81.07 |
| $HC_7H_5O_2$ (benzoic acid) | 122.12 | $Na_2SO_4$ | 142.04 |
| $HCl$ | 36.46 | $Na_2S_2O_3 \cdot 5H_2O$ | 248.18 |
| $HClO_4$ | 100.46 | $NH_4Cl$ | 53.49 |
| $H_2C_2O_4 \cdot 2H_2O$ | 126.07 | $(NH_4)_2C_2O_4 \cdot H_2O$ | 142.11 |
| $H_5IO_6$ | 227.94 | $NH_4NO_3$ | 80.04 |
| $HNO_3$ | 63.01 | $(NH_4)_2SO_4$ | 132.14 |
| $H_2O$ | 18.015 | $(NH_4)_2S_2O_8$ | 228.18 |
| $H_2O_2$ | 34.01 | $NH_4VO_3$ | 116.98 |
| $H_3PO_4$ | 98.00 | $Ni(C_4H_6O_2N_2)_2$ | 286.91 |
| $H_2S$ | 34.08 | $PbCrO_4$ | 323.18 |
| $H_2SO_3$ | 82.08 | $PbO$ | 223.19 |
| $H_2SO_4$ | 98.08 | $PbO_2$ | 239.19 |
| $HgO$ | 216.59 | $PbSO_4$ | 303.25 |
| $Hg_2Cl_2$ | 472.09 | $P_2O_5$ | 141.94 |
| $HgCl_2$ | 271.50 | $Sb_2S_3$ | 339.69 |
| $KBr$ | 119.01 | $SiO_2$ | 60.08 |
| $KBrO_3$ | 167.01 | $SnCl_2$ | 189.60 |
| $KCl$ | 74.56 | $SnO_2$ | 150.69 |
| $KClO_3$ | 122.55 | $SO_2$ | 64.06 |
| $KCN$ | 65.12 | $SO_3$ | 80.06 |
| $K_2CrO_4$ | 194.20 | $Zn_2P_2O_7$ | 304.68 |

# Analytical Chemistry: an introduction FOURTH EDITION

**Douglas A. Skoog**  Stanford University

**Donald M. West**  San Jose State University

*SAUNDERS GOLDEN SUNBURST SERIES*

 **Saunders College Publishing**
Philadelphia  New York  Chicago
San Francisco  Montreal  Toronto
London  Sydney  Tokyo  Mexico City
Rio de Janeiro  Madrid

*Address orders to:*
383 Madison Avenue
New York, NY 10017

*Address editorial correspondence to:*
West Washington Square
Philadelphia, PA 19105

Text Typeface: 10/12 Times Roman
Compositor: Progressive Typographers, Inc.
Acquisitions Editor: John Vondeling
Project Editors: Maureen R. Iannuzzi, Ellen Newman
Copyeditor: Ruth Melnick
Art Director: Carol Bleistine
Art/Design Assistant: Virginia A. Bollard
Text Design: Emily Harste
Cover Design: Lawrence R. Didona
Text Artwork: ANCO/Boston
Production Manager: Tim Frelick
Assistant Production Manager: JoAnn Melody

Cover Credit: Photo courtesy of the Perkin-Elmer Corporation

**Library of Congress Cataloging in Publication Data**

Skoog, Douglas Arvid.
    Analytical chemistry: an introduction

    (Saunders golden sunburst series)
    Includes index.

    1. Chemistry, Analytic — Quantitative.   I. West,
Donald M.  II. Title
QD101.2.S55    1985    545    85-14628
ISBN 0-03-002954-6

ANALYTICAL CHEMISTRY: AN INTRODUCTION        ISBN 0-03-002954-6

3456   016   987654321

**CBS COLLEGE PUBLISHING**
Saunders College Publishing
Holt, Rinehart and Winston
The Dryden Press

# Preface

The fourth edition of *Analytical Chemistry: an introduction,* like its predecessors, is a shortened version of the authors' other text, *Fundamentals of Analytical Chemistry*[1]. The abbreviated text is directed toward two types of courses in analytical chemistry. The first is a one-semester analytical course for students whose career goals are in such fields as medicine, biology, geology, and the physical sciences (other than chemistry). The second is a sophomore-level, one-semester course for chemistry majors where fewer laboratory experiments and a less detailed treatment of descriptive topics is desired. The latter course is generally followed by a senior-level instrumental analysis course.

It is our feeling that a book for either type of audience should present a balanced view of modern analytical chemistry and should thus include material concerned not only with the classical methods of analysis but also with methods based on spectroscopy, chromatography, and electrochemistry. In all probability, this course will be the nonchemistry student's only exposure to these important techniques; for the chemistry major, this instrumental material should be helpful as background for future course work, including a more advanced course dealing with instrumental analysis.

A major aim of the fourth edition of this text is to provide the student with a rigorous background in those chemical principles that are especially pertinent to the understanding of analytical chemistry. A second goal is to develop the student's appreciation for the importance of the frequently difficult task of judging the accuracy and precision of experimental data, as well as to provide the tools for sharpening these judgments. A third aim is to introduce the student to the wide range of techniques that are now available for performing chemical analyses. A final goal is to teach those laboratory skills that will give students confidence in their ability to obtain high-quality analytical data.

The fourth edition differs from the third edition in several regards. In general, we have increased the number of pages devoted to chromatography and spectroscopy at the expense of descriptive material dealing with the gravimetric and volumetric procedures. A full chapter is now devoted to chromatography,

[1] D. A. Skoog and D. M. West, *Fundamentals of Analytical Chemistry,* 4th ed. Philadelphia: Saunders College Publishing, 1982.

with much new material on the various types of high-performance methods. We have revised the presentation on flame spectroscopic methods, expanded the section dealing with atomic absorption, and added a section describing plasma sources. We have also introduced a brief new section on three-electrode polarography and have considerably revised the discussion of membrane electrodes, devoting more space to solid state and immobilized liquid membranes. Another addition is a section dealing with the least-squares method for deriving calibration curves. To accommodate the aforementioned additions, it was necessary to delete certain parts of the descriptive material in the third edition.

We have also made a number of changes in the chapters devoted to laboratory work, replacing the section on equal-arm balances with a section on modern electronic balances, introducing several new experiments and deleting some of the older ones. The additions include the following: a chromatographic procedure for determining the alcohol content of beverages; a method for the separation of a mixture of nickel and zinc based on an ion-exchange separation, followed by titration of the two ions with EDTA; atomic absorption methods for the determination of lead in brass and lead extracted from a pottery glaze; a method for the amperometric titration of lead by chromate ion; a polarographic method for determining copper and lead in a sample of brass; a coulometric method for determining cyclohexene; and methods for the determination of fluoride in tap water and in toothpaste by direct potentiometric measurements with a specific-ion electrode. We have deleted several of the more conventional methods involving gravimetric and volumetric determinations to make room for these new procedures.

The fourth edition also contains a new and expanded set of problems, with answers provided for approximately half of them. A solutions manual is also available.

We would like to acknowledge with gratitude the helpful comments and suggestions of the following people, who have used earlier editions of this book and have also read a preliminary draft of this edition: Professor J. N. Cooper of Bucknell University, Professor R. H. Hanson of the University of Arkansas at Little Rock, Professor J. J. Topping of Towson State University, Professor L. R. Sherman of the University of Scranton, Professor T. J. Bydalek of the University of Minnesota at Duluth, Professor D. M. King of Western Washington University, Professor A. M. Harper of the University of Texas at El Paso, and Professor R. D. Caton of the University of New Mexico.

*Douglas A. Skoog*
*Donald M. West*

# Contents Overview

v

# Contents

# Chapter 1
# Introduction

A quantitative analysis provides numerical information about the amount of some species (the *analyte*) that is contained in a measured quantity of matter (the *sample*). The results of a quantitative analysis are ordinarily expressed in relative terms, such as parts per hundred (that is, as a percentage), parts per thousand, parts per million, or perhaps parts per billion of the sample. Other ways of expressing results include the weight (or the volume) of analyte per unit volume of sample, as well as the mole fraction.

Chemistry in general, and analytical chemistry in particular, will only be peripheral to the intellectual goals of many readers of this book; indeed, the need for study in this area may be questioned. In 1984, Wilhelm Ostwald wrote,

> Analytical chemistry, or the art of recognizing different substances and determining their constituents, takes a prominent position among the applications of science, since the questions which it enables us to answer arise wherever chemical processes are employed for scientific or technical purposes. Its supreme importance has caused it to be assiduously cultivated from a very early period in the history of chemistry, and its records comprise a large part of the quantitative work which is spread over the whole domain of science.

The amount of our chemical knowledge has increased enormously since 1894; these words, nevertheless, are as apt today as they were then. The fact is that few (if any) areas of manufacture, and few (if any) of the physical sciences, escape the need for quantitative chemical information. The amount of protein in foodstuffs, for example, is commonly established by a quantitative analysis for nitrogen content. The analysis of soils and the needs of plants provide the farmer with the information that will maximize yields through the judicious application of fertilizers. The properties of alloys depend in large measure upon composition, which, in turn, is monitored by chemical analysis. The effectiveness of devices for the control of automotive and industrial pollution is likewise established by analysis. Physicians rely upon the chemical analysis of body fluids as an important aid in diagnosis. A listing such as this is virtually endless.

## 1A CLASSIFICATION OF QUANTITATIVE METHODS

Quantitative methods can be subdivided into several groups, based upon the nature of the final measurement in the analysis, the magnitude of which is proportional to the amount of analyte in the sample. The final measurement in a *gravimetric* analysis involves a determination of mass. Likewise, the final measurement of a typical *volumetric* analysis consists of measuring the volume of a solution that contains sufficient reagent to react completely with the analyte. *Electroanalytical* methods are based upon the measurement of such electrical quantities as volts, amperes, ohms, and coulombs. *Spectroscopic* methods are based upon measurements of the interaction of electromagnetic radiation (X-ray, ultraviolet, visible, infrared, and radio radiation) with analyte atoms or molecules or upon the production of such radiation by the analyte.

## 1B STEPS IN A CHEMICAL ANALYSIS

In most analyses, the measurement step just described must be preceded by several preliminary steps, which are often more time consuming and more formidable than the final measurement itself. Indeed, this final step is likely to be the most straightforward part of the entire analytical process. The preliminary steps that are common to most analytical procedures are described in this section.

**Definition of the Problem**    At the outset, the analyst must have a clear understanding of the accuracy that is needed in the results of an analysis. The selection of a method and the care with which it is performed are critically dependent upon the accuracy that is required. It is safe to state that an increase in the reliability of a measurement is likely to require an exponential increase in the time and effort needed to perform the measurement. An analysis that is incapable of providing the required reliability will clearly be a waste of time. Equally wasteful is the performance of an analysis at a level of accuracy that exceeds by very much the demands made upon the data. For example, numerous methods exist for the analysis of chlorine. A very simple (but not very accurate) method is sufficient to establish whether the chlorine content of swimming pool water is high enough to provide safety and yet not so high as to cause eye irritation. In contrast, investigation of a reaction mechanism in which chlorine is a participant is likely to require a method of measurement that possesses a far greater degree of reliability. A compromise may also be required between the accuracy that can be attained and the time available for performing the analysis. A physician with a critically ill patient cannot afford to wait hours (or perhaps days) for the results of an elegant analysis; here the demands of time may dictate the use of a method that will provide the needed guidance at the expense of the ultimate in accuracy.

**Sampling**[1]    Fundamental to any chemical analysis is the acquisition of a sample with a composition that is representative of a larger quantity of matter. In

---

[1] For a detailed discussion of sampling, see: C. A. Bicking, in *Treatise on Analytical Chemistry,* 2nd ed., I. M. Kolthoff and P. J. Elving, Eds., Part I, Vol. 1, Chapter 6. New York: John Wiley & Sons, 1978.

general, the more finely divided and homogeneous the material, the easier it is to obtain such a sample. At one extreme is a well mixed gas or solution where inhomogeneity exists only at the molecular level. Here, one can be confident that even the smallest sample will be truly representative of a much larger mass of material. An example of the other extreme would be a 25-ton shipment of silver ore where buyer and seller must come to agreement as to its average silver content. The ore, however, is inherently heterogeneous, consisting of particles that range in size from a few tenths of a millimeter to several centimeters. Furthermore, the silver content of these particles may range from zero to several percent depending upon their particle size. The assay of this shipment will be performed on a sample that weighs something on the order of 1 g. The composition of this sample must be the same as the average for the entire 25 tons (or 22,700,000 g) in the shipment. The task of isolating 1 g with any confidence that its composition truly reflects that of the nearly 23,000,000 g from which it was taken is a nontrivial undertaking that requires a systematic manipulation of the entire shipment.

Many sampling problems will be less formidable than the one just described. Regardless of difficulty, however, assurance is needed that the sample that is employed in the analysis is truly representative of the whole before proceeding further.

**Preparation of the Laboratory Sample for Analysis**   Most solid materials must be ground to decrease particle size and then thoroughly mixed to ensure homogeneity. Moreover, the adsorption or desorption of water from the atmosphere will cause the percent composition of the sample to depend upon the humidity of its surroundings. This source of difficulty is ordinarily overcome by putting the sample through some sort of drying cycle.

We have noted that quantitative analytical results are commonly reported in relative terms (such as percentage); it is therefore necessary to determine the weight or volume of the sample upon which the analysis is perfomed.

**Solution of the Sample**[2]   Most (but certainly not all) analyses are performed on solutions of the sample. Ideally, the solvent should dissolve the entire sample (not just the analyte) rapidly and under sufficiently mild conditions that loss of the analyte does not occur. Solvents with these properties simply do not exist for many of the materials that are of interest to the scientist — a silicate mineral, a high-molecular-weight polymer, or a sample of animal tissue, for example. For such materials, the conversion to a soluble form is often difficult and time consuming.

---

[2] For a detailed treatment of decomposition and solution of samples, see: R. Bock, *A Handbook of Decomposition Methods in Analytical Chemistry.* New York: John Wiley & Sons, 1979; and D. C. Bogen, in *Treatise on Analytical Chemistry,* 2nd ed., P. J. Elving, E. Grushka, and I. M. Kolthoff, Eds., Part I, Vol. 5, Chapter 1. New York: John Wiley & Sons, 1982.

**Separation of Potential Interferences** Few chemical or physical properties of analytical importance are unique to a single chemical species; instead, the reactions used and the properties measured are shared by several elements or compounds. This lack of specificity adds greatly to the difficulties faced by the analyst because a scheme must be devised for isolating the species of interest from all others in the sample that can influence the final measurement. Substances that prevent the direct measurement of the analyte concentration are called *interferences;* their elimination prior to the final measurement is an important step in most analyses. No hard and fast rules exist for the elimination of interferences; this problem is frequently the most demanding aspect of the analysis. Separation methods are treated in Chapters 17 and 18.

**Completion of the Analysis** All preliminary steps in an analysis are undertaken to ensure that the final measurement is a true gauge of the quantity of analyte in the sample. The foregoing survey of the steps in an analysis suggests, correctly, that this final measurement is frequently the least difficult step.

Many types of final measurements are discussed in the chapters that follow, along with the principles upon which such measurements are based.

## 1C CHOICE OF METHOD FOR AN ANALYSIS

The chemist or scientist who needs analytical data is frequently confronted with numerous methods that can, in principle, provide the desired information. The success or failure of an analysis can be critically dependent upon the choice of method. Speed, sample composition, convenience, accuracy, availability of equipment, number of analyses, amount of sample available for analysis, and likely concentration range of the analyte are all factors that will influence this choice. There is no substitute for experience in making the decision.

*Analytical Chemistry: An Introduction* deals with (1) the chemical principles upon which all analytical methods are based; (2) the accuracy and precision of quantitative analyses; (3) the principles of gravimetric, volumetric, electroanalytical, and certain spectroscopic and chromatographic methods; and (4) unit operations, such as weighing, measuring volumes, drying, and evaporating, which are part of all analytical methods. In addition, specific directions for several typical methods of analysis are provided. Mastery of this material will permit the student to perform useful chemical analyses and will also provide a background that will aid in the selection of procedures for solving analytical problems.

# Chapter 2
# Review of Some Elementary Concepts

The analyte in a typical quantitative analysis will ordinarily exist in solution at one stage or another. Thus, familiarity with solution chemistry as well as an understanding of mass relationships between reactants and products in solution is fundamental to the study of analytical chemistry. This chapter provides a brief review of these topics.

## 2A SOLUTIONS AND THEIR COMPOSITION

### 2A–1 Electrolytes

Solutes that dissociate to produce conducting solutions are classed as electrolytes. *Strong electrolytes* are ionized completely, or nearly so, while *weak electrolytes* are only partially ionized. Table 2–1 is a compilation of solutes that act as strong and as weak electrolytes in aqueous solution.

### 2A–2 Acids and Bases

The concept of acid-base behavior—as proposed independently by Brønsted and Lowry in 1923—is of particular importance to the analytical chemist.[1]

Table 2–1  Classification of Electrolytes

| Strong Electrolytes | Weak Electrolytes |
|---|---|
| 1. The inorganic acids $HNO_3$, $HClO_4$, $H_2SO_4$* $HCl$, $HI$, $HBr$, $HClO_3$, $HBrO_3$ <br> 2. Alkali and alkaline-earth hydroxides <br> 3. Most salts | 1. Many inorganic acids such as $H_2CO_3$, $H_3BO_3$, $H_3PO_4$, $H_2S$, $H_2SO_3$ <br> 2. Most organic acids <br> 3. Ammonia and most organic bases <br> 4. Halides, cyanides, and thiocyanates of Hg, Zn, and Cd |

* $H_2SO_4$ is completely dissociated into $HSO_4^-$ and $H_3O^+$ ions and for this reason is classified as a strong electrolyte. However, it should be noted that the $HSO_4^-$ ion is a weak electrolyte, being only partially dissociated.

---

[1] For a thorough treatment of the various acid-base concepts, see: I. M. Kolthoff, in *Treatise on Analytical Chemistry,* 2nd ed., I. M. Kolthoff and P. J. Elving, Eds., Part I, Vol. 2, Chapter 17. New York: John Wiley & Sons, 1979.

According to this view, *an acid is a substance capable of donating a proton; a base is a substance that can accept a proton.* It is important to appreciate that the proton-donating capacity of an acid will be observed only in the presence of a proton acceptor, or base. Similarly, a substance will only manifest its proton-accepting character in the presence of a proton donor, or acid.

A feature of the Brønsted-Lowry concept is that each acid has associated with it a *conjugate base,* that is, the entity that remains following the donation of a proton. Similarly, every base produces a *conjugate acid* as a result of accepting a proton. Examples of conjugate acid-base relationships are pointed out in Equations 2–1 to 2–4.

Many solvents are themselves proton donors or acceptors and can thus induce basic or acidic behavior on the part of solutes dissolved in them. For example, in an aqueous solution of ammonia, the solvent donates a proton and thus acts as an acid with respect to the solute:

$$NH_3 + H_2O \rightleftharpoons NH_4^+ + OH^- \qquad (2-1)$$
$$\text{base}_1 \quad \text{acid}_2 \quad \text{acid}_1 \quad \text{base}_2$$

Ammonium ion is the conjugate acid of the base $NH_3$ while $OH^-$ is the conjugate base of the acid water. In contrast, water acts as a proton acceptor, or base, in an aqueous solution of nitrous acid:

$$H_2O + HNO_2 \rightleftharpoons H_3O^+ + NO_2^- \qquad (2-2)$$
$$\text{base}_1 \quad \text{acid}_2 \quad \text{acid}_1 \quad \text{base}_2$$

Nitrite ion is the conjugate base of the acid $HNO_2$; $H_3O^+$ is the conjugate acid of the base water. Neither of these processes is complete because ammonia and nitrous acid are both weak electrolytes.

Water is the classic example of an *amphiprotic* solvent: depending upon the solute, it acts either as a donor (Equation 2–1) or as an acceptor (Equation 2–2) of protons. Other common amphiprotic solvents include methanol, ethanol, and anhydrous acetic acid. In methanol, for example, the equilibria that are analogous to those shown in Equations 2–1 and 2–2 are

$$NH_3 + CH_3OH \rightleftharpoons NH_4^+ + CH_3O^- \qquad (2-3)$$
$$\text{base}_1 \quad \text{acid}_2 \quad \text{acid}_1 \quad \text{base}_2$$

$$CH_3OH + HNO_2 \rightleftharpoons CH_3OH_2^+ + NO_2^- \qquad (2-4)$$
$$\text{base}_1 \quad \text{acid}_2 \quad \text{acid}_1 \quad \text{base}_2$$

It is important to recognize that an acid, having donated a proton, becomes a conjugate base that has the capability of accepting a proton to re-form the original acid; the converse holds equally well. Thus, nitrite ion, the species produced by the loss of a proton from nitrous acid, is a potential acceptor of a proton from a suitable donor. It is this reaction that causes an aqueous solution of nitrite ion to be slightly basic:

$$NO_2^- + H_2O \rightleftharpoons HNO_2 + OH^-$$
$$\text{base}_1 \quad \text{acid}_2 \quad \text{acid}_1 \quad \text{base}_2$$

## 2A-3 Autoprotolysis

Amphiprotic solvents undergo self-dissociation, or *autoprotolysis,* to form a pair of ionic species. Autoprotolysis is another example of an acid-base reaction, as illustrated by the following equations:

$$\text{base}_1 + \text{acid}_2 \quad \rightleftharpoons \quad \text{acid}_1 + \text{base}_2$$

$$H_2O + H_2O \quad \rightleftharpoons \quad H_3O^+ + OH^-$$

$$CH_3OH + CH_3OH \quad \rightleftharpoons \quad CH_3OH_2^+ + CH_3O^-$$

$$HCOOH + HCOOH \rightleftharpoons HCOOH_2^+ + HCOO^-$$

$$NH_3 + NH_3 \quad \rightleftharpoons \quad NH_4^+ + NH_2^-$$

The cation produced from the autoprotolysis of water is called the *hydronium ion;* the proton is covalently bound to the parent molecule by one of the unshared electron pairs of the oxygen. Other hydrates, such as $H_5O_2^+$ and $H_7O_3^+$, undoubtedly exist, but none possesses stability comparable to $H_3O^+$. The unhydrated proton does not appear to exist in aqueous solution.

Chemists frequently use the notation $H_3O^+$ in equations in recognition of the unique stability of that species in aqueous solution. The use of $H^+$ to symbolize the proton, whatever the degree of its hydration, has the advantage of simplifying the writing of equations that require inclusion of the proton for balance. The reader should be familiar with both representations; we shall use either, as convenient, in various sections of this text.

## 2A-4 Strength of Acids and Bases

Reactions of selected acids with water are shown in Figure 2-1. The first two are *strong acids* because reaction with this solvent is sufficiently complete as to leave no undissociated molecules in aqueous solution. The remainder are *weak acids* which react incompletely with water to give solutions that contain significant quantities of both the parent acid and its conjugate base. Note that acids can be cationic, anionic, or electrically neutral.

The acids in Figure 2-1 become progressively weaker from top to bottom.

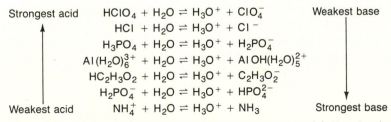

**Figure 2-1**   Relative strengths of some common weak acids and their conjugate bases.

Perchloric acid and hydrochloric acid are completely dissociated; in contrast, ammonium ion is dissociated only to a few thousandths of a percent. Another generality illustrated in Figure 2 – 1 is that the weakest acid forms the strongest conjugate base; that is, ammonia has a much stronger affinity for protons than any base above it.

The tendency of a solvent to accept or to donate protons in large measure determines the strength of a solute acid (or base) dissolved in it. For example, perchloric, hydrobromic, and hydrochloric acids all qualify as strong acids in water. If anhydrous acetic acid, a poorer proton acceptor than water, is substituted *as the solvent,* none of these acids undergoes complete dissociation; instead, equilibria such as the following are established:

$$\underset{\text{base}_1}{CH_3COOH} + \underset{\text{acid}_2}{HClO_4} \rightleftharpoons \underset{\text{acid}_1}{CH_3COOH_2^+} + \underset{\text{base}_2}{ClO_4^-}$$

It is noteworthy that perchloric acid is considerably stronger than the other two in this solvent; its dissociation is about 5000 times greater than that for hydrochloric acid. Acetic acid thus acts as a *differentiating* solvent toward these acids in the sense that its use reveals inherent differences in their acidities. Water, on the other hand, is a *leveling* solvent for these three acids in that all are completely ionized in this medium and thus exhibit no differences in strength.

## 2B  UNITS OF WEIGHT AND CONCENTRATION

The mass of a substance is ordinarily obtained in such metric units as kilograms (kg), grams (g), milligrams (mg), micrograms ($\mu$g), nanograms (ng), or picograms (pg).[2] For chemical calculations, however, it is advantageous to employ chemical units of mass that express the weight relationships (or *stoichiometry*) among reacting species in terms of small whole numbers. The *gram-formula weight,* the *gram-molecular weight,* and the *gram-equivalent weight* serve this purpose; these terms are frequently shortened to the formula weight, the molecular weight, and the equivalent weight, respectively.

### 2B – 1  Chemical Formulas, Formula Weights, and Molecular Weights

An *empirical formula* expresses the simplest combining ratio for the atoms in a substance. The *chemical formula,* on the other hand, specifies the number of atoms in a molecule. An empirical formula may be shared by more than one substance. For example, $CH_2O$ is both the empirical formula and the chemical formula for formaldehyde; it is also the empirical formula for such diverse

---

[2] The relationships among these is

$$10^{-3} \text{ kg} = \text{g} = 10^3 \text{ mg} = 10^6 \text{ }\mu\text{g} = 10^9 \text{ ng} = 10^{12} \text{ pg}$$

substances as acetic acid $C_2H_4O_2$, glyceraldehyde $C_3H_6O_3$, and glucose $C_6H_{12}O_6$, as well as more than 50 other substances containing six or fewer carbons. The empirical formula can be evaluated from percent composition data; the chemical formula additionally requires knowledge of the molecular weight.[3]

The entity expressed by the chemical formula may or may not exist. For example, $H_2$ represents the molecular formula for hydrogen, there being ample evidence to confirm that the gas consists of diatomic molecules under ordinary conditions. In contrast, molecules having the chemical formula NaCl do not exist either in solid sodium chloride or in solutions of the salt. Instead, this substance is made up of an array of sodium ions and chloride ions, no one of which is in simple combination with a specific ion of opposite charge. Nonetheless, for purposes of stoichiometric accounting, both $H_2$ and NaCl are equally useful as chemical formulas.

The chemical formula may represent only the principal form for a substance. Thus, for example, water in the liquid state contains small amounts of such entities as $H_3O^+$, $OH^-$, $H_4O_2$, and undoubtedly others. Here, the chemical formula $H_2O$ is that for the predominant species and is perfectly satisfactory from a stoichiometric standpoint; two hydrogens exist for each oxygen, regardless of dissociation or association.

The *gram-formula weight* (gfw) is the summation of the atomic weights (in grams) for all of the atoms in the chemical formula of a substance. The gram-formula weight for hydrogen is thus 2.016 ($2 \times 1.008$); that for sodium chloride is 58.44 ($22.99 + 35.45$). Note that the gram-formula weight carries with it no inference as to the existence or nonexistence of the substance for which it has been calculated.

The term *gram-molecular weight* (gmw) will be reserved for the summation of atomic weights (in grams) in the chemical formula of a real chemical species. Thus, the gram-molecular weight for hydrogen is the same as its gram-formula weight (2.016 g). We are not entitled, under this convention, to assign a molecular weight to sodium chloride, because the *species* NaCl does not exist. It would be perfectly proper to assign gram-molecular weights to $Na^+$ (22.99 g) and $Cl^-$ (35.45 g) because these are real chemical entities (these should probably be referred to as gram-ionic weights).

The value of making a distinction between the gram-formula weight and the gram-molecular weight may seem minor, and for stoichiometric problems it surely is. On the other hand, the convention eliminates possibilities for ambiguity in describing the concentration of solutions.

By definition, one gram-molecular weight of a species contains Avogadro's

---

[3] The *molecular formula* may also provide structural information about a substance. Thus, the chemically different ethanol $C_2H_5OH$ and diethyl ether $CH_3OCH_3$ share the same chemical (and empirical) formula $C_2H_6O$.

number $(6.02 \times 10^{23})$ of molecules, ions, or other chemical entities. This quantity is commonly referred to as the *mole* (abbreviation: mol).[4] Similarly, one gram-formula weight, called the *formula weight* (fw), represents $6.02 \times 10^{23}$ of the units expressed by the chemical formula, whether real or not.

---

**Example 2–1**    A 25.0-g sample of $H_2$ contains

$$25.0 \text{ g} \times \frac{\text{mol } H_2}{2.016 \text{ g}} = 12.40 \text{ mol } H_2$$

or

$$12.40 \text{ mol } H_2 \times \frac{6.02 \times 10^{23} \text{ molecules}}{\text{mol } H_2} = 7.47 \times 10^{24} \text{ molecules } H_2$$

The same weight of NaCl contains

$$25.0 \text{ g} \times \frac{\text{fw NaCl}}{58.44 \text{ g}} = 0.428 \text{ fw NaCl}$$

which corresponds to 0.428 mol $Na^+$ and 0.428 mol $Cl^-$. This amount of NaCl contains

$$0.428 \text{ fw } Na^+ \times \frac{6.02 \times 10^{23} \text{ ions}}{\text{mol } Na^+} = 2.58 \times 10^{23} \text{ ions } Na^+$$

and an equal number of $Cl^-$ ions.

---

The mole and the formula weight frequently represent inconveniently large amounts in terms of chemical analysis. The *millimole* (mmol) and the *milliformula weight* (mfw) are more useful; these represent 1/1000 of the gram-molecular weight and gram-formula weight, respectively.

## 2B–2 The Concentration of Solutions

**Formality, or Formal Concentration**    The formality, $F$, is equal to the number of formula weights of solute per liter of solution; it is also numerically identical to the number of milliformula weights of solute per milliliter of solution.

---

[4] In the International System of Units (SI), proposed by the International Bureau of Weights and Measures, the only chemical unit for an amount of substance is the *mole*. The mole is defined as the quantity of a material that contains as many elementary entities (these may be atoms, ions, electrons, ion-pairs, or molecules, and must be explicitly specified) as there are atoms of carbon in exactly 0.012 kg of carbon-12 (that is, Avogadro's number). It seems probable that a shift to SI units will ultimately occur. It is equally important to have an understanding of the units upon which the present chemical literature is based, even though these may ultimately disappear.

**Example 2–2**   Exactly 4.57 g of $BaCl_2 \cdot 2H_2O$ (gfw = 244) are dissolved in sufficient water to give 250 mL of solution. Calculate the formal concentration of $BaCl_2$ and of $Cl^-$ in this solution.

$$4.57 \text{ g} \times \frac{\text{mfw } BaCl_2 \cdot 2H_2O}{0.244 \text{ g}} = 18.73 \text{ mfw } BaCl_2 \cdot 2H_2O$$

$$\frac{18.73 \text{ mfw } BaCl_2 \cdot 2H_2O}{250 \text{ mL}} = \frac{0.0749 \text{ mfw } BaCl_2 \cdot 2H_2O}{\text{mL}}$$

$$= 0.0749 \; F \; BaCl_2 \cdot 2H_2O$$

Each $BaCl_2 \cdot 2H_2O$ supplies 2 $Cl^-$; thus,

$$\frac{0.0749 \text{ mfw } BaCl_2 \cdot 2H_2O}{\text{mL}} \times \frac{2 \text{mfw } Cl^-}{\text{mfw } BaCl_2 \cdot 2H_2O} = 0.150 \; F \; Cl^-$$

**Molarity, or Molar Concentration**   The molarity, $M$, expresses the number of moles of solute per liter of solution. In common with formality, the molarity is also equal to the number of millimoles of solute per milliliter of solution.

The following example illustrates that the formal and molar concentrations will be identical for some solutions and quite different for others.

**Example 2–3**   Calculate the formal and molar concentrations of

(a) an aqueous solution that contains 1.80 g of ethanol (gfw = 46.1) in 750 mL.

(b) an aqueous solution that contains 365 mg of iodic acid, $HIO_3$ (gfw = 176), in 20.0 mL (the acid is 71.0% ionized in this solution).

(a) $\dfrac{1.80 \text{ g } C_2H_5OH}{750 \text{ mL}} \times \dfrac{\text{mfw } C_2H_5OH}{0.0461 \text{ g}} = \dfrac{0.0521 \text{ mfw } C_2H_5OH}{\text{mL}}$

$$= 0.0521 \; F \; C_2H_5OH$$

The only solute species present in significant amount in an aqueous solution of ethanol is $C_2H_5OH$. Therefore,

$$M = F = 0.0521$$

(b) $\dfrac{365 \text{ mg } HIO_3}{20.0 \text{ ml}} \times \dfrac{\text{g}}{1000 \text{ mg}} \times \dfrac{\text{mfw } HIO_3}{0.176 \text{ g}} = \dfrac{0.104 \text{ mfw } HIO_3}{\text{mL}} = 0.104 \; F$

Here, only 29.0% (100% − 71.0%) of the solute exists as undissociated $HIO_3$. Thus, the molar concentration of this species will be

$$0.290 \times 0.104 = 0.0302 \; M \; HIO_3$$

The distinction between formal and molar concentrations is by no means universally accepted. The second part of Example 2–3 demonstrates the need to make a distinction between the concentration based on the recipe used to generate a solution (its formality) and the present concentration of a solute species within that solution (its molar concentration in terms of a particular species). Many chemists prefer to use *molar analytical concentration* in lieu of formal concentration and *molar species concentration* for moles per liter of the specific solute species. By this convention, the analytical iodic acid concentration in Example 2–3 is 0.104 $M$. Its species concentration is 0.0302 $M$.

Molar concentrations for specific species are also indicated by square brackets. Thus, the answers to the second part of Example 2–3 can be expressed as

$$[HIO_3] = 0.0302$$

$$[IO_3^-] = 0.104 \times 0.710 = 0.0738$$

---

**Example 2–4** Describe the preparation of 2.00 L of 0.108 $F$ $BaCl_2$ from $BaCl_2 \cdot 2H_2O$ (gfw = 244).

To produce this solution we will need

$$2.00 \text{ L} \times \frac{0.108 \text{ fw } BaCl_2 \cdot 2H_2O}{L} = 0.216 \text{ fw } BaCl_2 \cdot 2H_2O$$

The source of $BaCl_2$ is the solid dihydrate. Thus, we need

$$0.216 \text{ fw } BaCl_2 \cdot 2H_2O \times \frac{244 \text{ g}}{\text{fw } BaCl_2 \cdot 2H_2O} = 52.7 \text{ g } BaCl_2 \cdot 2H_2O$$

Dissolve 52.7 g of $BaCl_2 \cdot 2H_2O$ in water and dilute to 2.00 L.

---

**Example 2–5** Describe the preparation of 500 mL of 0.0740 $M$ $Cl^-$ solution from solid $BaCl_2 \cdot 2H_2O$.

Each $BaCl_2 \cdot 2H_2O$ supplies 2 $Cl^-$; thus, we need

$$500 \text{ mL} \times \frac{0.0740 \text{ mmol } Cl^-}{mL} \times \frac{\text{mfw } BaCl_2 \cdot 2H_2O}{2 \text{ mmol } Cl^-} = 18.5 \text{ mfw } BaCl_2 \cdot 2H_2O$$

$$18.5 \text{ mfw } BaCl_2 \cdot 2H_2O \times \frac{0.244 \text{ g}}{\text{mfw } BaCl_2 \cdot 2H_2O} = 4.51 \text{ g } BaCl_2 \cdot 2H_2O$$

Dissolve 4.51 g of $BaCl_2 \cdot 2H_2O$ in water and dilute to 500 mL.

---

**Normality or Normal Concentration** Normality expresses the number of equivalents of a solute that are contained in one liter of solution or the number of milliequivalents in one milliliter. The number of equivalents, in turn, is based upon the concept of the *gram-equivalent weight*. These topics are considered in Chapter 6.

**Titer**    Titer, also discussed in Chapter 6, defines concentration in terms of the weight of analyte with which 1.00 mL of solution reacts.

**p-Values**    It is frequently convenient to express concentration in terms of p-values or p-functions. The p-value for a particular species is defined as the negative logarithm (to the base 10) of the molar concentration of that species. Thus, for the species X:

$$pX = -\log [X] = \log \frac{1}{[X]} \qquad (2-5)$$

As shown by the following examples, p-values offer the advantage of providing concentration information in terms of small numbers that can be expressed in digital form.

---

**Example 2–6**    Calculate p-values for each ion in a solution that is $2.25 \times 10^{-3}$ *F* in NaCl and $7.65 \times 10^{-4}$ *F* in HCl.

$$[H_3O^+] = 7.65 \times 10^{-4}$$

$$pH = \log \frac{1}{7.65 \times 10^{-4}} = \log 1307 = 3.116$$

$$[Na^+] = 2.25 \times 10^{-3}$$

$$pNa = \log \frac{1}{2.25 \times 10^{-3}} = \log 444 = 2.648$$

Chloride ion is derived from both solutes; thus,

$$[Cl^-] = 7.65 \times 10^{-4} + 2.25 \times 10^{-3} = 3.015 \times 10^{-3}$$

$$pCl = \log \frac{1}{3.015 \times 10^{-3}} = \log 332 = 2.521$$

---

**Example 2–7**    A solution has a pAg of 6.372; calculate $[Ag^+]$.

$$pAg = -\log [Ag^+] = \log \frac{1}{[Ag^+]} = 6.372$$

$$\frac{1}{[Ag^+]} = \text{antilog } 6.372 = 2.36 \times 10^6$$

$$[Ag^+] = \frac{1}{2.36 \times 10^6} = 4.25 \times 10^{-7}$$

It is worth noting that the p-value becomes negative when the species concentration exceeds unity. Thus, in 3.00 $F$ HCl,

$$[H_3O^+] = 3.00$$

$$pH = \log \frac{1}{3.00} = \log 0.333 = -0.478$$

**Density and Specific Gravity**    The *density* of a substance measures the mass per unit volume; the *specific gravity* is the ratio of its mass to that of an equal volume of water at 4°C. In the metric system, density has the units of grams per milliliter, or kilograms per liter. Specific gravity is widely used in describing items of commerce because it is dimensionless and, therefore, not tied to any particular system of units. Because water at 4°C has a density of 1.00 g/mL, and because the metric system will be used throughout this text, we will usually be able to use density and specific gravity interchangeably.

**Parts Per Million; Parts Per Billion**    The concentration of a very dilute solution is conveniently expressed in terms of parts per million (ppm):

$$ppm = \frac{wt \ solute \times 10^6}{wt \ soln}$$

For even more dilute solutions, $1 \times 10^9$ rather than $1 \times 10^6$ is used in this equation, thereby giving the result in parts per billion (ppb).

The density of a very dilute aqueous solution does not differ significantly from 1.00 g/mL, or $10^6$ mg/L; thus,

$$ppm = \frac{mg \ solute}{10^6 \ mg \ water} = \frac{mg \ solute}{L \ soln}$$

**Example 2-8**    Calculate the molar concentration of $K^+$ ion in a solution that contains 75.0 ppm of $K_4Fe(CN)_6$ (gfw = 368).

This solution is so dilute that its density will be 1.00 g/mL, and

$$\frac{75.0 \ mg}{L} \times \frac{fw \ K_4Fe(CN)_6}{368 \times 10^3 \ mg} = 2.04 \times 10^{-4} \ F \ K_4Fe(CN)_6$$

$$\frac{2.04 \times 10^{-4} \ fw \ K_4Fe(CN)_6}{L} \times \frac{4 \ mol \ K^+}{fw \ K_4Fe(CN)_6} = 8.16 \times 10^{-4} \ M \ K^+$$

**Example 2-9**    Solutions containing $MnO_4^-$ ion are so intensely colored that a solution in which the $MnO_4^-$ concentration is $4.0 \times 10^{-6}$ $F$ has perceptible color to most observers. Calculate the parts per million of $MnO_4^-$ (gfw $= 119$) in such a solution.

$$\frac{4.0 \times 10^{-6} \text{ mfw } MnO_4^-}{mL} \times \frac{119 \text{ mg}}{\text{mfw } MnO_4^-} \times \frac{1000 \text{ mL}}{L} = \frac{0.476 \text{ mg } MnO_4^-}{L}$$

$$= 0.48 \text{ ppm}$$

**Percentage Concentration**    Chemists frequently express concentrations in terms of percentage (or parts per hundred). This practice is highly ambiguous unless the type of percentage is carefully specified. Common methods include

$$\text{weight percent (w/w)} = \frac{\text{wt solute}}{\text{wt soln}} \times 100$$

$$\text{volume percent (v/v)} = \frac{\text{volume solute}}{\text{volume soln}} \times 100$$

$$\text{weight-volume percent (w/v)} = \frac{\text{wt solute}}{\text{volume soln, mL}} \times 100$$

Note that the denominator in each of these expressions refers to the *solution* rather than the solvent. Moreover, the first two expressions are dimensionless (provided, of course, that the units in the numerator and denominator are consistent); units must be supplied for the third. Of the three expressions, only weight percentage has the virtue of being temperature independent.

Weight-percentage is frequently used to express the concentrations of commercial aqueous reagents. For example, nitric acid is sold as a 70% solution, which means that the reagent contains 70 g of $HNO_3$ per 100 g of solution.

---

**Example 2–10**    Calculate the formal concentration of $HNO_3$ (gfw $= 63.0$) in a solution that has a specific gravity of 1.42 and is 70% $HNO_3$ (w/w).

$$\frac{1.42 \text{ g soln}}{mL \text{ soln}} \times \frac{70 \text{ g } HNO_3}{100 \text{ g soln}} \times \frac{\text{mfw } HNO_3}{0.063 \text{ g } HNO_3} = \frac{15.8 \text{ mfw } HNO_3}{mL \text{ soln}}$$

$$= 15.8 = 16 \ F$$

---

**Example 2–11**    Describe the preparation of 100 mL of 6.0 $F$ HCl from a concentrated reagent that has a specific gravity of 1.18 and is 37% (w/w) HCl (gfw $= 36.5$).

We need

$$100 \text{ mL} \times \frac{6.0 \text{ mfw HCl}}{mL} = 600 \text{ mfw HCl}$$

The commercial reagent contains

$$\frac{1.18 \text{ g soln}}{\text{mL}} \times \frac{0.37 \text{ g HCl}}{\text{g soln}} \times \frac{\text{mfw HCl}}{0.0365 \text{ g HCl}} = 12.0 \; F$$

Thus, the volume of concentrated reagent to be taken should be

$$600 \text{ mfw HCl} \times \frac{\text{mL}}{12.0 \text{ mfw HCl}} = 50 \text{ mL}$$

**Solution-Diluent Ratios**   The composition of a dilute solution is sometimes specified as a ratio between the volume of a more concentrated reagent and the volume of solvent to be used to dilute it; the volume of the former is separated from the volume of the latter by a colon. According to this practice, a 1:4 HCl solution contains four volumes of water for each volume of acid. This method of expressing concentration is ambiguous because the concentration of the original acid is not specified. Futhermore, the second part of the ratio may be interpreted as being the relative volume of the diluted solution rather than the volume of solvent used. Virtually any other method of expressing concentration is preferable.

## 2C STOICHIOMETRIC RELATIONSHIPS

A balanced chemical equation is a statement of the combining ratios (in units of chemical mass) that exist between reacting substances and their products. Thus, the equation

$$2NaI(aq) + Pb(NO_3)_2(aq) = PbI_2(s) + 2NaNO_3(aq)$$

indicates that 2 fw of aqueous sodium iodide combine with 1 fw of lead nitrate to produce 1 fw of solid lead iodide and 2 fw of aqueous sodium nitrate.[5,6] A statement such as this is called the *stoichiometry* of the reaction.

Often, the chemist needs to convert an experimentally determined weight of an element or compound to the weight of some other species that is chemically equivalent to the original substance. Generally, the experimental and the cal-

---

[5] Here it is advantageous to depict the reaction in terms of chemical compounds. If we wish to focus on reacting species, the net ionic equation is preferable:

$$2I^-(aq) + Pb^{2+}(aq) \rightleftharpoons PbI_2(s)$$

The generation of net ionic equations is reviewed in Appendix 1.

[6] Chemists frequently include information about the physical state of substances in equations; thus, (g), (l), (s), and (aq) refer to gaseous, liquid, solid, and aqueous states, respectively. We shall follow this practice wherever the physical state of a reactant is pertinent to the discussion.

culated weights are in metric units such as grams or milligrams. A conversion of this type is a three-step process involving (1) transformation of the raw metric data into units of chemical mass (fw, mfw), (2) multiplication by a factor that accounts for the stoichiometry, and (3) reconversion of the chemical mass data to the metric units called for in the answer. The process can be summarized as follows:

$$
\begin{array}{c}
\text{quantity} \\
\text{measured} \\
\text{\scriptsize(metric} \\
\text{\scriptsize units)}
\end{array}
\times
\begin{array}{c}
\text{conversion} \\
\text{factor} \\
\text{\scriptsize(metric to} \\
\text{\scriptsize chemical)}
\end{array}
\times
\begin{array}{c}
\text{stoichiometric} \\
\text{relationship}
\end{array}
\times
\begin{array}{c}
\text{conversion} \\
\text{factor} \\
\text{\scriptsize(chemical} \\
\text{\scriptsize to metric)}
\end{array}
=
\begin{array}{c}
\text{quantity} \\
\text{sought} \\
\text{\scriptsize(metric} \\
\text{\scriptsize units)}
\end{array}
$$

The conversion factors in this sequence will be gram-formula weights or gram-molecular weights, as appropriate.

It is worthwhile to reemphasize the value of making certain that the units on both sides of the equal sign are the same by including them as part of the equation. Such agreement is the best proof that the correct relationship has been generated.

---

**Example 2–12**  What weight of $AgNO_3$ (gfw = 170) is needed to convert 2.33 g of $Na_2CO_3$ (gfw = 106) to $Ag_2CO_3$?

$$
\underset{\substack{\text{(metric}\\\text{quantity)}}}{2.33 \text{ g } Na_2CO_3} \times
\underset{\substack{\text{(conversion}\\\text{factor)}}}{\frac{\text{fw } Na_2CO_3}{106 \text{ g}}} \times
\underset{\substack{\text{(stoichiometry)}}}{\frac{2 \text{ fw } AgNO_3}{\text{fw } Na_2CO_3}} \times
\underset{\substack{\text{(conversion}\\\text{factor)}}}{\frac{170 \text{ g}}{\text{fw } AgNO_3}} =
\underset{\substack{\text{(metric}\\\text{quantity)}}}{7.47 \text{ g } AgNO_3}
$$

---

The consistent use of milliformula weights (i.e., fw/1000) is equally satisfactory, as illustrated in Example 2–13.

---

**Example 2–13**  What volume of 0.0750 *F* $AgNO_3$ will be needed to convert 0.214 g of pure $Na_2CO_3$ to $Ag_2CO_3$?

$$
\underset{\substack{\text{(metric}\\\text{quantity)}}}{0.214 \text{ g } Na_2CO_3} \times
\underset{\substack{\text{(conversion}\\\text{factor)}}}{\frac{\text{mfw } Na_2CO_3}{0.106 \text{ g}}} \times
\underset{\substack{\text{(stoichiometry)}}}{\frac{2 \text{ mfw } AgNO_3}{\text{mfw } Na_2CO_3}}
$$

$$
\times
\underset{\substack{\text{(conversion}\\\text{factor)}}}{\frac{\text{mL}}{0.0750 \text{ mfw } AgNO_3}} =
\underset{\substack{\text{(metric}\\\text{quantity)}}}{53.8 \text{ mL}}
$$

---

Here, reconversion to metric units required introduction of the formal $AgNO_3$ concentration in the denominator in order for the answer to have the units of milliliters, as called for in the problem.

---

**Example 2–14** What weight of $Ag_2CO_3$ (gfw = 276) is formed when 25.0 mL of 0.200 $F$ $AgNO_3$ are mixed with 50.0 mL of 0.0800 $F$ $Na_2CO_3$?

Mixing these two solutions will result in one (and only one) of three possible outcomes, specifically:

(a) an excess of $AgNO_3$ will remain after reaction is complete.
(b) an excess of $Na_2CO_3$ will remain after reaction is complete.
(c) an excess of neither reagent will exist.

As a first step, we must establish which of these situations applies by ascertaining the amounts of reactants (in chemical units) available at the outset.
    Initial amounts are

$$25.0 \text{ mL} \times \frac{0.200 \text{ mfw } AgNO_3}{\text{mL}} = 5.00 \text{ mfw } AgNO_3 = 5.00 \text{ mfw } Ag^+$$

$$50.0 \text{ mL} \times \frac{0.0800 \text{ mfw } Na_2CO_3}{\text{mL}} = 4.00 \text{ mfw } Na_2CO_3 = 4.00 \text{ mfw } CO_3^{2-}$$

Because each $CO_3^{2-}$ ion reacts with two $Ag^+$ ions, situation (b) prevails, and the amount of $Ag_2CO_3$ produced will be limited by the amount of $Ag^+$ available; thus,

$$5.00 \text{ mfw } Ag^+ \times \frac{\text{mfw } Ag_2CO_3}{2 \text{ mfw } Ag^+} \times \frac{0.276 \text{ g}}{\text{mfw } Ag_2CO_3} = 0.690 \text{ g } Ag_2CO_3$$

---

**Example 2–15** What will be the formal $CO_3^{2-}$ concentration in the solution produced when 25.0 mL of 0.200 $F$ $AgNO_3$ are mixed with 50.0 mL of 0.0800 $F$ $Na_2CO_3$?

We have seen in the previous example that formation of 5.00 mfw of $AgNO_3$ will require 2.50 mfw of $Ag_2CO_3$. The number of milliformula weights of unreacted $CO_3^{2-}$ is then given by

$$4.00 \text{ mfw } CO_3^{2-} - 5.00 \text{ mfw } AgNO_3 \times \frac{\text{mfw } CO_3^{2-}}{2 \text{ mfw } AgNO_3} = 1.50 \text{ mfw } CO_3^{2-}$$

$$\frac{1.50 \text{ mfw } CO_3^{2-}}{(50.0 + 25.0) \text{ mL}} = \frac{0.0200 \text{ mfw } CO_3^{2-}}{\text{mL}} = 0.0200 \text{ } F \text{ } CO_3^{2-}$$

## 2D  CHEMICAL EQUILIBRIUM

Throughout the field of analytical chemistry, few concepts are more pervasive than that of chemical equilibrium. The notion that a reaction is never entirely complete, but rather is characterized by a condition in which the ratio between reactants and products is fixed, will crop up time and again throughout this text.

Equilibrium constant expressions are *algebraic* equations that relate the molar concentrations of reactants and products to one another by means of a numerical quantity called an *equilibrium constant*. The ability to extract useful information from equilibrium constants is essential to the study of analytical chemistry.

### 2D – 1  The Equilibrium State

For purposes of discussion, consider the reaction

$$2Fe^{3+} + 3I^- \rightleftharpoons 2Fe^{2+} + I_3^-$$

At low concentrations, triiodide ion is the only species present that imparts color to a solution of the four ions. Consequently, the rate of the reaction can be judged by the rate at which the orange-red color of this ion appears or disappears. Equilibrium is signaled when the color becomes constant.

Equilibrium in this system can be approached in either of two general ways:

1. Solutions containing iron(III) and $I^-$ can be mixed; the rapid appearance of $I_3^-$ will be observed.
2. Solutions of iron(II) and $I_3^-$ can be mixed; here a decrease in intensity of color will be observed.

Through a suitable choice of initial reactant concentrations, the solutions that result will be indistinguishable from one another. Stated more generally, the concentration relationships among reactants and products upon attainment of equilibrium (that is, the *position of equilibrium*) are independent of the route by which the equilibrium state is achieved.

Application of a stress to a system in equilibrium will result in an alteration in the concentration relationships in a direction that tends to counteract that stress. These effects are predicted qualitatively by the *Le Châtelier principle*. Thus, an increase in temperature will alter the equilibrium relationship in favor of the direction that tends to absorb heat; an increase in pressure favors those participants that occupy the smaller volume. Of particular importance is the effect of introducing an additional amount of participating species to an equilibrium system. Here, the resulting stress is relieved by a shift in equilibrium in the direction that tends to use up the added substance. For example, introduction of iron(III) to the equilibrium we have been considering would cause an increase in color as more triiodide ion and iron(II) are formed; the addition of iron(II) would have the reverse effect. The equilibrium shift brought about by changing the amount of a participant is called the *mass action effect*.

The influence of concentration (or pressure if the species is a gas) on the position of equilibrium is quantitatively described by means of the equilibrium constant expression. Such expressions are of great practical importance because they permit the chemist to predict the direction and the completeness of a chemical reaction. They reveal no information, however, concerning the rate at which equilibrium will be attained.

Consider the generalized equation for a chemical equilibrium:

$$mM + nN \rightleftharpoons pP + qQ \qquad (2-6)$$

where the capital letters represent formulas for the participating species and the lower case letters are the integers required to balance the equation. Thus, Equation 2-6 states that reaction between $m$ formula weights of M and $n$ formula weights of N results in formation of $p$ formula weights of P and $q$ formula weights of Q. The approximate equilibrium constant, $K_{eq}$, for this system is[7]

$$K_{eq} = \frac{[P]^p[Q]^q}{[M]^m[N]^n}$$

where the letters in square brackets represent the molar concentrations of dissolved solutes or partial pressures (in atmospheres) if the species are gases.

Common types of equilibrium processes and their corresponding equilibrium constant expressions are shown in Table 2-2. Note that a term for the concentration of water does not appear in any of these expressions even though it is involved in the equilibrium. Water exists in enormous excess in the dilute aqueous solutions for which these expressions apply; its concentration is thus essentially constant and is incorporated in the numerical constant. Note also that in the solubility equilibrium expression, a term is not included for the concentration of lead iodide, which is only sparingly soluble in aqueous solutions. Its concentration in the solid phase is constant and is also included in the numerical constant $K_{sp}$. For this simpler form of the equilibrium constant expression to apply, it is only necessary that an excess of solid lead iodide be in contact and in equilibrium with the solution.

We will make extensive use of equilibrium concepts as they apply to precipitation, neutralization, complex formation, and oxidation-reduction systems. In these discussions, it is assumed that the reader is familiar with the generation of net ionic equations, the use of exponential notation, logarithms, and the solution of algebraic equations. Each of these topics is briefly reviewed in sections of the Appendix, along with a consideration of ways by which equations can be simplified through neglect of terms.

---

[7] It is assumed that each participant behaves totally independently of all other species, a situation that is approached in only the most dilute of solutions. The steps needed to correct equilibrium calculations for interaction effects are considered in Chapter 5.

**Table 2–2   Typical Equilibria and Their Equilibrium Constant Expressions**

*Dissociation of Water*

$$2H_2O \rightleftharpoons H_3O^+ + OH^- \qquad K_w = [H_3O^+][OH^-]$$

*Acid-base*

$$HCOOH + H_2O \rightleftharpoons H_3O^+ + HCOO^- \qquad K_a = \frac{[H_3O^+][HCOO^-]}{[HCOOH]}$$

$$NH_3 + H_2O \rightleftharpoons NH_4^+ + OH^- \qquad K_b = \frac{[NH_4^+][OH^-]}{[NH_3]}$$

*Solubility*

$$PbI_2(s) \rightleftharpoons Pb^{2+} + 2I^- \qquad K_{sp} = [Pb^{2+}][I^-]^2$$

*Complex formation*

$$Ni^{2+} + 4CN^- \rightleftharpoons Ni(CN)_4^{2-} \qquad K_f = \frac{[Ni(CN)_4^{2-}]}{[Ni^{2+}][CN^-]^4}$$

*Oxidation-reduction*

$$IO_3^- + 5I^- + 6H^+ \rightleftharpoons 3I_2(aq) + 3H_2O \qquad K_{eq} = \frac{[I_2]^3}{[IO_3^-][I^-]^5[H^+]^6}$$

$$Cl_2(g) + 2AgI(s) \rightleftharpoons 2AgCl(s) + I_2(g) \qquad K_{eq} = \frac{p_{I_2}*}{p_{Cl_2}}$$

*Distribution*

$$I_2(aq) \rightleftharpoons I_2(CCl_4) \qquad K_d = \frac{[I_2]_{CCl_4}}{[I_2]_{H_2O}}$$

---

\* Here $p$ refers to the partial pressure (atm) for the substance in the subscript.

## PROBLEMS

\*2–1. Indicate whether each of the following species acts as an acid or a base in aqueous solution. Give the formula for its conjugate base or acid.

    (a) $NH_3$                   (b) $H_3PO_4$

    (c) HCN                  (d) $HPO_4^{2-}$

    (e) NaOCl               (f) $CH_3CH_2NH_2$

2–2. Indicate whether each of the following species acts as an acid or a base in aqueous solution. Give the formula for its conjugate base or acid.

    (a) $H_2S$                   (b) $H_2CO_3$

    (c) $CO_3^{2-}$               (d) HOCl

    (e) $C_6H_5NH_3^+$         (f) $HCO_3^-$

2–3. Write autoprotolysis constants for the following amphiprotic solvents. Indicate the acid and the base formed as a result of autoprotolysis.

    (a) formic acid, HCOOH     (b) ethylamine, $CH_3CH_2NH_2$

    (c) methanol, $CH_3OH$        (d) hydrazine, $H_2NNH_2$

\*2–4. How many $Na^+$ ions are contained in 7.68 g of $Na_2C_2O_4$?

2–5. How may $K^+$ ions are contained in 5.45 mfw of $K_3PO_4$?

\*2–6. How many milliformula weights are contained in

    (a) 4.96 g of $B_2O_3$?

    (b) 313 mg of $Na_2B_4O_7 \cdot 10H_2O$?

     (c) 12.10 g of $Mn_3O_4$?

     (d) 87.9 mg of $CaC_2O_4$?

**2-7.** How many milliformula weights are contained in

     (a) 27.0 mg of $P_2O_5$?

     (b) 9.73 g of $CO_2$?

     (c) 40.0 g of $NaHCO_3$?

     (d) 584 mg of $MgNH_4PO_4$?

**\*2-8.** How many milliformula weights of solute are contained in

     (a) 2.00 L of $4.77 \times 10^{-3}$ $F$ $KMnO_4$?

     (b) 500 mL of 0.0276 $F$ KSCN?

     (c) 250 mL of a solution that contains 7.80 ppm of $CuSO_4$?

     (d) 3.50 L of 0.466 $F$ KCl?

**2-9.** How many milliformula weights of solute are contained in

     (a) 75.0 mL of 0.320 $F$ $HClO_4$?

     (b) 20.0 L of $8.05 \times 10^{-3}$ $F$ $K_2CrO_4$?

     (c) 5.00 L of an aqueous solution that contains 12.4 ppm of $AgNO_3$?

     (d) 684 mL of 0.0200 $F$ KOH?

**\*2-10.** How many milligrams are contained in

     (a) 0.405 mfw of $HNO_3$?

     (b) 325 mfw of MgO?

     (c) 16.0 fw of $NH_4NO_3$?

     (d) 4.79 mfw of $(NH_4)_2Ce(NO_3)_6$?

**2-11.** How many grams are contained in

     (a) 3.58 fw of KBr?

     (b) 9.50 mfw of PbO?

     (c) 5.76 fw of $MgSO_4$?

     (d) 12.8 mfw of $Fe(NH_4)_2(SO_4)_2 \cdot 6H_2O$?

**\*2-12.** How many milligrams of solute are contained in

     (a) 24.0 mL of 0.150 $F$ sucrose (gfw = 342)?

     (b) 4.80 L of $5.23 \times 10^{-3}$ $F$ $H_2SO_3$?

     (c) 527 mL of a solution that contains 7.38 ppm of $Pb(NO_3)_2$?

     (d) 4.09 mL of 0.0619 $F$ $KNO_3$?

**2-13.** How many grams of solute are contained in

     (a) 450 mL of 0.202 $F$ $H_2O_2$?

     (b) 61.0 mL of $8.75 \times 10^{-4}$ $F$ benzoic acid (gfw = 122)?

     (c) 6.50 L of a solution that contains 21.7 ppm of $SnCl_2$?

     (d) 13.8 mL of 0.0125 $F$ $KBrO_3$?

**2-14.** Calculate the p-value for each of the indicated ions in the following:

     \*(a) $Na^+$, $Cl^-$, and $OH^-$ in a solution that is 0.0820 $F$ in NaCl and 0.138 $F$ in NaOH

     (b) $Ba^{2+}$, $Mn^{2+}$, and $Cl^-$ in a solution that is $7.25 \times 10^{-3}$ $F$ in $BaCl_2$ and $2.38 \times 10^{-4}$ $F$ in $MnCl_2$

     \*(c) $H^+$, $Cl^-$, and $Zn^{2+}$ in a solution that is 1.50 $F$ in HCl and 0.120 $F$ in $ZnCl_2$

     (d) $Cu^{2+}$, $Zn^{2+}$, and $NO_3^-$ in a solution that is $8.40 \times 10^{-2}$ $F$ in $Cu(NO_3)_2$ and 0.175 $F$ in $Zn(NO_3)_2$

     \*(e) $K^+$, $OH^-$, and $Fe(CN)_6^{4-}$ in a solution that is $6.22 \times 10^{-6}$ $F$ in $K_4Fe(CN)_6$ and $3.18 \times 10^{-5}$ $F$ in KOH

     (f) $H^+$, $Ba^{2+}$, and $ClO_4^-$ in a solution that is $5.75 \times 10^{-4}$ $F$ in $Ba(ClO_4)_2$ and $3.61 \times 10^{-4}$ $F$ in $HClO_4$

**2-15.** Calculate the molar $H_3O^+$ ion concentration of a solution that has a pH of

    *(a) 10.64.                  (b) 3.58.

    *(c) 0.75.                    (d) 13.22.

    *(e) 6.49.                    (f) 8.32.

    *(g) −0.81.                 (h) −0.46.

**\*2-16.** Sea water contains an average of $1.08 \times 10^3$ ppm of $Na^+$ and 270 ppm of $SO_4^{2-}$. Calculate

    (a) the molar concentrations of $Na^+$ and $SO_4^{2-}$, given that the average density of sea water is 1.02 g/mL.

    (b) the pNa and pSO$_4$ for sea water.

**2-17.** Average human blood serum contains 18 mg of $K^+$ and 365 mg of $Cl^-$ per 100 mL. Calculate

    (a) the molar concentration for each of these species; use 1.00 g/mL for the density of serum.

    (b) pK and pCl for human serum.

**\*2-18.** A solution was prepared by dissolving 0.964 g of $KCl \cdot MgCl_2 \cdot 6H_2O$ (gfw = 278) in sufficient water to give 2.000 L. Calculate

    (a) the formal concentration of $KCl \cdot MgCl_2$ in this solution.

    (b) the molar concentration of $Mg^{2+}$.

    (c) the molar concentration of $Cl^-$.

    (d) the weight-volume percentage of $KCl \cdot MgCl_2 \cdot 6H_2O$.

    (e) the millimoles of $Cl^-$ in 25.0 mL of this solution.

    (f) ppm $K^+$.

    (g) pMg for the solution.

    (h) pCl for the solution.

**2-19.** A solution was prepared by dissolving 684 mg of $K_3Fe(CN)_6$ (gfw = 329) in sufficient water to give 500 mL. Calculate

    (a) the formal concentration of $K_3Fe(CN)_6$.

    (b) the molar concentration of $K^+$.

    (c) the molar concentration of $Fe(CN)_6^{3-}$.

    (d) the weight-volume percentage of $K_3Fe(CN)_6$.

    (e) the millimoles of $K^+$ in 50.0 mL of this solution.

    (f) ppm $Fe(CN)_6^{3-}$.

    (g) pK for the solution.

    (h) pFe(CN)$_6$ for the solution.

**\*2-20.** An 8.00% (w/w) $Fe(NO_3)_3$ solution has a density of 1.062 g/mL. Calculate

    (a) the formal concentration of $Fe(NO_3)_3$ in this solution.

    (b) the molar $NO_3^-$ concentration in the solution.

    (c) the grams of $Fe(NO_3)_3$ contained in each liter of this solution.

**2-21.** A 14.0% (w/w) $NiCl_2$ solution has a density of 1.143 g/mL. Calculate

    (a) the formal concentration of $NiCl_2$ in this solution.

    (b) the molar $Cl^-$ concentration of the solution.

    (c) the grams of $NiCl_2$ contained in each liter of this solution.

**\*2-22.** Describe the preparation of

    (a) 500 mL of 12.0% (w/v) aqueous ethanol ($C_2H_5OH$, gfw = 46.1).

    (b) 500 g of 12.0% (w/w) aqueous ethanol.

    (c) 500 mL of 12.0% (v/v) aqueous ethanol.

**2-23.** Describe the preparation of

    (a) 1.50 L of 24.0% (w/v) aqueous glycerol ($C_3H_8O_3$, gfw = 92.1).

      (b) 1.50 kg of 24.0% (w/w) aqueous glycerol.

      (c) 1.50 L of 24.0% (v/v) aqueous glycerol.

**\*2 – 24.** Describe the preparation of 250 mL of 6.00 $F$ $H_3PO_4$ from the commercial reagent that is 85% $H_3PO_4$ (w/w) and has a specific gravity of 1.69.

**2 – 25.** Describe the preparation of 750 mL of 3.00 $F$ $HNO_3$ from the commercial reagent that is 69% $HNO_3$ (w/w) and has a specific gravity of 1.42.

**\*2 – 26.** Describe the preparation of

      (a) 500 mL of 0.0800 $F$ $AgNO_3$ from the solid reagent.

      (b) 1.00 L of 0.200 $F$ HCl, starting with a 6.00 $F$ solution of the reagent.

      (c) 600 mL of a solution that is 0.0750 $M$ in $K^+$, starting with solid $K_4Fe(CN)_6$.

      (d) 750 mL of 3.00% (w/v) aqueous $BaCl_2$ from a 0.400 $F$ $BaCl_2$ solution.

      (e) 2.00 L of 0.120 $F$ $HClO_4$ from the commercial reagent [60% $HClO_4$ (w/w), sp gr 1.60].

      (f) 5.00 L of a solution that is 60.0 ppm in $Na^+$, starting with solid $Na_2SO_4$.

**2 – 27.** Describe the preparation of

      (a) 1.00 L of 0.0200 $F$ $KMnO_4$ from the solid reagent.

      (b) 2.50 L of 0.160 $F$ $HClO_4$, starting with an 8.00 $F$ solution of the reagent.

      (c) 750 mL of a solution that is 0.0400 $F$ in $I^-$, starting with $MgI_2$.

      (d) 200 mL of 5.00% (w/v) aqueous $CuSO_4$, from a 0.485 $F$ $CuSO_4$ solution.

      (e) 1.50 L of 0.100 $F$ NaOH from the concentrated commercial reagent [50% NaOH (w/w), sp gr 1.525].

      (f) 4.00 L of a solution that is 12 ppm in $K^+$, starting with solid $K_4Fe(CN)_6$.

**\*2 – 28.** Lead ion reacts with $I^-$ to form the sparingly soluble $PbI_2$ (gfw = 461).

      (a) What weight of KI will be needed to react completely with 2.41 g of $Pb(NO_3)_2$ (gfw = 331)?

      (b) What volume of 0.115 $F$ $Pb(NO_3)_2$ will be needed to react completely with 1.86 g of KI?

      (c) What volume of 0.0875 $F$ KI will be needed to react completely with 25.0 mL of 0.0628 $F$ $Pb(NO_3)_2$?

      (d) What weight of $PbI_2$ will be produced when 30.9 mL of 0.211 $F$ KI are mixed with 40.0 mL of 0.0939 $F$ $Pb(NO_3)_2$?

**2 – 29.** Trivalent Ce reacts with $IO_3^-$ to form the sparingly soluble $Ce(IO_3)_3$ (gfw = 701).

      (a) What weight of $CeCl_3$ will be needed to react completely with 2.86 g of $KIO_3$?

      (b) What volume of 0.0518 $F$ $CeCl_3$ will be needed to react completely with 40.0 mL of 0.0840 $F$ $KIO_3$?

      (c) What volume of 0.154 $F$ $Mg(IO_3)_2$ (gfw = 374) will be needed to react completely with 1.14 g of $CeCl_3$?

      (d) What weight of $Ce(IO_3)_3$ will be produced when 21.1 mL of 0.0671 $F$ $Mg(IO_3)_2$ and 43.2 mL of 0.0225 $F$ $CeCl_3$ are mixed?

**2 – 30.** Treatment of a solution containing $Mg^{2+}$ with base results in formation of the sparingly soluble $Mg(OH)_2$.

      (a) What weight of $Mg(OH)_2$ will be produced when 25.0 mL of 0.0774 $F$ $MgSO_4$ are mixed with an equal volume of 0.140 $F$ KOH?

      (b) What weight will remain undissolved after treatment of 0.366 g of solid $Mg(OH)_2$ with 46.3 mL of 0.162 $F$ HCl?

      (c) What will be the $Mg^{2+}$ concentration of the solution that results in (b)?

# Chapter 3
# The Evaluation of Analytical Data

Every physical measurement is subject to a degree of uncertainty that, at best, can only be decreased to an acceptable level. The determination of the magnitude of this uncertainty is often difficult and requires additional effort, ingenuity, and good judgment on the part of the observer. Nevertheless, evaluation of the uncertainty in analytical data is a task that cannot be neglected because a measurement of totally unknown reliability is worthless. On the other hand, a result that is not particularly accurate may be of great use if the limits of the probable error affecting it can be set with a high degree of certainty. Unfortunately, there exist no simple and generally applicable methods by which the quality of an experimental result can be assessed with absolute certainty; indeed, the work expended in evaluating the reliability of data is frequently comparable to the effort that went into obtaining them. Such an evaluation may take any of several courses, including search of the chemical literature to profit from the experiences of others, performance of further experiments that are designed to reveal potential sources of error, calibration of equipment used in the measurements, and application of statistical tests to the data that have been obtained. It must be recognized, however, that none of these recourses is infallible and that ultimately the scientist can only make a judgment as to the probable accuracy of a measurement; such judgments tend to become more pessimistic with increases in experience.

A direct relationship exists between the accuracy of an analytical result and the time required for its acquisition. A tenfold increase in reliability may involve hours, days, or perhaps weeks of additional labor. One of the first questions that must be considered at the outset of any analysis is the degree of reliability that is required; this consideration will in large measure determine the amount of time and effort that will be needed to perform the analysis. *It cannot be too strongly emphasized that a scientist cannot afford to waste time in the indiscriminate pursuit of great reliability where it is not needed.*

This chapter is devoted to a consideration of the errors that can affect an analysis, methods for their recognition, and techniques for estimating and reporting their magnitude.

## 3A DEFINITION OF TERMS

The chemist ordinarily performs a particular analysis on two to five samples. The individual results in such a set of measurements will seldom be identical; it thus becomes necessary to select a central "best" value for the set. Intuitively, the added effort of replication can be justified in two ways. First, the central value of the set ought to be more reliable than any individual result; second, the variations among the results should provide some measure of the reliability of the "best" value that has been chosen.

Either the *mean* or the *median* may serve as the central value for a set of measurements.

### 3A-1 The Mean and the Median

The *mean, arithmetic mean,* and *average* $(\bar{x})$ are synonymous terms for the numerical value that is obtained by dividing the sum of a set of replicate measurements by the number of individual results in the set.

The *median* of a set is that result about which all of the others are equally distributed, half being numerically greater and half numerically smaller. The median of a set consisting of an odd number of measurements can be evaluated directly. The mean of the central pair is used for a set with an even number of measurements.

---

**Example 3-1**    Calculate the mean and the median for 10.06, 10.20, 10.08, and 10.10.

$$\text{mean} = \bar{x} = \frac{10.06 + 10.20 + 10.08 + 10.10}{4} = 10.11$$

Because the set contains an even number of measurements, the median is the average of the middle pair:

$$\text{median} = \frac{10.08 + 10.10}{2} = 10.09$$

---

Ideally, the mean and the median will be numerically identical, but they frequently are not, particularly when the number of individual measurements in a set is small.

### 3A-2 Precision

The term *precision* is used to describe the reproducibility of results. It can be defined as the agreement between the numerical values for two or more measurements that have been made *in an identical fashion.* Precision can be measured in any of several ways.

**Absolute Methods for Expressing Precision** The *deviation from the mean* $(x_i - \bar{x})$ is a common method for describing precision and is simply the numerical difference, *without regard to sign,* between an experimental value and the mean of the set (occasionally, deviation from the median is encountered). To illustrate, suppose that the determination of chloride in a sample yielded the accompanying results:

| Sample | Percent Chloride | Deviation from the Mean $\lvert x_i - \bar{x} \rvert$ | Deviation from the Median |
|--------|------------------|------------------------------------|---------------------------|
| $x_1$ | 24.39 | 0.077 | 0.03 |
| $x_2$ | 24.19 | 0.123 | 0.17 |
| $x_3$ | 24.36 | 0.047 | 0.00 |
| | 3⟌72.94 | 3⟌0.247 | 3⟌0.20 |
| | $\bar{x} = 24.313 = 24.31$ | av $= 0.082 = 0.08$ | av $= 0.067 = 0.07$ |

The mean for this set is 24.31% Cl⁻. The deviation of the second result from the mean is 0.12; it is 0.17 from the median. The average deviation from the mean is 0.08% Cl⁻. Note that the data in the third column were calculated by using 24.313 as the mean instead of the rounded value of 24.31. This practice of carrying an extra figure through calculations should be followed generally in the interest of minimizing rounding errors in the final result.

The *spread* or *range* ($w$) in a set of data is also a measure of precision and is simply the numerical difference between the highest and the lowest result. In the present example the spread is 0.20% Cl⁻.

The *standard deviation* and the *variance* are more significant measures of precision; these terms are defined in Section $3D^{-2}$.

**Relative Methods for Expressing Precision** We have thus far calculated precision in absolute terms. It is frequently more informative to indicate the relative deviation from the mean (or median) in parts per thousand or as a percentage. Thus, for example, the relative deviation of sample $x_1$ from the mean in the present set is

$$\frac{0.077 \times 100}{24.313} = 0.32\% = 0.3\%$$

Note that we again rounded only at the end of the calculation. Similarly, the average deviation of the set from the median, in parts per thousand, is

$$\frac{0.067 \times 1000}{24.36} = 2.8 = 3 \text{ ppt}$$

## 3A – 3 Accuracy

The term *accuracy* denotes the nearness of a measurement to its accepted value and is expressed in terms of error. Note the fundamental difference that exists between accuracy and precision. Accuracy involves a comparison with a true or accepted value while precision measures internal agreement among measurements that have been made in the same way.

The *absolute error, E,* of a measurement is the difference between the observed value $x_i$ and the accepted value, $x_t$:

$$E = x_i - x_t \qquad\qquad (3-1)$$

In contrast to precision, where the only concern is a numerical difference, the sign associated with error is as important as the numerical value itself because the chemist needs to know whether the effect of error has caused the result (or results) to be high or low.

The accepted value may itself be subject to appreciable error; consequently, a realistic estimate regarding the accuracy of a measurement is frequently difficult. Returning to our example, suppose that the accepted value for the percentage of chloride in the sample is 24.35%. The absolute error of the mean is thus

$$24.31 - 24.35 = -0.04\% \ Cl^-$$

The absolute error of the median is

$$24.36 - 24.35 = +0.01\% \ Cl^-$$

We shall see that the very small error associated with the median for these data is probably fortuitous.

The *relative error,* which is frequently a more revealing measure of accuracy, is often expressed as a percentage (parts per hundred) or as parts per thousand. Thus, for the chloride analysis that we have been considering,

$$\text{relative error of the mean} = \frac{-0.04 \times 100}{24.35} = -0.2\%$$

$$\text{relative error of the median} = \frac{0.01 \times 1000}{24.35} = 0.4 \text{ ppt}$$

## 3B PRECISION AND ACCURACY OF EXPERIMENTAL DATA

The precision of a measurement is readily evaluated by performing replicate experiments under the same conditions. The same cannot be said for accuracy since this quantity requires knowledge of the true value, the very information that is being sought. It is tempting to ascribe a direct relationship between accuracy and precision. The danger of this practice is illustrated in Figure 3 – 1,

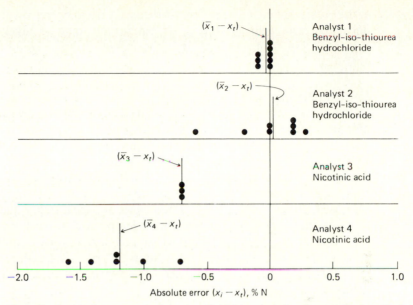

**Figure 3 – 1**   Absolute errors in the micro-Kjeldahl analysis for nitrogen. Each vertical line labeled $(x_i - x_t)$ is the absolute average deviation of the set from the true value. (Data from C. O. Willits and C. L. Ogg, *J. Assoc. Offic. Anal. Chem.*, **1949,** *32*, 561. With permission.)

which summarizes the results obtained by four analysts for the determination of nitrogen in two pure compounds. The dots represent the absolute errors in the individual data. Note that Analyst 1 achieved both high precision and high accuracy. In contrast, Analyst's 2 data were of considerably lower precision but still showed good accuracy. The results of Analyst 3 are not at all uncommon; here, the precision was excellent, but the accuracy was poor. The results by Analyst 4 were wanting in both accuracy and precision.

The effects illustrated in Figure 3 – 1 can be rationalized by postulating that two general types of error affect experimental data and that precision estimates provide no information concerning the existence of one type.

## 3B – 1 Classes of Errors

As a matter of convenience, the phenomena that are responsible for uncertainties in analytical measurements are subdivided into two broad categories: (1) *determinate errors* and (2) *indeterminate* or *random errors.*[1] It should be noted

---

[1] For a thorough treatment of sources of error, see: L. A. Currie, in *Treatise on Analytical Chemistry,* 2nd ed., I. M. Kolthoff and P. J. Elving, Eds., Part I, Vol. 1, Chapter 4. New York: John Wiley & Sons, 1978.

at the outset that no hard and fast rules exist for deciding the category in which a particular error belongs; indeed, the assignment may be largely a matter of subjective judgment. Nevertheless, the classification is useful in discussing analytical errors.

**Determinate Errors**   Determinate errors have assignable causes and definite values, which in principle can be measured and compensated for. A determinate error is often unidirectional in the sense that it causes all of the results from a set of replicate measurements to be high or low, but not both high and low. The data from Analysts 3 and 4 in Figure 3–1 appear to suffer from a negative determinate error the source of which is well understood. The first step in this analysis involves decomposition of the organic sample with concentrated sulfuric acid, a process which normally converts organic nitrogen quantitatively to ammonium ion, which is subsequently determined. It has been found, however, that unless special precautions are taken, this conversion is incomplete with compounds, such as nicotinic acid, which contain pyridine rings. Apparently, Analysts 3 and 4 failed to observe these special precautions and a negative determinate error resulted.

**Indeterminate Errors**   Indeterminate errors are encountered whenever a measuring system is extended to its maximum sensitivity. Under these circumstances, the results are not constant but instead fluctuate in a random manner about a mean value. The sources of these fluctuations can never be identified because they are made up of a myriad of uncertainties that are individually so small that they can never be detected. What is observed then is a summation of a very large number of minute unobservable uncertainties. If, by chance, the signs of the majority of the individual uncertainties are positive at any instant, a net positive fluctuation is observed. If, on the other hand, most are negative, their sum will be a detectable negative indeterminate error. The most likely situation arises when the number and size of the positive and negative uncertainties are about the same. Then, the net fluctuation approaches zero.

The scatter of individual errors in Figure 3–1 is the result of the operation of indeterminate errors. Larger indeterminate errors affect the results reported by Analysts 2 and 4 than are present in the data of Analysts 1 and 3.

## 3B–2 Sources of Errors

It is not possible to catalog all conceivable sources of error that can affect analytical measurements. We can recognize, however, that limitations on both accuracy and precision can be traced to three general sources: (1) *instrumental uncertainties* that are attributable to imperfections in measuring devices, (2) *method uncertainties* that are caused by nonideal chemical or physical behavior of analytical systems, and (3) *personal uncertainties* that result from physical or psychological limitations of the analyst.

## 3C DETERMINATE ERRORS: THEIR DETECTION, EFFECTS, AND CORRECTION

### 3C–1 Types of Determinate Errors

**Instrumental Uncertainties**    All measuring devices are potential sources of determinate error. For example, volumetric glassware will frequently deliver or contain volumes that are measurably different from those indicated by their graduations. These differences may have such origins as utilization of the glassware at temperatures that differ significantly from the temperature of calibration, distortions in the container walls resulting from excessive heat during drying, errors in the original calibration, or contamination on interior surfaces. Calibration will eliminate most determinate errors of this type.

**Method Uncertainties**    Determinate errors are often caused by the nonideal chemical or physical behavior of the reactions or reagents upon which an analysis is based. Causes of nonideality include the slowness of some reactions, the incompleteness of others, the instability of many species, the nonspecificity of most reagents, and the occurrence of side reactions that interfere with the measurement process. For example, the chemist performing a gravimetric analysis is confronted with the problem of isolating the species to be determined as a solid that possesses the highest possible purity. If it is not washed sufficiently, this precipitate will be contaminated with foreign substances and have a spuriously high weight. On the other hand, the amount of washing needed to eliminate these contaminants may cause a weighable quantity of the precipitate to be lost, owing to its finite solubility; a negative determinate error will result. In either event, the accuracy of the analysis is ultimately limited by inherent and unavoidable method errors.

A frequently encountered method error in volumetric analysis results from the need for a volume of reagent, in excess of the theoretical, to cause an indicator to undergo the color change that signals completion of a titration. As in the previous example, the ultimate accuracy that can be attained is limited by the very phenomenon upon which the analysis is based.

Errors inherent in a method are frequently difficult to detect and are thus the most serious type of determinate errors.

**Personal Uncertainties**    Many measurements require personal judgments. Examples include estimating the position of a pointer between two scale divisions, the level of a liquid with respect to a graduation mark, or the color of a solution at the end point of a titration. Judgments of this type are frequently subject to systematic, unidirectional uncertainties. A person may consistently read a pointer high, another may be uniformly slow in activating a switch, while a third may lack sensitivity to color changes and thus use excess titrant in a volumetric analysis. Color blindness and other physical handicaps increase the probability for determinate errors of this type.

A nearly universal source of personal error is prejudice, or *bias*. Most of us, no matter how objective we claim to be, have a tendency to make estimates in measurements that tend to improve the precision of a set of results, or to place greater credence on data that lie closer to a preconceived notion of the true value for a measurement. Number bias is yet another common source of personal error. The numbers 0 and 5 tend to be preferred over others in estimating the position of an indicator with respect to a scale; also prevalent is prejudice favoring small digits over large and even numbers over odd. A scientist must actively fight against bias; it is insufficient to hold that the problem occurs only in others.

Finally, there is the personal error that results from making a *gross mistake.* Errors in calculation, transposition of numbers in recording data, reading a scale backward, reversing a sign, or using an incorrect scale are common examples. A gross mistake can affect a single value or an entire series of measurements. Errors of this type can ordinarily be traced to carelessness and can be eliminated by self-discipline.

## 3C–2 Effect of Determinate Errors upon the Results of an Analysis

Determinate errors may be classified as being either *constant* or *proportional.* The magnitude of a constant error is independent of the size of the quantity measured. On the other hand, proportional errors increase or decrease in proportion to the size of the sample taken for analysis.

**Constant Errors**   For any given analysis a constant error will become more serious as the size of the quantity measured decreases. This problem is illustrated by the solubility losses that attend the washing of a precipitate.

---

**Example 3–2**   Suppose that 0.50 mg of precipitate is lost as a result of washing with 200 mL of wash liquid. If the precipitate weighs a total of 500 mg, the relative error due to this solubility loss will be $-(0.50 \times 100/500) = -0.1\%$; if the precipitate weighs only 50.0 mg, the relative error will increase to $-1.0\%$.

---

The amount of reagent required to bring about the color change in a volumetric analysis is another example of constant error. This volume, usually small, remains the same regardless of the total volume of reagent used in the titration. Again, the relative error from this source increases as the volume required for the titration becomes smaller.

Clearly, one way of minimizing the effect of constant error is to use as large a sample as is consistent with the method at hand.

**Proportional Errors**   The effect of interfering contaminants in a sample, if not eliminated somehow, will cause a proportional error. For example, a common method for the analysis of copper involves reaction of copper(II) ion with iodide; the quantity of iodine produced in the reaction is then measured.

Iron(III), if present, will also react with iodide, liberating iodine as well. Unless measures are taken to prevent this interference, the iodine produced will be a measure of the sum of copper and iron in the sample, and the analysis will yield a spuriously high result for the percentage of copper. The magnitude of this error will be fixed by the *fraction* of iron contamination and will produce the same relative effect regardless of sample size. If the sample size is doubled, for example, the amount of iodine liberated by both the copper and the iron contaminant will likewise be doubled. Thus, the reported percentage of copper will be independent of sample size.

### 3C-3  Detection and Elimination of Determinate Instrumental and Personal Errors

Determinate instrumental errors are usually found and corrected by calibration. The response of most instruments will undergo change with time owing to wear, corrosion, or mistreatment. The periodic recalibration of equipment is always worthwhile.

Personal errors can be minimized through care and self-discipline. Thus, most scientists develop the habit of systematically checking instrument readings, notebook entries, and calculations. Errors that result from a physical handicap can ordinarily be avoided through a judicious choice of method—provided, of course, that the handicap is recognized.

### 3C-4  Detection of Determinate Method Errors

Determinate method errors are particularly difficult to detect. Identification and compensation for such errors may take one or more of the approaches described in this section.

**Analysis of Standard Samples**   A method may be tested for determinate error by analysis of a synthetic sample whose overall composition is known and closely resembles that of the material for which the analysis is intended. Standard samples must be carefully prepared to ensure that the analyte concentration is known with a high degree of certainty. Unfortunately, it is frequently impossible to produce a sample whose composition truly resembles that of a complex natural substance; indeed, the problems associated with the preparation of standard samples may be so imposing as to prevent the use of this approach.

The National Bureau of Standards offers several hundred carefully analyzed common substances such as ores, alloys, glasses, oils containing polychlorinated biphenyls, and freeze-dried urine.[2] Standard samples of such diverse

---

[2] See the current edition of NBS Special Publication 260, *NBS Standard Reference Materials Catalog.* Washington: U.S. Government Printing Office, 1984. The Reference Material Program is described by R. Alvarez, S. D. Rasberry, and G. A. Uriano in: *Anal. Chem.,* **1982,** *54,* 1226A; see also: G. A. Uriano, *ASTM Standardization News,* **1979,** *7,* 8.

materials as coals, pesticides, pollutants, and petroleum products are also available from several commercial sources. These standards have been specifically designed for testing analytical methods for determinate errors.

**Independent Analysis**   Parallel analysis of a sample by a different method of established reliability is of particular value where samples of known composition are not available. In general, the independent method should not resemble the one under investigation in order to minimize the possibility that some factor in the sample will have a common effect on both methods.

**Blank Determinations**   Constant errors affecting physical measurements can frequently be evaluated with a blank determination in which all steps of the analysis are performed in the absence of a sample; the result is then applied as a correction to the actual measurement. Blank determinations are of particular value in exposing errors that are due to the introduction of interfering contaminants from reagents and vessels used in the method. Blanks also enable the analyst to correct titration data for the volume of reagent needed to cause an indicator to undergo its end-point color change.

**Variation in Sample Size**   As was demonstrated in Example 3–2, the fact that a constant error has a decreasing effect as the size of the measurement increases can be utilized to detect such errors in an analytical method. A series of analyses is performed in which the sample size is varied as widely as possible. The results are then examined for systematic increases or decreases that can be correlated with sample size.

## 3D INDETERMINATE ERROR

The presence of indeterminate error is revealed by the random fluctuation in results that are encountered when replicate experimental data are collected. As the name suggests, the specific causes of these fluctuations are unknown because they do not have a single source; instead, they result from an accumulation of individual uncertainties, no one of which is large enough to be detected.

Table 3–1 illustrates the effect of indeterminate error upon the relatively simple process of calibrating a pipet. The procedure involves determining the weight of water delivered by the pipet. The temperature of the water must be measured to establish its density. The experimentally determined weight can then be converted to the volume delivered by the pipet.

The data in Table 3–1 are typical of those that might be obtained by an experienced and competent worker who performs the weighings to the nearest milligram (which corresponds to 0.001 mL), with every effort being made to recognize and eliminate determinate errors. Even so, the average deviation from the mean of 24 measurements is ±0.005 mL, and the spread is 0.023 mL. This dispersion among the data is the direct result of indeterminate error.

Variations among replicate results such as those in Table 3–1 can be ration-

Table 3 – 1   Replicate Measurements from the Calibration of a 10-mL Pipet

| Trial | Volume of Water Delivered, mL | Trial | Volume of Water Delivered, mL | Trial | Volume of Water Delivered, mL |
|---|---|---|---|---|---|
| 1 | 9.990 | 9 | 9.988 | 17 | 9.977 |
| 2 | 9.993* | 10 | 9.976 | 18 | 9.982 |
| 3 | 9.973 | 11 | 9.981 | 19 | 9.974 |
| 4 | 9.980 | 12 | 9.974 | 20 | 9.985 |
| 5 | 9.982 | 13 | 9.970† | 21 | 9.987 |
| 6 | 9.988 | 14 | 9.989 | 22 | 9.982 |
| 7 | 9.985 | 15 | 9.981 | 23 | 9.979 |
| 8 | 9.970† | 16 | 9.985 | 24 | 9.988 |

Mean volume = 9.9816 = 9.982 mL
Median volume = 9.982 mL
Average deviation from mean = 0.0052 mL = 0.005 mL
Spread = 9.993 − 9.970 = 0.023 mL
Standard deviation = 0.0065 mL = 0.006 mL

* Maximum value.
† Minimum value.

alized by assuming that *any* measurement process is affected by numerous small and individually undetectable instrument, method, and personal uncertainties attributable to uncontrolled variables in the experiment. The cumulative effect of such uncertainties is likewise variable. Ordinarily, they tend to cancel one another and thus exert a minimal effect. Occasionally, however, they act in concert to produce a relatively large positive or negative error. Sources for such uncertainties in the calibration of a pipet include such visual judgments as the level of the water with respect to the marking etched on the pipet, the mercury level in the thermometer, and the position of an indicator with respect to a balance scale (all personal uncertainties). Other sources include variation in the drainage time, the angle of the pipet as it drains (both method uncertainties), and changes in temperature resulting from the manner in which the pipet is handled (an instrument uncertainty). Numerous other uncertainties undoubtedly exist in addition to the ones cited. It is clear that many small and uncontrolled variables accompany even as simple a process as a pipet calibration. Although we are unable to account for the influence of any one of these uncertainties, their cumulative effect is an indeterminate error that is responsible for the scatter of data about the mean.

## 3D – 1 The Distribution of Data from Replicate Measurements

In contrast to determinate errors, indeterminate errors cannot be eliminated from measurements. Moreover, the scientist cannot ignore their existence simply because they ordinarily tend to be small. For example, it would probably be safe to assume that the average value of the 24 measurements in Table 3–1 is closer to the true volume delivered by the pipet than any one of the individual data. Suppose, however, that only a duplicate calibration, consisting of trials 1 and 2, had been performed; the average of these two values, 9.992, differs from the mean of the 24 measurements by 0.010 mL. Note, also, that the average deviation of these two measurements *from their own mean* is only ±0.0015 mL. This figure would represent a highly optimistic estimate of the indeterminate error associated with the calibration process. Suppose, further, that the user of this pipet needed to deliver volumes known to, say, ±0.002 mL. Failure to recognize the true magnitude of the indeterminate error would create a totally false sense of security with respect to the performance of the pipet. It can be shown that, in 1000 measurements with this pipet, two to three transfers are likely to differ from the mean of 9.982 mL by as much as 0.02 mL; more than 100 would differ by 0.01 mL (or greater) despite every precaution on the part of the user.

A qualitative notion of the way small uncertainties affect the outcome of replicate measurements can be developed by considering an imaginary situation in which just four uncertainties are the cause of indeterminate error. Each of these uncertainties is considered to have an equal probability of occurrence and each can cause the final result to be in error by plus or minus a fixed amount $U$. Finally, we shall stipulate that the magnitude of $U$ is the same for each of the four uncertainties.

Table 3–2 shows all of the possible ways these uncertainties can combine to

**Table 3–2    Possible Ways Four Equal-Sized Uncertainties $U_1$, $U_2$, $U_3$, and $U_4$ Can Combine**

| Combinations of Uncertainties | Magnitude of Indeterminate Error | Relative Frequency of Error |
|---|:---:|:---:|
| $+U_1 + U_2 + U_3 + U_4$ | $+4U$ | 1 |
| $-U_1 + U_2 + U_3 + U_4$<br>$+U_1 - U_2 + U_3 + U_4$<br>$+U_1 + U_2 - U_3 + U_4$<br>$+U_1 + U_2 + U_3 - U_4$ | $+2U$ | 4 |
| $-U_1 - U_2 + U_3 + U_4$<br>$+U_1 + U_2 - U_3 - U_4$<br>$+U_1 - U_2 + U_3 - U_4$<br>$-U_1 + U_2 - U_3 + U_4$<br>$-U_1 + U_2 + U_3 - U_4$<br>$+U_1 - U_2 - U_3 + U_4$ | 0 | 6 |
| $+U_1 - U_2 - U_3 - U_4$<br>$-U_1 + U_2 - U_3 - U_4$<br>$-U_1 - U_2 + U_3 - U_4$<br>$-U_1 - U_2 - U_3 + U_4$ | $-2U$ | 4 |
| $-U_1 - U_2 - U_3 - U_4$ | $-4U$ | 1 |

give the indicated indeterminate errors. We note that there is only one way in which the maximum positive error of $4U$ can occur, compared with four combinations that lead to a positive error of $2U$, and six combinations that result in zero error. The same relationship exists for negative indeterminate errors. This ratio of $6:4:1$ is a measure of the probability for an error of each magnitude; if we were to make sufficient measurements, a frequency distribution such as that shown in Figure 3–2a would be expected. Figure 3–2b shows a similar distribution for ten equal-sized uncertainties; again we see that the most frequent occurrence is zero error, while a maximum error of $10U$ would be expected to occur only occasionally (about once in 500 measurements).

It can be demonstrated that extension of the foregoing arguments to a very large number of uncertainties will result in the continuous curve shown in Figure 3–2c. This bell-shaped curve is called a *Gaussian*, or *normal error*

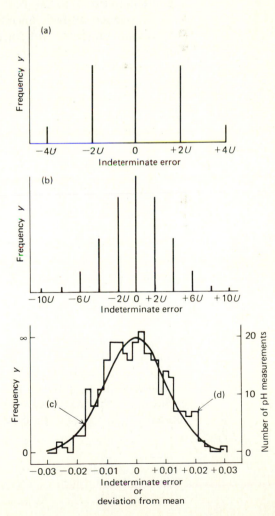

**Figure 3–2** Theoretical distribution of indeterminate error arising from (a) four uncertainties, (b) ten uncertainties, and (c) a very large number of uncertainties. Curve (c) represents a normal or Gaussian distribution. Curve (d) is an experimental distribution curve that might be obtained by plotting the deviations of 250 replicate pH measurements against the number of times each deviation was observed.

*curve.*[3] Its properties include: (1) a maximum frequency in the occurrence of zero indeterminate error, (2) a symmetry about this maximum, indicating that positive and negative errors occur with equal frequency, and (3) an exponential decrease in frequency as the magnitude of the error increases. Thus, small indeterminate errors will be observed much more frequently than a very large one.

Numerous *empirical* observations have confirmed that indeterminate errors in a typical chemical analysis will distribute themselves in a manner that approaches a Gaussian distribution. Thus, for example, if the deviations from the mean for hundreds of repetitive pH measurements of a single sample were plotted against the frequency with which each deviation occurred, one would expect to obtain a curve approximating that shown in Figure 3–2c.

The frequent experimental observation of Gaussian behavior lends credibility to the notion that the indeterminate error occurring in analytical measurements can be traced to the accumulation of a large number of small, independent, and uncontrolled uncertainties. Equally important, the Gaussian distribution associated with the precision of analytical data permits the use of statistical techniques to estimate the effects of indeterminate error on such measurements.

## 3D–2 Classical Statistics

A mathematical description of random processes, such as the effect of indeterminate error on the results of a chemical analysis, can be approached through the application of statistics. It is important to realize, however, that the techniques of classical statistics are strictly applicable to an *infinite* number of observations, a situation which is far removed from a typical set of analytical data; application of statistics to the two to five replicate measurements that the chemist can afford to make can lead to seriously incorrect and misleadingly optimistic conclusions regarding the effect of indeterminate error. The modifications needed to adapt these concepts to small sets of data is conveniently approached from relationships of classical statistics.

**Properties of the Normal Error Curve**    The upper two curves of Figure 3–3 are normal error curves for two different analytical methods. The uppermost curve represents data from the more precise of the two methods inasmuch as the results are distributed more closely about the central value.

Figure 3–3 demonstrates that normal error curves can be plotted in any of several ways. In each, the ordinate $y$ represents the frequency of occurrence for each value of $x$ along the abscissa. The observed values for the measurement are plotted as $x_i$ in curve (a); the central value is thus the mean, which is symbolized

---

[3] We have developed our argument for an example in which all of the uncertainties have the same magnitude. Such a restriction is not needed in deriving the equation for a Gaussian curve.

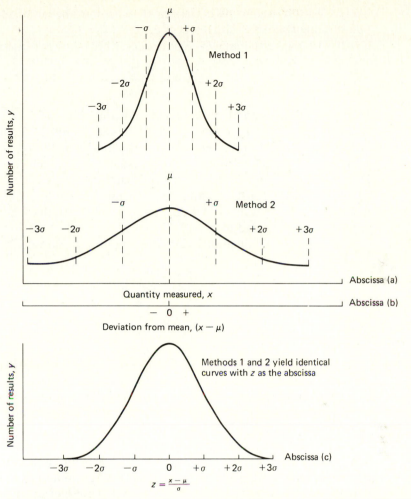

**Figure 3–3**   Normal error curves for measurement of the same quantity by two different methods. Method 1 is more reliable than method 2; thus, $\sigma$ is smaller. Note three types of abscissa: (a) measured quantity $x$, with maximum at $\mu$, (b) deviation from the mean, with maximum at 0, and (c) $z$, as defined in Equation 3–3. Abscissa (c) reduces data from methods 1 and 2 to a single curve.

by $\mu$. Abscissa (b) consists of individual deviations from the mean, $x_i - \mu$; here, the data are distributed about zero, the most frequently occurring deviation. Curve (c) represents yet another way of plotting the data and will be considered presently.

It is important to emphasize that the curves under discussion are theoretical in that they represent distributions that would be expected as the number of observations approaches infinity. For a physically realizable set of data, a discontinuous distribution, such as that shown in Figure 3–2d, would be more likely. Clearly, the concepts of classical statistics must be modified considerably before being applied to small sets of data.

All of the curves in Figure 3–3 can be described in terms of a single equation:

$$y = \frac{e^{-(x_i-\mu)^2/2\sigma^2}}{\sigma\sqrt{2\pi}} \qquad (3-2)$$

In this equation $x_i$ represents values of individual measurements, and $\mu$ is the arithmetic mean for an infinite set of such measurements. The quantity $(x_i - \mu)$ is thus the deviation from the mean; $y$ is the frequency of occurrence for each value of $(x_i - \mu)$. The symbol $\pi$ has its usual meaning, and $e$ is the base for Naperian logarithms, 2.718. . . . The parameter $\sigma$ is called the *standard deviation* and is a constant with a unique value for each individual set of data made up of a large number of measurements. The breadth of the normal error curve is directly related to $\sigma$.

The exponent in Equation 3–2 can be simplified by introducing the variable $z$, where

$$z = \frac{x_i - \mu}{\sigma} \qquad (3-3)$$

Equation 3–2 thus becomes

$$y = \frac{e^{-z^2/2}}{\sigma\sqrt{2\pi}} \qquad (3-4)$$

The quantity $z$ expresses the deviation of the mean in units of standard deviations. As shown in Figure 3–3c, the use of $z$ as the abscissa produces a single curve for all values of $\sigma$.

**The Standard Deviation**   Equation 3–2 indicates that a unique distribution curve exists for each value of the standard deviation, $\sigma$. Regardless of the magnitude of $\sigma$, however, it can be shown that 68.3% of the area beneath the curve lies within one standard deviation $(\pm 1\sigma)$ of the mean $\mu$. Approximately 95.5% of all values will be within $\pm 2\sigma$; 99.7% will be within $\pm 3\sigma$. Values of $(x_i - \mu)$ corresponding to $\pm 1\sigma, \pm 2\sigma$, and $\pm 3\sigma$ are indicated by broken vertical lines in curves (a) and (b) of Figure 3–3; the abscissa of curve (c) is already in terms of $\sigma$.

These percentages suggest that the standard deviation, if known, would be a useful predictive tool; knowledge of its value would permit one to assert that the chances are 68.3 in 100 that the indeterminate error associated with any single measurement is smaller than $\pm 1\sigma$. Similarly, the chances are 95.5 in 100 that the error is less than $\pm 2\sigma$, and so forth.

For a very large set of data, the standard deviation is given by

$$\sigma = \sqrt{\frac{\sum_{i=1}^{N}(x_i - \mu)^2}{N}} \qquad (3-5)$$

The individual deviations from the mean $(x_i - \mu)$ are squared, summed, and divided by the number of measurements in the set, $N$. Extraction of the square root of this quotient yields $\sigma$.

The square of the standard deviation, $\sigma^2$, is called the *variance*. The great virtue of the variance as a precision estimate is its additivity. Thus, if $n$ independent sources of indeterminate errors exist within a system (or a calculation), the total variance, $\sigma_T^2$, is simply the sum of the individual variances:

$$\sigma_T^2 = \sigma_1^2 + \sigma_2^2 + \cdots + \sigma_n^2$$

The standard deviation has the same units as the quantity measured, and is favored over the variance by most experimental scientists for this reason.

The direct application of classical statistics to small sets of data (2 to 20 results) tends to produce unrealistic conclusions concerning the probable magnitude of indeterminate error. Fortunately, modifications of the relationships have been developed to permit valid statements about the random error associated with as few as two or three values.

Equations 3–2 and 3–5 are not directly applicable to a small set of replicate measurements because $\mu$, the mean value of an infinitely large number of measurements (and the true value of the measurement, barring determinate error), is not known and cannot be known. In its stead, we are forced to make use of $\bar{x}$, the mean of our small set. More often than not, $\bar{x}$ will differ somewhat from $\mu$. This difference, of course, is the result of the indeterminate error whose magnitude we are attempting to assess. It is important to appreciate that any error in $\bar{x}$ causes a corresponding error in the standard deviation calculated with Equation 3–5. Thus, with a small set of data, not only is the mean $\bar{x}$ likely to differ from $\mu$, but, equally important, the deviations of individual values from this mean (and hence the estimate of the standard deviation) may also be misleading. In short, we have *two* uncertainties to cope with, the one residing in the mean and the other in the standard deviation.

A decrease in the number of individual measurements in a set of data has two effects on the calculated value of the standard deviation. First, there is a decrease in its reproducibility. Second, as a measure of precision, the calculated standard deviation develops a negative *bias;* that is, there is a tendency for the value calculated to be small rather than large as the number of measurements becomes smaller.

The negative bias in a standard deviation calculated from a small set of data is attributable to the fact that both a mean and a standard deviation must be extracted from the same small set. It can be shown that this bias can be largely eliminated by substituting the *degrees of freedom* $(N-1)$ for $N$ in Equation 3–5; that is, we define the standard deviation for a small number of measurements as

$$s = \sqrt{\frac{\sum_{i=1}^{N}(x_i - \bar{x})^2}{N-1}} \qquad (3-6)$$

Note that this equation differs from Equation 3–5 in two regards. First, the denominator is now $(N-1)$. Second, $\bar{x}$, the mean for the small set, replaces the true (and unknown) mean, $\mu$. To emphasize that it is but an approximation of the true value, the resulting standard deviation is given the symbol $s$ instead of $\sigma$.

The rationale for the use of $(N-1)$ in Equation 3–6 is as follows. When $\sigma$ is not known, we must extract two quantities, namely $\bar{x}$ and $s$ from our set of replicate data. The need to establish the mean $\bar{x}$ from the data removes one degree of freedom: that is, if their signs are retained, the sum of the individual deviations must total zero; once $(N-1)$ deviations have been established, the final one is necessarily known as well. Thus, only $(N-1)$ deviations provide independent measures of the precision for the set.

**Methods for Calculation of the Standard Deviation**    A value for $s$ can be obtained by substituting the appropriate quantities into Equation 3–6 or alternatively into its algebraic identity

$$s = \sqrt{\frac{\Sigma x_i^2 - (\Sigma x_i)^2/N}{N-1}} \tag{3-7}$$

Equation 3–7 is particularly useful for computing $s$ with a handheld calculator.

---

**Example 3–3**    Calculate the standard deviation $s$ for a subset consisting of the first five values in Table 3–1.

| $x_i$ | $x_i^2$ |
|---|---|
| 9.990 | 99.800100 |
| 9.993 | 99.860049 |
| 9.973 | 99.460729 |
| 9.980 | 99.600400 |
| 9.982 | 99.640324 |
| $\Sigma x_i = 49.918$ | $\Sigma x_i^2 = 498.361602$ |

$$\frac{(\Sigma x_i)^2}{N} = \frac{(49.918)^2}{5} = 498.361345$$

Substituting these values into Equation 3–7 gives

$$s = \sqrt{\frac{498.361602 - 498.361345}{5-1}} = \sqrt{\frac{0.0002572}{4}}$$

$$= \pm 0.00802 = \pm 0.008$$

Note that the values for $(\Sigma x_i)^2/N$ and $\Sigma x_i^2$ are nearly alike. Clearly, rounding must be postponed until the end in calculations of this kind. After completion of the computation, $s$ is rounded to one significant figure.

**Coefficient of Variation**    The *coefficient of variation* (CV) is a term that is sometimes used to describe the precision of analytical results. It is defined as

$$CV = \frac{\sigma \times 100}{\mu} \cong \frac{s \times 100}{\bar{x}}$$

Note that the coefficient of variation is simply the relative standard deviation expressed as a percentage. Thus, for the measurement in Example 3–3,

$$CV = \frac{0.008 \times 100}{9.984} = 0.082 = 0.08\%$$

## 3E APPLICATIONS OF STATISTICS TO SMALL SETS OF DATA

Experimentalists employ statistical calculations to sharpen their judgments concerning the effects of indeterminate error. We will consider four such applications:[4]

1. Definition of an interval around the experimental mean of a set within which the true mean can be expected to be found with a particular degree of probability.
2. Determination of the number of measurements that need to be made to have a given probability that the true mean will be found within a predetermined interval about the experimental mean.
3. Determination, with a given degree of probability, whether an outlying value in a set of replicate measurements is part of a normal distribution and should be retained (or perhaps rejected) in calculating the mean for the set.
4. The fitting of a straight line to a set of experimental points.

### 3E – 1 Confidence Intervals

The true mean, $\mu$, of a measurement is a constant that must always be unknown. With the aid of statistical theory, however, limits may be set about the experimentally determined mean, $\bar{x}$, within which we may expect to find the true mean with a given degree of probability. These bounds are called *confidence limits;* the interval defined by such limits is known as the *confidence interval.*

---

[4] For other applications of statistical calculations, see the following sources: J. C. Miller, *Statistics for Analytical Chemistry.* New York: John Wiley & Sons, 1985; J. Mandel, in *Treatise on Analytical Chemistry,* 2nd ed., I. M. Kolthoff and P. J. Elving, Eds., Part I, Vol. 1, Chapter 5. New York: John Wiley & Sons, 1978.

It is worthwhile to consider the properties of the confidence interval. For a given set of data, the size of the interval depends in part upon the odds for correctness desired. Clearly, for a prediction to be absolutely correct, we would have to choose an interval about the mean large enough to include all conceivable values that $x_i$ might take. Such an interval, of course, has no predictive value. On the other hand, the interval does not need to be this large if we are willing to accept the probability of being correct 99 times in 100; it can be even smaller if 95% probability is acceptable. In short, as the probability for making a correct prediction is made less favorable, the interval included by the confidence limits becomes smaller.

The confidence interval depends not only upon the standard deviation, $s$, for the measurement but also upon the certainty with which this quantity is known. In some situations, the chemist will have reason to believe that the experimental value for $s$ is an excellent approximation of $\sigma$; in others, $s$ may be known with considerably less certainty. A greater uncertainty in $s$ will require a larger confidence interval for the same probability of correctness.

**Methods for Obtaining a Good Approximation of $\sigma$**    Fluctuations in the calculated value for $s$ decrease as the number of measurements $N$ in Equation 3–6 increases; in fact, for all practical purposes it is safe to assume that $s$ and $\sigma$ are identical for a set that contains 20 or more individual values. It thus becomes feasible for the chemist to obtain a good approximation of $\sigma$ provided the method of measurement is not overly time consuming, and provided further that an adequate amount of sample is available. For example, if a particular investigation requires pH measurements for numerous solutions, it might prove worthwhile to evaluate $s$ in a preliminary experiment. This measurement is simple, requiring only that a pair of rinsed and dried electrodes be immersed in the test solution; the potential across the electrodes serves as a measure of the pH. To evaluate $s$, the pH of 20 to 30 portions of a solution with a fixed pH could be measured, following all steps of the procedure exactly. It would probably be safe to assume that the indeterminate error associated with this test would be the same as that in subsequent measurements and that the value for $s$ calculated by means of Equation 3–6 would be a valid and accurate measure of $\sigma$.

For analyses that are time consuming, the foregoing procedure is not ordinarily practical. Here, however, precision data from a series of samples can often be pooled to provide a value for $s$ that is superior to the value for any individual subset. It is necessary to assume that the sources of indeterminate error are the same for all samples. This assumption is usually valid, provided all the samples have similar composition and have been analyzed identically. To obtain a pooled estimate of $s$, deviations from the mean for each subset are squared; the squares for all of the subsets are then summed and divided by the appropriate number of degrees of freedom. Finally, the pooled $s$ is obtained by extracting the square root of the quotient.

One degree of freedom is lost for each subset. Thus, the number of degrees of freedom for the pooled $s$ is equal to the total number of measurements minus the number of subsets. An example of this calculation follows.

---

**Example 3–4**  The mercury in samples of seven fish taken from Lake Erie was determined by a method based upon the absorption of radiation by gaseous elemental mercury. Calculate a pooled estimate of the standard deviation for the method, based upon the accompanying data:

| Specimen Number | Number of Samples Measured | Hg Content, ppm | Mean, ppm Hg | Sum of Squares of Deviations from Means |
|---|---|---|---|---|
| 1 | 3 | 1.80, 1.58, 1.64 | 1.673 | 0.0259 |
| 2 | 4 | 0.96, 0.98, 1.02, 1.10 | 1.015 | 0.0115 |
| 3 | 2 | 3.13, 3.35 | 3.240 | 0.0242 |
| 4 | 6 | 2.06, 1.93, 2.12, 2.16, 1.89, 1.95 | 2.018 | 0.0611 |
| 5 | 4 | 0.57, 0.58, 0.64, 0.49 | 0.570 | 0.0114 |
| 6 | 5 | 2.35, 2.44, 2.70, 2.48, 2.44 | 2.482 | 0.0685 |
| 7 | 4 | 1.11, 1.15, 1.22, 1.04 | 1.130 | 0.0170 |
| Number of measurements | 28 | | Sum of squares = | 0.2196 |

The values for sample 1 in column 5 were derived as follows:

| $x_i$ | $\lvert(x_i - \bar{x})\rvert$ | $(x_i - \bar{x})^2$ |
|---|---|---|
| 1.80 | 0.127 | 0.0161 |
| 1.58 | 0.093 | 0.0087 |
| 1.64 | 0.033 | 0.0011 |
| 5.02 | Sum of squares = | 0.0259 |

$$\bar{x} = \frac{5.02}{3} = 1.673$$

The other data in column 5 were obtained similarly. Then

$$s_{\text{pooled}} = \sqrt{\frac{0.0259 + 0.0115 + 0.0242 + 0.0611 + 0.0114 + 0.0685 + 0.0170}{28 - 7}}$$

$$= 0.10 \text{ ppm Hg}$$

Since the number of degrees of freedom exceeds 20, this estimate of $s$ can be considered to be a good approximation of $\sigma$.

**Confidence Interval Where $s$ Represents a Good Approximation of $\sigma$**   As indicated earlier (p. 40), the breadth of the normal error curve is determined by $\sigma$. For any given value of $\sigma$, the area under a part of the normal error curve relative to the total area can be related to the parameter $z$ by means of Equation 3–3. This ratio of areas (usually expressed as a percentage) is called the confidence level, and is a measure of the probability for the absolute deviation $(x_i - \mu)$ being equal to or less than $z$. Thus, the area under the curve encompassed by $z = \pm 1.96\sigma$ corresponds to 95% of the total area. Here the confidence level is 95%, and we may state that for a large number of measurements, the chances are 95 in 100 that the calculated $(x_i - \mu)$ will be equal to or less than $\pm 1.96\ \sigma$. Table 3–3 lists confidence levels for various values of $z$.

The confidence limit for a single measurement can be obtained by rearranging Equation 3–3 and remembering that the sign for $z$ can be either plus or minus. Thus,

$$\text{confidence limit} = x_i \pm z\sigma \qquad (3-8)$$

Example 3–5 demonstrates the use of Equation 3–8.

**Example 3–5**   Calculate the 50% and the 95% confidence limits for the first entry (1.80 ppm Hg) in Example 3–4.

Here the pooled $s$ of 0.10 is based on sufficient data to permit the assumption that $s \rightarrow \sigma$. Table 3–3 reveals that $z = \pm 0.67$ and $\pm 1.96$ for the two confidence levels in question. Upon substituting these quantities in Equation 3–8, we find that

50% confidence limit $= 1.80 \pm 0.67 \times 0.10 = 1.80 \pm 0.07$ ppm Hg

95% confidence limit $= 1.80 \pm 1.96 \times 0.10 = 1.8 \pm 0.2$ ppm Hg

The chances are 50 in 100 that $\mu$, the true mean (and, in the absence of determinate error, the true value), is in the interval between 1.73 and 1.87 ppm Hg; the chances are 95 in 100 that $\mu$ is in the interval between 1.6 and 2.0 ppm Hg.

Equation 3–8 applies to the result of a single measurement. It can be shown that the confidence interval is decreased by $\sqrt{N}$ for the average of $N$ replicate measurements. Thus, a more general form of Equation 3–8 is

$$\text{confidence limit for } \mu = \bar{x} \pm \frac{z\sigma}{\sqrt{N}} \qquad (3-9)$$

**Table 3–3   Values of z for Various Confidence Levels**

| Confidence Level, % | z |
|---|---|
| 50 | 0.67 |
| 68 | 1.00 |
| 80 | 1.29 |
| 90 | 1.64 |
| 95 | 1.96 |
| 96 | 2.00 |
| 99 | 2.58 |
| 99.7 | 3.00 |
| 99.9 | 3.29 |

**Example 3–6**   Calculate 50% and 95% confidence limits, this time using the mean value (1.67 ppm HgO) for specimen 1 in Example 3–4. As before, $s \to \sigma = 0.10$; values for $z$ are still 0.67 and 1.96, respectively. Thus,

$$50\% \text{ confidence limit} = 1.67 \pm \frac{0.67 \times 0.10}{\sqrt{3}} = 1.67 \pm 0.04 \text{ ppm Hg}$$

$$95\% \text{ confidence limit} = 1.67 \pm \frac{1.96 \times 0.10}{\sqrt{3}} = 1.67 \pm 0.11 \text{ ppm Hg}$$

For the same odds, the confidence limits are now substantially smaller. There is a 50% probability that the true mean, $\mu$, is included in the interval between 1.63 and 1.71, and a 95% probability that it lies between 1.56 and 1.78.

**Example 3–7**   How many replicate measurements of specimen 1 in Example 3–4 would be needed to decrease the 95% confidence interval to 0.07 ppm Hg?

The pooled value for $s$ is a good estimate for $\sigma$. For a confidence interval of $\pm 0.07$ ppm Hg, then,

$$0.07 = \pm \frac{1.96 \times 0.10}{\sqrt{N}}$$

$$\sqrt{N} = \pm \frac{1.96 \times 0.10}{0.07} = 2.80$$

$$N = 7.8$$

Thus, 8 measurements would provide a slightly better than 95% chance of the true mean lying within $\pm 0.07$ ppm of the experimental mean.

A consideration of Equation 3–9 indicates that the confidence interval for an analysis can be halved by performing four measurements. Sixteen measurements would be required to narrow the limit by yet another factor of two. It is apparent that a point of diminishing return is rapidly reached in acquiring additional data. Consequently, the analyst ordinarily takes advantage of the relatively large gain provided by averaging two to four measurements but can seldom afford the time required for further increases in confidence.

**Confidence Limits Where $\sigma$ is Unknown**     A chemist is frequently required to make use of an unfamiliar method. Moreover, limitations in time or the amount of available sample may make it impossible to obtain a reliable estimate of $\sigma$. Here, a single set of replicate measurements must provide not only a mean value but also a precision estimate. We have noted (p. 41) that $s$ calculated from a limited set of data is likely to be subject to considerable uncertainty. Therefore, the confidence limits clearly must be broader under these circumstances.

To account for the potential variability in $s$, use is made of the statistical parameter $t$, which is defined as

$$t = \frac{\bar{x}_i - \mu}{s}$$

In contrast to $z$ in Equation 3–3, $t$ depends not only on the desired confidence level, but also upon the number of degrees of freedom available in the calculation of $s$. Table 3–4 provides values for $t$ for various degrees of freedom and

**Table 3–4     Values of t for Various Levels of Probability**

| Degrees of Freedom | Factor for Confidence Interval, % | | | | |
|---|---|---|---|---|---|
| | 80 | 90 | 95 | 99 | 99.9 |
| 1 | 3.08 | 6.31 | 12.7 | 63.7 | 637 |
| 2 | 1.89 | 2.92 | 4.30 | 9.92 | 31.6 |
| 3 | 1.64 | 2.35 | 3.18 | 5.84 | 12.9 |
| 4 | 1.53 | 2.13 | 2.78 | 4.60 | 8.60 |
| 5 | 1.48 | 2.02 | 2.57 | 4.03 | 6.86 |
| 6 | 1.44 | 1.94 | 2.45 | 3.71 | 5.96 |
| 7 | 1.42 | 1.90 | 2.36 | 3.50 | 5.40 |
| 8 | 1.40 | 1.86 | 2.31 | 3.36 | 5.04 |
| 9 | 1.38 | 1.83 | 2.26 | 3.25 | 4.78 |
| 10 | 1.37 | 1.81 | 2.23 | 3.17 | 4.59 |
| 11 | 1.36 | 1.80 | 2.20 | 3.11 | 4.44 |
| 12 | 1.36 | 1.78 | 2.18 | 3.06 | 4.32 |
| 13 | 1.35 | 1.77 | 2.16 | 3.01 | 4.22 |
| 14 | 1.34 | 1.76 | 2.14 | 2.98 | 4.14 |
| $\infty$ | 1.29 | 1.64 | 1.96 | 2.58 | 3.29 |

confidence levels; more extensive compilations are to be found in mathematical handbooks. Note that values for $t$ become equal to those for $z$ (Table 3–3) as the number of degrees of freedom becomes infinite.

The confidence limit can be derived from $t$ by a relationship that is analogous to Equation 3–9; that is,

$$\text{confidence limit} = \bar{x} \pm \frac{ts}{\sqrt{N}} \qquad (3-10)$$

---

**Example 3–8**  A chemist obtained the following data for the percentage of alcohol in a sample of blood: 0.084%, 0.089%, and 0.079%. Calculate the 95% confidence limit for the mean, assuming that

(a) no additional knowledge concerning precision of the method is available.

(b) $s \rightarrow \sigma = 0.005$, based on extensive past experience.

(a)
$$\Sigma x_i = 0.084 + 0.089 + 0.079 = 0.252$$

$$\bar{x} = \Sigma x_i / N = 0.252/3 = 0.084$$

$$(\Sigma x_i)^2 / N = (0.252)^2/3 = 0.021168$$

$$\Sigma x_i^2 = (0.084)^2 + (0.089)^2 + (0.079)^2 = 0.0212180$$

Substituting into Equation 3–7 gives

$$s = \sqrt{\frac{0.0212180 - 0.021168}{3-1}} = \sqrt{\frac{0.00005}{2}} = 0.005$$

Table 3–4 indicates that $t = \pm 4.30$ for two degrees of freedom and 95% confidence. Thus,

$$95\% \text{ confidence limit} = 0.084 \pm \frac{4.30 \times 0.0050}{\sqrt{3}}$$

$$= 0.084 \pm 0.012$$

(b) Since a good value for $\sigma$ is available,

$$95\% \text{ confidence limit} = 0.084 \pm \frac{z\sigma}{\sqrt{N}}$$

$$= 0.084 \pm \frac{1.96 \times 0.0050}{\sqrt{3}}$$

$$= 0.084 \pm 0.006$$

Note that sure knowledge of $\sigma$ decreased the confidence interval to half.

## 3E–2 Rejection of Data

When a set of data contains an outlying result that appears to differ excessively from the average, the decision must be made to retain or reject it. The choice of criterion for the rejection of a suspected result has its perils. If we set a stringent criterion that causes the rejection of a questionable result to be difficult, we run the risk of retaining results that are spurious and have an inordinate effect on the average of the data. On the other hand, if we set lenient limits on precision and make easy the rejection of a result, we are likely to discard measurements that rightfully belong in the set, thus introducing a bias into the data. It is an unfortunate fact that no universal rule can be invoked to settle the question of retention or rejection.

Of the numerous statistical criteria suggested to aid in deciding whether to retain or reject a measurement, the $Q$ test[5] is to be preferred.

To apply the $Q$ test, the difference between the questionable result and its nearest neighbor is divided by the spread for the entire set. The resulting ratio, $Q_{exp}$, is then compared with rejection values that are critical for a particular degree of confidence. If $Q_{exp}$ is smaller than $Q_{crit}$, retention of the suspected result is indicated; if $Q_{exp}$ is larger than $Q_{crit}$, a statistical basis for rejection exists. Table 3–5 provides critical values for $Q$ at several confidence levels.

---

**Example 3–9**   The analysis of a calcite sample yielded values of 55.95, 56.08, 56.04, 56.00, and 56.23, respectively. The last value appears to be anomalously high. Should it be retained or rejected at the 90% confidence level?

The difference between 56.23 and its nearest neighbor, 56.08, is 0.15%. The spread for the set $(56.23 - 55.95)$ is 0.28%. Thus,

$$Q_{exp} = \frac{0.15}{0.28} = 0.54$$

For five measurements and 90% confidence, $Q_{crit}$ is 0.64. Since 0.54 is smaller than 0.64, retention is indicated.

---

Despite its superiority over other criteria, the $Q$ test must be used with good judgment as well. For example, there will be situations in which the dispersion associated with a set of measurements will be fortuitously small, and the indiscriminate application of the $Q$ test will cause rejection of a value that should actually be retained; indeed, in a three-member set that contains a pair of identical values, the experimental value for $Q$ will inevitably exceed the critical value. On the other hand, it has been pointed out that the magnitudes of

---

[5] R. B. Dean and W. J. Dixon, *Anal. Chem.,* **1951,** *23,* 636.

**Table 3–5   Critical Values for Rejection Quotient** Q*

| Number of Observations | $Q_{crit}$ (Reject if $Q_{exp} > Q_{crit}$) | | |
|---|---|---|---|
| | 90% Confidence | 96% Confidence | 99% Confidence |
| 3 | 0.94 | 0.98 | 0.99 |
| 4 | 0.76 | 0.85 | 0.93 |
| 5 | 0.64 | 0.73 | 0.82 |
| 6 | 0.56 | 0.64 | 0.74 |
| 7 | 0.51 | 0.59 | 0.68 |
| 8 | 0.47 | 0.54 | 0.63 |
| 9 | 0.44 | 0.51 | 0.60 |
| 10 | 0.41 | 0.48 | 0.57 |

* From W. J. Dixon, *Ann. Math. Stat.*, **1951**, *68*, 22.

rejection quotients for small sets are likely to cause the retention of erroneous data.[6]

The blind application of statistical tests to the decision for retention or rejection of a suspected measurement in a small set of data is not likely to be much more fruitful than an arbitrary decision; indeed, the application of good judgment, based upon an estimate of the precision to be expected, may be a more sound approach—particularly if this estimate is based upon broad experience with the method that is involved. Ultimately, however, the only entirely valid reason for rejecting an experimental result from a small set is the certain knowledge that a mistake has been made in its acquisition. In the absence of such knowledge, the matter of rejection should be approached with great caution.

To summarize, several recommendations can be made for the treatment of a small set of results that contains an outlying result:

1. Re-examine carefully all data relating to the suspected value with the aim of uncovering a gross error. *A properly maintained notebook, containing careful notations of all observations, is essential if this recommendation is to be helpful.*
2. Attempt to estimate the precision that can reasonably be expected from the method; in short, be certain that the outlying result is in fact questionable.
3. Repeat the analysis if sufficient sample and time are available. Agreement of the newly acquired data with those that appear valid will lend weight to the notion that the outlying result should be rejected. Moreover, the questionable result will have a smaller effect on the mean of the enlarged set of data if retention is still indicated.

[6] R. B. Dean and W. J. Dixon, *Anal. Chem.*, **1951**, *23*, 636.

4. If additional data cannot be secured, apply the $Q$ test to establish whether a statistical basis exists for rejection of the doubtful result.
5. If the $Q$ test indicates retention, give consideration to use of the median as the "best" value for the set, rather than the mean. The median has the great virtue of allowing inclusion of all data in a set without undue influence from an outlying value.

### 3E – 3 Generation of Calibration Curves: The Least-Squares Method

Most analytical methods require a calibration step in which standards containing known amounts ($x$) of analyte are treated in the same way as the samples. The experimental quantity measured ($y$) is plotted against $x$ to give a calibration curve such as that shown in Figure 3 – 4. Such plots typically approximate a straight line. Seldom, however, do the data fall exactly on that line, owing to the existence of indeterminate error in the measuring process. The investigator is thus obligated to draw a "best" straight line through the experimental points. Statistical methods exist for the objective generation of such a line and for estimation of the uncertainties associated with its use. Statisticians refer to these techniques as *regression analyses.* We shall limit our consideration to the most straightforward regression procedure, called the *method of least squares,* and to the equations needed to perform a least-squares analysis.[7]

**Assumptions**    Application of the least-squares method to the generation of a calibration curve requires two assumptions. The first of these is that a linear relationship does in fact exist between the amount of analyte ($x$) and the magnitude of the measured variable ($y$); that is,

$$y = a + bx$$

where $a$ is the value for $y$ when $x$ is zero (the *intercept*) and $b$ is the *slope* of the line. A second assumption is that any deviation of individual points from a straight line is entirely the consequence of indeterminate error in the measurement of $y$; that is, no significant error exists in the composition of the standards.

**Derivation of a Least-Squares Line**    The line generated by a least-squares evaluation is the one that minimizes the squares of the individual vertical displacements, or *residuals,* from that line. In addition to providing the best fit between the experimental points and a straight line, the method provides the means of determining the intercept $a$ and the slope $b$ of the line.

For convenience, we shall define three quantities, $S_{xx}$, $S_{yy}$, and $S_{xy}$, as follows:

---

[7] For further information concerning regression analysis and the least-squares method, see the following sources: *Statistical Methods in Research and Production,* 4th ed., O. L. Davies and P. L. Goldsmith, Eds., Chapter 7. New York: Longman Group, Ltd., 1972. W. J. Dixon and F. J. Massey, Jr., *Introduction to Statistical Analysis,* 3rd ed. New York: McGraw-Hill Book Co., 1969.

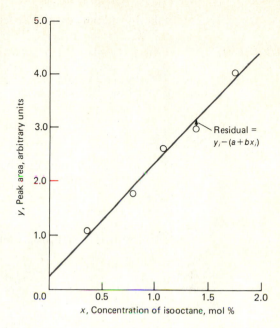

**Figure 3 – 4** Calibration curve for the determination of isooctane in a hydrocarbon mixture.

$$S_{xx} = \Sigma(x_i - \bar{x})^2 = \Sigma x_i^2 - (\Sigma x_i)^2/n \qquad (3-11)^8$$

$$S_{yy} = \Sigma(y_i - \bar{y})^2 = \Sigma y_i^2 - (\Sigma y_i)^2/n \qquad (3-12)^8$$

$$S_{xy} = \Sigma(x_i - \bar{x})(y_i - \bar{y}) = \Sigma x_i y_i - \Sigma x_i \Sigma y_i/n \qquad (3-13)$$

Here, $x_i$ and $y_i$ are individual pairs of values for $x$ and $y$ that are used to define points along the least-squares line. The quantity $n$ is the number of pairs of data used in preparation of the calibration curve, while $\bar{x}$ and $\bar{y}$ are the average values for the variables; that is,

$$\bar{x} = \frac{\Sigma x_i}{n}$$

$$\bar{y} = \frac{\Sigma y_i}{n}$$

Note that $S_{xx}$ and $S_{yy}$ in Equations 3 – 11 and 3 – 12 are simply the sums of the squares of the deviations from the means for individual values for $x$ and $y$. The equivalent expressions shown to the right are more convenient from the standpoint of computation with a handheld calculator.

Calculating $S_{xx}$, $S_{yy}$, and $S_{xy}$ permits evaluation of five useful quantities:

1. The slope of the line $b$:

$$b = S_{xy}/S_{xx} \qquad (3-14)$$

---

[8] The reader must be careful to distinguish between $\Sigma x_i^2$ and $(\Sigma x_i)^2$ as well as $\Sigma y_i^2$ and $(\Sigma y_i)^2$. The first quantity is obtained by squaring each value of $x_i$ or $y_i$ and then summing. The second involves summing $x_i$ and $y_i$; the sums are then squared.

2. The intercept $a$:

$$a = \bar{y} - b\bar{x} \qquad (3-15)$$

3. The standard deviation about the regression $s_r$, which is based upon deviations of the individual points from the line, rather than $\bar{y}$ (see Figure 3–4):

$$s_r = \sqrt{\frac{S_{yy} - b^2 S_{xx}}{n-2}} \qquad (3-16)$$

4. The standard deviation of the slope $s_b$:

$$s_b = \sqrt{\frac{s_r^2}{S_{xx}}} \qquad (3-17)$$

5. The standard deviation, $s_c$, for results based on the calibration curve that is generated:

$$s_c = \frac{s_r}{b} \times \sqrt{\frac{1}{m} + \frac{1}{n} + \frac{(\bar{y}_c - \bar{y})^2}{b^2 S_{xx}}} \qquad (3-18)$$

Equation 3–18 permits calculation of the standard deviation from the mean $\bar{y}_c$ of a set of $m$ replicate measurements from a calibration curve derived from $n$ points; recall that $\bar{y}$ is the mean value of $y$ for the $n$ calibration data.

---

**Example 3–10**  Table 3–6 contains calibration data for a chromatographic determination of isooctane in a hydrocarbon mixture. Here, $x$ is the mole percent isooctane in a series of standards, while $y$ is the measured area under the experimentally obtained chromatographic peaks of the analyte (see Section 18D–2). Carry out a least-squares analysis for these data.

Columns 3, 4, and 5 of Table 3–6 contain computed values for $x_i^2$, $y_i^2$, and $x_i y_i$; their sums appear as the last entry of each column. Note that the number of figures carried in the computed values are the maximum allowed by the calculator; that is, *rounding should not be performed until the end* in order to avoid rounding errors.

We now use Equations 3–11, 3–12, and 3–13 to obtain values for $S_{xx}$, $S_{yy}$, and $S_{xy}$:[9]

$$S_{xx} = \Sigma x_i^2 - (\Sigma x_i)^2/n = 6.90201 - (5.365)^2/5 = 1.145365$$

$$S_{yy} = \Sigma y_i^2 - (\Sigma y_i)^2/n = 36.3775 - (12.51)^2/5 = 5.07748$$

$$S_{xy} = \Sigma x_i y_i - \Sigma x_i \Sigma y_i/n = 15.81992 - 5.365 \times 12.51/5 = 2.39669$$

Substitution of these quantities into Equations 3–14 through 3–17 yields

$$b = \frac{2.39669}{1.145365} = 2.0925 = 2.09$$

$$a = \frac{12.51}{5} - 2.0925 \times \frac{5.365}{5} = 0.2567 = 0.26$$

Thus, the equation for the least-squares line is

$$y = 0.26 + 2.09x$$

The standard deviation about the regression is

$$s_r = \sqrt{\frac{S_{yy} - b^2 S_{xx}}{n-2}} = \sqrt{\frac{5.07748 - (2.0925)^2 \times 1.145365}{5-2}}$$

$$= \pm 0.144 \cong \pm 0.14$$

and the standard deviation for the slope is

$$s_b = \sqrt{\frac{s_r^2}{S_{xx}}} = \sqrt{\frac{(0.144)^2}{1.145365}} = \pm 0.13$$

**Table 3–6  Calibration Data for the Determination of Isooctane in a Hydrocarbon Mixture**

| Mole Percent Isooctane, $x_i$ | Peak Area $y_i$ | $x_i^2$ | $y_i^2$ | $x_i y_i$ |
|---|---|---|---|---|
| 0.352 | 1.09 | 0.12390 | 1.1881 | 0.38368 |
| 0.803 | 1.78 | 0.64481 | 3.1684 | 1.42934 |
| 1.08 | 2.60 | 1.16640 | 6.7600 | 2.80800 |
| 1.38 | 3.03 | 1.90140 | 9.1809 | 4.18140 |
| 1.75 | 4.01 | 3.06250 | 16.0801 | 7.01750 |
| 5.365 | 12.51 | 6.90201 | 36.3775 | 15.81992 |

**Example 3–11**  The relationship derived in Example 3–10 was used to determine the isooctane content of a hydrocarbon mixture. Calculate

(a) the mole percentage (mol %) of isooctane in the mixture if a peak area of 2.65 was obtained.

(b) the standard deviation of the result based on a single measurement ($y$) of 2.65.

(c) the standard deviation of the result where 2.65 represents the mean of four measurements.

(a) Substituting into the equation derived in the earlier example, we find

$$y = 2.65 = 0.26 + 2.09x$$

$$x = (2.65 - 0.26)/2.09 = 1.14 \text{ mol \%}$$

---

[9] See footnote 8, p. 53.

(b) Substitution into Equation 3–18 gives

$$s_c = \frac{0.144}{2.09} \sqrt{\frac{1}{1} + \frac{1}{5} + \frac{(2.65 - 12.51/5)^2}{(2.09)^2 \times 1.145}} = \pm 0.076 = \pm 0.08 \text{ mol } \%$$

(c) For the mean of four measurements,

$$s_c = \frac{0.144}{2.09} \sqrt{\frac{1}{4} + \frac{1}{5} + \frac{(2.65 - 12.51/5)^2}{(2.09)^2 \times 1.145}} = \pm 0.046 = \pm 0.05 \text{ mol } \%$$

## 3F PROPAGATION OF INDETERMINATE ERROR IN COMPUTATIONS

Scientists must frequently estimate the uncertainty in a result that has been computed from two or more data, each of which has an indeterminate error associated with it. The way in which the uncertainty in the result is derived depends upon the kind of arithmetic relationship that exists between the terms containing the errors and the quantity being computed.

**Propagation of Error in Sums and Differences**    Example 3–12 demonstrates how the standard deviation of a sum and a difference is obtained.

**Example 3–12**    Consider the summation

$$\begin{array}{r} +0.50 \ (\pm 0.02) \\ +4.10 \ (\pm 0.03) \\ \underline{-1.97 \ (\pm 0.05)} \\ 2.63 \ (\pm ?) \end{array}$$

where the numbers in parentheses are the absolute standard deviations. The uncertainty associated with the answer could be as large as 0.10 if by chance the signs of the three standard deviations were all positive or all negative. It is also possible that the three uncertainties could combine to yield an accumulated error of zero. Neither of these possibilities is as likely as a combination resulting in an uncertainty that is intermediate between these extremes. It can be shown statistically that the most probable standard deviation for a sum or a difference is given by the square root of the sum of the individual *absolute variances:*

$$s_y = \sqrt{s_a^2 + s_b^2 + s_c^2}$$

where $s_y$ is the standard deviation of the answer and $s_a$, $s_b$, and $s_c$ are the standard deviations in the three terms of the sum. Thus, in the present example,

$$s_y = \sqrt{(\pm 0.02)^2 + (\pm 0.03)^2 + (\pm 0.05)^2}$$
$$= \pm 0.06$$

and the sum could be reported as 2.63 ($\pm 0.06$).

**Propagation of Error in Products and Quotients**   Example 3–13 demonstrates how the standard deviation of a product or quotient can be calculated.

---

**Example 3–13**   Consider the following calculations:

$$\frac{4.10\ (\pm 0.02) \times 0.0050\ (\pm 0.0001)}{1.97\ (\pm 0.04)} = 0.0104\ (\pm ?)$$

Note that the standard deviations of two of the numbers in this calculation are larger than the result itself. Clearly, then, we cannot obtain the desired standard deviation by direct combination of the uncertainties, as in an addition or subtraction. For products or quotients, it is necessary to evaluate the *relative* standard deviation of the result from the relative standard deviations of the numbers involved in the computation. Thus,

$$(s_a)_r = \frac{\pm 0.02}{4.10} = \pm 0.0049$$

$$(s_b)_r = \frac{\pm 0.0001}{0.0050} = \pm 0.020$$

$$(s_c)_r = \frac{\pm 0.04}{1.97} = \pm 0.020$$

Here, the relative standard deviation of the result $(s_y)_r$ is equal to the square root of the sum of the squares of the relative standard deviations for the individual numbers. Thus,

$$(s_y)_r = \frac{s_y}{y} = \sqrt{(s_a)_r^2 + (s_b)_r^2 + (s_c)_r^2}$$
$$= \sqrt{(\pm 0.0049)^2 + (\pm 0.020)^2 + (\pm 0.020)^2}$$
$$= \pm 0.029$$

and to obtain the absolute standard deviation of the result,

$$s_y = y \times (\pm 0.029) = 0.0104 \times (\pm 0.029)$$
$$= \pm 0.0003$$

The uncertainty in the answer can be indicated by the notation $0.0104 \pm 0.0003$.

---

For calculations that involve both sums and differences as well as products and quotients, the uncertainties associated with the former are evaluated first.

**Example 3–14**   Evaluate the uncertainty in

$$\frac{[14.3\ (\pm 0.2) - 11.6\ (\pm 0.2)] \times 0.050\ (\pm 0.001)}{[820\ (\pm 10) + 1030\ (\pm 5)] \times 42.3\ (\pm 0.4)} = 1.725\ (\pm ?) \times 10^{-6}$$

The absolute uncertainty for the difference in the numerator is

$$s_a = \sqrt{(\pm 0.2)^2 + (\pm 0.2)^2} = \pm 0.28$$

and for the sum in the denominator

$$s_b = \sqrt{(\pm 10)^2 + (\pm 5)^2} = \pm 11$$

The equation may now be rewritten as

$$\frac{2.7\ (\pm 0.28) \times 0.050\ (\pm 0.001)}{1850\ (\pm 11) \times 42.3\ (\pm 0.4)} = 1.725 \times 10^{-6}$$

The equation now involves only products and a quotient. Thus,

$$(s_a)_r = \frac{\pm 0.28}{2.7} = \pm 0.104$$

$$(s_b)_r = \frac{\pm 0.001}{0.050} = \pm 0.020$$

$$(s_c)_r = \frac{\pm 11}{1850} = \pm 0.0059$$

$$(s_d)_r = \frac{\pm 0.4}{42.3} = \pm 0.0095$$

Finally,

$$(s_y)_r = \sqrt{(\pm 0.104)^2 + (\pm 0.020)^2 + (\pm 0.0059)^2 + (\pm 0.0095)^2}$$
$$= \pm 0.106$$

The absolute standard deviation of the result is

$$s_y = 1.725 \times 10^{-6} \times (\pm 0.106) = \pm 0.18 \times 10^{-6}$$

and the answer is rounded to

$$1.7\ (\pm 0.2) \times 10^{-6}$$

## 3G THE SIGNIFICANT FIGURE CONVENTION

A report of analytical results should always contain not only what the chemist believes to be the best value for the quantity measured (the mean or the median), but also an estimate of the uncertainty arising from indeterminate errors. The best measure of this uncertainty is the standard deviation because this parameter permits the user to derive such useful quantities as the confidence limit for the data. In the older literature, the precision of results was often reported in terms of average deviation from the mean or sometimes the spread because these quantities were much easier to calculate than the standard deviation. Now, however, the statistically more significant standard deviation is easily derived with handheld calculators, many of which have the standard deviation as a keyboard function.

In addition to providing a precision estimate in a report, it is also common practice to round data so that they contain only digits that are known with certainty plus the first uncertain digit. This practice is known as *the significant figure convention.* As we have mentioned earlier, rounding to the proper number of significant figures should be delayed until the final result has been computed. Rounding errors are avoided in this way.

To illustrate how data are rounded to include significant figures only, consider the four replicate results: 41.60, 41.46, 41.55, and 41.61. The mean and standard deviation for the data are 41.555 and $\pm 0.069$ respectively. The latter figure indicates that the digit in the second decimal place is uncertain and that the mean should be rounded accordingly. The question of reporting 41.55 or 41.56 must be considered, 41.555 being equally spaced between them. A good rule is to round a number ending in 5 to the nearest even number; by so doing, any tendency to round in a set direction is eliminated, there being an equal probability that the nearest even number will be the higher or the lower in any given situation. Accordingly, we would report the result as $41.56 \pm 0.07$. If we had reason to doubt that $\pm 0.07$ was a valid estimate of the precision, we might choose to present the result as $41.6 \pm 0.1$.

The significant figure convention is frequently used instead of a specific estimate of the precision in a result. Thus, by simply reporting 41.6 in this example, we would be implying that the 4 and the 1 are certain digits, but that doubt exists with regard to the 6. The disadvantage of this approach is that the extent of the uncertainty — here, greater than $\pm 0.05$ and smaller than $\pm 0.5$ — is not made clear to the reader.

In the significant figure convention the zero not only functions as a number, but also as an indicator for placement of the decimal in very large and very small numbers. Avogadro's number is a case in point. The first three digits, 6, 0, and 2, are known with certainty; the next number is uncertain, but is probably a 3. Since the digits that follow 6023 are not known, we introduce 19 zeros; these zeros indicate the order of magnitude of the number and have no other meaning. It is clear that a distinction must be made between zeros that have physical significance (that is, zeros that are *significant figures*) and those that are either unknown or are meaningless owing to the inadequacies of measurement.

Zeros bounded by digits only on the right are never significant; zeros bracketed by digits are always significant. Zeros bounded by digits only on the left may or may not be significant. Thus, use of 20.0 mg to report mass of an object carries the inference that its mass is known to three significant figures (expression of this mass as 0.0200 g does not alter the number of these figures, because the zero immediately following the decimal simply serves to indicate the order of magnitude). On the other hand, if we wish to express the volume of a 2-L beaker as 2000 mL, the zeros may or may not be significant. If it has been shown by experiment that the beaker holds 2.0 L, the first zero becomes significant while the other two in 2000 indicate order of magnitude. The use of exponential notation eliminates the possibility for ambiguity; we could express such a volume as $2.0 \times 10^3$ mL.

A certain amount of care is needed in determining the number of significant figures to carry in the result of an arithmetic combination of two or more numbers. For addition or subtraction, the number of significant figures can be established by inspection. Consider, for example,

$$3.4 + 0.02 + 1.31 = 4.7$$

Clearly, the second decimal place cannot be significant because uncertainty is introduced in the first decimal place by the 3.4.

For multiplication or division, it is frequently assumed that the number of significant figures in the result is the same as that of the component quantity that contains the fewest significant figures. Unfortunately, this assumption may not be valid. Consider the two calculations:

$$\frac{24 \times 4.52}{100.0} = 1.08$$

and

$$\frac{24 \times 4.02}{100.0} = 0.965$$

We can determine the way the answers should be rounded by assuming a unit uncertainty in the last digit of each number involved in the calculations. Here, the relative uncertainty of 1 part in 24 is substantially greater than that in any of the other numbers (1 part in 452 and 1 part in 1000 in the first example; note that the location of the decimal is of no concern in making this judgment). Thus, the absolute uncertainty in the two answers will be

$$1.08 \times 1/24 = 0.04$$

and

$$0.965 \times 1/24 = 0.04$$

The answers should therefore be rounded to 1.08 and 0.96, respectively. Note that one is rounded to three significant figures while the other is rounded to two because 4 parts in 108 are not significantly different from 4 parts in 96.

Particular care is needed in rounding logarithms and antilogarithms. Consider, for example, the following computations:

$$\log 1.124 \times 10^{13} = 13.05077 = 13.0508$$

$$\log 1.125 \times 10^{13} = 13.05112 = 13.0511$$

$$\log 1.126 \times 10^{13} = 13.05154 = 13.0515$$

Note that a change of $\pm 1$ in the *fourth* digit of the number whose logarithm is sought results in a change of $\pm 3$ to $\pm 4$ in the *sixth* digit of the corresponding logarithm. It turns out that this apparent gain in the number of significant figures is an artifact. The characteristic of the logarithm (here, 13) serves only to locate the decimal point in the original number; all of the information concerning 1.125, for example, is contained in the mantissa, 0.0511. Thus, the number of figures to the right of the decimal point in the logarithm should correspond to the number of significant figures in the original number. Note that *all zeros in a mantissa are significant, regardless of location.*

General rules in determining the number of significant figures in logarithms and antilogarithms are as follows:[10]

1. When converting numbers to logarithms, use as many decimal places in the mantissa as there are significant figures in the number.
2. In evaluating an antilogarithm, keep as many significant digits as there are decimal places in the mantissa. Thus, the antilogarithm of 20.60 should be reported as $4.0 \times 10^{20}$; the two decimal places in the mantissa limit to two the number of significant digits in the antilogarithm.

## PROBLEMS

**3–1.** Consider the following sets of data:

| *A | B | *C | D | *E | F |
|-----|-------|--------|-------|-------|-------|
| 624 | 12.34 | 0.0723 | 4.810 | 39.01 | 0.252 |
| 613 | 12.54 | 0.0703 | 4.806 | 38.74 | 0.246 |
| 596 | 12.50 | 0.0719 | 4.814 | 38.66 | 0.252 |
| 607 |       | 0.0724 | 4.843 |       | 0.275 |
| 618 |       |        | 4.799 |       | 0.250 |

Calculate
  (a) the mean.
  (b) the median.
  (c) the range.
  (d) the absolute average deviation from the mean.
  (e) the relative average deviation from the mean (in parts per thousand).
  (f) the absolute standard deviation.
  (g) the percent relative standard deviation.

**3–2.** Which of the following pairs of data from Problem 3–1 has
  (i)   the larger absolute deviation?
  (ii)  the larger relative deviation from the mean (in parts per thousand)?
  (iii) the larger standard deviation?
  (iv)  the larger percent relative standard deviation?
  (a) Set A and Set F         (b) Set B and Set E
  (c) Set C and Set D         (d) Set B and Set F
  (e) Set C and Set E         (f) Set D and Set F

---

[10] D. E. Jones, *J. Chem. Educ.,* **1972**, *49*, 753.

**3–3.** Accepted values for the sets of data in Problem 3–1 are as follows:

| | | | |
|---|---|---|---|
| *Set A | 614 | Set D | 4.828 |
| Set B | 12.41 | *Set E | 38.85 |
| *Set C | 0.0716 | Set F | 0.264 |

Calculate
   (a) the absolute error.
   (b) the percent relative error of the mean for each set.

**3–4.** A particular method of analysis yields weights for Co that are high by 0.4 mg. Calculate the percent relative error caused by this uncertainty if the weight of Co in the sample is
   *(a) 800 mg.                          (b) 500 mg.
   *(c) 100 mg.                          (d) 25 mg.

**3–5.** The method described in Problems 3–4 is to be used for the analysis of an ore that assays about 6.1% Co. What minimum sample weight should be taken if the relative error resulting from a 0.4-mg loss is not to exceed
   (a) 0.2%?                             (b) 0.5%?
   (c) 0.8%?                             (d) 1.2%?

**\*3–6.** Analysis of two standard samples yielded the accompanying values for the percentage of nitrogen:

| Set A | | Set B | |
|---|---|---|---|
| **Sought** | **Found** | **Sought** | **Found** |
| | 11.7 | | 0.46 |
| 11.9 | 12.4 | 0.54 | 0.46 |
| | 12.0 | | 0.51 |
| | 11.8 | | 0.44 |

   (a) Compare the precision of the two analyses in terms of absolute and relative standard deviations.
   (b) Compare the errors associated with the means of the two analyses in absolute and in relative terms.

**3–7.** Consider the accompanying data:

| Set A | Set B |
|---|---|
| 67.6 | 4.37 |
| 67.4 | 4.40 |
| 67.1 | 4.33 |
| 67.5 | 4.30 |
| Accepted value: 68.0 | Accepted value: 4.32 |

   (a) Which set has the better absolute precision? Which has the better relative precision?
   (b) Which set has the smaller absolute error? Which has the smaller relative error?

**\*3–8.** Analysis of several plant food preparations for $K^+$ yielded the accompanying data:

| Sample | Mean Percent K$^+$ | Number of Observations | Deviation of Individual Results from Mean Values |
|--------|--------------------|------------------------|-------------------------------------------------|
| 1 | 5.10 | 5 | 0.12, 0.10, 0.08, 0.06, 0.07 |
| 2 | 8.24 | 3 | 0.10, 0.07, 0.12 |
| 3 | 3.69 | 6 | 0.04, 0.08, 0.03, 0.14, 0.07, 0.09 |
| 4 | 4.07 | 4 | 0.11, 0.07, 0.05, 0.10 |
| 5 | 7.13 | 5 | 0.06, 0.07, 0.13, 0.10, 0.09 |

(a) Evaluate the standard deviation, $s$, for each sample.

(b) Obtain a pooled estimate for $s$.

**3–9.** Six bottles of wine were analyzed for residual sugar, with the following results:

| Bottle | Percent (w/v) Residual Sugar | Number of Observations | Deviation of Individual Results from Their Means |
|--------|------------------------------|------------------------|-------------------------------------------------|
| 1 | 0.94 | 3 | 0.04, 0.09, 0.07 |
| 2 | 1.08 | 4 | 0.08, 0.05, 0.05, 0.09 |
| 3 | 1.20 | 5 | 0.04, 0.11, 0.07, 0.00, 0.08 |
| 4 | 0.67 | 4 | 0.05, 0.08, 0.05, 0.10 |
| 5 | 0.83 | 3 | 0.07, 0.09, 0.10 |
| 6 | 0.76 | 4 | 0.06, 0.11, 0.04, 0.07 |

(a) Evaluate the standard deviation, $s$, for each analysis.

(b) Pool the data to establish an absolute standard deviation for the method.

**\*3–10.** Evaluate a pooled estimate for the standard deviation associated with an analysis for trace amounts of Zn, given the accompanying data:

| Sample | Concentration of Zn, ppb |
|--------|--------------------------|
| 1 | 61.7, 70.3, 66.9, 64.2 |
| 2 | 120, 110, 124, 118 |
| 3 | 47.7, 49.3, 54.1, 45.0 |
| 4 | 358, 349, 354, 361 |

**3–11.** A method for determining the particulate content of Pb in air samples is based upon drawing a measured volume of air through a filter and performing the analysis on a circle cut from the filter. Calculate individual values for $s$ as well as a pooled estimate for $s$ from the accompanying data:

| Sample | $\mu$g Pb/m$^3$ Air |
|--------|---------------------|
| 1 | 2.4, 2.1, 1.6, 2.2 |
| 2 | 1.3, 1.9, 1.7 |
| 3 | 2.5, 2.8, 2.0, 2.3 |
| 4 | 1.6, 1.9, 1.2, 1.7, 1.4 |
| 5 | 4.3, 3.7, 3.5, 3.9 |

**\*3–12.** Extensive past experience with a particular analysis for the percentage of Cu has provided a pooled standard deviation of 0.08 for the method. A triplicate analysis has yielded an average of 3.64% Cu. Calculate

(a) the 95% confidence limit.

(b) the 90% confidence limit.

(c) the 80% confidence limit for the analysis.

**3–13.** A pooled standard deviation of a method for the determination of atmospheric $SO_2$ is 0.72 ppm (that is, $s_{pooled} \rightarrow \sigma$). Calculate the 90% confidence limit for an analysis that yielded a value of 5.71 ppm $SO_2$, based on

(a) a single measurement.

(b) the average of three measurements.

(c) the average of five measurements.

**3–14.** How many measurements would be needed to lower the 90% confidence limit in Problem 3–13 to no more than ±0.50 ppm $SO_2$?

**\*3–15.** The method described in Problem 3–13 was modified and found to yield a standard deviation of 0.60 ppm, based upon 5 measurements. Establish the 90% confidence limit for the modified method.

**3–16.** Use the data in Problem 3–1 to evaluate

\*(a) the 80% confidence limit for Set A.

(b) the 90% confidence limit for Set B.

\*(c) the 95% confidence limit for Set C.

(d) the 99% confidence limit for Set D.

\*(e) the 90% confidence limit for Set E.

(f) the 90% confidence limit for Set F.

**3–17.** Apply the $Q$ test to the data in Problem 3–1 to establish whether statistical grounds exist for rejection of the outlying result in

\*(a) Set A at the 90% level.

(b) Set B at the 96% level.

\*(c) Set C at the 99% level.

(d) Set D at the 90% level.

\*(e) Set E at the 99% level.

(f) Set F at the 96% level.

**3–18.** An equation for the linear relation between $x_i$ and $y_i$ is to be evaluated by the least-squares method.

(a) Supply the missing numbers in the accompanying tabulation:

| $x_i$ | $y_i$ | $x_i^2$ | $y_i^2$ | $x_i y_i$ |
|-------|-------|---------|---------|-----------|
| 0.80 | 3.1 | 0.6400 | | |
| 1.65 | 9.8 | | 96.04 | |
| 2.70 | 15.4 | | | 41.580 |
| 3.80 | 19.5 | 14.440 | | |
| 4.75 | 24.5 | | 600.25 | |
| 5.35 | 31.0 | | | 165.850 |

$S_{xx} = 76.2775 - (19.05)^2/6 =$

$S_{yy} =$

$S_{xy} =$

(b) Evaluate the slope, $b$, and the intercept, $a$, for the line.

(c) Write the equation for the line.

(d) Calculate the standard deviation for the slope, $s_b$, and for the residuals, $s_r$, of this line.

**\*3–19.** The linear relationship between absorbance and concentration of $MnO_4^-$ can be used to determine the Mn content of steel samples. A calibration curve was generated by performing this analysis upon samples that contained known amounts of Mn.

| Amount of Mn Taken, mg | Instrument Response (arbitrary units) |
|---|---|
| 1.0 | 0.060 |
| 2.0 | 0.140 |
| 3.3 | 0.217 |
| 5.3 | 0.331 |
| 6.8 | 0.430 |
| 8.6 | 0.542 |

(a) Use these data to derive the equation for the calibration line.
(b) Calculate the standard deviation for the slope, $s_b$, and about regression, $s_r$, of this line.

**3–20.** Use the curve generated in Problem 3–19 and the accompanying data to calculate

| | Wt Sample Taken, g | Instrument Response (arbitrary units) |
|---|---|---|
| \*(i) | 0.962 | 0.438 |
| (ii) | 1.264 | 0.519 |
| \*(iii) | 0.897 | 0.602 |
| (iv) | 1.084 | 0.469 |

(a) the percentage of Mn in a series of steel samples.
(b) the standard deviation of the result, assuming that the recorded instrument response was a single measurement and that the uncertainty in the result is determined by the instrument response only.
(c) the standard deviation of the result, assuming that the recorded instrument response was the mean of 3 measurements and that the limiting uncertainty was in the instrument response.

**3–21.** The accompanying data were obtained in the calibration of a spectrophotometer for the determination of $K^+$ in mineral waters.

| mg $K^+$/100 mL | Instrument Response (arbitrary units) |
|---|---|
| 1.10 | 14.2 |
| 1.95 | 28.0 |
| 2.90 | 37.3 |
| 3.85 | 48.4 |
| 4.60 | 62.7 |
| 4.95 | 69.4 |

It has been established that a linear relationship exists between the $K^+$ concentration and the response of the instrument.

(a) Derive an equation for the best straight line through the experimental points.

(b) Calculate the standard deviation for the slope and residuals of this line.

**3-22.** Calculate the concentration of $K^+$ (in mg/100 mL) as well as the standard deviation in this quantity, based upon the accompanying data and the calibration data in Problem 3-21.

|       | Number of Measurements | Average Instrument Response (arbitrary units) |
|-------|:----------------------:|:---------------------------------------------:|
| *(a)  | 3                      | 46.1                                          |
| (b)   | 5                      | 36.4                                          |
| *(c)  | 4                      | 50.7                                          |
| (d)   | 4                      | 23.8                                          |

**3-23.** How many significant figures are there in

*(a) 41.94?    (b) 4.019?

*(c) 7357.02?    (d) $3.00 \times 10^{-6}$?

*(e) 21.22?    (f) 0.0075?

*(g) 1.0075?    (h) $8.8 \times 10^4$?

*(i) 6126.64?    (j) 94305?

**3-24.** Estimate the absolute and relative standard deviations in the results of the accompanying calculations (the numbers in parentheses are standard deviations associated with the individual data). Round the result to the proper number of significant figures.

*(a)  $41.49 \,(\pm 0.03) + 12.37 \,(\pm 0.04) = 53.86$

(b)  $8.575 \,(\pm 0.003) - 8.4128 \,(\pm 0.0007) = 0.1662$

*(c)  $19.47 \,(\pm 0.02) + 21.66 \,(\pm 0.04) - 4.67 \,(\pm 0.03) = 36.46$

(d)  $1.493 \,(\pm 0.004) + 12.41 \,(\pm 0.01) - 8.947 \,(\pm 0.002) = 4.956$

*(e)  $313.7 \,(\pm 0.1) + 97.0 \,(\pm 0.1) - 14.36 \,(\pm 0.08) = 396.34$

(f)  $2.04 \,(\pm 0.07) \times 10^{-2} - 8.61 \,(\pm 0.03) \times 10^{-3} = 1.179 \times 10^{-2}$

**3-25.** Estimate the absolute and relative standard deviations in the results of the accompanying calculations (the numbers in parentheses are standard deviations associated with the individual data). Round the result to the appropriate number of significant figures.

*(a)  $64.4 \,(\pm 0.2) \times 0.381 \,(\pm 0.007) = 24.5364$

(b)  $18.18 \,(\pm 0.03) \times 4.764 \,(\pm 0.009) = 86.60952$

*(c)  $26.94 \,(\pm 0.08) \div 0.0496 \,(\pm 0.0004) = 543.14516 \ldots$

(d)  $0.9194 \,(\pm 0.0008) \div 46.18 \,(\pm 0.03) = 0.01990905 \ldots$

**3-26.** Estimate the absolute and relative standard deviations in the results of the accompanying calculations (the numbers in parentheses are standard deviations associated with the individual measurements). Round each result to the appropriate number of significant figures.

*(a)  $\dfrac{29.67 \,(\pm 0.03) - 8.51 \,(\pm 0.01)}{6.36 \,(\pm 0.02) + 4.83 \,(\pm 0.02)} = 1.89097 \ldots$

(b)  $\dfrac{7.614 \,(\pm 0.008) - 6.923 \,(\pm 0.005)}{14.2468 \,(\pm 0.0002) - 13.6719 \,(\pm 0.0001)} = 1.2019 \ldots$

*(c) $\dfrac{3.44\,(\pm 0.01)\times 10^{-5}}{0.100\,(\pm 0.004)+0.250\,(\pm 0.001)} = 9.82857\ldots\times 10^{-5}$

(d) $\dfrac{[44.41\,(\pm 0.02)-3.12\,(\pm 0.01)]\times 0.2048\,(\pm 0.0006)}{12.6349\,(\pm 0.0001)-12.2775\,(\pm 0.0001)} = 23.6603\ldots$

*(e) $\dfrac{8.40\,(\pm 0.01)\times 1.697\,(\pm 0.002)}{186.1\,(\pm 0.3)\times 0.5843\,(\pm 0.0007)} = 0.13109\ldots$

(f) $\dfrac{765\,(\pm 1)\times 3.564\,(\pm 0.004)}{192.5\,(\pm 0.2)} = 14.1634\ldots$

**3-27.** Round the following results to the proper number of significant figures:
*(a) $\log(3.06\times 10^{-3}) = -2.51428\ldots$
(b) $\log 2.000 = 0.301030\ldots$
*(c) $\log(6.02\times 10^{23}) = 23.77960\ldots$
(d) $1/\log(6.02\times 10^{23}) = 0.0420529\ldots$
*(e) antilog $4.41 = 2.5704\ldots\times 10^{4}$
(f) antilog $12.3 = 1.9953\ldots\times 10^{12}$
*(g) antilog $(-14.62) = 2.39883\ldots\times 10^{-15}$
(h) antilog $(-1.7798) = 0.01660351\ldots$

# Chapter 4
# Gravimetric Methods of Analysis

A gravimetric analysis is based upon the measurement of the weight of a substance that has a known composition and is chemically related to the analyte.[1] In *precipitation methods,* the species to be determined is precipitated by a reagent that yields a sparingly soluble product that either has a known composition or can be converted to such a substance. This product is weighed following filtration, washing, and suitable heat treatment. In a *volatilization method,* the analyte or its decomposition products are volatilized at an appropriate temperature. The product is then weighed or, alternatively, the weight of the residue is determined. Precipitation methods are far more commonly encountered than volatilization procedures.

## 4A CALCULATION OF RESULTS FROM GRAVIMETRIC DATA

Two experimental measurements form the basis for a gravimetric analysis, specifically, the weight of sample taken, and the weight of solid produced from this amount of sample. The results of the analysis are ordinarily expressed in terms of the percentage of analyte, A:

$$\% \, A = \frac{\text{weight of A}}{\text{weight of sample}} \times 100 \qquad (4-1)$$

The denominator of Equation 4–1 is established at the outset by performing the analysis upon carefully weighed quantities of the sample; the numerator is then determined after suitable treatment of the sample. Occasionally, the product is A itself and its weight is measured directly. More commonly, the product that is actually isolated and weighed either contains A or is chemically related to A. In either event, a calculation identical to that described in Section 2C is used (1) to convert the weight of product from units of metric mass to chemical mass, (2) to account for the stoichiometry between the product and A, and finally (3) to convert the mass of A from chemical to metric units. The collection of

---

[1] For a thorough treatment of gravimetric methods, see: C. L. Rulfs, in *Treatise on Analytical Chemistry,* I. M. Kolthoff and P. J. Elving, Eds., Part I, Vol. 11, Chapter 113. New York: John Wiley & Sons, 1975.

constants associated with these transformations is called the *gravimetric factor;* the generation of this factor is conveniently demonstrated with examples.

---

**Example 4–1** What weight of Cl is contained in 0.204 g of AgCl?

$$0.204 \text{ g} \times \underbrace{\frac{\text{gfw AgCl}}{143.3 \text{ g}}}_{\substack{\text{metric mass} \\ \text{to chemical mass}}} \times \underbrace{\frac{\text{gfw Cl}}{\text{gfw AgCl}}}_{\text{(stoichiometry)}} \times \underbrace{\frac{35.45 \text{ g}}{\text{gfw Cl}}}_{\substack{\text{(chemical mass} \\ \text{(to metric mass)}}} = 0.0505 \text{ g Cl}$$

---

**Example 4–2** To what weight of $AlCl_3$ would 0.204 g of AgCl correspond?

Proceeding as in Example 4–1, we write

$$0.204 \text{ g} \times \frac{\text{gfw AgCl}}{143.3 \text{ g}} \times \frac{\text{gfw AlCl}_3}{3 \text{ gfw AgCl}} \times \frac{133.3 \text{ g}}{\text{gfw AlCl}_3} = 0.0633 \text{ g AlCl}_3$$

---

Note how these two calculations resemble one another. In both, the weight of one substance is converted into the corresponding weight of another substance through multiplication by a set of constant terms. The gravimetric factor (G.F.) is simply the collection of these terms; that is,

$$\text{G.F.} = \frac{\text{gfw Cl}}{\text{gfw AgCl}} \qquad \text{(Example 4–1)}$$

and

$$\text{G.F.} = \frac{\text{gfw AlCl}_3}{3 \times \text{gfw AgCl}} \qquad \text{(Example 4–2)}$$

Note further that the number of chlorides is the same in the numerator and denominator of each expression.

---

**Example 4–3** What weight of $Fe_2O_3$ can be obtained from 1.63 g of $Fe_3O_4$? What is the gravimetric factor for this conversion?

Here, it is necessary to assume that ample oxygen is available to accomplish this transformation; that is,

$$2Fe_3O_4 + [O] = 3Fe_2O_3$$

$$1.63 \text{ g} \times \frac{\text{gfw Fe}_3O_4}{231.5 \text{ g}} \times \frac{3 \text{ gfw Fe}_2O_3}{2 \text{ gfw Fe}_3O_4} \times \frac{159.7 \text{ g}}{\text{gfw Fe}_2O_3} = 1.69 \text{ g Fe}_2O_3$$

As before, collection of the constant terms in this transformation yields the gravimetric factor:

$$\text{G.F.} = \frac{3 \times \text{gfw Fe}_2\text{O}_3}{2 \times \text{gfw Fe}_3\text{O}_4} = \frac{3 \times 159.7}{2 \times 231.5} = 1.035$$

These examples suggest that the gravimetric factor takes the form

$$\text{G.F.} = \frac{a}{b} \times \frac{\text{gfw substance sought}}{\text{gfw substance weighed}}$$

where $a$ and $b$ are small integers that take whatever values are needed for *chemical equivalence* between the substances in the numerator and denominator; this condition is frequently achieved by balancing the number of atoms of an element (other than oxygen) that is common to both terms.

Equation 4–1 can now be expressed in the more useful form

$$\% \text{ A} = \frac{\text{wt ppt} \times \dfrac{a \times \text{gfw A}}{b \times \text{gfw ppt}}}{\text{wt of sample}} \times 100$$

Additional examples of gravimetric factors are provided in Table 4–1.

The situation will occasionally arise in which no element (other than oxygen) is common to both numerator and denominator of the gravimetric factor. Consider, for example, an indirect analysis for the iron in a sample of iron(III) sulfate, $\text{Fe}_2(\text{SO}_4)_3$, that involves the precipitation and weighing of barium sulfate, $\text{BaSO}_4$. Here, we must seek further for the means of establishing the chemical equivalence between the substance sought in the analysis (Fe) and the substance that has been weighed ($\text{BaSO}_4$). We note that

$$2 \text{ gfw Fe} \equiv 1 \text{ gfw Fe}_2(\text{SO}_4)_3 \equiv 3 \text{ gfw SO}_4^{2-} \equiv 3 \text{ gfw BaSO}_4$$

**Table 4–1  Typical Gravimetric Factors**

| Species Sought | Species Weighed | Gravimetric Factor |
|---|---|---|
| In | $\text{In}_2\text{O}_3$ | $\dfrac{2 \times \text{gfw In}}{\text{gfw In}_2\text{O}_3}$ |
| HgO | $\text{Hg}_5(\text{IO}_6)_2$ | $\dfrac{5 \times \text{gfw HgO}}{\text{gfw Hg}_5(\text{IO}_6)_2}$ |
| I | $\text{Hg}_5(\text{IO}_6)_2$ | $\dfrac{2 \times \text{gfw I}}{\text{gfw Hg}_5(\text{IO}_6)_2}$ |
| $\text{K}_3\text{PO}_4$ | $\text{K}_2\text{PtCl}_6$ | $\dfrac{2 \times \text{gfw K}_3\text{PO}_4}{3 \times \text{gfw K}_2\text{PtCl}_6}$ |
| $\text{K}_3\text{PO}_4$ | $\text{Mg}_2\text{P}_2\text{O}_7$ | $\dfrac{2 \times \text{gfw K}_3\text{PO}_4}{\text{gfw Mg}_2\text{P}_2\text{O}_7}$ |

The gravimetric factor for calculating the percentage of iron will thus be

$$\text{G.F.} = \frac{2 \times \text{gfw Fe}}{3 \times \text{gfw BaSO}_4}$$

The examples that follow will illustrate the use of the gravimetric factor in the calculation of results of an analysis.

---

**Example 4–4**   A 0.703-g sample of a commercial detergent was ignited at red heat to destroy its organic components. Treatment of the residue with hot HCl brought the phosphorus into solution as $H_3PO_4$. The phosphate was then precipitated as $MgNH_4PO_4 \cdot 6H_2O$ by the addition of $Mg^{2+}$ followed by aqueous $NH_3$. The precipitate was filtered, washed, and subsequently converted to $Mg_2P_2O_7$ by ignition at 1000°C. This residue was found to weigh 0.432 g. Calculate the percentage of phosphorus in the sample.

$$\% \text{ P} = \frac{\text{wt Mg}_2\text{P}_2\text{O}_7 \times \dfrac{2 \times \text{gfw P}}{\text{gfw Mg}_2\text{P}_2\text{O}_7}}{\text{wt sample}} \times 100$$

$$= \frac{0.432 \times \dfrac{2 \times 30.97}{222.6}}{0.703} \times 100 = 17.1$$

---

**Example 4–5**   At elevated temperatures, $Na_2C_2O_4$ is converted to $Na_2CO_3$ with the evolution of CO

$$\text{Na}_2\text{C}_2\text{O}_4 \rightarrow \text{Na}_2\text{CO}_3 + \text{CO(g)}$$

Ignition of a 1.1906-g sample of impure $Na_2C_2O_4$ yielded a residue that weighed 0.9859 g. Calculate the percentage purity of the sample.

Here, it must be assumed that the difference between the initial and final weights represents the weight of CO evolved during the ignition; it is this weight loss that forms the basis for the analysis. Since each formula weight of $Na_2C_2O_4$ yields one formula weight of CO,

$$\% \text{ Na}_2\text{C}_2\text{O}_4 = \frac{\text{wt CO} \times \dfrac{\text{gfw Na}_2\text{C}_2\text{O}_4}{\text{gfw CO}}}{\text{wt sample}} \times 100$$

$$= \frac{(1.1906 - 0.9859) \times \dfrac{134.0}{28.01}}{1.1906} \times 100 = 82.25$$

**Example 4–6**  A 0.2510-g sample of an alloy that contained only Mg and Zn was dissolved in acid. Treatment with $(NH_4)_3PO_4$ and $NH_3$, followed by filtration and ignition, resulted in the formation of $Mg_2P_2O_7$ (gfw = 222.57) and $Zn_2P_2O_7$ (gfw = 304.68) with a combined weight of 0.9513 g. Calculate the percent composition of the sample.

The problem contains two unknowns; two independent relationships are therefore needed. Since the sample contains only the two components, one such relationship is

$$0.2510 = \text{wt Mg} + \text{wt Zn}$$

therefore,

$$\text{wt Zn} = 0.2510 - \text{wt Mg}$$

Another is

$$0.9513 = \text{wt } Mg_2P_2O_7 + \text{wt } Zn_2P_2O_7$$

$$= \text{wt Mg} \times \frac{\text{gfw } Mg_2P_2O_7}{2 \times \text{gfw Mg}} + \text{wt Zn} \times \frac{\text{gfw } Zn_2P_2O_7}{2 \times \text{gfw Zn}}$$

Substitution for wt Zn in this expression gives

$$0.9513 = \text{wt Mg} \times \frac{222.57}{2 \times 24.312} + (0.2510 - \text{wt Mg}) \times \frac{304.68}{2 \times 65.37}$$

$$= 4.5774 \text{ wt Mg} + 0.5849 - 2.3304 \text{ wt Mg}$$

Rearrangement gives

$$\text{wt Mg} = \frac{0.9513 - 0.5849}{4.5774 - 2.3304} = \frac{0.3664}{2.2470} = 0.1631 \text{ g}$$

Thus,

$$\text{wt Zn} = 0.2510 - 0.1631 = 0.0879 \text{ g}$$

Finally, then,

$$\% \text{ Mg} = \frac{0.1631}{0.2510} \times 100 = 64.98$$

$$\% \text{ Zn} = \frac{0.0879}{0.2510} \times 100 = 35.02$$

**Example 4–7**  A 1.878-g sample containing $NH_4Cl$, $NaIO_3$, and inert materials was dissolved in sufficient water to give exactly 250.0 mL of solution. Treatment of a 50.00-mL aliquot of the diluted solution with an excess of $AgNO_3$ yielded a mixture of AgCl and $AgIO_3$ that weighed 0.6020 g; treatment of a second 50.00-mL aliquot with an excess of $Ba(NO_3)_2$ resulted in the formation of 0.1974 g of $Ba(IO_3)_2$. Calculate the percentages of $NH_4Cl$ and $NaIO_3$ in the sample.

This problem resembles Example 4–6 in that there are two unknowns, and therefore, two independent relationships are required. Here, however, the sample contains inert materials in addition to the two analytes; the second precipitating agent is therefore needed.

We can write that

$$\text{wt AgIO}_3 = \text{wt Ba(IO}_3)_2 \times \frac{2 \times \text{gfw AgIO}_3}{\text{gfw Ba(IO}_3)_2}$$

$$= 0.1974 \times \frac{2 \times 282.8}{487.2} = 0.2292 \text{ g}$$

We also know that

$$\text{wt AgCl} + \text{wt AgIO}_3 = 0.6020 \text{ g}$$

or

$$\text{wt AgCl} = 0.6020 - 0.2292 = 0.3728 \text{ g}$$

Each analysis is performed on 50/250 of the sample; thus,

$$\% \text{ NH}_4\text{Cl} = \frac{\text{wt AgCl} \times \dfrac{\text{gfw NH}_4\text{Cl}}{\text{gfw AgCl}}}{\dfrac{50.00}{250.0} \times \text{wt sample}} \times 100$$

$$= \frac{0.3728 \times \dfrac{53.49}{143.3}}{0.3756} \times 100 = 37.05$$

Similarly, then,

$$\% \text{ NaIO}_3 = \frac{\text{wt Ba(IO}_3)_2 \times \dfrac{2 \times \text{gfw NaIO}_3}{\text{gfw Ba(IO}_3)_2}}{0.3756} \times 100$$

$$= \frac{0.1974 \times \dfrac{2 \times 197.9}{487.2}}{0.3756} \times 100 = 42.70$$

## 4B PROPERTIES OF PRECIPITATES

The ideal precipitating reagent for a gravimetric analysis would react with one and only one analyte to produce a solid that (1) possesses a sufficiently low solubility that losses from that source are negligible, (2) is readily filtered and washed free of contaminants, (3) is unreactive with constituents of the atmosphere, and (4) is of known composition after drying or, if necessary, ignition. Few precipitates or reagents meet this ideal. Reagents tend to yield sparingly soluble products with more than one analyte; many precipitates are difficult to manipulate and purify. The chemist is frequently obliged to perform analyses with products and reactions that leave much to be desired.

### 4B – 1 Filterability and Purity of Precipitates[2]

Particle size determines the ease with which a precipitate is filtered and purified. The relationship between particle size and ease of filtration is straightforward; coarse solids are readily retained upon porous media and are thus rapidly filtered. Finely divided precipitates require dense filters, which lead to low filtration rates. The relation between particle size and purity is more complex. In general, however, decreases in soluble contaminants tend to accompany increases in particle size.

Precipitates are frequently found to be contaminated by soluble compounds that would not be expected to precipitate under the conditions extant in the solution. This carrying down of *normally soluble* species is called *coprecipitation.* It should be emphasized that the solution is *under saturated* with respect to the substances that are coprecipitated and that the *simultaneous precipitation of a second substance whose solubility has been exceeded does not constitute coprecipitation.*

**Factors Affecting the Particle Size of Precipitates**   Particle size depends not only on the chemical composition of a precipitate, but also upon the conditions that exist at the time of its formation. The range of particle sizes encountered is enormous. At one extreme are *colloidal suspensions,* the individual particles of which are so small as to be invisible to the naked eye (diameters of $10^{-6}$ to $10^{-4}$ mm). These particles show no tendency to settle out of solution nor are they retained by common filtering media. At the other extreme are particles with dimensions on the order of several tenths of a millimeter. The temporary dispersion of such particles in the liquid phase is called a *crystalline suspension.* The particles of a crystalline suspension settle out rapidly and spontaneously to give precipitates that are readily filtered.

No sharp discontinuities in physical properties exist as the dimensions of the particles in the solid phase increase from colloidal to those typical of crystals. Indeed, some precipitates possess characteristics intermediate between these defined extremes. Nevertheless, most precipitates are easily recognizable as being predominately colloidal or predominately crystalline in character; the classification, while imperfect, can be successfully applied to most solid phases.

---

[2] For a detailed treatment of the properties of precipitates, see: A. E. Nielsen, in *Treatise on Analytical Chemistry,* 2nd ed., I. M. Kolthoff and P. J. Elving, Eds., Part I, Vol. 3, Chapter 27. New York: John Wiley & Sons, 1983.

Although the phenomenon of precipitation has long attracted the interest of chemists, fundamental information concerning the mechanism of the process remains incompletely understood. It is certain, however, that the particle size of the solid that forms is influenced in part by such experimental variables as the temperature, the solubility of the precipitate in the medium in which it is formed, reactant concentrations, and the rate at which the reactants are brought together. The influence of these variables on the particle size of precipitates can be accounted for by a single property of the system called its *relative supersaturation,*[3]

$$\text{relative supersaturation} = \frac{Q - S}{S} \qquad (4-2)$$

where $Q$ is the concentration of the solute at any instant, and $S$ is its equilibrium solubility.

Each addition of precipitating reagent to the solution containing the analyte presumably causes a condition of momentary supersaturation (that is, $Q > S$). This unstable condition is ordinarily and quickly relieved by precipitate formation. Experimental evidence suggests that the particle size of the resulting solid varies inversely with the average degree of relative supersaturation that exists after each addition of reagent. Thus, precipitates tend to form as colloids when $(Q - S)/S$ is large, and as crystals under conditions where $(Q - S)/S$ is small.

**Mechanism of Precipitate Formation**   The influence of relative supersaturation on particle size can be rationalized by postulating that the precipitation process involves two mechanisms, *nucleation* and *particle growth*. The particle size of a freshly formed precipitate is governed by the extent to which one of these processes predominates over the other.

During nucleus formation, some minimum number of ions or molecules unite to form a stable second phase. The process may take place at the surface of a solid contaminant in the solution *(heterogeneous nucleation)* or else may involve the fortuitous arrangement of the component species in an orientation that permits bonding forces to form the solid *(homogeneous nucleation)*. Further precipitation can occur either by the generation of additional nuclei or by deposition of solid on the nuclei that have already been produced (particle growth). If nucleation predominates, the precipitate will consist of an enormous number of very small particles; if growth predominates, a smaller number of larger particles will be produced.

The rate of homogeneous nucleation is believed to increase exponentially

---

[3] P. P. von Weimarn first proposed the concept of relative supersaturation and described its effect on particle size. An account of von Weimarn's work is to be found in *Chem. Rev.,* **1925,** *2,* 217. The von Weimarn viewpoint adequately suggests the general conditions for obtaining satisfactory particle size during precipitation. Other theories are superior in accounting for the details of the process. See, for example, A. E. Nielsen, *The Kinetics of Precipitation.* New York: The Macmillan Company, 1964.

with increases in the relative supersaturation, while the rate of particle growth is approximately linear. Thus, when the supersaturation is high, the nucleation rate far exceeds that for particle growth and is the predominant precipitation mechanism. On the other hand, the condition of low supersaturation may favor particle growth; here, deposition of solid on particles that already exist can occur to the exclusion of further nucleation.

**Experimental Control of Particle Size**    Experimental variables that minimize supersaturation and thus favor the production of crystalline precipitates include elevated temperatures (to increase $S$), the use of dilute solutions (to minimize $Q$), and the slow addition of precipitating reagent with good stirring (also to minimize the average value of $Q$).

An additional variable that may influence relative supersaturation and thus particle size is pH. For example, large and easily filtered crystals of calcium oxalate can be obtained by forming the bulk of the solid in a somewhat acidic environment in which the salt is moderately soluble. The precipitation is then completed by the slow addition of aqueous ammonia until the acidity is decreased to the level needed for the quantitative formation of the calcium oxalate; the additional precipitate generated during this process forms on the solid.

A crystalline precipitate is much easier to manipulate than a solid that forms as a colloid. Particle growth is thus preferable to additional nucleation during the formation of a precipitate. Nevertheless, if the solubility of a precipitate is very small, it becomes essentially impossible to avoid momentary large relative supersaturations as the solutions are brought together; in this case, $Q$ is inevitably enormous with respect to $S$, and a colloidal suspension is the unavoidable result. Many hydrous oxides [iron(III), chromium(III) and aluminum, among others] and the sulfides of most heavy metal ions form only as colloids owing to their very low solubilities.[4]

## 4B–2 Colloidal Precipitates

Individual colloidal particles are so small that they are not retained by ordinary filtering media; moreover, Brownian motion prevents their settling out of solution under the influence of gravity. Fortunately, however, the individual particles of most colloids can be caused to coagulate (or agglomerate) to give a filterable, noncrystalline mass that will settle out of solution.

**Coagulation of Colloids**    The coagulation process can be hastened through the use of heating, stirring, and the addition of an electrolyte to the medium. To understand the effectiveness of these measures, we need to look into the causes for the stability of a colloidal suspension.

---

[4] Silver chloride illustrates that the relative supersaturation concept is imperfect. It ordinarily forms as a colloid, yet the formal solubility of AgCl is not significantly different from other compounds, such as $BaSO_4$, which generally form as crystals.

The individual particles of a typical colloid bear either a positive or a negative charge as a result of *adsorption,* a phenomenon by which cations or anions are bound to the surface of the particles. The existence of charge is readily demonstrated by observing the migration of colloidal particles under the influence of an electric field.

The adsorption of ions upon an ionic solid has, as its origin, the normal bonding forces that are responsible for crystal growth. For example, a silver ion at the surface of a silver chloride particle has a partially unsatisfied bonding capacity owing to its surface location. Negative ions are attracted to this site by the same forces that hold chloride ions in the silver chloride lattice. Chloride ions at the surface exert an analogous attraction for cations dissolved in the solvent.

The nature and magnitude of the charge on the particles of a colloidal suspension depend, in a complex way, on several variables. For a suspension produced in the course of a gravimetric analysis, however, the species adsorbed, and hence the charge on the particles, can be readily predicted by the empirical observation that lattice ions are generally more strongly held than others. For example, when silver nitrate is first added to a solution containing chloride ion, the colloidal particles of the precipitate are negatively charged due to adsorption of chloride ions. This charge, however, becomes positive when enough silver nitrate has been added to provide an excess of silver ions. Not surprisingly, the surface charge is at a minimum when the supernatant liquid contains equivalent amounts of the two ions.

The extent of adsorption, and thus the charge on a given particle, increases rapidly with the first additions of a common ion to a solution containing a colloid. As the surface of the particles becomes covered with the adsorbed ion, however, the charge becomes constant and independent of concentration.

Figure 4–1 depicts schematically a colloidal silver chloride particle in a solution that contains an excess of silver nitrate. Attached directly to the solid surface is the *primary adsorption layer,* consisting mainly of adsorbed silver ions. Surrounding the charged particle is a *volume of solution, called the counter-ion layer,* which contains a large enough excess of negative ions (principally nitrate) to just balance the charge on the surface of the particle. The primarily adsorbed silver ions and the negative counter-ion layer constitute an *electrical double layer* that imparts stability to the colloidal suspension. As colloidal particles approach one another, this double layer exerts an electrostatic repulsive force that prevents particles from colliding and adhering.

The influence of the double layer on the stability of a colloid is illustrated in the precipitation of silver chloride by the slow addition of silver nitrate to a solution of sodium chloride. At the outset, the newly formed silver chloride particles carry a high negative charge because of the large concentration of chloride ions in their environment. As a consequence of their high charge, the volume of the counter-ion layers surrounding each particle must also be large in order to contain sufficient sodium and hydrogen ions to match the large number of chloride ions adsorbed on the particles. Under these circumstances, close

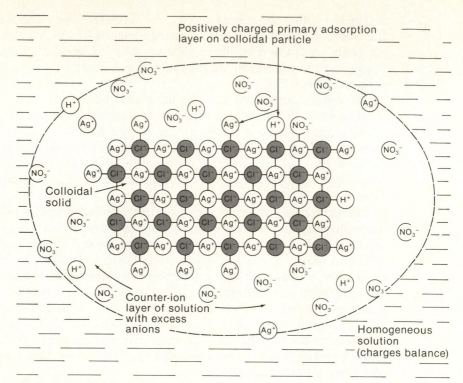

**Figure 4–1**  A colloidal silver chloride particle suspended in a solution of silver nitrate.

approach of particles is prevented, and the suspension is stable. With further additions of silver nitrate, more particles are formed and the concentration of chloride ion is lowered. The consequence is a smaller charge per particle and a concomitant decrease of volume of the counter-ion layer. As chemical equivalence is approached, shrinkage of the double layer is sufficient to permit particles to approach one another closely enough for agglomeration to take place. A coagulated mass of silver chloride particles makes its appearance at this point.

It is of interest that the coagulation process can be reversed by addition of an excess of either chloride ion or of silver ion. The resulting particles will, of course, carry different charges depending on which of these reagents is used.

Coagulation of a colloidal suspension can often be brought about by a short period of heating, particularly if accompanied by stirring. Heating brings about coagulation by reducing the number of adsorbed ions and thus the volume of the double layer. The particles may also gain sufficient kinetic energy at the higher temperature to overcome the barrier to close approach posed by the double layer.

An even more effective way to coagulate a colloid is to increase the electrolyte concentration of the solution. Upon the addition of a suitable ionic compound, the volume of solution that contains sufficient counter ions to balance the charge of the primary adsorption layer decreases; the net effect of adding an electrolyte is thus a shrinkage of the counter-ion layer, with the result that the surface charge on the particles is more completely neutralized. The particles can then approach one another more closely.

**Coprecipitation in Coagulated Colloids**    A coagulated colloid consists of irregularly arranged particles that form a loosely packed, porous mass. Within this mass, large internal surface areas remain in contact with the solvent phase (see Figure 4–2). These surfaces will retain most of the primarily adsorbed ions that were on the uncoagulated particles. Even though the counter-ion layer surrounding the original colloidal particle is part of the solution, sufficient counter ions to impart electrical neutrality must accompany the particle (in the film of liquid surrounding the particle) through the processes of coagulation and filtration. *The net effect of surface adsorption is, therefore, the carrying down of an otherwise soluble compound as a surface contaminant.*

**Peptization of Colloids**    *Peptization* refers to the process by which a coagulated colloid reverts in whole or in part to its original dispersed state. When a coagulated colloid is washed, some of the electrolyte responsible for its coagulation is leached from the internal liquid that contacts the solid particles. Removal of this electrolyte has the effect of increasing the volume of the counter-ion layer. The repulsive forces responsible for the original colloidal state are reestablished as a consequence, and particles detach themselves from the coagulated mass. The washings become cloudy as the freshly dispersed particles pass through the filter.

 The chemist is thus faced with a dilemma in working with coagulated colloids. On the one hand, washing is needed to minimize contamination; on the other, there is the risk of losses resulting from peptization if pure water is used. The problem is commonly resolved by washing the precipitate with a solution containing an electrolyte that will volatilize during the subsequent drying or ignition step. For example, silver chloride is ordinarily washed with a dilute solution of nitric acid. While the precipitate undoubtedly becomes contami-

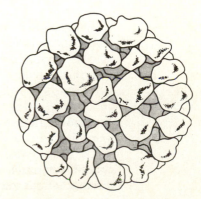

*Figure 4–2*   Coagulated colloidal particles.

nated with the acid, no harm results, since the nitric acid is volatilized during the ensuing drying step.

**Practical Treatment of Colloidal Precipitates**   Colloids are best precipitated from hot, stirred solutions to which sufficient electrolyte has been added to ensure coagulation. The filterability of a coagulated colloid frequently improves after it is allowed to stand for an hour or more in contact with the hot solution from which it was formed. During this process, which is known as *digestion,* weakly bound water appears to be lost from the precipitate; the result is a denser mass that is easier to filter.

As previously noted, a dilute solution of a volatile electrolyte is used to wash the filtered precipitate. Washing does not appreciably affect primarily adsorbed ions because the attraction between these species and the solid is too strong. Some exchange, however, may occur between the existing counter ions and one of the ions of the wash solution. Under any circumstances, it must be expected that the precipitate will still be contaminated to some degree, even after extensive washing. The error introduced into the analysis from this source can range from 1 to 2 ppt (as in the coprecipitation of silver nitrate upon silver chloride) to an intolerable level [as in the coprecipitation of heavy metal hydroxides upon the hydrous oxides of iron(III) or aluminum].

A drastic, but very effective, way to improve the purity of a coagulated colloid is *reprecipitation,* in which the filtered solid is redissolved and again precipitated. The original precipitate ordinarily carries down only a fraction of the total contaminant that exists in the sample. Thus, the solution containing the redissolved precipitate will have a significantly smaller contaminant concentration than the original; as a result, a smaller amount of contaminant will be carried down by the second precipitate. Reprecipitation adds substantially to the time required for an analysis; nevertheless, it is almost a necessity for such precipitates as the hydrous oxides of iron(III) and aluminum, which possess extraordinary tendencies to adsorb the hydroxides of such cations as zinc, cadmium, and manganese.

### 4B-3 Crystalline Precipitates

Crystalline precipitates tend to be more easily manipulated than coagulated colloids. The size of the individual particles is determined to a degree by experimental conditions; the physical properties and the purity of the solid are thus determined by experimental variables over which the chemist has some control.

**Methods of Improving Particle Size and Purity**   The particle size of a crystalline solid is frequently improved by keeping the relative supersaturation at a low level while the precipitate is being formed. Equation 4-2 suggests that this condition can be realized by minimizing $Q$, maximizing $S$, or both. The use of dilute solutions and the slow addition of precipitating reagent with efficient stirring will minimize momentary local supersaturation within the solution.

Moreover, *S* can usually be increased by generating the precipitate at elevated temperatures. Significant improvements in particle size can be obtained with these simple measures.

**Purity of Crystalline Precipitates**   In contrast to colloidal solids, the surface area exposed by a crystalline precipitate is small; contamination as a result of surface adsorption is thus negligibly small. However, other forms of coprecipitation, which involve entrainment of contaminants within the interior of the crystals, may cause serious errors.

Two types of coprecipitation affect crystalline precipitates, namely *inclusion* and *occlusion*. In both, the normally soluble foreign matter is entrapped within the crystal as the precipitate forms. Included impurities distribute themselves randomly throughout the crystalline structure without causing significant lattice distortion. In one type of inclusion, ions of the contaminant replace ions of like charge in the lattice structure. In a second type of inclusion, the impurity behaves as if it were in solid solution in the precipitate.

Occlusion differs from inclusion in that the impurities are not randomly distributed but instead occur in pockets within the crystal. Occlusion takes place when entire droplets of solution are trapped and surrounded by the rapidly growing crystal.

Washing has no effect on either inclusion or occlusion because the entrapped contaminant cannot be reached by the wash liquid. Digestion decreases occlusion but is often of little help for included impurities.

**Digestion of Crystalline Precipitates**   The heating of a crystalline precipitate in contact with the solution from which it was formed will frequently improve the purity as well as the particle size of the product. The improvement in purity undoubtedly results from the solution and recrystallization of the solid that occur continuously and at an enhanced rate at elevated temperatures. Contaminants that were occluded become exposed and are able to return to the solution; a more perfect solid is thus obtained.

Solution and recrystallization are probably responsible for the improvement in filterability of many crystalline precipitates after digestion. Bridging between adjacent particles yields larger crystalline aggregates that are more easily filtered. The observation that little or no improvement in filterability is obtained if stirring accompanies digestion tends to confirm this view.

## 4B-4 Direction of Coprecipitation Errors

Coprecipitated impurities can cause the results of an analysis to be high as well as low. Positive errors will always result from contamination by a species that does not involve the analyte. For example, adsorption of silver nitrate on colloidal silver chloride will cause the results of a chloride analysis to be high. On the other hand, the results of an analysis may be either high or low if the impurity contains the ion that is being determined. Thus, the determination of barium by precipitation as barium sulfate is subject to error owing to the occlusion of barium salts. If the occluded contaminant is barium nitrate, the error will be positive, since this compound has a larger formula weight than the barium sulfate that would have been produced had coprecipitation not occurred. By the same reasoning, occlusion of barium chloride would result in a negative error, its formula weight being smaller than that for barium sulfate.

### 4B–5 Precipitation from Homogeneous Solution

Precipitation from homogeneous solution involves chemical generation of the precipitating reagent. Local reagent excesses do not occur because the precipitating agent is produced slowly and uniformly throughout the solution containing the analyte; the relative supersaturation is thus kept low. In general, precipitates formed in this manner are better suited for gravimetric analysis than solids formed by direct addition of a reagent.

Urea is often utilized as a source of hydroxide ion. The reaction can be expressed by the equation

$$(H_2N)_2CO + 3H_2O \rightarrow CO_2(g) + 2NH_4^+ + 2OH^-$$

The reaction proceeds slowly at temperatures just below boiling; typically, one to two hours are needed for production of sufficient hydroxide ion to complete a precipitation. This method is particularly valuable for the precipitation of hydrous oxides or basic salts. For example, when formed by the direct addition of base, the hydrous oxides of iron(III) and aluminum are bulky, gelatinous masses that are heavily contaminated and are difficult to filter. In contrast, when produced by the homogeneous generation of hydroxide ion, the same products have significantly greater density and are thus more readily filtered; in addition, their purity is improved. The homogeneous formation of crystalline precipitates also results in marked increases in crystal size as well as improvements in purity.

Representative methods based upon the homogeneous generation of precipitating reagents are listed in Table 4–2.

### 4B–6 Drying and Ignition of Precipitates

Some sort of heat treatment is needed to free a filtered precipitate from solvent as well as any volatile electrolytes that have been coprecipitated with the solid. In addition, some precipitates must be thermally decomposed to obtain a product that has a known composition.

The temperature needed to produce a solid of known composition varies from precipitate to precipitate. Figure 4–3 shows the effect of heat treatment on several common analytical precipitates. These data were recorded by an automatic thermobalance,[5] an instrument that continuously measures the weight of

---

[5] For a review of thermobalances, see: W. W. Wendlandt, *Thermal Methods of Analysis,* 2nd ed. New York: John Wiley & Sons, 1974.

Table 4 – 2   Methods for the Homogeneous Generation of Precipitants

| Precipitant | Reagent | Generation Reaction | Elements Precipitated |
|---|---|---|---|
| $OH^-$ | Urea | $(NH_2)_2CO + 3H_2O = CO_2 + 2NH_4^+ + 2OH^-$ | Al, Ga, Th, Bi, Fe, Sn |
| $PO_4^{3-}$ | Trimethyl phosphate | $(CH_3O)_3PO + 3H_2O = 3CH_3OH + H_3PO_4$ | Zr, Hf |
| $C_2O_4^{2-}$ | Ethyl oxalate | $(C_2H_5)_2C_2O_4 + 2H_2O = 2C_2H_5OH + H_2C_2O_4$ | Mg, Zn, Ca |
| $SO_4^{2-}$ | Dimethyl sulfate | $(CH_3O)_2SO_2 + 4H_2O = 2CH_3OH + SO_4^{2-} + 2H_3O^+$ | Ba, Ca, Sr, Pb |
| $CO_3^{2-}$ | Trichloroacetic acid | $Cl_3CCOOH + 2OH^- = CHCl_3 + CO_3^{2-} + H_2O$ | La, Ba, Ra |
| $S^{2-}$ | Thioacetamide | $CH_3\overset{\|\|}{\underset{S}{C}}NH_2 + H_2O = CH_3\overset{\|\|}{\underset{O}{C}}NH_2 + H_2S$ | Sb, Mo, Cu, Cd |
| 8-Hydroxyquinoline | 8-Acetoxyquinoline | | Al, U, Mg, Zn |

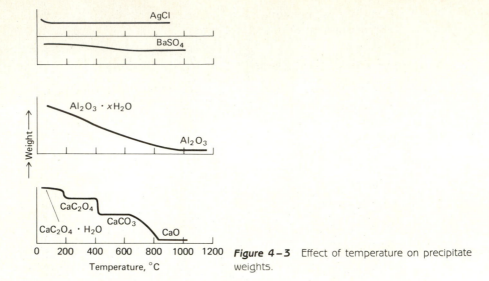

**Figure 4-3**   Effect of temperature on precipitate weights.

a substance as its temperature is raised. The heating of three precipitates—silver chloride, barium sulfate, and aluminum oxide—simply eliminates water and perhaps volatile electrolytes that were carried down during the precipitation process. To be sure, the temperature needed to bring about this removal differs enormously among these precipitates. Thus, moisture is completely removed from silver chloride at temperatures between 110 and 120°C; in contrast, dehydration of aluminum oxide requires temperatures in excess of 1000°C. It is of interest to note that aluminum oxide, formed homogeneously with urea, is completely dehydrated at about 650°C.

The thermal curve for calcium oxalate is considerably more complex than the others shown in Figure 4-3. At temperatures less than 135°C, unbound water is eliminated yielding the monohydrate $CaC_2O_4 \cdot H_2O$. Removal of the hydrate water requires a temperature of about 225°C. The abrupt decrease in weight at about 450°C results from the decomposition of calcium oxalate to calcium carbonate and carbon monoxide. The final step in the curve, which occurs between 700 and 800°C, involves conversion of the carbonate to calcium oxide and carbon dioxide. Clearly, ignition conditions will dictate the nature of the solid that is ultimately weighed in a gravimetric analysis that is based upon the isolation of calcium as the oxalate.

## 4C APPLICATIONS OF GRAVIMETRIC ANALYSIS

Gravimetric methods have been developed for most, if not all, inorganic anions and cations as well as for such neutral species as water, sulfur dioxide, carbon dioxide, and iodine. Numerous organic compounds can also be determined gravimetrically, including lactose in milk products, salicylates in drug preparations, phenolphthalein in laxatives, nicotine in pesticides, cholesterol in cereals, and benzaldehyde in almond extracts.

## 4C–1 Inorganic Precipitating Reagents

Table 4–3 lists common inorganic precipitating reagents; these typically cause the formation of a slightly soluble salt or hydrous oxide. The lack of specificity of most inorganic reagents is demonstrated by the numerous entries in the second column of that table.

Detailed instructions for the gravimetric determination of two inorganic species are to be found in Chapter 20, Section 20A.

## 4C–2 Reducing Reagents

Table 4–4 lists several reagents that convert the analyte to its elemental form for weighing.

**Table 4–3    Some Inorganic Precipitating Agents†**

| Precipitating Agent | Element Precipitated‡ |
|---|---|
| $NH_3(aq)$ | **Be** (BeO), **Al** ($Al_2O_3$), **Sc** ($Sc_2O_3$), Cr ($Cr_2O_3$),* **Fe** ($Fe_2O_3$), Ga ($Ga_2O_3$), **Zr** ($ZrO_2$), **In** ($In_2O_3$), Sn ($SnO_2$), U ($U_3O_8$) |
| $H_2S$ | Cu (CuO),* **Zn** (ZnO, or $ZnSO_4$), **Ge** ($GeO_2$), As ($\underline{As_2O_3}$, or $As_2O_5$), Mo ($MoO_3$), Sn ($SnO_2$),* Sb ($\underline{Sb_2O_3}$, or $Sb_2O_5$), $\underline{Bi}$ ($Bi_2S_3$) |
| $(NH_4)_2S$ | Hg ($\underline{HgS}$), Co ($Co_3O_4$) |
| $(NH_4)_2HPO_4$ | **Mg** ($Mg_2P_2O_7$), Al ($AlPO_4$), Mn ($Mn_2P_2O_7$), Zn ($Zn_2P_2O_7$), Zr ($Zr_2P_2O_7$), Cd ($Cd_2P_2O_7$), Bi ($BiPO_4$) |
| $H_2SO_4$ | Li, Mn, **Sr, Cd, Pb, Ba** (all as sulfates) |
| $H_2PtCl_6$ | K ($K_2PtCl_6$, or Pt), Rb ($Rb_2PtCl_6$), Cs ($\underline{Cs_2PtCl_6}$) |
| $H_2C_2O_4$ | Ca (CaO), Sr (SrO), **Th** ($\overline{ThO_2}$) |
| $(NH_4)_2MoO_4$ | Cd ($CdMoO_4$),* Pb ($\underline{PbMoO_4}$) |
| HCl | **Ag** (AgCl), Hg ($\underline{Hg_2Cl_2}$), Na (as NaCl from butyl alcohol), Si ($SiO_2$) |
| $AgNO_3$ | **Cl** (AgCl), Br ($\underline{AgBr}$), I($\underline{AgI}$) |
| $(NH_4)_2CO_3$ | **Bi** ($Bi_2O_3$) |
| $NH_4SCN$ | Cu ($Cu_2(SCN)_2$) |
| $NaHCO_3$ | Ru, Os, Ir (precipitated as hydrous oxides; reduced with $H_2$ to metallic state) |
| $HNO_3$ | Sn ($SnO_2$) |
| $H_5IO_6$ | Hg ($Hg_5(IO_6)_2$) |
| NaCl, $Pb(NO_3)_2$ | F (PbClF) |
| $BaCl_2$ | $SO_4^{2-}$ ($BaSO_4$) |
| $MgCl_2$, $NH_4Cl$ | $PO_4^{3-}$ ($Mg_2P_2O_7$) |

† From W. F. Hillebrand, G. E. F. Lundell, H. A. Bright, and J. I. Hoffman, *Applied Inorganic Analysis*. New York: John Wiley & Sons, 1953.

‡ Boldface type indicates that gravimetric analysis is the preferred method for the element or ion. The weighed form is indicated in parentheses. An asterisk indicates that the gravimetric method is seldom used. An underscored entry indicates the most reliable gravimetric method.

Table 4 – 4  Some Reducing Reagents Employed in Gravimetric Methods

| Reducing Agent | Analyte |
|---|---|
| $SO_2$ | Se, Au |
| $SO_2 + H_2NOH$ | Te |
| $H_2NOH$ | Se |
| $H_2C_2O_4$ | Au |
| $H_2$ | Re, Ir |
| HCOOH | Pt |
| $NaNO_2$ | Au |
| $TiCl_2$ | Rh |
| $SnCl_2$ | Hg |
| Electrolytic reduction | Co, Ni, Cu, Zn, Ag, In, Sn, Sb, Cd, Re, Bi |

## 4C – 3 Organic Precipitating Reagents

A number of organic reagents have been developed for the gravimetric determination of inorganic species. Some of these yield products in which bonding between the reagent and the analyte is primarily ionic. Others form sparingly soluble, nonionic complexes, called *coordination compounds*. Organic reagents tend to be more selective than the inorganic reagents listed in Table 4 – 3; as a result, they are particularly useful for gravimetric analysis. We shall limit our discussion to three typical examples of organic reagents.[6]

**Sodium Tetraphenylboron**  Sodium tetraphenylboron, $(C_6H_5)_4B^-Na^+$, is an important example of organic precipitating reagents that form saltlike precipitates. In cold mineral acid solutions, it is a near-specific precipitating reagent for potassium and ammonium ions. The products are stoichiometric, corresponding to the salts of the respective cations; they can be isolated by vacuum filtration and can be brought to constant weight at 105 to 110°C. Mercury(II), rubidium, and cesium are the only common sources of interference that must be removed by prior treatment.

**Dimethylglyoxime**  Dimethylglyoxime is an organic precipitating reagent of unparalleled specificity. It has the formula

$$CH_3-\underset{\underset{OH}{\overset{\|}{N}}}{C}-\underset{\underset{OH}{\overset{\|}{N}}}{C}-CH_3$$

---

[6] For further information on organic reagents, see: K. L. Cheng, K. Ueno, and T. Imamura, *Organic Analytical Reagents*. Boca Raton, Florida: CRC Press, Inc., 1982.

Its coordination compound with palladium is the only one that is sparingly soluble in acid solution. Similarly, only the nickel compound precipitates from a weakly alkaline environment. Nickel dimethylglyoxime is bright red and has the structure

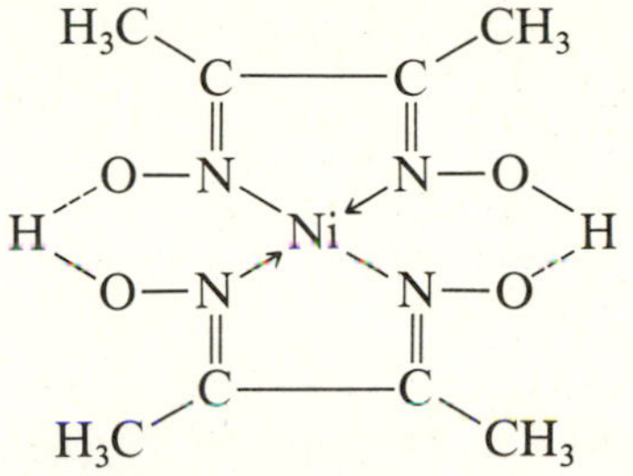

The precipitate is bulky and has an exasperating tendency to creep as it is filtered and washed; these properties place an upper limit on the amount of nickel that can be conveniently isolated with the reagent. Nickel dimethylglyoxime is readily dried at 110°C; the dried product has the composition shown by its formula.

**8-Hydroxyquinoline**    Approximately two dozen cations form sparingly soluble coordination compounds with 8-hydroxyquinoline, which is also known as *oxine:*

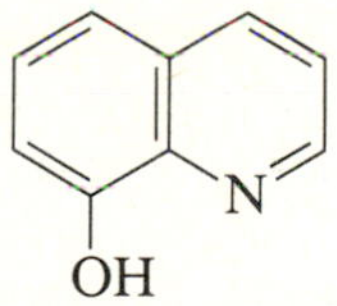

Typical of these is the product with magnesium:

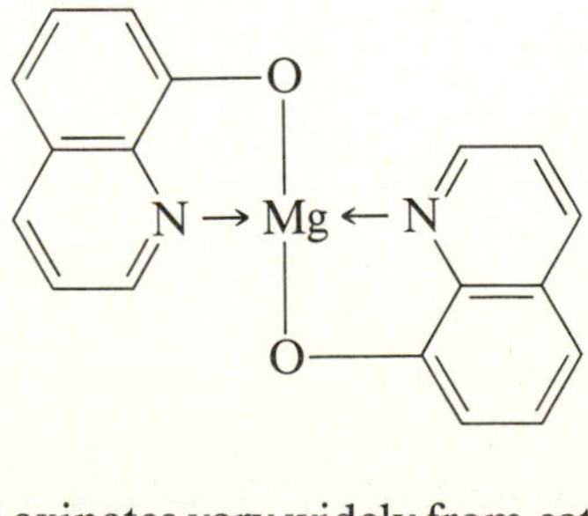

The solubilities of metal oxinates vary widely from cation to cation and, moreover, are pH dependent because proton formation always accompanies the chelation reaction. Therefore, a considerable degree of selectivity can be imparted to the action of 8-hydroxyquinoline through pH control.

Table 4 – 5  *Gravimetric Methods for Organic Functional Groups*

| Functional Group | Basis for Method | Reaction and Product Weighed* |
|---|---|---|
| Carbonyl | Weight of precipitate with 2,4-dinitrophenylhydrazine | $RCHO + H_2NNHC_6H_3(NO_2)_2 \rightarrow$ $R\!-\!CH\!=\!NN\underline{HC_6H_3(NO_2)_2(s)} + H_2O$ (RCOR' reacts similarly) |
| Aromatic carbonyl | Weight of $CO_2$ formed at 230°C in quinoline; $CO_2$ distilled, absorbed, and weighed | $ArCHO \xrightarrow[CuCO_3]{230°C} Ar + \underline{CO_2}(g)$ |
| Methoxyl and ethoxyl | Weight of AgI formed after distillation and decomposition of $CH_3I$ or $C_2H_5I$ | $\left.\begin{array}{l} ROCH_3 \quad + HI \rightarrow ROH \quad\ + CH_3I \\ RCOOCH_3 + HI \rightarrow RCOOH + CH_3I \\ ROC_2H_5 \ + HI \rightarrow ROH \quad\ + C_2H_5I \end{array}\right\} CH_3I + Ag^+ + H_2O \rightarrow$ $\underline{AgI}(s) + CH_3OH$ |
| Aromatic nitro | Weight loss of Sn | $RNO_2 + \tfrac{3}{2}\underline{Sn}(s) + 6H^+ \rightarrow RNH_2 + \tfrac{3}{2}Sn^{4+} + 2H_2O$ |
| Azo | Weight loss of Cu | $RN = NR' + 2\underline{Cu}(s) + 4H^+ \rightarrow RNH_2 + R'NH_2 + 2Cu^{2+}$ |
| Phosphate | Weight of Ba salt | $RO\overset{\overset{\textstyle O}{\|}}{P}(OH)_2 + Ba^{2+} \rightarrow \underline{ROPO_2Ba}(s) + 2H^+$ |
| Sulfamic acid | Weight of $BaSO_4$ after oxidation with $HNO_2$ | $RNHSO_3H + HNO_2 + Ba^{2+} \rightarrow ROH + \underline{BaSO_4}(s) + N_2 + 2H^+$ |
| Sulfinic acid | Weight of $Fe_2O_3$ after ignition of $Fe^{3+}$ sulfinate | $3ROSOH + Fe^{3+} \rightarrow (ROSO)_3Fe(s) + 3H^+$ $(ROSO)_3Fe \xrightarrow{O_2} CO_2 + H_2O + SO_2 + \underline{Fe_2O_3}(s)$ |

* The substance weighed is underlined.

## 4C–4 Gravimetric Organic Functional Group Analysis

Several reagents have been shown to react selectively with certain organic functional groups and thus can be used for the determination of most compounds containing these groups. A list of gravimetric functional group reagents is given in Table 4–5. Many of the reactions shown can also be used for volumetric and spectrophotometric determinations. For the occasional analysis, the gravimetric procedure will often be the method of choice since no calibration or standardization is required.

## PROBLEMS

**4–1.** Use chemical symbols to generate gravimetric factors for the following analytes:

| Analyte | Weighed | | Analyte | Weighed |
|---|---|---|---|---|
| *(a) $ZnCl_2$ | $AgCl$ | | (f) $Fe_6S_{17}$ | $CdSO_4$ |
| (b) $ZnCl_2$ | $Zn_2P_2O_7$ | | *(g) $Fe_6S_{17}$ | $Fe_2O_3$ |
| *(c) $(NH_4)_2SO_4$ | $BaSO_4$ | | (h) $Cu_2HgI_4$ | $AgI$ |
| (d) $(NH_4)_2SO_4$ | $(C_6H_5)_4BNH_4$ | | *(i) $Cu_2HgI_4$ | $HgO$ |
| *(e) $Mn_3O_4$ | $MnO_2$ | | (j) $Cu_2HgI_4$ | $Cu$ |

**4–2.** Use chemical symbols to generate the gravimetric factor for an analysis in which the results are to be expressed in terms of percent $(NH_4)_3Fe(C_2O_4)_3 \cdot 3H_2O$ and the species weighed is
*(a) $Fe_2O_3$.    *(c) $(C_6H_5)_4BNH_4$.    *(e) $CO_2$.
(b) $H_2O$.    (d) $CaC_2O_4$.    (f) $NH_3$.

**4–3.** Use chemical symbols to generate the gravimetric factor for an analysis in which the results are to be reported in terms of percent $MgSnCl_6 \cdot 6H_2O$, and the species weighed is
*(a) $MgO$.    *(c) $AgCl$.    *(e) $Mg_2P_2O_7$.
(b) $SnO_2$.    (d) $PbClF$.    (f) $H_2O$.

**4–4.** Which of the weighing forms in Problem 4–2 will yield the smallest weight of precipitate from a given weight of $(NH_4)_3Fe(C_2O_4)_3 \cdot 3H_2O$ (gfw = 428.1)?

**4–5.** Which of the weighing forms in Problem 4–3 will yield the largest weight of precipitate from a given weight of $MgSnCl_6 \cdot 6H_2O$ (gfw = 463.8)?

**4–6.** Use chemical formulas to express the gravimetric factor needed to calculate the percentage of $CaC_2O_4 \cdot 2H_2O$ based on
*(a) the weight for the anhydrous salt after removal of hydrate water at 250°C (see Figure 4–3).
*(b) the loss of weight that results from heating $CaC_2O_4 \cdot 2H_2O$ at 250°C.
(c) the weight of $CaCO_3$ produced as the result of heating at 600°C.
(d) the loss in weight that results when $CaC_2O_4 \cdot 2H_2O$ is heated at 600°C.
*(e) the weight of $CaO$ produced as the result of igniting $CaC_2O_4 \cdot 2H_2O$ at 1000°C.
(f) the loss in weight that results from ignition of $CaC_2O_4 \cdot 2H_2O$ at 1000°C.

**4–7.** What weight of $Ag_2Cr_2O_7$ can be produced from 2.05 g of
(a) $AgNO_3$?    (b) $K_2CrO_4$?    (c) $K_2Cr_2O_7$?

**4-8.** What weight of $MgNH_4PO_4$ can be produced from a 1.76-g sample that is 49.5%
*(a) $MgSO_4$?               *(c) $(NH_4)_2SO_4$?               *(e) $Cu(NH_3)_4SO_4$?
 (b) $H_4P_2O_7$?             (d) $H_3PO_4$?                    (f) $P_2O_5$?

*4-9. How many grams of $H_3PO_4$ are needed to produce 6.55 g of $Mg_2P_2O_7$?

4-10. What weight of $KIO_3$ is needed to produce 3.12 g of $Ba(IO_3)_2$?

*4-11. Calculate the weight of AgCl produced when 0.525 g of AgI is heated in a stream of $Cl_2$. Reaction:

$$2AgI(s) + Cl_2(g) \rightarrow 2AgCl(s) + I_2(g)$$

4-12. Calculate the weight of MgO produced when 7.44 g of $MgC_2O_4$ are heated at 500°C. Reaction:

$$MgC_2O_4(s) \rightarrow MgO(s) + CO(g) + CO_2(g)$$

4-13. Calculate the minimum weight of urea (gfw = 60.1) needed to precipitate the Mg contained in a 0.472-g sample that is 89.5% $MgCl_2$. Reaction:

$$Mg^{2+} + (H_2N)_2CO + 3H_2O \rightarrow Mg(OH)_2(s) + CO_2(g) + 2NH_4^+$$

4-14. Treatment of a 0.5024-g sample with an excess of $BaCl_2$ yielded 0.2986 g of $BaSO_4$. Express the results of this analysis in terms of percent
*(a) $Na_2SO_4$.                      (d) $Fe_2(SO_4)_3$.
 (b) $Fe(NH_4)_2(SO_4)_2 \cdot 6H_2O$.    *(e) $Na_2S_2O_5$.
*(c) $SO_2$.                          (f) $(NH_4)_2SO_4 \cdot MgSO_4 \cdot 6H_2O$.

4-15. The Fe in a 0.8504-g sample was isolated as $Fe_2O_3 \cdot xH_2O$ by treatment with excess $NH_3$ and was subsequently converted to 0.3895 g of $Fe_2O_3$ by ignition at 1000°C. Calculate the percentage of $FeSO_4$ in the sample.

4-16. The Hg in a 0.6833-g mineral specimen was brought into solution, converted to the $+2$ state, and precipitated with an excess of paraperiodic acid, $H_5IO_6$. Reaction:

$$5Hg^{2+} + 2H_5IO_6 \rightarrow Hg_5(IO_6)_2(s) + 10H^+$$

The precipitated $Hg_5(IO_6)_2$ was dried to a constant weight of 0.5249 g. Calculate the percentage of HgO in the mineral.

*4-17. The As in a 12.75-g sample of ant poison was converted to the $+5$ state and precipitated as $Ag_3AsO_4$, 0.0916 g being recovered. Express the results of this analysis in terms of percent $As_2O_3$.

4-18. Calculate the percentage composition of an alloy if a 0.4315-g sample yielded 0.3808 g of $Cu_2(SCN)_2$, 0.3409 g of $Zn_2P_2O_7$, and 0.4248 g of nickel dimethylglyoxime.

4-19. The S in seven saccharin ($C_7H_5NO_3S$, gfw = 183) tablets that weighed a total of 0.2996 g was oxidized to $SO_4^{2-}$ and then precipitated as $BaSO_4$.
*(a) Calculate the average number of milligrams of saccharin in these tablets if 0.2895 g of $BaSO_4$ was recovered.
 (b) This analysis contains an unnecessary step; identify it.

*4-20. Calculate the formal concentration of an HCl solution if a 25.00-mL aliquot yielded 0.3107 g of AgCl when treated with an excess of $AgNO_3$.

4-21. Calculate the formal concentration of an $AgNO_3$ solution if a 50.0-mL aliquot yielded 0.2621 g of $Ag_2CrO_4$ when treated with an excess of $K_2CrO_4$.

*4-22. Ignition of a 1.045-g steel sample in a stream of $O_2$ resulted in conversion of its C to $CO_2$, which was collected in a tube containing an absorbent for the gas. The tube weighed 15.9733 g at the outset and 16.0087 g at the completion of the analysis. Calculate the percentage of C in the steel.

**4–23.** Combustion of a 0.3826-g sample of a purified organic compound containing only C, H, and O resulted in the formation of 0.7823 g of $CO_2$ and 0.2402 g of $H_2O$. Calculate

    (a) the percentages of C, H, and O in the sample.

    (b) the empirical formula for the compound.

**\*4–24.** Chloride ion is quantitatively oxidized to $Cl_2$ by a known excess of $MnO_2$ in an acidic medium. Reaction:

$$MnO_2(s) + 2Cl^- + 4H^+ \rightarrow Mn^{2+} + Cl_2(g) + 2H_2O$$

The weight of the unused $MnO_2$ is determined after the reaction is complete. Use the accompanying data to calculate the percentage of $NH_4Cl \cdot NiCl_2 \cdot 6H_2O$ (gfw = 291.2) in an impure sample.

| | |
|---|---|
| wt sample taken | 0.9471 g |
| wt $MnO_2$ added | 0.5382 g |
| wt residual $MnO_2$ | 0.2619 g |

**4–25.** W. W. White and P. J. Murphy (*Anal. Chem., 1979, 51,* 1864) reported that the anionic complex between Ag(I) and thiosulfate, $S_2O_3^{2-}$, is quantitatively precipitated with hexamminecobalt(III) trichloride. Reaction:

$$Co(NH_3)_6^{3+} + Ag(S_2O_3)_2^{3-} \rightarrow [Co(NH_3)_6][Ag(S_2O_3)_2](s)$$

The product (gfw = 493.2) is brought to constant weight at $95\,^\circ C$. Analysis of a 50.0-mL aliquot of a photographic fixing solution by this method results in the formation of 0.5954 g of precipitate. Calculate the number of grams of Ag contained in each liter of this solution; sufficient thiosulfate ion, $S_2O_3^{2-}$, is present to ensure essentially complete complexation of $Ag^+$.

**\*4–26.** Addition of an excess of $AgNO_3$ to a 0.5012-g sample yielded a mixture of AgCl and AgI that weighed 0.4715 g. The precipitate was then heated in a stream of $Cl_2$ to convert the AgI to AgCl

$$2AgI(s) + Cl_2(g) \rightarrow 2AgCl(s) + I_2(g)$$

The precipitate was found to weigh 0.3922 g following this treatment. Calculate the percentages of KI and $NH_4Cl$ in the sample.

**4–27.** A 1.461-g sample containing $K_2SO_4$, $NH_4NO_3$, and inert materials was dissolved in sufficient water to give exactly 250 mL of solution. A 25.0-mL aliquot of this solution yielded 0.2999 g of precipitate that consisted of $(C_6H_5)_4BK$ (gfw = 358) and $(C_6H_5)_4BNH_4$ (gfw = 337) upon being treated with an excess of sodium tetraphenylboron $(C_6H_5)_4BNa$. A 50.0-mL aliquot of the sample was made alkaline and warmed to drive off the ammonia

$$NH_4^+ + OH^- \rightarrow NH_3(g) + H_2O$$

following which treatment with $(C_6H_5)_4BNa$ gave 0.3230 of $(C_5H_5)_4BK$. Calculate the percentages of $K_2SO_4$ and $NH_4NO_3$ in the sample.

**\*4–28.** A 1.008-g sample that contained $NH_4NO_3$, $(NH_4)_2SO_4$, and inert materials was dissolved in sufficient water to give exactly 500 mL of solution. A 50.0-mL aliquot of this solution yielded 0.3819 g of $(C_6H_5)_4BNH_4$ (gfw = 337) when treated with an excess of sodium tetraphylboron $(C_6H_5)_4BNa$. A second 50.0-mL aliquot was made alkaline and heated with Devarda's alloy (50% Cu, 45% Al, 5% Zn) to convert the $NO_3^-$ to $NH_3$:

$$NO_3^- + 6H_2O + 8e \rightarrow NH_3 + 9OH^-$$

The $NH_3$ produced in this reaction, as well as that derived from $NH_4^+$, was distilled and collected in dilute acid. Treatment of the distillate with $(C_6H_5)_4BNa$ yielded 0.5996 g of $(C_6H_5)_4BNH_4$. Calculate the percentages of $(NH_4)_2SO_4$ and $NH_4NO_3$ in the sample.

**4-29.** At 500°C, $CaC_2O_4$ is converted to $CaCO_3$, while $MgC_2O_4$ is converted to MgO:

$$CaC_2O_4(s) \rightarrow CaCO_3(s) + CO(g)$$

$$MgC_2O_4(s) \rightarrow MgO(s) + CO(g) + CO_2(g)$$

Heating at 900°C causes further decomposition of $CaCO_3$ to CaO

$$CaCO_3(s) \rightarrow CaO + CO_2(g)$$

The Ca and Mg in weighed samples were isolated as oxalates; after filtration, these oxalate mixtures were brought to constant weight at 500°C and again at 900°C. Use this information and the accompanying data to establish the percentage composition of Ca and Mg in the following:

| | Wt of Sample, g | Wt of Precipitate at | |
|---|---|---|---|
| | | 500°C | 900°C |
| (a) | 0.6397 | 0.3100 | 0.2361 |
| (b) | 0.7438 | 0.5096 | 0.3585 |
| (c) | 0.8142 | 0.2439 | 0.2441 |
| (d) | 0.8236 | 0.5351 | 0.3556 |
| (e) | 0.6044 | 0.3195 | 0.1789 |

**4-30.** Several alloys that contained only Ag and Cu were analyzed by dissolving weighed quantities in $HNO_3$, introducing an excess of $IO_3^-$, and bringing the filtered mixture of $AgIO_3$ and $Cu(IO_3)_2$ to constant weight. Use the accompanying data to calculate the percentage composition of the alloys.

| | Wt of Sample, g | Wt of Precipitate, g |
|---|---|---|
| *(a) | 0.2175 | 0.7391 |
| (b) | 0.1948 | 0.7225 |
| *(c) | 0.2473 | 0.7443 |
| (d) | 0.2386 | 0.9962 |
| *(e) | 0.1864 | 0.8506 |

**4-31.** The routine gravimetric determination of the $Cl^-$ content of nickel(II) ammonium chloride is being planned. From past experience it is known that the product will range between 82% and 93% $NiCl_2 \cdot NH_4Cl \cdot 6H_2O$ (gfw = 291.2).
  (a) What sample weight should be taken to ensure that the AgCl precipitated will weigh no less than 0.400 g?

(b) If the sample weight calculated in (a) is used, what will be the maximum weight of AgCl to be expected?

(c) What volume of $0.100\ F$ $AgNO_3$ should be used with 0.350-g samples to ensure—at a minimum—a 5% excess of $Ag^+$?

(d) It is desired that 50.0 mL of a $AgNO_3$ solution provide a 5% excess over the maximum needed to precipitate $Cl^-$ from 0.350-g samples. What should be the formal concentration of this solution?

(e) What sample weight should be taken in order to have the percentage of $NiCl_2 \cdot NH_4Cl \cdot 6H_2O$ exceed the weight of AgCl produced by a factor of 150?

# Chapter 5
# The Solubility of Precipitates

Reactions that yield products of limited solubility find application in three important analytical processes: (1) gravimetric analysis, in which the weight of a solid that has been formed is chemically related to the amount of analyte; (2) isolation of an analyte from potential interferences; and (3) titrimetric methods based on the volume of a standard reagent that is needed to bring about essentially complete precipitation of the analyte. Each of these applications has the common need for a solid that is reasonably pure, has a suitable particle size, and possesses low solubility. Factors that affect the particle size and purity of precipitates were considered in Chapter 4; here, we shall be concerned with the variables that influence the solubility.[1]

## 5A THE SOLUBILITY PRODUCT CONSTANT

To review, a condition of equilibrium exists between a sparingly soluble compound and its ions in solution. Using silver iodate as an example, equilibrium is described by the equation

$$AgIO_3(s) \rightleftharpoons Ag^+(aq) + IO_3^-(aq)$$

This is a *dynamic* equilibrium in the sense that solid silver iodate is continuously undergoing solution as well as formation. Because the rates of the two processes are equal in the equilibrium state, the system experiences no *net* change in composition.

As described in Chapter 2, Section 2D–1, the qualitative effects of stress upon a system in equilibrium can be predicted by the LeChâtelier principle. Furthermore, and in common with all other types of chemical equilibrium, the law of mass action describes the condition of equilibrium involving a sparingly soluble solid in contact with its saturated solution. Thus, equilibrium between

---

[1] For a detailed treatment of solubility, see: E. S. Amis, in *Treatise on Analytical Chemistry,* 2nd ed., I. M. Kolthoff and P. J. Elving, Eds., Part I, Vol. 3, Chapter 26. New York: John Wiley & Sons, 1983.

solid silver iodate and its ions in aqueous solution is described by the expression

$$K_{eq} = \frac{[Ag^+][IO_3^-]}{[AgIO_3(s)]}$$

where the bracketed terms represent molar concentrations.[2]

Now, this expression can be simplified by taking advantage of the fact that the position of equilibrium is unaffected by the amount of solid, so long as some is present; that is,

$$[AgIO_3(s)] = \text{a constant}$$

Combination of $[AgIO_3(s)]$ with $K_{eq}$ yields $K_{sp}$, the *solubility product constant:*

$$K_{eq}[AgIO_3(s)] = K_{sp} = [Ag^+][IO_3^-] \tag{5-1}$$

More generally, for the sparingly soluble substance $A_xB_y$,

$$A_xB_y(s) \rightleftharpoons xA(aq) + yB(aq) \tag{5-2}$$

$$K_{sp} = [A]^x[B]^y$$

where $x$ and $y$ are small whole numbers and A and B represent the cation and anion of the precipitate; their charges have been omitted as a matter of convenience. Note that the solubility product constant defines the condition of equilibrium in terms of the concentrations (raised to appropriate powers) of the ions derived from the solid. It follows that precipitation can be expected in solutions where the product of these concentrations (raised to the same powers) is numerically larger than $K_{sp}$, and that no precipitation will occur in solutions where this product is smaller than $K_{sp}$.

It is of the utmost importance to appreciate that a solubility product constant, such as that shown in Equation 5-1 or 5-2, applies *only* to a saturated solution that is in contact with an excess of undissolved solid. In common with other types of equilibrium constant expressions, numerical values for $K_{sp}$ are temperature dependent.

### 5A-1 Calculations Involving the Solubility Product Constant

Solubility product constants for numerous sparingly soluble inorganic species are tabulated in Appendix 9. Typical uses of these constants are illustrated in the accompanying examples.

---

[2] The use of *activities* instead of molar concentrations in this expression will yield results that are in better conformity with experimental observation. The concept of activity is described in Section 5D.

**Example 5-1** What weight of $Ba(IO_3)_2$ (gfw = 487) can be dissolved in 500 mL of water?

At 25°C, the solubility product constant for $Ba(IO_3)_2$ is $1.57 \times 10^{-9}$ (Appendix 9). Thus,

$$Ba(IO_3)_2(s) \rightleftharpoons Ba^{2+} + 2IO_3^-$$

and

$$K_{sp} = 1.57 \times 10^{-9} = [Ba^{2+}][IO_3^-]^2$$

Each formula weight of dissolved $Ba(IO_3)_2$ contributes 1 mol of $Ba^{2+}$ and 2 mol of $IO_3^-$. Thus, if we symbolize the formal solubility of $Ba(IO_3)_2$ as $s$, then

$$[Ba^{2+}] = s$$

$$[IO_3^-] = 2s$$

Substitution of these quantities into the solubility product expression gives

$$1.57 \times 10^{-9} = (s)(2s)^2 = 4s^3$$

$$s^3 = \frac{1.57 \times 10^{-9}}{4} = 3.92 \times 10^{-10}$$

$$s = 7.32 \times 10^{-4} \text{ gfw/L}$$

Thus, a saturated solution of $Ba(IO_3)_2$ is $7.32 \times 10^{-4}$ $F$; the weight of $Ba(IO_3)_2$ that is dissolved in 500 mL of such a solution is

$$\text{weight } Ba(IO_3)_2 = \frac{7.32 \times 10^{-4} \text{ mfw } Ba(IO_3)_2}{mL} \times \frac{0.487 \text{ g}}{\text{mfw } Ba(IO_3)_2} \times 500 \text{ mL}$$

$$= 0.178 \text{ g}$$

Examples 5-2 and 5-3 demonstrate the effect of a common ion on solubility.

**Example 5-2** Calculate the formal solubility of $Ba(IO_3)_2$ in a solution that is $0.0200$ $F$ in $Ba(NO_3)_2$.

The $Ba^{2+}$ concentration in this solution has two sources, specifically, $Ba(NO_3)_2$ and $Ba(IO_3)_2$. If we again let $s$ equal the formal solubility of $Ba(IO_3)_2$, then

$$[Ba^{2+}] = 0.0200 + s$$

$$[IO_3^-] = 2s$$

Substitution of these quantities into the solubility product expression yields

$$1.57 \times 10^{-9} = (0.0200 + s)(2s)^2$$

Since the exact solution for $s$ will involve a cubic equation, it is worthwhile to seek an approximation that will simplify the algebra (see Appendix 6). The small numerical value for $K_{sp}$ suggests that the solubility of $Ba(IO_3)_2$ is not large. It is reasonable, therefore, to calculate a provisional answer to the problem based on the assumption that the contribution of $Ba^{2+}$ from the solubility of $Ba(IO_3)_2$ is small with respect to that from the $Ba(NO_3)_2$ (that is, $0.0200 \gg s$). The original equation then simplifies to

$$0.0200(2s)^2 = 1.57 \times 10^{-9}$$

$$s = 1.40 \times 10^{-4} \, F$$

The assumption that

$$(0.0200 + s) = 0.0200$$

will not introduce a serious error, because $s$ is only about 0.7% of 0.0200; uncertainties of 5 to 10% are ordinarily tolerable in calculations such as this. Note that the addition of a common ion has lowered the solubility of $Ba(IO_3)_2$ over that in pure water (Example 5–1) by a factor of about 5.

If mixing of two solutions is likely to cause formation of a sparingly soluble product, one of four possible situations will result:

1. No precipitate will form, because the solution is not saturated with respect to the sparingly soluble product.
2. The solution will be saturated with respect to the precipitate and will additionally contain an excess of its cation.
3. The solution will be saturated with respect to the precipitate and will additionally contain an excess of its anion.
4. The solution will be saturated with respect to the precipitate and will not contain an excess of either ion.

It is necessary to know which of these situations prevails before further calculations are undertaken. Example 5–3 illustrates how this information is obtained.

**Example 5–3**    Calculate the formal solubility of $Ba(IO_3)_2$ in the solution that results when 200 mL of 0.0100 $F$ $Ba(NO_3)_2$ are mixed with 100 mL of 0.100 $F$ $NaIO_3$.

Mixing will result in a solution in which the initial $Ba^{2+}$ concentration multiplied by the square of the $IO_3^-$ concentration ($7.41 \times 10^{-6}$) is numerically larger than $K_{sp}$ for $Ba(IO_3)_2$ ($1.57 \times 10^{-9}$); we can therefore eliminate possibility 1 and can assert that precipitation will occur. The amounts of reactants involved are

$$200 \text{ mL} \times \frac{0.0100 \text{ mfw Ba}^{2+}}{\text{mL}} = 2.00 \text{ mfw Ba}^{2+}$$

$$100 \text{ mL} \times \frac{0.100 \text{ mfw IO}_3^-}{\text{mL}} = 10.00 \text{ mfw IO}_3^-$$

The solution will clearly contain an excess of $IO_3^-$. If formation of $Ba(IO_3)_2$ is complete,

$$\text{excess } IO_3^- = 10.00 - 2 \times 2.00 = 6.00 \text{ mfw}$$

Thus, the formal concentration of $IO_3^-$ will be

$$\frac{6.00 \text{ mfw IO}_3^-}{300 \text{ mL}} = 0.0200 \ F \ IO_3^-$$

If we again use $s$ to symbolize the formal solubility of $Ba(IO_3)_2$, then

$$[Ba^{2+}] = s$$

The molar concentration of $IO_3^-$ will be

$$[IO_3^-] = 0.0200 + 2s$$

where $2s$ represents the contribution of $IO_3^-$ owing to the finite solubility of the precipitate. A provisional answer to the problem can be obtained by making the assumption that $2s$ is negligible compared with 0.0200. If such is indeed the case, then

$$[IO_3^-] \cong 0.0200$$

and

$$\text{solubility of Ba(IO}_3)_2 = s = \frac{K_{sp}}{[IO_3^-]^2}$$

$$= \frac{1.57 \times 10^{-9}}{(0.0200)^2} = 3.93 \times 10^{-6} \ F$$

The assumption that $2s \ll 0.0200$ is clearly reasonable.

Examples 5-2 and 5-3 demonstrate that the presence of excess iodate is more effective in decreasing the solubility of barium iodate than an equal formal excess of barium ion.

**Example 5-4**    Iodide is slowly added to a solution that is 0.050 $F$ in $Ag^+$ and 0.045 $F$ in $Bi^{3+}$. Calculate

(a) the $I^-$ concentration needed to start precipitation of each cation.

(b)  the formal concentration of the less soluble iodide as precipitation of the more soluble iodide commences. For AgI, $K_{sp} = 8.3 \times 10^{-17}$; for $BiI_3$, $K_{sp} = 8.1 \times 10^{-19}$.

(a)  The $I^-$ concentrations at the outset of precipitation of each cation will be

$$[I^-] = \frac{K_{sp}}{[Ag^+]} = \frac{8.3 \times 10^{-17}}{5.0 \times 10^{-2}} = 1.66 \times 10^{-15} = 1.7 \times 10^{-15}$$

$$[I^-]^3 = \frac{K_{sp}}{[Bi^{3+}]} = \frac{8.1 \times 10^{-19}}{4.5 \times 10^{-2}} = 1.8 \times 10^{-17}$$

$$[I^-] = 2.6 \times 10^{-6}$$

The $I^-$ concentration needed to exceed the ion product for AgI is smaller than that needed for $BiI_3$; AgI will thus precipitate first. Note that the compound with the numerically larger $K_{sp}$ is actually the less soluble of the two; this apparent anomaly is a consequence of the exponential term for $[I^-]$ in $K_{sp}$ for $BiI_3$.

(b)  When $[I^-] = 2.6 \times 10^{-6}$, the formal solubility, $s$, for AgI will be

$$s = [Ag^+] = \frac{K_{sp}}{[I^-]} = \frac{8.3 \times 10^{-17}}{2.6 \times 10^{-6}} = 3.2 \times 10^{-11}$$

## 5B  THE EFFECT OF COMPETING EQUILIBRIA ON THE SOLUBILITY OF PRECIPITATES

The solubility of a precipitate is increased by solutes that compete for one (or more) of the ions of the solid. Typical examples of this effect are shown in Table 5–1. In the first example, the solubility of barium sulfate is increased by the introduction of strong acid. Sulfate ion, the conjugate base of the weak acid

Table 5 – 1   *Increases in Solubility Brought About by Completing Equilibria*

| Precipitate | Species Causing Solubility Increases | Equilibria |
|---|---|---|
| $BaSO_4$ | $H^+$ | $BaSO_4 \rightleftharpoons Ba^{2+} + SO_4^{2-}$<br>$+$<br>$H_3O^+$<br>$\updownarrow$<br>$HSO_4^- + H_2O$ |
| AgBr | $NH_3$ | $AgBr \rightleftharpoons Ag^+ + Br^-$<br>$+$<br>$2NH_3$<br>$\updownarrow$<br>$Ag(NH_3)_2^+$ |

$HSO_4^-$, reacts with hydronium ions of the strong acid; it is evident (recall the principle of Le Châtelier) that the added acid will cause an increase in the hydrogen sulfate concentration. The consequent decrease in the sulfate ion concentration is partially offset by a shift of the first equilibrium to the right; the net effect is an increase in the solubility of barium sulfate.

The second example illustrates how the solubility of silver bromide is increased through the introduction of ammonia, which combines with silver ion to form an ammine complex. Ammonia molecules tend to lower the silver ion concentration; the consequent shift in the solubility equilibrium to offset this change results in an increase in solubility.

*The reader should appreciate that the existence of competing equilibria in no way alters the validity of the solubility product expression or the numerical value of the solubility product constant.* Regardless of the existence of strong acid in a saturated solution of barium sulfate or of ammonia in a saturated silver bromide solution, the solubilities of these solutes are still described by

$$BaSO_4(s) \rightleftharpoons Ba^{2+} + SO_4^{2-} \qquad K_{sp} = [Ba^{2+}][SO_4^{2-}]$$

and

$$AgBr(s) \rightleftharpoons Ag^+ + Br^- \qquad K_{sp} = [Ag^+][Br^-]$$

To be sure, the competing equilibria have caused changes in the concentrations needed to satisfy the respective solubility product constant expressions.

Accounting for the effect of competing equilibria requires additional knowledge concerning the concentration(s) of the competing species as well as numerical values for all pertinent equilibrium constants. Typically, several algebraic expressions are needed to describe completely the concentration relationships that exist in such a solution, and the solubility calculation requires consideration of several simultaneous equations. The algebra associated with solution of these equations is potentially more formidable than the task of generating them.

A systematic approach by which any problem involving several equilibria can be attacked is presented in the following section. This approach is then illustrated with typical examples involving sparingly soluble compounds.

### 5B – 1 Systematic Method for Solving Problems that Involve Multiple Equilibria

1. Write balanced net-ionic equations for all equilibria that appear to have any bearing on the problem.
2. State the quantity being sought in the problem in terms of equilibrium concentration(s) of participating species.
3. Write equilibrium constant expressions for all of the equilibria produced in step 1; locate numerical values for the constants in appropriate tables.
4. Write mass-balance equations for the system. These are algebraic expressions that relate the equilibrium concentrations of the various participants

to one another and to the formal concentrations of species used in preparing the solution.

5. Write a charge-balance equation. The concentrations of anions and cations in any solution must be such that electrical neutrality is maintained; the charge-balance equation expresses this relationship.[3]

6. Count the number of unknown quantities in the equations generated in steps 3, 4, and 5, and compare with the number of independent equations. The problem can be solved exactly by suitable algebraic manipulation, provided the number of equations is equal to the number of unknown quantities. An exact solution is not possible if the number of unknowns exceeds the number of equations; under these circumstances, one must either attempt to generate additional independent equations or else conclude that an exact solution is not possible with the data at hand. To be sure, an approximate solution may still be possible.

7. Make suitable approximations to simplify the algebra or to decrease the number of unknowns so that the problem can be solved.

8. Solve for the equilibrium concentrations that are needed to obtain a provisional answer as defined in step 2.

9. Check the approximations that were made in step 7 for validity, using the concentrations obtained in step 8.

Step 6 of this scheme is particularly important because it indicates whether an exact solution for the problem is feasible. If the number of independent equations is as great as the number of unknowns, the problem becomes purely algebraic, involving the solution of several simultaneous equations. On the other hand, if the number of equations is fewer than the number of unknowns, a

---

[3] To illustrate the generation of a charge-balance relationship, consider the solution that results when 0.10 fw of NaCl is dissolved in a liter of water. This solution has a net charge of zero even though it contains positive ions and negative ions. Electrical neutrality is a consequence of the relationship

$$[Na^+] + [H_3O^+] = [Cl^-] + [OH^-]$$

Note that the concentrations of some species appearing in a charge-balance expression may be negligible with respect to others. Substitution of numerical values in the present example illustrates this point. A solution of sodium chloride is neutral; thus,

$$0.10 + 1.0 \times 10^{-7} = 0.10 + 1.0 \times 10^{-7}$$

Now, consider a $0.10\ F$ $MgCl_2$ solution. Here, we must write

$$2[Mg^{2+}] + [H_3O^+] = [Cl^-] + [OH^-]$$

It is necessary to multiply the magnesium ion concentration by 2 to account for the two units of charge contributed by this ion. Since a $0.10\ F$ solution of $MgCl_2$ will be $0.10\ F$ with respect to $Mg^{2+}$ and $0.20\ F$ with respect to $Cl^-$ (that is, $[Cl^-] = 2[Mg^{2+}]$), it follows that

$$2 \times 0.10 + 1.0 \times 10^{-7} = 0.20 + 1.0 \times 10^{-7}$$

and electrical neutrality is preserved. The concentration of a triply charged species, if present, would have to be multiplied by 3. Thus, the charge-balance equation for an aqueous solution containing $Al_2(SO_4)_3$ would be

$$3[Al^{3+}] + [H_3O^+] = 2[SO_4^{2-}] + [HSO_4^-] + [OH^-]$$

search for additional relationships, or for approximations that will decrease the number of unknowns, is essential. Time expended in attempting to solve a problem for which there are insufficient data is time wasted.

### 5B–2 The Effect of pH on Solubility

A precipitate that contains an anion with basic properties, a cation with acidic properties, or both, will have a solubility that is pH dependent. In the context of analytical chemistry, this property can be used to bring about separations that are based upon differences in solubility.

**Solubility Calculations Where the Hydronium Ion Concentration Is Fixed and Known**    Analytical precipitations are frequently performed under conditions of a fixed and predetermined hydronium ion concentration. Example 5–5 demonstrates how the recommended systematic approach is applied to such a system.

---

**Example 5–5**    Calculate the formal solubility of $CaC_2O_4$ in a solution with a fixed hydronium ion concentration of $1.00 \times 10^{-4}\ M$.

*Step 1: Chemical equations.* Foremost among these will be the equilibrium between the solid and its ions in solution

$$CaC_2O_4(s) \rightleftharpoons Ca^{2+} + C_2O_4^{2-} \qquad (5-3)$$

Three oxalate-containing species ($C_2O_4^{2-}$, $HC_2O_4^-$, and $H_2C_2O_4$) will exist in the solution; dissociation equations for the parent acid should therefore be included

$$H_2C_2O_4 + H_2O \rightleftharpoons H_3O^+ + HC_2O_4^- \qquad (5-4)$$

$$HC_2O_4^- + H_2O \rightleftharpoons H_3O^+ + C_2O_4^{2-} \qquad (5-5)$$

Because hydronium ions are involved in both of these equilibria, an additional chemical system (a buffer) will be needed to satisfy the requirement of a fixed pH.

*Step 2: Definition of the unknown.* What is sought? We wish to calculate the formal solubility of $CaC_2O_4$. Since $CaC_2O_4$ is ionic, its formal concentration will be numerically equal to the molar concentration of calcium ion. It will also be equal to the sum of the equilibrium concentrations of the oxalate species; that is,

$$\text{solubility, } s = [Ca^{2+}]$$

$$= [H_2C_2O_4] + [HC_2O_4^-] + [C_2O_4^{2-}]$$

Thus, if we can evaluate either the calcium ion concentration or the sum of the concentrations of oxalate species, we shall have obtained a solution to the problem.

*Step 3. Equilibrium constant expressions.* The equilibrium constant expressions for Equations 5–3, 5–4, and 5–5 are

$$K_{sp} = [Ca^{2+}][C_2O_4^{2-}] = 2.3 \times 10^{-9} \tag{5–6}$$

$$K_1 = \frac{[H_3O^+][HC_2O_4^-]}{[H_2C_2O_4]} = 5.36 \times 10^{-2} \tag{5–7}$$

$$K_2 = \frac{[H_3O^+][C_2O_4^{2-}]}{[HC_2O_4^-]} = 5.42 \times 10^{-5} \tag{5–8}$$

*Step 4. Mass-balance equations.* Dissolved calcium oxalate is the only source for calcium ion as well as for the various oxalate-containing species. It follows, then, that

$$[Ca^{2+}] = [H_2C_2O_4] + [HC_2O_4^-] + [C_2O_4^{2-}] \tag{5–9}$$

Furthermore, the problem stipulates that

$$[H_3O^+] = 1.00 \times 10^{-4} \tag{5–10}$$

*Step 5. Charge-balance equations.* We do not have sufficient information to permit the writing of a charge-balance equation. The condition of a fixed hydronium ion concentration is maintained by means of an auxiliary chemical system (consisting, most likely, of a weak acid HX and its conjugate base X⁻). Since information regarding this auxiliary system has not been provided, an equation based on the electrical neutrality of the system cannot be written. As it turns out, this problem can be solved without the need for a charge-balance equation.

*Step 6. Comparison of equations and unknowns.* We have four unknowns ($[Ca^{2+}]$, $[C_2O_4^{2-}]$, $[HC_2O_4^-]$, and $[H_2C_2O_4]$). We also have four independent algebraic relationships (Equations 5–6, 5–7, 5–8, and 5–9). Therefore, an exact solution is possible, and the problem has now become one of algebra.

*Step 7. Approximations.* None will be needed, since sufficient information is available for an exact solution.

*Step 8. Solution of the equations.* A convenient way to solve the four equations is to express mass balance (Equation 5–9) in terms of $[Ca^{2+}]$, $[H_3O^+]$, and the several equilibrium constants. Rearrangement of Equation 5–8 makes it possible to express $[HC_2O_4^-]$ in terms of $[C_2O_4^{2-}]$:

$$[HC_2O_4^-] = \frac{[H_3O^+][C_2O_4^{2-}]}{K_2}$$

Multiplication of Equation 5–7 by Equation 5–8 permits expression of $[H_2C_2O_4]$ in the same terms:

$$K_1K_2 = \frac{[H_3O^+][\cancel{HC_2O_4^-}]}{[H_2C_2O_4]} \times \frac{[H_3O^+][C_2O_4^{2-}]}{[\cancel{HC_2O_4^-}]}$$

Rearrangement gives

$$[H_2C_2O_4] = \frac{[H_3O^+]^2[C_2O_4^{2-}]}{K_1 K_2}$$

Equation 5–9 can now be written as

$$[Ca^{2+}] = \frac{[H_3O^+]^2[C_2O_4^{2-}]}{K_1 K_2} + \frac{[H_3O^+][C_2O_4^{2-}]}{K_2} + [C_2O_4^{2-}]$$

or

$$[Ca^{2+}] = [C_2O_4^{2-}] \times \left(\frac{[H_3O^+]^2}{K_1 K_2} + \frac{[H_3O^+]}{K_2} + 1\right)$$

Substitution of $K_{sp}/[Ca^{2+}]$ for $[C_2O_4^{2-}]$ gives

$$[Ca^{2+}] = \frac{K_{sp}}{[Ca^{2+}]} \times \left(\frac{[H_3O^+]^2}{K_1 K_2} + \frac{[H_3O^+]}{K_2} + 1\right)$$

Finally, then

$$[Ca^{2+}]^2 = 2.3 \times 10^{-9} \times \left(\frac{(1.00 \times 10^{-4})^2}{(5.36 \times 10^{-2})(5.42 \times 10^{-5})} + \frac{1.00 \times 10^{-4}}{5.42 \times 10^{-5}} + 1\right)$$

$$= 6.54 \times 10^{-9}$$

$$[Ca^{2+}] = s = 8.1 \times 10^{-5}$$

**Solubility Calculations Where the Hydronium Ion Concentration Is Variable**
Solutes containing basic anions (such as calcium oxalate) or acidic cations (such as ammonium chloride) influence the hydronium ion concentration of their aqueous solutions. If an auxiliary reagent is not available to maintain the pH at a fixed value, the hydronium ion concentration will depend upon the extent to which such solutes dissolve. For example, a saturated aqueous solution of calcium oxalate will be basic as a consequence of the reactions

$$CaC_2O_4(s) \rightleftharpoons Ca^{2+} + C_2O_4^{2-}$$

$$C_2O_4^{2-} + H_2O \rightleftharpoons HC_2O_4^- + OH^-$$

$$HC_2O_4^- + H_2O \rightleftharpoons H_2C_2O_4 + OH^-$$

In contrast to the example just considered, the hydroxide ion now becomes an additional variable, and another equation must be sought.

The reaction between a precipitate and water is frequently the major factor that determines the pH of the solution; failure to take account of such reactions can cause serious errors in solubility product calculations. The magnitude of this error depends upon the solubility of the precipitate in question as well as the dissociation constant of the conjugate acid (or conjugate base) from which the ion of the precipitate is derived.

Sufficient algebraic equations can be generated to permit computation of the solubility of a precipitate in pure water by the systematic approach (p. 100). The algebra involved in an exact solution to these equations is difficult unless suitable approximations are made.[4]

**Solubility of Metal Hydroxides**   Two equilibria should be considered in calculating solubility of a metal hydroxide. Using $M(OH)_2$ as an example, these equilibria are

$$M(OH)_2(s) \rightleftharpoons M^{2+} + 2OH^- \qquad K_{sp} = [M^{2+}][OH^-]^2$$

$$2H_2O \rightleftharpoons H_3O^+ + OH^- \qquad K_w = [H_3O^+][OH^-]$$

Charge balance requires that

$$2[M^{2+}] + [H_3O^+] = [OH^-] \qquad\qquad (5-11)$$

If the formal solubility, $s$, of $M(OH)_2$ is reasonably large, $[H_3O^+]$ will necessarily be small; Equation 5–11 thus simplifies to

$$2[M^{2+}] = [OH^-] = 2s$$

Substitution into the solubility product constant for $M(OH)_2$ yields an expression that is identical with that in Example 5–1:

$$K_{sp} = (s)(2s)^2 = 4s^3$$

$$s = \left(\frac{K_{sp}}{4}\right)^{1/3}$$

On the other hand, if the formal solubility of $M(OH)_2$ is very low, $2[M^{2+}]$ becomes much smaller than $[H_3O^+]$, and Equation 5–11 becomes

$$[H_3O^+] = [OH^-] = 1.00 \times 10^{-7}$$

and

$$s = [M^{2+}] = \frac{K_{sp}}{[OH^-]^2} = \frac{K_{sp}}{1.00 \times 10^{-14}}$$

---

**Example 5–6**   Calculate the solubility of $Fe(OH)_3$ in water.

Charge balance in this system requires that

$$3[Fe^{3+}] + [H_3O^+] = [OH^-] \qquad\qquad (5-12)$$

As a hypothesis, let us assume that the contribution of $H_3O^+$ to the positive

---

[4] See, for example, D. A. Skoog and D. M. West, *Fundamentals of Analytical Chemistry,* 4th ed., pp. 97–102. Philadelphia: Saunders College Publishing, 1982.

charge in this system is vanishingly small; Equation 5–12 then simplifies to

$$3[Fe^{3+}] = [OH^-]$$

In terms of the formal solubility, $s$, of $Fe(OH)_3$,

$$[Fe^{3+}] = s$$

$$[OH^-] = 3s$$

and

$$(s)(3s)^3 = K_{sp} = 4 \times 10^{-38}$$

$$s = \left(\frac{4 \times 10^{-38}}{27}\right)^{1/4} = 2 \times 10^{-10}$$

Since $[OH^-] = 3s = 6 \times 10^{-10}$, then

$$[H_3O^+] = \frac{1.00 \times 10^{-14}}{6 \times 10^{-10}} = 1.7 \times 10^{-5}$$

Clearly, $[H_3O^+]$ is *not* much smaller than $3[Fe^{3+}]$, as assumed; indeed, the reverse appears to be the case:

$$3[Fe^{3+}] \ll [H_3O^+]$$

Equation 5–12 thus simplifies to

$$[H_3O^+] = [OH^-] = 1.00 \times 10^{-7}$$

and substitution for $[OH^-]$ in the solubility product expression yields

$$[Fe^{3+}] = s = \frac{4 \times 10^{-38}}{(1.00 \times 10^{-7})^3} = 4 \times 10^{-17}$$

The assumption that $3[Fe^{3+}] \ll [H_3O^+]$ is clearly valid. Note the very large error in the first calculation that resulted from use of the incorrect assumption.

## 5B–3 Complex Ion Formation and Solubility

The solubility of a precipitate may be greatly altered by the presence of some species that will react with the anion or the cation of the precipitate to form a stable complex. For example, the precipitation of aluminum ion with base is never complete in the presence of fluoride ion, even though aluminum hydroxide has an extremely low solubility; the fluoride complexes of aluminum(III) are sufficiently stable to prevent the quantitative removal of aluminum from solution. The equilibria involved can be represented by

$$Al(OH)_3 \rightleftharpoons Al^{3+} + 3OH^-$$
$$+$$
$$6F^-$$
$$\updownarrow$$
$$AlF_6^-$$

The equilibrium constants for the two reactions are such that fluoride ion competes successfully with hydroxide ion for aluminum(III). Thus, the soluble fluoroaluminate complex forms at the expense of the solid.

**Quantitative Treatment of the Effect of Complex Formation on Solubility**    Solubility calculations for a precipitate in the presence of a complexing reagent are similar in principle to those discussed in the previous section. Formation constants for the complexes involved are, of course, required.[5]

**Complex Formation with an Ion that Is Common to the Precipitate**    Many precipitates react with the precipitating reagent to form soluble complexes. In a gravimetric analysis, this tendency may have the unfortunate effect of reducing the recovery of precipitate if too large an excess of reagent is used. For example, in solutions containing high concentrations of chloride, silver chloride forms chloro complexes such as $AgCl_2^-$ and $AgCl_3^{2-}$. The effect of these complexes is illustrated in Figure 5–1, in which the experimentally determined solubility of silver chloride is plotted against the logarithm of the potassium chloride concentration. For low anion concentrations, the experimental solubilities do not differ greatly from those calculated with the solubility product constant for silver chloride; beyond a chloride ion concentration of about $10^{-3}$, however, the calculated solubilities approach zero while the measured values rise steeply. Note that the solubility of silver chloride is about the same in 0.3 $F$ KCl as in pure water, and is about eight times that figure in a 1 $F$ solution. A quantitative description of these effects can be achieved if the compositions of the complexes and their formation constants are known.

Increases in solubility caused by large excesses of a common ion are by no means rare. Of particular interest are amphoteric hydroxides, which are sparingly soluble in dilute base but are redissolved by excess hydroxide ion. The hydroxides of zinc and aluminum, for example, are converted to the soluble zincate and aluminate ions upon treatment with excess base. For zinc, the equilibria can be represented as

$$Zn^{2+} + 2OH^- \rightleftharpoons Zn(OH)_2(s)$$
$$Zn(OH)_2(s) + 2OH^- \rightleftharpoons Zn(OH)_4^{2-}$$

As with silver chloride, the solubilities of amphoteric hydroxides pass through minima and then increase rapidly with increasing concentrations of base. The hydroxide ion concentration at which the solubility is a minimum can be calculated, provided equilibrium constants for the reactions are available.

---

[5] For an example of such a calculation, see: D. A. Skoog and D. M. West, *Fundamentals of Analytical Chemistry,* 4th ed., p. 106. Philadelphia: Saunders College Publishing, 1982.

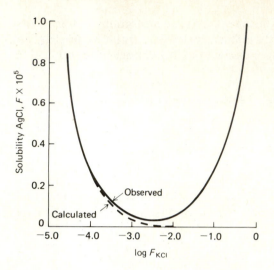

**Figure 5 – 1**   Solubility of silver chloride in potassium chloride solutions. The dashed curve is calculated from $K_{sp}$; the solid curve is plotted from experimental data of A. Pinkus and A. M. Timmermans, *Bull. Soc. Belges*, **1937**, *46*, 46 – 73.

## 5C SEPARATIONS BASED UPON DIFFERENCES IN SOLUBILITY

Slow addition of an ion that forms sparingly soluble compounds with each of two species will first cause formation of the less soluble product. If the solubilities are sufficiently different, quantitative removal of the first ion from solution may be achieved without interference from the second. A successful separation requires careful control of the precipitant concentration at some suitable predetermined level. Many important analytical separations, notably those involving sulfide ion, hydroxide ion, and organic reagents, are based upon solubility differences.

### 5C – 1 Calculation of the Feasibility of Separations

An important application of solubility product calculations involves determining the feasibility and the optimum conditions for separations based on the control of reagent concentration. Example 5 – 7 illustrates such an application.

**Example 5 – 7**   Establish whether the difference in the solubilities of $Fe(OH)_3$ and $Mg(OH)_2$ is sufficient to permit the quantitative separation of $Fe^{3+}$ from $Mg^{2+}$ in a solution that is originally $0.100\ F$ in each cation. If such a separation is possible, calculate the concentration range within which the $OH^-$ concentration must be held. Solubility product constants for the two compounds are

$$[Fe^{3+}][OH^-]^3 = 4 \times 10^{-38}$$

$$[Mg^{2+}][OH^-]^2 = 1.8 \times 10^{-11}$$

The numerical values for their solubility product constants suggest that

$Fe(OH)_3$ will form at a lower $OH^-$ concentration than $Mg(OH)_2$.[6] We can answer the questions posed in this problem by (1) calculating the $OH^-$ concentration needed to bring about the quantitative removal of $Fe^{3+}$ from this solution and (2) determining the $OH^-$ concentration needed to initiate precipitation of $Mg(OH)_2$. If (1) is smaller than (2), a separation is feasible, and the range within which the $OH^-$ concentration must be maintained will be defined by the two values.

We must first decide what constitutes quantitative removal of iron(III) from this solution. It is impossible to remove every $Fe^{3+}$ ion so that we must *arbitrarily* set some limit below which, for all practical purposes, the further presence of this ion can be neglected. When its concentration has been decreased to $10^{-6}$ *M*, the iron remaining in solution will be only 1/100,000 of the original amount. For most purposes, removal of all but this fraction can be considered a quantitative separation.

Substitution of $1.0 \times 10^{-6}$ for $[Fe^{3+}]$ in the solubility product constant permits evaluation of the required $OH^-$ concentration:

$$[OH^-]^3 = \frac{K_{sp}}{[Fe^{3+}]} = \frac{4 \times 10^{-38}}{1.0 \times 10^{-6}} = 4 \times 10^{-32}$$

$$[OH^-] = 3.4 \times 10^{-11}$$

Thus, an $OH^-$ ion concentration of $3.4 \times 10^{-11}$ will lower $[Fe^{3+}]$ to $1 \times 10^{-6}$ *M*. It is of interest to note that the quantitative precipitation of $Fe(OH)_3$ is achieved in a distinctly acidic solution.

We must now perform a similar calculation to establish the $OH^-$ concentration needed to initiate precipitation of $Mg(OH)_2$. Substitution of 0.10 for $[Mg^{2+}]$ into the solubility product expression will permit calculation of the *maximum*[7] $OH^-$ concentration that can be tolerated without formation of $Mg(OH)_2$:

$$[OH^-]^2 = \frac{K_{sp}}{[Mg^{2+}]} = \frac{1.8 \times 10^{-11}}{1.0 \times 10^{-1}} = 1.8 \times 10^{-10}$$

$$[OH^-] = 1.3 \times 10^{-5}$$

These calculations suggest that quantitative separation of $Fe(OH)_3$ can be achieved if the hydroxide ion concentration is greater than $3.4 \times 10^{-11}$ *M*, and that $Mg(OH)_2$ will not precipitate until $[OH^-]$ is about $1.3 \times 10^{-5}$ *M*. It should

---

[6] The reader should be reminded that it is only the enormous difference between the two constants that permits this judgment. The units of $K_{sp}$ for $Mg(OH)_2$ ($mol^3/L^3$) are not the same as those for $Fe(OH)_3$ ($mol^4/L^4$); strictly, then, the two constants are not comparable.

[7] Addition of base to precipitate the iron(III) will cause the concentration of magnesium ion to be less than 0.10, owing to dilution. Use of 0.10 in the calculation will provide a conservative estimate of the $[OH^-]$ that will cause the ion product for $Mg(OH)_2$ to exceed $K_{sp}$.

therefore be possible to separate $Fe^{3+}$ from $Mg^{2+}$ by maintaining the hydroxide ion concentration within this range.

### 5C – 2 Sulfide Separations

Many important separations of cations are based on controlling the concentration of the precipitating anion. This control, in turn, is achieved by regulating the hydronium ion concentration of the solution through use of a suitable buffer.[8] Perhaps the best known of these methods makes use of hydrogen sulfide as the precipitating reagent. Hydrogen sulfide is a very weak acid that undergoes stepwise dissociation:

$$H_2S + H_2O \rightleftharpoons H_3O^+ + HS^- \qquad K_1 = \frac{[H_3O^+][HS^-]}{[H_2S]} = 5.7 \times 10^{-8}$$

$$HS^- + H_2O \rightleftharpoons H_3O^+ + S^{2-} \qquad K_2 = \frac{[H_3O^+][S^{2-}]}{[HS^-]} = 1.2 \times 10^{-15}$$

These equations can be combined to give an expression for the overall dissociation of hydrogen sulfide:

$$H_2S + 2H_2O \rightleftharpoons 2H_3O^+ + S^{2-} \qquad K_1K_2 = \frac{[H_3O^+]^2[S^{2-}]}{[H_2S]} = 6.8 \times 10^{-23}$$

The constant for the overall equilibrium is simply the product of $K_1$ and $K_2$.

The formal concentration of hydrogen sulfide is essentially constant during a typical sulfide separation because the solution is continuously saturated with the gas. Since it is such a weak acid, the actual molar concentration of the species hydrogen sulfide will correspond closely to the solubility of the gas in water, which is about $0.1\ F$; thus, it is permissible to state that

$$[H_2S] \cong 0.1$$

Substitution of this value into the overall dissociation constant expression yields

$$\frac{[H_3O^+]^2[S^{2-}]}{0.1} = 6.8 \times 10^{-23}$$

which rearranges to

$$[S^{2-}] = \frac{6.8 \times 10^{-24}}{[H_3O^+]^2} \qquad\qquad (5-13)$$

Thus, the molar concentration of the sulfide ion varies inversely with the square

---

[8] The preparation and properties of buffer solutions are considered in Chapter 8 (Section 8C–3). A particularly important property of a buffer is its ability to maintain the hydronium ion concentration of a solution at an approximately fixed and predetermined level.

of the hydronium ion concentration of the solution. This relationship makes it possible to establish optimum conditions for the separation of cations based on differences in the solubility of their sulfides.

---

**Example 5-8** Determine the conditions under which $Pb^{2+}$ and $Tl^{+}$ can be separated as sulfides from a solution that is originally 0.1 $F$ with respect to each cation. Solubility data for the two sulfides are

$$PbS(s) \rightleftharpoons Pb^{2+} + S^{2-} \qquad [Pb^{2+}][S^{2-}] = 7 \times 10^{-28}$$

$$Tl_2S(s) \rightleftharpoons 2Tl^{+} + S^{2-} \qquad [Tl^{+}]^2[S^{2-}] = 1 \times 10^{-22}$$

Lead forms the less soluble sulfide and will precipitate first. Assuming again that lowering $[Pb^{2+}]$ to $10^{-6}$ or less constitutes quantitative removal, we can evaluate the required $S^{2-}$ concentration

$$[S^{2-}] = \frac{K_{sp}}{[Pb^{2+}]} = \frac{7 \times 10^{-28}}{1 \times 10^{-6}} = 7 \times 10^{-22}$$

This value can then be compared with the $S^{2-}$ concentration needed to initiate precipitation of $Tl_2S$ from a 0.1 $F$ solution:

$$[S^{2-}] = \frac{K_{sp}}{[Tl^{+}]^2} = \frac{1 \times 10^{-22}}{(1 \times 10^{-1})^2} = 1 \times 10^{-20}$$

Thus, a separation should be feasible if the sulfide ion concentration can be maintained between $7 \times 10^{-22}$ and $1 \times 10^{-20}$ $M$. We may now substitute these limiting values for $[S^{2-}]$ into Equation 5-13 to calculate the corresponding hydronium ion concentrations. To lower $[Pb^{2+}]$ to $1 \times 10^{-6}$ $M$,

$$[H_3O^{+}]^2 = \frac{6.8 \times 10^{-24}}{7 \times 10^{-22}} = 0.99 \times 10^{-2}$$

$$[H_3O^{+}] = 0.099 \cong 0.1$$

and to initiate precipitation of $Tl_2S$,

$$[H_3O^{+}]^2 = \frac{6.8 \times 10^{-24}}{1 \times 10^{-20}} = 6.8 \times 10^{-4}$$

$$[H_3O^{+}] = 0.026 \cong 0.03$$

---

These calculations indicate that PbS can be precipitated without interference from $Tl_2S$ by maintaining $[H_3O^{+}]$ between 0.03 and 0.1. From a practical standpoint, however, it is doubtful that experimental conditions could be controlled sufficiently to give a clean separation.

## 5D EFFECT OF ELECTROLYTE CONCENTRATION ON SOLUBILITY

It is an experimental fact that an electrolyte with no ion common to a sparingly soluble solid causes the solubility of that solid to be greater than it is in water alone. Figure 5–2 demonstrates this effect for barium sulfate, barium iodate, and silver chloride. It is seen that an increase in the potassium nitrate concentration from 0 to 0.02 $F$ causes the solubility for $BaSO_4$ to increase by a factor of 2, for $Ba(IO_3)_2$ by 1.25, and for AgCl by 1.20.

The effect of electrolyte concentration upon solubility has its origins in the electrostatic attractions that exist between foreign ions and the ions of opposite charge in the precipitate. Such interactions shift the position of equilibrium. It is important to realize that this effect is not peculiar to solubility equilibria but is observed with all other types as well. For example, the data in Table 5–2 show that the degree of dissociation for acetic acid increases significantly with increases in the concentration of sodium chloride. The experimental dissociation constants were obtained by measuring the equilibrium concentrations of hydronium and acetate ions in solutions containing the indicated salt concentrations. As with equilibria involving sparingly soluble precipitates, the shift in equilibrium can be attributed to the attraction of the sodium and chloride ions from the electrolyte for the acetate and hydronium ions derived from acetic acid.

These findings, along with myriad others, demonstrate that the equilibrium law, as we have thus far encountered it, is a *limiting law* in the sense that it applies strictly to dilute solutions in which the electrolyte concentration is vanishingly small (that is, ideal solutions). Only we must now consider a more rigorous form of the law which can be applied to nonideal solutions.

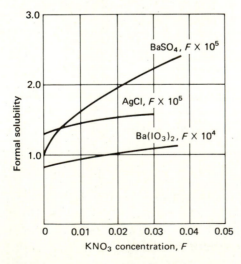

**Figure 5–2**  Effect of electrolyte concentration on the solubility of some salts.

Table 5–2  Dissociation Constants for Acetic Acid
in Solutions of Sodium Chloride at 25°C*

| Concentration of NaCl, $F$ | Apparent Dissociation Constant, $K'_a$ |
|---|---|
| 0.00 | $1.75 \times 10^{-5}$ |
| 0.02 | $2.29 \times 10^{-5}$ |
| 0.11 | $2.85 \times 10^{-5}$ |
| 0.51 | $3.31 \times 10^{-5}$ |
| 1.01 | $3.16 \times 10^{-5}$ |

* From H. S. Harned and C. F. Hickey, *J. Am. Chem. Soc.,* **1937,** *59,* 1289. Courtesy of the American Chemical Society.

## 5D–1 Some Empirical Observations

Extensive studies concerned with the influence of electrolyte concentration upon chemical equilibrium have led to a number of important observations. One is that the magnitude of the effect is highly dependent upon the charges carried by the species involved in the equilibrium. Where all are neutral particles, little or no variation in the position of equilibrium occurs; on the other hand, the effects become greater with increasing charge on the participants of the equilibrium. Thus, for example, for the two equilibria

$$AgCl(s) \rightleftharpoons Ag^+ + Cl^-$$

$$BaSO_4(s) \rightleftharpoons Ba^{2+} + SO_4^{2-}$$

the same concentration of potassium nitrate shifts the second farther to the right than it does the first (Figure 5–2).

A second important generality is that while the effects are essentially independent of the nature of the indifferent electrolyte, they do depend upon a concentration parameter of the solution, called its *ionic strength.* This quantity is defined by the equation

$$\text{ionic strength} = \mu = \tfrac{1}{2}\,(m_1 Z_1^2 + m_2 Z_2^2 + m_3 Z_3^2 + \cdots) \qquad (5-14)$$

where $m_1$, $m_2$, $m_3$, . . . , represent the molar concentrations of the various ions in the solution, and $Z_1$, $Z_2$, $Z_3$, . . . , are their charges. Example 5–9 illustrates how the ionic strength of a solution is calculated.

**Example 5–9**  Calculate the ionic strength of a solution that is

(a)  0.10 $F$ in $KNO_3$.
(b)  0.10 $F$ in $Na_2SO_4$.
(c)  0.050 $F$ in $KNO_3$ and 0.10 $F$ in $Na_2SO_4$.

(a)  Since $m_{K^+} = m_{NO_3^-} = 0.10$,

$$\mu = \tfrac{1}{2}(0.10 \times 1^2 + 0.10 \times 1^2) = 0.10$$

(b)  Here, $m_{Na^+} = 2 \times 0.10$ and $m_{SO_4^{2-}} = 0.10$; thus,

$$\mu = \tfrac{1}{2}(0.20 \times 1^2 + 0.10 \times 2^2) = 0.30$$

(c)  Similarly,

$$\mu = \tfrac{1}{2}(0.050 \times 1^2 + 0.050 \times 1^2 + 0.20 \times 1^2 + 0.10 \times 2^2)$$
$$= 0.35$$

Note that the ionic strength is identical to the formal concentration of a strong electrolyte solution consisting solely of singly charged ions [Example 5–9(a)]. Solutions containing ions that carry multiple charges will possess ionic strengths that are greater than their formal concentrations [Examples 5–9(b) and (c)].

For solutions with ionic strengths that are 0.1 or less, the effect of electrolyte concentration on equilibria is dependent upon the ionic strength and is *independent* of the kinds of ions that are present. Thus, the degree of dissociation for acetic acid is the same in solutions that contain potassium nitrate, sodium chloride, or barium iodide, provided that the concentration of each solute is such that the ionic strengths are identical. It should also be noted that this independence with respect to electrolyte species disappears in solutions with high ionic strengths.

## 5D–2 Activity and Activity Coefficients[9]

The effect of ionic strength on chemical equilibria can be described by substituting concentration parameters called *activities* for molar concentrations in equilibrium constant expressions. The activity for a species is given by

$$a_A = [A]f_A \qquad (5-15)$$

where $a_A$ is the activity for A, [A] is its molar concentration, and $f_A$ is a dimensionless quantity called the *activity coefficient*. The activity coefficient (and thus the activity) of A varies with the ionic strength such that employment of $a_A$ instead of [A] in an equilibrium constant expression frees the numerical value of that constant from dependence upon the ionic strength. Using the dissociation of acetic acid as an example,

$$K_a = \frac{a_{H_3O^+} \cdot a_{OAc^-}}{a_{HOAc}} = \frac{[H_3O^+][OAc^-]}{[HOAc]} \times \frac{f_{H_3O^+} \cdot f_{OAc^-}}{f_{HOAc}}$$

[9] For a detailed treatment of activity, see: T. S. Lee and O. Popovych, in *Treatise on Analytical Chemistry,* 2nd ed., I. M. Kolthoff and P. J. Elving, Eds., Part I, Vol. 1, pp. 527–550. New York: John Wiley & Sons, 1978.

where $f_{H_3O^+}$, $f_{OAc^-}$, and $f_{HOAc}$ vary with the ionic strength in such a way as to keep $K_a$ numerically constant over a wide range of ionic strengths (in contrast to the *apparent $K'_a$* shown in Table 5–2).

**Properties of Activity Coefficients** Activity coefficients have the following properties:

1. The activity coefficient for a species can be thought of as a measure of the effectiveness with which that species influences an equilibrium in which it is a participant. In very dilute solutions, where the ionic strength is low, this effectiveness becomes constant; the activity coefficient then acquires a value of unity, and the activity is identical with the molar concentration. With increasing ionic strength, the ion loses some of its effectiveness, and its activity coefficient decreases. In terms of Equation 5–15, $f_A < 1$ at moderate ionic strengths; as the solution under consideration approaches infinite dilution, $f_A \rightarrow 1$, and $a_A \rightarrow [A]$.

   Activity coefficients for some species become greater than 1.00 at high ionic strengths. Interpretation of such behavior is difficult; we shall confine our discussion to solutions in which $\mu < 0.1$.

   Figure 5–3 illustrates how activity coefficients of some common cations vary as a function of ionic strength.

2. In dilute solutions, the activity coefficient for a given species depends only upon the ionic strength and is independent of the kind of electrolyte(s) contributing to the ionic strength.

3. The activity coefficient of an ion with a multiple charge is smaller at a given ionic strength than the coefficient of an ion of lesser charge (Table 5–3). The activity coefficient for an uncharged molecule is approximately one at all ionic strengths.

4. The activity coefficients for ions of like charge are approximately the same at any given ionic strength. Such small variations as do exist among their activity coefficients can be correlated with their effective diameters in solution.

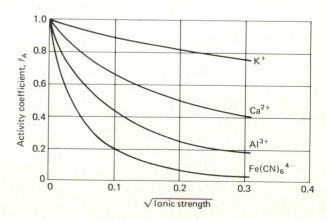

**Figure 5–3** Effect of ionic strength on activity coefficients.

5. The activity coefficient for a given ion describes its effective behavior in all equilibria in which it participates. Thus, for a given ionic strength, a single activity coefficient describes the influence of cyanide ion upon any of the following equilibria:

$$HCN + H_2O \rightleftharpoons H_3O^+ + CN^-$$

$$AgCN(s) \rightleftharpoons Ag^+ + CN^-$$

$$Ni^{2+} + 4CN^- \rightleftharpoons Ni(CN)_4^{2-}$$

$$2MnO_4^- + CN^- + 2OH^- \rightleftharpoons 2MnO_4^{2-} + CNO^- + H_2O$$

**Evaluation of Activity Coefficients**   In 1923 P. Debye and E. Hückel[10] derived a theoretical equation to permit calculation of activity coefficients for ions:

$$-\log f_A = \frac{0.5085\ Z_A^2\ \sqrt{\mu}}{1 + 0.3281\ \alpha_A\ \sqrt{\mu}} \qquad (5-16)$$

where

$f_A$ = activity coefficient for the ionic species A

$Z_A$ = charge carried by A

$\mu$ = ionic strength of the solution

$\alpha_A$ = the effective diameter of the hydrated ion A in ångstrom units (1 ångstrom unit = $10^{-8}$ cm)

The numerical constants in Equation 5–16 are applicable to solutions at 25°C; different values must be used at temperatures other than 25°C.

The least reliable quantity in Equation 5–16 is the magnitude of $\alpha_A$. It appears to be approximately 3 Å for most singly charged ions; for these species, then, the denominator of the Debye-Hückel equation simplifies to approximately $(1 + \sqrt{\mu})$. For ions with higher charge, $\alpha_A$ may be as large as 10 Å. It should be noted that the second term of the denominator becomes negligible for ionic strengths less than 0.01; under these circumstances, uncertainties in $\alpha_A$ have little effect upon activity coefficient calculations.

Kielland[11] has calculated values of $\alpha_A$ for numerous ions, based upon a variety of chemical evidence. His "best" values for effective diameters are given in Table 5–3. Also provided are activity coefficients calculated with Equation 5–16, using the tabulated values for the size parameter.

---

[10] P. Debye and E. Hückel, *Physik Z.,* **1923,** *24,* 185.

[11] J. Kielland, *J. Am. Chem. Soc.,* **1937,** *59,* 1675; see also: T. S. Lee and O. Popovych, in *Treatise on Analytical Chemistry,* 2nd ed., I. M. Kolthoff and P. J. Elving, Eds., Part I, Vol. 1, pp. 538–539. New York: John Wiley & Sons, 1978.

Unfortunately, experimental verification of individual activity coefficients such as those shown in Table 5–3 is impossible; all experimental methods yield only a mean activity coefficient for the positively and negatively charged participants in an equilibrium.[12] It should be pointed out, however, that *mean* activity coefficients calculated from the data in Table 5–3 agree satisfactorily with experimental values.

The Debye-Hückel equation and the data in Table 5–3 give satisfactory values for activity coefficients in solutions with ionic strengths up to about 0.1. For higher concentrations, the equation no longer applies; it thus becomes necessary to evaluate mean activity coefficients experimentally.

**Solubility Calculations Involving Activity Coefficients**  The use of activities instead of molar concentrations in equilibrium constant calculations yields information that is more in accord with experimental fact. Unless otherwise specified, tabulated values for solubility product constants are based upon activities (activity-based constants are called *thermodynamic equilibrium constants*). Thus, for the precipitate $A_mB_n$, we may write

$$K_{sp} = a_A^m \cdot a_B^n = [A]^m[B]^n \cdot f_A^m \cdot f_B^n$$

or

$$[A]^m[B]^n = \frac{K_{sp}}{f_A^m \cdot f_B^n} = K'_{sp}$$

where the bracketed terms are the molar concentrations of A and B. Division of the thermodynamic constant, $K_{sp}$, by the product of the activity coefficients for A and B (or else the mean activity coefficient), produces a concentration constant, $K'_{sp}$ that is applicable to a solution with the ionic strength in question. This constant is then employed in precisely the same way as the thermodynamic constant.

---

[12] The mean activity coefficient $f_\pm$ for the electrolyte $A_mB_n$ is defined as follows:

$$f_\pm = (f_A^m \cdot f_B^n)^{1/(m+n)}$$

Although the mean activity coefficient can be measured in any of several ways, it is experimentally impossible to resolve this term into the individual activity coefficients $f_A$ and $f_B$. If, for example $A_mB_n$ is a sparingly soluble solid, we can write

$$K_{sp} = [A]^m[B]^n \cdot f_A^m \cdot f_B^n = [A]^m[B]^n \cdot f_\pm^{(m+n)}$$

A value for $K_{sp}$ can be obtained by measuring the solubility of $A_mB_n$ in a solution that has an electrolyte concentration that approaches zero (that is, where $f_A$ and $f_B \rightarrow 1$). A second solubility measurement at some ionic strength $\mu_1$ would provide values for [A] and [B]. These data would then permit evaluation of the mean activity coefficient $f_A^m \cdot f_B^n = f_\pm^{(m+n)}$ for ionic strength $\mu_1$. It is important to appreciate that there are insufficient experimental data to allow calculation of the individual activity coefficients $f_A$ and $f_B$ and that there appears to be no additional experimental information that would permit evaluation of these quantities. This situation is general; the *experimental* determination of individual activity coefficients appears to be impossible.

**Table 5–3    Activity Coefficients for Ions at 25°C***

| Ion | $\alpha_A$ Effective Diameter, Å | Activity Coefficient at Indicated Ionic Strengths | | | | |
|---|---|---|---|---|---|---|
| | | 0.001 | 0.005 | 0.01 | 0.05 | 0.1 |
| $H_3O^+$ | 9 | 0.967 | 0.933 | 0.914 | 0.86 | 0.83 |
| $Li^+$, $C_6H_5COO^-$ | 6 | 0.965 | 0.929 | 0.907 | 0.84 | 0.80 |
| $Na^+$, $IO_3^-$, $HSO_3^-$, $HCO_3^-$, $H_2PO_4^-$, $H_2AsO_4^-$, $OAc^-$ | 4–4.5 | 0.964 | 0.928 | 0.902 | 0.82 | 0.78 |
| $OH^-$, $F^-$, $SCN^-$, $HS^-$, $ClO_3^-$, $ClO_4^-$, $BrO_3^-$, $IO_4^-$, $MnO_4^-$ | 3.5 | 0.964 | 0.926 | 0.900 | 0.81 | 0.76 |
| $K^+$, $Cl^-$, $Br^-$, $I^-$, $CN^-$, $NO_2^-$, $NO_3^-$, $HCOO^-$ | 3 | 0.964 | 0.925 | 0.899 | 0.80 | 0.76 |
| $Rb^+$, $Cs^+$, $Tl^+$, $Ag^+$, $NH_4^+$ | 2.5 | 0.964 | 0.924 | 0.898 | 0.80 | 0.75 |
| $Mg^{2+}$, $Be^{2+}$ | 8 | 0.872 | 0.755 | 0.69 | 0.52 | 0.45 |
| $Ca^{2+}$, $Cu^{2+}$, $Zn^{2+}$, $Sn^{2+}$, $Mn^{2+}$, $Fe^{2+}$, $Ni^{2+}$, $Co^{2+}$, $Phthalate^{2-}$ | 6 | 0.870 | 0.749 | 0.675 | 0.48 | 0.40 |
| $Sr^{2+}$, $Ba^{2+}$, $Cd^{2+}$, $Hg^{2+}$, $S^{2-}$ | 5 | 0.868 | 0.744 | 0.67 | 0.46 | 0.38 |
| $Pb^{2+}$, $CO_3^{2-}$, $SO_3^{2-}$, $C_2O_4^{2-}$ | 4.5 | 0.868 | 0.742 | 0.665 | 0.46 | 0.37 |
| $Hg_2^{2+}$, $SO_4^{2-}$, $S_2O_3^{2-}$, $CrO_4^{2-}$, $HPO_4^{2-}$ | 4.0 | 0.867 | 0.740 | 0.660 | 0.44 | 0.36 |
| $Al^{3+}$, $Fe^{3+}$, $Cr^{3+}$, $La^{3+}$, $Ce^{3+}$ | 9 | 0.738 | 0.54 | 0.44 | 0.24 | 0.18 |
| $PO_4^{3-}$, $Fe(CN)_6^{3-}$ | 4 | 0.725 | 0.50 | 0.40 | 0.16 | 0.095 |
| $Th^{4+}$, $Zr^{4+}$, $Ce^{4+}$, $Sn^{4+}$ | 11 | 0.588 | 0.35 | 0.255 | 0.10 | 0.065 |
| $Fe(CN)_6^{4-}$ | 5 | 0.57 | 0.31 | 0.20 | 0.048 | 0.021 |

\* From J. Kielland, *J. Am. Chem. Soc.,* **1937,** *59,* 1675. Courtesy of the American Chemical Society.

**Example 5–10**    Use activities to calculate the solubility of $Ba(IO_3)_2$ in a solution that is 0.033 $F$ in $Mg(IO_3)_2$; the thermodynamic solubility product constant for $Ba(IO_3)_2$ is $1.57 \times 10^{-9}$ (Appendix 9).

At the outset, we may write

$$K'_{sp} = \frac{1.57 \times 10^{-9}}{f_{Ba^{2+}} f^2_{IO_3^-}} = [Ba^{2+}][IO_3^-]^2$$

We must next evaluate the activity coefficients for $Ba^{2+}$ and $IO_3^-$; this calculation, in turn, requires computation of the ionic strength for the solution. Thus,

$$\mu = \tfrac{1}{2}(m_{Mg^{2+}} \times 2^2 + m_{IO_3^-} \times 1^2)$$
$$= \tfrac{1}{2}(0.033 \times 4 + 0.066 \times 1) = 0.099 \cong 0.10$$

This computation is based on the assumption that the solubility of $Ba(IO_3^-)_2$ makes no significant contribution to the ionic strength of the solution. Owing to the low solubility of the solid, this assumption seems reasonable; nevertheless, verification will be necessary after a provisional value for the solubility of $Ba(IO_3^-)_2$ has been calculated (in situations where this assumption is not justified, an improved value for $\mu$ can be obtained by calculating approximate concentrations of the ions based on the notion that the activities and concentrations are identical; these molar concentrations can then be used in Equation 5–14 to obtain a better estimate of $\mu$).

Table 5–3 indicates that, at an ionic strength of 0.10,

$$f_{Ba^{2+}} = 0.38 \qquad f_{IO_3^-} = 0.78$$

Equation 5–16 could have been used to calculate these quantities if the ionic strength had not matched one of the columns in Table 5–3.

The concentration constant for $Ba(IO_3^-)_2$ at an ionic strength of 0.10 is thus

$$K'_{sp} = [Ba^{2+}][IO_3^-]^2 = \frac{1.57 \times 10^{-9}}{(0.38)(0.78)^2} = 6.8 \times 10^{-9}$$

The remainder of the calculation is identical with that shown in Example 5–3; that is,

$$\text{solubility}, \; s = [Ba^{2+}]$$

$$[IO_3^-] = 2 \times 0.033 + 2s \cong 0.066$$

$$s(0.066)^2 = 6.8 \times 10^{-9}$$

$$s = \frac{6.8 \times 10^{-9}}{4.36 \times 10^{-3}} = 1.56 \times 10^{-6} \; \text{fw/L}$$

The assumption that the solubility of $Ba(IO_3)_2$ makes a negligible contribution to the ionic strength is clearly justified. Note also that neglect of ionic strength effects would have led to a value of $3.60 \times 10^{-7}$ for $s$.

We shall ordinarily use molar concentrations instead of activities in equilibrium calculations in the interest of keeping the calculations simpler. The errors introduced by the assumption of unity for the activity coefficient will not ordinarily be large enough to lead to false conclusions. Nevertheless, Example 5–10 clearly demonstrates that disregard of activity coefficients may introduce a significant numerical error into calculations of this kind; relative errors on the order of 100% are not uncommon.

The reader should be alert to the conditions under which the substitution of concentration for activity is likely to cause appreciable errors. Significant discrepancies can be expected under conditions of high ionic strength (0.01 or larger) or where the ions involved have multiple charges (Table 5–3). The substitution of concentrations for activities in an equilibrium constant will cause the least error in dilute solutions ($\mu < 0.01$) of nonelectrolytes or singly charged ions.

It is also important to note that the decrease in solubility resulting from the introduction of an ion common to the precipitate is in part counterbalanced by the effects of increased ionic strength associated with the presence of the species containing the common ion.

## 5E  ADDITIONAL VARIABLES THAT AFFECT SOLUBILITY

The solubility of a precipitate is influenced by temperature and the presence of organic solvents. Most solids absorb heat as they dissolve and are thus more soluble at higher temperatures. It follows, then, that solubility product constants tend to become numerically larger at higher temperatures.

The addition of an organic solvent tends to cause the solubility of most inorganic substances to be smaller than in pure aqueous solutions. The data for calcium sulfate in Table 5–4 are typical of this effect.

## 5F  RATE OF PRECIPITATE FORMATION

It is important to stress that the numerical value of an equilibrium constant provides no clues concerning the rate at which the condition of equilibrium is achieved. Many reactions with favorable equilibrium constants approach equilibrium at a slow and sometimes imperceptible rate.

**Table 5–4    Solubility of Calcium Sulfate in Aqueous Ethanol Solutions***

| Concentration of Ethanol (w/w %) | Solubility, g CaSO$_4$/100 g Solvent |
|---|---|
| 0 | 0.208 |
| 6.2 | 0.100 |
| 13.6 | 0.044 |
| 23.8 | 0.014 |
| 33.0 | 0.0052 |
| 41.0 | 0.0029 |

* From T. Yamamoto, *Bull. Inst. Phys. Chem. Res.* (Tokyo) **1930**, *9*, 352.

Precipitation reactions are often quite slow with the time required for attainment of equilibrium often being minutes or even hours. The chemist can occasionally take advantage of a slow reaction rate to accomplish separations that would not be feasible if equilibrium were approached rapidly. For example, it would appear impossible to separate calcium and magnesium based upon the difference in solubilities of their oxalates. Nevertheless, this separation is feasible because the rate at which magnesium oxalate is formed is much slower than that for calcium oxalate. If the calcium oxalate is filtered shortly after it has been formed, a solid that is essentially free of contamination by magnesium is obtained; if the precipitate is permitted to remain in extended contact with the solution from which it was formed, contamination by magnesium oxalate will occur.

## PROBLEMS

See Appendix 9 for additional solubility product data. Unless specified otherwise, consider that activity coefficients $\rightarrow 1.0$.

**5-1.** Write a net ionic equation for the equilibrium that is established between the following sparingly soluble substances and their ions in solution:

- *(a) AgSCN
- (b) $RaSO_4$
- *(c) $Ag_2S$
- (d) $Cd(OH)_2$
- *(e) $Ag_3AsO_4$
- (f) $La(IO_3)_3$
- *(g) PbClF
- (h) $MgNH_4PO_4$
- *(i) $Ca_3(AsO_4)_2$
- (j) $La_2(C_2O_4)_3$

**\*5-2.** Generate the solubility product expression for each of the equilibria in Problem 5-1.

**5-3.** Information concerning saturated solutions is provided in the accompanying table. Supply the missing data.

| Substance (gfw) | Solubility, mg/L | Molar Concentration of Cation | of Anion | $K_{sp}$ |
|---|---|---|---|---|
| *(a) AgI (234) | | $9.1 \times 10^{-9}$ | | |
| (b) $RaSO_4$ (322) | 2.11 | | | |
| *(c) $Pb(IO_3)_2$ (557) | | | | $3.2 \times 10^{-13}$ |
| (d) $Hg_2I_2$ (655) | | $2.2 \times 10^{-10}$ | $4.4 \times 10^{-10}$ | |
| *(e) $BiI_3$ (590) | | | | $8.1 \times 10^{-19}$ |
| (f) $La(OH)_3$ (190) | 1.76 | | | |
| (g) $Zn_2Fe(CN)_6$ (343) | | | | $4.1 \times 10^{-16}$ |
| (h) $Ag_3Fe(CN)_6$ (535) | 0.66 | | | |
| *(i) $In_4[Fe(CN)_6]_3$ (1109) | | | $4.8 \times 10^{-7}$ | |

**5-4.** Calculate the formal solubility of AgSCN in
- (a) water.
- (b) 0.0200 $F$ $AgNO_3$.
- (c) 0.0200 $F$ KSCN.

**\*5–5.** Calculate the formal solubility of $Mg(OH)_2$ in
    (a) water.
    (b) 0.0300 $F$ $MgSO_4$.
    (c) 0.0300 $F$ KOH.

**5–6.** Calculate the formal solubility of the following in (i) water, (ii) a 0.0800 $F$ solution of the cation, and (iii) a 0.0800 $F$ solution of the anion:
    \*(a) CuBr
    (b) $SrSO_4$
    \*(c) $PbI_2$
    (d) $Ag_2CrO_4$
    \*(e) $Cd(IO_3)_2$ $[K_{sp} = 4.3 \times 10^{-8}]$
    (f) $La(IO_3)_3$
    \*(g) $Ag_3Fe(CN)_6$ $[K_{sp} = 6.2 \times 10^{-23}]$
    (h) $Cd_2Fe(CN)_6$ $[K_{sp} = 3.2 \times 10^{-17}]$

**5–7.** Calculate the molar concentration of $IO_3^-$ in a solution that is 0.050 $F$ in
    \*(a) $Ag^+$ and saturated with $AgIO_3$.
    (b) $Sr^{2+}$ and saturated with $Sr(IO_3)_2$ $[K_{sp} = 3.3 \times 10^{-7}]$.
    \*(c) $In^{3+}$ and saturated with $In(IO_3)_3$ $[K_{sp} = 1.3 \times 10^{-11}]$.
    (d) $Ce^{4+}$ and saturated with $Ce(IO_3)_4$ $[K_{sp} = 4.7 \times 10^{-17}]$.

**5–8.** Calculate the molar cation concentration of a solution that is 0.040 $F$ in $IO_3^-$ and saturated with
    \*(a) $AgIO_3$.         (b) $Sr(IO_3)_2$.
    \*(c) $In(IO_3)_3$.       (d) $Ce(IO_3)_4$.
    (See Problem 5–7 for solubility product data.)

**5–9.** Establish whether a precipitate can be expected when the indicated volumes of solutions A and B are mixed.

| Solution A | Solution B |
| --- | --- |
| \*(a) 40.0 mL 0.0060 $F$ HCl | 40.0 mL 0.0400 $F$ $TlNO_3$ |
| (b) 60.0 mL 0.0150 $F$ KCl | 40.0 mL 0.0400 $F$ $TlNO_3$ |
| \*(c) 60.0 mL 0.0150 $F$ $CaCl_2$ | 40.0 mL 0.0400 $F$ $TlNO_3$ |
| (d) 40.0 mL 0.0400 $F$ KCl | 80.0 mL 0.0600 $F$ $Pb(NO_3)_2$ |
| \*(e) 40.0 mL 0.0400 $F$ $CaCl_2$ | 80.0 mL 0.0600 $F$ $Pb(NO_3)_2$ |
| (f) 40.0 mL 0.0400 $F$ $Pb(NO_3)_2$ | 80.0 mL 0.0600 $F$ KCl |

**5–10.** Establish whether a precipitate can be expected when 25.0 mL of 0.030 $F$ $HIO_3$ are mixed with 35.0 mL of a solution that is $3.0 \times 10^{-3}$ $F$ in
    (a) KCl $[K_{sp}\ KIO_3 = 5 \times 10^{-2}]$.
    (b) $CuCl_2$ $[K_{sp}\ Cu(IO_3)_2 = 7.4 \times 10^{-8}]$.
    (c) $CaCl_2$ $[K_{sp}\ Ca(IO_3)_2 = 7.1 \times 10^{-7}]$.
    (d) $CeCl_3$ $[K_{sp}\ Ce(IO_3)_3 = 3.2 \times 10^{-10}]$.

**\*5–11.** Calculate the concentration of each cation in the solution that results when 60.0 mL of 0.0400 $F$ KI are mixed with 40.0 mL of a solution that is 0.025 $F$ in
    (a) $Ag^+$.
    (b) $Hg_2^{2+}$.
    (c) $Bi^{3+}$ $[K_{sp}\ BiI_3 = 8.1 \times 10^{-19}]$.

**5–12.** Calculate the concentration of each cation in the solution that results when

80.0 mL of 0.050 $F$ $KIO_3$ are mixed with 40.0 mL of a solution that is 0.030 $F$ in

(a) $AgNO_3$.        (b) $SrCl_2$.

(c) $InCl_3$.        (d) $Ce(SO_4)_2$.

(See Problem 5–7 for solubility product data.)

**5–13.** Calculate the concentration of the cation in a saturated solution of

*(a) BiOOH $[K_{sp} = 4.0 \times 10^{-10}]$.

(b) $Co(OH)_2$ $[K_{sp} = 2.0 \times 10^{-16}]$.

*(c) $Zr(OH)_4$ $[K_{sp} = 1 \times 10^{-52}]$.

(d) $La(OH)_3$ $[K_{sp} = 2 \times 10^{-19}]$.

*(e) $TiO(OH)_2$ $[K_{sp} = 1 \times 10^{-29}]$.

(f) $Ce(OH)_3$ $[K_{sp} = 2 \times 10^{-20}]$.

*(g) $Co(OH)_3$ $[K_{sp} = 1 \times 10^{-43}]$.

(h) $Hf(OH)_4$ $[K_{sp} = 4 \times 10^{-26}]$.

**5–14.** Write material-balance equations and charge-balance equations for a solution that is

*(a) 0.050 $F$ in KCl.

(b) 0.090 $F$ in $Mg(NO_3)_2$.

*(c) 0.075 $F$ in HOCl.

(d) 0.040 $F$ in HOCl and 0.060 $F$ in NaOCl.

*(e) 0.020 $F$ in $Ba(OH)_2$ and 0.100 $F$ in $BaCl_2$.

(f) 0.060 $F$ in $H_2SO_3$.

*(g) 0.050 $F$ in $Na_2SO_4$ and 0.040 $F$ in $MgCl_2$.

(h) 0.025 $F$ in $H_2SO_3$ and 0.060 $F$ in $NaHSO_3$.

*(i) 0.060 $F$ in $NaHSO_3$ and 0.050 $F$ in $Na_2SO_3$.

**5–15.** A 0.0800 $F$ solution of acetic acid is saturated with the sparingly soluble silver acetate. Write material-balance and charge-balance equations that will aid in the calculation of the $Ag^+$ concentration of the solution.

**5–16.** Write material-balance and charge-balance equations for the following systems. If such is needed, $s$ should be used to indicate the solubility of a sparingly soluble species.

(a) a saturated solution of AgI

*(b) a saturated solution of $PbI_2$

(c) a 0.020 $F$ $MgI_2$ solution that is saturated with AgI

*(d) 100 mL of 0.100 $F$ NaCN, to which 2.00 mfw of AgCl have been added [product: $Ag(CN)_2^-$]

**5–17.** Calculate the solubility of AgCN in a solution that has a fixed $H_3O^+$ concentration of

*(a) $1.00 \times 10^{-5}$.        *(b) $1.00 \times 10^{-7}$.

(c) $1.00 \times 10^{-9}$.        (d) $1.00 \times 10^{-11}$.

[Formation of $Ag(CN)_2^-$ will be negligible in these solutions.]

**5–18.** Calculate the solubility of $BaCO_3$ in a solution that has a fixed $H_3O^+$ concentration of

*(a) $1.00 \times 10^{-6}$.        *(b) $1.00 \times 10^{-8}$.

(c) $1.00 \times 10^{-10}$.        (d) $1.00 \times 10^{-12}$.

**5–19.** Calculate the solubility of $Ag_2CO_3$ in a solution that has a fixed hydronium ion concentration of

*(a) $1.00 \times 10^{-6}$.        (b) $1.00 \times 10^{-8}$.

(c) $1.00 \times 10^{-10}$.        (d) $1.00 \times 10^{-12}$.

**5-20.** The equilibrium constant for the formation of $CuCl_2^-$ is given by

$$Cu^+ + 2Cl^- \rightleftharpoons CuCl_2^- \qquad K_f = \frac{[CuCl_2^-]}{[Cu^+][Cl^-]^2} = 8.7 \times 10^4$$

Calculate the solubility of CuCl in a solution that has a formal NaCl concentration of

*(a)  $1.00 \times 10^{-1}$.    (b)  $1.00 \times 10^{-2}$.

(c)  $1.00 \times 10^{-3}$.    (d)  $1.00 \times 10^{-4}$.

**5-21.** Potassium iodate is slowly introduced to a solution that is 0.040 $F$ in $Ag^+$ and 0.050 $F$ in $Ba^{2+}$.

(a)  Which iodate will precipitate first?

(b)  What will be the concentration of the ion that forms the less soluble iodate at the onset of precipitation by the more soluble iodate?

**5-22.** Determine whether the following separations are theoretically feasible. Indicate which species will precipitate first; use a residual concentration of $1.0 \times 10^{-6}$ $M$ as the criterion for quantitative removal. Where separation is feasible, calculate the range within which the concentration of the precipitating reagent should be maintained.

| Species to Be Separated | Precipitant |
|---|---|
| *(a)  0.040 $F$ $Ag^+$, 0.075 $F$ $Hg_2^{2+}$ | $SCN^-$ |
| $\quad$ [$K_{sp}$ $Hg_2(SCN)_2 = 3.0 \times 10^{-20}$] | |
| *(b)  0.060 $F$ $Hg_2^{2+}$, 0.080 $F$ $Bi^{3+}$ | $I^-$ |
| $\quad$ [$K_{sp}$ $BiI_3 = 8.1 \times 10^{-19}$] | |
| (c)  0.060 $F$ $Pb^{2+}$, 0.050 $F$ $Bi^{3+}$ | $I^-$ |
| (d)  0.0090 $F$ $Mg^{2+}$, 0.100 $F$ $Cu^{2+}$ | $OH^-$ |
| *(e)  0.025 $F$ $Zn^{2+}$, 0.0080 $F$ $In^{3+}$ | $OH^-$ |
| $\quad$ [$K_{sp}$ $In(OH)_3 = 6 \times 10^{-34}$] | |
| (f)  0.015 $F$ $La^{3+}$, 0.075 $F$ $Ag^+$ | $IO_3^-$ |

**5-23.** Determine which of the following cations can be separated as sulfides through control of the $H_3O^+$ concentration in a saturated $H_2S$ solution. Indicate which sulfide will precipitate first; use a residual concentration of $1.0 \times 10^{-6}$ $M$ as the criterion for quantitative removal. Where separation is feasible, calculate the range within which the hydronium ion concentration should be maintained.

*(a)  0.050 $F$ $Fe^{2+}$ and 0.080 $F$ $Cd^{2+}$

(b)  0.075 $F$ $Mn^{2+}$ and 0.060 $F$ $Ce^{3+}$ [$K_{sp}$ $Ce_2S_3 = 6 \times 10^{-11}$]

*(c)  0.025 $F$ $Fe^{2+}$ and 0.040 $F$ $Tl^+$

(d)  0.030 $F$ $Zn^{2+}$ and 0.045 $F$ $Fe^{2+}$

*(e)  0.050 $F$ $Zn^{2+}$ and 0.080 $F$ $Tl^+$

**5-24.** Calculate the ionic strength of a solution that is

*(a)  $1.67 \times 10^{-3}$ $F$ in $Mg(NO_3)_2$.

(b)  $3.33 \times 10^{-3}$ $F$ in $K_2SO_4$.

*(c)  $1.33 \times 10^{-3}$ $F$ in $BaCl_2$ and $6.01 \times 10^{-3}$ $F$ in $KNO_3$.

(d)  $1.25 \times 10^{-3}$ $F$ in $MgNH_4Cl_3$.

*(e)  $1.00 \times 10^{-2}$ $F$ in $Ca(NO_3)_2$ and 0.070 $F$ in $NaNO_3$.

(f)  $8.00 \times 10^{-3}$ $F$ in $K_4Fe(CN)_6$ and $2.0 \times 10^{-2}$ $F$ in $NaCl$.

*(g)  $1.00 \times 10^{-2}$ $F$ in $MgSO_4$ and $5.00 \times 10^{-3}$ $F$ in $Mg_2Fe(CN)_6$.

**5-25.** Calculate the concentration constant $K'_{sp}$ for
  *(a) $Ba(IO_3)_2$ in the solution described in Problem 5-24(a).
  (b) $MgC_2O_4$ in the solution described in Problem 5-24(b).
  *(c) $Fe(OH)_2$ in the solution described in Problem 5-24(c).
  (d) $Cd_2Fe(CN)_6$ $(K_{sp} = 3.2 \times 10^{-17})$ in the solution described in Problem 5-24(d).

**\*5-26.** Calculate the ionic strength of the solution that results when 1.50 mfw of $Mg(OH)_2$ are mixed with 100 mL of 0.0350 $F$ HCl.

**5-27.** A 0.232-g specimen of $Mg(OH)_2$ is treated with 100 mL of 0.0250 $F$ $H_2SO_4$.
  (a) Calculate the ionic strength of the resulting solution.
  (b) Calculate the concentration constant, $K'_{sp}$, for $Mg(OH)_2$ in this solution.
  (c) Compare values for the $OH^-$ concentration computed with $K_{sp}$ and with $K'_{sp}$. Calculate the percent relative error associated with neglect of activities in this system.

# Chapter 6
# An Introduction to Titrimetric Methods of Analysis

In a *titrimetric method* of analysis, the concentration of an analyte is determined by measuring its combining capacity for a standard reagent.[1] The most common standard reagent is a solution containing a known concentration of a species that reacts more or less completely with the analyte. The volume of the standard solution required to complete this reaction serves as the analytical parameter. Such a procedure is often called a *volumetric method*. A *coulometric method* is a second type of titrimetric procedure in which the standard reagent is an electric current of accurately known magnitude. Here, the quantity of electricity (that is, the number of moles of electrons) to react completely with the analyte is determined. Coulometric methods are discussed in Chapter 14, Section 14C. This chapter is concerned with volumetric methods.

## 6A TERMINOLOGY ASSOCIATED WITH VOLUMETRIC METHODS

A volumetric method makes use of one or more *standard solutions,* which are reagents whose concentrations are exactly known. A *titration* involves determination of the volume of standard required to react completely with the analyte contained in a known weight or volume of a sample. In some instances, it is convenient or necessary to add an excess of the standard solution and then determine the excess by *back-titration* with a second standard reagent.

Great care is taken to establish the concentration of a standard solution accurately because the ultimate reliability of a volumetric method depends directly on the quality of this parameter. The concentration of a standard solution is established either

1. directly, by dissolving a carefully weighed quantity of a highly pure compound and diluting to an exactly known volume; or
2. indirectly, by titrating a solution containing a weighed quantity of a pure compound with the standard solution.

---

[1] For a detailed discussion of titrimetric methods, see: J. I. Watters in *Treatise on Analytical Chemistry,* I. M. Kolthoff and P. J. Elving, Eds., Part I, Vol. 11, Chapter 114. New York: John Wiley & Sons, 1975.

The highly pure substance required for either of these approaches is called a *primary standard.* The process by which the concentration of a standard solution is determined by titration, as in (2), is called a *standardization.*

The goal of every titration is the introduction of standard solution in an amount that is chemically equivalent to the substance with which it reacts. This condition is achieved at the *equivalence point.* For example, the equivalence point in the titration of sodium chloride with silver nitrate occurs after exactly one formula weight of silver ion has been introduced for each formula weight of chloride ion in the sample. Likewise, the equivalence point in the titration of sulfuric acid with sodium hydroxide involves the introduction of two formula weights of base for each formula weight of the acid.

The equivalence point for a titration is a theoretical concept; in fact, this point can only be estimated by observing some physical change associated with the condition of equivalence. Such changes occur at the *end point* of the titration. It is to be hoped that any volume difference between the equivalence point and the end point will be small. Such differences do exist, however, owing to inadequacies in the physical changes and our ability to observe them; the result is a *titration error.*

A common method of detecting the end point in a volumetric analysis involves the use of a supplementary substance that undergoes a change in color as a result of the concentration changes that occur in the vicinity of the equivalence point; such a substance is called an *indicator.*

## 6B REACTIONS AND REAGENTS USED IN VOLUMETRIC ANALYSIS

Volumetric methods are based upon precipitation, neutralization (acid-base), complex formation, and oxidation-reduction reactions. The equivalent weight is defined differently for each of these reaction types; additional differences among them include the equilibria involved, as well as the reagents, primary standards, and indicators that are available.

### 6B – 1 Primary Standards

The accuracy of a volumetric analysis is critically dependent upon the *primary standard* that is used — directly or indirectly — to establish the concentration of the standard solution. Important requirements for primary standard substances include the following:

1. Highest purity; established methods should also exist to confirm its purity.
2. Stability; it should not be attacked by constituents of the atmosphere.
3. Absence of hydrate water; the possibility of deliquescence or efflorescence introduces uncertainty regarding the composition of the substance.
4. Ready availability at modest cost.
5. A reasonably large equivalent weight to minimize the relative error associated with the weighing operation.

Few substances meet or even approach these criteria; as a result, the number of primary-standard substances available to the chemist is limited.

It may occasionally become necessary to base an analysis on a substance that fails to meet all of the requirements of a primary standard. The assay (that is, the percentage purity) of such a *secondary standard* must be established by careful analysis.

### 6B – 2 Standard Solutions

The ideal standard solution for a titrimetric analysis should possess the following properties:

1. Its concentration should remain unchanged indefinitely to eliminate the need for periodic restandardization.
2. Its reaction with the analyte should be sufficiently rapid so that there is no need for a waiting period after each addition of reagent.
3. Its reaction with the analyte should be reasonably complete, because only then are satisfactory end points realized.
4. It should react only with the analyte, and this reaction should be describable by a balanced chemical equation.

Few volumetric reagents meet all these requirements perfectly.

## 6C CALCULATIONS ASSOCIATED WITH VOLUMETRIC METHODS OF ANALYSIS

In preparation for the discussion that follows, it may be helpful to review the material in Chapter 2, Section 2B on the chemical units of mass and concentration.

The terms *equivalent weight* (or *milliequivalent weight*) and *normality* are frequently used by chemists in conjunction with volumetric calculations. The definition of these terms depends upon the type of reaction that serves as the basis for the titration — that is, whether neutralization, oxidation-reduction, precipitation, or complex formation is involved. It should be noted that a compound can often enter into more than one type of reaction and can therefore possess more than one equivalent weight. *Thus, the definition of equivalent weight or milliequivalent weight for a substance is always based on its behavior in a specific chemical reaction (or sequence of reactions). This reaction (or sequence of reactions) must be unambiguously stated before the equivalent weight can be defined. Likewise, the concentration of a solution cannot be expressed in terms of normality without this information.*

### 6C – 1 Definition of Equivalent Weight for Neutralization Reactions

The equivalent weight (eq wt) of a substance participating in a neutralization reaction is the weight in grams that will react with, or supply, one gram-formula weight of hydrogen ion *in a particular reaction.* The milliequivalent weight (meq wt) is equal to the equivalent weight divided by 1000.

The relationship between equivalent weight and gram-formula weight is straightforward for acids or bases that contain a single reactive hydrogen or hydroxide ion. For example, the equivalent weights for hydrochloric acid and potassium hydroxide must necessarily be identical with their formula weights. Similarly, we know that only one hydrogen in acetic acid, $HC_2H_3O_2$, is acidic; the equivalent weight and the formula weight for this acid must also be identical. Barium hydroxide, $Ba(OH)_2$, is a strong base with two hydroxides that are indistinguishable. This base will react with two hydrogen ions in any acid-base reaction; its equivalent weight will thus be equal to one-half of its gram-formula weight. For aqueous solutions of sulfuric acid, dissociation of the second hydrogen ion is not complete. The hydrogen sulfate ion, however, is a sufficiently strong acid so that both hydrogens participate in all aqueous neutralization reactions. The equivalent weight for $H_2SO_4$ as an acid, therefore, is always one-half of its gram-formula weight.

The situation becomes more complex for acids that contain two (or more) hydrogen ions with differing tendencies to dissociate. For example, it is possible to select an indicator that undergoes a color change when the first proton of phosphoric acid has been neutralized; that is,

$$H_3PO_4 + OH^- \rightarrow H_2PO_4^- + H_2O$$

Other indicators undergo color change after two of the protons have reacted:

$$H_3PO_4 + 2OH^- \rightarrow HPO_4^{2-} + 2H_2O$$

The equivalent weight for phosphoric acid will be equal to its formula weight in the first of these reactions, and one-half of its formula weight in the second. Here, definition of the equivalent weight requires knowledge as to which of these two stoichiometries is involved (since it is not practical to titrate the third proton of phosphoric acid, an equivalent weight that is one-third of its formula weight is not encountered in the context of a neutralization titration).

### 6C – 2 Definition of Equivalent Weight for Oxidation-Reduction Reactions

One equivalent of a participant in an oxidation-reduction reaction is the weight in grams that is directly or indirectly involved in the transfer of one mole of electrons. The numerical value for the equivalent weight is conveniently established by dividing the gram-formula weight of the substance of interest by the change in oxidation number that is associated with its reaction. For example, consider the oxidation of oxalic acid by permanganate:

$$5H_2C_2O_4 + 2MnO_4^- + 6H^+ \rightarrow 10CO_2 + 2Mn^{2+} + 8H_2O$$

Manganese undergoes a change from the $+7$ to the $+2$ state; the equivalent weight for $MnO_4^-$ will therefore be one-fifth of its formula weight. Each carbon in oxalic acid is oxidized from the $+3$ to the $+4$ state. A two-electron change is associated with each oxalic acid because it contains two carbons; the equivalent weight of the acid will therefore be equal to one-half of its formula weight. By the same reasoning, the equivalent weight of carbon dioxide is equal to its formula weight because it contains a single carbon.

Table 6–1 contains several additional examples demonstrating the derivation of equivalent weights. It is important to appreciate that the equivalent weight for a substance is evaluated from the change in oxidation state that occurs *in the titration* and not necessarily from its oxidation state in the sample. For example, the oxidation state for manganese is $+3$ in $Mn_2O_3$. If the reaction we have been considering were to be used in an analysis for $Mn_2O_3$, a preliminary step would be needed to convert all of the manganese in the sample to permanganate. Each manganese would then undergo a five-electron change during the titration, and the equivalent weight of $Mn_2O_3$ would thus be equal to one-tenth of its formula weight.

In common with neutralization reactions, the equivalent weight for a given oxidizing or reducing agent may vary from reaction to reaction. Potassium permanganate, for example, undergoes four different reactions with reducing

**Table 6–1    Equivalent Weights for Manganese and Carbon Species, Based on the Reaction:**

$$5H_2C_2O_4 + 2MnO_4^- + 6H^+ \rightarrow 10CO_2 + 2Mn^{2+} + 8H_2O$$

| Substance | Equivalent Weight |
|---|---|
| $Mn$ | $\dfrac{\text{gfw } Mn}{5}$ |
| $KMnO_4$ | $\dfrac{\text{gfw } KMnO_4}{5}$ |
| $Ca(MnO_4)_2 \cdot 4H_2O$ | $\dfrac{\text{gfw } Ca(MnO_4)_2 \cdot 4H_2O}{2 \times 5}$ |
| $Mn_2O_3$ | $\dfrac{\text{gfw } Mn_2O_3}{2 \times 5}$ |
| $Mn_3O_4$ | $\dfrac{\text{gfw } Mn_3O_4}{3 \times 5}$ |
| $CO_2$ | $\dfrac{\text{gfw } CO_2}{1}$ |
| $CaC_2O_4$ | $\dfrac{\text{gfw } CaC_2O_4}{2 \times 1}$ |
| $Al_2(C_2O_4)_3$ | $\dfrac{\text{gfw } Al_2(C_2O_4)_3}{6 \times 1}$ |

agents, depending upon the conditions that exist in the solutions. The half-reactions are

$$MnO_4^- + e \rightarrow MnO_4^{2-}$$

$$MnO_4^- + 3e + 2H_2O \rightarrow MnO_2(s) + 4OH^-$$

$$MnO_4^- + 4e + 3H_2P_2O_7^{2-} + 8H^+ \rightarrow Mn(H_2P_2O_7)_3^{3-} + 4H_2O$$

$$MnO_4^- + 5e + 8H^+ \rightarrow Mn^{2+} + 4H_2O$$

The changes in oxidation state for manganese are 1, 3, 4, and 5, respectively; the equivalent weights for potassium permanganate will therefore be equal to the gram-formula weight for the first reaction, and one-third, one-fourth, and one-fifth of the gram-formula weight for the others.

### 6C-3 Definition of Equivalent Weight for Precipitation and Complex Formation Reactions

An unambiguous definition of the equivalent weight for a substance involved in a precipitation or a complex formation reaction is awkward at best. For this reason, many chemists prefer to avoid the term for reactions of this type and instead use the gram-formula weight exclusively. We are in agreement with this practice and will follow it throughout this text. Nevertheless, the reader may encounter situations where equivalent or milliequivalent weights are specified for substances involved in precipitation or complex formation; it is thus important to understand how these quantities are defined.

The equivalent weight of a participant in a precipitation or a complex formation reaction is the weight that reacts with or provides one gram-formula weight of the *reacting* cation if it is univalent, one-half of a gram-formula weight if it is divalent, one-third of a gram-formula weight if it is trivalent, and so on. Our earlier definitions of equivalent weight were based either on one mole of hydrogen ions or on one mole of electrons. Here, one mole of a univalent cation (or the equivalent thereof) is used. *The cation referred to in this definition is always the cation directly involved in the reaction* and not necessarily the cation contained in the compound whose equivalent weight is being defined.

A participant in a complex formation reaction sometimes has an equivalent weight that is a multiple of its formula weight. For example, in the titration of potassium cyanide with silver ion, the volumetric reaction is

$$Ag^+ + 2CN^- \rightarrow Ag(CN)_2^-$$

The reacting cation, $Ag^+$, is clearly univalent, and two cyanide ions combine with it. Thus, the equivalent weight of KCN will by definition be twice its gram-formula weight.

## 6C – 4  Definition of Equivalent Weight for Species that Do Not Participate Directly in a Volumetric Reaction

Occasionally, an equivalent weight may be needed for an analyte that is related only indirectly to the actual participants in the titration. To illustrate, lead(II) can be determined by precipitation as lead chromate from a slightly acidic solution:

$$Pb^{2+} + CrO_4^{2-} \rightarrow PbCrO_4(s)$$

The precipitate is then filtered, washed free of excess precipitant, and redissolved in dilute hydrochloric acid:

$$2PbCrO_4(s) + 2H^+ \rightarrow 2Pb^{2+} + Cr_2O_7^{2-} + H_2O$$

Finally, the dichromate ion that is produced is titrated with a standard solution of iron(II):

$$Cr_2O_7^{2-} + 6Fe^{2+} + 14H^+ \rightarrow 2Cr^{3+} + 6Fe^{3+} + 7H_2O$$

Because the titration involves oxidation-reduction, the equivalent weight of lead must be based upon a change in oxidation state. Clearly, lead exhibits no such change in this sequence. On the other hand, lead combines with chromium in a 1 : 1 ratio and each chromium undergoes a change from the $+6$ to the $+3$ state in the titration. Therefore, a change in oxidation state of 3 is *associated* with each lead, and its equivalent weight is one-third of its gram-formula weight in the context of the overall process.

It often helps to make an inventory of the chemical relationships that exist between the substance whose equivalent weight is sought and the various participants in the analytical process. The reaction type for the titration is first identified. The various processes are then scanned for the desired relationships. Thus, in the present example,

$$2Pb^{2+} \equiv 2CrO_4^{2-} \equiv Cr_2O_7^{2-} \equiv 6Fe^{2+} \equiv 6e$$

This accounting shows that 2 gfw of lead are indirectly associated with the transfer of 6 mol of electrons. The weight of lead that is involved in the transfer of 1 mol of electrons is thus

$$\frac{2 \times gfw\ Pb}{6} = \frac{gfw\ Pb}{3}$$

The equivalent weight of any other participant in this process can be deduced in the same way.

## 6C – 5  Calculation of Number of Equivalents and Number of Milliequivalents

The *number of equivalents* (eq) is obtained by dividing the weight of a substance in grams by its equivalent weight. This operation is entirely analogous to that shown in Chapter 2, Section 2B – 1 for calculating the number of formula weights and serves the same purpose — the conversion of a mass from metric to chemical units. The *number of milliequivalents* (meq) of a substance is simply 1000 times the number of equivalents.

---

**Example 6–1** Calculate the number of equivalents in 2.84 g of $KMnO_4$ with respect to the half-reaction

$$MnO_4^- + 5e + 8H^+ \rightarrow Mn^{2+} + 4H_2O$$

$$2.84 \text{ g} \times \frac{5 \text{ eq } KMnO_4}{158.04 \text{ g}} = 0.0899 \text{ eq } KMnO_4$$

---

**Example 6–2** Calculate the number of milliequivalents in 2.84 g of $KMnO_4$ with respect to the half-reaction

$$MnO_4^- + 3e + 2H_2O \rightarrow MnO_2(s) + 4OH^-$$

$$2.84 \text{ g} \times \frac{3 \text{ eq } KMnO_4}{158.04 \text{ g}} \times \frac{1000 \text{ meq}}{\text{eq}} = 53.9 \text{ meq } KMnO_4$$

---

### 6C–6 Concentration Units Used in Volumetric Calculations

Many of the ways by which chemists express the concentration of solutions were reviewed in Chapter 2, Section 2B–2; included among these were formal and molar concentrations, p-values, and various types of percent composition. We must now introduce two additional terms, normality and titer, which are commonly used to define the concentration of solutions for volumetric analysis.

**Normality** *Normality, N,* expresses the number of equivalents (eq) of solute that are contained in one liter of solution or the number of milliequivalents (meq) of solute per milliliter of solution. The latter definition is often more convenient to use. Note that normality is a specialized way of expressing concentration that requires a knowledge of the stoichiometry of the reaction between the standard and the analyte. Little possibility for ambiguity exists where the concentration of a hydrochloric solution is given as, say, 0.02 *N* because this solute reacts in one and only one way as a volumetric reagent. In contrast, a 0.02 *N* $KMnO_4$ solution will have uncertain composition unless an equation that defines the stoichiometry of its reaction with an analyte is also provided (see p. 131).

**Example 6–3**  A solution was prepared by dissolving 3.380 g of primary standard $K_2Cr_2O_7$ (gfw = 294.2) in sufficient water to give a total volume of 500.0 mL. Calculate the normality of this solution in terms of the half-reaction

$$Cr_2O_7^{2-} + 6e + 14H^+ \rightarrow 2Cr^{3+} + 7H_2O$$

As in Example 6–2,

$$3.380 \text{ g} \times \frac{6 \text{ eq } K_2Cr_2O_7}{294.2 \text{ g}} \times \frac{1000 \text{ meq}}{\text{eq}} = 68.93 \text{ meq}$$

Thus,

$$\frac{68.93 \text{ meq } K_2Cr_2O_7}{500 \text{ mL}} = 0.1379 \text{ } N$$

**Example 6–4**  How many milliequivalents are involved in a titration that required 43.50 mL of 0.1379 $N$ $K_2Cr_2O_7$?

$$43.50 \text{ mL} \times \frac{0.1379 \text{ meq } K_2Cr_2O_7}{\text{mL}} = 5.9987 = 6.000 \text{ meq } K_2Cr_2O_7$$

**Titer**   This method of expressing concentration is even more specialized than normality in that it refers to a single analyte. *Titer* is defined as the weight of a substance (ordinarily in milligrams) that reacts with exactly 1.00 mL of the reagent. Thus, a silver nitrate solution with a titer of 1.00 mg of chloride contains just enough $AgNO_3$ in each milliliter to react completely with 1.00 mg of chloride ion. The titer of the same solution could also be expressed in terms of milligrams of potassium chloride, barium chloride, sodium iodide, or any other substance with which silver nitrate reacts. Titer is a useful way of expressing the concentration of a reagent that is to be used for the routine analysis of a particular analyte in a large number of samples.

**Example 6–5**  Calculate the $BaCl_2$ titer of a 0.125 $F$ $AgNO_3$ solution.

$$\text{titer} = \frac{0.125 \text{ mfw } AgNO_3}{\text{mL } AgNO_3} \times \frac{\text{mfw } BaCl_2}{2 \text{ mfw } AgNO_3} \times \frac{208.2 \text{ mg}}{\text{mfw } BaCl_2} = \frac{13.0 \text{ mg } BaCl_2}{\text{mL } AgNO_3}$$

**Example 6-6**   The label on a bottle containing dilute $AgNO_3$ states that the solution has a titer of 8.78 mg NaCl. Calculate the formal concentration of this solution.

$$F_{AgNO_3} = \frac{8.78 \text{ mg NaCl}}{\text{mL } AgNO_3} \times \frac{\text{mfw}}{58.44 \text{ mg NaCl}} \times \frac{\text{mfw } AgNO_3}{\text{mfw NaCl}}$$

$$= \frac{0.1502 \text{ mfw } AgNO_3}{\text{mL}} = 0.1502 \ F$$

---

**Example 6-7**   Calculate the percentage of NaCl in a 0.245-g sample if titration with the silver nitrate solution in Example 6-5 required 19.87 mL.

$$\frac{19.87 \text{ mL } AgNO_3}{245 \text{ mg sample}} \times \frac{8.78 \text{ mg NaCl}}{\text{mL } AgNO_3} \times 100 = 71.2\% \text{ NaCl}$$

### 6C-7 A Fundamental Relation Between Quantities of Reacting Substances

As a consequence of the way equivalent and milliequivalent weight is defined, it is possible to state that *at the equivalence point in a titration, the number of milliequivalents of standard are exactly equal to the number of milliequivalents of analyte.* For example, 1 meq of an acid analyte contributes 1 meq of hydronium ion to a solution regardless of whether the acid is $H_2SO_4$ or HCl; 1 meq of standard base, regardless of kind, consumes just this number of milliequivalents of hydronium ions. Similarly, 1 meq of an oxidizing agent reacts with exactly 1 meq of reductant since the milliequivalent weight of each is based upon transfer of 1 mmol of electrons. The use of the 1:1 relationship between milliequivalents of analyte and of titrant is demonstrated in the examples that follow.

**Standardization of Solutions**   The normality of a standard solution is computed either from the data related to its preparation or from a standardization titration.

---

**Example 6-8**   Calculate the normality of the solution that results from diluting 21.65 mL of 0.0443 $N$ $Ba(OH)_2$ to 50.00 mL.

The solution will contain the same number of milliequivalents of $Ba(OH)_2$ after dilution as it did before. Since the normality of a solution expresses the number of milliequivalents of solute per milliliter, the product of volume and normality provides the number of milliequivalents of $Ba(OH)_2$ taken initially. Thus,

$$mL_{final} \times N_{final} = mL_{initial} \times N_{initial}$$

$$N_{final} = \frac{21.65 \text{ mL} \times 0.0443 \ N}{50.00 \text{ mL}} = 0.0192$$

**Example 6-9**  Calculate the normality of a $Ba(OH)_2$ solution if 31.76 mL were needed to neutralize 46.25 mL of 0.1280 $N$ HCl.

At the equivalence point,

$$\text{meq } Ba(OH)_2 = \text{meq HCl}$$

Substitution of equivalent quantities into this relationship gives

$$mL \ Ba(OH)_2 \times \frac{\text{meq } Ba(OH)_2}{mL \ Ba(OH)_2} = mL \ HCl \times \frac{\text{meq HCl}}{mL \ HCl}$$

$$mL \ Ba(OH)_2 \times N_{Ba(OH)_2} = mL \ HCl \times N_{HCl}$$

$$31.76 \text{ mL} \times N_{Ba(OH)_2} = 46.25 \text{ mL} \times 0.1280 \text{ meq HCl/mL}$$

$$N_{Ba(OH)_2} = \frac{46.25 \times 0.1280}{31.76} = 0.1864 \text{ meq } Ba(OH)_2/mL$$

**Example 6-10**  Calculate the normality of an $I_2$ solution if 37.34 mL were needed to titrate 0.2040 g of primary standard $As_2O_3$ (gfw = 197.8). Reaction:

$$I_2 + H_2AsO_3^- + H_2O \rightarrow 2I^- + H_2AsO_4^- + 2H^+$$

At the equivalence point,

$$\text{meq } I_2 = \text{meq } As_2O_3$$

The number of milliequivalents of $I_2$ can be expressed in terms of the volume and the normality of the solution; the number of milliequivalents of $As_2O_3$ can be calculated from the weight taken for titration. Thus,

$$mL \ I_2 \times N_{I_2} = \frac{\text{wt } As_2O_3}{\text{meq wt } As_2O_3}$$

Arsenic is oxidized from the $+3$ to the $+5$ state in this reaction. Since two atoms of As are contained in each $As_2O_3$, the equivalent weight of the solid must be equal to one-fourth of its gram-formula weight; that is,

$$As_2O_3 \equiv 2H_2AsO_3^- \equiv 2I_2 \equiv 4e$$

Finally, then

$$37.34 \text{ mL} \times N_{I_2} = 0.2040 \text{ g As}_2\text{O}_3 \times \frac{4 \text{ meq As}_2\text{O}_3}{0.1978 \text{ g As}_2\text{O}_3}$$

$$N_{I_2} = \frac{0.2040}{37.34 \times 0.04945} = 0.1105$$

**Calculation of Results from Titrations**   Examples 6–11, 6–12, and 6–13 are typical calculations encountered in working up the results of volumetric analyses.

**Example 6–11**   A 0.804-g sample of iron ore was dissolved in acid. The Fe was reduced to the +2 state and titrated with 47.2 mL of a 0.112 $N$ KMnO$_4$ solution. Reaction:

$$5\text{Fe}^{2+} + \text{MnO}_4^- + 8\text{H}^+ \rightarrow 5\text{Fe}^{3+} + \text{Mn}^{2+} + 4\text{H}_2\text{O}$$

Express the results of this analysis in terms of

(a) the percentage of Fe (gfw = 55.85).
(b) the percentage of Fe$_3$O$_4$ (gfw = 231.5).

(a) At the equivalence point

$$\text{meq Fe} = \text{meq KMnO}_4$$

$$= 47.2 \text{ mL} \times \frac{0.112 \text{ meq}}{\text{mL}}$$

Each Fe(II) loses one electron in this reaction; its milliequivalent weight is therefore identical with its milliformula weight. Thus,

$$\text{wt Fe} = 47.2 \text{ mL} \times \frac{0.112 \text{ meq}}{\text{mL}} \times \frac{55.85 \text{ g Fe}}{1000 \text{ meq}}$$

Finally, then

$$\frac{47.2 \times 0.112 \times 0.05585}{0.804} \times 100 = 36.7\% \text{ Fe}$$

(b) At the equivalence point

$$\text{meq Fe}_3\text{O}_4 = \text{meq KMnO}_4$$

Since each Fe$_3$O$_4$ contains three Fe, and since each Fe loses an electron in the reaction, the equivalent weight for Fe$_3$O$_4$ must be equal to its gram-formula weight divided by 3. By the same arguments, then

$$\frac{47.2 \times 0.112 \times 0.2315/3}{0.804} \times 100 = 50.7\% \ Fe_3O_4$$

Note that the oxidation state of Fe in $Fe_3O_4$ was of no concern; only the change from the $+2$ state to the $+3$ state in the titration was important.

---

**Example 6–12** A 0.475-g sample of impure $(NH_4)_2SO_4$ was dissolved in water and made alkaline with NaOH. The liberated $NH_3$ was distilled into exactly 50.0 mL of 0.100 $N$ HCl. The excess HCl required a back-titration with 11.1 mL of 0.121 $N$ NaOH. Express the results of this analysis in terms of

(a) the percentage of $NH_3$ (gfw = 17.03).
(b) the percentage of $(NH_4)_2SO_4$ (gfw = 132.1).

(a) As always, the number of milliequivalents of acid and base are equal at the equivalence point. In this titration, however, both $NH_3$ and NaOH act as bases. Thus, at the equivalence point

$$meq \ HCl = meq \ NH_3 + meq \ NaOH$$

or

$$meq \ NH_3 = meq \ HCl - meq \ NaOH$$
$$= 50.0 \times 0.100 - 11.1 \times 0.121$$

$$\frac{(50.0 \times 0.100 - 11.1 \times 0.121) \times 0.01703}{0.475} \times 100 = 13.1\% \ NH_3$$

(b) The milliequivalent weight of $(NH_4)_2SO_4$ will be one-half of its gram-formula weight because

$$(NH_4)_2SO_4 \equiv 2NH_3 \equiv 2H^+$$

Thus,

$$\frac{(50.0 \times 0.100 - 11.1 \times 0.121) \times 0.1321/2}{0.475} \times 100 = 50.8\% \ (NH_4)_2SO_4$$

---

**Example 6–13** The organic matter in a 3.77-g sample of a mercurial ointment was decomposed with $HNO_3$. After dilution, the Hg(II) was titrated with 21.3 mL of 0.114 $F$ $NH_4SCN$. Reaction:

$$Hg^{2+} + 2SCN^- \rightarrow Hg(SCN)_2(aq)$$

Calculate

   (a)  the percentage of Hg (gfw = 200.6) in the ointment.

   (b)  the percentage of $Hg(NO_3)_2$ (gfw = 324.6) in the ointment.

Because the reaction involves formation of a complex, we will perform the required calculations in terms of milliformula weights rather than milliequivalent weights (see p. 131).

(a)  The weight of Hg is given by

$$21.3 \text{ mL} \times \frac{0.114 \text{ mfw } NH_4SCN}{\text{mL}} \times \frac{\text{mfw Hg}}{2 \text{ mfw } NH_4SCN}$$

$$\times \frac{0.2006 \text{ g}}{\text{mfw Hg}} = 0.244 \text{ g Hg}$$

and the percentage of Hg is thus

$$\frac{0.244 \text{ g Hg}}{3.77 \text{ g sample}} \times 100 = 6.46\% \text{ Hg}$$

(b)  By the same argument,

$$21.3 \text{ mL} \times \frac{0.114 \text{ mfw } NH_4SCN}{\text{mL}} \times \frac{\text{mfw } Hg(NO_3)_2}{2 \text{ mfw } NH_4SCN}$$

$$\times \frac{0.3246 \text{ g}}{\text{mfw } Hg(NO_3)_2} = 0.394 \text{ g } Hg(NO_3)_2$$

The percentage of $Hg(NO_3)_2$ in the sample is thus

$$\frac{0.394 \text{ g } Hg(NO_3)_2}{3.77 \text{ g sample}} \times 100 = 10.5\%$$

## 6D END POINTS FOR VOLUMETRIC ANALYSIS

End points are based upon observable physical changes that occur in solution during the course of a titration. Methods for visual detection include changes in color, turbidity, and light scattering. Electrical methods include changes in potential, current, and conductivity. End points based upon temperature change are also encountered. The two most widely used end points involve either a change in color due to the reagent, the analyte, or an indicator, or a change in the potential of an electrode that is sensitive to the concentration of the analyte or the reagent (or perhaps both).

    Two general methods exist for the detection of an end point. In one, observations made in the immediate vicinity of the equivalence point form the basis for detection; in the other, the observations are made in regions well removed from that point. The first method offers the advantages of speed and convenience; the second is potentially more sensitive and is applicable to reactions in which the equilibrium constant is not particularly favorable.

### 6D – 1  End Points Based upon Observations Near the Equivalence Point

End points of this type are the result of marked changes in relative reactant or product concentrations in the equivalence point region. To illustrate, the second column of Table 6 – 2 shows changes in the hydronium ion concentration during the titration of 50.00 mL of 0.1000 $F$ HCl with standard 0.1000 $F$ NaOH. To emphasize the *relative* changes that are occurring, the volumes selected for tabulation correspond to the increments needed to cause the concentrations of the reacting species to decrease or increase by an order of magnitude. For example, 40.91 mL of base are needed to decrease the hydronium ion concentration from its initial value of 0.1000 to 0.01000 $F$. Subsequent tenfold decreases require 8.10 and 0.89 mL additions of NaOH. Analogous increases in the hydroxide ion concentration occur simultaneously.

The data in Table 6 – 2 illustrate that a maximum in the rate of change in the concentrations of reacting species occurs in the immediate vicinity of the equivalence point. Owing to the enormous concentration changes that are involved, a plot of the hydronium ion concentration against the volume of base added (columns 1 and 2) yields a curve that clearly shows changes in the early stages of the titration but is not at all informative in depicting events in the equivalence point region. A clearer representation of the changes that occur in this region is obtained by plotting p-values (Chapter 2, Section 2B – 2) rather than concentrations as the ordinate. Thus, pH or pOH would be used for the titration under

**Table 6 – 2   Changes in pH and pOH During Titration of 50.00 mL of 0.1000 F HCl with 0.1000 F NaOH**

| Volume of 0.1000 $F$ NaOH added, mL* | Concentration of $H_3O^+$, mol/L | Concentration of $OH^-$, mol/L | pH | pOH | Volume (mL) of NaOH Needed to Cause a Tenfold change in pH |
|---|---|---|---|---|---|
| 0.00 | $1.0 \times 10^{-1}$ | $1.0 \times 10^{-13}$ | 1.00 | 13.00 | — |
| 40.91 | $1.0 \times 10^{-2}$ | $1.0 \times 10^{-12}$ | 2.00 | 12.00 | 40.91 |
| 49.01 | $1.0 \times 10^{-3}$ | $1.0 \times 10^{-11}$ | 3.00 | 11.00 | 8.10 |
| 49.90 | $1.0 \times 10^{-4}$ | $1.0 \times 10^{-10}$ | 4.00 | 10.00 | 0.89 |
| 49.990 | $1.0 \times 10^{-5}$ | $1.0 \times 10^{-9}$ | 5.00 | 9.00 | 0.09 |
| 49.9990 | $1.0 \times 10^{-6}$ | $1.0 \times 10^{-8}$ | 6.00 | 8.00 | 0.009 |
| 50.0000 | $1.0 \times 10^{-7}$ | $1.0 \times 10^{-7}$ | 7.00 | 7.00 | 0.001 |
| 50.0010 | $1.0 \times 10^{-8}$ | $1.0 \times 10^{-6}$ | 8.00 | 6.00 | 0.001 |
| 50.010 | $1.0 \times 10^{-9}$ | $1.0 \times 10^{-5}$ | 9.00 | 5.00 | 0.009 |
| 50.10 | $1.0 \times 10^{-10}$ | $1.0 \times 10^{-4}$ | 10.00 | 4.00 | 0.09 |
| 51.01 | $1.0 \times 10^{-11}$ | $1.0 \times 10^{-3}$ | 11.00 | 3.00 | 0.91 |
| 61.1 | $1.0 \times 10^{-12}$ | $1.0 \times 10^{-2}$ | 12.00 | 2.00 | 10.09 |

* These volume data are useful for purposes of illustration only. It should be understood that the measurement of volumes and normalities to six significant figures is ordinarily impossible.

discussion; columns 3 and 4 list p-values after each addition of base. Figure 6 – 1 illustrates that the equivalence point region is characterized by a major change in p-values — the pH undergoing a pronounced increase, the pOH experiencing a corresponding decrease.

Curves such as those shown in Figure 6 – 1 are called *titration curves.* Plots of analogous data for precipitation, complex formation, and oxidation-reduction titrations have the same general characteristics. From a practical standpoint, these curves are valuable aids in the selection of an indicator for a particular titration. We shall make frequent use of titration curves in subsequent discussions.

**Visual End Points**   End points based upon visual observations of changes in color or turbidity offer the considerable advantages of speed and simplicity of equipment. The applicability of visual end points is critically dependent, however, upon the magnitude of the relative concentration changes that occur in the region surrounding the equivalence point. The sensitivity of the human eye is such that, typically, a one- to two-unit change in p-value is needed to provide a detectable change in an indicator. For an end point to be useful, then, a change of this magnitude must occur within a volume range that corresponds to the allowable titration error. Ideally, this change will occur symmetrically around the true equivalence point for the titration. Note that a change of four pH units occurs over a volume range corresponding to ±0.01 mL of the equivalence point in the titration illustrated in Table 6 – 2 and Figure 6 – 1. This titration is thus admirably suited for visual end-point detection. Several indicators that exhibit sharp color changes in this region are available, and the titration can be performed with a minimal volumetric error.

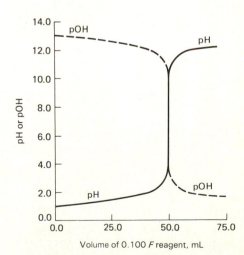

**Figure 6 – 1**   Curves for the titration of 50.00 mL of 0.1000 *F* HCl with 0.1000 *F* NaOH.

**Potentiometric End Points**   The course of a titration in which the change in p-value is too small for practical visual end-point detection can often be followed successfully by measuring the potential that develops between a pair of electrodes, one of which is sensitive to changes in the composition of the solution, the other of which is not. Potentiometric titrations are considered in Chapter 13.

### 6D – 2 End Points Based upon Observations Made Well Away from Equivalence

Visual, and to a lesser extent, potentiometric end points are limited to reactions with favorable equilibrium constants. For less complete reactions, detection must be based upon measurements made in regions where the common ion effect forces a linear relationship between the quantity measured and the concentration of a participant in the titration; the end point is established by graphical extrapolation.

### PROBLEMS

**6–1.** Classify the accompanying reactions according to type, and express the equivalent weight of each listed substance as a fraction or a multiple of its gram-formula weight.

*(a) $CO_3^{2-} + 2H^+ \rightarrow CO_2 + H_2O$  $\qquad$ $HClO_4$, $CO_2$, $Fe_2(CO_3)_3$

(b) $CO_3^{2-} + H^+ \rightarrow HCO_3^-$  $\qquad$ $Na_2CO_3$, $HBr$, $C_2O_3$, $Fe(CO)_5$

*(c) $2MnO_4^- + 5Sn^{2+} + 16H^+ \rightarrow 2Mn^{2+} + 5Sn^{4+} + 8H_2O$
$\qquad\qquad\qquad\qquad MnSO_4 \cdot 7H_2O$, $SnF_2$, $SnO \cdot Sn(NO_3)_2$, $Mn_2O_3$

(d) $B_4O_7^{2-} + 2H^+ + 5H_2O \rightarrow 4H_3BO_3$  $\qquad$ $Na_2B_4O_7 \cdot 10H_2O$, $B_2O_3$, $B$, $B_5H_9$

*(e) $V(OH)_4^+ + Fe^{2+} + 2H^+ \rightarrow VO^{2+} + Fe^{3+} + 3H_2O$
$\qquad\qquad\qquad\qquad\qquad\qquad Na_3VO_4$, $H_2V_4O_{11}$, $Fe_2O_3$

(f) $HONH_3^+ + OH^- \rightarrow HONH_2 + H_2O$
$\qquad\qquad\qquad\qquad NaOH$, $HONH_2$, $(HONH_3)_3PO_4$, $Ba(OH)_2$

*(g) $I_2 + 2S_2O_3^{2-} \rightarrow 2I^- + S_4O_6^{2-}$  $\qquad$ $I_2$, $Na_2S_2O_3 \cdot 5H_2O$, $Na_3Ag(S_2O_3)_2$

**6–2.** Organic compounds containing sulfur can be determined by combustion in a stream of $O_2$ to $CO_2$, $H_2O$, and $SO_2$. The combustion products are passed through a dilute solution of $H_2O_2$; the $SO_2$ is retained as a result of the reaction:

$$H_2O_2 + SO_2 \rightarrow H_2SO_4$$

The $H_2SO_4$ is subsequently titrated with standard base. Use chemical formulas to express the equivalent weight of the following as fractions or multiples of their gram-formula weights when determined by this method.

*(a) thioacetamide, $C_2H_5NS$  $\qquad$ (b) thioctic acid, $C_8H_{14}O_2S_2$

*(c) allyl disulfide, $C_6H_{10}S_2$  $\qquad$ (d) thioacetaldehyde, $C_6H_{12}S_3$

*(e) lenthionine, $C_2H_4S_5$  $\qquad$ (f) merthiolate, $C_9H_9HgNaO_2S$

*(g) cystamine, $C_4H_{12}N_2S_2$  $\qquad$ (h) sulthiamine, $C_{10}H_{14}N_2O_4S_2$

**6–3.** Analysis for nitrogen in organic compounds frequently involves conversion to $NH_4^+$ by digestion in boiling $H_2SO_4$. The digested mixture is cooled, and then rendered alkaline; the liberated $NH_3$ is collected by distillation and subsequently titrated with standard acid. Use chemical formulas to express the equivalent weight of the following compounds as fractions or multiples of their respective gram-formula weights in terms of this analysis.

(a) glutamic acid, $C_5H_9NO_4$

(b) vitamin $K_6$, $C_{11}H_{12}N_2$

(c) thiamiprine, $C_9H_8N_8O_2S$

(d) valinomycin, $C_{54}H_{90}N_6O_{18}$

(e) cyamelide, $C_3H_3N_3O_3$

(f) sulfanilurea, $C_7H_9N_3O_3S$

(g) viomycin, $C_{25}H_{43}N_{13}O_{10}$

(h) guanidine nitrate, $CH_5N_3 \cdot HNO_3$

**6-4.** In each of the following analyses, the cation is first isolated as a sparingly soluble precipitate. The precipitate is filtered, washed free of excess precipitating reagent, and then redissolved; the resulting solution is directly or indirectly titrated with a standard reagent.

Calculate the equivalent weight attributable to the indicated substances in the accompanying analyses; the last equation in each sequence is the titration upon which the analysis is based.

*(a) $Ca^{2+} + C_2O_4^{2-} = CaC_2O_4(s)$

$CaC_2O_4(s) + 2H^+ = H_2C_2O_4 + Ca^{2+}$

$5H_2C_2O_4 + 2MnO_4^- + 6H^+ = 10CO_2(g) + 2Mn^{2+} + 8H_2O$

What is the equivalent weight of Ca, $CaCl_2$, $Ca_3Al_2O_6$, $Ca_2Fe(CN)_6$?

(b) $2Tl^+ + CrO_4^{2-} = Tl_2CrO_4(s)$

$Tl_2CrO_4(s) + 2H^+ = 4Tl^+ + H_2O + Cr_2O_7^{2-}$

$6Fe^{2+} + Cr_2O_7^{2-} + 14H^+ = 6Fe^{3+} + 2Cr^{3+} + 7H_2O$

What is the equivalent weight of Tl, $Tl_2SO_4$, $Tl_4P_2O_7$?

(c) $Hg_2^{2+} + 2IO_3^- = Hg_2(IO_3)_2(s)$

$Hg_2(IO_3)_2(s) + 2H^+ = Hg_2^{2+} + 2HIO_3$

$HIO_3 + 5I^- + 6H^+ = 3I_2 + 3H_2O$

$I_2 + 2S_2O_3^{2-} = S_4O_6^{2-} + 2I^-$

What is the equivalent weight of Hg, $Hg_2(NO_3)_2$?

(d) $Cd^{2+} + S^{2-} = CdS(s)$

$CdS(s) + 2H^+ = Cd^{2+} + H_2S$

$H_2S + I_2 = S(s) + 2I^- + 2H^+$

What is the equivalent weight of Cd, $Cd_3(PO_4)_2$, $Cd_2P_2O_7$?

*6-5. Potassium hydrogen iodate is a versatile primary standard. In addition to being a fairly strong oxidizing agent, it has an acidic hydrogen. A solution is prepared by dissolving 19.46 g of $KH(IO_3)_2$ in sufficient water to give a total volume of 1.000 L. Express the concentration of this solution in terms of its

(a) formality.

(b) normality as an acid.

(c) normality as an oxidizing agent in the reaction:

$$IO_3^- + H_2NNH_2 + 2H^+ + 2Cl^- = ICl_2^- + N_2 + 3H_2O$$

(d) titer, as milligrams $H_2SO_3$, based on the reaction:

$$2H_2SO_3 + IO_3^- + 2Cl^- = 2SO_4^{2-} + ICl_2^- + 2H^+ + H_2O$$

**6-6.** An aqueous solution contains 13.79 g of $K_4Fe(CN)_6$ per liter. Express the concentration of this solution in terms of its

(a) formality.

(b) normality with respect to the reaction:

$$5Fe(CN)_6^{4-} + MnO_4^- + 8H^+ = 5Fe(CN)_6^{3-} + Mn^{2+} + 4H_2O$$

(c) titer, as milligrams Pb per milliliter with respect to the reaction:

$$Fe(CN)_6^{4-} + 2Pb^{2+} = Pb_2Fe(CN)_6(s)$$

(d) titer, as milligrams Zn per milliliter with respect to the reaction:

$$3Zn^{2+} + 2Fe(CN)_6^{4-} + 2K^+ = K_2Zn_3[Fe(CN)_6](s)$$

*6–7. A solution contains 6.50 g of KI in 500 mL. Express the concentration of this solution in terms of its
   (a) formality.
   (b) normality, based on the reaction:

$$BrO_3^- + 3I^- + 3HCN + 3H^+ = Br^- + ICN + 3H_2O$$

   (c) normality, based on the reaction:

$$I^- + 8MnO_4^- + 8OH^- = 8MnO_4^{2-} + IO_4^- + 4H_2O$$

   (d) titer, as milligrams HgO per milliliter based on the reaction:

$$HgO(s) + 4I^- + H_2O = HgI_4^{2-} + 2OH^-$$

6–8. How many milliequivalents of solute are contained in 25.0 mL of
   *(a) 0.0187 $F$ KMnO$_4$ (product: MnF$_6^{3-}$)?
   (b) 0.0187 $F$ KMnO$_4$ (product: MnO$_4^{2-}$)?
   *(c) 0.0244 $N$ I$_2$ (product: I$^-$)?
   (d) 0.0244 $F$ I$_2$ (product: ICl$_2^-$)?
   *(e) 0.0316 $F$ Na$_2$S$_2$O$_3$ (product: S$_4$O$_6^{2-}$)?
   (f) 0.0316 $N$ Na$_2$S$_2$O$_3$ (product: SO$_4^{2-}$)?
6–9. How many grams are contained in 45.0 mL of
   *(a) a K$_2$CrO$_4$ solution that has a titer of 1.20 mg Pb/mL?
   (b) a KCN solution that has a titer of 1.00 mg Ni/mL [product: Ni(CN)$_4^{2-}$]?
   *(c) an HClO$_4$ solution that has a titer of 2.50 mg Na$_3$PO$_4$/mL (product: H$_2$PO$_4^-$)?
   (d) an I$_2$ solution that has a titer of 3.58 mg Na$_2$S$_2$O$_3$/mL (products: I$^-$, S$_4$O$_6^{2-}$)?
   *(e) an Na$_2$S$_2$O$_3$ solution that has a titer of 3.58 mg I$_2$/mL [see Problem 6–8(d)]?
6–10. Calculate the number of milliequivalents in
   *(a) 0.2167 g of Fe as a primary standard for Ce$^{4+}$:

$$Fe^{2+} + Ce^{4+} = Fe^{3+} + Ce^{3+}$$

   (b) 0.2027 g of Na$_3$PO$_4$, based on the reaction:

$$PO_4^{3-} + 2H_3O^+ = H_2PO_4^- + 2H_2O$$

   *(c) 0.6322 g of HgO as a primary standard for acids:

$$HgO(s) + 2H_3O^+ + 4Br^- = HgBr_4^{2-} + 3H_2O$$

   (d) 0.1450 g of Cu, as a primary standard for Na$_2$S$_2$O$_3$:

$$2Cu^{2+} + 4I^- = Cu_2I_2(s) + I_2$$

$$I_2 + 2S_2O_3^{2-} = 2I^- + S_4O_6^{2-}$$

*(e) 0.1137 g of $As_2O_3$, based on the reaction:

$$HAsO_3^{2-} + I_2 + 2HCO_3^- = HAsO_4^{2-} + 2I^- + 2CO_2 + H_2O$$

**6-11.** Based on the reaction

$$MnO_4^- + 5Fe^{2+} + 8H^+ = Mn^{2+} + 5Fe^{3+} + 8H_2O$$

calculate

*(a) the $FeSO_4$ titer of a 0.1090 *F* $KMnO_4$ solution.
(b) the $FeSO_4$ titer of a 0.1090 *N* $KMnO_4$ solution.
*(c) the $Fe_2O_3$ titer of a 0.1090 *N* $KMnO_4$ solution.
(d) the normality of a $KMnO_4$ solution that has a titer of 4.00 mg $Fe(NH_4)_2(SO_4)_2 \cdot 6H_2O$/mL.
*(e) the normality of a $KMnO_4$ solution that a titer of 3.61 mg $Fe_2O_3$/mL.
(f) the normality of an $FeSO_4$ solution, 27.44 mL of which react with 16.84 mL of 0.1090 *F* $KMnO_4$.
*(g) the normality of an $FeSO_4$ solution, 27.44 mL of which react with 16.84 mL of 0.1090 *N* $KMnO_4$.

**6-12.** Calculate the normality of a 0.0250 *F* solution of HI

*(a) as an acid.
(b) as a reducing agent for $IO_3^-$ in strong HCl (product: $ICl_2^-$).
*(c) as a reducing agent for $O_2$ (products: $I_2$, $H_2O$).
(d) as a reducing agent for $Cl_2$ (products: $IO_3^-$, $Cl^-$).

**6-13.** What volume of 0.0886 *N* KOH will be needed to react with

*(a) 0.3144 g of benzoic acid, $C_6H_5COOH$?
(b) 0.6138 g of potassium hydrogen phthalate, $KHC_8H_4O_4$?
*(c) 0.1974 g of $H_3PO_4$ (product: $HPO_4^{2-}$)?
(d) 17.41 mL of 0.1015 *N* HBr?
*(e) 17.41 mL of 0.1015 *N* $H_2SO_4$?

**6-14.** Describe how 1.00 L of the following solutions should be prepared for use in each of the indicated reactions.

*(a) 0.120 *N* NaOH from a concentrated solution that is 50% NaOH (w/w) and has a density of 1.525 g/mL. Reaction:

$$H_3O^+ + OH^- = 2H_2O$$

(b) 0.200 *N* $H_3PO_4$ from a concentrated solution that is 85% (w/w) $H_3PO_4$ and has a density of 1.69 g/mL. Reaction:

$$H_3PO_4 + 2OH^- = HPO_4^{2-}$$

*(c) 0.0500 *N* disodium hydrogen phosphate from $Na_2HPO_4 \cdot 7H_2O$. Reaction:

$$HPO_4^{2-} + H_3O^+ = H_2PO_4^-$$

(d) 0.0800 *N* oxalic acid from $H_2C_2O_4 \cdot 2H_2O$. Reaction:

$$MnO_4^- + 5H_2C_2O_4 + 6H^+ = Mn^{2+} + 10CO_2(g) + 8H_2O$$

*(e) a 0.0500 *N* $IO_3^-$ solution from $KH(IO_3)_2$. Reaction:

$$IO_3^- + 2I_2 + 10Cl^- + 6H^+ = 5ICl_2^- + 3H_2O$$

(f) a 0.0500 $N$ $H_3O^+$ solution from $KH(IO_3)_2$. Reaction:

$$H(IO_3)_2^- + OH^- = 2IO_3^- + H_2O$$

**6–15.** Calculate the normality of the solution in column A if the volume indicated is needed to titrate the amount of primary standard (or volume of standard solution) shown in Column B. The products of the titration are shown in parentheses.

| A | B |
|---|---|
| *(a)  36.74 mL $HClO_4$ | 0.2444 g $Na_2CO_3$ ($CO_2$, $H_2O$) |
| (b)  39.51 mL $Ba(OH)_2$ | 0.2866 g $KHC_8H_4O_4$ ($C_8H_4O_4^{2-}$) |
| *(c)  46.89 mL $KMnO_4$ | 0.2913 g $Na_2C_2O_4$ ($Mn^{2+}$, $CO_2$) |
| (d)  41.62 mL $I_2$ | 0.1552 g $KIO_3$ ($ICl_2^-$) |
| *(e)  19.25 mL $NaHCO_3$ | 10.00 mL of the $HClO_4$ solution in (a) |
| (f)  11.46 mL HBr | 20.00 mL of the $Ba(OH)_2$ solution in (b) |
| *(g)  21.84 mL $Fe(NH_4)_2(SO_4)_2$ | 25.00 mL of the $KMnO_4$ solution in (c) |
| (h)  33.64 mL $Na_2S_2O_3$ | 50.00 mL of the $I_2$ solution in (d) |

**\*6–16.** Calculate the normality of a dilute HCl solution if a 50.00-mL aliquot yielded 0.8620 g of AgCl upon being treated with an excess of $AgNO_3$.

**6–17.** Calculate the normality of an $H_2SO_4$ solution if 0.3084 g of $BaSO_4$ was obtained by treatment of a 25.00-mL aliquot with an excess of $BaCl_2$.

**\*6–18.** Calculate the normality of an $Fe(NH_4)_2(SO_4)_2$ solution if a 25.00-mL aliquot required 29.29 mL of 0.1004 $N$ $K_2Cr_2O_7$. Reaction:

$$6Fe^{2+} + Cr_2O_7^{2-} + 14H^+ = 6Fe^{3+} + 2Cr^{3+} + 7H_2O$$

**6–19.** Calculate the normality of a $KMnO_4$ solution if 38.71 mL were needed to titrate a 50.00-mL aliquot of 0.05251 $N$ $Na_2C_2O_4$. Reaction:

$$2MnO_4^- + 5H_2C_2O_4 + 6H^+ = 2Mn^{2+} + 10CO_2(g) + 8H_2O$$

**\*6–20.** Calculate the normality of an $Na_2S_2O_3$ solution if 41.27 mL were needed to titrate the $I_2$ derived from 0.1238 g of $KIO_3$. Reactions:

$$IO_3^- + 5I^- + 6H^+ = 3I_2 + 3H_2O$$

$$I_2 + 2S_2O_3^{2-} = 2I^- + S_4O_6^{2-}$$

**6–21.** Calculate the normality of a $Ba(OH)_2$ solution if 47.21 mL were needed to titrate a 25.00-mL aliquot of 0.08370 $N$ $HClO_4$.

**\*6–22.** A 0.1884-g sample of impure $Na_2CO_3$ required 31.56 mL of 0.1056 $N$ HCl. Reaction:

$$CO_3^{2-} + 2H^+ = H_2O + CO_2(g)$$

Calculate the percent of $Na_2CO_3$ in the sample.

**6–23.** The sulfur in a 0.5073-g sample of organic matter was burned in a stream of $O_2$; the combustion products were bubbled through $H_2O_2$ where $SO_2$ was converted to $H_2SO_4$ (see Problem 6–2). Titration of the $H_2SO_4$ required 33.29 mL of 0.1115 $N$ NaOH. Calculate the percent S in the sample.

**\*6–24.** The carbonate in a 0.3063-g mineral specimen was converted to $CO_2$ by treatment with 50.00 mL of 0.1270 $N$ HCl. The solution was boiled to expel the $CO_2$, following which the excess HCl was titrated with 4.15 mL of 0.1283 $N$ NaOH. Express the results of this analysis in terms of percent $MgCO_3$.

**6–25.** A 0.4347-g sample of impure $KClO_3$ was treated with 50.00 mL of 0.1026 $N$ $Fe(NH_4)_2(SO_4)_2$. Reaction:

$$ClO_3^- + 6Fe^{2+} + 6H^+ = Cl^- + 6Fe^{3+} + 3H_2O$$

When reaction was complete the excess iron(II) was back-titrated with 9.11 mL of 0.1004 $N$ $K_2Cr_2O_7$. Calculate the percentage of $KClO_3$ in the sample.

**\*6–26.** The phosphate in a 0.7104-g sample of plant food was converted to $H_2PO_4^-$ and subsequently titrated with 35.02 mL of an $AgNO_3$ solution with a titer of 2.13 mg $P_2O_5$/mL. Reaction:

$$3Ag^+ + H_2PO_4^- = Ag_3PO_4(s) + 2H^+$$

Express the results of this analysis as percent $P_2O_5$.

**6–27.** The $CN^-$ in a 0.6546-g sample was oxidized to cyanate by $KMnO_4$ in a strongly alkaline solution:

$$MnO_4^- + CN^- + 2OH^- = MnO_4^{2-} + CNO^- + H_2O$$

Calculate the percentage of KCN in the sample if the titration required 41.36 mL of a $KMnO_4$ solution with a titer of 5.97 mg KCN/mL.

# Chapter 7
# Precipitation Titrations

Volumetric methods based on the formation of sparingly soluble silver salts are among the oldest known; these procedures were and still are routinely used for the analysis of silver, as well as for such ions as chloride, bromide, iodide, and thiocyanate. Volumetric precipitation methods that do not involve silver as one of the reactants are relatively rare in modern analytical laboratories.[1]

## 7A CURVES FOR PRECIPITATION REACTIONS

Titration curves are useful for selecting indicators and for estimating the titration error associated with the use of a given indicator. Ordinarily the p-value for the reacting cation is plotted as a function of volume of titrant. Precipitation titration curves are readily derived from solubility product data and agree closely with curves obtained in the laboratory. The calculations needed to derive such curves are identical with those discussed in Chapter 5. In the region short of the end point, an excess of the analyte exists; here, it is ordinarily safe to assume that the contribution from the small solubility of the precipitate is negligible and that the equilibrium concentration of the analyte is equal to its formal concentration. A comparable situation exists beyond the equivalence point, where the titrant is now in excess. Finally, at the equivalence point, the only source of the two reactant species is the sparingly soluble product; the concentrations of these ions are readily calculated from the solubility product constant.

---

**Example 7–1** Generate a curve for the titration of 50.00 mL of 0.00500 $F$ NaBr with 0.01000 $F$ AgNO$_3$.

We shall calculate both pAg and pBr, even though only one of these is actually needed.

---

[1] For a review of precipitation titration, see: J. F. Coetzee, in *Treatise on Analytical Chemistry,* 2nd ed., I. M. Kolthoff and P. J. Elving, Eds., Part I, Vol. 2, Chapter 28. New York: John Wiley & Sons, 1979.

(a) *0.00 mL of AgNO₃ added.* At the outset, the solution is 0.00500 $F$ in $Br^-$. Thus,

$$pBr = -\log(5.00 \times 10^{-3}) = \log \frac{1}{5.00 \times 10^{-3}} = \log \frac{10^3}{5.00} = 3 - \log 5.00$$

$$= 2.301 = 2.30$$

Since $[Ag^+] = 0$, pAg is indeterminate.

(b) *5.00 mL of AgNO₃ added.* Here, the $Br^-$ concentration will have been decreased by precipitation as well as dilution. Thus,

$$F_{NaBr} = \frac{(50.00 \times 0.00500) - (5.00 \times 0.01000)}{50.00 + 5.00} = 3.64 \times 10^{-3}$$

The first term in the numerator is the number of milliformula weights of NaBr that were originally present in the sample; the second term represents the milliformula weights of $AgNO_3$ added and hence the number of milliformula weights of AgBr produced. The increase in volume is accounted for in the denominator.

The $Br^-$ concentration of the solution will be

$$[Br^-] = 3.64 \times 10^{-3} + [Ag^+]$$

The first term on the right represents the $Br^-$ from the unreacted NaBr. The second term accounts for the contribution from dissolved AgBr, one $Ag^+$ ion being introduced for each $Br^-$ ion; the second term can ordinarily be neglected as being vanishingly small. Here,

$$[Ag^+] \ll 3.64 \times 10^{-3}$$

Thus,

$$[Br^-] = 3.64 \times 10^{-3}$$

and

$$pBr = -\log(3.64 \times 10^{-3}) = 3 - \log 3.64 = 2.44$$

A convenient way to evaluate pAg is to take the negative logarithm of the solubility product expression for AgBr; that is,

$$-\log([Ag^+][Br^-]) = -\log K_{sp} = -\log(5.2 \times 10^{-13})$$

$$-\log[Ag^+] - \log[Br^-] = -\log K_{sp} = 12.28$$

$$pAg + pBr = pK_{sp} = 12.28$$

This relationship applies to any saturated solution of AgBr. In the present example, then,

$$pAg = 12.28 - 2.44 = 9.84$$

(c) *25.00 mL of AgNO₃ added.* At the equivalence point, neither NaBr nor

$AgNO_3$ is in excess. The only source of $Ag^+$ as well as $Br^-$ is from $AgBr$; necessarily, then,

$$[Ag^+] = [Br^-] = \sqrt{5.2 \times 10^{-13}} = 7.21 \times 10^{-7}$$

$$pAg = pBr = 7 - \log 7.21 = 6.14$$

(d) *25.10 mL of AgNO₃ added.* The solution now contains an excess of $AgNO_3$; thus,

$$F_{AgNO_3} = \frac{25.10 \times 0.0100 - 50.00 \times 0.0050}{50.00 + 25.10} = 1.33 \times 10^{-5}$$

and the equilibrium $Ag^+$ concentration will be given by

$$[Ag^+] = 1.33 \times 10^{-5} + [Br^-] \cong 1.33 \times 10^{-5}$$

The second term on the right-hand side of the equation accounts for the $Ag^+$ due to the finite solubility of $AgBr$, which can ordinarily be neglected. Thus,

$$pAg = 5 - \log 1.33 = 4.876 = 4.88$$

$$pBr = 12.28 - 4.88 = 7.40$$

This type of calculation permits evaluation of additional points in the region beyond equivalence.

## 7A – 1 Significant Figures in Titration Curve Calculations

Calculated concentration values in the equivalence-point region are necessarily uncertain because they are based on small differences between large numbers. Thus, the silver ion concentration after addition of 25.10 mL of $AgNO_3$ in the foregoing example contains only two significant figures as a result of the small difference between the two quantities in the numerator $(0.2510 - 0.2500 = 0.0010)$; at best, then, $F_{AgNO_3}$ is known to two significant figures as well. Note, however, that in the interests of minimizing rounding errors, we retained three digits throughout the calculation. In rounding the value for pAg, it is important to recall (Chapter 3, Section 3G) that *the significant figure convention applies only to the mantissa of a logarithm* (that is, the numbers to the right of the decimal point) because the characteristic merely serves to locate the decimal point. Thus, in the example under consideration, pAg was rounded to 4.88 to give a mantissa with just two figures.

Note that justification exists for retaining an additional figure in p-values well removed from the equivalence point. For example, the initial pBr could have been reported as 2.301. Little is gained by so doing, however, because that region of the curve is not of primary interest. Here, and in other titration curve calculations, p-values will ordinarily be rounded to two significant digits to the right of the decimal. The large changes in p-functions that typify most equivalence points are not obscured by this limited precision in calculations.

## 7A – 2 Factors Influencing the Sharpness of End Points

A sharp and easily located end point is observed when small additions of titrant cause large changes in p-values. It is therefore of interest to examine those variables that influence the magnitude of change in p-function during a titration.

**Reagent Concentration**    Figure 7 – 1 illustrates curves for the titration of bromide ion with silver nitrate at three reactant concentrations. Also shown is the pAg range required to produce a color change of the indicator In⁻. The three titration curves were derived from calculations similar to those in Example 7 – 1. It is clear that increases in the analyte and titrant concentrations enhance the change in pAg in the equivalence-point region; the same conclusion would be drawn from a plot of pBr as a function of titrant volume.

These effects have practical significance for the titration of bromide ion. Easily detected end points and small titration errors can be expected provided the analyte concentration is sufficient to permit the use of silver nitrate solutions that are 0.1 $F$ or stronger. Conversely, with solutions that are 0.001 $F$ or less, the change in pAg (or pBr) is so small that end-point detection becomes difficult, and a large titration error must be expected.

These observations are in no way limited to precipitation titrations. We shall see that they are equally applicable to titrations that are based upon other types of reactions as well.

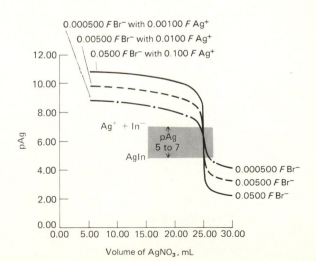

**Figure 7 – 1**   Effect of reagent concentration on titration curves. Each titration involves 50.00 mL of NaBr.

**Completeness of Reaction**    Figure 7–2 shows how the solubility of the species produced influences curves for titrations in which 0.1 $F$ silver nitrate serves as the titrant. The greatest change in pAg clearly occurs in the titration of iodide ion, which, of all the anions considered, forms the least soluble silver salt and hence reacts most completely with silver ion. The smallest change in pAg is observed for the titration with bromate ion, where the reaction is least favorable. Reactions that produce silver salts with solubilities that are intermediate between these extremes yield titration curves with end-point breaks that are also intermediate. Again, we shall see that this effect is common to all reaction types.

## 7A–3 Titration Curves for Mixtures

The methods developed in Example 7–1 can be extended to the titration of mixtures that form precipitates with differing solubilities. Consider, for example, the titration of a solution containing both iodide and chloride ions with standard silver nitrate. Initial additions of reagent will result in the formation of silver iodide exclusively because the solubility of silver iodide is substantially smaller than that for silver chloride. The first portion of the curve will thus resemble the preequivalence region of the curve for iodide in Figure 7–1. With the first appearance of silver chloride, essentially all of the iodide will have been precipitated.[2] Consequently, the remainder of the curve will be indistinguishable from one for the titration of chloride alone. Curve $A$ in Figure 7–3 is for the titration of the mixture of chloride and iodide ions, while curve $B$ is for a solution containing similar amounts of bromide and chloride ions. The solubility of silver bromide is greater than that for silver iodide, which accounts for the lower pAg in the initial portion of the latter titration and the smaller change in pAg at the equivalence point.

Curves that resemble those in Figure 7–3 can be obtained experimentally by measuring the potential of a silver electrode immersed in the solution. Hence, halide mixtures can be analyzed by this method (see Chapter 13).

---

[2] For a solution that is saturated with both silver iodide and silver chloride, the silver ion concentration must be such as to satisfy both solubility product expressions; that is,

$$\frac{K_{sp_{AgCl}}}{[Cl^-]} = [Ag^+] = \frac{K_{sp_{AgI}}}{[I^-]}$$

or

$$[I^-] = \frac{8.3 \times 10^{-17}}{1.82 \times 10^{-10}} \times [Cl^-] = 4.6 \times 10^{-7}\,[Cl^-]$$

The iodide ion concentration is thus a minuscule fraction of the chloride ion concentration throughout the remainder of the titration.

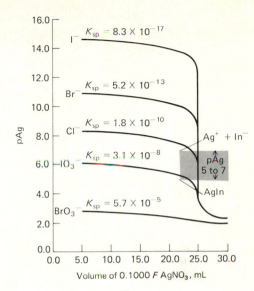

**Figure 7–2** Effect of completeness of reaction on titration curves. Each curve represents the titration of 50.00 mL of solution that is 0.0500 $F$ in the anion with 0.1000 $F$ AgNO$_3$.

# 7B INDICATORS FOR PRECIPITATION TITRATIONS

## 7B–1 Theory of Indicator Behavior

A *chemical indicator* produces a visually detectable signal—ordinarily a change in color or turbidity—in the solution being titrated. The indicator functions by reacting competitively with one of the participants of the titration. For example, consider the titration of the species X with reagent R in the presence of an indicator In that reacts with R to give a colored product InR. This titration system can be described chemically by the equations

$$X + R \rightleftharpoons XR(s)$$

$$In + R \rightleftharpoons InR$$

For satisfactory indicator behavior, InR must impart a significantly different color to the solution than In. Moreover, the color of InR should be so intense that a negligible amount of R will be needed to produce the color change. Finally, the equilibrium constant for the indicator reaction must be such that the ratio [InR]/[In] is shifted from a small to a large value as a consequence of changes in the concentration of R (or pR) in the equivalence-point region. The last condition is most likely to be realized in systems where the change in pR is large.

We can expand on these notions by considering the application of a hypothetical indicator, In, which changes color as the pAg changes from 7 to 5, to the bromide titrations illustrated in Figure 7–1. Recalling that a 1- to 2-unit change in p-function is required in order for the average person to detect the color

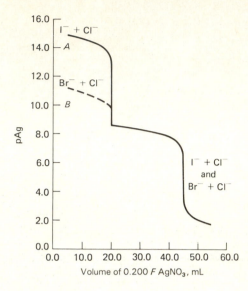

**Figure 7–3** Curves for the titration of halide mixtures with 0.200 $F$ AgNO$_3$. Curve A: 50.0 mL of solution that is 0.0800 $F$ in I$^-$ and 0.100 $F$ in Cl$^-$. Curve B: 50.0 mL of solution that is 0.800 $F$ in Br$^-$ and 0.100 $F$ in Cl$^-$.

change of a typical indicator (Chapter 6, Section 6D–1), it is clear that each titration requires a different volume of titrant to produce an observable end point. Thus, for the most concentrated solution, less than 0.10 mL of 0.1 $F$ silver nitrate is needed; the change in color will begin at about 24.95 mL and will be complete at 25.05 mL. The result will be an abrupt change in color and a minimal titration error. The indicator would fail for the titration involving the least concentrated reactants; here, the onset of color change will occur after about 24.5 mL of 0.001 $F$ AgNO$_3$ have been added and will be complete at about 25.8 mL. The exact location of such an end point is impossible. Somewhat less than 0.2 mL of 0.01 $F$ AgNO$_3$ would be needed to cause the indicator to exhibit its color change in the titration of 0.005 $F$ bromide ion. Here, the use of the indicator would be possible, but the titration error would likely be substantial.

Consider now the applicability and effectiveness of the same indicator in terms of the titrations illustrated by the curves in Figure 7–2. The indicator will exist largely as AgIn throughout the titration of bromate or iodate; thus, no color change would be observed following the first addition of silver nitrate. In contrast, the solubilities of silver bromide and silver iodide are small enough to prevent formation of significant amounts of AgIn until the equivalence-point region is reached. Note that pAg remains above 7 until just before the equivalence point in the bromide titration and just beyond the equivalence point of the iodide titration. The excess of silver ion needed to complete the color change is less than 0.01 mL for both titrations. The error associated with the functioning of the indicator would thus be negligible.

An indicator with a pAg range of 5 to 7 would be unsatisfactory for a chloride titration because AgIn would start to form approximately 1 mL short of the equivalence point and would continue to form over a range of approximately 1 mL; exact location of an end point would be impossible. On the other hand, an indicator with a transition pAg range between 4 and 6 would be entirely satisfactory. Chemical indicators do not exist for the titration of either bromate or iodate with silver ion because the changes in pAg in the equivalence-point region are too small.

The underlying arguments developed in this section are equally applicable to titrations that are based upon other types of chemical reaction.

## 7B–2 Examples of Indicators for Precipitation Titrations

This section describes three widely used indicators for precipitation titrations.

**Formation of a Second Precipitate: The Mohr Method**    The formation of a second, highly colored precipitate is the basis for end-point detection in the *Mohr method*. This titration has been used extensively for the determination of chloride ion and bromide ion with standard silver nitrate. Chromate ion is the indicator, and the end point is signaled by the appearance of the brick-red silver chromate precipitate, $Ag_2CrO_4$.

The formal solubility of silver chromate is several times greater than that for silver chloride. It should be possible, therefore, to select a chromate ion concentration such that silver chromate will appear at a silver ion concentration that corresponds to the equivalence point for the titration (see Problem 7–19); in fact, however, this theoretical chromate ion concentration imparts sufficient color of its own to the solution to obscure the color of the silver chromate. A smaller concentration results in the need for a silver ion concentration that is somewhat greater than hypothetical. In addition, a finite amount of silver nitrate must be added to cause formation of a perceptible amount of silver chromate. Both of these factors cause an overconsumption of reagent, which can be appreciable with dilute solutions but is vanishingly small at the 0.1 $F$ level. A correction can be made by titrating a similar volume of a chloride-free calcium carbonate suspension containing the same quantity of indicator as the sample. This *indicator blank* also serves as a convenient color standard for subsequent titrations. A better approach, which largely eliminates the indicator error, is to use the Mohr method to standardize the silver nitrate solution against pure sodium chloride. The "working concentration" that is obtained will not only compensate for the overconsumption of reagent but also for the visual acuity of the analyst in detecting the color change.

The successful determination of chloride by the Mohr method requires reasonable control over the acidity of the solution. The equilibrium

$$2CrO_4^{2-} + 2H^+ \rightleftharpoons Cr_2O_7^{2-} + H_2O$$

is displaced to the right with increases in the hydrogen ion concentration. Because silver dichromate is considerably more soluble than silver chromate, the indicator reaction in acid solution requires substantially larger silver ion concentrations, if indeed it occurs at all. In an alkaline environment the possibility exists that silver will precipitate as its oxide:

$$2Ag^+ + 2OH^- \rightarrow 2AgOH(s) \rightarrow Ag_2O(s) + H_2O$$

Thus, the determination of chloride by the Mohr method must be performed in a medium that is neutral or nearly so (pH 7 to 10). The introduction of either sodium hydrogen carbonate or borax keeps the hydrogen ion concentration of the titration medium within acceptable bounds.

**Formation of a Colored Complex: The Volhard Method**    A standard solution of potassium thiocyanate is used to titrate silver ion by the *Volhard method:*

$$Ag^+ + SCN^- \rightarrow AgSCN(s)$$

The indicator is iron(III), which imparts a red color to the titration mixture with the first slight excess of thiocyanate:

$$Fe^{3+} + SCN^- \rightarrow \underset{\text{red}}{FeSCN^{2+}}$$

The titration must be performed in acid solution to prevent the precipitation of iron(III) as the hydrous oxide. The indicator concentration is not critical for a successful Volhard titration as long as the iron(III) concentration is kept between 0.002 and 0.2 $F$.

The most important application of the Volhard method is for the indirect determination of chloride ion. A measured excess of standard silver nitrate is added to the sample, and the excess is determined by back-titration with a standard thiocyanate solution. A distinct advantage of the Volhard method for the determination of halides is the need for an acidic medium, which prevents such anions as carbonate, oxalate, and arsenate from precipitating as silver salts and thus interfering with the analysis.

In contrast to the other silver halides, silver chloride is more soluble than silver thiocyanate. As a consequence, the reaction

$$AgCl(s) + SCN^- \rightarrow AgSCN(s) + Cl^-$$

occurs to a significant extent near the end point during back-titration of the excess silver ion. The resulting decrease in the thiocyanate ion concentration causes the indicator equilibrium to be shifted to the left and the end point to fade. Additional thiocyanate must then be added to produce a permanent end point, which causes the results of a chloride analysis to be low. Any of several general methods can be used to circumvent this problem. The most widespread of these involves isolation of the precipitated silver chloride before back-titration with the thiocyanate. Filtration, followed by titration of an aliquot of the filtrate, will yield excellent results provided the silver chloride is digested briefly before the filtration step. Probably the most widely encountered modification is that of Caldwell and Moyer,[3] which involves coating the silver chloride precipitate with nitrobenzene, thereby effectively removing the solid from contact with the solution. The coating is accomplished by shaking the titration mixture with a few milliliters of the organic liquid prior to back-titration.

**Adsorption Indicators: The Fajans Method**   An *adsorption indicator* is an organic dye that is adsorbed on or desorbed from the surface of the solid that is produced during a precipitation titration. The adsorption or desorption ideally occurs in the equivalence-point region and results not only in a color change but also in a transfer of color from the solution to the solid (or the reverse). Precipitation titrations based upon adsorption indicators are known as *Fajans titrations* in honor of the chemist who was active in their development.

The organic dye fluorescein is a typical adsorption indicator that is useful for the titration of chloride ions with silver nitrate. Fluorescein is a weak acid in aqueous solution and partially dissociates to give fluoresceinate ions, which impart a yellowish green color to the medium. Solid silver fluoresceinate is red and possesses limited solubility. *It is important to appreciate, however, that the concentration of the dye in a Fajans titration is such that the solubility product of silver fluoresceinate is never exceeded.*

In the initial stages of a titration with the indicator, the negatively charged fluoresceinate ions are electrostatically repelled from the surface of the silver chloride particles, which also carry a negative charge owing to the adsorbed chloride ions (p. 77). Beyond the equivalence point, however, the particles of the precipitate acquire a positive charge because of their strong tendency to adsorb silver ions; under these conditions, fluoresceinate ions are attracted to the counter-ion layer surrounding the particles. The net result is the appearance of the red color of silver fluoresceinate on the surface of the precipitate. *It is important to stress that this color change is an adsorption rather than a precipitation process because at no time is the solubility product constant for silver fluoresceinate exceeded by the product of silver ion and fluoresceinate ion concentrations in the solution.* The adsorption process is reversible, the dye being desorbed upon back-titration with chloride ion.

The successful application of the Fajans method to a precipitation titration depends upon chemical and physical properties possessed by the solid as well as the indicator.

1. The particles of the precipitate must have colloidal dimensions to provide maximum surface at which the dye can be adsorbed. A small amount of the high-molecular-weight carbohydrate, dextrin, can be added to the analyte solution to prevent coagulation of the colloidal particles.
2. The precipitate must strongly adsorb its own ions. We have seen (Chapter 4, Section 4B–2) that this property is characteristic of colloidal precipitates.
3. The indicator dye must be strongly held in the counter-ion layer by a pri-

---

[3] J. R. Caldwell and H. V. Moyer, *Ind. Eng. Chem., Anal. Ed.,* **1935,** *7,* 38.

marily adsorbed ion. In general, this type of adsorption correlates with a low solubility for the salt that is formed between the dye and the lattice ion. Nevertheless, this species must have sufficient solubility to remain in solution at all times.

4. The solution must be maintained at a pH in which the indicator will exist predominantly in its ionic form. Activity of most adsorption indicators resides in the ion that is the conjugate acid or base of the parent dye molecule. For a given formal dye concentration, then, the concentration of the active species will be pH dependent.

Titrations involving adsorption indicators are rapid, accurate, and reliable. Their application, however, is limited to a relatively small number of reactions in which a colloidal precipitate is rapidly produced. End points with adsorption indicators tend to be less satisfactory in media with high electrolyte concentrations, where coagulation decreases the surface area at which dye activity can occur.

Most adsorption indicators are weak acids. Their use is thus confined to neutral and slightly alkaline solutions in which the conjugate base form predominates. A few weak bases also act as adsorption indicators and are suitable

**Table 7–1  Typical Argentometric Precipitation Methods**

| Substance Determined | End Point | Remarks |
|---|---|---|
| $AsO_4^{3-}$, $Br^-$, $I^-$, $CNO^-$, $SCN^-$ | Volhard | Removal of silver salt not required |
| $CO_3^{2-}$, $CrO_4^{2-}$, $CN^-$, $Cl^-$, $C_2O_4^{2-}$, $PO_4^{3-}$, $S^{2-}$ | Volhard | Removal of silver salt required before back-titration of excess $Ag^+$ |
| $BH_4^-$ | Modified Volhard | Titrate excess $Ag^+$ following: $BH_4^- + 8Ag^+ + 8OH^- \rightleftarrows 8Ag(s) + H_2BO_3^- + 5H_2O$ |
| Epoxide | Volhard | Titrate excess $Cl^-$ following hydrohalogenation |
| $K^+$ | Modified Volhard | Precipitate $K^+$ with known excess of $B(C_6H_5)_4^-$; add excess $Ag^+$, which precipitates $AgB(C_6H_5)_4$; back-titrate excess $Ag^+$ |
| $Br^-$, $Cl^-$ | Mohr | |
| $Br^-$, $Cl^-$, $I^-$, $SeO_3^{2-}$ | Adsorption indicator | |
| $V(OH)_4^+$, fatty acids, mercaptans | Electroanalytical | Direct titration with $Ag^+$ |
| $Zn^{2+}$ | Modified Volhard | Precipitate as $ZnHg(SCN)_4$; filter, dissolve in acid, add excess $Ag^+$; back-titrate excess $Ag^+$ |
| $F^-$ | Modified Volhard | Precipitate as $PbClF$; filter, dissolve in acid, add excess $Ag^+$; back-titrate excess $Ag^+$ |

for titrations in acid solution. Indicator action here is between the conjugate acid form of the indicator and negatively charged particles of the precipitate.

Finally, many adsorption indicators enhance the tendency of silver salts to undergo photodecomposition. Therefore, titrations are best performed in subdued light.

Directions for the determination of chloride by the Fajans method are to be found in Chapter 20, Section 20B–1.

**Potentiometric End Points** End points for precipitation titrations involving silver ion are readily obtained by measuring the potential between a silver electrode and a reference electrode when the two are immersed in the analyte solution. Potentiometric end points for precipitation titrations are discussed in detail in Chapter 13; specific directions for the titration of halides are given in Chapter 20.

## 7C APPLICATIONS OF PRECIPITATION TITRATIONS

Most precipitation titrations make use of a standard silver nitrate solution and are known as *argentometric methods*. Table 7–1 lists typical applications of argentometry. Note that many of these analyses are based upon precipitation of the analyte with a measured excess of silver nitrate followed by a Volhard titration of the excess silver ion with potassium thiocyanate. Both reagents are obtainable in primary standard quality, and solutions of both are indefinitely stable. Potassium thiocyanate is, however, somewhat hygroscopic and its solutions are ordinarily standardized against silver nitrate.

The high cost of silver nitrate is a distinct disadvantage to argentometry. Methods for the recovery of silver from its solutions and its sparingly soluble salts have appeared in the literature.[4]

Table 7–2 lists miscellaneous volumetric precipitation titrations based on reagents other than silver nitrate.

**Table 7–2** *Miscellaneous Volumetric Precipitation Methods*

| Reagent | Ion Determined | Reaction Product | End Point |
|---|---|---|---|
| $K_4Fe(CN)_6$ | $Zn^{2+}$ | $K_2Zn_3[Fe(CN)_6]_2$ | Diphenylamine |
| $Pb(NO_3)_2$ | $SO_4^{2-}$ | $PbSO_4$ | Erythrosin B |
| | $MoO_4^{2-}$ | $PbMoO_4$ | Eosin A |
| $Pb(OAc)_2$ | $PO_4^{3-}$ | $Pb_3(PO_4)_2$ | Dibromofluorescein |
| | $C_2O_4^{2-}$ | $PbC_2O_4$ | Fluorescein |
| $Th(NO_3)_4$ | $F^-$ | $ThF_4$ | Alizarin red |
| $Hg_2(NO_3)_2$ | $Cl^-, Br^-$ | $Hg_2X_2$ | Bromophenol blue |
| $NaCl$ | $Hg_2^{2+}$ | $Hg_2X_2$ | Bromophenol blue |

[4] See, for example, K. J. Bush and H. Diehl, *J. Chem. Educ.,* **1979**, *56,* 54.

## PROBLEMS

**7-1.** Calculate the formal concentration of an $AgNO_3$ solution if 36.2 mL are needed to react with
   *(a)  0.207 g of NaCl.
    (b)  0.612 g of $BaCl_2 \cdot 2H_2O$.
   *(c)  42.5 mL of 0.0291 $F$ $FeCl_3$.
    (d)  29.7 mL of 0.1044 $F$ $Ba(SCN)_2$.

*\**7-2.** Calculate the formal $Ag^+$ concentration of a solution if 1.000 mL reacts with 2.250 mg of
   *(a)  KI.                          (b)  KSCN.
   *(c)  $K_3Fe(CN)_6$.                (d)  $K_2CrO_4$.
   *(e)  $H_2S$.                       (f)  $Na_3PO_4$.

**7-3.** Express the titer of a 0.0496 $F$ $AgNO_3$ solution in terms of
   *(a)  mg KI/mL.                    (b)  mg KSCN/mL.
   *(c)  mg $K_3Fe(CN)_6$/mL.          (d)  mg $K_2CrO_4$/mL.
   *(e)  mg $H_2S$/mL.                 (f)  mg $Na_3PO_4$/mL.

*\**7-4.** Lead(II) can be titrated with standard $K_2CrO_4$ [product: $PbCrO_4(s)$]. A 0.712-g mineral specimen consisting principally of cerussite, $PbCO_3$, was dissolved and subsequently titrated with 28.68 mL of 0.0681 $F$ $K_2CrO_4$. Calculate the percentage of cerussite in the sample.

**7-5.** A Fajans titration of a 0.7908-g sample required 45.32 mL of 0.1046 $F$ $AgNO_3$. Express the results of this analysis in terms of the percentage of
   (a)  $Cl^-$.
   (b)  $BaCl_2 \cdot 2H_2O$.
   (c)  $ZnCl_2 \cdot 2NH_4Cl$ (gfw = 243.3).

*\**7-6.** The sulfide in a sample of brackish water was determined by making a 100.0-mL sample ammoniacal and titrating with 7.04 mL of 0.0150 $F$ $AgNO_3$. Reaction:

$$2Ag^+ + S^{2-} \rightarrow Ag_2S(s)$$

Express the results of this analysis in terms of ppm $H_2S$.

**7-7.** The phosphate in a 4.258-g sample of plant food was precipitated as $Ag_3PO_4$ through the addition of 50.00 mL of 0.0820 $F$ $AgNO_3$:

$$3Ag^+ + HPO_4^{2-} \rightarrow Ag_3PO_4(s) + H^+$$

The solid was filtered and washed, following which the filtrate and washings were diluted to exactly 250.0 mL. Titration of a 50.00-mL aliquot of this solution required a 4.64-mL back-titration with 0.0625 $F$ KSCN. Express the results of this analysis in terms of the percentage of $P_2O_5$.

**7-8.** The Association of Official Analytical Chemists recommends a Volhard titration for the analysis of heptachlor, $C_{10}H_5Cl_7$; the percentage of heptachlor is given by the equation

$$\% \text{ heptachlor} = \frac{(mL_{Ag^+} \times F_{Ag^+} - mL_{SCN^-} \times F_{SCN^-}) \times 37.33}{\text{wt sample}}$$

What does this calculation reveal concerning the stoichiometry of the reaction?

**7-9.** The theobromine ($C_7H_8N_4O_2$) in a 2.95-g sample of ground cocoa beans was converted to the sparingly soluble silver salt $C_7H_7N_4O_2Ag$ by warming in an ammoniacal solution containing 25.0 mL of 0.0100 $F$ AgNO$_3$. After reaction was complete, all solids were removed by filtration. Calculate the percentage of theobromine (gfw = 180.1) in the sample if the combined filtrate and washings required a back titration with 7.69 mL of 0.0108 $F$ KSCN.

*7–10. Borohydride ion can be determined through reaction with a measured excess of Ag$^+$:

$$BH_4^- + 8Ag^+ + 8OH^- \rightarrow H_2BO_3^- + 8Ag(s) + 5H_2O$$

The purity of a quantity of KBH$_4$ to be used in organic synthesis was established by diluting a 0.3127-g sample to exactly 250 mL, treating a 25.0-mL aliquot with 50.0 mL of 0.1008 $F$ AgNO$_3$, and titrating the excess Ag$^+$ with 8.96 mL of 0.0936 $F$ KSCN. Calculate the percentage purity of the KBH$_4$ (gfw = 53.94).

7–11. What volume of 0.0936 $F$ KSCN would have been needed if the analysis in Problem 7–10 had been completed by filtering off the precipitated Ag, dissolving it in acid, diluting the solution to 100.0 mL, and titrating a 25.0-mL aliquot?

*7–12. A 0.1750-g sample of pure Zn was dissolved in acid and titrated with 38.79 mL of a K$_4$Fe(CN)$_6$ solution:

$$3Zn^{2+} + 2Fe(CN)_6^{4-} + 2K^+ \rightarrow K_2Zn_3[Fe(CN)_6](s)$$

The same reaction was involved when 26.35 mL of this Fe(CN)$_6^{4-}$ solution was used to titrate the Zn$^{2+}$ derived from a 0.2829-g mineral specimen. Express the results of this analysis in terms of the percentage of willemite, Zn$_2$SiO$_4$ (gfw = 222.8).

7–13. A 2.00-L sample of mineral water was evaporated to a small volume, following which K$^+$ was precipitated with excess sodium tetraphenylboron:

$$K^+ + NaB(C_6H_5)_4 \rightarrow KB(C_6H_5)_4(s)$$

The precipitate was filtered, washed, and redissolved in acetone. The analysis was completed by a Mohr titration that required 43.85 mL of 0.03941 $F$ AgNO$_3$:

$$KB(C_6H_5)_4 + Ag^+ \rightarrow AgB(C_6H_5)_4(s) + K^+$$

Calculate the K$^+$ concentration (in ppm) of the water sample.

*7–14. The action of an alkaline I$_2$ solution upon the rodenticide warfarin, C$_{19}$H$_{16}$O$_4$ (gfw = 308.3), results in the formation of one formula weight of iodoform, CHI$_3$, from each formula weight of the parent compound. The analysis for warfarin can then be based upon the reaction between CHI$_3$ and Ag$^+$:

$$CHI_3 + 3Ag^+ + H_2O \rightarrow 3AgI(s) + 3H^+ + CO(g)$$

The CHI$_3$ produced from a 14.82-g sample was treated with 25.00 mL of 0.0227 $F$ AgNO$_3$, following which the excess Ag$^+$ was titrated with 8.83 mL of 0.0359 $F$ KSCN. Calculate the percentage of warfarin in the sample.

7–15. Each formula weight of acetone (CH$_3$COCH$_3$) yields one formula weight of iodoform (CHI$_3$) when treated with excess alkaline I$_2$. The CHI$_3$ produced from the acetone in a 100-mL urine specimen was treated with 20.00 mL of 0.0232 $F$ AgNO$_3$ (see Problem 7–14 for reaction). Calculate the weight (mg) of acetone in the sample if 0.83 mL of 0.0209 $F$ KSCN was needed to titrate the excess Ag$^+$.

*7–16. A 3.095-g sample containing KCl, KClO$_4$, and inert materials was dissolved in sufficient water to give 250.0 mL of solution. A 50.00-mL aliquot required 38.32 mL of 0.0637 $F$ $AgNO_3$ in a Mohr titration. A 25.00-mL aliquot was then treated with $V_2(SO_4)_3$ to reduce $ClO_4^-$ to $Cl^-$:

$$ClO_4^- + 8V^{3+} + 4H_2O \rightarrow Cl^- + 8VO^{2+} + 8H^+$$

following which titration required 39.63 mL of the same $AgNO_3$ solution. Calculate the respective percentages of KCl and $KClO_4$ in the sample.

**7–17.** A 4.269-g sample containing $NH_4Cl$, $(NH_4)_2SO_4$, and inert materials was diluted to exactly 500.0 mL. The $Cl^-$ in a 50.00-mL aliquot of this solution required 24.04 mL of 0.0682 $F$ $AgNO_3$. The $NH_4^+$ in a 25.00-mL aliquot was converted to $NH_3$ and collected in 100.0 mL of a 0.03070 $F$ sodium tetraphenylboron solution

$$NH_3(g) + NaB(C_6H_5)_4 + H^+ \rightarrow NH_4B(C_6H_5)_4(s) + Na^+$$

After the solid had been removed by filtration, 7.50 mL of the $AgNO_3$ solution was needed to titrate the filtrate and washings:

$$Ag^+ + NaB(C_6H_5)_4 \rightarrow AgB(C_6H_5)_4(s) + Na^+$$

Calculate the percentages of $NH_4Cl$ and $(NH_4)_2SO_4$ in the sample.

**7–18.** Calculate the $Ag^+$ concentration after the addition of 5.00, 20.0, 30.0, 35.0, 39.0, 40.0, 41.0, 45.0, and 50.0 mL of 0.100 $F$ $AgNO_3$ to 50.0 mL of
\*(a)  0.080 $F$ KI.
 (b)  0.080 $F$ KSCN.
\*(c)  0.080 $F$ KCl.
 (d)  0.040 $F$ $K_2CrO_4$ (under conditions where $[HCrO_4^-]$ and $[H_2CrO_4]$ are negligibly small).

**\*7–19** Calculate the $CrO_4^{2-}$ concentration required to initiate formation of $Ag_2CrO_4$ at the equivalence point in a Mohr titration for $Cl^-$.

**7–20.** Because the yellow color of $CrO_4^{2-}$ tends to obscure the first appearance of the red $Ag_2CrO_4$, it is common practice to hold the $CrO_4^{2-}$ concentration to about $2.5 \times 10^{-3}$ $M$. Calculate the relative titration error (neglecting the volume of $AgNO_3$ needed to produce a detectable amount of $Ag_2CrO_4$) in the titration of
\*(a)  50.0 mL of 0.0500 $F$ NaCl with 0.1000 $F$ $AgNO_3$.
 (b)  50.0 mL of 0.0100 $F$ NaCl with 0.0200 $F$ $AgNO_3$.

**7–21.** The reaction between AgCl and $SCN^-$ is shown on page 156.
\*(a)  Calculate the equilibrium constant for this reaction.
 (b)  Assuming that the $SCN^-$ concentration needed to produce a detectable amount of $FeSCN^-$ is $2 \times 10^{-7}$ $M$, calculate the relative error resulting from the reaction between AgCl and $SCN^-$ in a titration involving the introduction of 30.0 mL of 0.100 $F$ $AgNO_3$ to 50.0 mL of 0.0500 $F$ NaCl, followed by a back-titration with 5.00 mL of 0.100 $F$ KSCN.

# Chapter 8
# Theory of Neutralization Titrations

End points in neutralization titrations are based upon the abrupt changes in pH that occur in the vicinity of equivalence points. The pH range for such changes varies from titration to titration and is determined by the nature and concentration of both the analyte and the titrant. Generally, the largest pH changes occur when one or both of the reactants are completely ionized. For this reason, the standard solutions used in neutralization titrations are always strong acids or strong bases.[1]

As with precipitation titrations, the selection of indicators and the estimation of titration errors are conveniently based upon titration curves. Thus, much of this chapter is devoted to the derivation of such curves for various kinds of acids and bases. Consideration is also given to the behavior of acid-base indicators.

## 8A ACID-BASE INDICATORS

Many substances, both naturally occurring and synthetic, display colors that depend upon the pH of the solutions in which they are dissolved. Some of these substances have been used for centuries to indicate the acidity or alkalinity of water and are important to the modern chemist for detecting the end points of acid-base titrations.

The theory of indicator behavior is considered in Section 8C–4. For the present, it is sufficient to note that the typical acid-base indicator is an organic compound that exhibits an "acid color" or a "base color," depending upon the pH of the environment, and that the transition from the pure acid color to the pure base color typically requires a pH change on the order of two units. The transition range of an indicator should correspond to the sharp change in pH (or pOH) that is characteristic of the equivalence-point region of a neutralization titration to minimize the volume of titrant needed to bring about this two-unit change.

---

[1] For a thorough treatment of neutralization titrations, see: D. Rosenthal and P. Zuman, in *Treatise on Analytical Chemistry,* 2nd ed., I. M. Kolthoff and P. J. Elving, Eds., Part I, Vol. 2, Chapter 18. New York: John Wiley & Sons, 1979.

## 8A–1 Types of Acid-Base Indicators

Substances that possess acid-base indicator properties include a variety of organic structures. An indicator exists for just about any desired pH range. A number of commonly encountered indicators are listed in Table 8–1.

## 8A–2 Titration Errors Associated with Acid-Base Indicators

The use of acid-base indicators is subject to determinate as well as indeterminate error. The former occurs when the transition range of the indicator fails to encompass the equivalence-point pH. This type of error can be minimized through prudent indicator selection; a blank will frequently provide an appropriate correction if such is necessary.

The limited ability of the human eye to distinguish reproducibly the point at which the color change of an indicator occurs is a source of indeterminate error. The magnitude of this random error will depend upon the change in pH per milliliter of titrant in the equivalence-point region, the concentration of the indicator in the titration medium, and the sensitivity of the eye to the two indicator colors. For most individuals, this uncertainty amounts to about

*Table 8–1  Some Important Acid-Base Indicators\**

| Common Name | Transition pH Range | $pK_a$† | Color Change‡ | Indicator Type§ |
|---|---|---|---|---|
| Thymol blue | 1.2–2.8 | 1.65 | R–Y | 1 |
|  | 8.0–9.6 | 8.90 | Y–B |  |
| Quinaldine red | 1.3–3.2 | 2.75 | C–R | 2 |
| Bromophenol blue | 3.0–4.6 | 4.10 | Y–B | 1 |
| Methyl orange | 3.1–4.4 | 3.46¶ | R–O | 2 |
| Bromocresol green | 4.0–5.6 | 4.66 | Y–B | 1 |
| Methyl red | 4.4–6.2 | 5.00¶ | R–Y | 2 |
| Bromocresol purple | 5.2–6.8 | 6.12 | Y–P | 1 |
| Bromothymol blue | 6.2–7.6 | 7.10 | Y–B | 1 |
| Phenol red | 6.4–8.0 | 7.81 | Y–R | 1 |
| Cresol purple | 7.6–9.2 | 8.32 | Y–P | 1 |
| Phenolphthalein | 8.0–10.0 | 9.7 | C–P | 1 |
| Thymolphthalein | 9.4–10.6 | 10 | C–B | 1 |
| Alizarin yellow GG | 10–12 | 11.23 | C–Y | 2 |

\* Data principally from R. G. Bates, in *Treatise on Analytical Chemistry,* 2nd ed., I. M. Kolthoff and P. J. Elving, Eds., Part I, Vol. 2, p. 854. New York: John Wiley & Sons, 1978; and C. A. Streuli, in *Handbook of Analytical Chemistry,* L. Meites, Ed., pp. **3-35, 3-36.** New York: McGraw-Hill Book Co., 1963. With permission.
† Ionic strength = 0.1.
‡ B = blue; C = colorless; O = orange; P = purple; R = red; Y = yellow.
§ (1) Acid type: $HIn + H_2O \rightleftharpoons H_3O^+ + In^-$
  (2) Base type: $In + H_2O \rightleftharpoons InH^+ + OH^-$
¶ Reaction type: $InH^+ + H_2O \rightleftharpoons H_3O^+ + In$

$\pm 0.5$ pH unit in an ordinary titration, although with the use of comparison standards, uncertainties on the order of $\pm 0.1$ pH unit can often be achieved. It must be emphasized that these magnitudes are rough estimates that will vary from indicator to indicator as well as from person to person.

## 8B CURVES FOR THE TITRATION OF STRONG ACIDS OR STRONG BASES

When both analyte and titrant are completely dissociated, the net-ionic equation for a neutralization reaction is

$$H_3O^+ + OH^- \rightleftharpoons 2H_2O$$

and derivation of a curve for such a titration is analogous to that for a precipitation titration (Chapter 7, Section 7A).

### 8B–1 Titration of a Strong Acid with a Strong Base

An aqueous solution of a strong acid has two sources of hydronium ion, specifically, that derived from the solute, and that from the dissociation of water itself. The ion-product constant for water is, however, so very small that its contribution to the acidity of a solution of a strong acid is inconsequential except when the acid is extremely dilute (say $1 \times 10^{-6}F$ or less). Normally, then, in a solution of a strong acid, such as hydrochloric acid, the hydronium ion concentration is numerically equal to the formal concentration of the acid; that is,

$$[H_3O^+] = F_{HCl}$$

A comparable situation exists in solutions of strong bases. Both sodium hydroxide and barium hydroxide are completely dissociated; thus,

$$[OH^-] = F_{NaOH}$$

and

$$[OH^-] = 2 \times F_{Ba(OH)_2}$$

A relationship that is useful in calculating the pH of basic solutions is obtained by taking the negative logarithm of each side of the ion product constant for water; that is,

$$-\log K_w = -\log ([H_3O^+][OH^-]) = -\log [H_3O^+] - \log [OH^-]$$

$$pK_w = pH + pOH$$

At 25°C, $pK_w$ is numerically equal to 14.00.

The following example illustrates that derivation of a curve for the titration of a strong acid with a strong base is straightforward because the hydronium ion or hydroxide ion concentration is evaluated directly from the formal concentration of the acid or base that is present in excess. The pH (rather than the pOH) is customarily plotted as the ordinate.

Table 8 – 2   *Changes in pH During the Titration of a Strong Acid with a Strong Base*

| Volume of NaOH, mL | pH | |
|---|---|---|
| | 50.00 mL of 0.0500 *F* HCl with 0.1000 *F* NaOH | 50.00 mL of 0.000500 *F* HCl with 0.001000 *F* NaOH |
| 0.00 | 1.30 | 3.30 |
| 10.00 | 1.60 | 3.60 |
| 20.00 | 2.15 | 4.15 |
| 24.00 | 2.87 | 4.87 |
| 24.90 | 3.87 | 5.87 |
| 25.00 | 7.00 | 7.00 |
| 25.10 | 10.12 | 8.12 |
| 26.00 | 11.12 | 9.12 |
| 30.00 | 11.80 | 9.80 |

**Example 8 – 1**   Generate a curve for the titration of 50.00 mL of 0.0500 *F* HCl with 0.1000 *F* NaOH. Round pH data to two places beyond the decimal point.

(a) *Initial pH.* The solution is $5.00 \times 10^{-2}$ *F* in HCl. Since HCl is completely dissociated,

$$[H_3O^+] = 5.00 \times 10^{-2}$$

$$\begin{aligned} pH &= -\log (5.00 \times 10^{-2}) = -\log 5.00 - \log 10^{-2} \\ &= 2 - \log 5.00 \\ &= 2 - 0.699 = 1.301 = 1.30 \end{aligned}$$

(b) *pH after addition of 10.00 mL NaOH.* The volume of the solution is now 60.00 mL and the HCl has been partially neutralized. Thus,

$$[H_3O^+] = \frac{50.00 \times 0.0500 - 10.00 \times 0.1000}{50.00 + 10.00} = 2.50 \times 10^{-2}$$

$$pH = 2 - \log 2.50 = 1.60$$

Additional data to define the curve short of the equivalence point are calculated in the same way. The results of such calculations are to be found in column 2 of Table 8 – 2.

(c) *pH after addition of 25.00 mL of NaOH.* This volume corresponds to the equivalence point for the titration, where neither HCl nor NaOH is in excess. Here, the dissociation of water is now the only source for hydronium and hydroxide ions, and

$$[H_3O^+] = [OH^-] = \sqrt{K_w} = 1.00 \times 10^{-7}$$

$$pH = 7.00$$

(d) *pH after addition of 25.10 mL of NaOH.* The solution now contains an excess of base; the formal concentration of NaOH will be given by

$$F_{\text{NaOH}} = \frac{25.10 \times 0.1000 - 50.00 \times 0.0500}{50.00 + 25.10} = 1.33 \times 10^{-4}$$

Provided the contribution of water to the hydroxide ion concentration is negligible with respect to that from the excess of base,

$$[\text{OH}^-] = -\log 1.33 \times 10^{-4}$$

$$\text{pOH} = 4 - \log 1.33 = 3.88$$

Finally, then,

$$\text{pH} = 14.00 - 3.88 = 10.12$$

Additional data for this titration, calculated in the same way, are to be found in column 2 of Table 8–2.

**Effect of Concentration**    The effects of reagent concentration and analyte concentration on neutralization titration curves are shown by the two sets of data in Table 8–2. These data are plotted in Figure 8–1; shown also are the transition pH ranges for three common acid-base indicators. Note the enormous pH change in the equivalence-point region of a titration involving the 0.0500 *F* acid (curve *A*). This change is markedly less for the titration of 0.000500 *F* hydrochloric acid (curve *B*) but is still pronounced.

Figure 8–1 demonstrates that indicator selection is not at all critical for a titration involving strong acids and strong bases with concentrations on the order of 0.1 *F*. Here, the volume differences among titrations with the three indicators are of the same magnitude as the uncertainties associated with the reading of the buret and are thus negligible. On the other hand, bromocresol

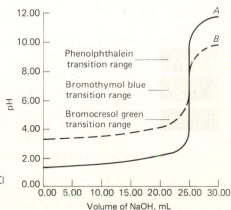

**Figure 8–1**    Curves for the titration of HCl with NaOH. Curve *A*: 50.0 mL of 0.0500 *F* HCl with 0.1000 *F* NaOH. Curve *B*: 50.0 mL of 0.000500 *F* HCl with 0.001000 *F* NaOH.

green is clearly unsuited for a titration with the 0.001 *F* reagent because the color change would take place continuously over a substantial range of titrant volumes. In addition, transition to the alkaline form would be essentially complete before the equivalence point was reached: a significant titration error would result. The same objections would attend the use of phenolphthalein. Of the three indicators, only bromothymol blue would provide an end point with a minimum titration error.

### 8B–2 Titration of a Strong Base with a Strong Acid

An analogous situation exists for the titration of strong bases with strong acids. In such titrations the region short of the equivalence point is highly alkaline, the molar hydroxide ion concentration being numerically equal to the normality of the base. The solution is neutral at the equivalence point for precisely the same reason noted in Example 8–1. Finally, the solution becomes acidic in the region beyond the equivalence point; here, the pH is determined by the formal concentration of the excess strong acid that has been introduced. A curve for the titration of a strong base is shown in Figure 8–6 (p. 191). The selection of an indicator is subject to the same considerations that were noted for the titration of a strong acid with a strong base.

## 8C EQUILIBRIA INVOLVING WEAK ACIDS AND WEAK BASES

Calculation of the pH of a solution of a weak acid or a weak base requires a knowledge of not only the formal concentration of the acid or base but its dissociation constant as well in order to take into account the incomplete dissociation of the species.[2]

### 8C–1 Dissociation Equilibria for Weak Acids and Bases

In contrast to their strong counterparts, weak acids and weak bases only undergo partial dissociation in solution. Thus, a solution of nitrous acid is described by the equilibrium

$$HNO_2 + H_2O \rightleftharpoons H_3O^+ + NO_2^- \qquad K_a = \frac{[H_3O^+][NO_2^-]}{[HNO_2]}$$

where $K_a$ is the *acid dissociation constant* for nitrous acid. The corresponding equilibrium for a solution of ammonia is

$$NH_3 + H_2O \rightleftharpoons NH_4^+ + OH^- \qquad K_b = \frac{[NH_4^+][OH^-]}{[NH_3]}$$

Similarly, $K_b$ is the *basic dissociation constant* for ammonia.

---

[2] We shall restrict our discussion to systems in which water is the solvent. The reader should be aware that acid-base behavior occurs in other solvents as well.

Note that a concentration term for water ($[H_2O]$) does not appear in the denominator of either equilibrium constant expression because the solvent concentration is so large that it is unaffected by the introduction of the acid or base. Thus, the concentration of water has been incorporated in the equilibrium constants $K_a$ and $K_b$. Numerical values for the dissociation constants of weak acids and weak bases are tabulated in Appendixes 10 and 11.

**Relationship Between the Dissociation Constants for Conjugate Acid-Base Pairs**   Consider the equilibria for the dissociation of ammonia as well as its conjugate acid, ammonium ion:

$$NH_3 + H_2O \rightleftharpoons NH_4^+ + OH^- \qquad K_b = \frac{[NH_4^+][OH^-]}{[NH_3]}$$

and

$$NH_4^+ + H_2O \rightleftharpoons NH_3 + H_3O^+ \qquad K_a = \frac{[NH_3][H_3O^+]}{[NH_4^+]}$$

Multiplication of one equilibrium constant expression by the other gives

$$K_a K_b = \frac{[NH_3][H_3O^+]}{[NH_4^+]} \times \frac{[NH_4^+][OH^-]}{[NH_3]} = [H_3O^+][OH^-]$$

Since

$$[H_3O^+][OH^-] = K_w$$

it follows that

$$K_a K_b = K_w \qquad\qquad (8-1)$$

This is a general relationship for all conjugate acid-base pairs. Tabulations of dissociation constant data seldom include both the acid and base dissociation constants for conjugate pairs, since it is so easy to calculate one from the other by means of Equation 8-1.

---

**Example 8-2**   Calculate $K_b$ for the equilibrium

$$CN^- + H_2O \rightleftharpoons HCN + OH^-$$

Examination of the table of basic dissociation constants (Appendix 11) reveals no entry for $CN^-$. The table of acid dissociation constants (Appendix 10), however, does show a value of $2.1 \times 10^{-9}$ for HCN. Thus,

$$K_b = \frac{[HCN][OH^-]}{[CN^-]} = \frac{1.00 \times 10^{-14}}{2.1 \times 10^{-9}} = 4.8 \times 10^{-6}$$

## 8C–2 Equilibrium Calculations Involving Weak Acids and Weak Bases

The dissociation constant expression for a weak acid (or a weak base) can be used to account for the phenomenon of buffering, the behavior of acid-base indicators, and the shape of titration curves. A convenient approach to a problem that requires the use of such a constant has, as a first step, identification of the principal solute species that will influence the pH of the solution after equilibrium has been established. For the present, we shall be concerned with just three possibilities:

1. The principal solute species is the weak acid, HA.
2. The solution contains both HA and its conjugate base, $A^-$, as principal solute species.
3. The principal solute species is the conjugate base, $A^-$.

These three possibilities also apply to a solution of a weak base and its conjugate acid.

Each of these situations is considered in the sections that follow. Also discussed are the approximations, which — where applicable — greatly ease the burden of calculation.

**Calculation of the Hydronium Ion Concentration for a Solution Containing the Weak Acid HA**    Consider the equilibrium that is established when the weak acid HA is dissolved in water to give a solution with a formal acid concentration of $F_{HA}$. Both solvent and solute contribute to the hydronium ion concentration of the resulting solution

$$HA + H_2O \rightleftharpoons H_3O^+ + A^- \qquad K_a = \frac{[H_3O^+][A^-]}{[HA]} \qquad (8-2)$$

$$2H_2O \rightleftharpoons H_3O^+ + OH^- \qquad K_w = [H_3O^+][OH^-] \qquad (8-3)$$

This solution contains $A^-$ as well as HA as a result of dissociation. Since the only source of these two species is the original acid, material balance requires that

$$F_{HA} = [HA] + [A^-] \qquad (8-4)$$

Note that the fraction of A-containing species that exists as $[A^-]$ will increase as the dissociation constant for the weak acid under consideration *increases*.

The solution can also be described in terms of charge balance (footnote 3, p. 101); that is,

$$[A^-] + [OH^-] = [H_3O^+]$$

Dissociation of the weak acid will ordinarily repress the dissociation of water to such an extent that the formation of hydroxide ion is negligible. Thus for all but most dilute solutions of very weak acids it is possible to assume that $[OH^-]$ is much smaller than $[A^-]$. The foregoing equation then becomes

$$[A^-] \cong [H_3O^+] \qquad (8-5)$$

Substitution of Equation 8–5 into 8–4 and rearrangement gives

$$[HA] = F_{HA} - [H_3O^+] \tag{8-6}$$

Substitution of Equations 8–5 and 8–6 into Equation 8–2 gives

$$K_a = \frac{[H_3O^+]^2}{F_{HA} - [H_3O^+]} \tag{8-7}$$

which in turn rearranges to

$$[H_3O^+]^2 + K_a[H_3O^+] - K_a F_{HA} = 0 \tag{8-8}$$

The positive root to this quadratic equation is

$$[H_3O^+] = \frac{-K_a + \sqrt{(K_a)^2 + 4K_a F_{HA}}}{2}$$

It is frequently possible to make the further approximation that $[H_3O^+]$ is much smaller than $F_{HA}$. Under this circumstance, Equation 8–7 can be rearranged to

$$[H_3O^+] \cong \sqrt{K_a F_{HA}} \tag{8-9}$$

The magnitude of the error introduced by the assumption that $[H_3O^+] \ll F_{HA}$ will increase as the formal concentration of the acid becomes smaller and as the numerical value for $K_a$ becomes larger. This statement is supported by the data in Table 8–3. Note that the error is about 0.5% when the ratio $F_{HA}/K_a$ is

**Table 8–3  Errors Introduced by Assuming H₃O⁺ Concentration Small Relative to $F_{HA}$ in Equation 8–7**

| Value of $K_a$ | Value of $F_{HA}$ | Value for $[H_3O^+]$ Using Assumption | Value for $[H_3O^+]$ by More Exact Equation | Percent Error |
|---|---|---|---|---|
| $1.00 \times 10^{-2}$ | $1.00 \times 10^{-3}$ | $3.16 \times 10^{-3}$ | $0.92 \times 10^{-3}$ | 244 |
| | $1.00 \times 10^{-2}$ | $1.00 \times 10^{-2}$ | $0.62 \times 10^{-2}$ | 61 |
| | $1.00 \times 10^{-1}$ | $3.16 \times 10^{-2}$ | $2.70 \times 10^{-2}$ | 17 |
| $1.00 \times 10^{-4}$ | $1.00 \times 10^{-4}$ | $1.00 \times 10^{-4}$ | $0.62 \times 10^{-4}$ | 61 |
| | $1.00 \times 10^{-3}$ | $3.16 \times 10^{-4}$ | $2.70 \times 10^{-4}$ | 17 |
| | $1.00 \times 10^{-2}$ | $1.00 \times 10^{-3}$ | $0.95 \times 10^{-3}$ | 5.3 |
| | $1.00 \times 10^{-1}$ | $3.16 \times 10^{-3}$ | $3.11 \times 10^{-3}$ | 1.6 |
| $1.00 \times 10^{-6}$ | $1.00 \times 10^{-5}$ | $3.16 \times 10^{-6}$ | $2.70 \times 10^{-6}$ | 17 |
| | $1.00 \times 10^{-4}$ | $1.00 \times 10^{-5}$ | $0.95 \times 10^{-5}$ | 5.3 |
| | $1.00 \times 10^{-3}$ | $3.16 \times 10^{-5}$ | $3.11 \times 10^{-5}$ | 1.6 |
| | $1.00 \times 10^{-2}$ | $1.00 \times 10^{-4}$ | $9.95 \times 10^{-5}$ | 0.5 |
| | $1.00 \times 10^{-1}$ | $3.16 \times 10^{-4}$ | $3.16 \times 10^{-4}$ | 0.0 |

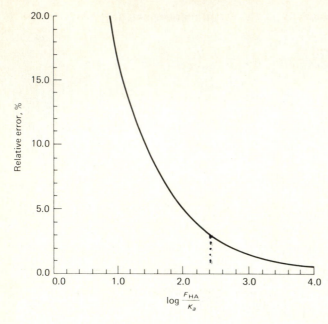

Figure 8-2  Relative error resulting from the assumption that $F_{HA} - [H_3O^+] \cong F_{HA}$ in Equation 8-7.

$10^4$ and increases to 1.6%, 5%, and 17%, respectively, as the ratio decreases by successive orders of magnitude. Figure 8-2 depicts this effect graphically. It is noteworthy that the hydronium ion concentration becomes equal to or greater than the formal concentration for the acid itself where $F_{HA}/K_a$ is 1 or smaller; clearly, the approximation leads to meaningless results under these circumstances.

It is ordinarily good practice to use Equation 8-9 to obtain a provisional value for $[H_3O^+]$; this result can then be used to judge the validity of the approximation. If an assumption does not appear to be justified, it will be necessary to use the quadratic equation.

---

**Example 8-3**  Calculate the $H_3O^+$ concentration of a $0.120\ F$ $HNO_2$ solution.

The principal equilibrium in this solution is

$$HNO_2 + H_2O \rightleftharpoons H_3O^+ + NO_2^-$$

for which

$$K_a = \frac{[H_3O^+][NO_2^-]}{[HNO_2]} = 5.1 \times 10^{-4} \qquad \text{(Appendix 10)}$$

Equations 8-5 and 8-6 become

$$[H_3O^+] = [NO_2^-]$$

$$[HNO_2] = 0.120 - [H_3O^+] \cong 0.120$$

Substitution of these quantities into the dissociation constant expression for $HNO_2$ gives, provisionally,

$$[H_3O^+]^2 = (5.1 \times 10^{-4})(0.120) = 6.12 \times 10^{-5}$$

$$[H_3O^+] = 7.8 \times 10^{-3}$$

$$pH = 3 - \log 7.8 = 2.11$$

The magnitude of the relative error in $[H_3O^+]$ resulting from the assumption that $[H_3O^+] \ll [HA]$ can be approximated by the use of Figure 8-2. That is,

$$\log \frac{F_{HNO_2}}{K_a} = \frac{0.120}{5.1 \times 10^{-4}} = 2.4$$

It is seen that a relative error between 2 and 3% was incurred. Use of the quadratic equation for this example gives $[H_3O^+] = 7.6 \times 10^{-3}$.

---

**Example 8-4** Calculate the hydronium ion concentration for a solution that is $1.00 \times 10^{-2}$ $F$ in chloroacetic acid. For $ClCH_2COOH$,

$$K_a = \frac{[H_3O^+][ClCH_2COO^-]}{[ClCH_2COOH]} = 1.36 \times 10^{-3}$$

Proceeding as in the previous example,

$$[H_3O^+] \cong [ClCH_2COO^-]$$

$$[ClCH_2COOH] = 1.00 \times 10^{-2} - [H_3O^+] \cong 1.00 \times 10^{-2}$$

Substitution into the dissociation constant expression gives

$$[H_3O^+]^2 = (1.36 \times 10^{-3})(1.00 \times 10^{-2}) = 1.36 \times 10^{-5}$$

$$[H_3O^+] = 3.69 \times 10^{-3}$$

It is clear that $3.69 \times 10^{-3}$ is *not* negligible with respect to $1.00 \times 10^{-2}$, as we assumed in calculating the provisional answer. It thus becomes necessary to use $1.00 \times 10^{-2} - [H_3O^+]$ for $[ClCH_2COOH]$ in the denominator of the equilibrium constant expression:

$$1.36 \times 10^{-3} = \frac{[H_3O^+]^2}{1.00 \times 10^{-2} - [H_3O^+]}$$

which rearranges to

$$[H_3O^+]^2 + 1.36 \times 10^{-3}[H_3O^+] - 1.36 \times 10^{-5} = 0$$

$$[H_3O^+] = \frac{-1.36 \times 10^{-3} + \sqrt{(1.36 \times 10^{-3})^2 + 4 \times 1.36 \times 10^{-5}}}{2}$$

$$= 3.07 \times 10^{-3}$$

**Calculation of $[H_3O^+]$ for a Solution Containing the Conjugate Base of the Weak Acid HA** A solution of NaA will be alkaline, as a result of the equilibrium

$$A^- + H_2O \rightleftharpoons HA + OH^- \qquad K_b = \frac{[HA][OH^-]}{[A^-]}$$

and the process of solving for $[OH^-]$ is analogous to the calculation of $[H_3O^+]$ for a solution of HA.

Consider a solution in which the formal concentration of NaA is $F_{A^-}$. Material balance requires that

$$\begin{aligned} F_{A^-} &= [Na^+] \\ &= [HA] + [A^-] \end{aligned} \qquad (8-10)$$

Since $F_A = [Na^+]$, the charge-balance relationship can be written as

$$F_{A^-} + [H_3O^+] = [OH^-] + [A^-] \qquad (8-11)$$

Subtraction of Equation 8–10 from Equation 8–11 and rearrangement yields

$$[OH^-] = [H_3O^+] + [HA] \qquad (8-12)$$

Since the solution is alkaline, it is permissible to simplify Equation 8–12 to

$$[OH^-] \cong [HA] \qquad (8-13)$$

This approximation is generally valid because the hydroxide ions produced by the reaction of $A^-$ with water will repress the dissociation of water to the point where $[H_3O^+]$ is minuscule. Substitution of Equation 8–13 into Equation 8–10 gives

$$F_{A^-} = [OH^-] + [A^-]$$

or

$$[A^-] = F_{A^-} - [OH^-] \qquad (8-14)$$

Where dissociation of the base is not extensive, it is possible to simplify this equation to

$$F_{A^-} \cong [A^-]$$

This approximation is most likely to be justified where the basic dissociation constant is numerically small and the formal concentration of NaA is large. Substitution of this equation and Equation 8–13 into the equilibrium constant expression yields

$$K_b = \frac{[OH^-]^2}{F_{A^-}}$$

$$[OH^-] = \sqrt{K_b F_{A^-}}$$

If the provisional value for $[OH^-]$ is appreciable with respect to the concentration of $A^-$, $(F_{A^-} - [OH^-])$ must be used in the denominator of the dissociation constant expression and it becomes necessary to solve a quadratic equation; that is,

$$K_b = \frac{[OH^-]^2}{F_{A^-} - [OH^-]}$$

$$[OH^-]^2 + K_b[OH^-] - K_b F_{A^-} = 0 \qquad (8-15)$$

**Example 8-5** Calculate the hydronium ion concentration of a 0.0100 $F$ NaOCl solution.

The equilibrium between $OCl^-$ and water is

$$OCl^- + H_2O \rightleftharpoons HOCl + OH^-$$

for which (Equation 8-1)

$$K_b = \frac{[HOCl][OH^-]}{[OCl^-]} = \frac{1.00 \times 10^{-14}}{3.0 \times 10^{-8}} = 3.3 \times 10^{-7}$$

Substitution into Equations 8-13 and 8-14 gives

$$[HOCl] = [OH^-]$$

$$[OCl^-] = F_{OCl^-} - [OH^-] \cong 1.00 \times 10^{-2}$$

A provisional value for $[OH^-]$ is obtained by substituting these quantities into the basic dissociation constant for hypochlorite ion:

$$[OH^-]^2 = K_b F_{OCl^-} = (3.3 \times 10^{-7})(1.00 \times 10^{-2}) = 3.3 \times 10^{-9}$$

$$[OH^-] = 5.7 \times 10^{-5}$$

The error introduced by the approximations is negligible. Finally, then,

$$[H_3O^+] = \frac{1.00 \times 10^{-14}}{5.7 \times 10^{-5}} = 1.7 \times 10^{-10}$$

**Calculation of the pH of a Solution Containing a Weak Acid, HA, and Its Conjugate Base, NaA**     Consider a solution in which the formal concentration of HA is $F_{HA}$ while that for NaA is $F_{A^-}$. We have noted that each solute enters into equilibrium with the solvent:

$$HA + H_2O \rightleftharpoons H_3O^+ + A^-$$

$$A^- + H_2O \rightleftharpoons OH^- + HA$$

The first of these equilibria will decrease the concentration of HA by an amount equal to $[H_3O^+]$, while the second equilibrium will result in an increase in HA in an amount equal to $[OH^-]$. Thus, the molar concentration of HA is given by

$$[HA] = F_{HA} - [H_3O^+] + [OH^-] \qquad (8-16)$$

Similarly, the first equilibrium will be responsible for an increase in the concentration of $A^-$ by an amount equal to $[H_3O^+]$, while the second equilibrium will cause a decrease in an amount equal to $[OH^-]$. The equilibrium concentration of $A^-$ will thus be given by

$$[A^-] = F_{A^-} + [H_3O^+] - [OH^-] \qquad (8-17)$$

It will always be possible to eliminate either $[OH^-]$ or $[H_3O^+]$ from Equations 8-16 and 8-17. If the solution is appreciably acidic, $[OH^-]$ will necessarily be small and can be neglected. Similarly, $[H_3O^+]$ will be negligibly small in an alkaline solution. Moreover, it is frequently true that the *difference* in concentration between $[H_3O^+]$ and $[OH^-]$ is small relative to the molar concentrations of either HA or $A^-$; under such circumstances, Equations 8-16 and 8-17 further simplify to

$$[HA] \cong F_{HA} \qquad (8-18)$$

$$[A^-] \cong F_{A^-} \qquad (8-19)$$

Substitution of these quantities into the dissociation constant expression for HA and rearrangement gives

$$[H_3O^+] = K_a \frac{F_{HA}}{F_{A^-}} \qquad (8-20)^3$$

Within the limits imposed by the assumptions made in its derivation, Equation 8-20 states that the hydronium ion concentration of a solution containing a weak acid and its conjugate base is dependent only upon the *ratio* between the molar concentrations of these two solutes. Furthermore, this ratio remains *independent of dilution* because the concentration of each component changes proportionately as the result of a volume change. Thus, the hydronium ion concentration of a solution containing appreciable quantities of a weak acid and its conjugate base tends to be independent of dilution and depends only upon the ratio of molar concentrations between the two solutes. This independence of pH from dilution is one manifestation of the *buffering* properties of such solutions. Buffer solutions are considered in greater detail in Section 8C-3.

---

[3] An alternative form of Equation 8-20 is frequently encountered in the biological and biochemical literature; it is obtained by taking the negative logarithm of each term. Thus,

$$-\log [H_3O^+] = -\log K_a - \log \frac{F_{HA}}{F_A^-}$$

Upon inverting the concentration ratio to keep all terms positive, we obtain

$$pH = pk_a + \log \frac{F_A^-}{F_{HA}}$$

This expression is known as the *Henderson-Hasselbalch equation.*

The assumption that both $[H_3O^+]$ and $[OH^-]$ can be eliminated from Equations 8–16 and 8–17 will fail for solutions involving acids or bases with dissociation constants equal to or greater than about $10^{-3}$ or where the concentration of either the acid or its conjugate base (or both) is small. Retention of $[H_3O^+]$ will be required if the solution is acidic; alternatively, $[OH^-]$ will be required if the solution is basic. A quadratic equation will result under either circumstance.

As in earlier examples, it is always prudent to obtain a provisional answer through use of Equation 8–20; the calculated value can then be used to test the assumptions.

**Example 8–6** Calculate the pH of a solution that is $0.120\ F$ in $HNO_2$ and $0.200\ F$ in $NaNO_2$.

The equilibrium between $HNO_2$ and water is

$$HNO_2 + H_2O \rightleftharpoons H_3O^+ + NO_2^-$$

for which

$$K_a = \frac{[H_3O^+][NO_2^-]}{[HNO_2]} = 5.1 \times 10^{-4}$$

Application of Equations 8–18 and 8–19 gives

$$[HNO_2] \cong F_{HNO_2} = 0.120$$

$$[NO_2^-] \cong F_{NO_2^-} = 0.200$$

Substitution of these quantities into Equation 8–20 yields

$$[H_3O^+] = 5.1 \times 10^{-4} \times \frac{0.120}{0.200} = 3.1 \times 10^{-4}$$

Note that the assumption that $[H_3O^+] \ll [HNO_2]$ and $[NO_2^-]$ is justified. Thus,

$$pH = 4 - \log 3.1 = 3.51$$

Comparison of the hydronium ion concentration of this solution with that calculated for $0.120\ F\ HNO_2$ (Example 8–3) reveals that introduction of sodium nitrite has caused a decrease in $[H_3O^+]$ by a factor of about 25.

**Example 8–7** Calculate the pH of a solution that is $2.00 \times 10^{-2}\ F$ in chloroacetic acid and $1.00 \times 10^{-2}\ F$ in sodium chloroacetate.

For $ClCH_2COOH$,

$$K_a = \frac{[H_3O^+][ClCH_2COO^-]}{[ClCH_2COOH]} = 1.36 \times 10^{-3}$$

Proceeding as in the previous example,

$$[ClCH_2COOH] \cong F_{ClCH_2COOH} = 2.00 \times 10^{-2}$$

$$[ClCH_2COO^-] \cong F_{ClCH_2COO^-} = 1.00 \times 10^{-2}$$

and

$$[H_3O^+] = 1.36 \times 10^{-3} \times \frac{2.00 \times 10^{-2}}{1.00 \times 10^{-2}} = 2.72 \times 10^{-3}$$

This provisional value for $[H_3O^+]$ is unsatisfactory because it is not small with respect to the formal solute concentrations. Since the solution is clearly acidic, Equations 8–16 and 8–17 can be simplified through elimination of $[OH^-]$; thus,

$$[ClCH_2COOH] = 2.00 \times 10^{-2} - [H_3O^+]$$

$$[ClCH_2COO^-] = 1.00 \times 10^{-2} + [H_3O^+]$$

Substitution of these quantities into the equilibrium constant expression gives

$$1.36 \times 10^{-3} = \frac{[H_3O^+](1.00 \times 10^{-2} + [H_3O^+])}{2.00 \times 10^{-2} - [H_3O^+]}$$

which rearranges to

$$[H_3O^+]^2 + (1.00 \times 10^{-2} + 1.36 \times 10^{-3})[H_3O^+]$$
$$- (1.36 \times 10^{-3})(2.00 \times 10^{-2}) = 0$$

$$[H_3O^+]^2 + 1.14 \times 10^{-2}[H_3O^+] - 2.72 \times 10^{-5} = 0$$

The positive root to this quadratic equation is

$$[H_3O^+] = 2.03 \times 10^{-3}$$

$$pH = 3 - \log 2.03 = 2.69$$

Note that, as in Example 8–6, introduction of chloroacetate ion has had the effect of raising the pH significantly.

**Calculation of the pH for a Solution Containing a Weak Base and Its Conjugate Acid**   The calculation of the hydroxide ion concentration for a solution consisting of a weak base and its conjugate acid is entirely analogous to that developed in the preceding section.

**Example 8–8**   Calculate the pH of a solution that is $0.0700\ F$ in $NH_3$ and $0.280\ F$ in $NH_4Cl$. The basic dissociation constant for $NH_3$ (Appendix 11) is $1.76 \times 10^{-5}$.

The equilibria of interest are

$$NH_3 + H_2O \rightleftharpoons NH_4^+ + OH^-$$

$$NH_4^+ + H_2O \rightleftharpoons NH_3 + H_3O^+$$

Using the arguments in the previous examples, the molar concentrations for $NH_3$ and $NH_4^+$ are

$$[NH_3] = 0.0700 - [OH^-] + [H_3O^+] = 0.0700 - [OH^-] \cong 0.0700$$

$$[NH_4^+] = 0.280 + [OH^-] - [H_3O^+] = 0.280 + [OH^-] \cong 0.280$$

A provisional value for $[OH^-]$ is obtained by substituting the approximate values for $[NH_3]$ and $[NH_4^+]$ into the rearranged form of the equilibrium constant expression for $NH_3$:

$$[OH^-] = K_b \times \frac{[NH_3]}{[NH_4^+]} = 1.76 \times 10^{-5} \times \frac{0.0700}{0.280} = 4.40 \times 10^{-6}$$

The approximations are clearly justified; thus,

$$pOH = 5 - \log 4.40 = 5.36$$

$$pH = 14.00 - 5.36 = 8.64$$

## 8C-3 Buffer Solutions

A *buffer solution* is defined as a solution that resists changes in pH as a result of (1) dilution or (2) small additions of acids or bases. The most versatile buffer solutions contain large and approximately equal concentrations of a conjugate acid-base pair.

**Effect of Dilution** The pH of a buffer solution remains essentially independent of dilution until its concentrations are decreased to the point where the approximations used to develop Equations 8-18 and 8-19 fail. Example 8-9 illustrates effects of dilution upon the pH of a typical buffer.

**Example 8-9** Calculate the pH of the nitrous acid–nitrite buffer in Example 8-6 after dilution by a factor of (a) 10 and (b) 1000. The pH of the undiluted buffer was 3.51.

(a) After a 10-fold dilution,

$$[HNO_2] = \frac{0.120}{10} - [H_3O^+] \cong 0.0120$$

$$[NO_2^-] = \frac{0.200}{10} + [H_3O^+] \cong 0.0200$$

If we again assume that $[H_3O^+]$ is small with respect to $[HNO_2]$ and $[NO_2^-]$ (p. 176), we obtain

$$[H_3O^+] = 5.1 \times 10^{-4} \times \frac{0.0120}{0.0200} = 3.1 \times 10^{-4}$$

which is identical to the hydronium ion concentration calculated in Example 8–6. This value is still small with respect to the formal solute concentrations; the approximations are thus reasonably good, and the calculated pH remains 3.51 (a quadratic solution would yield a value of 3.53).

(b) After a 1000-fold dilution,

$$[HNO_2] = \frac{0.120}{1000} - [H_3O^+] \cong 1.20 \times 10^{-4}$$

$$[NO_2^-] = \frac{0.200}{1000} + [H_3O^+] \cong 2.00 \times 10^{-4}$$

If, as before, we were to substitute these values into $K_a$ for $HNO_2$, we would again obtain a value of $3.1 \times 10^{-4}$ for $[H_3O^+]$. Now, however, this provisional value is actually larger than $[HNO_2]$ and $[NO_2^-]$. It thus becomes necessary to make use of the more exact Equations 8–16 and 8–17; that is,

$$[HNO_2] = F_{HNO_2} - [H_3O^+] + [OH^-]$$

$$[NO_2^-] = F_{NO_2^-} + [H_3O^+] - [OH^-]$$

The solution is acidic; further simplification to

$$[HNO_2] \cong F_{HNO_2} - [H_3O^+] = 1.20 \times 10^{-4} - [H_3O^+]$$

$$[NO_2^-] \cong F_{NO_2^-} + [H_3O^+] = 2.00 \times 10^{-4} + [H_3O^+]$$

is justified. Substitution of these quantities into $K_a$ yields

$$5.1 \times 10^{-4} = \frac{[H_3O^+](2.00 \times 10^{-4} + [H_3O^+])}{1.20 \times 10^{-4} - [H_3O^+]}$$

and rearrangement gives the quadratic equation

$$[H_3O^+]^2 + 7.1 \times 10^{-4}[H_3O^+] - 6.12 \times 10^{-8} = 0$$

$$[H_3O^+] = 7.8 \times 10^{-5}; \qquad pH = 5 - \log 7.8 = 4.11$$

We see that a 1000-fold dilution has caused the pH to change by about 0.6 unit, while the 10-fold dilution had essentially no effect.

Figure 8–3 shows how buffered and unbuffered solutions are affected by dilution. The resistance of the buffered solution to changes in pH is clear.

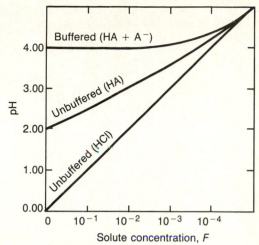

**Figure 8-3**  Effect of dilution on the pH of buffered and unbuffered solutions. Initial solute concentrations are 1.00 $F$; $K_{HA} = 1.00 \times 10^{-4}$.

**Effect of Added Acids and Bases upon Buffers**  Examples 8-10 and 8-11 demonstrate the capability of buffers to maintain a nearly unchanged pH after small amounts of acid or base have been introduced.

**Example 8-10**  Calculate the pH change that occurs when 100 mL of (a) 0.0500 $F$ NaOH and (b) 0.0500 $F$ HCl are added to 400 mL of a buffer solution that is 0.200 $F$ in $NH_3$ and 0.300 $F$ in $NH_4Cl$.

This solution will be alkaline. Thus, initially,

$$[NH_3] = 0.200 - [OH^-] \cong 0.200$$

$$[NH_4^+] = 0.300 + [OH^-] \cong 0.300$$

Substitution of these values into the dissociation contant for $NH_3$ and rearrangement gives

$$[OH^-] = 1.76 \times 10^{-5} \times \frac{0.200}{0.300} = 1.17 \times 10^{-5}$$

$$pOH = 5 - \log 1.17 = 4.93$$

$$pH = 14.00 - 4.93 = 9.07$$

(a)  Addition of NaOH converts part of the $NH_4^+$ to $NH_3$:

$$NH_4^+ + OH^- \rightarrow NH_3 + H_2O$$

The formal solute concentrations thus become

$$F_{NH_3} = \frac{400 \times 0.200 + 100 \times 0.0500}{500} = \frac{85.0}{500} \cong [NH_3]$$

$$F_{NH_4^+} = \frac{400 \times 0.300 - 100 \times 0.0500}{500} = \frac{115}{500} \cong [NH_4^+]$$

Substitution into the rearranged form of the dissociation constant for $NH_3$ gives

$$[OH^-] = 1.76 \times 10^{-5} \times \frac{85.0/500}{115/500} = 1.30 \times 10^{-5}$$

$$pOH = 5 - \log 1.30 = 4.89$$

$$pH = 14.00 - 4.89 = 9.11$$

$$\Delta pH = 9.11 - 9.07 = 0.04$$

(b)  Addition of HCl converts part of the $NH_3$ to $NH_4^+$:

$$NH_3 + H_3O^+ \rightarrow NH_4^+ + H_2O$$

Thus,

$$F_{NH_3} = \frac{400 \times 0.200 - 100 \times 0.0500}{500} = \frac{75}{500} \cong [NH_3]$$

$$F_{NH_4^+} = \frac{400 \times 0.300 + 100 \times 0.0500}{500} = \frac{125}{500} \cong [NH_4^+]$$

Substitution as before gives

$$[OH^-] = 1.76 \times 10^{-5} \times \frac{75/500}{125/500} = 1.06 \times 10^{-5}$$

$$pOH = 5 - \log 1.06 = 4.98$$

$$pH = 14.00 - 4.98 = 9.02$$

$$\Delta pH = 9.02 - 9.07 = -0.05$$

**Example 8–11**  Contrast the behavior of an unbuffered solution with a pH of 9.07 ($[OH^-] = 1.17 \times 10^{-5}$) when the same volumes of acid and base are added as were introduced to the $NH_3/NH_4^+$ buffer in Example 8–10.

(a) After the addition of 100 mL of 0.0500 $F$ NaOH,

$$[OH^-] = \frac{400 \times 1.17 \times 10^{-5} + 100 \times 0.0500}{500} = \frac{5.00}{500}$$

$$= 1.00 \times 10^{-2}$$

$$pOH = 2 - \log 1.00 = 2.00$$

$$pH = 14.00 - 2.00 = 12.00$$

$$\Delta pH = 12.00 - 9.07 = 2.93$$

(b) After the addition of 100 mL of 0.0500 $F$ HCl,

$$[H_3O^+] = \frac{100 \times 0.0500 - 400 \times 1.17 \times 10^{-5}}{500} = \frac{5.00}{500}$$

$$= 1.00 \times 10^{-2}$$

$$pH = 2 - \log 1.00 = 2.00$$

$$\Delta pH = 2.00 - 9.07 = -7.07$$

**Buffer Capacity** The foregoing examples demonstrate that a solution containing a conjugate acid-base pair shows remarkable resistance to changes in pH resulting from the addition of acids or bases. The ability of a buffer to resist such changes is directly related to the total concentration of the buffering species as well as their concentration ratios. For example, 400 mL of a solution produced by a tenfold dilution of the $NH_3/NH_4^+$ buffer in Example 8 – 10 would change by about 0.5 pH unit when treated with the same amounts of sodium hydroxide or hydrochloric acid; recall that the change for the undiluted buffer was only about 0.04 pH unit.

*Buffer capacity* is defined as the number of equivalents of strong acid or strong base needed to cause 1.00 L of a solution to experience a 1.00-unit change in pH. The maximum buffer capacity is associated with a concentration ratio of 1.00 between a weak acid and its conjugate base.[4]

**Preparation of Buffers** In principle, a buffer solution of any desired pH can be prepared by combining calculated quantities of a suitable conjugate acid-base pair. As a practical matter, however, the pH of such a buffer is likely to differ somewhat from the predicted values. These differences are the result of uncertainties in dissociation constants and the approximations that were used in the pH calculations. Moreover, the ionic strength of a typical buffer is likely to be

---

[4] See, for example: D. Rosenthal and P. Zuman, in *Treatise on Analytical Chemistry,* 2nd ed., I. M. Kolthoff and P. J. Elving, Eds., Part I, Vol. 2, p. 208. New York: John Wiley & Sons, 1979.

well beyond the range within which the Debye-Hückel expression applies. Thus, the ionic strength of the $NH_3/NH_4^+$ buffer in Example 8 – 10 is about 0.30; the concentration constant $K_b$ (p. 112) would therefore be significantly larger than $1.76 \times 10^{-5}$ and quite uncertain (about $4 \times 10^{-5}$).

Empirically derived recipes for the preparation of buffer solutions are available in chemical handbooks and reference works.[5] Two such systems deserve specific mention because of the frequency with which they are encountered. McIlvaine buffers cover a pH range from about 2 to 8 and are prepared by mixing solutions of citric acid and disodium hydrogen phosphate. Clark and Lubs buffers, which encompass a pH range from 2 to 10, make use of phthalic acid – potassium hydrogen phthalate, potassium dihydrogen phosphate – dipotassium hydrogen phosphate, and boric acid – sodium borate systems.

## 8C – 4 Theory of Indicator Behavior

Acid-base indicators are themselves weak acids or bases that dissociate in water to give hydronium or hydroxide ions. The property that sets indicators apart from other acids and bases is the internal structural changes that accompany dissociation and lead to changes in color. The dissociation of an acidic type indicator, HIn, or a basic type indicator, In, can be represented by equations such as

$$HIn + H_2O \rightleftharpoons H_3O^+ + In^- \qquad K_a = \frac{[H_3O^+][In^-]}{[HIn]}$$

$$\text{acid} \qquad\qquad\qquad\qquad \text{base}$$
$$\text{color} \qquad\qquad\qquad\qquad \text{color}$$

or

$$In + H_2O \rightleftharpoons InH^+ + OH^- \qquad K_b = \frac{[InH^+][OH^-]}{[In]}$$

$$\text{base} \qquad\qquad \text{acid}$$
$$\text{color} \qquad\qquad \text{color}$$

The species HIn will predominate in a strongly acidic environment and will be responsible for the "acid color" of the first indicator; $In^-$ will predominate in a strongly alkaline solution and will be the source for the "base color." Similarly, In and $InH^+$ will provide the "base color" and the "acid color," respectively, in alkaline and acidic media.

The equilibrium constant expressions can be rearranged to give

$$\frac{[In^-]}{[HIn]} = \frac{K_a}{[H_3O^+]}$$

and

$$\frac{[InH^+]}{[In]} = \frac{K_b}{[OH^-]} = \frac{K_b[H_3O^+]}{K_w}$$

[5] See, for example, P. J. Elving, as well as R. G. Bates and V. E. Bower, in *Handbook of Analytical Chemistry*, L. Meites, Ed., pp. **1**-112, **11**-5–**11**-7. New York: McGraw-Hill Book Co., 1963.

The human eye is relatively insensitive to small changes in the color of an indicator. Typically, a five- to tenfold excess of one form of the indicator is required before the color of that species appears predominant to the observer; further increases in the ratio have no detectable effect. Thus, the subjective "color change" involves a major alteration in the equilibrium position of the indicator. Using HIn as an example, we may write that the indicator exhibits its pure acid color to the average observer when

$$\frac{[In^-]}{[HIn]} \le \frac{1}{10}$$

and its pure base color when

$$\frac{[In^-]}{[HIn]} \ge \frac{10}{1}$$

The color appears to be intermediate for ratios between these two values. These figures, of course, are averages with some indicators requiring smaller ratio changes and others larger. Furthermore, individuals vary considerably in their ability to judge color differences; indeed, a color-blind person may be unable to discern the color change of certain indicators.

Substitution of these two concentration ratios into the dissociation constant expression for HIn permits evaluation of the range of hydronium ion concentrations needed to bring about color change. Thus, for observation of the full acid color,

$$\frac{[H_3O^+][In^-]}{[HIn]} = \frac{[H_3O^+] \times 1}{10} = K_a$$

which rearranges to

$$[H_3O^+] = 10K_a$$

Similarly, for the full base color,

$$\frac{[H_3O^+] \times 10}{1} = K_a$$

or

$$[H_3O^+] = \frac{1}{10} K_a$$

The pH range for the full color transition can be obtained by taking the negative logarithm of the two ratios; that is,

$$\text{pH range} = -\log 10K_a \text{ to } -\log \frac{1}{10} K_a$$

$$= -1 + pK_a \text{ to } -(-1) + pK_a$$
$$= pK_a \pm 1$$

As an example, an indicator with acid dissociation constant of $1 \times 10^{-5}$ will undergo a full color change when the solution in which it is dissolved changes pH from 4 to 6.

## 8D TITRATION CURVES FOR WEAK ACIDS

Before undertaking the derivation of a curve for the titration of a weak acid, it is worthwhile to deduce its qualitative features, based upon our discussion thus far. First, we know that the initial pH will be higher than that for a strong acid with the same formal concentration. Furthermore, the ratio $[HA]/[A^-]$ at the outset will be numerically large; that is, $[HA] \gg [A^-]$. The first additions of base may cause the concentration of $A^-$ to increase by an order of magnitude or more; the *relative* decrease in the concentration of HA, on the other hand, will be but a few percent of its original concentration. The consequence is that the ratio $[HA]/[A^-]$ will undergo a marked decrease. A proportionate increase in pH is thus to be expected.

After this initial surge in pH, the titration enters the buffer region, in which the formal concentrations of both HA and $A^-$ are significant. We have demonstrated that such solutions exhibit remarkable resistance to pH changes with additions of strong base. Thus, the curve will be relatively flat until almost all of the weak acid has been titrated.

Additions of base in the region just short of the equivalence point will cause significantly greater relative changes in $[HA]$, which is now small, than in $[A^-]$, which is now large. As a result, the rate of decrease in the ratio $[HA]/[A^-]$ will again accelerate, as will the increase in pH.

The principal solute species at the equivalence point is $A^-$, a weak base; the solution will therefore be alkaline.

Four different calculations are involved in generating a curve for the titration of a weak acid, HA, with a strong base:

1. At the outset, the principal solute species that determines the pH is HA; the pH is calculated from the formal concentration of that solute (see Examples 8–3 and 8–4).
2. With the addition of base, the solution will consist of a series of buffers, with HA and $A^-$ as principal solute species (see Examples 8–6 and 8–7).
3. At the equivalence point, the principal solute species will be $A^-$; the pH is calculated from the formal concentration of $A^-$ (see Example 8–5).
4. Beyond the equivalence point the solution will contain the weak base $A^-$ as well as an excess of strong base; the pH will be governed largely by the concentration of excess titrant.

The first three situations have been considered in earlier sections. Beyond the equivalence point the inherently small dissociation of the weak base $A^-$ is repressed even more by the introduction of excess hydroxide ion from the titrant. Thus, the pH can be calculated from the formal concentration of excess base.

**Table 8-4** *Changes in pH During the Titration of a Weak Acid with a Strong Base*

| Volume of NaOH, mL | pH | |
|---|---|---|
| | 50.00 mL of 0.1000 $F$ HOAc Titrated with 0.1000 $F$ NaOH | 50.00 mL of 0.001000 $F$ HOAc Titrated with 0.001000 $F$ NaOH |
| 0.00 | 2.88 | 3.91 |
| 10.00 | 4.16 | 4.30 |
| 25.00 | 4.76 | 4.80 |
| 40.00 | 5.36 | 5.38 |
| 49.00 | 6.45 | 6.46 |
| 49.90 | 7.46 | 7.47 |
| 50.00 | 8.73 | 7.73 |
| 50.10 | 10.00 | 8.09 |
| 51.00 | 11.00 | 9.00 |
| 60.00 | 11.96 | 9.96 |
| 75.00 | 12.30 | 10.30 |

**Example 8-12** Derive a curve for the titration of 50.0 mL of 0.1000 $F$ acetic acid ($K_a = 1.75 \times 10^{-5}$) with 0.1000 $F$ NaOH.

(a) *Initial pH.* The principal solute species at this point is undissociated HOAc. A calculation identical with that in Example 8-3 yields a pH of 2.88.

(b) *pH after addition of 10.00 mL of base.* A buffer system consisting of HOAc and NaOAc has now been produced. The formal concentration of the two species will be given by

$$F_{HOAc} = \frac{50.00 \times 0.1000 - 10.00 \times 0.1000}{50.00 + 10.00} = \frac{4.000}{60.00}$$

$$F_{NaOAc} = \frac{10.00 \times 0.1000}{50.00 \times 10.00} = \frac{1.000}{60.00}$$

Upon substituting these concentrations into the dissociation constant for HOAc and rearranging, we obtain

$$[H_3O^+] = 1.75 \times 10^{-5} \times \frac{4.000/60.00}{1.000/60.00} = 7.00 \times 10^{-5}$$

$$pH = 5 - \log 7.00 = 4.16$$

Calculations similar to this will define the curve throughout the buffer region. Data from such calculations are given in column 2 of Table 8-4.

(c) *Equivalence-point pH.* The principal solute species at the equivalence point in this titration will be acetate ion. The solution is undistinguishable from one produced by dissolving an appropriate amount of NaOAc in water. The pH calculation here is identical to that shown in Example 8–5 for a weak base; that is,

$$[OH^-] = [HOAc]$$

$$[OAc^-] = \frac{50.00 \times 0.1000}{100.0} - [OH^-] \cong 0.0500$$

Thus,

$$[OH^-] = \sqrt{\frac{K_w}{K_a} \times 0.0500} = \sqrt{\frac{1.00 \times 10^{-14} \times 0.0500}{1.75 \times 10^{-5}}}$$

$$= 5.34 \times 10^{-6}$$

$$pOH = 6 - \log 5.34 = 5.27$$

$$pH = 14.00 - 5.34 = 8.73$$

(d) *pH after addition of 50.10 mL of base.* After 50.10 mL of base have been introduced, the excess NaOH as well as the equilibrium between $OAc^-$ and $H_2O$ are sources of $OH^-$. The contribution of the latter is small enough to be neglected, however, since the excess of strong base will tend to repress this inherently unfavorable equilibrium even further. Recall that the $OH^-$ concentration was only $5.34 \times 10^{-6}$ *M* at the equivalence point; once an excess of strong base has been added, the contribution from the equilibrium between $OAc^-$ and $H_2O$ will be even less. Thus,

$$[OH^-] = F_{NaOH} = \frac{50.10 \times 0.1000 - 50.00 \times 0.1000}{100.1} = 1.00 \times 10^{-4}$$

$$pOH = 4 - \log 1.00 = 4.00$$

$$pH = 14.00 - 4.00 = 10.00$$

Note that slightly past the end point and beyond, the curve for the titration of a weak acid is identical to that for a strong acid with the same concentration.

Of special interest in Table 8–4 is the point at which the acid has been 50% neutralized (after addition of 25.00 mL of base in this particular titration). Here, the formal concentrations of acid and conjugate base are identical. Within the limits of the usual approximations, so also are their molar concentrations. Thus, the two terms cancel in the equilibrium constant expression, and the hydronium ion concentration is numerically equal to the dissociation constant. Likewise, the hydroxide ion concentration is equal to the dissociation constant for a weak base at the midpoint in its titration with a strong acid.

**Effect of Concentration**    The second and third columns of Table 8–4 contain data for the titration of 0.1000 *F* and 0.001000 *F* acetic acid solutions with sodium hydroxide solutions of the same concentrations. It was necessary to use the quadratic equation to calculate pH values for the more dilute solution, inasmuch as the formal solute concentrations were so small; see Examples 8–4 and 8–7.

Figure 8–4 is a plot of the data in Table 8–4. Note that the initial pH values are somewhat larger and the equivalence pH is somewhat smaller for the 0.001000 $F$ titration. At intermediate titrant volumes, however, the pH values differ only slightly because acetic acid–sodium acetate buffers exist within this region. Figure 8–4 is graphical confirmation that the pH of buffers is largely independent of dilution.

Curves for the titration of 0.1000 $F$ acids with differing dissociation constants are shown in Figure 8–5. Note that the pH change in the vicinity of the equivalence point becomes smaller with decreasing values for $K_a$—that is, as reaction between the acid and the base becomes less complete. The relation between completeness of reaction and reagent concentrations illustrated in Figures 8–4 and 8–5 is analogous to that seen in precipitation titration curves (p. 151).

**Indicator Choice: Feasibility of Titration**    Figures 8–4 and 8–5 clearly demonstrate that the choice of indicator for the titration of a weak acid is more limited than that for a stronger acid. For example, bromocresol green is totally unsuited for the titration of 0.1000 $F$ acetic acid; nor would bromothymol blue be satisfactory, since its full color change would occur over the range between about 47 and 50 mL of 0.100 $F$ base. Bromothymol blue might still be useful if the titration were carried to the full basic color of the indicator; a comparison standard containing the same concentration of the indicator as the solution would, however, be necessary. An indicator such as phenolphthalein, which changes color in the basic region, would be most satisfactory for this titration.

The equivalence-point pH change shown by curve $B$ in Figure 8–4 is small enough to cause a significant titration error at this low analyte concentration.

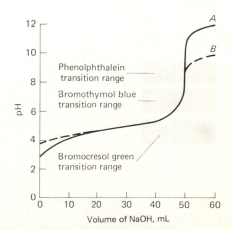

**Figure 8–4**   Curves for the titration of acetic acid with sodium hydroxide. Curve A: 0.1000 $F$ acid with 0.1000 $F$ base. Curve B: 0.001000 $F$ acid with 0.01000 $F$ base.

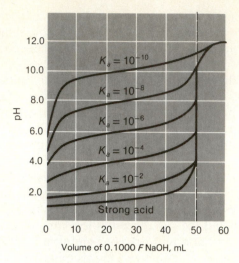

**Figure 8-5**  Influence of acid strength upon titration curves. Each curve represents the titration of 50.00 mL of 0.1000 $F$ acid with 0.1000 $F$ base.

By careful choice of indicator and with a color comparison standard, it might be possible to establish this end point with a reproducibility of a few percent.

Figure 8–5 shows titration curves for acids having different dissociation constants. Clearly the indeterminate error associated with a titration becomes greater as the dissociation constant for an acid becomes smaller. It has been found that a precision on the order of $\pm 2$ ppt can be realized in the titration of 0.1 $F$ solutions of an acid having a dissociation constant of $10^{-8}$ provided a suitable color comparison standard is used. With more concentrated solutions, somewhat weaker acids can be titrated with reasonable precision.

## 8E TITRATION CURVES FOR WEAK BASES

Figure 8–6 shows theoretical curves for a series of bases with differing strengths. These curves were derived by methods analogous to those described for the generation of the curves for weak acids. It is clear that *acid range indicators* will be needed for the successful titration of the weaker bases.

## 8F TITRATION CURVES FOR MIXTURES OF STRONG AND WEAK ACIDS

The hydronium ion concentration of a solution containing, say, hydrochloric acid and the weak acid HA, is given by the equation

$$[H_3O^+] = F_{HCl} + [A^-] \qquad (8-21)$$

The contribution from the strong acid is numerically equal to its formal concentration, while that from the weak acid is equal to the concentration of its conjugate base, $[A^-]$. Strictly, the dissociation of water also provides hydro-

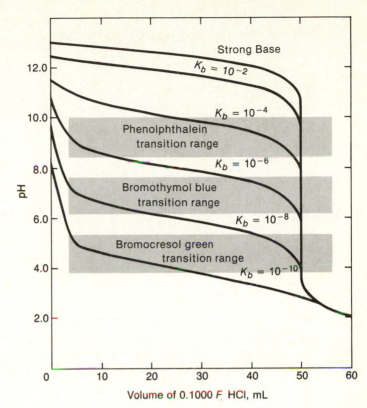

**Figure 8–6** Influence of base strength upon titration curves. Each curve represents titration of 50.00 mL of 0.1000 $F$ base with 0.1000 $F$ HCl.

nium ions, but its contribution is vanishingly small in the presence of the strong acid and can be neglected.

If $K_{HA}$ is less than about $10^{-4}$, the presence of the strong acid will repress dissociation of HA to the point that $[A^-]$ is negligibly small with respect to $F_{HCl}$ in Equation 8–21; the pH of such a mixture is thus determined solely by the formal concentration of the strong acid. As a result, the initial portions of a curve for the titration of such a mixture with sodium hydroxide will be indistinguishable from that for the strong acid alone. At the equivalence point for the titration of the strong acid, the solution will contain sodium chloride, which will not directly contribute to the pH, plus the as yet untitrated weak acid, which most assuredly will. The remainder of the titration will thus be described by a curve that is typical for the weak acid. The shape of the curve, and hence the information obtainable from it, will depend upon the strength of the weak acid and its concentration. Figure 8–7 depicts curves for the titration of hydrochloric acid–weak acid mixtures with sodium hydroxide. It can be seen that a break in the titration curve at the first end point will be essentially nonexistent if the weak acid has a relatively large dissociation constant (curve $A$); for such mix-

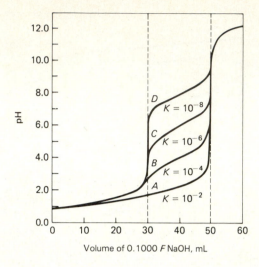

**Figure 8 – 7**  Curves for the titration of strong acid – weak acid mixtures with 0.1000 $F$ NaOH. Each titration involves 25.00 mL of a solution that is 0.1200 $F$ with respect to HCl and 0.0800 $F$ with respect to HA.

tures, only the total acidity (i.e., $F_{HCl} + F_{HA}$) is obtainable from the titration. Conversely, in a mixture involving a very weak acid, only the strong acid content can be determined. Two useful end points may exist in curves for the titration of mixtures containing weak acids of intermediate strength.

## 8G TITRATION CURVES FOR POLYFUNCTIONAL ACIDS AND BASES

Examination of Appendix 10 reveals that many acids possess more than one ionizable proton. For example, three equilibria are established in an aqueous solution of phosphoric acid:

$$H_3PO_4 + H_2O \rightleftharpoons H_2PO_4^- + H_3O^+ \qquad K_1 = \frac{[H_3O^+][H_2PO_4^-]}{[H_3PO_4]}$$
$$= 7.11 \times 10^{-3}$$

$$H_2PO_4^- + H_2O \rightleftharpoons HPO_4^{2-} + H_3O^+ \qquad K_2 = \frac{[H_3O^+][HPO_4^{2-}]}{[H_2PO_4^-]}$$
$$= 6.34 \times 10^{-8}$$

$$HPO_4^{2-} + H_2O \rightleftharpoons PO_4^{3-} + H_3O^+ \qquad K_3 = \frac{[H_3O^+][PO_4^{3-}]}{[HPO_4^{2-}]}$$
$$= 4.2 \times 10^{-13}$$

With this acid, as with others, $K_1 > K_2 > K_3$.

Polyfunctional bases are also common, an example being sodium carbonate.

Carbonate ion, the conjugate base of the hydrogen carbonate ion, is involved in the stepwise equilibria:

$$CO_3^{2-} + H_2O \rightleftharpoons HCO_3^- + OH^- \qquad (K_b)_1 = \frac{[HCO_3^-][OH^-]}{[CO_3^{2-}]}$$

$$\frac{K_w}{K_2} = 2.13 \times 10^{-4}$$

$$HCO_3^- + H_2O \rightleftharpoons H_2CO_3 + OH^- \qquad (K_b)_2 = \frac{[H_2CO_3][OH^-]}{[HCO_3^-]}$$

$$\frac{K_w}{K_1} = 2.25 \times 10^{-8}$$

where $K_1$ and $K_2$ are the first and second dissociation constants for carbonic acid (see Appendix 10).

The pH of a solution containing a polyfunctional acid or a buffer mixture involving such an acid can be evaluated rigorously through use of the systematic approach to multiequilibrium problems described in Chapter 5. Solution of the several simultaneous equations that are involved can be difficult and time consuming, however. Fortunately, a simplifying assumption can be invoked where the successive equilibrium constants for the acid (or base) differ by a factor of about $10^3$ (or more). We will restrict our consideration to such acids and bases.

### 8G – 1 pH Calculations for Salts of the Type NaHA

One additional calculation is needed to establish the pH of a solution in which the principal solute species is an "acid salt" with the general formula NaHA. This species is different from species that we have considered thus far in that it is not only an acid in its own right but it is also the conjugate base of a stronger acid; it is thus necessary to take account of two equilibria involving NaHA and water:

$$HA^- + H_2O \begin{cases} A^{2-} + H_3O^+ \qquad K_2 = \dfrac{[H_3O^+][A^{2-}]}{[HA^-]} \\[2em] H_2A + OH^- \qquad K_b = \dfrac{K_w}{K_1} = \dfrac{[H_2A][OH^-]}{[HA^-]} \end{cases}$$

The solution will be basic if $K_b$ is greater than $K_2$, while it will be acidic if the reverse is true. A solute, such as NaHA, that can behave either as an acid or as a base is said to be *amphiprotic*.

A solution of NaHA can be described in terms of material balance

$$F_{HA^-} = [Na^+] = [H_2A] + [HA^-] + [A^{2-}] \tag{8-22}$$

as well as charge balance

$$[Na^+] + [H_3O^+] = [OH^-] + [HA^-] + 2[A^{2-}]$$

This equation can be rewritten as

$$F_{HA^-} + [H_3O^+] = [OH^-] + [HA^-] + 2[A^{2-}] \tag{8-23}$$

The system has five unknowns, namely, $[H_3O^+]$, $[OH^-]$, $[H_2A]$, $[HA^-]$, and $[A^{2-}]$, and five independent algebraic relationships. The latter include the ion product constant expression for water, the two dissociation constant expressions for the acid, the material balance equation (Equation 8–22), and the charge-balance equation (Equation 8–23). Thus, a rigorous solution to the problem is feasible but somewhat tedious. Fortunately, one approximation can be made in most instances that simplifies the algebra greatly.

Subtraction of Equation 8–22 from Equation 8–23 yields

$$[H_3O^+] = [OH^-] + [A^{2-}] - [H_2A]$$

This relationship can be expressed in terms of $[H_3O^+]$ and the principal solute species $[HA^-]$:

$$[H_3O^+] = \frac{K_w}{[H_3O^+]} + \frac{K_2[HA^-]}{[H_3O^+]} - \frac{[H_3O^+][HA^-]}{K_1}$$

Multiplication of both sides of this expression by $[H_3O^+]$ gives

$$[H_3O^+]^2 = K_w + K_2[HA^-] - \frac{[H_3O^+]^2[HA^-]}{K_1}$$

After collecting the two terms that contain $[H_3O^+]^2$, we obtain

$$[H_3O^+]^2 \left( 1 + \frac{[HA^-]}{K_1} \right) = K_w + K_2[HA^-]$$

which rearranges to

$$[H_3O^+]^2 = \frac{K_w + K_2[HA^-]}{1 + [HA^-]/K_1}$$

Taking the square root of both sides gives

$$[H_3O^+] = \sqrt{\frac{K_w + K_2[HA^-]}{1 + [HA^-]/K_1}} \tag{8-24}$$

Finally, for most acid salts it is permissible to make the approximation that

$[HA^-] \cong F_{HA^-}$. Equation 8–24 then becomes

$$[H_3O^+] \cong \sqrt{\frac{K_w + K_2 F_{HA^-}}{1 + F_{HA^-}/K_1}} \qquad (8\text{–}25)^6$$

Equation 8–25 is a reasonable approximation provided the two equilibrium constants that contain $[HA^-]$ are small and the formal concentration of NaHA is fairly large.

It is worthwhile to consider the possibilities for further simplification of Equation 8–25. The ion product for water is often small with respect to $K_2 F_{HA^-}$ and can be neglected (the converse situation, while possible, is seldom encountered). A similar comparison may suggest elimination of one of the two terms in the denominator. Where both $K_2 F_{HA^-} \gg K_w$ and $F_{HA^-}/K_1 \gg 1$, Equation 8–25 simplifies to

$$[H_3O^+] = \sqrt{K_1 K_2} \qquad (8\text{–}26)$$

Note that a solution for which Equation 8–26 is applicable will tend to have a constant pH over a substantial concentration range.

---

**Example 8–13**  Calculate the hydronium ion concentration of a 0.100 $F$ solution of $NaHCO_3$. For $H_2CO_3$, $K_1 = 4.45 \times 10^{-7}$ and $K_2 = 4.7 \times 10^{-11}$.

We must first examine the assumptions implicit in the use of Equation 8–26. The quantity $0.100/4.45 \times 10^{-7}$ is enormous with respect to 1; simplification of the denominator is clearly justified. Likewise, $K_2 F_{HA^-}$ is substantially larger than $K_w$. We are therefore entitled to use the simplified equation, and

$$[H_3O^+] = \sqrt{(4.45 \times 10^{-7})(4.7 \times 10^{-11})} = 4.6 \times 10^{-9}$$

---

**Example 8–14**  Calculate the hydronium ion concentration of a solution that is $1.00 \times 10^{-3}$ $F$ with respect to $Na_2HPO_4$.

Here, the pertinent constants (i.e., the two that contain a term for the principal solute species $[HPO_4^{2-}]$) are $K_2$ and $K_3$ for $H_3PO_4$, which have values of $6.34 \times 10^{-8}$ and $4.2 \times 10^{-13}$. Considering again the assumptions that might be applied to Equation 8–25, it turns out that the denominator can be simplified, since

---

[6] Note that $K_1$ and $K_2$ in Equation 8–25 are the first and second dissociation constants for $H_2A$. Both of these expressions contain a term for the principal solute species in the solution, namely $HA^-$.

$1.00 \times 10^{-3}/6.34 \times 10^{-8} \gg 1$. On the other hand, $K_w$ and $K_3 F_{HPO_4^{2-}}[(4.2 \times 10^{-13})(1.00 \times 10^{-3})]$ are of similar magnitude. Thus,

$$[H_3O^+] = \sqrt{\frac{1.00 \times 10^{-14} + (4.2 \times 10^{-13})(1.00 \times 10^{-3})}{1.00 \times 10^{-3}/6.34 \times 10^{-8}}}$$

$$= 8.1 \times 10^{-10}$$

The reader is encouraged to use the same reasoning to verify that the hydronium ion concentration of a $1.00 \times 10^{-2}\ F$ solution of $NaH_2PO_4$ is $1.62 \times 10^{-5}\ M$.

### 8G–2 Curves for the Titration of Polyfunctional Acids

Titration curves for solutions of polyfunctional acids or bases will have multiple inflection points provided the acidic or basic groups differ sufficiently in strength; such curves may have more than one useful end point.

Our discussion will be limited to polyfunctional systems in which successive equilibrium constants differ from one another by a factor of 1000 or more because reasonably accurate theoretical curves for such systems can be derived readily with the methods we have described in this chapter. Where the ratio between the dissociation constants is smaller than $10^3$, the approximate methods fail, particularly in the vicinity of the first equivalence point. Treatment of the equilibrium relationships with greater rigor, as is required in this case, is mathematically more involved.

A curve for the titration of maleic acid is derived in Example 8–15. Maleic acid is a weak dibasic acid with the formula HOOCCH=CHCOOH. The two dissociation equilibria are

$$H_2M + H_2O \rightleftharpoons H_3O^+ + HM^- \qquad K_1 = 1.20 \times 10^{-2}$$

$$HM^- + H_2O \rightleftharpoons H_3O^+ + M^{2-} \qquad K_2 = 5.96 \times 10^{-7}$$

where $H_2M$ symbolizes the undissociated acid. The ratio $K_1/K_2$ is sufficiently large ($2 \times 10^4$) so that it is permissible to neglect the influence of the second dissociation in deriving points short of the first equivalence point. By so doing, we are making the assumption that $[M^{2-}]$ is vanishingly small with respect to $[H_2M]$ and $[HM^-]$ in this region; it can be shown that this assumption does not cause appreciable error to within a few hundredths of a milliliter of the first equivalence point. Shortly beyond this point, the second equilibrium is sufficiently dominant so that the basic dissociation of $HM^-$,

$$HM^- + H_2O \rightleftharpoons H_2M + OH^-$$

does not significantly influence the pH. In this region, we can assume that $[H_2M] \ll [HM^-]$ or $[M^{2-}]$.

**Example 8-15**    Derive a curve for the titration of 25.00 mL of 0.1000 $F$ maleic acid with 0.1000 $F$ NaOH.

(a) *Initial pH.* Only the first dissociation makes an appreciable contribution to $[H_3O^+]$; thus,

$$[H_3O^+] = [HM^-]$$

$$[H_2M] = 0.1000 - [HM^-] - [M^{2-}]$$
$$\cong 0.1000 - [HM^-] \qquad (8-27)^7$$

Substituting these relationships into the expression for $K_1$ gives

$$1.20 \times 10^{-2} = \frac{[H_3O^+]^2}{0.1000 - [H_3O^+]}$$

Calculation of $[H_3O^+]$ will require solution of a quadratic equation because $K_1$ for maleic acid is so large; that is,

$$[H_3O^+]^2 + 1.20 \times 10^{-2}[H_3O^+] - 1.20 \times 10^{-3} = 0$$

$$[H_3O^+] = 2.92 \times 10^{-2}$$

$$pH = 2 - \log 2.92 = 1.54$$

(b) *First buffer region.* Addition of NaOH initially results in the formation of a buffer consisting of $H_2M$ and its conjugate base $HM^-$. Because the dissociation of $HM^-$ to give $M^{2-}$ is negligible, the solution can be treated as a simple buffer system. Thus, after the addition of 5.00 mL of 0.1000 $F$ NaOH,

$$[H_2M] = \frac{25.00 \times 0.1000 - 5.00 \times 0.1000}{25.00 + 5.00} - [H_3O^+] \cong 6.67 \times 10^{-2}$$

$$[HM^-] = \frac{5.00 \times 0.1000}{30.00} + [H_3O^+] \cong 1.67 \times 10^{-2}$$

Substitution of these approximate values for $[H_2M]$ and $[HM^-]$ into $K_1$ yields a provisional value of $4.8 \times 10^{-2}$ for $[H_3O^+]$, which is *not* negligible with respect to $[H_2M]$ and $[HM^-]$. It is thus necessary to use $(6.67 \times 10^{-2} - [H_3O^+])$ for $[H_2M]$ and $(1.67 \times 10^{-2} + [H_3O^+])$ for $[HM^-]$. Inser-

---

[7] The propriety of neglecting the second dissociation reaction can be confirmed by substituting the equality $[H_3O^+] = [HM^-]$ into $K_2$:

$$K_2 = \frac{[H_3O^+][M^{2-}]}{[HM^-]} = 5.96 \times 10^{-7}$$

Clearly, $[M^{2-}] \ll [H_2M]$ or $[HM^-]$.

tion of these quantities into $K_1$ and rearrangement results in the quadratic equation

$$[H_3O^+]^2 + 2.87 \times 10^{-2} [H_3O^+] - 8.00 \times 10^{-4} = 0$$

$$[H_3O^+] = 1.74 \times 10^{-2}$$

$$pH = 2 - \log 1.74 = 1.76$$

Additional points short of the first equivalence point can be calculated similarly. Because $K_1$ is so large, calculations will require use of the quadratic equation throughout the first buffer region.

(c) *First equivalence point.* At the first equivalence point, the volume is 50.00 mL; thus,

$$F_{HM^-} = \frac{25.00 \times 0.1000}{50.00} = 5.00 \times 10^{-2}$$

Simplification of the numerator of Equation 8–25 is clearly justified. On the other hand, the terms in the denominator are of the same order of magnitude. Thus,

$$[H_3O^+] = \sqrt{\frac{(5.96 \times 10^{-7})(5.00 \times 10^{-2})}{1 + 5.00 \times 10^{-2}/1.20 \times 10^{-2}}}$$

$$= 7.60 \times 10^{-5}$$

$$pH = 5 - \log 7.60 = 4.12$$

(d) *Second buffer region.* The remainder of the titration curve is obtained by treating the system as 50.00 mL of a 0.05000 $F$ solution of a weak acid $HM^-$ with a dissociation constant of $5.96 \times 10^{-7}$. For example, after addition of 25.50 mL of base, the formality of $Na_2M$ formed will be

$$F_{Na_2M} = \frac{(25.50 - 25.00) \times 0.1000}{50.50} = 0.050/50.50$$

Here, the first 25.00 mL of reagent was consumed in converting $H_2M$ to $HM^-$ and the remaining 0.50 mL went toward formation of $M^{2-}$. Furthermore,

$$F_{NaHM} = \frac{25.00 \times 0.1000 - 0.50 \times 0.1000}{50.50} = \frac{2.450}{50.50}$$

Here, we are assuming that $[H_3O^+] \ll F_{Na_2M}$ and $F_{NaHM}$. Therefore, $[M^{2-}] \cong F_{Na_2M}$ and $[HM^-] \cong F_{NaHM}$. Substituting these values into the second ionization constant expression gives

$$\frac{[H_3O^+] \, 0.050/50.50}{2.450/50.50} = 5.96 \times 10^{-7}$$

and

$$[H_3O^+] = 2.92 \times 10^{-5}$$

$$pH = 5 - \log 2.92 = 4.54$$

The assumption that $[H_3O^+]$ is small with respect to the two formal concentrations is valid.

(e) *Second equivalence point.* After 50.00 mL of NaOH have been added, the solution will be 0.0333 $F$ with respect to Na$_2$M. Reaction of the base M$^{2-}$ with water is the predominant equilibrium in the system and the only one that must be taken into account. Thus,

$$M^{2-} + H_2O \rightleftharpoons HM^- + OH^-$$

$$K_b = \frac{K_w}{K_2} = \frac{[OH^-][HM^-]}{[M^{2-}]} = \frac{1.00 \times 10^{-14}}{5.96 \times 10^{-7}} = 1.68 \times 10^{-8}$$

$$[OH^-] = [HM^-]$$

$$[M^{2-}] = 0.0333 - [OH^-] \cong 0.0333$$

$$[OH^-] = \sqrt{(1.68 \times 10^{-8})(0.0333)} = 2.37 \times 10^{-5}$$

$$pOH = 5 - \log 2.37 = 4.63$$

$$pH = 14.00 - 4.63 = 9.37$$

(f) *pH beyond the second equivalence point.* Further additions of sodium hydroxide will repress the basic dissociation of M$^{2-}$. The pH is calculated from the concentration of NaOH in excess of that required for the complete neutralization of H$_2$M. An example of this calculation appears on page 188.

Figure 8–8a depicts the titration of 0.1000 $F$ maleic acid with 0.1000 $F$ sodium hydroxide. (The significance of Figure 8–8b will be described in the next section.) The calculations shown in Example 8–15 were used to generate this curve. Major features include the existence of two buffer regions as well as two end points. In principle, the judicious choice of indicator would permit titration of either the first or both of the acidic hydrogens. The pH change associated with the second end point is the more pronounced and is clearly the better choice.

Figure 8–9 shows titration curves for three other polybasic acids. These curves illustrate that a well-defined end point corresponding to the first equivalence point will be observed only if the difference between successive dissociation constants is sufficiently large. Thus, curve $B$ for titration of the dibasic oxalic acid shows inflections corresponding to the titration of each hydrogen. The ratio between $K_1$ and $K_2$ (about 1000) is so small, however, that the inflection corresponding to titration of the first hydrogen is insufficient to permit

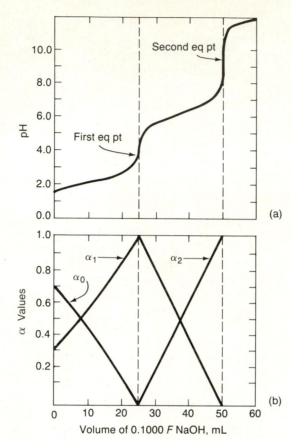

(a)

(b)

**Figure 8 – 8** Titration of 25.00 mL of 0.1000 $F$ maleic acid with 0.1000 $F$ sodium hydroxide. Plot of (a) pH vs. titrant volume and (b) mole fraction of maleate-containing species vs. titrant volume.

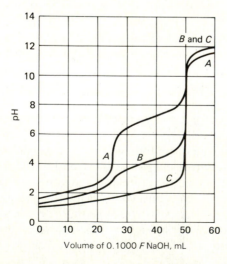

**Figure 8 – 9** Curves for the titration of polybasic acids. A 0.1000 $F$ sodium hydroxide solution is used to titrate 25.00 mL of 0.1000 $F$ phosphoric acid (curve A), 0.1000 $F$ oxalic acid (curve B), and 0.1000 $F$ sulfuric acid (curve C).

location of this end point with a chemical indicator. Note that the second end point would be conveniently detected with an alkaline range indicator.

Curve $A$ is a theoretical curve for the titration of the tribasic phosphoric acid. The ratio between $K_1$ and $K_2$ is approximately $10^5$ (or about 100 times larger than that for oxalic acid). Two well-defined end points are apparent, either of which is satisfactory for analytical purposes; an acid range indicator would be needed for detection of the first end point, while an alkaline range indicator would permit detection of the second. The third hydrogen of phosphoric acid is so slightly dissociated ($K_3 = 4.2 \times 10^{-13}$) that it does not yield an end point of any practical value. Nevertheless, the buffering action of the third dissociation is discernible, with the result that the pH for curve $A$ is somewhat lower than the other two in the region beyond the second equivalence point.

In general, the titration of a polyfunctional acid or base will yield multiple end points of practical value only if the ratio between the two dissociation constants is at least $10^4$. Otherwise, the change associated with the first equivalence point will be too small for accurate detection with a chemical indicator; then, only the second end point will be satisfactory for analytical purposes.

### 8G – 3  Titration Curves for Weak Polyfunctional Bases

The derivation of a curve for the titration of a polyfunctional weak base involves no new principles. In fact, such a derivation for the conjugate base of a polybasic acid can be thought of as being simply the reverse of that used in generating a curve for the acid (Table 8 – 5). For example, consider the titration of sodium carbonate with hydrochloric acid. For carbonic acid,

$$H_2CO_3 + H_2O \rightleftharpoons H_3O^+ + HCO_3^- \qquad K_1 = \frac{[H_3O^+][HCO_3^-]}{[H_2CO_3]}$$

$$HCO_3^- + H_2O \rightleftharpoons H_3O^+ + CO_3^{2-} \qquad K_2 = \frac{[H_3O^+][CO_3^{2-}]}{[HCO_3^-]}$$

At the outset of this titration, the reaction between carbonate ion and water governs the pH of the solution. Note that this is completely analogous to the

**Table 8 – 5**  *Comparison of Titrations for the Weak Acid $H_2A$ with Strong Base and for the Related Weak Base $A^{2-}$ with Strong Acid*

| Principal A-Containing Species in Solution | Stage of the Titration of Polyfunctional | | Calculate pH with Indicated Dissociation Constant for $H_2A$ |
|:---:|:---|:---|:---:|
| | Weak Acid, $H_2A$ | Weak Base, $A^{2-}$ | |
| $H_2A$ | At outset | At 2nd eq point | $K_1$ |
| $H_2A$, $HA^-$ | In 1st buffer region | In 2nd buffer region | $K_1$ |
| $HA^-$ | At 1st end point | At 1st end point | $K_1, K_2$ |
| $HA^-$, $A^{2-}$ | In 2nd buffer region | In 1st buffer region | $K_2$ |
| $A^{2-}$ | At 2nd end point | At outset | $K_w/K_2$ |

situation that existed at the second end point in the titration of maleic acid; clearly, the calculation shown in Example 8 – 15(e) will be applicable here as well. With the first additions of acid a carbonate – hydrogen carbonate buffer is established. These are the carbonate-containing species that appear in the expression for $K_2$; that equilibrium constant expression will therefore provide the means of calculating the hydronium ion concentration. Note that this calculation is analogous to that shown in Example 8 – 15(d). Sodium hydrogen carbonate is the principal species at the first equivalence point; calculation of the hydronium ion concentration will differ from Example 8 – 15(c) only in that it will be possible to simplify the denominator here (see Example 8 – 13). With further additions of acid a hydrogen carbonate – carbonic acid buffer will govern the pH; these are the species that appear in $K_1$ for carbonic acid. The calculation in this region will differ from that shown in Example 8 – 15(b) only in that there is no need to solve a quadratic equation. Because the principal solute species at the second equivalence point is carbonic acid, the hydronium ion concentration can be calculated as in Example 8 – 15(a), although a quadratic solution will not be necessary. Finally, addition of the first excess of hydrochloric acid will repress the dissociation of $H_2CO_3$ to the point where the hydronium ion concentration is essentially that of the formal concentration of the strong acid.

Figure 8 – 10 is a curve for the titration of a sodium carbonate solution. Two end points are discernible, the second being appreciably sharper than the first. This curve suggests that resolution of a carbonate – hydrogen carbonate mixture by titration with standard acid should be possible. Two indicators, one changing in the alkaline range, the other in the acid range, would be needed. This titration is among those discussed in Chapter 9.

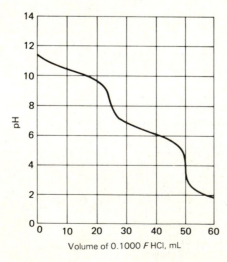

**Figure 8 – 10**   Curve for the titration of 25.00 mL of 0.1000 $F$ Na$_2$CO$_3$ with 0.1000 $F$ HCl.

## 8H COMPOSITION OF POLYBASIC ACID SOLUTIONS AS A FUNCTION OF pH

The titration curve is useful in accounting for the pH changes that occur in a solution as a function of titrant volume. In the interest of depicting the changes in the concentration of the various analyte species as a function of reagent volume and hence pH, a second type of plot is sometimes encountered. Here the *relative* amounts of each solute species are plotted as a function of pH. Such relative amounts are expressed in terms of $\alpha$-values, which are the fractions of the total concentration of species derived from a weak acid that exists in one particular form. Turning again to the maleic acid system, we will define $C_T$ as the sum of the molar concentrations of maleate-containing species in a solution; that is,

$$C_T = [H_2M] + [HM^-] + [M^{2+}] \tag{8-28}$$

By definition, then,

$$\alpha_0 = \frac{[H_2M]}{C_T}$$

$$\alpha_1 = \frac{[HM^-]}{C_T}$$

$$\alpha_2 = \frac{[M^{2-}]}{C_T}$$

The sum of the individual $\alpha$-values must equal unity so that

$$\alpha_0 + \alpha_1 + \alpha_2 = 1.00$$

Equation 8–28 can be expressed in terms of $[H_2M]$, $[H_3O^+]$, $K_1$, and $K_2$:

$$[HM^-] = \frac{K_1[H_2M]}{[H_3O^+]}$$

$$[M^{2-}] = \frac{K_1K_2[H_2M]}{[H_3O^+]^2}$$

Substitution gives

$$C_T = [H_2M] + \frac{K_1[H_2M]}{[H_3O^+]} + \frac{K_1K_2[H_2M]}{[H_3O^+]^2}$$

$$= [H_2M] \left( \frac{[H_3O^+]^2 + K_1[H_3O^+] + K_1K_2}{[H_3O^+]^2} \right)$$

Substitution into the equation that defines $\alpha_0$ and rearrangement yields

$$\alpha_0 = \frac{[H_2M]}{C_T} = \left( \frac{[H_3O^+]^2}{[H_3O^+]^2 + K_1[H_3O^+] + K_1K_2} \right) \tag{8-29}$$

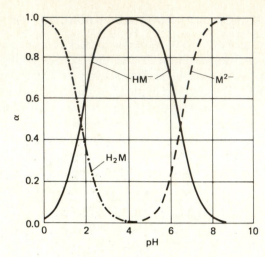

**Figure 8 – 11** Composition of $H_2M$ solutions as a function of pH.

Similar substitutions lead to values for $\alpha_1$ and $\alpha_2$:

$$\alpha_1 = \frac{[HM^-]}{C_T} = \left( \frac{K_1[H_3O^+]}{[H_3O^+]^2 + K_1[H_3O^+] + K_1K_2} \right) \qquad (8-30)$$

$$\alpha_2 = \frac{[M^{2-}]}{C_T} = \left( \frac{K_1K_2}{[H_3O^+]^2 + K_1[H_3O^+] + K_1K_2} \right) \qquad (8-31)$$

Note that the denominator is the same for each expression, that the numerator for $\alpha_0$ is the first term in the denominator, and that each successive $\alpha$-value has in its numerator the next term from the denominator. Generation of expressions for $\alpha$-values is thus a simple matter,[8] as is the calculation of their numerical values at any given pH. Note that the fractional amount of each species is fixed at any given pH and is independent of $C_T$.

Figure 8 – 11 illustrates how $\alpha$-values for each maleate-containing species change as a function of pH. In contrast, Figure 8 – 8b depicts the changes in the same $\alpha$-values but now as a function of volume of sodium hydroxide as the acid is titrated. A consideration of this figure in conjunction with the titration curve in Figure 8 – 8a gives a clear picture of the concentration changes that occur during the titration. Thus, initially, when the pH is 1.5, $\alpha_0$ is about 0.7, while $\alpha_1$ has a value of approximately 0.3; that is, about 70% of the solute is in the form of the undissociated acid and 30% is present as the acid salt $HM^-$. The concentration of $M^{2-}$ is for all practical purposes zero. At the first equivalence point (pH = 4.12), essentially all of the solute exists as $HM^-$ ($\alpha_1 \rightarrow 1$). Beyond the

---

[8] The weak acid $H_nA$ has $n$ hydrogens as well as $n$ dissociation constant expressions. The first term in the denominator for $\alpha$-values will be $[H_3O^+]^n$, the second will be $K_1[H_3O^+]^{n-1}$, the third $K_1K_2[H_3O^+]^{n-2}$, and so forth. The final term will contain only equilibrium constants.

first equivalence point, $\alpha_2$ increases at the expense of $\alpha_1$; here, $\alpha_0$ is vanishingly small. At the second equivalence point (pH $= 9.37$), essentially all of the solute exists as the maleate ion $M^{2-}$.

## PROBLEMS

Round all pH or pOH calculations to two places beyond the decimal.

*8-1. Supply the missing information for the following aqueous solutions.

| Solute | Percent Composition (w/w) | Density, g/mL | $[H_3O^+]$ | $[OH^-]$ |
|---|---|---|---|---|
| (a) HCl | 20.0 | 1.100 | | |
| (b) $HNO_3$ | 14.0 | 1.078 | | |
| (c) KOH | | 1.081 | | 1.73 |
| (d) $HNO_3$ | | 1.023 | 0.731 | |
| (e) $Ba(OH)_2$ | 9.90 | | | 1.89 |

8-2. Calculate the pH of the solution that results when 20.0 mL of 0.0900 $F$ HCl are mixed with 30.0 mL of
 *(a) water.      (b) 0.0500 $F$ KOH.
 *(c) 0.0500 $F$ Ba(OH)$_2$.      (d) 0.0600 $F$ NaOH.
 *(e) 0.0200 $F$ AgNO$_3$.      (f) 0.0200 $F$ HCl.

8-3. Calculate the pH of the solution that results when 40.0 mL of 0.0360 $F$ NaOH are mixed with 60.0 mL of
 (a) water.      *(b) 0.0120 $F$ HCl.
 (c) 0.0120 $F$ H$_2$SO$_4$.      *(d) 0.0300 $F$ HCl.
 (e) 0.0180 $F$ MgCl$_2$.      *(f) 0.0180 $F$ NaOH.

8-4. Calculate the pH of the solution that results when 40.0 mL of 0.0250 $F$ Ba(OH)$_2$ are mixed with 10.0 mL of
 *(a) 0.100 $F$ HCl.      (b) 0.100 $F$ H$_2$SO$_4$.
 *(c) 0.100 $F$ Na$_2$SO$_4$.      (d) 0.100 $F$ MgCl$_2$.
 *(e) 0.0500 $F$ MgCl$_2$.      (f) 0.150 $F$ MgCl$_2$.

8-5. Calculate the pH of a 0.100 $F$ solution of
 *(a) hydrogen cyanide.      (b) hypochlorous acid.
 *(c) acetic acid.      (d) lactic acid.
 *(e) chloroacetic acid.      (f) picric acid.

8-6. Calculate the pH of a 0.0100 $F$ solution of
 *(a) hydrogen cyanide.      (b) hypochlorous acid.
 *(c) acetic acid.      (d) lactic acid.
 *(e) chloroacetic acid.      (f) picric acid.

8-7. Calculate the pH of a 1.00 × 10$^{-4}$ $F$ solution of
 *(a) hydrogen cyanide.      *(b) hypochlorous acid.
 *(c) acetic acid.      (d) lactic acid.
 (e) chloroacetic acid.      (f) picric acid.

*8-8. Calculate the pH of a 0.100 $F$ solution of
 *(a) piperidine.      (b) sodium phenolate.

        *(c) sodium cyanide.            (d) hydroxylamine.
        *(e) hydrazine.               (f) sodium hypochlorite.

**8-9.** Calculate the pH of a 0.0100 $F$ solution of
        *(a) piperidine.              (b) sodium phenolate.
        *(c) sodium cyanide.            (d) hydroxylamine.
        *(e) hydrazine.               (f) sodium hypochlorite.

**8-10.** Calculate the pH of a $1.00 \times 10^{-4}$ $F$ solution of
        *(a) piperidine.              (b) sodium phenolate.
        *(c) sodium cyanide.            (d) hydroxylamine.
        *(e) hydrazine.               (f) sodium hypochlorite.

**8-11.** Calculate the pH of the solution that results when 20.0 mL of 0.150 $F$ nitrous acid are mixed with 30.0 mL of
        *(a) 0.0100 $F$ NaOH.          (b) 0.0500 $F$ NaOH.
        *(c) 0.0500 $F$ Ba(OH)$_2$.      (d) 0.0500 $F$ NaNO$_2$.
        *(e) 0.100 $F$ NaNO$_2$.        (f) 0.150 $F$ NaNO$_2$.

**8-12.** Calculate the pH of the solution that results when 25.0 mL of 0.160 $F$ NH$_3$ are mixed with 50.0 mL of
        *(a) 0.0200 $F$ HCl.           (b) 0.0500 $F$ HCl.
        *(c) 0.0800 $F$ HCl.           (d) 0.0200 $F$ NH$_4$NO$_3$.
        *(e) 0.0500 $F$ NH$_4$NO$_3$.     (f) 0.0800 $F$ NH$_4$NO$_3$.

**8-13.** Calculate the pH of the solution that results when 40.00 mL of 0.0600 $F$ HClO$_4$ are mixed with 30.00 mL of
        *(a) 0.100 $F$ NaOCl.        (b) 0.0800 $F$ NH$_3$.
        *(c) 0.0500 $F$ Ca(HCOO)$_2$.   (d) 0.0400 $F$ Ca(C$_2$H$_3$O$_2$)$_2$.
        (e) 0.120 $F$ sodium lactate.   (f) 0.100 $F$ Mg(NO$_2$)$_2$.
        (g) 0.0500 $F$ Ce(C$_2$H$_3$O$_2$)$_3$.   (h) 0.125 $F$ (CH$_3$)$_2$NH.

**8-14.** Calculate the pH of the solution that results when 25.00 mL of 0.0400 $F$ NaOH are mixed with 20.00 mL of
        *(a) 0.0600 $F$ salicylic acid.   (b) 0.150 $F$ CH$_3$NH$_3^+$Cl$^-$.
        *(c) 0.100 $F$ hydrazoic acid.    (d) 0.100 $F$ picric acid.

**\*8-15.** Calculate the pH of a solution
        (a) containing 14.68 g of mandelic acid and 20.59 g of sodium mandelate in 400 mL of solution.
        (b) prepared by adding 74.4 mL of 0.0968 $F$ NaOH to 95.8 mL of 0.216 $F$ mandelic acid.
        (c) prepared by dissolving 1.886 g of sodium mandelate in 40.0 mL of 0.1460 $F$ HCl.
        (d) prepared by dissolving 2.006 g of mandelic acid in 50.0 mL of 0.0894 $F$ NaOH.

**8-16.** Calculate the pH of a solution
        (a) containing 1.578 g of salicylic acid and 2.460 g of sodium salicylate in 250 mL of solution.
        (b) prepared by adding 45.0 mL of 0.0250 $F$ HCl to 100.0 mL of 0.0300 $F$ sodium salicylate.
        (c) prepared by dissolving 2.42 g of salicylic acid in 75.0 mL of 0.0804 $F$ NaOH.
        (d) prepared by dissolving 5.11 g of sodium salicylate in 140.0 mL of 0.1649 $F$ HCl.

**8–17.** Calculate the pH of the accompanying solutions:

| | Formal Concentration of | |
|---|---|---|
| Weak Acid | Conjugate Base | |
| *(a) 0.100 $F$ HNO$_2$ | 0.0500 $F$ NaNO$_2$ | |
| (b) 0.0500 $F$ NH$_4$Cl | 0.0600 $F$ NH$_3$ | |
| *(c) 0.0475 $F$ HCOOH | 0.100 $F$ HCOONa | |
| *(d) 0.0200 $F$ HCOOH | 0.0167 $F$ Ca(HCOO)$_2$ | |

**8–18.** Calculate the molar ratio of acid to conjugate base in a solution that has a pH of 5.40 and contains
   *(a) lactic acid and sodium lactate.
   (b) benzoic acid and sodium benzoate.
   *(c) hypochlorous acid and potassium hypochlorite.
   (d) formic acid and sodium formate.
   *(e) hydroxylamine hydrochloride and hydroxylamine.
   (f) aniline hydrochloride and aniline.

**\*8–19.** What weight of sodium lactate must be added to 600 mL of 0.269 $F$ lactic acid to produce a buffer solution with a pH of 4.10?

**8–20.** What weight of NH$_4$Cl must be added to 250 mL of 0.300 $F$ NH$_3$ to produce a buffer solution with a pH of 9.45?

**\*8–21.** Calculate the change in pH that results from a 50-fold dilution of a solution that is
   (a) 0.145 $F$ in lactic acid.
   (b) 0.259 $F$ in sodium lactate.
   (c) 0.145 $F$ in lactic acid and 0.259 $F$ in sodium lactate.
   (d) 0.0145 $F$ in lactic acid and 0.0259 $F$ in sodium lactate.

**8–22.** Calculate the effect of a 100-fold dilution upon a solution that is
   (a) 0.165 $F$ in NH$_3$.
   (b) 0.137 $F$ in NH$_4$Cl.
   (c) 0.165 $F$ in NH$_3$ and 0.137 $F$ in NH$_4$Cl.
   (d) 0.0165 $F$ in NH$_3$ and 0.0137 $F$ in NH$_4$Cl.

**\*8–23.** Calculate the change in pH that occurs when 25.0 mL of 0.0400 $F$ HCl are added to 100 mL of the undiluted solutions described in Problem 8–21.

**8–24.** Calculate the change in pH that occurs when 25.0 mL of 0.0400 $F$ NaOH are added to 100 mL of the undiluted solutions described in Problem 8–22.

**\*8–25.** What weight of calcium acetate should be added to 200 mL of 0.425 $F$ acetic acid in order to produce a buffer solution with a pH of 4.90?

**8–26.** What weight of NH$_4$Cl should be added to 400 mL of 0.304 $F$ NH$_3$ in order to produce a buffer solution with a pH of 9.00?

**\*8–27.** What volume of 0.307 $F$ NaOH must be added to 200 mL of 0.540 $F$ acetic acid to produce a buffer solution with a pH of 4.25?

**8–28.** What volume of 0.500 $F$ HCl must be added to 400 mL of 0.304 $F$ NH$_3$ to produce a buffer solution with a pH of 9.00?

**\*8–29.** Calculate the pH of a 0.0500 $F$ solution of
   (a) sulfurous acid.

     (b) carbonic acid.

     (c) phosphorous acid.

**8-30.** Calculate the pH of a 0.0750 $F$ solution of

     (a) maleic acid.

     (b) ethylenediamine.

     (c) hydrogen sulfide.

**\*8-31.** Calculate the pH of a solution that is 0.0500 $F$ in

     (a) sodium hydrogen sulfite.

     (b) sodium hydrogen carbonate.

     (c) disodium hydrogen phosphate.

**8-32.** Calculate the pH of a solution that is 0.0400 $F$ in

     (a) sodium hydrogen sulfide.

     (b) disodium hydrogen arsenate.

     (c) sodium dihydrogen arsenate.

**\*8-33.** Calculate the pH of a solution that is 0.0500 $F$ in

     (a) sodium sulfite.

     (b) sodium carbonate.

     (c) sodium malonate.

**8-34.** Calculate the pH of a solution that is 0.0360 $F$ in

     (a) ethylenediamine dihydrochloride.

     (b) potassium phthalate.

     (c) sodium sulfide.

**8-35.** Calculate the pH of a solution that is

     \*(a) 0.1095 $F$ in $Na_2SO_3$ and 0.1440 $F$ in $NaHSO_3$.

     (b) 0.0485 $F$ in $H_2C_2O_4$ and 0.0765 $F$ in $NaHC_2O_4$.

     \*(c) 0.0800 $F$ in $H_3PO_4$ and 0.0460 $F$ in $NaH_2PO_4$.

     \*(d) 0.0573 $F$ in $H_2SO_3$ and 0.0812 $F$ in $NaHSO_3$.

     (e) 0.0500 $F$ in $Na_2C_2O_4$ and 0.1020 $F$ in $NaHC_2O_4$.

     \*(f) 0.1342 $F$ in $Na_2HPO_4$ and 0.0739 $F$ in $NaH_2PO_4$.

     (g) 0.1127 $F$ in $Na_2HPO_4$ and 0.1067 $F$ in $Na_3PO_4$.

**\*8-36.** Identify the principal conjugate acid-base pair and calculate the ratio between them in a solution buffered to a pH of 4.00 with species derived from

     (a) malic acid.

     (b) oxalic acid.

     (c) tartaric acid.

**8-37.** Identify the principal conjugate acid-base pair and calculate the ratio between them in a solution buffered to a pH of 7.00 with species derived from

     (a) arsenic acid.

     (b) citric acid.

     (c) phosphoric acid.

**8-38.** Calculate the pH of the solution that results when 40.0 mL of 0.300 $F$ HCl are added to 20.0 mL of $Na_3PO_4$ with a concentration of

     \*(a) 0.400 $F$.               (b) 1.20 $F$.

     \*(c) 0.300 $F$.               (d) 0.200 $F$.

     (e) 0.600 $F$.               \*(f) 0.240 $F$.

**8-39.** Calculate the pH of the solution that results when 30.0 mL of 0.400 $F$ NaOH are added to 20.0 mL of phosphoric acid with a concentration of

     \*(a) 0.200 $F$.               (b) 0.600 $F$.

*(c) 0.750 *F*.          (d) 0.400 *F*.
*(e) 0.300 *F*.         (f) 0.350 *F*.

**8–40.** Describe the preparation of 1.00 L of the following buffer solutions:

| | pH Sought | Start with | Add |
|---|---|---|---|
| *(a) | 10.00 | 75.0 g of $NaHCO_3$ | $Na_2CO_3$ |
| (b) | 7.00 | 50.0 g of $Na_2HPO_4$ | $NaH_2PO_4$ |
| *(c) | 9.80 | 750 mL of 0.450 $F$ $Na_2CO_3$ | $NaHCO_3$ |
| (d) | 7.00 | 800 mL of 0.300 $F$ $NaH_2PO_4$ | $Na_2HPO_4$ |
| *(e) | 7.00 | 600 mL of 0.0900 $F$ $H_3PO_4$ | $Na_3PO_4$ |
| (f) | 10.00 | 500 mL of 0.400 $F$ $Na_2CO_3$ | 0.350 $F$ HCl |

**8–41.** Calculate $\alpha$-values for the following acids or bases at the indicated pH values:

| | Solute | pH |
|---|---|---|
| *(a) | tartaric acid | 2.00, 4.00, 6.00 |
| (b) | oxalic acid | 1.00, 3.00, 5.00 |
| *(c) | carbonic acid | 6.00, 8.00, 10.00 |
| (d) | malic acid | 2.00, 3.00, 4.00 |
| *(e) | ethylenediamine | 6.00, 8.00, 10.00 |
| (f) | citric acid | 2.00, 4.00, 6.00 |

**8–42.** Use data from Problem 8–41 to calculate the molar concentration of the indicated species in solutions that had been brought to the indicated pH.

| | Formal Solute Concentration | pH | Molar Concentration Sought |
|---|---|---|---|
| *(a) | 0.0500 $F$ $NaHCO_3$ | 10.00 | $H_2CO_3$ |
| (b) | 0.0800 $F$ citric acid, $H_3C$ | 6.00 | $H_2C^-$ |
| *(c) | 0.0400 $F$ tartaric acid, $H_2T$ | 4.00 | $HT^-$ |
| (d) | 0.0300 $F$ $H_3C$ | 4.00 | $H_2C^-$ |

**8–43.** Calculate $\alpha$-values for each of the following:
*(a) sulfide-containing species in a solution with a pH of 7.64
(b) *o*-phthalate-containing species in a solution with a pH of 2.30
*(c) phosphate-containing species in solutions with a pH of 10.70 and a pH of 2.40
(d) tartrate-containing species in a solution with a pH of 5.80

*8–44. Derive a curve for the titration of 50.0 mL of 0.100 $F$ NaOH with 0.100 $F$ HCl. Calculate the pH of the solution after the addition of 0.0, 10.0, 25.0, 40.0, 45.0, 49.0, 50.0, 51.0, and 60.0 mL of acid; prepare a titration curve from the data.

8–45. Calculate the pH after the addition of 0.0, 10.0, 25.0, 40.0, 45.0, 49.0, 50.0, 51.0, 55.0, and 60.0 mL of the titrant listed in column B have been introduced to 50.0 mL of the analyte solution listed in column A; prepare a titration curve from the data.

| Column A: Analyte | Column B: Titrant |
|---|---|
| *(a)  0.0800 $F$ HNO$_2$ | 0.0800 $F$ NaOH |
| (b)  0.1200 $F$ lactic acid | 0.1200 $F$ KOH |
| *(c)  0.1000 $F$ NH$_3$ | 0.1000 $F$ HCl |
| (d)  0.0900 $F$ CH$_3$NH$_2$ | 0.0900 $F$ HCl |

**8–46.** Calculate the pH after the addition of 0.00, 5.00, 10.0, 19.0, 20.0, 21.0, 30.0, 39.0, 40.0, 41.0, and 50.0 mL of 0.200 $F$ titrant listed in column B have been added to 50.0 mL of a solution that is 0.0800 $F$ with respect to the analyte listed in column A.

| Column A: Analyte | Column B: Titrant |
|---|---|
| *(a)          H$_2$SO$_3$ | NaOH |
| (b)          Na$_2$CO$_3$ | HCl |
| *(c)          H$_3$PO$_3$ | NaOH |
| (d)          Na$_3$PO$_4$ | HClO$_4$ |

# Chapter 9
# Applications of Neutralization Titrations

Volumetric methods based upon neutralization involve the titration of hydronium or hydroxide ions produced directly or indirectly by the analyte. Neutralization methods find extensive application in chemical analysis.[1] Water serves as the solvent in most titrations. It should be recalled, however, that the acidic or basic nature of a solute is determined in part by the solvent in which it is dissolved; substitution of another solvent often permits a titration that cannot be successfully performed in an aqueous environment.[2] We shall restrict our discussion to aqueous systems.

## 9A REAGENTS FOR NEUTRALIZATION REACTIONS

The standard solutions employed for neutralization titrations are always strong acids or strong bases because these completely ionized reagents produce the most pronounced changes in pH at the equivalence point of titrations.

### 9A–1 Preparation of Standard Acid Solutions

Hydrochloric acid is the most common acid reagent for volumetric analyses. Dilute solutions of the reagent are indefinitely stable and can be used in the presence of most cations without complicating precipitation reactions. It has been reported that 0.1 $N$ solutions can be boiled for as long as an hour without loss of solute, provided that water lost by evaporation is periodically replaced; 0.5 $N$ solutions can be boiled for at least 10 min without significant loss.

Solutions of perchloric acid and sulfuric acid are also stable and are used in titrations where the presence of chloride ion would interfere. Standard solutions of nitric acid are seldom encountered, owing to the oxidizing properties of this solute.

---

[1] For a review of applications of neutralization titrations, see: D. Rosenthal and P. Zuman, in *Treatise on Analytical Chemistry,* 2nd ed., I. M. Kolthoff and P. J. Elving, Eds., Part I, Vol. 2, Chapter 18. New York: John Wiley & Sons, 1979.

[2] For a review of nonaqueous acid-base titrimetry, see: *Treatise on Analytical Chemistry,* 2nd ed., I. M. Kolthoff and P. J. Elving, Eds., Part I, Vol. 2, Chapters 19A–19E. New York: John Wiley & Sons, 1979.

Standard acid solutions are usually prepared by diluting an approximate volume of concentrated reagent and subsequently standardizing the diluted solution against a primary standard base. Less frequently, the composition of the concentrated acid is established through careful density measurement, following which a weighed quantity is diluted to an exact volume (tables that relate the density of reagents to composition can be found in many chemistry and chemical engineering handbooks). A stock hydrochloric acid solution of exactly known solute concentration can also be prepared by distillation of the concentrated reagent; the final quarter of the distillate, known as *constant boiling* HCl, has a fixed and known composition that depends only upon the atmospheric pressure. For ambient pressures between 670 and 780 torr, the weight of distillate (in air) that contains exactly one equivalent of hydrochloric acid is given by[3]

$$\frac{\text{wt constant boiling acid, g}}{\text{equivalent}} = 164.673 + 0.02039 \ P \qquad (9-1)$$

where $P$ is the pressure in torr. A standard solution is prepared by diluting a weighed quantity of this acid to an exactly known volume.

### 9A–2 Primary Standards for Acids

**Sodium Carbonate**    Sodium carbonate is a convenient and frequently used standard for acid solutions. Primary-standard sodium carbonate is available from commercial suppliers; it can also be prepared by heating purified sodium hydrogen carbonate between 270 and 300°C for an hour:

$$2NaHCO_3(s) \rightarrow Na_2CO_3(s) + H_2O(g) + CO_2(g)$$

We have noted (p. 202) that two end points are observed in the titration of sodium carbonate. The first, corresponding to conversion to hydrogen carbonate, occurs at a pH of about 8.3; the second, involving the formation of carbonic acid, is observed at a pH of about 3.8. This second end point is preferred for standardizations because the pH change associated with it is greater. An even sharper end point can be achieved by eliminating carbonic acid, the reaction product. The sample is first titrated to the color change of an acid range indicator (such as bromocresol green or methyl orange). At this point the solution contains a large amount of carbonic acid and a small amount of unreacted hydrogen carbonate. Boiling eliminates the carbonic acid

$$H_2CO_3(aq) \rightarrow CO_2(g) + H_2O(l)$$

and effectively destroys this buffer. As a result, the solution acquires an alkaline pH owing to the residual hydrogen carbonate ion. The titration is completed

---

[3] *Official Methods of Analysis of the AOAC,* 14th ed., p. 1004. Washington, D.C.: Association of Official Analytical Chemists, 1984.

after the solution has been cooled. Now, however, a substantially larger pH decrease attends the final additions of acid; a sharper color change is observed.

An alternative method of standardization involves addition of the acid in an amount that is slightly greater than that needed to convert the sodium carbonate to carbonic acid. The solution is boiled to eliminate carbon dioxide, as before; after cooling, the excess acid is back-titrated with a dilute solution of base. Since this back-titration involves a strong acid and a strong base, the choice of indicator is not critical (see Figure 8 – 1). An independent titration is needed to establish the volume ratio that exists between the acid and the base (see Chapter 20, Section 20C – 4).

Directions for the preparation and standardization of hydrochloric acid solutions against sodium carbonate are given in Chapter 20, Sections 20C – 2 and 20C – 5.

**Other Primary Standards**    *Tris*-(hydroxymethyl)aminomethane (also known as THAM, or TRIS), $(HOCH_2)_3CNH_2$, is available in primary-standard purity from commercial sources. The reagent possesses the advantage of a substantially larger equivalent weight (121.14) than sodium carbonate (53.00).

Other standards for acids include sodium tetraborate, mercury(II) oxide, and calcium oxalate; details concerning their use can be found in standard reference works.[4]

## 9A – 3 Preparation of Standard Solutions of Base

Sodium hydroxide is by far the most common basic reagent, although potassium hydroxide and barium hydroxide are also encountered. None of these is obtainable in primary-standard purity; after being prepared to approximately the desired concentration, their solutions must be standardized.

**Effect of Carbon Dioxide upon Standard Base Solutions**    In solution, as well as in the solid state, the hydroxides of sodium, potassium, and barium avidly react with atmospheric carbon dioxide to produce carbonate ion:

$$CO_2(g) + 2OH^- \rightarrow CO_3^{2-} + H_2O$$

Even though the absorption of carbon dioxide involves the loss of hydroxide ions, the acid titer of a standardized base solution may not necessarily suffer a change. For example, if potassium or sodium hydroxide is to be used in a titration for which an acid range indicator is appropriate, each carbonate ion in the reagent will have reacted with two hydronium ions of the analyte at the end point (see Figure 8 – 10); that is,

$$CO_3^{2-} + 2H_3O^+ \rightarrow H_2CO_3 + 2H_2O$$

[4] See for example, C. A. Streuli with L. Meites, *Handbook of Analytical Chemistry*, L. Meites, Ed., p. **3** – 34. New York: McGraw-Hill Book Co., 1963.

Since the hydronium ion consumed by this reaction is identical to the amount of sodium hydroxide lost during the formation of carbonate ion, the absorption of carbon dioxide will cause no error.

Unfortunately, most titrations involving standard base solutions require the use of indicators with basic transition ranges (phenolphthalein, for example). The color change of such an indicator will occur at a pH that corresponds to the formation of hydrogen carbonate ion

$$CO_3^{2-} + OH^- \rightarrow HCO_3^- + H_2O$$

Because each carbonate ion will have reacted with only one hydronium ion, the effective normality of the base is diminished, and a determinate *carbonate error* will result.

The absorption of carbon dioxide by barium hydroxide results in the precipitation of barium carbonate

$$CO_2(g) + Ba^{2+} + 2OH^- \rightarrow BaCO_3(s) + H_2O$$

and the acid titer is thus decreased regardless of the indicator used in the titration; a carbonate error is the inevitable consequence.

Absorption of atmospheric carbon dioxide can be the source of extensive contamination in the solid hydroxides used to prepare standard base solutions. As a result, even freshly prepared solutions of base are likely to contain significant quantities of carbonate. Its presence will not cause a carbonate error, provided the titrations are performed with the same indicator as was used for standardization; this restriction, however, limits the versatility of the reagent.

Several methods exist for the preparation of carbonate-free hydroxide solutions. The use of barium hydroxide is one such possibility, particularly if barium carbonate is made even less soluble through the introduction of a neutral barium salt, such as the chloride or the nitrate. A barium salt can also be used to precipitate carbonate from solutions of sodium or potassium hydroxide. The presence of barium ion is frequently undesirable, however, owing to its tendency to form slightly soluble barium salts with anions that may be present in the sample.

The preferred method for the preparation of sodium hydroxide solutions takes advantage of the very low solubility of sodium carbonate in concentrated solutions of the alkali. An approximately 50% aqueous solution of sodium hydroxide is prepared (or purchased). The sodium carbonate is either allowed to settle or else is removed by vacuum filtration to give a clear liquid that is then diluted to give the desired concentration. Details for this preparation are given in Chapter 20, Section 20C – 3.

A carbonate-free base solution must be prepared from water that has been freed of carbon dioxide. Distilled water, which is sometimes supersaturated with $CO_2$, should be boiled briefly to eliminate the gas. The water is cooled before the base is added because hot alkali solutions rapidly absorb carbon dioxide.

Standard solutions of base are reasonably stable as long as they are protected

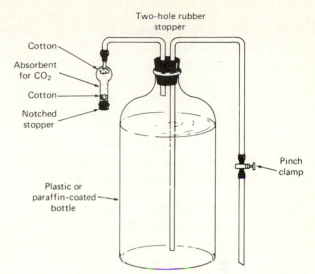

**Figure 9–1** Arrangement for the long-term storage of standard base solutions.

from the atmosphere. Figure 9–1 shows an arrangement for preventing the uptake of carbon dioxide during storage. Air entering the bottle passes over a solid absorbent for $CO_2$, such as soda lime or Ascarite II.[5] Any contamination that occurs as the solution is transferred from this storage bottle to the buret is negligible. Absorption during a titration can be minimized by covering the open end of the buret with a small test tube or beaker.

A tightly capped polyethylene bottle will usually provide sufficient short-term protection against the uptake of atmospheric carbon dioxide by standard base solutions. Care should be taken to keep the bottle capped at all times except during the brief periods when the contents are being transferred to a buret. The bottle should also be squeezed before capping to minimize the interior air space.

Sodium hydroxide solutions react slowly with glass to form silicates. Because of this reaction, the parts of any ground-glass fitting will ultimately freeze upon prolonged exposure to alkaline solutions. For this reason, standard solutions of base should not be stored in glass-stoppered containers. Similarly, a buret equipped with a glass stopcock should be promptly drained and thoroughly rinsed after being used to dispense these reagents. Use of a buret equipped with a Teflon stopcock is a practical alternative.

## 9A–4 Primary Standards for Bases

Several excellent primary standards are available for the standardization of bases. Most are weak organic acids that require use of an indicator with a basic transition range for their titration.

---

[5] Ascarite II is a registered trademark of Arthur H. Thomas Company, Philadelphia, PA. Ascarite II consists of sodium hydroxide deposited on a nonfibrous silicate matrix.

**Potassium Hydrogen Phthalate, KHC₈H₄O₄**   Potassium hydrogen phthalate (KHP) possesses many of the qualities of the ideal primary standard. It is a nonhygroscopic solid with a high equivalent weight. The commercial analytical-grade reagent can be used without further purification for most purposes. Potassium hydrogen phthalate of certified purity is available from the National Bureau of Standards for the most exacting work. Directions for the standardization of sodium hydroxide solutions are given in Chapter 20, Section 20C–6.

**Other Primary Standards for Bases**   Benzoic acid is obtainable in a highly pure state and can be used as a primary standard for bases. The reagent has limited solubility and is usually dissolved in ethanol before being diluted with water.

Potassium hydrogen iodate, $KH(IO_3)_2$, is an excellent primary standard with a large equivalent weight. It is also a strong acid that can be titrated using virtually any indicator with a pH transition range between 4 and 10.

## 9B TYPICAL APPLICATIONS OF NEUTRALIZATION TITRATIONS

Neutralization titrations are used for determination of the innumerable inorganic, organic, and biochemical species that possess inherent acidic or basic properties. Equally important, however, are the many applications that involve some chemical step that converts the analyte to an acid or a base which is then titrated with standard base or acid.

Two major types of end points are available for neutralization titrations. The first, based on the color change of an indicator, has been discussed in Chapter 8. The second involves the experimental generation of the titration curve through measurement of the pH as a function of titrant volume. A glass–calomel electrode system immersed in a solution of the analyte develops a potential that is directly proportional to pH; the end point is then determined graphically. Potentiometric titrations are described in Chapter 13.

### 9B–1 Elemental Analysis

A number of elements that occur in important organic and biological systems are conveniently determined by methods that involve a neutralization titration as the final step. The elements that are susceptible to this type of analysis are for the most part nonmetallic; principal among these are carbon, nitrogen, chlorine, bromine, sulfur, phosphorus, and fluorine. The analysis for each of these elements involves a preliminary step (or steps) that brings about conversion to an inorganic acid or base that can then be titrated. Typical examples are described in this section.

**Nitrogen**   Nitrogen occurs in many important materials such as proteins, synthetic drugs, fertilizers, explosives, and potable water supplies. The analysis for nitrogen is thus of singular importance to research and to industry.

Organic nitrogen can be determined in either of two ways. The *Dumas* method is suitable for virtually all organic nitrogen compounds. The sample is mixed with powdered copper(II) oxide and ignited in a combustion tube to carbon dioxide, water, nitrogen, and perhaps nitrogen oxides. A stream of carbon dioxide forces the ignition products through a packing of hot copper (to reduce any nitrogen oxides to elemental nitrogen) and thence into a gas buret filled with a highly concentrated solution of potassium hydroxide. The only component that is not absorbed by the base is nitrogen; its volume is measured directly.

The *Kjeldahl* determination of nitrogen was first described in 1883 and to this day is one of the most widely used methods of chemical analysis. It requires no special equipment and is readily adapted to the routine analysis of large numbers of samples. The Kjeldahl method (or one of its modifications) is the standard means for determining the protein content of grains, meats, and other biological materials.

In the Kjeldahl method, the sample is decomposed in hot concentrated sulfuric acid, which converts the nitrogen in most organic functional groups to ammonium sulfate. After the decomposition is complete, the solution is cooled, diluted, and made basic with strong sodium hydroxide. The liberated ammonia is distilled, collected, and subsequently titrated.

The critical step in the Kjeldahl method is the decomposition with sulfuric acid, which oxidizes the carbon and hydrogen in the sample to carbon dioxide and water. The fate of the nitrogen, however, depends upon its state of combination in the original sample. Amine and amide nitrogen are quantitatively converted to ammonium ion. In contrast, nitro-, azo-, or azoxy nitrogen is likely to yield elemental nitrogen or nitrogen oxides, which distill from the decomposition medium. When these oxidized functional groups are present, the sample is often pretreated with a reducing agent to form reduced products that behave as amide or amine nitrogen. In one such prereduction scheme, salicylic acid and sodium thiosulfate are added to the sulfuric acid along with the sample; the digestion is then performed in the usual way.

Some aromatic heterocyclic compounds, such as pyridine and its derivatives, are particularly resistant to complete decomposition by sulfuric acid. Such compounds yield low results unless special precautions are taken (see Figure 3–1, p. 29).

Digestion of the sample is the most time-consuming step in a Kjeldahl analysis; an hour (or more) may be needed for refractory samples. Attempts to hasten the process through the introduction of such stronger oxidizing agents as perchloric acid, potassium permanganate, and hydrogen peroxide have largely failed, owing to the partial oxidation of ammonium ion to volatile products.

Of the many schemes aimed at improving the kinetics of the digestion process, the modification proposed by Gunning[6] has been the most satisfactory.

[6] J. W. Gunning, *Z. Anal. Chem.,* **1889,** *28,* 188.

Gunning added a neutral salt, such as potassium sulfate, to increase the boiling point of the sulfuric acid and thus the temperature at which digestion of the sample occurs. The amount of salt that is added must be controlled to eliminate the possibility of oxidation of ammonium ion. This difficulty is encountered if excessive amounts of acid evaporate during digestion.

Many substances catalyze the decomposition of organic compounds by sulfuric acid. Mercury, copper, and selenium, either combined or in the elemental state, are effective. Mercury(II), if not precipitated with hydrogen sulfide prior to the distillation step, will retain some ammonia as an ammine complex.

Figure 9–2 illustrates typical arrangements for Kjeldahl distillations. The long-necked container, which is used for both decomposition and distillation, is called a *Kjeldahl flask.* After the digestion is judged complete, the flask and its contents are cooled, diluted with water, and rendered alkaline to convert the ammonium ion to ammonia

$$NH_4^+ + OH^- \rightarrow NH_3(g) + H_2O$$

The arrangement shown in Figure 9–2a permits the addition of base by partially opening the stopcock of the storage vessel; the liberated ammonia is carried to the receiving flask by steam distillation. As an alternative, dense sodium hydroxide solution is carefully poured down the side of the flask to form a second lower layer. The flask is immediately connected to a spray trap and condenser (Figure 9–2b) before loss of ammonia can occur. Only then are the two layers mixed by gently swirling the flask. The end of the condenser extends

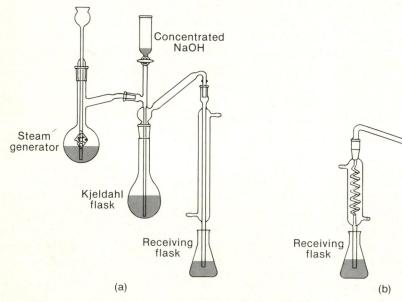

Concentrated NaOH

Steam generator

Kjeldahl flask

Receiving flask

(a)

Spray trap

Receiving flask

Kjeldahl flask

(b)

**Figure 9–2**  Arrangements for Kjeldahl distillations.

into the liquid in the receiving flask during the distillation process. It must be removed as heating is discontinued to prevent the contents of the flask from being drawn back into the condenser.

Two methods are used for collecting and determining the ammonia liberated from the sample solution. In one, the ammonia is distilled into a solution containing a measured excess of standard acid. The excess is then titrated with standard base; an indicator with an acidic transition range must be used because of the acidity of the ammonium ion present at the equivalence point. A convenient alternative, which requires only one standard solution, involves collection of the distilled ammonia in an unmeasured excess of boric acid, which retains the ammonia by the reaction

$$H_3BO_3 + NH_3 \rightarrow NH_4^+ + H_2BO_3^-$$

The dihydrogen borate ion is a reasonably strong base that can be titrated with a standard solution of acid:

$$H_2BO_3^- + H_3O^+ \rightarrow H_3BO_3 + H_2O$$

At equivalence, the solution contains boric acid and ammonium ion; an indicator with an acidic transition range is again required.

**Sulfur**   Sulfur in organic and biological specimens can be conveniently determined by burning a weighed sample in a stream of oxygen. Sulfur dioxide (and perhaps sulfur trioxide) produced during the oxidation is collected in a dilute solution of hydrogen peroxide:

$$SO_2(g) + H_2O_2 \rightarrow H_2SO_4$$

The sulfuric acid is then titrated with standard base.

**Other Elements**   Table 9–1 lists other elements that can be determined by neutralization methods.

Table 9–1   *Elemental Analyses Based On Neutralization Titrations*

| Element | Converted to | Absorption or Precipitation Products | Titration |
|---------|--------------|--------------------------------------|-----------|
| N | $NH_3$ | $NH_3(g) + H_3O^+ \rightarrow NH_4^+ + H_2O$ | Excess HCl with NaOH |
| S | $SO_2$ | $SO_2(g) + H_2O_2 \rightarrow H_2SO_4$ | NaOH |
| C | $CO_2$ | $CO_2(g) + Ba(OH)_2 \rightarrow BaCO_3(s) + H_2O$ | Excess $Ba(OH)_2$ with HCl |
| Cl(Br) | HCl | $HCl(g) + H_2O \rightarrow Cl^- + H_3O^+$ | NaOH |
| F | $SiF_4$ | $SiF_4(g) + H_2O \rightarrow H_2SiF_6$ | NaOH |
| P | $H_3PO_4$ | $12H_2MoO_4 + 3NH_4^+ + H_3PO_4 \rightarrow$ $(NH_4)_3PO_4 \cdot 12MoO_3(s) + 12H_2O + 3H^+$ $(NH_4)_3PO_4 \cdot 12MoO_3(s) + 26OH^- \rightarrow$ $HPO_4^{2-} + 12MoO_4^{2-} + 14H_2O + 3NH_3(g)$ | Excess NaOH with HCl |

## 9B – 2 Determination of Inorganic Species

Numerous inorganic species can be determined by titration with strong acids and bases. Typical among these are ammonium ion, nitrates and nitrites, and species derived from carbonic acid.

**Ammonium Salts**   Ammonium salts can be conveniently determined by conversion to ammonia with strong base and distillation. The liberated ammonia is collected and titrated as in the Kjeldahl method.

**Nitrates and Nitrites**   The method just described can be extended to the determination of inorganic nitrate or nitrite. These species are reduced to ammonium ion by Devarda's alloy (50% Cu, 45% Al, 5% Zn). Granules of the alloy are introduced to a strongly alkaline solution of the sample in a Kjeldahl flask; the ammonia is distilled after reaction is complete. Arnd's alloy (60% Cu, 40% Mg) is also effective as a reducing agent.

**Carbonate and Carbonate Mixtures**   The qualitative and quantitative determination of constituents in a solution containing sodium carbonate, sodium hydrogen carbonate, and sodium hydroxide, alone or in combination, provide interesting examples of the applications of neutralization titrations. No more than two of these three constituents can exist in appreciable amounts in any solution because reaction will eliminate the third. Thus, mixing solutions containing sodium hydroxide and sodium hydrogen carbonate will result in the formation of sodium carbonate until one or the other (or perhaps both) of the original constituents is exhausted. If the sodium hydroxide is used up, the solution will contain sodium carbonate and sodium hydrogen carbonate; if sodium hydrogen carbonate is used up, sodium carbonate and sodium hydroxide will remain. Finally, if chemically equivalent amounts of the two reactants are mixed, the principal solute species will be sodium carbonate.

Analysis of such mixtures requires two titrations: the one involving an alkaline range indicator (phenolphthalein, for example), and the other an acid range indicator (such as bromocresol green). The composition of the solution can be deduced from the relative volumes of standard acid needed to titrate equal volumes of the sample (see Table 9 – 2 and Figure 8 – 10). Once the composition has been established, the volume data can be used to determine the amount of each component in the sample.

---

**Example 9 – 1**   A solution to be analyzed may contain $NaHCO_3$, $Na_2CO_3$, NaOH, or any permissible combination of these solutes. Titration of a 50.0-mL aliquot to a phenolphthalein end point required 12.6 mL of 0.100 $F$ HCl. A second 50.0-mL aliquot required 48.4 mL of the HCl when titrated to a bromocresol green end point. Deduce the composition and the formal solute concentrations of the original solution.

A solution that contained only sodium hydroxide would use the same volume of acid, regardless of indicator (that is, $V_{phth} = V_{bcg}$). Similarly, if the solution contained only sodium carbonate, the volume of acid needed to titrate to a bromocresol green end point would be just twice that needed for a phenolphthalein end point. Neither of these situations fits the experimental facts, so we can rule out both as possibilities. Because $V_{bcg} > 2 \times V_{phth}$, the solution must have contained some $NaHCO_3$ in addition to $Na_2CO_3$. We can now calculate the concentration of each constituent. Titration to the phenolphthalein end point involved conversion of $CO_3^{2-}$ to $HCO_3^-$; thus,

$$12.6 \text{ mL} \times \frac{0.100 \text{ mfw HCl}}{\text{mL}} \times \frac{\text{mfw Na}_2\text{CO}_3}{\text{mfw HCl}} = 1.26 \text{ mfw Na}_2\text{CO}_3$$

The difference between the volumes needed to observe the two end points ($48.4 - 12.6 = 35.8$ mL) involved titration of the hydrogen carbonate that was originally present as well as that formed in the titration of the carbonate. Thus,

$$\text{mfw HCO}_3^- + \text{mfw CO}_3^{2-} = 35.8 \times 0.100 = 3.58$$

and

$$3.58 - 1.26 = 2.32 \text{ mfw HCO}_3^-$$

Finally, then,

$$\frac{1.26 \text{ mfw Na}_2\text{CO}_3}{50.0 \text{ mL}} = 0.0252 \ F \text{ Na}_2\text{CO}_3$$

$$\frac{2.32 \text{ mfw NaHCO}_3}{50.0 \text{ mL}} = 0.0464 \ F \text{ NaHCO}_3$$

The titration described in Example 9−1 is not entirely satisfactory because the pH change corresponding to the hydrogen carbonate end point is insufficient to give a sharp change with a chemical indicator (see Figure 8−10).

**Table 9−2** *Volume Relationships in the Analysis of Mixtures Containing Hydroxide, Carbonate, and Hydrogen Carbonate Ions*

| Constituent(s) in Sample | Relationship Between $V_{phth}$ and $V_{bcg}$ in the Titration of an Equal Volume of Sample* |
|---|---|
| NaOH | $V_{phth} = V_{bcg}$ |
| $Na_2CO_3$ | $V_{phth} = \frac{1}{2} V_{bcg}$ |
| $NaHCO_3$ | $V_{phth} = 0; V_{bcg} > 0$ |
| NaOH, $Na_2CO_3$ | $V_{phth} > \frac{1}{2} V_{bcg}$ |
| $Na_2CO_3$, $NaHCO_3$ | $V_{phth} < \frac{1}{2} V_{bcg}$ |

* $V_{phth}$ = volume of acid needed for a phenolphthalein end point; $V_{bcg}$ = volume of acid needed for a bromocresol green end point.

Although titration to a color match with a solution containing an approximately equivalent amount of $NaHCO_3$ is helpful, titration errors on the order of 1% must be expected.

The limited solubility of barium carbonate can be used to improve the titration of hydroxide–carbonate and carbonate–hydrogen carbonate mixtures. The *Winkler* method for the determination of hydroxide–carbonate mixtures involves titration of components in an aliquot with an acid range indicator (after boiling to eliminate $CO_2$). An unmeasured excess of neutral barium chloride is then added to a second aliquot to precipitate the carbonate ion, following which the hydroxide ion is titrated to a phenolphthalein end point. The sparingly soluble barium carbonate will not interfere with the titration as long as the concentration of the excess barium ion is about 0.1 $F$.

An accurate analysis of a carbonate–hydrogen carbonate mixture can be achieved by establishing the total equivalents through titration of an aliquot to the color change of an acid range indicator. The hydrogen carbonate in a second aliquot is converted to carbonate by a measured excess of sodium hydroxide. A large excess of barium chloride is introduced, following which the excess hydroxide is titrated to a phenolphthalein end point. The presence of solid barium carbonate does not hamper end-point detection in either of these methods.

Directions for the determination of sodium carbonate in an impure sample are given in Chapter 20, Section 20C–10; a method for the analysis of a carbonate–hydrogen carbonate mixture is found in Section 20C–11.

### 9B–3 Determination of Organic Functional Groups

A number of organic functional groups can be determined—directly or indirectly—by neutralization titrations. Examples of typical methods are briefly described in this section.

**Carboxylic and Sulfonic Acid Groups**   Carboxylic and sulfonic acid groups are the two most common structures that impart acidity to organic compounds. Most carboxylic acids have dissociation constants that range between $10^{-4}$ and $10^{-6}$ and are thus readily titrated. An alkaline range indicator is needed for such titrations.

Many carboxylic acids are not sufficiently soluble in water to permit a direct titration in this medium. Substitution of ethanol as solvent for the acid is a useful way of circumventing this problem. Alternatively, the acid can be dissolved in a measured excess of standard base and the unreacted base back-titrated with standard acid.

Sulfonic acids tend to be strong and are readily soluble in water; their titration with base is therefore straightforward.

**Amine Groups**   Aliphatic amines ordinarily have basic dissociation constants on the order of $10^{-5}$ and can thus be titrated directly with a standard acid. Aromatic amines, such as aniline and its derivatives, are usually too weak for titration in an aqueous medium ($K_b \sim 10^{-10}$); the same holds true for cyclic amines with aromatic character, such as pyridine and its derivatives. Saturated cyclic amines, such as piperidine, tend to resemble aliphatic amines.

Many amines that are too weak to be titrated as bases in water are readily determined in a nonaqueous solvent, such as anhydrous acetic acid, which enhances their basicity.

**Ester Groups** Esters are commonly determined by saponification with a measured excess of standard base

$$R_1COOR_2 + OH^- \rightarrow R_1COO^- + HOR_2$$

The excess base is then titrated with standard acid. The saponification of some esters is sufficiently rapid to permit direct titration with standard base while refluxing with excess base for several hours may be required for others.

**Hydroxyl Groups** Organic hydroxyl groups can be determined by esterification with carboxylic acid anhydrides or chlorides. The two most common reagents are probably acetic anhydride and phthalic anhydride; the reaction with the former is

$$(CH_3CO)_2O + ROH \rightarrow CH_3COOR + CH_3COOH$$

The acetylation step involves mixing the sample with a measured excess of the anhydride in pyridine and heating. Water is subsequently added to destroy the excess anhydride

$$(CH_3CO)_2O + H_2O \rightarrow 2CH_3COOH$$

The acetic acid is then titrated with a standard solution of sodium hydroxide or potassium hydroxide in alcohol. A blank is carried through the analysis to establish the original amount of anhydride.

**Carbonyl Groups** Many aldehydes and ketones can be quantitatively converted to oximes by reaction with hydroxylamine hydrochloride:

$$\begin{array}{c} R_1 \\ \diagdown \\ \diagup \\ R_2 \end{array} C{=}O + HONH_2 \cdot HCl \longrightarrow \begin{array}{c} R_1 \\ \diagdown \\ \diagup \\ R_2 \end{array} C{=}NOH + HCl + H_2O$$

where $R_2$ may be hydrogen. The liberated hydrochloric acid is titrated with standard base. Oxime formation with aldehydes is frequently complete after a 30-min heating period; reaction with ketones may require a reflux period of an hour or more.

## PROBLEMS

9-1. Identify the acid on the left side of each equation and express its equivalent weight as a fraction or a multiple of its gram-formula weight.

*(a) $Na_2CO_3 + H_2SO_4 \rightarrow Na_2SO_4 + 2H_2O + CO_2$

(b) $H_4P_2O_7 + 2Ba(OH)_2 \rightarrow Ba_2P_2O_7(s) + 4H_2O$

*(c) $H_5IO_6 + KOH \rightarrow KH_4IO_6 + H_2O$

(d) $Ba(OH)_2 + 2KH(IO_3)_2 \rightarrow Ba(IO_3)_2(s) + 2H_2O + 2KIO_3$

*(e) $2HONH_2 + H_2SO_4 \rightarrow 2HONH_3^+ + SO_4^{2-}$

(f) $HClO_4 + C_6H_5ONa \rightarrow C_6H_5OH + NaClO_4$

*(g) $Na_3PO_4 + 2HCl \rightarrow NaH_2PO_4 + 2NaCl$

(h) $2HCl + Na_2B_4O_7 + 5H_2O \rightarrow 4H_3BO_3 + 2NaCl$

*9–2. Express the equivalent weight of the base on the left side of each equation in Problem 9–1 as a fraction or a multiple of its gram-formula weight.

9–3. Calculate the number of milliequivalents in

*(a) 0.142 g of $Na_2CO_3$ in terms of reaction 9–1(a).

(b) 23.4 mL of 0.0359 $F$ $H_4P_2O_7$ in terms of reaction 9–1(b).

*(c) 50.0 mL of 0.0566 $N$ $H_5IO_6$ in terms of reaction 9–1(c).

(d) 0.367 g of $KH(IO_3)_2$ in terms of reaction 9–1(d).

*(e) 0.100 g of $HONH_2$ in terms of reaction 9–1(e).

(f) 46.7 mL of 0.0846 $F$ $C_6H_5ONa$ in terms of reaction 9–1(f).

*(g) 0.738 g of $Na_3PO_4$ in terms of reaction 9–1(g).

(h) 0.163 g of $Na_2B_4O_7 \cdot 10H_2O$ in terms of reaction 9–1(h).

9–4. Calculate the volume of

*(a) 0.0606 $F$ $H_2SO_4$ needed to titrate the $Na_2CO_3$ in Problem 9–3(a).

(b) 0.0689 $N$ $Ba(OH)_2$ needed to titrate the $H_4P_2O_7$ in Problem 9–3(b).

*(c) 0.1450 $N$ KOH needed to titrate the $H_5IO_6$ in Problem 9–3(c).

(d) 0.0224 $F$ $Ba(OH)_2$ needed to titrate the $KH(IO_3)_2$ in Problem 9–3(d).

*(e) 0.1070 $N$ $H_2SO_4$ needed to titrate the $HONH_2$ in Problem 9–3(e).

(f) 0.1040 $N$ $HClO_4$ needed to titrate the $C_6H_5ONa$ in Problem 9–3(f).

*(g) 0.1964 $N$ HCl needed to titrate the $Na_3PO_4$ in Problem 9–3(g).

(h) 0.0567 $N$ HCl needed to titrate the $Na_2B_4O_7 \cdot 10H_2O$ in Problem 9–3(h).

*9–5. Describe the preparation of

(a) 2.0 L of approximately 0.15 $N$ HCl from the concentrated reagent [sp gr 1.19, 37% HCl (w/w)].

(b) 2.000 L of 0.0800 $N$ reagent from constant-boiling HCl which was distilled at 765 torr.

(c) 500 mL of 0.0800 $N$ reagent from 0.965 $N$ HCl.

(d) 1.000 L of 0.1260 $N$ $HIO_3$ from the pure solid.

9–6. Describe the preparation of

(a) 4.0 L of approximately 0.12 $N$ NaOH from the concentrated reagent [sp gr 1.525, 50% NaOH (w/w)].

(b) 200 mL of approximately 0.30 $N$ NaOH from the solid.

(c) 1.000 L of 0.0200 $N$ NaOH from a 0.400 $N$ solution.

(d) 500 mL of 0.120 $N$ $Na_2CO_3$ in terms of the reaction

$$CO_3^{2-} + 2H^+ \rightarrow CO_2(g) + H_2O$$

*9–7. Calculate the normality of an HCl solution if

(a) a 25.00-mL aliquot yielded 0.314 g of AgCl.

(b) 42.71 mL were needed to titrate 1.2875 g of pure $Na_2B_4O_7 \cdot 10H_2O$ [see Problem 9–1(h) for reaction].

(c) 39.69 mL were needed to titrate a 50.00-mL aliquot of 0.1086 $N$ NaOH.

(d) 28.65 mL were needed to titrate the $Na_2CO_3$ (product: $CO_2$) produced by ignition of 0.2204 g of primary standard $Na_2C_2O_4$. Reaction:

$$Na_2C_2O_4 \rightarrow Na_2CO_3 + CO(g)$$

**9–8.** Calculate the normality of a $Ba(OH)_2$ solution if
   (a) a 50.00-mL aliquot yielded 0.2088 g of $BaSO_4$.
   (b) 41.28 mL were needed to titrate a 0.4013-g sample of primary-standard potassium hydrogen phthalate.
   (c) 38.76 mL were needed to titrate a 25.00-mL aliquot of 0.03184 $N$ $HClO_4$.
   (d) after the addition of a 50.00-mL aliquot to a 0.2571-g sample of benzoic acid, a 1.64-mL back-titration with 0.0584 $N$ HCl was required.

**\*9–9.** The accompanying data were recorded for the standardization of an NaOH solution against weighed portions of primary-standard potassium hydrogen phthalate (KHP):

| Weight KHP, g | 0.8609 | 0.8135 | 0.8327 | 0.8443 |
|---|---|---|---|---|
| Volume of NaOH, mL | 38.89 | 36.60 | 37.87 | 38.11 |

   (a) Calculate the mean value for the normality of the NaOH.
   (b) Calculate the standard deviation for the data.
   (c) Apply the $Q$ test to determine whether statistical grounds exist for disregarding the outlying result at the 90% confidence level.

**9–10.** Standardization of an NaOH solution against weighed quantities of primary-standard $KH(IO_3)_2$ yielded the accompanying results:

| Weight $KH(IO_3)_2$, g | 1.5602 | 1.6797 | 1.4934 | 1.5183 |
|---|---|---|---|---|
| Volume of NaOH, mL | 38.33 | 41.41 | 36.59 | 37.22 |

Calculate
   (a) the mean normality of the NaOH solution.
   (b) the percent relative deviation of the set from the mean.
   (c) the relative standard deviation (in parts per thousand) for the set.
   (d) the 90% confidence limit for the mean normality, based solely upon these data.

**\*9–11.** Calculate the $H_3PO_4$ titer of a 0.1429 $N$ NaOH solution if the product is
   (a) $Na_2HPO_4$.
   (b) $NaH_2PO_4$.

**9–12.** Calculate the $Na_3PO_4$ titer of a 0.1376 $N$ HCl solution if the product is
   (a) $Na_2HPO_4$.
   (b) $NaH_2PO_4$.

**9–13.** Suggest a range of sample weights that would yield a 35- to 45-mL titration in the standardization of
   \*(a) 0.12 $N$ KOH against primary standard potassium hydrogen phthalate.
   (b) 0.050 $N$ HCl against the $OH^-$ produced by treatment of primary standard HgO with excess $Br^-$. Reaction:

$$HgO(s) + H_2O + 4Br^- \rightarrow HgBr_4^{2-} + 2OH^-$$

   \*(c) 0.040 $N$ $Ba(OH)_2$ against primary standard $KH(IO_3)_2$.

(d) 0.080 $N$ HCl against pure $Na_2B_4O_7 \cdot 10H_2O$. Reaction:

$$B_4O_7^{2-} + 2H^+ + 5H_2O \rightarrow 4H_3BO_3$$

*(e) 0.10 $N$ $HClO_4$ against primary standard sodium oxalate. Reactions:

$$Na_2C_2O_4 \xrightarrow{heat} Na_2CO_3 + CO(g)$$

$$CO_3^{2-} + 2H^+ \longrightarrow CO_2(g) + H_2O$$

(f) 0.060 $N$ NaOH against primary-standard benzoic acid.

**9–14.** Briefly explain why $KH(IO_3)_2$ would be preferred over benzoic acid as a primary standard for a 0.010 $N$ NaOH solution.

**\*9–15.** Select an indicator from Table 8–1 that would be suitable for the titration of
  (a) 0.100 $N$ $HClO_4$ with 0.040 $N$ $Ba(OH)_2$.
  (b) 0.100 $N$ $(CH_3)_2NH$ with 0.100 $N$ HCl.
  (c) 0.100 $N$ $C_6H_5ONa$ with 0.080 $N$ HCl.
  (d) 0.150 $F$ $H_3PO_4$ to the $HPO_4^{2-}$ end point, with 0.100 $N$ NaOH.
  (e) the HCl in a solution that is 0.080 $N$ in HCl and 0.050 $N$ in propanoic acid, with 0.100 $N$ NaOH.
  (f) the total acidity of the solution in (e).

**9–16.** Select an indicator from Table 8–1 that would be suitable for the titration of
  (a) 0.0050 $N$ KOH with 0.0070 $N$ HCl.
  (b) 0.100 $N$ formic acid with 0.100 $N$ NaOH.
  (c) 0.080 $N$ aniline hydrochloride with 0.100 $N$ NaOH.
  (d) 0.100 $F$ $H_3AsO_4$ to the $H_2AsO_4^-$ end point, with 0.100 $N$ NaOH.
  (e) the NaOH in a solution that is 0.075 $N$ in NaOH and 0.040 $N$ in ethanol-amine, with 0.100 $N$ HCl.

**\*9–17.** Calculate the equivalent weight of a weak dibasic acid if 37.38 mL of 0.1057 $N$ NaOH were needed to titrate a 0.4034-g sample to a phenolphthalein end point.

**9–18.** A dilute solution of an unknown weak acid required a 28.62-mL titration with 0.1084 $N$ NaOH to reach a phenolphthalein end point. The titrated solution was evaporated to dryness, following which the sodium salt was collected and found to weigh 0.2110 g. Calculate the equivalent weight of the acid.

**\*9–19.** A 25.0-mL aliquot of vinegar was diluted to 250 mL in a volumetric flask. Titration of 50.0-mL aliquots of the diluted solution required an average of 34.88 mL of 0.09600 $N$ NaOH. Express the acidity of the vinegar in terms of the percentage (w/v) of acetic acid.

**9–20.** The acids in a 10.0-mL aliquot of white table wine were titrated to a phenolphthalein end point with 25.01 mL of 0.04930 $N$ NaOH. Express the acidity in terms of grams tartaric acid ($H_2C_4H_4O_6$, gfw = 150) per 100 mL of the wine, assuming that both of the acidic hydrogens are titrated.

**9–21.** Titration of a 0.7439-g sample of impure $Na_2B_4O_7$ required 31.64 mL of 0.1081 $N$ NaOH [see Problem 9–1(h) for reaction]. Express the results of this analysis in terms of the percentage of
  (a) $Na_2B_4O_7$.                          (b) $Na_2B_4O_7 \cdot 10H_2O$.
  (c) $B_2O_3$.                              (d) B.

**\*9–22.** The HgO in an impure 0.5214-g sample of the mineral montroydite was brought into solution with an excess of KI. Reaction:

$$HgO(s) + 4I^- + H_2O \rightarrow HgI_4^{2-} + 2OH^-$$

Calculate the percentage of HgO in the sample if 38.79 mL of 0.0968 $N$ HCl were needed to titrate the liberated base.

**9–23.** Three capsules containing chloral hydrate [$Cl_3CCH(OH)_2$ (gfw = 165.1)] were treated with 50.0 mL of 0.1048 $N$ NaOH. Reaction:

$$Cl_3CCH(OH)_2 + OH^- \rightarrow CHCl_3 + HCOO^- + H_2O$$

Calculate the average weight (in milligrams) of chloral hydrate per capsule if 9.30 mL of 0.0750 $N$ HCl were needed to titrate the excess base.

**9–24.** The benzoic acid extracted from a 106.3-g sample of catsup required a 14.76-mL titration with 0.0514 $N$ NaOH. Express the results of this analysis in terms of the percent sodium benzoate (gfw = 144) in the sample.

**9–25.** A 0.5843-g sample of a plant food preparation was analyzed for its N content by the Kjeldahl method, the liberated $NH_3$ being collected in 50.00 mL of 0.1062 $N$ HCl. The excess acid required an 11.89 mL back-titration with 0.0925 $N$ NaOH. Express the results of this analysis in terms of percent
*(a) N.                                    (b) urea, $H_2NCONH_2$.
*(c) $(NH_4)_2SO_4$.                       (d) $(NH_4)_3PO_4$.

**\*9–26.** A 1.378-g sample of canned meat was analyzed by the Kjeldahl method. The liberated $NH_3$ was collected in an excess of 4% $H_3BO_3$ and subsequently titrated with 30.27 mL of 0.1300 $N$ HCl. Express the results of this analysis in terms of
(a) percentage of N.
(b) percentage of protein, based upon 16% as an average value for the N content in protein.

**9–27.** A six-tablet sample of the diuretic hydrochlorothiazide, $C_7H_8ClN_3O_4S_2$ (gfw = 297.7), was analyzed by the Kjeldahl method, the ammonia produced being collected in an unmeasured excess of $H_3BO_3$ and subsequently titrated with 31.26 mL of 0.0967 $N$ HCl. Calculate the average weight (in milligrams) hydrochlorothiazide in these tablets.

**\*9–28.** A 2.513-g mixture containing $NH_4Cl$, $NaNO_3$, and inert materials was dissolved in sufficient water to give 250 mL. A 50.00-mL aliquot of this solution was rendered strongly alkaline, following which the liberated $NH_3$ was distilled into a $H_3BO_3$ solution. Titration of the $NH_4H_2BO_3$ produced required 23.66 mL of 0.1039 $N$ HCl. The $NH_3$ produced as a result of treating a second 50.00-mL aliquot with Devarda's alloy (p. 220) was again collected in $H_3BO_3$ and subsequently titrated with 41.71 mL of the standard HCl. Calculate the percent $NH_4Cl$ and $NaNO_3$ in the sample.

**9–29.** The method described in Problem 9–28 was used to determine the $NH_4NO_3$ and $(NH_4)_2SO_4$ in a 3.661-g sample that had been dissolved in sufficient water to give 250 mL of solution. The $NH_3$ evolved when a 50.00-mL aliquot was made basic required 45.42 mL of 0.1039 $N$ HCl. Treatment of a 25.00-mL aliquot with Devarda's alloy and distillation produced an amount of $NH_3$ that required a 30.40-mL titration with the HCl. Calculate the percentages of the two salts in the original sample.

**\*9–30.** Federal regulations set an upper limit of 50 ppm for the $NH_3$ ($d = 0.771$ g/L) in the air of a work environment (i.e., 50 mL $NH_3/10^6$ mL air). Air from a manufacturing operation was passed at the rate of 10.0 L/min through a trap that contained 100.0 mL of 0.01038 $N$ HCl for a total of 12.0 min. Titration of the excess acid in the trap required 13.06 mL of 0.0657 $N$ NaOH. Do the results of this analysis indicate that the manufacturer is in compliance with regulations?

**9–31.** The atmosphere within a paper mill was tested for compliance with the federal limit of less than 5.0 ppm $SO_2$ (i.e., 5.0 mL $SO_2/10^6$ mL air). Air was drawn at the rate of 9.00 L/min through a trap containing 100.0 mL of 1% $H_2O_2$. Reaction:

$$H_2O_2 + SO_2(g) \rightarrow 2H^+ + SO_4^{2-}$$

The $H_2SO_4$ produced during a 30.0-min test required 5.97 mL of 0.00977 $N$ NaOH. Calculate the parts per million of $SO_2$ ($d = 2.85$ g/L) in the air.

**\*9–32.** The formaldehyde, HCHO, in a 0.8610-g sample of a pesticide preparation was treated with a neutralized solution of $Na_2SO_3$. Reaction:

$$HCHO + Na_2SO_3 + H_2O \rightarrow \overset{\overset{\displaystyle H}{|}}{\underset{\underset{\displaystyle OH}{|}}{H-C-SO_3Na}} + OH^-$$

Calculate the percentage of HCHO in the sample if 38.60 mL of 0.2025 $N$ HCl were needed to titrate the liberated base.

**9–33.** Esters are saponified by treatment with an excess of base:

$$R-\overset{\overset{\displaystyle O}{\|}}{C}-OR' + OH^- \rightarrow R-\overset{\overset{\displaystyle O}{\|}}{C}-O^- + R'OH$$

A 2.259-g sample of a purified ester was heated with 50.00 mL of 0.4627 $N$ NaOH. Calculate the equivalent weight of the ester if 9.08 mL of 0.4063 $N$ HCl were needed to titrate the excess base.

**9–34.** Calculate the volume of 0.1364 $N$ HCl needed to titrate
  \*(a) 12.10 mL of 0.1238 $F$ NaOH to a thymolphthalein end point.
  (b) 18.74 mL of 0.1069 $F$ $Na_3PO_4$ to a thymolphthalein end point.
  \*(c) 12.10 mL of 0.1238 $F$ NaOH to a bromocresol green end point.
  (d) 18.74 mL of 0.1069 $F$ $Na_3PO_4$ to a bromocresol green end point.
  \*(e) the solution that results from mixing 12.10 mL of 0.1238 $F$ NaOH and 18.74 mL of 0.1069 $F$ $Na_3PO_4$ to a thymolphthalein end point.
  (f) the solution in (e), titrated to a bromocresol green end point.

**9–35.** Calculate the volume of 0.1110 $N$ NaOH needed to titrate
  \*(a) 13.88 mL of 0.1041 $F$ HCl to a bromocresol green end point.
  (b) 15.43 mL of 0.1200 $F$ $H_3PO_4$ to a bromocresol green end point.
  \*(c) 13.22 mL of 0.0988 $F$ $NaH_2PO_4$ to a thymolphthalein end point.
  (d) 15.43 mL of 0.1200 $F$ $H_3PO_4$ to a thymolphthalein end point.
  \*(e) the solution that results from mixing 13.88 mL of 0.1041 $F$ HCl and 15.43 mL of 0.1200 $F$ $H_3PO_4$, to a thymolphthalein end point.
  (f) the solution that results from mixing 15.43 mL of 0.1200 $F$ $H_3PO_4$ and 13.22 mL of 0.0988 $F$ $NaH_2PO_4$, to a bromocresol green end point.
  \*(g) the solution in (f), titrated to a thymolphthalein end point.

**\*9–36.** A 50.00-mL aliquot containing $H_3PO_4$ and $NaH_2PO_4$ required 23.71 mL of 0.1046 $N$ NaOH to reach a bromocresol green end point. A 25.00-mL aliquot, titrated to a phenolphthalein end point, required 43.68 mL of the base. Calculate the formal concentrations of $H_3PO_4$ and $NaH_2PO_4$ in the sample.

**9-37.** A 5.63-g sample containing $Na_2CO_3$, $NaHCO_3$, and inert materials was dissolved in sufficient water to give 250 mL of solution. A 50.00-mL aliquot required 14.91 mL of 0.1175 $N$ HCl when titrated to a phenolphthalein end point. A second 50.00-mL aliquot, titrated to a bromocresol green end point, required 44.78 mL of the acid. Calculate the percentages of $Na_2CO_3$ and $NaHCO_3$ in the sample.

**\*9-38.** Solutions containing NaOH, $Na_2CO_3$, and $NaHCO_3$, alone or in any compatible combination, were titrated with standard HCl. Use the accompanying data to establish the composition of each solution and the weight (in milligrams) of solute(s) in each milliliter.

**Volume of 0.1364 $N$ HCl Needed to Titrate
25.00-mL Aliquots to**

| Solution | Bromocresol Green End Point | Phenolphthalein End Point |
|---|---|---|
| (a) | 43.07 | 21.54 |
| (b) | 19.82 | 0.00 |
| (c) | 29.97 | 30.01 |
| (d) | 43.96 | 13.51 |
| (e) | 46.51 | 35.68 |

**9-39.** A series of solutions can contain HCl, $H_3PO_4$, and $NaH_2PO_4$, alone or in any compatible combination. Use the accompanying titration data to establish the composition of each solution and the weight (in milligrams) of solute(s) in each milliliter.

**Volume of 0.1203 $N$ NaOH Needed to Titrate
50.00-mL Aliquots to**

| Solution | Bromocresol Green End Point | Phenolphthalein End Point |
|---|---|---|
| (a) | 31.84 | 49.02 |
| (b) | 24.73 | 24.70 |
| (c) | 0.00 | 40.88 |
| (d) | 22.01 | 43.97 |
| (e) | 10.45 | 39.86 |

# Chapter 10
# Complex Formation Titrations

Many metal ions react with electron-pair donors to form coordination compounds or complex ions. The donor species, or *ligand,* must have at least one pair of unshared electrons; the water molecule, ammonia, and the halide ions are typical ligands.

Although exceptions exist, a specific cation will coordinate with a maximum of two, four, or six ligands; its *coordination number* is a statement of this maximum. The species that is formed as a result of coordination can be electrically positive, neutral, or negative. Thus, for example, copper(II) (with a coordination number of four) forms a cationic ammine complex with ammonia $[Cu(NH_3)_4^{2+}]$, a neutral complex with glycine $[Cu(NH_2CH_2COO)_2]$, and an anionic complex with chloride ion $[CuCl_4^{2-}]$.

Complex formation has been used for analyses for well over a century. The truly remarkable growth in analytical applications is of fairly recent origin and is based upon a particular class of coordination compounds called *chelates.* A chelate is produced by the coordination of a cation and a ligand that possesses more than one unshared electron pair. The copper complex with glycine, mentioned above, is an example. Here, copper is bonded by the oxygen of the carboxylate groups as well as the nitrogen of the amine groups. A chelating agent with two available donor groups is called *bidentate;* one with three groups is called *tridentate.* Tetra-, penta-, and hexadentate ligands exist as well.[1]

A majority of complex formation or *complexometric* titrations are based on chelating reagents because reactions of these species with cations is usually a one-step process. In contrast, simpler complexing agents, such as ammonia or cyanide ion, tend to react with cations in a stepwise manner, yielding a series of complexes. The advantage of formation of a single complex can be seen by considering the reaction of a tetracoordinate metal ion M with a tetradentate, a

---

[1] For a detailed discussion of complex-formation reactions in analytical chemistry, see: A. Ringbom and E. Wänninen, in *Treatise on Analytical Chemistry,* 2nd ed., I. M. Kolthoff and P. J. Elving, Eds., Part I, Vol. 2, Chapter 20. New York: John Wiley & Sons, 1979; and R. Přibil, *Applied Complexometry.* New York: Pergamon Press, 1982.

bidentate, and a monodentate ligand. The equilibrium with the tetradentate ligand D is shown by the equation[2]

$$M + D \rightleftharpoons MD$$

The equilibrium constant for this process is

$$K_f = \frac{[MD]}{[M][D]}$$

where $K_f$ is the *formation constant* for MD.

The equilibrium between M and the bidentate ligand B can be expressed as

$$M + 2B \rightleftharpoons MB_2$$

This equation, however, is the summation of a two-step process that involves formation of the intermediate MB

$$M + B \rightleftharpoons MB \qquad K_1 = \frac{[MB]}{[M][B]}$$

$$MB + B \rightleftharpoons MB_2 \qquad K_2 = \frac{[MB_2]}{[MB][B]}$$

The product of $K_1$ and $K_2$ yields the equilibrium constant, $\beta_2$, for the overall process[3]

$$\beta_2 = K_1 K_2 = \frac{[\cancel{MB}]}{[M][B]} \times \frac{[MB_2]}{[B][\cancel{MB}]} = \frac{[MB_2]}{[M][B]^2}$$

In a like manner, the reaction between M and the monodentate ligand A involves the overall equilibrium

$$M + 4A \rightleftharpoons MA_4$$

and the equilibrium constant, $\beta_4$, for the formation of $MA_4$ from M and A is numerically equal to the product of the equilibrium constants for the four subordinate processes.

Each of the titration curves depicted in Figure 10–1 is based on a reaction with an overall equilibrium constant of $1.0 \times 10^{20}$. Curve *A* illustrates the formation of MD in a single step ($K_f = 10^{20}$). Curve *B* involves the formation of $MB_2$ in a two-step process in which $K_1 = 10^{12}$ and $K_2 = 10^8$. Curve *C* is for the formation of $MA_4$, for which the constants for the successive equilibria are $10^8$, $10^6$, $10^4$, and $10^2$, respectively. Large changes in p-values in the equivalence-point region of a titration have been shown to cause sharp color changes in chemical indicators (see Chapters 7 and 8). The curves in Figure 10–1 demon-

---

[2] The charges carried by M and D are of no consequence in this discussion and have been omitted.

[3] It is customary to use the symbol $\beta_i$ to indicate an overall equilibrium constant; thus, for example, $\beta_2 = K_1 K_2$, $\beta_3 = K_1 K_2 K_3$, $\beta_4 = K_1 K_2 K_3 K_4$, and so forth.

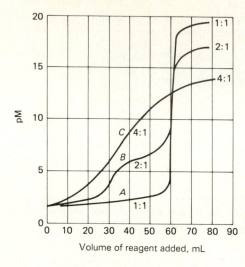

**Figure 10-1** Curves for complex formation titrations. Titrations of 60.0 mL of 0.020 F M with (curve A) a 0.020 F solution of the tetradentate ligand D to produce MD, (curve B) a 0.040 F solution of the bidentate ligand B to give $MB_2$, and (curve C) a 0.080 F solution of the monodentate ligand A to give $MA_4$. The overall equilibrium constant for each product is $1.0 \times 10^{20}$.

strate the clear superiority of a ligand that combines with a cation in a 1 : 1 ratio because the change in pM in the equivalence-point region is a maximum in such a system. It is for this reason that polydentate ligands are ordinarily superior reagents for complex formation titrations.

## 10A TITRATIONS WITH INORGANIC COMPLEXING REAGENTS

Complexometric titrations are among the oldest of volumetric methods.[4] For example, the titration of iodide ion with mercury(II)

$$Hg^{2+} + 4I^- \rightarrow HgI_4^{2-}$$

was first reported in 1834 while the determination of cyanide based on the formation of the dicyanoargentate(I) ion, $Ag(CN)_2^-$, was described by Liebig in 1851. Table 10-1 lists typical nonchelating complexing reagents as well as some of the uses to which they have been put.[5]

## 10B TITRATIONS WITH AMINOPOLYCARBOXYLIC ACIDS

Several tertiary amines that also contain carboxylic acid groups form remarkably stable chelates with numerous metal ions. Their potential as analytical reagents was first exploited by Schwarzenbach in 1945; they have been exhaus-

---

[4] See I. M. Kolthoff and V. A. Stenger, *Volumetric Analysis,* Vol. II, pp. 282, 331. New York: Interscience Publishers, 1947.

[5] Derivation of a theoretical curve for the titration of chloride by mercury(II) can be found in D. A. Skoog and D. M. West, *Fundamentals of Analytical Chemistry,* 4th ed., pp. 280–282. Philadelphia: Saunders College Publishing, 1982.

Table 10–1    Typical Inorganic Complex Formation Titrations*

| Titrant | Analyte | Remarks |
|---|---|---|
| $Hg(NO_3)_2$ | $Br^-$, $Cl^-$, $SCN^-$, $CN^-$, thiourea | Neutral mercury(II) complexes produced; various indicators used |
| $AgNO_3$ | $CN^-$ | Product is $Ag(CN)_2^-$, indicator is $I^-$; titrate to first turbidity of AgI |
| $NiSO_4$ | $CN^-$ | Product is $Ni(CN)_4^{2-}$, indicator is AgI; titrate to first turbidity of AgI |
| KCN | $Cu^{2+}$, $Hg^{2+}$, $Ni^{2+}$ | Products are $Cu(CN)_4^{2-}$, $Hg(CN)_2$, $Ni(CN)_4^{2-}$; various indicators used |

\* For further applications and selected references, see R. Püschel in *Handbook of Analytical Chemistry*, L. Meites, Ed., pp. 3-226–3-234. New York: McGraw-Hill Book Co., 1963.

tively investigated since that time. The revival of interest in complexometric titrations is attributable to these reagents.[6]

The discussion that follows is based on ethylenediaminetetraacetic acid, which is undoubtedly the most widely used and most versatile chelating reagent.

## 10B–1 Ethylenediaminetetraacetic Acid

Ethylenediaminetetraacetic acid [also called (ethylenedinitrilo) tetraacetic acid and often abbreviated EDTA] has the structure

$$HOOC-CH_2 \atop HOOC-CH_2 \!\!\!\diagdown \!\! N-CH_2-CH_2-N \!\! \diagup \!\!\! {CH_2-COOH \atop CH_2-COOH}$$

The successive equilibrium constants for dissociation of the four acidic hydrogens of EDTA are

$$K_1 = 1.02 \times 10^{-2} \qquad K_2 = 2.14 \times 10^{-3}$$

$$K_3 = 6.92 \times 10^{-7} \qquad K_4 = 5.50 \times 10^{-11}$$

These values indicate that the first two protons are lost much more readily than the remaining two. In addition to these four acidic hydrogens, each nitrogen has

---

[6] Several excellent monographs are devoted to these reagents and their applications. See, for example, G. Schwarzenbach and H. Flashka, *Complexometric Titrations*, 2nd ed. (trans. by H. M. N. H. Irving). London: Methuen & Co., 1969; A. Ringbom and E. Wänninen, in *Treatise on Analytical Chemistry*, 2nd ed., I. M. Kolthoff and P. J. Elving, Eds., Part I, Vol. 2, Chapter 20. New York: John Wiley & Sons, 1979. For annotated summaries of applications, see C. N. Reilley and A. J. Barnard, Jr., in *Handbook of Analytical Chemistry*, L. Meites, Ed., pp. 3-167–3-234. New York: McGraw-Hill Book Co., 1963.

an unshared electron pair; the molecule thus has six potential sites for bonding with a metal and is therefore classed as a hexadentate ligand. The abbreviations $H_4Y$, $H_3Y^-$, $H_2Y^{2-}$, $HY^{3-}$ and $Y^{4-}$ are encountered in discussions that pertain to the equilibrium aspects of EDTA and its coordination compounds.

Both the free acid $H_4Y$ and the disodium salt $Na_2H_2Y$ are available in reagent grade quality. The former can serve as a primary standard after being dried at $105\,°C$; it is then dissolved in the minimum amount of base needed for complete solution. The disodium salt $Na_2H_2Y \cdot 2H_2O$ contains 0.3% moisture in excess of the stoichiometric amount under ordinary atmospheric conditions; for all but the most exacting work, this excess is sufficiently reproducible to permit use of a corrected weight for the salt in the direct preparation of standard solutions. If necessary, the pure dihydrate can be obtained by heating the solid at $80\,°C$ for several days in an atmosphere with a relative humidity of 50%.

### 10B–2 Complexes of EDTA and Metal Ions

EDTA combines with metal ions in a 1:1 ratio regardless of the charge on the cation. Thus, complex formation with silver(I) and aluminum(III) can be described by the equations

$$Ag^+ + Y^{4-} \rightleftharpoons AgY^{3-}$$

$$Al^{3+} + Y^{4-} \rightleftharpoons AlY^-$$

EDTA forms complexes with virtually all cations, most of which are sufficiently stable to form the basis for a volumetric analysis. The stability of these complexes undoubtedly results from the six sites within the EDTA molecule that are available for coordination bonding. A structure for an EDTA complex with a divalent cation is shown in Figure 10–2; note that all six ligand groups of the parent molecule are involved in bonding with the cation.

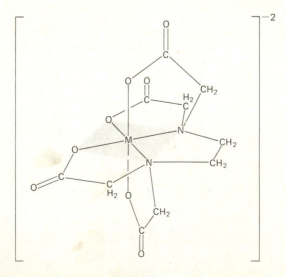

**Figure 10–2** Structure of an EDTA chelate with a divalent cation (bond lengths are not to scale).

Table 10–2   Formation Constants for EDTA Complexes*

| Cation | $K_{MY}$ | $\log K_{MY}$ | Cation | $K_{MY}$ | $\log K_{MY}$ |
|--------|----------|---------------|--------|----------|---------------|
| $Ag^+$ | $2.1 \times 10^7$ | 7.32 | $Cu^{2+}$ | $6.3 \times 10^{18}$ | 18.80 |
| $Mg^{2+}$ | $4.9 \times 10^8$ | 8.69 | $Zn^{2+}$ | $3.2 \times 10^{16}$ | 16.50 |
| $Ca^{2+}$ | $5.0 \times 10^{10}$ | 10.70 | $Cd^{2+}$ | $2.9 \times 10^{16}$ | 16.46 |
| $Sr^{2+}$ | $4.3 \times 10^8$ | 8.63 | $Hg^{2+}$ | $6.3 \times 10^{21}$ | 21.80 |
| $Ba^{2+}$ | $5.8 \times 10^7$ | 7.76 | $Pb^{2+}$ | $1.1 \times 10^{18}$ | 18.04 |
| $Mn^{2+}$ | $6.2 \times 10^{13}$ | 13.79 | $Al^{3+}$ | $1.3 \times 10^{16}$ | 16.13 |
| $Fe^{2+}$ | $2.1 \times 10^{14}$ | 14.33 | $Fe^{3+}$ | $1 \times 10^{25}$ | 25.1 |
| $Co^{2+}$ | $2.0 \times 10^{16}$ | 16.31 | $V^{3+}$ | $8 \times 10^{25}$ | 25.9 |
| $Ni^{2+}$ | $4.2 \times 10^{18}$ | 18.62 | $Th^{4+}$ | $2 \times 10^{23}$ | 23.2 |

* Data from G. Schwarzenbach, *Complexometric Titrations,* p. 8. New York: Interscience (London: Chapman & Hall), 1957. With permission. (Constants valid at 20°C and an ionic strength of 0.1.)

Table 10–2 contains numerical values for the formation constants, $K_{MY}$, of a number of important EDTA–cation complexes. Note that these constants refer to equilibria between the completely deprotonated $Y^{4-}$ and the cation; that is,

$$M^{n+} + Y^{4-} \rightleftharpoons MY^{(n-4)+}$$

for which

$$K_{MY} = \frac{[MY^{(n-4)+}]}{[M^{n+}][Y^{4-}]} \qquad (10-1)$$

### 10B–3 Equilibrium Calculations Involving EDTA

Most indicators used in EDTA titrations respond to changes in the concentration of the uncomplexed cation $M^{n+}$. The ordinate of a titration curve with EDTA will thus be pM. In order to calculate the concentration of $M^{n+}$, and thus pM, at and beyond the equivalence point, it is necessary to derive the molar concentration of the species $Y^{4-}$. This calculation requires that the pH of the medium be taken into account.

**Effect of pH on the Composition of EDTA Solutions**   Application of Equation 10–1 to the computation of $[M^{n+}]$ is complicated by the fact that EDTA solutions contain not only $Y^{4-}$ but also $HY^{3-}$, $H_2Y^{2-}$, $H_3Y^-$, and $H_4Y$. Clearly, the relative concentrations of these five species is strongly pH dependent. Generally, however, EDTA titrations are performed in well-buffered solutions of known pH because control of pH permits selective titration of certain cations in the presence of others. Because pH is known, the concentration of the various EDTA species can be readily derived by means of their $\alpha$-values (see Chapter 8, Section 8H). Thus, if we define $C_T$ as the sum of concentrations of the *uncom-*

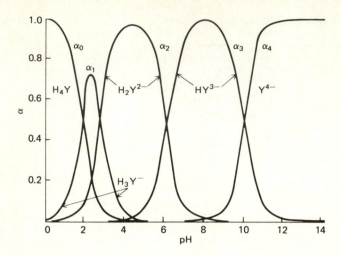

**Figure 10–3** Composition of EDTA solutions as a function of pH.

*plexed* EDTA species, we define the $\alpha$-value for $Y^{4-}$ as

$$\alpha_4 = \frac{[Y^{4-}]}{C_T} \tag{10-2}$$

where

$$C_T = [Y^{4-}] + [HY^{3-}] + [H_2Y^{2-}] + [H_3Y^-] + [H_4Y]$$

An expression for calculating $\alpha_4$ for a known hydronium ion concentration is readily derived by the method described in Chapter 8, Section 8H. Thus,

$$\alpha_4 = \frac{K_1K_2K_3K_4}{[H^+]^4 + K_1[H^+]^3 + K_1K_2[H^+]^2 + K_1K_2K_3[H^+] + K_1K_2K_3K_4} \tag{10-3}$$

where $K_1$, $K_2$, $K_3$, and $K_4$ are the four dissociation constants for $H_4Y$.[7] Values for the other species are similarly obtained; that is,

$$\alpha_0 = [H^+]^4/D \qquad \alpha_1 = K_1[H^+]^3/D$$

$$\alpha_2 = K_1K_2[H^+]^2/D \qquad \alpha_3 = K_1K_2K_3[H^+]/D$$

where $D$ is the denominator of Equation 10–3.

Figure 10–3 illustrates how the mole fractions for the five EDTA-containing species vary as a function of pH. It is apparent that $H_2Y^{2-}$ predominates in moderately acidic media (pH 3 to 6). Only at pH values greater than 10 does $Y^{4-}$ become a major component of an EDTA solution.

Table 10–3 lists values of $\alpha_4$ for integer pH values. Note that only about

---

[7] As a matter of convenience, we shall revert to use of $H^+$ as a shorthand notation for $H_3O^+$ and from time to time shall refer to the species represented by $H^+$ as the hydrogen ion.

Table 10–3   Values of $\alpha_4$ for EDTA in Solutions of Various pH

| pH | $\alpha_4$ | pH | $\alpha_4$ |
|---|---|---|---|
| 2.0 | $3.7 \times 10^{-14}$ | 7.0 | $4.8 \times 10^{-4}$ |
| 3.0 | $2.5 \times 10^{-11}$ | 8.0 | $5.4 \times 10^{-3}$ |
| 4.0 | $3.6 \times 10^{-9}$ | 9.0 | $5.2 \times 10^{-2}$ |
| 5.0 | $3.5 \times 10^{-7}$ | 10.0 | $3.5 \times 10^{-1}$ |
| 6.0 | $2.2 \times 10^{-5}$ | 11.0 | $8.5 \times 10^{-1}$ |
|  |  | 12.0 | $9.8 \times 10^{-1}$ |

$4 \times 10^{-12}$ percent of EDTA exists as $Y^{4-}$ at a pH of 2.00. Example 10–1 illustrates the calculation for $[Y^{4-}]$ in a solution of known pH.

**Example 10–1**   Calculate the molar $Y^{4-}$ concentration in a 0.0200 $F$ EDTA solution that has been buffered to a pH of 10.00.

At pH 10.00, $\alpha_4$ is 0.35 (Table 10–3). Thus,

$$[Y^{4-}] = \alpha_4 C_T = (0.35)(0.0200) = 7.0 \times 10^{-3} \, M$$

**Calculation of the Cation Concentration in a Solution of an EDTA Chelate**   Examples 10–2 and 10–3 illustrate how the concentration of uncomplexed cation is evaluated in a solution of $MY^{2-}$ as well as in a solution that, in addition, contains an excess of the chelating reagent.

**Example 10–2**   Calculate the molar concentration of $Ni^{2+}$ in a 0.0150 $F$ solution of $NiY^{2-}$ that has been buffered to a pH of (a) 3.00 and (b) 8.00.

The formation constant for $NiY^{2-}$ is given by (Table 10–2)

$$Ni^{2+} + Y^{4-} \rightleftharpoons NiY^{2-} \qquad K_{NiY} = \frac{[NiY^{2-}]}{[Ni^{2+}][Y^{4-}]}$$
$$= 4.2 \times 10^{18}$$

The molar concentration of the chelate will be equal to its formal concentration less any that has been lost as the result of dissociation; that is,

$$[NiY^{2-}] = 0.0150 - [Ni^{2+}]$$

The magnitude of $K_{NiY}$ suggests that $[Ni^{2+}]$ will be vanishingly small with respect to 0.0150 and that $[NiY^{2-}] \cong 0.0150$.

Since the complex is the only source of $Ni^{2+}$ and EDTA species in this solution, it follows that

$$[Ni^{2+}] = [Y^{4-}] + [HY^{3-}] + [H_2Y^{2-}] + [H_3Y^-] + [H_4Y] = C_T$$

According to Equation 10-2,

$$\alpha_4 = \frac{[Y^{4-}]}{C_T}$$

Thus,

$$[Y^{4-}] = \alpha_4 C_T$$

Substitution of this relationship into the formation constant expression gives

$$\frac{[NiY^{2-}]}{[Ni^{2+}]\alpha_4 C_T} = K_{NiY}$$

Combination of the two constants in this equation gives

$$\frac{[NiY^{2-}]}{[Ni^{2+}]C_T} = \alpha_4 K_{NiY} = K'_{NiY} \qquad (10-4)$$

The new constant $K'_{NiY}$ is a *conditional constant*, $K'_{MY}$, which is applicable only at the pH for which it has been calculated. We have noted that $[Ni^{2+}] = C_T$. Thus,

$$\frac{[NiY^{2-}]}{[Ni^{2+}]^2} = K'_{NiY} \qquad (10-5)$$

(a) Table 10-3 indicates that $\alpha_4$ is $2.5 \times 10^{-11}$ at pH 3.00. Substitution of this value into Equation 10-4 gives

$$K'_{NiY} = (2.5 \times 10^{-11})(4.2 \times 10^{18}) = 1.05 \times 10^8$$

and rearrangement of Equation 10-5 permits calculation of $[Ni^{2+}]$; thus,

$$[Ni^{2+}] = \sqrt{\frac{[NiY^{2-}]}{K'_{NiY}}} = \sqrt{\frac{0.0150}{1.05 \times 10^8}} = 1.2 \times 10^{-5}$$

The assumption that $0.0150 \gg [Ni^{2+}]$ is indeed justified.

(b) At pH 8.00, $\alpha_4$ is $5.4 \times 10^{-3}$; similarly, then

$$K'_{NiY} = (5.4 \times 10^{-3})(4.2 \times 10^{18}) = 2.27 \times 10^{16}$$

$$[Ni^{2+}] = \sqrt{\frac{0.0150}{2.27 \times 10^{16}}} = 8.1 \times 10^{-10}$$

Note that the concentration of $Ni^{2+}$ is significantly lower at pH 8.0 owing to the greater fraction of Y-containing species that exists as $Y^{4-}$ at the higher pH.

**Calculation of the Cation Concentration in the Presence of an Excess of EDTA**   In a solution containing an excess of the chelating reagent, the only source for the free cation will be from dissociation of the complex. Example 10–3 illustrates how the cation concentration is calculated under this circumstance.

---

**Example 10–3**   Calculate the concentration of nickel(II) in a solution that was prepared by mixing 50.0 mL of 0.0300 $F$ $Ni^{2+}$ with 50.0 mL of 0.0500 $F$ EDTA, both of which were buffered to a pH of 3.00.

The solution produced contains an excess of EDTA, and the concentration of the complex is determined by the amount of $Ni^{2+}$ that was originally available. Thus,

$$F_{NiY^{2-}} = \frac{50.0 \times 0.0300}{100.0} = 0.0150$$

$$F_{EDTA} = \frac{50.0 \times 0.0500 - 50.0 \times 0.0300}{100.0} = 0.0100 = C_T$$

The quantities needed for Equation 10–4 are

$$K'_{NiY} = 1.05 \times 10^8 \quad [\text{Example } 10\text{–}2(a)]$$

$$C_T = 0.0100 + [Ni^{2+}] \cong 0.0100$$

$$[NiY^{2-}] = 0.0150 - [Ni^{2+}]$$

It is worthwhile to obtain a provisional value for $[Ni^{2+}]$ with the assumption that $0.0150 - [Ni^{2+}] \cong 0.0150$. Thus, we may write Equation 10–4 as

$$\frac{[NiY^{2-}]}{[Ni^{2+}] \, C_T} = K'_{NiY} = 1.05 \times 10^8$$

or

$$[Ni^{2+}] = \frac{0.0150}{(1.05 \times 10^8)(0.0100)} = 1.4 \times 10^{-8}$$

Note that both assumptions concerning the magnitude of $[Ni^{2+}]$ are justified.

---

Conditional constants are readily computed and provide a simple means by which the equilibrium concentrations of the metal ion and its chelate can be calculated at any point in a titration curve. Note that the expression for the conditional constant differs from that for the formation constant only in that the term $C_T$ replaces the molar concentration of the completely dissociated anion $[Y^{4-}]$. This difference is significant, however, because $C_T$ is readily evaluated from stoichiometric considerations while $[Y^{4-}]$ is not.

## 10B–4 Derivation of a Curve for an EDTA Titration

**Example 10–4**  Derive a curve relating pCa to the volume of EDTA added in the titration of 50.0 mL of 0.00500 $F$ $Ca^{2+}$ with 0.0100 $F$ EDTA in a solution buffered to a pH of 10.00.

(a) *Calculation of the conditional constant.* The conditional constant for the calcium–EDTA chelate is obtained by substituting the formation constant for the complex (Table 10–2) and the value of $\alpha_4$ for pH 10 (Table 10–3) into an expression similar to Equation 10–4; thus,

$$K'_{CaY} = \alpha_4 K_{CaY} = (3.5 \times 10^{-1})(5.0 \times 10^{10}) = \frac{[CaY^{2-}]}{[Ca^{2+}]C_T}$$
$$= 1.75 \times 10^{10}$$

(b) *Calculation of pCa prior to the equivalence point.* Short of the equivalence point, the molar $Ca^{2+}$ concentration will be equal to the sum of the contributions from the as yet untitrated excess of the cation and that from the dissociation of the chelate, the latter being numerically equal to $C_T$. Here, it is ordinarily permissible to make the assumption that $C_T$ is small with respect to the formal concentration of the uncomplexed $Ca^{2+}$. Thus, for example, after 10.0 mL of EDTA have been added,

$$[Ca^{2+}] = \frac{50.0 \times 0.00500 - 10.0 \times 0.0100}{50.0 + 10.0} + C_T \cong 2.50 \times 10^{-3}$$

and

$$pCa = 3 - \log 2.50 = 2.60$$

(c) *Equivalence-point pCa.* A situation identical to Example 10–2 will exist after 25.0 mL of the EDTA solution have been added. The formal concentration of the chelate is given by

$$F_{CaY^{2-}} = \frac{50.0 \times 0.00500}{50.0 + 25.0} = 3.33 \times 10^{-3}$$

The only source of $Ca^{2+}$ will be from dissociation of the chelate. It also follows that the $Ca^{2+}$ concentration must necessarily be identical to the sum of the concentrations of uncomplexed EDTA ions, $C_T$. Thus,

$$[Ca^{2+}] = C_T$$

$$[CaY^{2-}] = 0.00333 - [Ca^{2+}] \cong 0.00333$$

Substitution of these quantities into the conditional formation constant expression (see Equation 10–5) gives

$$\frac{0.00333}{[Ca^{2+}]^2} = 1.75 \times 10^{10}$$

$$[Ca^{2+}] = 4.36 \times 10^{-7}$$

$$pCa = 7 - \log 4.36 = 6.36$$

(d) *Postequivalence values for pCa.* Beyond the equivalence point, the formal concentrations of $CaY^{2-}$ and EDTA are available from the stoichiometry of the reaction; computation of $[Ca^{2+}]$ involves use of a calculation similar to that shown in Example 10–3. Thus, after 35.0 mL of reagent have been added, approximate values for these two concentrations will be

$$[CaY^{2-}] = \frac{50.0 \times 0.00500}{85.0} - [Ca^{2+}] \cong \frac{0.25}{85.0} = 2.94 \times 10^{-3}$$

$$C_T = \frac{35.0 \times 0.0100 - 0.25}{85.0} + [Ca^{2+}] \cong \frac{0.10}{85.0} = 1.18 \times 10^{-3}$$

Substitution into the conditional formation constant expression and rearrangement yields

$$[Ca^{2+}] = \frac{2.94 \times 10^{-3}}{(1.75 \times 10^{10})(1.18 \times 10^{-3})} = 1.42 \times 10^{-10}$$

$$pCa = 10 - \log 1.42 = 9.85$$

Note that neglect of the term for $[Ca^{2+}]$ was justified in arriving at values for $[CaY^{2-}]$ and $C_T$.

---

Figure 10–4 depicts the curve for the titration of calcium ion at pH 10, as well as at several other pH levels. It is apparent that appreciable changes in pCa will occur only if the pH of the solution is maintained at about 8 or greater. Figure 10–5 demonstrates that cations with larger formation constants provide good end points even in acidic solutions. Figure 10–6 shows the minimum permissible pH at which satisfactory titrations of various metal ions can be expected in the absence of auxiliary complexing equilibria. Note that many

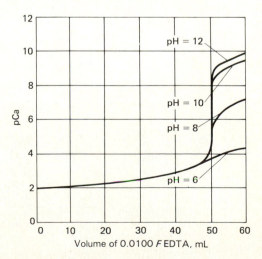

**Figure 10–4** Influence of pH upon the titration of 0.0100 F $Ca^{2+}$ with 0.0100 F EDTA.

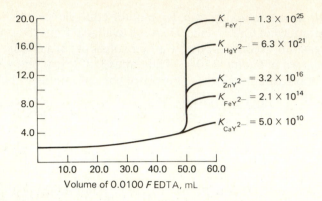

**Figure 10–5** Titration curves for 50.0-mL aliquots of 0.0100 $F$ cation solutions with EDTA at pH 6.0.

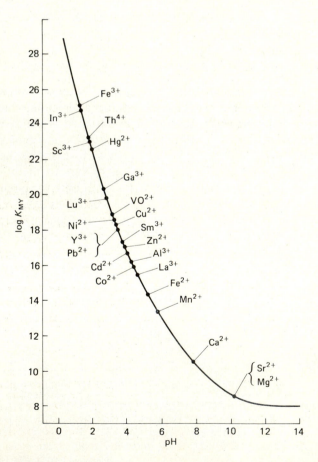

**Figure 10–6** Minimum pH needed for satisfactory titration of various cations with EDTA. (From C. N. Reilley and R. W. Schmidt, *Anal. Chem.*, **1958**, *30*, 947. With permission of the American Chemical Society).

divalent heavy metal cations can be titrated in a moderately acidic solution and that iron(III) provides a useful end point in a strongly acidic environment.

**Effect of Other Complexing Agents on EDTA Titrations**    Many EDTA titrations are complicated by the tendency on the part of the cation being titrated to form a hydrous oxide or hydroxide at the pH needed for a satisfactory end point. Here, an auxiliary complexing agent is needed to keep the cation in solution, particularly in the early stages of the titration. Thus, for example, zinc(II) is ordinarily titrated in a medium that has fairly high concentrations of ammonia and ammonium chloride. These species buffer the solution to a pH that will ensure a complete reaction between cation and titrant; in addition, ammonia forms ammine complexes with zinc(II) and prevents formation of the sparingly soluble zinc hydroxide, particularly in the early stages of the titration. A somewhat more realistic equation for the titration would thus be

$$Zn(NH_3)_4^{2+} + HY^{3-} \rightarrow ZnY^{2-} + 3NH_3 + NH_4^+$$

The solution also contains such other zinc–ammonia species as $Zn(NH_3)_3^{2+}$, $Zn(NH_3)_2^{2+}$, and $Zn(NH_3)^{2+}$. Calculation of pZn in a solution that contains ammonia must take these species into account.[8] Qualitatively, the presence of an auxiliary complexing reagent causes pM values to be larger than in a solution with no such reagent. Figure 10–7 shows two theoretical curves for the titration of zinc(II) with EDTA at pH 9.00. The concentration of ammonia was 0.100 $F$ for the one titration and 0.0100 $F$ for the other. Note that the larger concentration of ammonia has the effect of decreasing the change in pZn in the equivalence-point region. Thus, it is desirable to keep the concentration of any auxiliary complexing reagent at the minimum needed to prevent hydroxide formation. Note also that the postequivalence pZn values are not influenced by the concentration of the auxiliary reagent; to be sure, $\alpha_4$ (and thus pH) are important in defining this region of the titration curve (see Figure 10–4).

### 10B–5 Indicators for EDTA Titrations

Reilley and Barnard[9] have cataloged nearly 200 organic compounds that have been suggested as indicators for metal ions in EDTA titrations. In general, these indicators are organic dyes that form chelates with metal ions that are so intensely colored as to be discernible to the eye in the range of $10^{-6}$ to $10^{-7}$ $M$.

Most of the dyes that serve as metal ion indicators also function as acid-base indicators and display colors that resemble those of their metal chelates. Such dyes are useful only in pH ranges where competition with the proton does not mask reaction with the analyte cation.

---

[8] See, for example, D. A. Skoog and D. M. West, *Fundamentals of Analytical Chemistry,* 4th ed., p. 291. Philadelphia, PA: Saunders College Publishing, 1982.

[9] C. N. Reilley and A. J. Barnard, Jr., in *Handbook of Analytical Chemistry,* L. Meites, Ed., pp. 3-100–3-165. New York: McGraw-Hill Book Co., 1963.

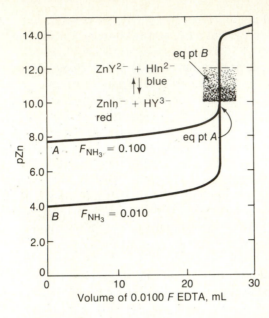

**Figure 10–7** Influence of an auxiliary complexing reagent upon an EDTA titration: titration of 50.0 mL of 0.00500 $F$ $Zn^{2+}$ with 0.100 $F$ EDTA. The shaded area shows the transition range for Eriochrome Black T.

These properties are exhibited by *Eriochrome Black T* (sometimes shortened to Erio T), a widely used metal ion indicator. The predominant acid-base equilibrium for the indicator in acidic and moderately alkaline solution is

$$\underset{\text{red}}{H_2In^-} + H_2O \rightleftharpoons \underset{\text{blue}}{HIn^{2-}} + H_3O^+$$

At very high pH levels, $HIn^{2-}$ dissociates further to $In^{3-}$, which is orange.

The metal complexes of Eriochrome Black T are generally red. A pH of 7 (or above) is needed to cause the unchelated dye to exist predominantly as the blue $HIn^{2-}$. The color change at the end point of an EDTA titration thus involves removal of the cation from the dye by the titrant

$$\underset{\text{red}}{MIn^-} + HY^{3-} \rightleftharpoons \underset{\text{blue}}{HIn^{2-}} + MY^{2-}$$

Eriochrome Black T forms red complexes with more than two dozen cations, but only a few possess stabilities that are appropriate for end-point detection. To be suitable, the conditional constant for the cation–indicator chelate should be less than one-tenth that of the cation–EDTA chelate; otherwise, a premature end point will be observed. On the other hand, if this ratio becomes too small, as with the calcium–Erio T complex, late end points will be observed. The applicability of a particular indicator to an EDTA titration can be determined from the change in pM in the equivalence-point region, provided the formation constant for the cation–indicator chelate is known.[10]

---

[10] For a discussion of the principles governing choice of indicators for complex formation titrations, see: A. Ringbom and E. Wänninen, in *Treatise on Analytical Chemistry,* 2nd ed., I. M. Kolthoff and P. J. Elving, Eds., Part I, Vol. 2, Chapter 20. New York: John Wiley & Sons, 1979.

A limitation of Eriochrome Black T is that its solutions decompose slowly on standing. Refrigeration slows the process. It is claimed that solutions of Calmagite, an indicator with properties similar to Eriochrome Black T, do not suffer from this disadvantage.

## 10B–6 Types of EDTA Titrations

Solutions of EDTA can be used to titrate metal ions in several different ways. The most common of these are considered here.

**Direct Titration**  Reilley and Barnard[11] list 40 elements that can be determined by direct titration with EDTA, using metal ion indicators for end-point detection. Direct titrations are limited to those reactions for which a method of end-point detection exists and to those metal ions that react rapidly with EDTA. Where direct methods fail, a back-titration or a displacement titration may still permit an analysis.

**Back Titration**  Back titrations are useful for the analysis of cations that form very stable EDTA complexes and for which a satisfactory indicator is not available. In such an analysis, the excess EDTA is determined by back-titration with a standard magnesium solution, with Eriochrome Black T serving as indicator. The cation–EDTA chelate must be more stable than the magnesium–EDTA complex to prevent displacement of the analyte cation by magnesium. Back-titrations are also useful where samples contain anions that would otherwise form slightly soluble precipitates with the analyte cation under the conditions of the analysis; the excess EDTA keeps the cation in solution.

**Displacement Titration**  In a displacement titration, the sample is first treated with an unmeasured excess of an Mg–EDTA (or Zn–EDTA) solution. If the analyte cation forms a more stable complex than that of magnesium (or zinc), the following reaction occurs:

$$MgY^{2-} + M^{2+} \rightarrow MY^{2-} + Mg^{2+}$$

The liberated magnesium is then titrated with a standard EDTA solution. Displacement titrations are valuable where a satisfactory indicator for the analyte cation is not available.

**Alkalimetric Titration**  An alkalimetric titration involves adding an excess of $Na_2H_2Y$ to a neutral solution of the analyte cation:

$$M^{2+} + H_2Y^{2-} \rightarrow MY^{2-} + 2H^+$$

The liberated hydrogen ions are then titrated with a standard solution of base.

[11] C. N. Reilley and A. J. Barnard, Jr., in *Handbook of Analytical Chemistry*, L. Meites, Ed., pp. 3-166–3-200. New York: McGraw-Hill Book Co., 1963.

Directions for the preparation and for typical applications of EDTA solutions are found in Chapter 20, Section 20D.

### 10B–7 Scope of EDTA Titrations

EDTA forms chelates with virtually every cation, and methods based upon this property have been developed for the analysis of most cations. At first glance it might appear that the reagent would be highly nonspecific; in fact, however, considerable control can be achieved through regulation of the pH of the reaction medium and through the use of auxiliary complexing reagents. For example, it is generally possible to titrate trivalent cations without interference from divalent species by maintaining a pH of about one (see Figure 10–6); under these circumstances, the less stable divalent chelates do not form to any significant extent, while the trivalent ions are quantitatively complexed. Similarly, the chelates of ions such as cadmium and zinc are sufficiently stable to permit titration in a solution buffered to a pH of 7, with Eriochrome Black T serving as indicator. Magnesium ion, if present, does not interfere because formation of the Mg–EDTA chelate is negligible at this pH. Finally, interference from a particular cation can frequently be eliminated through use of a *masking agent,* an auxiliary ligand that preferentially forms highly stable complexes with the potential interference.[12] For example, cyanide ion is a useful masking agent that will permit the titration of magnesium and calcium ions in the presence of other ions such as cadmium, cobalt, copper, nickel, zinc, and palladium. All of the latter form sufficiently stable cyanides to prevent reaction with EDTA.

### 10B–8 Determination of Hardness in Water

Historically, "hardness" was defined in terms of the capacity of cations in a water sample to replace the sodium or potassium ions in soaps and form sparingly soluble products. This undesirable property is shared by most cations that possess multiple charges; in natural waters, however, the concentrations of calcium and magnesium ions generally far exceed those of any other metal ion. *Hardness* has thus come to mean the concentration of calcium carbonate that is chemically equivalent to the concentration of multivalent cations (principally calcium and magnesium) in the sample.

The determination of hardness is a useful analytical measure of water quality. Hardness is of particular concern to industrial users owing to the fact that heating of hard water causes the precipitation of calcium carbonate, which then clogs boilers and pipes. In addition, the precipitates formed between soaps and the cations responsible for hardness are a nuisance to domestic users.

---

[12] For further information, see: D. D. Perrin, in *Treatise on Analytical Chemistry,* 2nd ed., I. M. Kolthoff and P. J. Elving, Eds., Part I, Vol. 2, Chapter 21. New York: John Wiley & Sons, 1979.

Test kits are available to permit determination of hardness in household water. These consist of a vessel calibrated to contain a known volume of water, a measuring scoop to deliver an appropriate amount of a buffer mixture, an indicator solution, and a dropper bottle containing a standard solution of EDTA. The drops of EDTA needed to cause a color change by the indicator is a measure of hardness in the sample; the concentration of the EDTA solution is ordinarily such that one drop corresponds to one grain (about 0.065 g) of calcium carbonate per gallon of water. A procedure for the determination of the hardness of a water sample is found in Chapter 20, Section 20D–5.

## PROBLEMS

**\*10–1.** Describe the preparation of 1.000 L of 0.0500 $F$ EDTA from $Na_2H_2Y \cdot 2H_2O$; recall that the solid has 0.3% excess moisture.

**10–2.** Express the concentration of a 0.0500 $F$ EDTA solution in terms of its titer as
\*(a) mg CaO/mL.  (b) mg $MgCO_3$/mL.
\*(c) mg $Zn_2P_2O_7$/mL.  (d) mg $Fe_3O_4$/mL.

**10–3.** Calculate the volume of 0.0500 $F$ EDTA needed to titrate
\*(a) 26.37 mL of 0.0741 $F$ $Mg(NO_3)_2$.
(b) the Ca in 0.2145 g of $CaCO_3$.
\*(c) the Ca in a 0.4397-g mineral specimen that is 81.4% brushite, $CaHPO_4 \cdot 2H_2O$ (gfw = 172.1).
(d) the Mg in a 0.2080-g sample of the mineral hydromagnesite, $3MgCO_3 \cdot Mg(OH)_2 \cdot 3H_2O$ (gfw = 365.3).
\*(e) the Ca and Mg in a 0.1557-g sample that is 92.5% dolomite, $CaCO_3 \cdot MgCO_3$ (gfw = 184.4).

**10–4.** An EDTA solution has a titer of 3.18 mg HgO/mL. Calculate
(a) the formal concentration of this solution.
(b) the percentage of HgO if 28.33 mL of this solution are needed to titrate a 0.1149-g sample.
(c) the Zn titer of this solution.
(d) the volume of this solution that would be needed to titrate a 25.00-mL aliquot of 0.01881 $F$ $HgCl_2$.

**\*10–5.** A 1.471-g sample of a mercurial ointment was suspended in $CHCl_3$ and shaken vigorously with dilute $HNO_3$ to extract the Hg(II). Titration of the extract required 20.78 mL of 0.05144 $F$ KSCN, with Fe(III) acting to indicate the first excess of titrant. Reactions:

$$Hg^{2+} + 2SCN^- \rightarrow Hg(SCN)_2$$

$$SCN^- + Fe^{3+} \rightarrow FeSCN^{2+} \quad (red)$$

Calculate the percentage of HgO in the ointment.

**10–6.** Addition of Hg(II) to a solution that contains $CN^-$ results in formation of the soluble, weakly dissociated $Hg(CN)_2$

$$Hg^{2+} + 2CN^- \rightarrow Hg(CN)_2$$

The excess titrant is titrated with a standard KSCN solution, with Fe(III) acting as indicator (see Problem 10–5 for reactions). A 50.00-mL aliquot from an electroplating bath was treated with 28.34 mL of 0.0818 $F$ Hg(NO$_3$)$_2$ solution, following which the excess Hg(II) was back titrated with 1.79 mL of 0.1006 $F$ KSCN. Express the results of this analysis in terms of grams of KCN per liter of solution.

*10–7. The Hg in 5.00 mL of a solution containing the sodium salt of the diuretic meralluride, C$_{16}$H$_{21}$HgN$_6$NaO$_7$ (gfw = 633.0), was titrated with 41.10 mL of 0.0473 $F$ KSCN (see Problem 10–5). Calculate the concentration of this solution in terms of
   (a) mg Hg/mL.
   (b) mg C$_{16}$H$_{21}$HgN$_6$NaO$_7$/mL.

10–8. An ammoniacal solution of Ni(II) can be titrated with a standard solution of NaCN. The end point is signaled by the disappearance of turbidity due to a trace of AgI suspended in the solution. Reactions:

$$Ni(NH_3)_4^{2+} + 4CN^- \rightarrow Ni(CN)_4^{2-} + 4NH_3$$

$$AgI(s) + 2CN^- \rightarrow Ag(CN)_2^- + I^-$$

The Ni in a 0.1484-g mineral specimen of the mineral heazlewoodite, Ni$_3$S$_2$, was brought into solution and pretreated to eliminate interferences. Aqueous NH$_3$ was introduced, along with about 1 mL of 25% KI. The solution was titrated with KCN until the blue-green color of Ni(NH$_3$)$_4^{2+}$ was faint. A few drops of AgNO$_3$ solution were added; the titration with KCN was continued until the opalescence of the AgI just disappeared. Express the results of this analysis in terms of percentage of Ni$_3$S$_2$ (gfw = 240.3) if a total of 45.72 mL of 0.1010 $F$ KCN were used in the titration.

*10–9. A solution contains 1.694 mg of CoSO$_4$ (gfw = 155.0) per milliliter. Calculate
   (a) the volume of 0.008640 $F$ EDTA needed to titrate a 25.00-mL aliquot of this solution.
   (b) the volume of 0.009450 $F$ Zn$^{2+}$ needed to titrate the excess reagent after addition of 50.00 mL of 0.008640 $F$ EDTA to a 25.00-mL aliquot of this solution.
   (c) the volume of 0.008640 $F$ EDTA needed to titrate the Zn$^{2+}$ displaced by Co$^{2+}$ following addition of an unmeasured excess of ZnY$^{2-}$ to a 25.00-mL aliquot of the CoSO$_4$ solution. Reaction:

$$Co^{2+} + ZnY^{2-} \rightarrow CoY^{2-} + Zn^{2+}$$

10–10. The Zn in a 0.7556-g sample of foot powder was titrated with 21.27 mL of 0.01645 $F$ EDTA. Calculate the percent Zn in this sample.

*10–11. The Cr plating on a surface that measured 3.00 × 4.00 cm was dissolved in HCl. The pH was suitably adjusted, following which 15.00 mL of 0.01768 $F$ EDTA were introduced. The excess reagent required a 4.30-mL back-titration with 0.008120 $F$ Cu$^{2+}$. Calculate the average weight of Cr on each square centimeter of surface.

10–12. The Tl in a 9.76-g sample of rodenticide was oxidized to the trivalent state and treated with an unmeasured excess of Mg–EDTA solution. Reaction:

$$Tl^{3+} + MgY^{2-} \rightarrow TlY^- + Mg^{2+}$$

Titration of the liberated Mg$^{2+}$ required 13.34 mL of 0.03560 $F$ EDTA. Calculate the percentage of Tl$_2$SO$_4$ (gfw = 504.8) in the sample.

*10–13. The Fe in 6.875-g of an indoor plant food was oxidized to the +3 state, following which the sample was diluted to exactly 100.0 mL in a volumetric flask. Calculate the percentage of Fe in the sample if 44.87 mL of the diluted solution were needed to titrate a 15.00-mL aliquot of 0.00360 $F$ EDTA.

**10–14.** Bismuth(III) can be determined indirectly by titration of the $Zn^{2+}$ produced in the reaction:

$$3Zn(Hg) + 2Bi^{3+} \rightarrow 3Zn^{2+} + 2Bi(Hg)$$

with a standard EDTA solution. A 1.100-g tablet for the treatment of peptic ulcers was dissolved, treated with zinc amalgam, and diluted to 250.0 mL in a volumetric flask. Calculate the percent of $BiONO_3$ (gfw = 287.0) in the tablet if a 50.00-mL aliquot of the diluted solution required 41.73 mL of 0.004383 $F$ EDTA.

**\*10–15.** A 24-hr urine specimen was diluted to exactly 2.000 L in a volumetric flask. The $Ca^{2+}$ and $Mg^{2+}$ in a 100.0-mL aliquot of the diluted solution required 20.81 mL of 0.00830 $F$ EDTA after being buffered to a pH of 10.0. The $Ca^{2+}$ in a second 100.0-mL aliquot was precipitated as $CaC_2O_4(s)$, which was filtered and washed. The filtrate and washings were titrated with 5.98 mL of the EDTA solution. Calculate the milligrams of $Ca^{2+}$ and $Mg^{2+}$ in the 24-hr sample.

**10–16.** Titration of the $Ca^{2+}$ and $Mg^{2+}$ in a 100.0-mL aliquot of hard water required 44.75 mL of 0.01115 $F$ EDTA. A second 100.0-mL aliquot was made strongly alkaline with NaOH [to remove Mg(II) as $Mg(OH)_2$], following which the supernatant liquid was titrated with 31.38 mL of the EDTA solution. Calculate

(a) the total hardness of the sample, expressed as ppm of $CaCO_3$.

(b) the ppm of $CaCO_3$ in the sample.

(c) the ppm of $MgCO_3$ in the sample.

**\*10–17.** The $SO_4^{2-}$ in a 0.2057-g sample was precipitated as $BaSO_4(s)$ by treatment with an excess of a solution that contained the Ba complex of EDTA, $BaY^{2-}$. The liberated EDTA in the filtrate and washings required 31.79 mL of 0.02644 $F$ $Mg^{2+}$ solution. Express the results of this analysis in terms of percent $Na_2SO_4$.

**10–18.** A 1.509-g sample of a Pb–Cd alloy was dissolved in acid and diluted to exactly 250.0 mL in a volumetric flask. A 50.00-mL aliquot of the diluted solution was brought to a pH of 10.0 with an $NH_4^+–NH_3$ buffer; the subsequent titration involved both cations and required 28.89 mL of 0.06950 $F$ EDTA. A second 50.00-mL aliquot was brought to a pH of 10.0 with an HCN–NaCN buffer, which also served to mask the $Cd^{2+}$; 11.56 mL of the EDTA solution were needed to titrate the $Pb^{2+}$. Calculate the percent Pb and Cd in the sample.

**10–19.** A 0.6004-g sample of Ni–Cu condenser tubing was dissolved in acid and diluted to 100.0 mL in a volumetric flask. Titration of both cations in a 25.00-mL aliquot of this solution required 45.81 mL of 0.05285 $F$ EDTA. Mercaptoacetic acid and $NH_3$ were then introduced; production of the Cu complex with the former resulted in the release of an equivalent amount of EDTA, which required a 22.85-mL titration with 0.07238 $F$ $Mg^{2+}$. Calculate the percent Cu and Ni in the alloy.

**10–20.** Calculate the conditional constant, $K'_{MY}$, for the formation of the EDTA complex of

\*(a) $Fe^{3+}$ at pH 4.00.

\*(b) $Cu^{2+}$ at pH 6.00.

(c) $Zn^{2+}$ at pH 10.00.

(d) $Cd^{2+}$ at pH 9.00.

(e) $Co^{2+}$ at pH 7.00.

**10–21.** Taking $10^{10}$ as the minimum value of $K'_{MY}$ for a successful EDTA titration, calculate the minimum integer pH value permissible for the titration of

*(a) $Zn^{2+}$.                  (b) $Mn^{2+}$.

*(c) $Hg^{2+}$.                  (d) $Ni^{2+}$.

**\*10–22.** Generate a curve for the titration of 50.00 mL of 0.0100 $F$ $Cu^{2+}$ with 0.0200 $F$ EDTA at a pH of 6.00. Calculate pCu after the addition of 0.00, 10.00, 24.00, 24.90, 25.00, 25.10, 26.00, and 30.00 mL of titrant.

**10–23.** Generate a curve for the titration of 0.0100 $F$ $Co^{2+}$ with 0.0100 $F$ EDTA in a solution that is buffered to a pH of 7.00. Calculate pCo after the addition of 0.00, 10.00, 24.00, 24.90, 25.00, 25.10, 26.00, and 30.00 mL of titrant to 25.00 mL of the analyte solution.

# Chapter 11
# Theory of Oxidation-Reduction Titrations

Chemical reactions in which electrons are transferred from one reactant to another are known as *oxidation-reduction,* or *redox,* reactions. Volumetric methods involving oxidation-reduction are more numerous and more diverse than those based on any other reaction type.

The first four sections of this chapter are concerned with the principles of electrochemistry that are needed to understand how oxidation-reduction titration curves are derived and how indicators for these titrations are selected.[1] This introductory material also provides a theoretical basis for the subject matter in Chapters 12 through 14.

## 11A OXIDATION-REDUCTION PROCESSES

*Oxidation* involves the loss of electrons by a substance, while *reduction* refers to the gain of electrons. A balanced oxidation-reduction equation requires a molar ratio between reactants such that the number of electrons lost by one species is equal to the number acquired by the other.

### 11A–1 Oxidizing and Reducing Agents

*Oxidizing agents,* or *oxidants,* are capable of abstracting electrons from other species and thus are responsible for the oxidation of those species. *Reducing agents,* or *reductants,* readily give up electrons and cause the reduction of other species. Note that an oxidizing agent, having been reduced as a result of acquiring one or more electrons, becomes a potential electron donor, or reducing agent. Similarly, a reducing agent, having lost an electron (or electrons), is now a potential electron acceptor. In its simplest form, then, an oxidation-reduction reaction can be written as

$$Red_1 + Ox_2 \rightleftharpoons Ox_1 + Red_2$$

---

[1] For further reading on oxidation-reduction equilibria and titrations, see: J. A. Goldman, in *Treatise on Analytical Chemistry,* 2nd ed., I. M. Kolthoff and P. J. Elving, Eds., Part I, Vol. 3, Chapter 24. New York: John Wiley & Sons, 1983.

If equilibrium in this process favors the products, we can state that $Ox_2$ is a more effective electron acceptor (or oxidizing agent) than is $Ox_1$, the species that results from the loss of electrons from $Red_1$. By the same reasoning, $Red_1$ is a more effective electron donor (and hence, reducing agent) than is $Red_2$. The mixing of a strong oxidizing agent with a strong reducing agent will result in an equilibrium in which the products are overwhelmingly favored; reactants that are less strong yield correspondingly less favorable equilibria. Note the analogy between these observations, involving the transfer of electrons, and the Brønsted-Lowry concepts of proton transfer (Chapter 2, Section 2A–2).

## 11A–2 Half-Reactions

It is often helpful to separate an oxidation-reduction equation into two *half-reactions,* one of which describes the oxidation process and the other the reduction process. Consider, for example, the equation for the oxidation of iron(II) ions by permanganate ions in an acidic solution:

$$5Fe^{2+} + MnO_4^- + 8H^+ \rightleftharpoons 5Fe^{3+} + Mn^{2+} + 4H_2O$$

The two half-reactions are

$$MnO_4^- + 8H^+ + 5e \rightleftharpoons Mn^{2+} + 4H_2O$$

$$5Fe^{2+} \rightleftharpoons 5Fe^{3+} + 5e$$

These equations show clearly that the permanganate ion is the electron acceptor, while the iron(II) ion is the electron donor. Note that before combining the two half-reactions, it is necessary to multiply the second by five in order to eliminate electrons from the final equation.[2]

## 11A–3 Oxidation-Reduction Reactions in Electrochemical Cells

Oxidation-reduction reactions may be the result of a direct transfer of electrons from the donor to the acceptor. Thus, if metallic zinc is immersed in a solution containing copper sulfate, copper(II) ions migrate to, and are reduced at, the surface of the zinc

$$Cu^{2+} + 2e \rightleftharpoons Cu(s)$$

and a chemically equivalent quantity of zinc is oxidized

$$Zn(s) \rightleftharpoons Zn^{2+} + 2e$$

The equation for the overall process is obtained by adding the equations for the two half-reactions, which gives

$$\underset{Red_1}{Zn(s)} + \underset{Ox_2}{Cu^{2+}} \rightleftharpoons \underset{Ox_1}{Zn^{2+}} + \underset{Red_2}{Cu(s)}$$

---

[2] The experimental observation that this reaction proceeds to essential completion makes it possible to state that iron(II) is a more effective electron donor than is manganese(II), and that permanganate ion is a more effective electron acceptor than is iron(III).

An interesting aspect of many oxidation-reduction reactions is that the two constituent half-reactions can be made to take place in regions that are physically isolated from one another. For example, Figure 11 – 1 illustrates how the half-reactions we have just considered can occur in separate compartments of a container or cell. The compartment on the left contains a piece of zinc immersed in a solution of zinc sulfate; the one on the right consists of a strip of copper in a copper sulfate solution. An external conductor provides a means by which electrons are transferred from the zinc to the copper. This flow of electrons cannot occur to any significant extent, however, unless means are provided by which the charge imbalance created by the movement of electrons can be offset; that is, movement of electrons from the zinc to the copper would create an excess of positive ions on the surface of the zinc that would, by electrostatic attraction, prevent further migration of electrons. Similarly, the solution immediately adjacent to the copper would become negatively charged owing to the removal of positive copper ions. Such charge imbalances do not develop, however, because the porous disk (which prevents extensive mixing of the two solutions) provides a pathway by which zinc ions can move away from the zinc surface toward the compartment containing copper ions, and the excess sulfate ions can move from the copper surface toward the zinc surface.

The equilibrium that is established in this cell is identical in every respect to that described earlier. Here, however, electrons are transferred from one species to the other as an electric current. Electron transfer will continue until the concentrations of copper(II) and zinc(II) achieve levels corresponding to equilibrium for the reaction

$$Zn(s) + Cu^{2+} \rightleftharpoons Zn^{2+} + Cu(s)$$

When this condition is reached, no further net flow of electrons occurs, and the

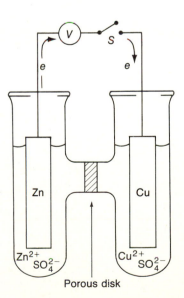

**Figure 11 – 1** A galvanic cell with liquid junction. Arrows show direction of electron flow when switch 5 is closed.

current drops to zero. It is essential to recognize that the overall process and the concentrations that exist at equilibrium are totally independent of the route that led to the condition of equilibrium, be it by direct contact between the reactants or by indirect reaction as in Figure 11 – 1. It is equally important to recognize that *the equation for the reaction and the numerical constant associated with it are also independent of the route by which the condition of equilibrium was reached.*

## 11B ELECTROCHEMICAL CELLS

The device illustrated in Figure 11 – 1 is a typical *electrochemical cell;* it consists of a pair of electrodes, ordinarily (but not necessarily) metallic, each of which is immersed in an electrolyte. The current that develops in this cell when the switch *S* is closed is due to the strong tendency for zinc to be oxidized at the surface of one electrode and for copper(II) to be reduced at the surface of the other. The magnitude of the potential that develops between the two electrodes provides a measure of the tendency of the two half-reactions to proceed toward equilibrium. As will be shown subsequently, this potential — which is readily measured by the voltage-measuring device, *V*, shown in Figure 11 – 1 — is directly related to the equilibrium constant for the particular oxidation-reduction process involved as well as to the extent to which the existing concentrations of the participants differ from their equilibrium values. Such potentials, in fact, are important experimental sources for numerical equilibrium constant data.

### 11B – 1 Galvanic and Electrolytic Cells

A *galvanic,* or *voltaic,* cell provides electrical energy. In contrast, operation of an *electrolytic cell* requires an external source of electricity. The cell shown in Figure 11 – 1 provides a spontaneous flow of electrons from the zinc electrode to the copper electrode through the external circuit and is thus a galvanic cell.

The same cell could be operated as an electrolytic cell. Here, it would be necessary to introduce a battery or some other source of electrical energy in the external circuit to force electrons to flow in the opposite direction through the external conductor. Under these circumstances, zinc would deposit and copper would dissolve. These nonspontaneous processes would consume energy from the external source.

**Reversible and Irreversible Cells**    The cell shown in Figure 11 – 1 is an example of an electrochemically *reversible* cell; that is, reversing the current simply results in a reversal of the chemical processes that take place at the two electrodes. Alteration in the direction of the current in an irreversible cell causes entirely different reactions at one or both electrodes. For example, the cell shown in Figure 11 – 1 becomes irreversible if the solution in the zinc compartment is made slightly acidic. Under this circumstance, zinc will not deposit; instead, hydrogen preferentially forms at the electrode

$$2H^+ + 2e \rightleftharpoons H_2(g)$$

and the overall process now becomes

$$Cu(s) + 2H^+ \rightarrow Cu^{2+} + H_2(g)$$

## 11B-2 Conduction in Electrochemical Cells

Three separate mechanisms are involved in the transport of electricity through an electrochemical cell. Electrons serve as carriers in the electrodes as well as in the external conductor. Anionic and cationic migration is responsible for transport within solutions. In Figure 11-1, copper ions, hydronium ions, and other positively charged species tend to move toward the copper electrode; likewise, negatively charged ions (here, $SO_4^{2-}$, $HSO_4^-$, and $OH^-$) are attracted toward the zinc electrode by the excess of positive ions generated in the electrochemical process. Finally, an oxidation at one electrode surface and a reduction at the other provide the mechanism whereby the ionic conduction of the solution is coupled with the electron conduction within the electrodes.

**Liquid Junctions**   The porous disk in Figure 11-1 prevents direct reaction between the components of the two half-cells. If mixing were allowed, copper would deposit directly on the zinc electrode; the result would be a decrease in cell efficiency. The two electrolyte solutions do contact one another within the disk and diffusion of solute species occurs across this interface. Because some ions diffuse more rapidly than others, a small *liquid junction potential* develops. The effects of liquid junction potentials can be minimized by interposing a *salt bridge* between the two cell compartments. This device consists of a U-shaped tube that contains a concentrated solution of a salt such as potassium chloride. The introduction of a salt bridge has the effect of creating two liquid junction potentials, one at either end. The effects of these two potentials are largely self-canceling.

## 11B-3 Cell Components

*The anode of an electrochemical cell is the electrode at which oxidation occurs, while the cathode is the electrode at which reduction takes place.*

Typical cathodic processes are illustrated by the following equations:

$$Ag^+ + e \rightleftharpoons Ag(s)$$

$$2H^+ + 2e \rightleftharpoons H_2(g)$$

$$Fe^{3+} + e \rightleftharpoons Fe^{2+}$$

$$IO_4^- + 2H^+ + 2e \rightleftharpoons IO_3^- + H_2O$$

The last three of these reactions take place at the surface of an inert cathode, such as platinum. Note that anions as well as cations can undergo reduction at the surface of a platinum electrode. Hydrogen is frequently the cathode product in solutions that contain no other more easily reduced species.

Typical anodic reactions include

$$Zn(s) \rightleftharpoons Zn^{2+} + 2e$$

$$2Cl^- \rightleftharpoons Cl_2(g) + 2e$$

$$Fe^{2+} \rightleftharpoons Fe^{3+} + e$$

$$2H_2O \rightleftharpoons O_2(g) + 4H^+ + 4e$$

The first reaction requires a zinc electrode, while the others occur at an inert electrode. The last reaction commonly occurs in solutions that do not contain more easily oxidized species.

## 11C ELECTRODE POTENTIALS

A precise definition of an *electrode potential* is to be found in Section 11C–2; for the present, it is sufficient to note that the term will always refer to the potential of a half-cell process that has been *written as a reduction.*

It is frequently convenient to consider a cell potential as being the difference between two electrode potentials, one being associated with the cathode and the other with the anode. Thus, for the cell shown in Figure 11–1, we may write

$$E_{cell} = E_{Cu} - E_{Zn}$$

where $E_{Cu}$ is the electrode potential for the copper electrode (which is acting as cathode), and $E_{Zn}$ is the electrode potential for the zinc electrode (here, the anode). A more general statement for the potential of a cell is

$$E_{cell} = E_{cathode} - E_{anode}$$

It must be emphasized at the outset that there is no way by which the absolute value for the potential of a single electrode can be determined, since all voltage-measuring devices measure *differences* in potential. One conductor from such a device is connected to the electrode in question. A second conductor must make contact—directly or indirectly—with the electrolyte solution of the half-cell. The latter contact inevitably involves a solid–solution interface that acts as a second half-cell at which chemical change also must occur if electricity is to flow. A potential will be associated with this second reaction. Thus, an absolute value for the desired half-cell potential is not realized; instead, the measured potential is a combination of the potential of interest and the one that develops between the second contact and the solution.

Our inability to measure absolute potentials for half-cell processes is not a serious drawback because relative half-cell potentials, measured against some reproducible reference half-cell, are just as useful. These relative potentials can be combined to give cell potentials and, in addition, are useful for calculating equilibrium constants for oxidation-reduction processes.

### 11C-1 Reference Electrodes

The standard hydrogen electrode (SHE) is the universal reference for reporting relative half-cell potentials. It is a carefully defined version of a hydrogen electrode.

**The Hydrogen Electrode**   The hydrogen electrode, which was used extensively in early electrochemical studies, consists of a platinum electrode immersed in an electrolyte solution that is kept saturated with hydrogen gas. As shown in Figure 11-2, a saturated condition is maintained by bubbling hydrogen continuously through the solution that is immediately adjacent to the electrode surface. In order to ensure that the half reaction

$$2H^+ + 2e \rightleftharpoons H_2(g)$$

proceeds rapidly and reversibly, it is necessary that the platinum be *platinized* — that is, coated with a layer of finely divided platinum to maximize its surface area.

The hydrogen electrode may serve as an anode or a cathode, depending upon the half-cell with which it is coupled. Hydrogen is oxidized to hydrogen ions when the electrode acts as an anode; the reverse occurs when it acts as a cathode. The purpose of the gas stream shown in Figure 11-2 is to ensure that the solution is continuously saturated with molecular hydrogen. Thus, when a

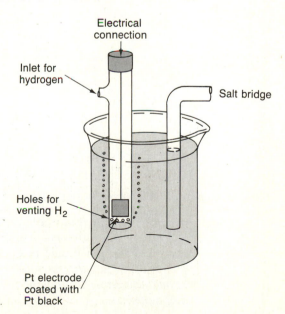

**Figure 11-2**   A hydrogen electrode.

Electrical connection

Inlet for hydrogen

Salt bridge

Holes for venting $H_2$

Pt electrode coated with Pt black

hydrogen electrode operates as an anode, the half-cell process is the sum of two reactions:

$$H_2(g) \rightleftharpoons H_2(aq \text{ sat'd})$$

$$H_2(aq \text{ sat'd}) \rightleftharpoons 2H^+(aq) + 2e$$

The overall reaction is

$$H_2(g) \rightleftharpoons 2H^+(aq) + 2e$$

it being understood that sufficient $H_2(g)$ is available to maintain saturation.

**The Standard Hydrogen Electrode (SHE)**   The potential of a hydrogen electrode depends upon the temperature, the hydrogen ion concentration (or, more correctly, its activity), and the pressure of the hydrogen at the surface of the electrode. Values for these parameters must be carefully defined in order for the half-cell process to serve as a reference. Specifications for the *standard hydrogen electrode* call for a hydrogen ion activity of unity, and a partial pressure of exactly 1.00 atm for hydrogen. *By convention, the potential of this electrode is assigned a value of exactly zero volt at all temperatures.*

**Other Reference Electrodes**   The flowing stream of gas needed for the operation of a hydrogen electrode is inconvenient as well as hazardous. Ordinarily, then, secondary reference electrodes are used in place of the hydrogen electrode. Common secondary references include calomel as well as silver–silver chloride electrodes. Calomel electrodes are based upon the half-reaction

$$Hg_2Cl_2(s) + 2e \rightleftharpoons Hg(1) + 2Cl^-$$

while the silver–silver chloride electrode involves the process

$$AgCl(s) + e \rightleftharpoons Ag(s) + Cl^-$$

The saturated calomel electrode (SCE) is particularly useful for electrochemical work; its potential ($E_{SCE}$) is $+0.244$ V relative to the standard hydrogen electrode. The numerical value for an electrode potential will thus be more negative (by 0.244 V) with respect to the saturated calomel electrode than its value with respect to the standard hydrogen electrode.

Practical aspects of calomel and silver–silver chloride electrodes are discussed in Chapter 13.

### 11C-2 Definition of Electrode Potentials

An *electrode potential* is defined as the potential for a cell consisting of the electrode in question *acting as cathode* and the standard hydrogen electrode *acting as anode*. It should be emphasized that, despite its name, *an electrode potential is in fact the potential of an electrochemical cell involving a carefully defined reference electrode.* It could be more properly called a "relative electrode potential."

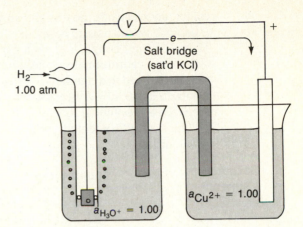

**Figure 11-3** Hypothetical cell illustrating the definition of electrode potential for the half-reaction

$$Cu^{2+} + 2e \rightleftharpoons Cu(s)$$

The cell in Figure 11-3 illustrates the definition of the electrode potential for the half-reaction

$$Cu^{2+} + 2e \rightleftharpoons Cu(s)$$

Here, the half-cell on the right consists of a piece of pure copper in contact with an $xM$ solution of copper(II); the electrode on the left is the standard hydrogen electrode.

If the copper ion activity in this cell is unity, a potential of 0.334 V develops, with the copper electrode functioning *as the cathode;* that is, the spontaneous cell reaction is

$$Cu^{2+} + H_2(g) \rightleftharpoons Cu(s) + 2H^+$$

Because the copper electrode serves as the cathode in this cell, the measured potential is, *by definition,* the electrode potential for the copper half-reaction (or the copper couple). Note that the copper electrode bears a positive charge with respect to the hydrogen electrode. The electrode is therefore assigned a positive sign, and we may write

$$Cu^{2+} + 2e \rightleftharpoons Cu(s) \qquad E_{Cu^{2+}} = +0.334 \text{ V}$$

Replacement of the $Cu-Cu^{2+}$ half-cell with a zinc electrode immersed in a solution with a zinc ion activity of unity results in a potential of 0.763 V. In contrast to the previous cell, however, the zinc electrode acts as the anode, and the spontaneous cell reaction is

$$Zn(s) + 2H^+ \rightleftharpoons H_2(g) + Zn^{2+}$$

that is, the zinc electrode bears a negative charge with respect to the standard hydrogen electrode.

Because the zinc electrode acts as an anode in this galvanic cell, 0.763 V is an *oxidation potential* rather than a reduction potential. In order to cause the zinc electrode to behave as a cathode (as it must if the measured potential is to be called an electrode potential), an external potential more negative than $-0.763$ V would have to be applied to the cell. Therefore the *electrode potential* of the Zn-$Zn^{2+}$ couple is given a negative sign and is equal to $-0.763$ V.

When a cadmium electrode, immersed in a $1.00\ M$ solution of cadmium ions, is coupled to a standard hydrogen electrode, a galvanic cell is formed with a potential of $0.403$ V. Because the cadmium electrode is the anode in this cell, the electrode potential for the Cd-$Cd^{2+}$ system is $-0.403$ V.

The electrode potentials for the four half-cells just described can be arranged in the order

| Half-Reaction | Electrode Potential |
|---|---|
| $Cu^{2+} + 2e \rightleftharpoons Cu(s)$ | 0.334 V |
| $2H^+ + 2e \rightleftharpoons H_2(g)$ | 0.000 V |
| $Cd^{2+} + 2e \rightleftharpoons Cd(s)$ | $-0.403$ V |
| $Zn^{2+} + 2e \rightleftharpoons Zn(s)$ | $-0.763$ V |

The magnitudes of these electrode potentials indicate the relative strength of the four species on the left as electron acceptors (or oxidizing agents); that is, in decreasing strength, $Cu^{2+} > H^+ > Cd^{2+} > Zn^{2+}$.

**Sign Conventions for Electrode Potentials** Historically, electrochemists did not always make use of the sign convention just described. Indeed, disagreements regarding the conventions to be used in specifying signs for half-cell processes caused much controversy and confusion in the development of electrochemistry. The International Union of Pure and Applied Chemistry (IUPAC) addressed itself to this problem at its 1953 meeting in Stockholm. The usages adopted at this meeting are collectively referred to as the Stockholm convention and are by now generally accepted. The sign convention described in the previous section is based upon the IUPAC recommendations.

Any sign convention must be based upon the expression of half-cell processes in a single way — that is, either as oxidations or as reductions. According to the IUPAC convention, the term *electrode potential* (or, more exactly, the *relative electrode potential*) *is reserved exclusively to describe half-reactions written as reductions.* There is no objection to the use of the term *oxidation potential* to connote a process written in the opposite sense, but it is not proper to refer to such a potential as an electrode potential.

The sign of an electrode potential is determined from the charge of the electrode system in the galvanic cell formed between it and a standard hydrogen electrode. Thus, as we have shown, a zinc electrode or a cadmium electrode will act as the anode in such a cell, electrons flowing through the external circuit to the standard hydrogen electrode. Because the two metal electrodes are more negative than the platinum conductor of the hydrogen electrode, their electrode potentials are given negative signs. By the same reasoning, the electrode potential for the copper half-cell is given a positive sign because a copper electrode acts as the cathode in the galvanic cell formed with the standard hydrogen electrode. Here, electrons flow from the hydrogen electrode through the external circuit toward the copper electrode.

It is important to emphasize that the electrode potential refers to the half-cell process written as a reduction. For the zinc and cadmium electrodes, which we have been considering, the spontaneous reactions are oxidations. *It is evident, then, that the sign of an electrode potential will indicate whether the reduction is spontaneous with respect to the standard hydrogen electrode.* The positive sign associated with the electrode potential for copper indicates that the process

$$Cu^{2+} + H_2(g) \rightleftharpoons Cu(s) + 2H^+$$

favors the products under ordinary conditions. Similarly, the negative sign of the electrode potential for zinc means that the analogous reaction

$$Zn^{2+} + H_2(g) \rightleftharpoons Zn(s) + 2H^+$$

is nonspontaneous; indeed, equilibrium favors the reactants over the products.

Reference works, particularly those published before 1953, may contain tabulations of electrode potential data that are not in accord with the IUPAC recommendations. For example, in a classic source of oxidation-potential data compiled by Latimer,[3] one finds

$$Zn(s) \rightleftharpoons Zn^{2+} + 2e \qquad E = +0.76 \text{ V}$$

$$Cu(s) \rightleftharpoons Cu^{2+} + 2e \qquad E = -0.34 \text{ V}$$

In converting these oxidation potentials to electrode potentials as defined by the IUPAC convention, one must mentally (1) express the half-reactions as reductions, and (2) change the signs of the potentials.

The sign convention used in a tabulation of electrode potential data may not be explicitly stated. This information can be readily determined, however, by noting the direction and the sign of the potential for a half-reaction with which one is familiar; whatever changes, if any, that may be needed to convert to the IUPAC convention are then applied to the remainder of the data in the table. For example, under the IUPAC convention, electrode potentials for strong oxidizing agents are numerically large and carry positive signs. Thus, the reaction

$$O_2(g) + 4H^+ + 4e \rightleftharpoons 2H_2O \qquad E = +1.229 \text{ V}$$

occurs spontaneously with respect to the standard hydrogen electrode. The sign and direction of this reaction can thus serve as a key to any changes that may be needed to convert all data to the IUPAC convention.

---

[3] W. M. Latimer, *The Oxidation States of the Elements and Their Potentials in Aqueous Solutions*, 2nd ed. Englewood Cliffs, N.J.: Prentice-Hall, 1952.

### 11C – 3  Effect of Concentration on Electrode Potentials: The Nernst Equation

We have noted that the magnitude of an electrode potential is a measure of the force that is driving a half-reaction to the condition of equilibrium (with respect to the standard hydrogen electrode). It follows, then, that numerical values for electrode potentials are concentration dependent. Thus, a concentrated solution of copper(II) tends to be more readily reduced to the elemental state than is a more dilute solution; the electrode potential must be correspondingly greater for the more concentrated solution.

The relationship between the electrode potential and concentration was first enunciated by Walther Nernst, a German chemist; it is fitting that his name is associated with this important relationship.

Consider the generalized reversible half-reaction

$$a\text{A} + b\text{B} + \cdots + ne \rightleftharpoons c\text{C} + d\text{D} + \cdots$$

where the capital letters represent formulas for species (whether charged or uncharged) participating in the electron transfer process, $e$ is the electron, and $a$, $b$, $c$, $d$, and $n$ are the number of moles of participants involved in the half-cell process as it has been written. It can be shown theoretically as well as experimentally that the potential, $E$, for this process is described by the relation

$$E = E^0 - \frac{RT}{nF} \ln \frac{[\text{C}]^c[\text{D}]^d \cdots}{[\text{A}]^a[\text{B}]^b \cdots} \qquad (11-1)$$

where

$E^0$ = a constant, called the *standard electrode potential,* which is characteristic for each half-reaction

$R$ = the gas constant, $8.316 \text{ J } °\text{K}^{-1} \text{ mol}^{-1}$

$T$ = temperature, $°\text{K}$

$n$ = number of moles of electrons that appear in the half-reaction for the electrode process, as it has been written

$F$ = the faraday = 96487 coulombs

$\ln$ = the base for natural logarithms = $2.303 \log_{10}$

Substitution of numerical values for the several constants, conversion to base 10 logarithms, and specification of 25°C for the temperature gives

$$E = E^0 - \frac{0.0591}{n} \log \frac{[\text{C}]^c[\text{D}]^d \cdots}{[\text{A}]^a[\text{B}]^b \cdots} \qquad (11-2)$$

The letters in brackets strictly represent the activities of the various species. Substitution of concentrations for activities in many calculations does not introduce an excessive error. Thus, if some participating species X is a solute,

$$[\text{X}] = \text{concentration, in moles per liter}$$

If X is a gas,

$$[X] = \text{partial pressure, in atmospheres}$$

If X exists in a second phase *as a pure liquid or solid,* then, by definition,

$$[X] = 1.00$$

The basis for the last statement is the same as that described in Chapter 5, Section 5A, specifically, that a pure liquid or solid which exists as a second phase has a constant effect upon a chemical equilibrium and is independent of amount. Similarly, although it may appear as a participant in a half-cell process, water in its capacity as a solvent has a concentration that is ordinarily enormous with respect to the other reactants; for all practical purposes, its contribution is also constant. Thus, if water appears in Equation 11–2,

$$[H_2O] = 1.00$$

Example 11–1 illustrates how the Nernst equation is applied to typical half-cell processes.

**Example 11–1**  Generate the Nernst expression for the following half-reactions.

(a) $Zn^{2+} + 2e \rightleftharpoons Zn(s)$

$$E = E^0 - \frac{0.0591}{2} \log \frac{1}{[Zn^{2+}]}$$

The activity of elemental Zn is unity, by definition, and the electrode potential varies with the logarithm of the reciprocal of the molar $Zn^{2+}$ concentration.

(b) $Fe^{3+} + e \rightleftharpoons Fe^{2+}$

$$E = E^0 - \frac{0.0591}{1} \log \frac{[Fe^{2+}]}{[Fe^{3+}]}$$

The potential for this half-reaction can be measured with an inert electrode immersed in a solution containing Fe(II) and Fe(III) ions. The potential that develops is dependent upon the logarithm of the ratio between the molar concentrations of these ions.

(c) $2H^+ + 2e \rightleftharpoons H_2(g)$

$$E = E^0 - \frac{0.0591}{2} \log \frac{p_{H_2}}{[H^+]^2}$$

The term $p_{H_2}$ represents the partial pressure of $H_2$, expressed in atmospheres, at the surface of the electrode. Ordinarily, $p_{H_2}$ will be very close to atmospheric pressure.

(d) $Cr_2O_7^{2-} + 6e + 14H^+ \rightleftharpoons 2Cr^{3+} + 7H_2O$

$$E = E^0 - \frac{0.0591}{6} \log \frac{[Cr^{3+}]^2}{[Cr_2O_7^{2-}][H^+]^{14}}$$

Note that the potential of this half-cell depends not only upon the concentrations of the two chromium-containing species but also upon the pH of the solution.

(e) $AgCl(s) + e \rightleftharpoons Ag(s) + Cl^-$

$$E = E^0 - \frac{0.0591}{1} \log [Cl^-]$$

This half-reaction describes the behavior of an Ag electrode immersed in a chloride solution that has been saturated with AgCl; it is the sum of two reactions, namely,

$AgCl(s) \rightleftharpoons Ag^+ + Cl^-$

$Ag^+ + e \rightleftharpoons Ag(s)$

The activities of both metallic Ag and AgCl are equal to unity, as long as both are present; therefore, the Nernst equation will contain only a logarithmic term for the $Cl^-$ concentration.

## 11C – 4 The Standard Electrode Potential, $E^0$

Examination of Equation 11–2 reveals that the constant $E^0$ is equal to the half-cell potential when the logarithmic term has a value of zero. This condition prevails whenever the activity quotient becomes unity; one such circumstance is when the activities of all reactants and products are equal to one. Thus, *the standard electrode potential may be defined as the electrode potential of a half-cell reaction (vs. the standard hydrogen electrode) when all reactants and products exist at unit activity.*

The standard electrode potential is an important physical constant that provides a quantitative description of the relative driving force for a half-cell reaction.[4] A number of facts concerning this constant, and half-cell potentials calculated from it, should be kept in mind.

1. The standard electrode potential is a relative quantity in the sense that it is actually the potential of an electrochemical cell in which the anode is the

---

[4] For further reading on standard electrode potentials, see: R. G. Bates, in *Treatise on Analytical Chemistry,* 2nd ed., I. M. Kolthoff and P. J. Elving, Eds., Part I, Vol. 1, Chapter 13. New York: John Wiley & Sons, 1978.

standard hydrogen electrode, whose potential has been arbitrarily assigned a value of zero volt.

2. Standard electrode potentials refer exclusively to half-cell processes that are written as reductions; that is, they are relative reduction potentials.

3. The standard electrode potential measures the relative intensity of the driving force for a half-reaction; as such, its numerical value is independent of the notation used to express the half-reaction. Thus, the potential for the process

$$Ag^+ + e \rightleftharpoons Ag(s) \qquad E^0 = +0.799 \text{ V}$$

does not change if for some reason we find it convenient to express this half-reaction as

$$5Ag^+ + 5e \rightleftharpoons 5Ag(s) \qquad E^0 = +0.799 \text{ V}$$

To be sure, the corresponding Nernst expression must be consistent with the half-reaction *as it has been written.* For the first of these, it will be

$$E = 0.799 - \frac{0.0591}{1} \log \frac{1}{[Ag^+]}$$

and for the second

$$E = 0.799 - \frac{0.0591}{5} \log \frac{1}{[Ag^+]^5}$$

4. The sign of a standard electrode potential is based upon the driving force for the reduction reaction relative to the reduction of hydrogen ion at unit activity. A positive sign indicates that the electrode (with all reactants and products at unit activity) acts as the cathode when coupled with the standard hydrogen electrode. Conversely, a negative sign indicates that the electrode will behave as the anode in a galvanic cell with respect to the standard hydrogen electrode.

5. The magnitude of an electrode potential is temperature dependent; care must be taken to specify the temperature to which measurements refer.

Compilations of standard electrode potentials provide the chemist with qualitative information regarding the extent and direction of electron transfer reactions between the tabulated species. Table 11–1 contains data needed for the examples that follow; a more extensive listing is given in Appendix 12.[5]

---

[5] Current sources for standard electrode potential data include G. Milazzo and S. Caroli, *Tables of Standard Electrode Potentials.* New York: Wiley-Interscience, 1977; M. S. Antelman and F. J. Harris, *Chemical Electrode Potentials.* New York: Plenum Press, 1982. Some compilations are arranged alphabetically by element; others are tabulated according to the numerical value of $E^0$. Slight variations are encountered among published values of standard electrode potentials.

**Table 11 – 1    Standard Electrode Potentials***

| Half-Reaction | at 25°C, V $E^0$ |
|---|---|
| $Cl_2(g) + 2e \rightleftharpoons 2Cl^-$ | +1.359 |
| $O_2(g) + 4H^+ + 4e \rightleftharpoons 2H_2O$ | +1.229 |
| $Br_2(aq) + 2e \rightleftharpoons 2Br^-$ | +1.087 |
| $Br_2(l) + 2e \rightleftharpoons 2Br^-$ | +1.065 |
| $Ag^+ + e \rightleftharpoons Ag(s)$ | +0.799 |
| $Fe^{3+} + e \rightleftharpoons Fe^{2+}$ | +0.771 |
| $I_3^- + 2e \rightleftharpoons 3I^-$ | +0.536 |
| $Cu^{2+} + 2e \rightleftharpoons Cu(s)$ | +0.337 |
| $Ag(S_2O_3)_2^{3-} + e \rightleftharpoons Ag(s) + 2S_2O_3^{2-}$ | +0.010 |
| $2H^+ + 2e \rightleftharpoons H_2(g)$ | 0.000 |
| $AgI(s) + e \rightleftharpoons Ag(s) + I^-$ | -0.151 |
| $Cd^{2+} + 2e \rightleftharpoons Cd(s)$ | -0.403 |
| $Zn^{2+} + 2e \rightleftharpoons Zn(s)$ | -0.763 |

\* See Appendix 12 for a more extensive listing.

Entries in both of these tabulations are arranged according to numerical values. Proceeding down the left side of such a tabulation, each succeeding species is a less effective electron acceptor than the one above it. The half-reactions at the bottom of the table have little tendency to take place, as written. On the other hand, these do tend to occur in the opposite sense, as electron donors. The most effective reducing agents, then, are species that appear in the lower right-hand side of the equations in a table of standard electrode potentials.

Inspection of Table 11 – 1 indicates that zinc is more readily oxidized than cadmium; it is possible to conclude, therefore, that the reaction

$$Zn(s) + Cd^{2+} \rightleftharpoons Zn^{2+} + Cd(s)$$
$$\text{Red}_1 \qquad \text{Ox}_2 \qquad \text{Ox}_1 \qquad \text{Red}_2$$

is spontaneous, as written, and nonspontaneous in the opposite sense. Similarly, iron(III) is seen to be a more effective electron acceptor than is triiodide ion; it is therefore possible to predict that iron(II) and triiodide ions will predominate in the equilibrium

$$3I^- + 2Fe^{3+} \rightleftharpoons I_3^- + 2Fe^{2+}$$
$$\text{Red}_1 \qquad \text{Ox}_2 \qquad \text{Ox}_1 \qquad \text{Red}_2$$

## 11C – 5  Calculation of Electrode Potentials from Standard Electrode Potential Data

Typical applications of the Nernst equation are illustrated in Examples 11 – 2 through 11 – 5.

**Example 11-2** Calculate the potential for a Cd electrode that is immersed in a 0.0100 $F$ $Cd^{2+}$ solution.

The standard electrode potential for the process

$$Cd^{2+} + 2e \rightleftharpoons Cd(s)$$

is $-0.403$ V (Table 11-1). Thus,

$$E = E^0 - \frac{0.0591}{2} \log \frac{1}{[Cd^{2+}]}$$

Substitution of 0.0100 for $[Cd^{2+}]$ and $-0.403$ for $E^0$ in this equation gives

$$E = -0.403 - \frac{0.0591}{2} \log \frac{1}{0.0100}$$

$$= -0.403 - \frac{0.0591}{2} \log 100$$

$$= -0.403 - \frac{0.0591}{2} \times 2$$

$$= -0.462 \text{ V}$$

The sign for an electrode potential indicates the direction of the corresponding half-cell when it is coupled with a standard hydrogen electrode. The negative sign here indicates that the reduction of cadmium ion is nonspontaneous with respect to this reference. Note also that the potential for the half-cell in Example 11-2 is more negative than the standard electrode potential itself. This follows from mass law considerations, since the half-reaction, *as written,* has less tendency to occur with the smaller cadmium ion concentration.

**Example 11-3** Calculate the potential of a Pt electrode immersed in a solution prepared by saturating a 0.0100 $F$ solution of KBr with $Br_2$.

The pertinent half-reaction is

$$Br_2(l) + 2e \rightleftharpoons 2Br^- \qquad E^0 = 1.065 \text{ V}$$

Note that the symbol (l) in this half-reaction indicates that liquid $Br_2$ is present and that the aqueous solution of KBr is at all times saturated with $Br_2$; then, by definition, the activity of $Br_2$ is 1.00. The overall process can be considered the sum of the two equilibria

$$Br_2(l) \rightleftharpoons Br_2(aq)$$

$$Br_2(aq) + 2e \rightleftharpoons 2Br^-$$

The Nernst equation for the overall process is therefore

$$E = 1.065 - \frac{0.0591}{2} \log \frac{[Br^-]^2}{1.00}$$

$$= 1.065 - \frac{0.0591}{2} \log \frac{(1.00 \times 10^{-2})^2}{1.00}$$

$$= 1.065 - \frac{0.0591}{2}(-4.00)$$

$$= 1.183 \text{ V}$$

**Example 11-4**  Calculate the potential of a Pt electrode immersed in a solution that is 0.0100 $F$ in KBr and $1.00 \times 10^{-3}$ $F$ in $Br_2$.

The standard electrode potential used in Example 11-2 is not applicable here because *the solution is no longer saturated with* $Br_2$. Table 11-1, however, has an entry for the half-reaction

$$Br_2(aq) + 2e \rightleftharpoons 2Br^- \qquad E^0 = 1.087 \text{ V}$$

The term (aq) refers to elemental $Br_2$ *in solution,* and 1.087 V is the potential for this half-cell when both $Br_2$ and $Br^-$ have unit activity. It turns out, however, that the solubility of $Br_2$ in water is only about 0.18 $M$ at 25°C. Thus, the recorded potential of 1.087 V is based on a system that—in terms of our definition for $E^0$—cannot be realized experimentally. Nevertheless, this potential is useful because it permits calculation of potentials for half-cells involving solutions that are undersaturated with respect to elemental bromine. Thus,

$$E = 1.087 - \frac{0.0591}{2} \log \frac{[Br^-]^2}{[Br_2]}$$

$$= 1.087 - \frac{0.0591}{2} \log \frac{(1.00 \times 10^{-2})^2}{1.00 \times 10^{-3}}$$

$$= 1.087 - \frac{0.0591}{2} \log 0.100$$

$$= 1.117 \text{ V}$$

Note that the activity for $Br_2$ here is taken as $1.00 \times 10^{-3}$ rather than 1.00, as was the situation in Example 11-2, where the solution was saturated and in contact with elemental bromine.

### 11C–6 Standard Electrode Potentials for Half-Reactions Involving Precipitation or Complex Formation

Example 11–5 demonstrates that reagents which compete for a participant of an electrode process can markedly influence the potential for that process.

---

**Example 11–5**  Calculate the potential of an Ag electrode in a solution that is saturated with AgI and has an $I^-$ activity of precisely 1.00.

The standard electrode potential for the half-cell process occurring at the silver electrode (Table 11–1) is

$$Ag^+ + e \rightleftharpoons Ag(s) \qquad E^0 = 0.799 \text{ V}$$

and,

$$E = 0.799 - \frac{0.0591}{1} \log \frac{1}{[Ag^+]}$$

Replacement of $[Ag^+]$ with $K_{sp}/[I^-]$ in the Nernst equation gives

$$E = 0.799 - \frac{0.0591}{1} \log \frac{[I^-]}{K_{sp}}$$

This expression can be rewritten as

$$E = 0.799 + 0.0591 \log K_{sp} - 0.0591 \log [I^-] \qquad (11-3)$$

Substitution of $8.3 \times 10^{-17}$ for $K_{sp}$ (Appendix 9) and 1.00 for $I^-$ gives

$$E = 0.799 + 0.0591 \log 8.3 \times 10^{-17} - 0.0591 \log 1.00$$
$$= 0.799 - 0.0591 \, (-16.08) - 0 = -0.151 \text{ V}$$

---

Note that the half-cell potential for a silver electrode is profoundly altered by the presence of iodide ion, which lowers the silver ion concentration and thus diminishes the tendency for reduction of that ion.

Equation 11–3 relates the potential of a silver electrode to the iodide ion concentration of a solution that is also saturated with silver iodide. The overall process can be considered the resultant of two reactions

$$AgI(s) \rightleftharpoons Ag^+ + I^-$$

$$Ag^+ + e \rightleftharpoons Ag(s)$$

and the potential is related to the constants associated with these processes. When the iodide ion activity is unity, the second logarithmic term in Equation

11 – 3 becomes zero, and the resulting potential is the standard electrode potential for the process

$$AgI(s) + e \rightleftharpoons Ag(s) + I^- \qquad E^0_{AgI} = -0.151 \text{ V}$$

where

$$E^0_{AgI} = 0.799 + 0.0591 \log K_{sp} = -0.151 \text{ V} \qquad (11-4)$$

The potential for a silver electrode that is immersed in a solution that is saturated with silver iodide can thus be described *either* in terms of its silver ion concentration (with $E^0$ for the reduction of silver ion) *or* in terms of the iodide ion concentration (with $E^0$ for the silver–silver iodide half-reaction).

The potential for a silver electrode in a solution containing an ion that forms a soluble complex with silver ion can be treated analogously. For example, in a solution containing thiosulfate ion, the half-reaction can be written as

$$Ag(S_2O_3)_2^{3-} + e \rightleftharpoons Ag(s) + 2S_2O_3^{2-}$$

A silver electrode immersed in a solution in which the activities of the complex and thiosulfate ion are both unity will have a potential that is numerically equal to the standard electrode potential for this half-cell process; here,

$$E^0 = +0.799 - 0.0591 \log K_f$$

where $K_f$ is the formation constant for the complex.

Data for the potential of a silver electrode in contact with solutions that contain ions that form complexes or sparingly soluble compounds with silver ion can be found in the literature (see Appendix 12). Similar information is available for other electrode systems. Such data simplify the calculation of many half-cell potentials.

## 11D CELLS AND CELL POTENTIALS

### 11D – 1 Schematic Representation of Cells

Chemists frequently use a shorthand notation to describe electrochemical cells. The cell shown in Figure 11 – 1 can be represented as

$$Zn|ZnSO_4(C_1)|CuSO_4(C_2)|Cu$$

where $C_1$ and $C_2$ are the formal concentrations of zinc sulfate and copper sulfate, respectively. *By convention, the anode process is displayed to the left.* Single vertical lines signify phase boundaries within the cell at which potentials develop. For example, the interface between the zinc electrode and the zinc sulfate solution with which it is in contact is the source for part of the potential of this cell. A small potential also develops at liquid junctions; thus, another vertical line is inserted between symbols for the zinc sulfate and copper sulfate solutions. The phrase boundary between the cathode and the copper sulfate solution is symbolized by yet another vertical line.

Two vertical lines are used to symbolize the presence of a salt bridge in an electrochemical cell. For example, the cell shown in Figure 11–3 (p. 259) would be represented as

$$Pt, H_2(p = 1.00 \text{ atm})|H^+(a = 1.00)\|Cu^{2+}(xF)|Cu$$

The junction potentials at each interface of a salt bridge tend to cancel one another; we shall ordinarily neglect their effects.

### 11D – 2 Calculation of Cell Potentials

A valuable application of standard electrode potentials is the calculation of the potential obtainable from a galvanic cell or the potential required to operate an electrolytic cell. These calculated potentials (sometimes called *reversible,* or *thermodynamic, potentials*) are theoretical in the sense that they apply to cells in which there is essentially no current; additional factors must be taken into account where a current is involved.

The potential of an electrochemical cell is obtained by combining the *electrode potentials* for the two half-cell processes as follows:

$$E_{cell} = E_{cathode} - E_{anode}$$

It should be stressed that *both* $E_{cathode}$ and $E_{anode}$ are electrode potentials and are therefore the potentials of the half-reactions written as reductions (p. 256).

Consider the hypothetical cell

$$Zn|ZnSO_4(a_{Zn^{2+}} = 1.00)\|CuSO_4(a_{Cu^{2+}} = 1.00)|Cu$$

This notation indicates that elemental zinc is being oxidized to Zn(II) at the anode, and that Cu(II) is being reduced to the elemental state at the cathode. Because the activities of the two ions are specified as unity, the electrode potentials in this cell are numerically identical with the standard potentials. Thus, using $E^0$ data from Table 11–1,

$$E_{cell} = +0.337 - (-0.763) = +1.100 \text{ V}$$

The positive sign for the cell potential indicates that the reaction

$$Zn(s) + Cu^{2+} \rightleftharpoons Zn^{2+} + Cu(s)$$

is spontaneous; the cell, therefore, is galvanic.

Representation of this cell with the diagram

$$Cu|CuSO_4(a_{Cu^{2+}} = 1.00) \| ZnSO_4(a_{Zn^{2+}} = 1.00)|Zn$$

implies that the copper electrode is now the anode. The potential for this cell is

$$E_{cell} = -0.763 - 0.337 = -1.100 \text{ V}$$

The negative sign signifies that the process

$$Cu(s) + Zn^{2+} \rightleftharpoons Cu^{2+} + Zn(s)$$

is nonspontaneous, and that application of an external potential that is greater than 1.100 V would be needed to cause this reaction to occur.

---

**Example 11–6**  Calculate the theoretical potential for the cell

$$Ag|AgI(sat'd), HI(0.0200 \ F)|H_2(0.800 \text{ atm}), Pt$$

Note that this cell does not require two compartments (and a salt bridge) because molecular $H_2$ has little tendency to react directly with the low concentration of $Ag^+$ in the electrolyte solution.

The two half-reactions and their corresponding standard electrode potentials are (Table 11–1)

$$2H^+ + 2e \rightleftharpoons H_2(g) \qquad E^0 = 0.000 \text{ V}$$

$$AgI(s) + e \rightleftharpoons Ag(s) + I^- \qquad E^0 = -0.151 \text{ V}$$

Provided the molar concentrations are reasonable approximations of activities, the two electrode potentials are given by

$$E_{Ag,AgI} = -0.151 - \frac{0.0591}{1} \log 0.0200 = -0.0506 = -0.051 \text{ V}$$

and

$$E_{Pt,H_2} = 0.000 - \frac{0.0591}{2} \log \frac{0.800}{(0.0200)^2} = -0.098 \text{ V}$$

The cell diagram specifies the Ag electrode as the anode and the $H_2$ electrode as the cathode. Thus,

$$E_{cell} = -0.098 - (-0.051) = -0.047 \text{ V}$$

The negative sign indicates that the cell reaction

$$2H^+ + 2Ag(s) + 2I^- \rightleftharpoons H_2(g) + 2AgI(s)$$

is nonspontaneous and the cell is electrolytic.

**Example 11-7**  The shorthand notation for the cell produced by substitution of a saturated calomel electrode (p. 258) for the hydrogen electrode in Example 11-6 would be[6]

$$Ag|AgI(sat'd), HI(0.0200\ F)\|KCl(sat'd), Hg_2Cl_2(sat'd)|Hg(l)$$

Note that a salt bridge has now become a necessity. The potential for this cell will be

$$E_{cell} = 0.244 - (-0.051) = 0.295\ V$$

The reaction

$$2Ag(s) + Hg_2Cl_2(s) + 2I^- \rightleftharpoons 2Hg(l) + 2AgI(s) + 2Cl^-$$

is spontaneous, and the cell is galvanic.

---

**Example 11-8**  Calculate the potential required to initiate deposition of Cu from a solution that is $0.010\ F$ in $CuSO_4$ and contains sufficient $H_2SO_4$ to give an $H^+$ concentration of $1.00 \times 10^{-4}\ M$.

The deposition of copper necessarily occurs at the cathode. Since there is no more easily oxidizable substance in the system, the oxidation of water, with the evolution of $O_2$, will occur at the anode. The required standard electrode potentials are (Appendix 12)

$$O_2(g) + 4H^+ + 4e \rightleftharpoons 2H_2O \qquad E^0 = +1.229\ V$$
$$Cu^{2+} + 2e \rightleftharpoons Cu(s) \qquad E^0 = +0.337\ V$$

The potential for the Cu electrode is given by

$$E = +0.337 - \frac{0.0591}{2} \log \frac{1}{0.010} = +0.278\ V$$

If $O_2$ is evolved at 1.00 atm, the potential for the oxygen electrode will be

$$E = +1.229 - \frac{0.0591}{4} \log \frac{1}{(1.00)(1.00 \times 10^{-4})^4} = +0.993\ V$$

and the cell potential will be

$$E_{cell} = +0.278 - 0.993 = -0.715\ V$$

---

[6] We shall sometimes abbreviate the cell diagram even further; that is,

$$Ag|AgI(sat'd), HI\ (0.0200\ F)\|SCE$$

where SCE is a shorthand representation for the saturated calomel electrode.

Thus, initiation of the reaction

$$2Cu^{2+} + 2H_2O \rightleftharpoons O_2(g) + 4H^+ + 2Cu(s)$$

would require the application of a potential greater than 0.715 V.

### 11D–3 Calculation of Equilibrium Constants for Oxidation-Reduction Reactions from Standard Electrode Potentials

Consider a galvanic cell based upon the reaction

$$Cu(s) + 2Ag^+ \rightleftharpoons Cu^{2+} + 2Ag(s)$$

The equilibrium constant expression for this reaction is

$$K_{eq} = \frac{[Cu^{2+}]}{[Ag^+]^2}$$

The cell potential is at all times given by

$$E_{cell} = E_{cathode} - E_{anode}$$
$$= E_{Ag} - E_{Cu}$$

and at all times,

$$E_{Ag} = E^0_{Ag} - \frac{0.0591}{1} \log \frac{1}{[Ag^+]}$$

$$E_{Cu} = E^0_{Cu} - \frac{0.0591}{2} \log \frac{1}{[Cu^{2+}]}$$

As electricity is drawn from this cell, the concentration of Cu(II) increases while that for Ag(I) decreases. These changes have the effect of rendering the potential of the copper electrode more positive and the potential for the silver electrode less positive. The net effect, then, is a decrease in the potential of the cell as it is discharged. The concentrations of Cu(II) and Ag(I) will ultimately attain values such that there remains no further tendency for the transfer of electrons to occur. Under these conditions, the potential of the cell becomes zero, and the system is in equilibrium. Thus,

$$E_{cell} = 0 = E_{cathode} - E_{anode}$$

which is tantamount to the statement that, at equilibrium,

$$E_{cathode} = E_{anode} \tag{11-5}$$

Equation 11–5 is an important and general relationship; *the electrode potentials for all half-reactions of an oxidation-reduction system will be equal when that system is in equilibrium.* Note that this generalization applies regardless of the number of half-reactions involved in the system because interactions among *all* must take place until all of the electrode potentials are identical. In terms of the galvanic cell we have been considering, equilibrium will have been attained when

$$E_{Ag} = E_{Cu}$$

This equality can be expressed in terms of the Nernst equations for the two half-cell processes:

$$E^0_{Ag} - \frac{0.0591}{2} \log \frac{1}{[Ag^+]^2} = E^0_{Cu} - \frac{0.0591}{2} \log \frac{1}{[Cu^{2+}]} \qquad (11-6)$$

It is important to note that the Nernst equation was applied to the silver half-reaction as it appears in the balanced equation

$$2Ag^+ + 2e \rightleftharpoons 2Ag(s) \qquad E^0 = 0.799 \text{ V}$$

Rearrangement gives

$$E^0_{Ag} - E^0_{Cu} = \frac{0.0591}{2} \log \frac{1}{[Ag^+]^2} - \frac{0.0591}{2} \log \frac{1}{[Cu^{2+}]}$$

$$= \frac{0.0591}{2} \log \frac{1}{[Ag^+]^2} + \frac{0.0591}{2} \log \frac{[Cu^{2+}]}{1}$$

Finally, then,

$$\frac{2(E^0_{Ag} - E^0_{Cu})}{0.0591} = \log \frac{[Cu^{2+}]}{[Ag^+]^2} = \log K_{eq}$$

Note that the concentration terms in this particular relationship are *equilibrium concentrations; the quotient in the logarithmic term is thus the equilibrium constant for the reaction.*

---

**Example 11-9** A piece of Cu is placed in 0.050 *F* $AgNO_3$. Calculate the equilibrium composition of this solution.

The required standard electrode potentials are (Appendix 12)

$$Ag^+ + e \rightleftharpoons Ag(s) \qquad E^0 = +0.799 \text{ V}$$

$$Cu^{2+} + 2e \rightleftharpoons Cu(s) \qquad E^0 = +0.337 \text{ V}$$

and, as just shown,

$$\log \frac{[Cu^{2+}]}{[Ag^+]^2} = \frac{2(E^0_{Ag^+} - E^0_{Cu^{2+}})}{0.0591}$$

$$= \frac{2(0.799 - 0.337)}{0.0591} = 15.63$$

Thus,

$$\frac{[Cu^{2+}]}{[Ag^+]^2} = K_{eq} = 4.3 \times 10^{15}$$

The magnitude of the equilibrium constant suggests that nearly all of the $Ag^+$ has been reduced. The concentration of $Cu^{2+}$ is therefore

$$[Cu^{2+}] = \tfrac{1}{2}(0.050 - [Ag^+]) = 0.025 - \tfrac{1}{2}[Ag^+]$$

It is worthwhile to obtain a provisional solution based on the assumption that $[Ag^+]$ is small with respect to 0.025; that is,

$$[Cu^{2+}] \cong 0.025$$

and

$$\frac{0.025}{[Ag^+]^2} = 4.3 \times 10^{15}$$

$$[Ag^+] = 2.4 \times 10^{-9}$$

The assumption is clearly justified.

---

A more general relationship for the calculation of equilibrium constants can be obtained by considering an equilibrium involving oxidation of the species $A_{red}$ to $A_{ox}$ and the reduction of $B_{ox}$ to $B_{red}$:

$$a A_{red} + b B_{ox} \rightleftharpoons a A_{ox} + b B_{red}$$

where $a$ and $b$ are the integers needed to balance the equation. Half-reactions for the two processes are

$$a A_{ox} + ne \rightleftharpoons a A_{red}$$
$$b B_{ox} + ne \rightleftharpoons b B_{red}$$

When this system is in equilibrium, the two electrode potentials $E_A$ and $E_B$ will be numerically identical; that is,

$$E_A = E_B$$

Substitution of the Nernst expressions into this equation reveals that, *at equilibrium*,

$$E_A^0 - \frac{0.0591}{n} \log \frac{[A_{red}]^a}{[A_{ox}]^a} = E_B^0 - \frac{0.0591}{n} \log \frac{[B_{red}]^b}{[B_{ox}]^b}$$

which can be rearranged to

$$E_B^0 - E_A^0 = \frac{0.0591}{n} \log \frac{[A_{ox}]^a [B_{red}]^b}{[A_{red}]^a [B_{ox}]^b} = \frac{0.0591}{n} \log K_{eq}$$

Finally, then,

$$\log K_{eq} = \frac{n(E_B^0 - E_A^0)}{0.0591} \tag{11-7}$$

---

**Example 11-10** Calculate the equilibrium constant for the reaction

$$MnO_4^- + 5Fe^{2+} + 8H^+ \rightleftharpoons Mn^{2+} + 5Fe^{3+} + 4H_2O$$

The standard electrode potentials are (Appendix 12)

$$MnO_4^- + 5e + 8H^+ \rightleftharpoons Mn^{2+} + 4H_2O \qquad E_{MnO_4^-}^0 = +1.51 \text{ V}$$

$$Fe^{3+} + e \rightleftharpoons Fe^{2+} \qquad E_{Fe^{3+}}^0 = +0.771 \text{ V}$$

The balanced net-ionic equation for this reaction involves 5 fw of Fe for each formula weight of Mn; it is therefore necessary to write the half-reaction for the $Fe^{3+} - Fe^{2+}$ couple as

$$5Fe^{3+} + 5e \rightleftharpoons 5Fe^{2+} \qquad E_{Fe^{3+}}^0 = +0.771 \text{ V}$$

*Note that multiplication of this half-reaction by 5 has not altered the value for $E^0$ (p. 265).*

When the system is in equilibrium,

$$E_{Fe^{3+}} = E_{MnO_4^-}$$

or

$$E_{Fe^{3+}}^0 - \frac{0.0591}{5} \log \frac{[Fe^{2+}]^5}{[Fe^{3+}]^5} = E_{MnO_4^-}^0 - \frac{0.0591}{5} \log \frac{[Mn^{2+}]}{[MnO_4^-][H^+]^8}$$

This equality can be rearranged to provide the logarithm of the equilibrium constant expression; that is,

$$\frac{0.0591}{5} \log \frac{[Mn^{2+}][Fe^{3+}]^5}{[MnO_4^-][Fe^{2+}]^5[H^+]^8} = E_{MnO_4^-}^0 - E_{Fe^{3+}}^0$$

Finally, then,

$$\log K_{eq} = \frac{5(1.51 - 0.771)}{0.0591} = 62.52$$

$$K_{eq} = 10^{0.52} \times 10^{62} = 3 \times 10^{62}$$

---

Note that it was necessary to round the result to a single significant figure. The first two integers of the logarithm were required to set the decimal point for the antilogarithm; only the third integer (5) was available for the result itself.

### 11D–4 Evaluation of Equilibrium Constants from the Measurement of Cell Potentials

Numerical values for solubility product constants, dissociation constants, and formation constants are conveniently evaluated through the measurement of cell potentials. One important virtue of this technique is that the measurement can be made without affecting appreciably any equilibria that may exist in the solution. For example, the potential of a silver electrode in an argentocyanide solution depends upon the activities of the silver ion, the cyanide ion, and the complex formed between them. It is possible to measure this potential with negligible passage of electricity. Since the activities of the participants are not sensibly altered during the measurement, the position of the equilibrium

$$Ag^+ + 2CN^- \rightleftharpoons Ag(CN)_2^-$$

is likewise undisturbed.

Examples 11–11 and 11–12, demonstrate how numerical values for equilibrium constants can be obtained from standard electrode potential data.

---

**Example 11–11**  Calculate the formation constant, $K_f$, for $Ag(CN)_2^-$

$$Ag^+ + 2CN^- \rightleftharpoons Ag(CN)_2^-$$

if the cell

$$Ag|Ag(CN)_2^-(7.50 \times 10^{-3}\ M),\ CN^-(0.0250\ M)\| SCE$$

develops a potential of 0.625 V.

Proceeding as in the earlier examples,

$$Ag^+ + e \rightleftharpoons Ag(s) \qquad E^0 = +0.799\ V$$

$$0.625 = E_{cathode} - E_{anode} = 0.244 - E_{Ag}$$

$$E_{Ag} = -0.381\ V$$

$$-0.381 = 0.799 - \frac{0.0591}{1} \log \frac{1}{[Ag^+]}$$

$$\log [Ag^+] = \frac{-0.381 - 0.799}{0.0591} = -20.0$$

$$[Ag^+] = 1 \times 10^{-20}$$

The result must be rounded to one significant figure:

$$K_f = \frac{[Ag(CN)_2^-]}{[Ag^+][CN^-]^2}$$

$$= \frac{7.50 \times 10^{-3}}{(1 \times 10^{-20})(2.50 \times 10^{-2})^2}$$

$$= 1.2 \times 10^{21} = 1 \times 10^{21}$$

In theory, any electrode system that includes a term for the hydrogen ion concentration can be used to evaluate dissociation constants for acids and bases. To be sure, relatively few of these systems have been used for determining such constants. Example 11–12 makes use of a hydrogen electrode and a saturated calomel electrode (p. 258) to obtain a numerical value for an acid dissociation constant.

---

**Example 11–12**   Calculate the dissociation constant for the weak acid HP if the cell

$$\text{Pt, H}_2(1.00 \text{ atm}) \,|\, \text{HP}(0.010 \; M), \text{NaP}(0.040 \; M) \,\|\, \text{SCE}$$

develops a potential of 0.591 V.

The diagram for this cell indicates that the saturated calomel electrode is the cathode. Thus,

$$E_{\text{cell}} = 0.244 - E_{\text{anode}}$$

or

$$E_{\text{anode}} = 0.244 - 0.591 = -0.347 \text{ V}$$

Application of the Nernst equation for the hydrogen electrode gives

$$-0.347 = 0.000 - \frac{0.0591}{2} \log \frac{1.00}{[\text{H}^+]^2}$$

$$= 0.000 + \frac{\cancel{2} \times 0.0591}{\cancel{2}} \log [\text{H}^+]$$

$$\log [\text{H}^+] = \frac{(-0.347 - 0.000)}{0.0591} = -5.87$$

$$[\text{H}^+] = 1.34 \times 10^{-6}$$

Substitution of this value for $[\text{H}^+]$ as well as the concentrations of the weak acid and its conjugate base into the dissociation constant expression gives

$$K_{\text{HP}} = \frac{[\text{H}^+][\text{P}^-]}{[\text{HP}]} = \frac{(1.34 \times 10^{-6})(0.040)}{0.010} = 5.4 \times 10^{-6}$$

---

## 11D–5 Limitations to the Use of Standard Electrode Potentials

The preceding examples suggest that the Nernst equation can provide answers to questions that are of importance to the analytical chemist. Significant differences between calculated and experimental potentials are, however, sometimes encountered; the reader should be aware of the sources of these differences and when they are likely to occur.

**Use of Concentrations Instead of Activities**   It is ordinarily more convenient to make use of molar concentrations instead of activities in calculations even though differences in these quantities tend to increase with increasing ionic strength (Chapter 5, Section 5D–2). Because the ionic strength of the typical electrochemical cell is large, potentials based on molar concentrations are likely to differ appreciably from those obtained by direct measurement.

**Effect of Other Equilibria**   The application of electrode potential data to many systems of interest is further complicated by solvolysis, dissociation, association, and complex formation equilibria that involve species appearing in the Nernst equation. Accounting for these phenomena requires knowledge of their existence and availability of the appropriate equilibrium constant data.

**Formal Potentials**   Swift[7] proposed substitution of *formal potentials* (also called *conditional potentials*) in place of standard electrode potentials to compensate for activity effects as well as errors due to the existence of competing equilibria. The formal potential of a system is the potential of the half-cell (with respect to the standard hydrogen electrode) when the concentration of each reactant and product is 1 $F$ and the concentrations of all other constituents of the solution are carefully specified.

Formal potentials for many half-reactions are listed in Appendix 12; note that large differences exist between the formal and standard potentials for some half-reactions. For example, the standard potential for the reduction of iron(III) to iron(II) is 0.771 V. In 1 $F$ perchloric acid the formal potential for the same half-reaction is 0.731 V. This difference is attributable to the fact that the activity coefficient of iron(III) is considerably smaller than that for iron(II) at the high ionic strength of the medium. As a consequence, the ratio of activities of the two species is less than one, which leads to a decrease in the electrode potential. In 1 $F$ hydrochloric acid, the formal potential for this couple is only 0.700 V. Here, the ratio of activity of iron(III) to iron(II) is even smaller because the chloro complexes of the former are more stable than the chloro complexes of the latter; a yet larger decrease in potential is the consequence.

Substitution of formal potentials for standard electrode potentials in the Nernst equation will result in better agreement between calculated and experimental results—provided, of course, that the electrolyte concentration of the solution approximates that for which the formal potential is applicable. Not surprisingly, attempts to apply formal potentials to systems that differ substantially in type and in concentration of electrolyte can result in errors that are larger than those associated with the use of standard electrode potentials. Henceforth, we shall use whichever is the more appropriate.

---

[7] E. H. Swift, *A System of Chemical Analysis,* p. 50. San Francisco: W. H. Freeman and Company, 1939.

**Reaction Rates** Standard potentials will reveal whether or not reaction is sufficiently complete for application to a particular analytical problem. Potential data, however, provide no information concerning the rate at which the equilibrium state will be approached; thus, a reaction that appears extremely favorable from equilibrium considerations may be totally unacceptable from a kinetic standpoint. The oxidation of arsenic(III) with cerium(IV) in dilute sulfuric acid is a typical example. The reaction is

$$H_3AsO_3 + 2Ce^{4+} + H_2O \rightleftharpoons H_3AsO_4 + 2Ce^{3+} + 2H^+$$

The formal potentials, $E^f$, for these two systems are

$$Ce^{4+} + e \rightleftharpoons Ce^{3+} \qquad\qquad E^f = +1.4 \text{ V}$$

$$H_3AsO_4 + 2H^+ + 2e \rightleftharpoons H_3AsO_3 + H_2O \qquad E^f = +0.56 \text{ V}$$

and an equilibrium constant of about $10^{28}$ can be deduced from these data. Even though this reaction is highly favorable from an equilibrium standpoint, solutions of arsenic(III) cannot be titrated with cerium(IV) because several hours are needed for the attainment of equilibrium. To be sure, this reaction can be used as the basis for a titration through the introduction of a suitable catalyst.

## 11E OXIDATION-REDUCTION TITRATIONS

The successful application of an oxidation-reduction reaction to volumetric analysis requires, among other things, the means for equivalence-point detection. As in other types of titrations, detection is based upon pronounced changes in the chemical system that occur in the region of the equivalence point.

### 11E – 1 Derivation of Titration Curves

For the titration curves considered thus far, the negative logarithm of the concentration of a participant in the titration (that is, a p-value) has been plotted as a function of titrant volume, the species selected being one to which the indicator for the reaction is sensitive. Most of the indicators used for oxidation-reduction titrations are themselves oxidizing or reducing agents that respond to changes in potential rather than to changes in the concentration of any particular reactant or product. It is thus customary to plot the *electrode potential for the system* as the ordinate of the curve for an oxidation-reduction titration rather than the p-function for a participant.

In order to show what is meant by the term electrode potential of the system, let us consider the titration of iron(II) with cerium(IV). Here, the reaction is

$$Fe^{2+} + Ce^{4+} \rightleftharpoons Fe^{3+} + Ce^{3+}$$

The reaction between these two species is rapid, so that the system can be considered to be at equilibrium at all times. We have noted that the electrode potentials for participants in an oxidation-reduction reaction are equal when the reaction is in equilibrium (p. 274). Therefore, for the reaction under consideration we may write

$$E_{Ce^{4+}} = E_{Fe^+} = E_{system}$$

where $E_{system}$ is the potential of the system. If a redox indicator is present in this solution, the ratio between its oxidized and reduced forms must adjust so that its potential is also the system potential; that is,

$$E_{In} = E_{Ce^{4+}} = E_{Fe^{3+}} = E_{system}$$

The theoretical potential of a system can be derived from standard potential data. Thus, for the reaction between iron(II) with cerium(IV), the titration mixture can be considered part of the hypothetical cell

$$SHE \| Ce^{4+}, Ce^{3+}, Fe^{3+}, Fe^{2+} | Pt$$

The potential of the half-cell on the right (with respect to the standard hydrogen electrode) is determined by the affinity of both iron(III) and cerium(IV) for electrons

$$Fe^{3+} + e \rightleftharpoons Fe^{2+}$$

$$Ce^{4+} + e \rightleftharpoons Ce^{3+}$$

The concentration ratios of the oxidized and reduced forms of each species at equilibrium are such that these two affinities (and thus their electrode potentials) are identical. Note that these concentration ratios vary continuously throughout the course of the titration; so also must $E_{system}$. It is this characteristic variation in $E_{system}$ that forms the basis for end-point detection.

It is clear that $E_{system}$ data for a titration curve can be generated through the use of the Nernst equation for either the cerium(IV) half-reaction or the iron(III) half-reaction. It turns out that one or the other will be the more convenient, depending upon the stage of the titration. For example, the concentrations of iron(II), iron(III), and cerium(III) are readily deduced from stoichiometric considerations in the region short of the equivalence point; the electrode potential for the system can thus be directly evaluated from the Nernst equation for the iron(III)–iron(II) couple. Use of the corresponding expression for the cerium(IV)–cerium(III) half-reaction would provide the same answer. However, at this stage in the titration, an additional calculation involving the equilibrium constant for the reaction (Section 11D–4) would be needed to establish a numerical value for cerium(IV). This situation is reversed after an excess of titrant has been introduced. Now, the concentrations of cerium(IV), cerium(III), and iron(III) are immediately available, while that for iron(II) would require a preliminary calculation involving the equilibrium constant. The cerium(IV)–cerium(III) couple will therefore provide the most direct means of evaluating the electrode potential of the system.

**Equivalence-Point Potential** The potential of the system at the equivalence point is of particular importance for indicator selection. It is not, however, possible to compute this potential directly from the Nernst equation, as it is for points before or after the equivalence point, because the concentration of neither cerium(IV) nor iron(II) can be obtained from the stoichiometry of the reaction. We can, however, take advantage of our knowledge that at the equivalence point, the concentrations of cerium(III) and iron(III) are equal, as well as the concentrations of cerium(IV) and iron(II). We also know that the equivalence point potential, $E_{eq}$, is given by

$$E_{eq} = E^0_{Ce^{4+}} - \frac{0.0591}{1} \log \frac{[Ce^{3+}]}{[Ce^{4+}]}$$

and also by

$$E_{eq} = E^0_{Fe^{3+}} - \frac{0.0591}{1} \log \frac{[Fe^{2+}]}{[Fe^{3+}]}$$

These expressions can be added to give

$$2E_{eq} = E^0_{Ce^{4+}} + E^0_{Fe^{3+}} - \frac{0.0591}{1} \log \frac{[Ce^{3+}][Fe^{2+}]}{[Ce^{4+}][Fe^{3+}]} \qquad (11-8)$$

*Note that the concentration quotient in this expression does not have the same form as the equilibrium constant expression for the reaction.*

The stoichiometry of the reaction requires that *at the equivalence point,*

$$[Fe^{3+}] = [Ce^{3+}]$$

$$[Fe^{2+}] = [Ce^{4+}]$$

Application of these equalities to Equation 11–7 results in elimination of the logarithmic term:

$$2E_{eq} = E^0_{Ce^{4+}} + E^0_{Fe^{3+}} - \frac{0.0591}{1} \log \frac{\cancel{[Ce^{3+}]}\cancel{[Fe^{3+}]}}{\cancel{[Ce^{4+}]}\cancel{[Fe^{2+}]}}$$

and,

$$E_{eq} = \frac{E^0_{Ce^{4+}} + E^0_{Fe^{3+}}}{2}$$

Example 11–13 demonstrates how the equilibrium concentrations of the reacting species can be calculated from the equivalence-point potential.

---

**Example 11–13** Calculate the concentration of each reactant and product at the equivalence point in the titration of 0.100 $N$ Fe(II) with 0.100 $N$ Ce(IV) in a solution that is 1.0 $F$ in $H_2SO_4$.

Formal potentials for the two half-reactions are available (Appendix 12) and will be used to calculate the equivalence-point potential; that is,

$$E_{eq} = \frac{+1.44 + 0.68}{2} = 1.06 \text{ V}$$

The molar ratio between, say, $Fe^{2+}$ and $Fe^{3+}$ at the equivalence point can be evaluated from the Nernst equation

$$+1.06 = +0.68 - \frac{0.0591}{1} \log \frac{[Fe^{2+}]}{[Fe^{3+}]}$$

$$\log \frac{[Fe^{2+}]}{[Fe^{3+}]} = \frac{-1.06 + 0.68}{0.0591} = -6.4$$

$$\frac{[Fe^{2+}]}{[Fe^{3+}]} = 4 \times 10^{-7}$$

It is clear that most of the $Fe^{2+}$ has been oxidized to $Fe^{3+}$ at the equivalence point. As a consequence of dilution, then, $[Fe^{3+}]$ at equivalence will be essentially equal to one-half of the original Fe(II) concentration; that is,

$$[Fe^{3+}] = \frac{0.100}{2} - [Fe^{2+}] \cong 0.050$$

Thus,

$$[Fe^{2+}] = 4 \times 10^{-7}[Fe^{3+}] = (4 \times 10^{-7})(0.050)$$
$$= 2 \times 10^{-8}$$

Finally, the stoichiometry of the reaction requires that

$$[Ce^{4+}] = [Fe^{2+}] = 2 \times 10^{-8}$$

$$[Ce^{3+}] = [Fe^{3+}] \cong 0.050$$

Note that identical results would have been obtained through the use of the Nernst equation for the cerium half-reaction.

Calculation of the equivalence-point potential in Example 11–14 is slightly more involved but embraces no additional principles.

**Example 11–14**  Derive an equation for the potential of the system at the equivalence point in the titration of $Fe^{2+}$ with $MnO_4^-$. Reaction:

$$5Fe^{2+} + MnO_4^- + 8H^+ \rightarrow 5Fe^{3+} + Mn^{2+} + 4H_2O$$

The required half-reactions are

$$MnO_4^- + 5e + 8H^+ \rightleftharpoons Mn^{2+} + 4H_2O \qquad E^0 = +1.51 \text{ V}$$
$$Fe^{3+} + e \rightleftharpoons Fe^{2+} \qquad\qquad\qquad E^0 = +0.771 \text{ V}$$

The equivalence-point potential of this system is given either by

$$E_{eq} = E^0_{Fe^{3+}} - \frac{0.0591}{1} \log \frac{[Fe^{2+}]}{[Fe^{3+}]}$$

or by

$$E_{eq} = E^0_{MnO_4^-} - \frac{0.0591}{5} \log \frac{[Mn^{2+}]}{[MnO_4^-][H^+]^8}$$

Before the two equations can be combined, the second must be multiplied through by 5, which gives

$$5E_{eq} = 5E^0_{MnO_4^-} - \frac{\cancel{5} \times 0.0591}{\cancel{5}} \log \frac{[Mn^{2+}]}{[MnO_4^-][H^+]^8}$$

Addition of this expression with that for $Fe^{3+}$ gives

$$6E_{eq} = E^0_{Fe^{3+}} + 5E^0_{MnO_4^-} - \frac{0.0591}{1} \log \frac{[Fe^{2+}][Mn^{2+}]}{[Fe^{3+}][MnO_4^-][H^+]^8}$$

At the equivalence point,

$$[Fe^{2+}] = 5[MnO_4^-]$$

$$[Fe^{3+}] = 5[Mn^{2+}]$$

Substitution and rearrangement gives

$$
\begin{aligned}
E_{eq} &= \frac{E^0_{Fe^{3+}} + 5E^0_{MnO_4^-}}{6} - \frac{0.0591}{6} \log \frac{\cancel{5}[MnO_4^-][Mn^{2+}]}{\cancel{5}[Mn^{2+}][MnO_4^-][H^+]^8} \\
&= \frac{0.771 + 5 \times 1.51}{6} + \frac{0.0591}{6} \log [H^+]^8 \\
&= 1.39 + \frac{8 \times 0.0591}{6} \log [H^+] = 1.39 - 0.0788 \text{ pH}
\end{aligned}
$$

Note that the equivalence-point potential for this titration is pH dependent.

**Variation in Potential as a Function of Titrant Volume**    The variables that affect the shape of the curve for an oxidation-reduction titration are conveniently illustrated with examples.

**Example 11-15**    Generate a curve for the titration of 0.0500 $N$ $Fe^{2+}$ with 0.1000 $N$ $Ce^{4+}$ in a medium that is 1.0 $F$ in $H_2SO_4$ at all times. Formal potential data for both half-cell processes are available in Appendix 12 and will be used for these calculations.

(a) *Initial potential.* The solution contains no Ce species at the outset. There will be, in all likelihood, a small but unknown amount of $Fe^{3+}$ owing to air oxidation of $Fe^{2+}$. In any event, we lack sufficient information to calculate an initial potential.

(b) *Potential after addition of 5.00 mL of cerium(IV).* With the introduction of oxidant, the solution acquires appreciable and known concentrations of three participants; that for the fourth, $Ce^{4+}$, will be vanishingly small. Therefore, it will be more convenient to use the concentrations of the two Fe species to calculate the electrode potential of the system.

The molar concentration of Fe(III) will be equal to its normal concentration less the equilibrium concentration of the unreacted Ce(IV); that is,

$$[Fe^{3+}] = \frac{5.00 \times 0.100}{50.00 + 5.000} - [Ce^{4+}] \cong \frac{0.500}{55.00}$$

Similarly, the $Fe^{2+}$ concentration is given by its normality plus $[Ce^{4+}]$:

$$[Fe^{2+}] = \frac{50.00 \times 0.0500 - 5.00 \times 0.1000}{55.00} + [Ce^{4+}] \cong \frac{2.00}{55.00}$$

The validity of the assumptions shown in the two equations is readily confirmed by computing the equilibrium constant for the reaction between Fe(II) and Ce(IV) using the technique described in Section 11D–3. The large value for this constant ($7 \times 10^{12}$) indicates clearly that the concentration of Ce(IV) must indeed be vanishingly small when compared with the concentrations of the two Fe species.

Substitution for $[Fe^{2+}]$ and $[Fe^{3+}]$ into the Nernst equation gives

$$E = +0.68 - \frac{0.0591}{1} \log \frac{2.00/55.00}{0.500/55.00} = +0.64 \text{ V}$$

Note that the volumes in the numerator and denominator cancel, which indicates that the potential is independent of dilution. This independence will persist until the solution becomes so dilute that it can no longer be assumed that the molar and formal concentrations of the Fe species are the same.

It is worth emphasizing again that the use of the Nernst equation for the Ce(IV)–Ce(III) system would have yielded the same values for $E$.

Additional values for potentials needed to define the titration curve short of the equivalence point can be obtained similarly. Such data are given in Table 11–2; the reader is encouraged to confirm one or two of these values.

(c) *Equivalence-point potential.* The equivalence-point potential has been shown to be $+1.06$ V (Example 11–13).

(d) *Potential after addition of 25.10 mL of cerium(IV).* The normal concentrations of Ce(III), Ce(IV), and Fe(III) are readily computed at this point, but that for iron(II) is not. Therefore, computations based on the Ce half-reaction will be more convenient. Here,

$$[Ce^{3+}] = \frac{25.00 \times 0.100}{75.10} - [Fe^{2+}] \cong \frac{2.500}{75.10}$$

$$[Ce^{4+}] = \frac{25.10 \times 0.1000 - 50.00 \times 0.0500}{75.10} + [Fe^{2+}] \cong \frac{0.010}{75.10}$$

The indicated approximations should be reasonable in view of the favorable equilibrium constant. Substitution of these concentrations into the Nernst equation gives

$$E = +1.44 - \frac{0.0591}{1} \log \frac{[Ce^{3+}]}{[Ce^{4+}]}$$

$$= +1.44 - \frac{0.0591}{1} \log \frac{2.500/75.10}{0.010/75.10} = +1.30 \text{ V}$$

The additional post-equivalence-point potentials in Table 11–2 were derived in a similar fashion.

The titration of iron(II) with cerium(IV) appears as curve $A$ in Figure 11–4. Its general shape resembles the curves that were encountered in neutralization, precipitation, and complex formation titrations, the equivalence point being signaled by a large change in the ordinate function. A titration involving 0.00500 $F$ iron(II) will yield a curve that is, for all practical purposes, identical to the one derived, since the electrode potential of the system is independent of dilution. Note that the cerium(IV) titration is symmetric about the equivalence

**Table 11–2  Electrode Potentials (vs. SHE) During Titrations of 50.0 mL of 0.0500 N Iron(II) Solutions\***

| Volume of 0.1000 $N$ Reagent, mL | Potential, V | |
|---|---|---|
| | Titration with 0.1000 $N$ Ce$^{4+}$ | Titration with 0.1000 $N$ MnO$_4^-$ |
| 5.00 | 0.64 | 0.64 |
| 15.00 | 0.69 | 0.69 |
| 20.00 | 0.72 | 0.72 |
| 24.00 | 0.76 | 0.76 |
| 24.90 | 0.82 | 0.82 |
| 25.00 | 1.06 ← Equivalence point → | 1.37 |
| 25.10 | 1.30 | 1.48 |
| 26.00 | 1.36 | 1.49 |
| 30.00 | 1.40 | 1.50 |

\* $[H^+] = 1.0$ $M$ throughout.

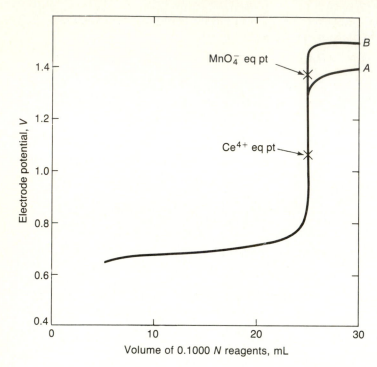

**Figure 11-4** Titration curves for 50.00 mL of 0.0500 N Fe(II) with 0.1000 N Ce(IV) (curve A) and with 0.1000 N KMnO₄ (curve B).

point, a consequence of the equiformal combining ratio that exists between oxidant and reductant. Example 11–16 demonstrates that an asymmetric curve results when this ratio differs from unity.

---

**Example 11–16**   Derive a curve for the titration of 50.00 mL of 0.0500 $N$ $Fe^{2+}$ with 0.1000 $N$ $KMnO_4$. For convenience, consider that the concentration of $H_2SO_4$ in the solution is sufficient to make $[H^+] = 1.00$. The reaction in this titration is

$$5Fe^{3+} + MnO_4^- + 8H^+ \rightleftharpoons 5Fe^{3+} + Mn^{2+} + 4H_2O$$

Since no formal potential is available for the reduction of $MnO_4^-$, we will use the standard potential of 1.51 V in its stead. It should be noted that the normal concentrations for the two Mn species must be divided by 5 to convert them to molar concentrations before substitution into the Nernst equation.

(a) *Pre-equivalence-point potentials.* The potential of the system short of the equivalence point is most readily derived from the concentrations of Fe(II) and Fe(III) in the solution. Their values will be identical to those computed in Example 11–15 for the titration with Ce(IV).

(b) *Equivalence-point potential.* The potential at the equivalence point for this reaction is given by (see Example 11–14)

$$E_{eq} = 1.39 - 0.0788 \text{ pH} = 1.39 - 0.0788 \, (-\log 1.00) = 1.39 \text{ V}$$

(c) *Post-equivalence-point potentials.* After 25.10 mL of 0.1000 $N$ KMnO$_4$ have been added, stoichiometry requires that

$$[Fe^{3+}] = \frac{50.00 \times 0.0500}{50.00 + 25.10} - [Fe^{2+}] \cong \frac{2.500}{75.10}$$

$$[Mn^{2+}] = \frac{1}{5} \left( \frac{50.00 \times 0.0500}{75.10} - [Fe^{2+}] \right) \cong \frac{0.500}{75.10}$$

$$[MnO_4^-] = \frac{1}{5} \left( \frac{25.10 \times 0.1000 - 50.00 \times 0.0500}{75.10} + [Fe^{2+}] \right)$$

$$\cong \frac{2.0 \times 10^{-3}}{75.10}$$

It is now advantageous to calculate the potential of the system with the Nernst expression for the Mn(VII)–Mn(II) half-reaction; that is,

$$E = 1.51 - \frac{0.0591}{5} \log \frac{[Mn^{2+}]}{[H^+]^8[MnO_4^-]}$$

$$= 1.51 - \frac{0.0591}{5} \log \frac{0.500/75.10}{(1.0)^8(2.0 \times 10^{-3})/75.10} = 1.48 \text{ V}$$

Additional data for this titration are tabulated in Table 11–2 and are plotted as curve *B* in Figure 11–4. The curves for the titration of iron(II) with the two oxidants are indistinguishable to within about 99.9% of the titrant volume needed for equivalence. Note, however, that the equivalence-point potentials are quite different. Moreover, the curve involving permanganate is markedly asymmetric, the potential increasing only slightly beyond the equivalence point. Finally, note that the change in potential associated with the equivalence-point region is somewhat greater with permanganate as titrant because the equilibrium constant for this reaction is more favorable.

## 11E–2 Effect of Concentration on Titration Curves

It is important to appreciate that the quantity plotted as the ordinate for an oxidation-reduction titration is the electrode potential of the system, which, in turn, is determined by the logarithm of a concentration ratio that is ordinarily unaffected by dilution. Consequently, titration curves for oxidation-reduction reactions are usually independent of analyte and reagent concentrations.[8] This behavior is in distinct contrast to the other types of titration curves we have encountered.

### 11E-3 Effect of Completeness of Reaction

The change in ordinate function in the equivalence-point region of an oxidation-reduction titration becomes larger as the reaction becomes more complete. This effect is demonstrated in Figure 11-5, which depicts curves for the titration of a hypothetical analyte having a standard electrode potential of 0.20 V with several hypothetical titrants with standard potentials ranging from 1.20 to 0.40 V; the corresponding equilibrium constants take values between about $8 \times 10^{16}$ and $2 \times 10^{3}$. Clearly, the greatest change in the potential of the system is associated with the reaction that is most complete; in this respect, then, oxidation-reduction titrations are no different from titrations involving other reaction types.

The curves in Figure 11-5 were derived for reactions in which both oxidant and reductant undergo a one-electron change; if one of the participants were involved in a two-electron change, the corresponding change in potential in the region between 24.9 and 25.1 mL would be larger by about 0.14 V.

## 11F OXIDATION-REDUCTION INDICATORS

We have demonstrated that the equivalence point in an oxidation-reduction titration is characterized by a marked change in the electrode potential of the system. Several methods exist for detecting such a change, any one of which can serve to signal the end point of the titration.

### 11F-1 Chemical Indicators

Chemical indicators for oxidation-reduction titrations are of two types. *Specific indicators* owe their behavior to a reaction with one of the participants in the

---

[8] Electrode potentials become dependent upon dilution when the number of moles of reactant and product of a half-reaction differ. An example is the reaction

$$I_3^- + 2e \rightleftarrows 3I^-$$

Application of the Nernst equation gives

$$E = E^0 - \frac{0.0591}{2} \log \frac{[I^-]^3}{[I_3]}$$

Here, the concentration term in the numerator bears the unit of $(mol/L)^3$, while the unit in the denominator is mol/L. Consequently, the ratio of the two terms is related to the square of the iodide concentration. The result is that potentials at and beyond the equivalence point in a titration involving the triiodide-iodide couple will depend upon dilution.

As mentioned earlier, electrode potentials also become concentration dependent when it can no longer be assumed that the molar concentrations of the various species are equal to the concentrations derived from stoichiometry.

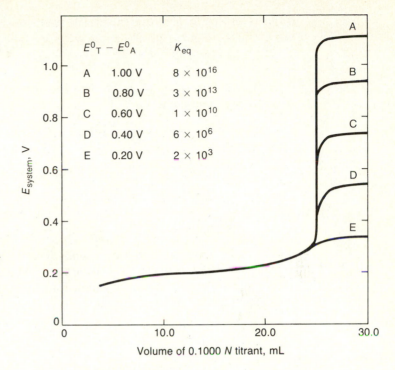

**Figure 11–5** Effect of titrant electrode potential upon completeness of reaction. The standard electrode potential for the analyte ($E_A^0$) is 0.200 V; starting with curve A, standard electrode potentials for the titrant ($E_T^0$) are 1.20, 1.00, 0.80, 0.60, and 0.40 V, respectively. Both analyte and titrant undergo a one-electron change.

titration. *True oxidation-reduction indicators,* on the other hand, respond to the potential of the system rather than to the appearance or disappearance of a particular species during the course of the titration.

**Specific Indicators**    Perhaps the best known specific indicator for oxidation-reduction titrations is starch, which forms a deep blue complex with iodine. The appearance or disappearance of this complex signals the end point in titrations involving either the production or the consumption of iodine.

Another common specific indicator is thiocyanate ion, which forms a red complex with iron(III). For example, thiocyanate ion can serve as an indicator in the titration of iron(III) with titanium(III). At the equivalence point in this titration, the concentration of iron(III) becomes vanishingly small, and the red color of the complex disappears, which serves to indicate the end point.

**True Oxidation-Reduction Indicators**    True oxidation-reduction indicators respond to the electrode potential of the system. These indicators are substantially more versatile than their specific counterparts and thus enjoy wider application.

The half-reaction responsible for color change of a typical true oxidation-reduction indicator can be written as

$$In_{ox} + ne \rightleftharpoons In_{red}$$

If this is an electrochemically reversible process, we may write

$$E = E^0 - \frac{0.0591}{n} \log \frac{[In_{red}]}{[In_{ox}]} \qquad (11-9)$$

In common with other types of chemical indicators, the color of $In_{ox}$ will typically predominate when

$$\frac{[In_{red}]}{[In_{ox}]} > \frac{1}{10}$$

whereas the color of $In_{red}$ will be perceived when

$$\frac{[In_{red}]}{[In_{ox}]} < \frac{10}{1}$$

Thus, the transition of an oxidation-reduction indicator from the color of its oxidized form to its reduced form usually requires a shift of about 100 in the concentration ratio between the two forms. The potentials associated with the full color change of the indicator can be found by substituting these values for the concentration ratio into Equation 11–9:[9]

$$E = E^0 \pm \frac{0.0591}{n} \qquad (11-10)$$

Equation 11–10 suggests that a typical oxidation-reduction indicator will undergo a detectable color change when the titrant causes a shift from $(E^0 + 0.0591/n)$ V to $(E^0 - 0.0591/n)$ V, which corresponds to a change of about $0.118/n$ V in the potential of the system. For many indicators, $n = 2$; a change of 0.059 V is thus sufficient.

The potential at which a color transition will occur depends upon the standard potential for the particular indicator system. As shown in Table 11–3, indicators with transition potentials as large as $+1.25$ V exist. Reference to this table as well as Figure 11–5 reveals that all of the indicators, save the first and the last, could be used to indicate an end point in which reagent A is titrant. In contrast, the use of reagent D would succeed only if indigo tetrasulfonate (or an indicator with a similar transition range) were used.

### 11F–2 Potentiometric End Points

End points for many oxidation-reduction titrations are readily observed by making the solution part of the cell

reference electrode $\parallel$ analyte solution $|$ Pt

and measuring the potential that develops between the two electrodes as a

---

[9] It should be pointed out that the proton is involved in the reduction of many oxidation-reduction indicators; the transition potentials for such indicators will thus be pH dependent.

Table 11–3    A Selected List of Oxidation-Reduction Indicators*

| | Color | | Transition Potential, | |
|---|---|---|---|---|
| Indicator | Oxidized | Reduced | V | Conditions |
| 5-Nitro-1, 10-phenanthroline iron(II) complex | Pale blue | Red-violet | +1.25 | 1 $F$ $H_2SO_4$ |
| 2,3'-Diphenylamine dicarboxylic acid | Blue-violet | Colorless | +1.12 | 7–10 $F$ $H_2SO_4$ |
| 1,10-Phenanthroline iron(II) complex | Pale blue | Red | +1.11 | 1 $F$ $H_2SO_4$ |
| Erioglaucin A | Bluish red | Yellow-green | +0.98 | 0.5 $F$ $H_2SO_4$ |
| Diphenylamine sulfonic acid | Red-violet | Colorless | +0.85 | Dilute acid |
| Diphenylamine | Violet | Colorless | +0.76 | Dilute acid |
| p-Ethoxychrysoidine | Yellow | Red | +0.76 | Dilute acid |
| Methylene blue | Blue | Colorless | +0.53 | 1 $F$ acid |
| Indigo tetrasulfonate | Blue | Colorless | +0.36 | 1 $F$ acid |
| Phenosafranine | Red | Colorless | +0.28 | 1 $F$ acid |

* Data taken in part from I. M. Kolthoff and V. A. Stenger, *Volumetric Analysis,* 2nd ed., Vol. 1, p. 140. New York: Interscience Publishers, 1942.

function of titrant volume. The curves that are produced are similar to those shown in Figures 11–4 and 11–5.

The reference electrode in this cell could be, but seldom is, a hydrogen electrode; it is ordinarily more convenient to make use of a secondary reference, such as the saturated calomel electrode. This substitution will cause the curves to be displaced along the vertical axis by an amount that corresponds to the difference between the potential of the secondary reference and the standard hydrogen electrode. The potentiometric end point is discussed in Chapter 13.

## 11G SUMMARY

Conclusions based upon the calculations in this chapter guide the chemist in the choice of reaction conditions and indicators for oxidation-reduction titrations. For example, the curves shown in Figures 11–4 and 11–5 clearly define a range of potentials within which an indicator must exhibit a color change for a successful titration. Nevertheless, it must be emphasized that these calculations are theoretical and that they may not necessarily take into account all factors that actually determine the applicability of an oxidation-reduction reaction. Included among such factors are the rates at which both the principal and the indicator reactions occur, the effects of electrolyte concentration, pH, and the existence of complexing species in solution, the presence of colored components other than the indicator, and variations in color perception among individuals. Our knowledge is frequently insufficient to treat the effects of these variables mathematically. Theoretical calculations can and will eliminate useless experiments and act as guides to the ones that are most likely to be successful. The final test must always come in the laboratory.

## PROBLEMS

*11-1. Write balanced net-ionic equations for the following (it may be necessary to introduce $H_2O$, $H^+$, or $OH^-$ to achieve balance):

| Reactants | Products |
|---|---|
| (a) $Fe^{3+}$, $Sn^{2+}$ | $Fe^{2+}$, $Sn^{4+}$ |
| (b) $Cr(s)$, $Ag^+$ | $Cr^{3+}$, $Ag(s)$ |
| (c) $NO_3^-$, $Cu(s)$ | $NO_2(g)$, $Cu^{2+}$ |
| (d) $MnO_4^-$, $H_2SO_3$ | $Mn^{2+}$, $SO_4^{2-}$ |
| (e) $Ti^{3+}$, $Fe(CN)_6^{3-}$ | $TiO^{2+}$, $Fe(CN)_6^{4-}$ |
| (f) $Ce^{4+}$, $H_2O_2$ | $Ce^{3+}$, $O_2$ |
| (g) $Ag(s)$, $I^-$, $Sn^{4+}$ | $AgI(s)$, $Sn^{2+}$ |
| (h) $UO_2^{2+}$, $Zn(s)$ | $U^{4+}$, $Zn^{2+}$ |
| (i) $HNO_2$, $MnO_4^-$ | $NO_3^-$, $Mn^{2+}$ |
| (j) $H_2NNH_2$, $IO_3^-$, $Cl^-$ | $ICl_2^-$, $N_2(g)$ |

11-2. Write balanced net-ionic equations for the following (it may be necessary to introduce $H_2O$, $H^+$, or $OH^-$ to achieve balance):

| Reactants | Products |
|---|---|
| (a) $MnO_4^-$, $VO^{2+}$ | $Mn^{2+}$, $V(OH)_4^+$ |
| (b) $I_2(aq)$, $H_2S(g)$ | $I^-$, $S(s)$ |
| (c) $Cr_2O_7^{2-}$, $U^{4+}$ | $Cr^{3+}$, $UO_2^{2+}$ |
| (d) $MnO_2(s)$, $Cl^-$ | $Mn^{2+}$, $Cl_2(g)$ |
| (e) $IO_3^-$, $I^-$ | $I_2(aq)$ |
| (f) $IO_3^-$, $I^-$, $Cl^-$ | $ICl_2^-$ |
| (g) $HPO_3^{2-}$, $MnO_4^-$ | $PO_4^{3-}$, $MnO_4^{2-}$ |
| (h) $SCN^-$, $BrO_3^-$ | $Br^-$, $SO_4^{2-}$, $HCN$ |
| (i) $V(OH)_4^+$, $V^{2+}$ | $VO^{2+}$ |
| (j) $MnO_4^-$, $Mn^{2+}$, $OH^-$ | $MnO_2(s)$ |

*11-3. Identify the reactant that acts as the oxidizing agent in each reaction of Problem 11-1; write a balanced equation for its half-reaction.

*11-4. Identify the reactant that acts as the reducing agent in each reaction of Problem 11-1; write a balanced equation for its half-reaction.

11-5. Identify the reactant that acts as the oxidizing agent in each reaction of Problem 11-2; write a balanced equation for its half-reaction.

11-6. Identify the reactant that acts as the reducing agent in each reaction of Problem 11-2; write a balanced equation for its half-reaction.

*11-7. Express the equivalent weight for each oxidant and each reductant in Problem 11-1 as a fraction or a multiple of its gram-formula weight.

11-8. Express the equivalent weight for each oxidant and each reductant in Problem 11-2 as a fraction or a multiple of its gram-formula weight.

**\*11–9.** Consider the following oxidation-reduction reactions:

$$2H^+ + Sn(s) \rightarrow H_2(g) + Sn^{2+}$$

$$Ag^+ + Fe^{2+} \rightarrow Ag(s) + Fe^{3+}$$

$$Sn^{4+} + H_2(g) \rightarrow Sn^{2+} + 2H^+$$

$$2Fe^{3+} + Sn^{2+} \rightarrow 2Fe^{2+} + Sn^{4+}$$

$$Sn^{2+} + Co(s) \rightarrow Sn(s) + Co^{2+}$$

(a) Write each net process in terms of two balanced half-reactions.
(b) Express each half-reaction as a reduction.
(c) Insofar as the overall reactions proceed spontaneously to the right, as written, arrange the half-reactions in (b) in order of decreasing effectiveness as electron acceptors.

**11–10.** Consider the following oxidation-reduction reactions:

$$AgBr(s) + V^{2+} \rightarrow Ag(s) + V^{3+} + Br^-$$

$$Tl^{3+} + 2Fe(CN)_6^{4-} \rightarrow Tl^+ + 2Fe(CN)_6^{3-}$$

$$2V^{3+} + Zn(s) \rightarrow 2V^{2+} + Zn^{2+}$$

$$Fe(CN)_6^{3-} + Ag(s) + Br^- \rightarrow Fe(CN)_6^{4-} + AgBr(s)$$

$$S_2O_8^{2-} + Tl^+ \rightarrow 2SO_4^{2-} + Tl^{3+}$$

(a) Write each net process in terms of two balanced half-reactions.
(b) Express each half-reaction as a reduction.
(c) Insofar as the net reactions proceed spontaneously to the right, as written, arrange the half-reactions in (b) in order of decreasing effectiveness as electron acceptors.

**\*11–11.** Calculate the half-cell potential of a Cu electrode that is immersed in
(a) $0.0800\ F\ CuSO_4$.
(b) $0.0500\ F\ NaCl$ and saturated with CuCl.
(c) $0.0250\ F$ in NaOH and saturated with $Cu(OH)_2$.
(d) a solution that has a pH of 3.00 and is saturated with both $H_2S$ and CuS.
(e) a solution that is $1.15 \times 10^{-3}\ F$ in $CuCl_2^-$ and $0.600\ F$ in $Cl^-$. Reaction:

$$Cu^+ + 2Cl^- \rightleftharpoons CuCl_2^- \qquad K_f = 8.7 \times 10^4$$

**11–12.** Calculate the half-cell potential of an Ag electrode immersed in
(a) $0.0400\ F\ AgNO_3$.
(b) $0.0500\ F$ in KBr and saturated with AgBr.
(c) $0.0412\ F$ in $MgI_2$ and saturated with AgI.
(d) $0.0200\ F\ K_2CrO_4$ and saturated with $Ag_2CrO_4$.
(e) $0.200\ F$ in KCN and $3.60 \times 10^{-3}\ F$ in $Ag(CN)_2^-$.

**\*11–13.** Calculate the potential of a Pt electrode that is immersed in a solution that is
(a) $0.120\ F$ in $Sn(SO_4)_2$ and $2.50 \times 10^{-3}\ F$ in $SnSO_4$.
(b) $6.75 \times 10^{-3}\ F$ in $K_4Fe(CN)_6$ and $8.00 \times 10^{-2}\ F$ in $K_3Fe(CN)_6$.
(c) $0.0150\ F$ in $K_2PtCl_4$ and $0.0700\ F$ in KCl.
(d) $0.0400\ F$ in $U^{4+}$, $0.0250\ F$ in $UO_2^{2+}$, and $0.0800\ F$ in $HClO_4$.
(e) $0.0250\ F$ in $VOSO_4$, $0.0100\ F$ in $V_2(SO_4)_3$, and $0.100\ F$ in $HClO_4$.
(f) buffered to a pH of 8.70 and saturated with $H_2(g)$ at 1.00 atm.

(g) 0.0100 $F$ in $KMnO_4$, saturated with $MnO_2$, and buffered to a pH of 6.00.

(h) 0.0480 $F$ in $KBr$ and saturated with $Br_2(l)$.

(i) prepared by mixing 50.0 mL of 0.0680 $F$ $Ce(SO_4)_2$ with an equal volume of 0.100 $F$ $SnCl_2$.

(j) prepared by mixing 25.0 mL of 0.0120 $F$ $K_2Cr_2O_7$ with 20.0 mL of 0.150 $F$ $FeSO_4$ and has a pH of 2.00.

**11–14.** Calculate the potential of a Pt electrode that is immersed in a solution that is

(a) 0.180 $F$ in $VCl_3$ and 0.0425 $F$ in $VCl_2$.

(b) 0.00615 $F$ in $I_3^-$ and 0.288 $F$ in $KI$.

(c) 0.0150 $F$ in $FeSO_4$ and 0.0480 $F$ in $Fe_2(SO_4)_3$.

(d) $2.14 \times 10^{-2}$ $F$ in $SbO^+$, saturated with $Sb_2O_5$, and buffered to a pH of 3.00.

(e) 0.0960 $F$ in $K_2Cr_2O_7$, 0.0408 $F$ in $Cr_2(SO_4)_3$, and 0.200 $F$ in $HClO_4$.

(f) 0.0500 $F$ in $H_2O_2$, buffered to a pH of 6.00, and saturated with $O_2(g)$ at 1.00 atm.

(g) 0.0275 $F$ in $HNO_3$ and 0.0140 $F$ in $HNO_2$.

(h) 0.300 $F$ in $KI$ and saturated with $CuI(s)$.

(i) prepared by mixing 25.0 mL of 0.100 $F$ $SnCl_2$ and an equal volume of 0.120 $F$ $FeCl_3$.

(j) prepared by mixing 30.0 mL of 0.0100 $F$ $V_2(SO_4)_3$ with 25.0 mL of 0.0360 $F$ $V(OH)_4^+$ and has a pH of 1.00.

**\*11–15.** Indicate whether each of the accompanying half-cells will act as the anode or as the cathode when coupled with a standard hydrogen electrode in a galvanic cell.

(a) $Sn|SnCl_2(0.0850\ F)$

(b) $Ag|AgBr(sat'd),\ KBr(0.0937\ F)$

(c) $Cu|Cu(OH)_2(sat'd),\ H_3O^+(1.00 \times 10^{-10}\ F)$

(d) $Pt,\ H_2(0.964\ atm)|HCl(6.33 \times 10^{-3}\ F)$

(e) $Pt,\ O_2(740\ torr)|HClO_4(1.88 \times 10^{-2}\ F)$

(f) $Ag|Ag(S_2O_3)_2^{3-}(4.72 \times 10^{-3}\ F),\ S_2O_3^{2-}(0.106\ F)$

(g) $Hg|HgY^{2-}(3.25 \times 10^{-3}\ F),\ HY^{3-}(0.0500\ F),\ H_3O^+(1.00 \times 10^{-8}\ F)$

**11–16.** Indicate whether each of the accompanying half-cells will act as the anode or as the cathode when coupled with a standard hydrogen electrode in a galvanic cell.

(a) $Cu|CuCl_2(0.0444\ F)$

(b) $Ag|AgI(sat'd),\ KI(0.127\ F)$

(c) $Hg|Hg_2I_2(sat'd),\ KI(3.75 \times 10^{-2}\ F)$

(d) $Pt,\ O_2(1.04\ atm)|HClO_4(0.225\ F)$

(e) $Pt,\ H_2(775\ torr)|HCl(3.00 \times 10^{-2}\ F)$

(f) $Ag|Ag(CN)_2^-(0.0841\ F),\ KCN(0.0337\ F)$

(g) $Cu|CuY^{2-}(0.00714\ F),\ HY^{3-}(0.0362\ F),\ H_3O^+(1.00 \times 10^{-9}\ F)$

**\*11–17.** Calculate theoretical potentials for the following cells. Indicate whether each cell, as written, is galvanic or electrolytic.

(a) $Ni|Ni^{2+}(7.50 \times 10^{-3}\ M)||Cu^{2+}(9.00 \times 10^{-3}\ M)|Cu$

(b) $Co|Co^{2+}(6.94 \times 10^{-3}\ M)||Zn^{2+}(4.60 \times 10^{-3}\ M)|Zn$

(c) $Sn|Sn^{2+}(3.87 \times 10^{-3}\ M)||AgI(sat'd),\ I^-(0.0800\ M)|Ag$

(d) $Pt,\ H_2(740\ torr)|H^+(5.67 \times 10^{-7}\ M)||Pd^{2+}(0.0120\ M)|Pd$

(e) $Pt|Tl^{3+}(4.23 \times 10^{-3}\ M),\ Tl^+(1.66 \times 10^{-1}\ M)||H^+(2.17 \times 10^{-3}\ M),\ O_2(1.00\ atm)|Pt$

(f) $Hg|Hg^{2+}(4.24 \times 10^{-3} M)\|Fe^{3+}(1.00 \times 10^{-1} M)$, $Fe^{2+}(5.00 \times 10^{-3} M)$ $|Pt$

(g) $Bi|BiO^+(6.24 \times 10^{-3} M)$, $H^+(1.50 \times 10^{-2} M)\|Cl^-(5.18 \times 10^{-2} M)$, $BiCl_4^-(3.87 \times 10^{-2} M)|Bi$

(h) $Pt, H_2(1.00 \text{ atm})|NaCN(0.0866 F)\|SCE$

(i) $Pt, H_2(1.00 \text{ atm})|HCN(0.125 F), NaCN(0.0866 F)\|SCE$

(j) $Pt|TiO^{2+}(4.18 \times 10^{-3} M)$, $Ti^{3+}(7.26 \times 10^{-3} M)$, $H^+(6.93 \times 10^{-2} M)$ $\|Sn^{4+}(2.40 \times 10^{-1} M)$, $Sn^{2+}(8.28 \times 10^{-3} M)|Pt$

**11–18.** Calculate the theoretical potentials for the following cells. Indicate whether each cell, as written, is galvanic or electrolytic.

(a) $Pb|Pb^{2+}(5.00 \times 10^{-2} M)\|Fe^{2+}(6.41 \times 10^{-3} M)|Fe$

(b) $Cu|Cu^{2+}(1.44 \times 10^{-3} M)\|Ag^+(0.100 M)|Ag$

(c) $Hg|Hg^{2+}(0.0915 M)\|Cl^-(2.50 \times 10^{-4} M), AgCl(sat'd)|Ag$

(d) $Pt, H_2(1.00 \text{ atm})|H^+(1.00 \times 10^{-6} M)\|Pb^{2+}(0.223)|Pb$

(e) $Pt, O_2(798 \text{ torr})|H^+(5.00 \times 10^{-9} M)\|I_3^-(2.76 \times 10^{-3} M), I^-(0.0839 M)$ $|Pt$

(f) $Pt|Fe^{3+}(5.00 \times 10^{-3} M), Fe^{2+}(0.100 M)\|Hg^{2+}(4.24 \times 10^{-3} M)|Hg$

(g) $Pb|PbSO_4(sat'd), SO_4^{2-}(7.44 \times 10^{-2} M)\|Pb^{2+}(1.40 \times 10^{-1} M)|Pb$

(h) $Pt, H_2(1.00 \text{ atm})|HCN(0.125 F)\|SCE$

(i) $Pt, H_2(1.00 \text{ atm})|Na_2CO_3(0.0618 F), NaHCO_3(0.247 F)\|SCE$

(j) $Pt|UO_2^{2+}(2.23 \times 10^{-4} M)$, $U^{4+}(9.75 \times 10^{-3} M)$, $H^+(1.50 \times 10^{-3} M)$ $\|I^-(5.65 \times 10^{-4} M), CuI(sat'd)|Cu$

**11–19.** Generate equilibrium constant expressions for the following reactions. Calculate numerical values for $K_{eq}$.

*(a) $Pd^{2+} + Hg(l) \rightleftharpoons Pd(s) + Hg^{2+}$

(b) $Cu^{2+} + Pb(s) \rightleftharpoons Cu(s) + Pb^{2+}$

*(c) $Cd(s) + Sn^{2+} \rightleftharpoons Cd^{2+} + Sn(s)$

(d) $Cu^{2+} + Sn^{2+} \rightleftharpoons Cu(s) + Sn^{4+}$

*(e) $Sn^{2+} + 2Cr^{2+} \rightleftharpoons Sn(s) + 2Cr^{3+}$

(f) $HO_2^- + 2MnO_4^{2-} + H_2O \rightleftharpoons 3OH^- + 2MnO_4^-$

(g) $PbO_2(s) + H_2O_2 + 2H^+ \rightleftharpoons Pb^{2+} + O_2(g) + 2H_2O$

*(h) $VO^{2+} + V^{2+} + 2H^+ \rightleftharpoons 2V^{3+} + H_2O$

(i) $TiO^{2+} + Ti^{2+} + 2H^+ \rightleftharpoons 2Ti^{3+} + H_2O$

*(j) $2MnO_4^- + 5Tl^+ + 16H^+ \rightleftharpoons 2Mn^{2+} + 5Tl^{3+} + 8H_2O$

(k) $3Sn^{2+} + 2BiO^+ + 4H^+ \rightleftharpoons 3Sn^{4+} + 2Bi(s) + 2H_2O$

*(l) $2V(OH)_4^+ + U^{4+} \rightleftharpoons 2VO^{2+} + UO_2^{2+} + 4H_2O$

**11–20.** Calculate $E^0$ for the following processes.

(a) $Zn(OH)_2(s) + 2e \rightleftharpoons Zn(s) + 2OH^-$

(b) $AgSCN(s) + e \rightleftharpoons Ag(s) + SCN^-$

(c) $BiOCl(s) + 2H^+ + 3e \rightleftharpoons Bi(s) + H_2O + Cl^-$

(d) $CdCO_3(s) + 2e \rightleftharpoons Cd(s) + CO_3^{2-}$

(e) $Cu_2P_2O_7(s) + 4e \rightleftharpoons 2Cu(s) + P_2O_7^{4-}$
    (for $Cu_2P_2O_7$, $K_{sp} = 8.4 \times 10^{-16}$)

(f) $Ag_2C_2O_4(s) + 2e \rightleftharpoons 2Ag(s) + C_2O_4^{2-}$
    (for $Ag_2C_2O_4$, $K_{sp} = 3.5 \times 10^{-11}$)

**\*11–21.** The formation constant for $Zn(NH_3)_4^{2+}$ is $2.5 \times 10^9$. Calculate $E^0$ for the process

$$Zn(NH_3)_4^{2+} + 2e \rightleftharpoons Zn(s) + 4NH_3$$

**11–22.** Calculate $E^0$ for the process

$$Ni(CN)_4^{2-} + 2e \rightleftharpoons Ni(s) + 4CN^-$$

given that $K_f$ for $Ni(CN)_4^{2-}$ is $1.0 \times 10^{22}$.

**\*11–23.** Given the formation constants

$$Fe^{3+} + Y^{4-} \rightleftharpoons FeY^- \qquad K_f = 1.0 \times 10^{25}$$
$$Fe^{2+} + Y^{4-} \rightleftharpoons FeY^{2-} \qquad K_f = 2.1 \times 10^{14}$$

calculate $E^0$ for the process

$$FeY^- + e \rightleftharpoons FeY^{2-}$$

**11–24.** Calculate $E^0$ for the process

$$Cu(NH_3)_4^{2+} + e \rightleftharpoons Cu(NH_3)_2^+ + 2NH_3$$

given that

$$Cu^+ + 2NH_3 \rightleftharpoons Cu(NH_3)_2^+ \qquad K_f = 7.2 \times 10^{10}$$
$$Cu^{2+} + 4NH_3 \rightleftharpoons Cu(NH_3)_4^{2+} \qquad K_f = 2.0 \times 10^{13}$$

**\*11–25.** Calculate $K_{sp}$ for $ZnCO_3$, given that

$$Zn^{2+} + 2e \rightleftharpoons Zn(s) \qquad E^0 = -0.763 \text{ V}$$
$$ZnCO_3(s) + 2e \rightleftharpoons Zn(s) + CO_3^{2-} \qquad E^0 = -1.060 \text{ V}$$

**11–26.** Use the accompanying standard electrode potentials to calculate $K_{sp}$ for copper(II) acetate, $Cu(OAc)_2$.

$$Cu^{2+} + 2e \rightleftharpoons Cu(s) \qquad E^0 = 0.337 \text{ V}$$
$$Cu(OAc)_2(s) + 2e \rightleftharpoons Cu(s) + 2OAc^- \qquad E^0 = 0.276 \text{ V}$$

**\*11–27.** The standard electrode potential for the process

$$Cu(S_2O_3)_2^{2-} + 2e \rightleftharpoons Cu(s) + 2S_2O_3^{2-}$$

is $-0.020$ V. Use this information to calculate the formation constant for $Cu(S_2O_3)_2^{2-}$.

**11–28.** Calculate $K_f$ for $HgI_4^{2-}$, given that the standard electrode potential for the process

$$HgI_4^{2-} + 2e \rightleftharpoons Hg(l) + 4I^-$$

is $-0.033$ V.

**\*11–29.** The hypothetical cell

$$Zn \mid Zn(NH_3)_4^{2+}(1.20 \times 10^{-3} \, M), NH_3(0.150 \, M) \parallel SHE$$

develops a potential of 1.029 V.
 (a) Calculate $K_f$ for $Zn(NH_3)_4^{2+}$.
 (b) If a saturated calomel electrode ($E_{SCE} = +0.244$ V) is substituted for the standard hydrogen electrode, what will be the potential of the resulting cell?

**11–30.** A Pd electrode in contact with a solution that is 0.0235 $M$ in $PdCl_4^{2-}$ and 0.720 $M$ in $Cl^-$ acts as the anode when coupled with a saturated calomel electrode ($E_{SCE} = +0.244$ V).
  (a) Use the shorthand notation to describe this cell.
  (b) Calculate $K_f$ for $PdCl_4^{2-}$, given that this cell develops a potential of −0.318 V.

**\*11–31.** Calculate $K_a$ for sulfanilic acid (HA), given that the cell

$$Pt, H_2(1.00\ atm)|HA(0.0285\ F), NaA(0.0633\ F)\|SCE$$

has a potential of 0.543 V.

**11–32.** Calculate $K_a$ for picric acid (HP), given that the potential of the cell

$$Pt, H_2(1.00\ atm)|HP(0.0960\ F), NaP(0.120\ F)\|SCE$$

is 0.257 V.

**\*11–33.** The cell

$$SCE\|Ag_2C_2O_4(sat'd), C_2O_4^{2-}(0.108\ M)|Ag$$

develops a potential of 0.277 V. Use this information to calculate $K_{sp}$ for $Ag_2C_2O_4$.

**11–34.** Calculate $K_{sp}$ for $Pb(SCN)_2$, given that the cell

$$Pb|Pb(SCN)_2(sat'd), SCN^-(0.0920\ M)\|SCE$$

has a potential of 0.406 V.

**\*11–35.** Calculate the equivalence-point potential for the following titrations. Where necessary, use 0.0500 $N$ for the concentrations of analyte and titrant at the outset of the titration, 0.1000 $M$ for [H$^+$] at the equivalence point, and a partial pressure of 1.00 atm for any gas.
  (a) $Fe^{3+} + Cr^{2+} \rightleftharpoons Fe^{2+} + Cr^{3+}$
  (b) $V(OH)_4^+ + Fe(CN)_6^{4-} + 2H^+ \rightleftharpoons VO^{2+} + Fe(CN)_6^{3-} + 3H_2O$
  (c) $H_2SO_3 + 2Ce^{4+} + H_2O \rightleftharpoons SO_4^{2-} + 2Ce^{3+} + 4H^+$ (1 $F$ HNO$_3$)
  (d) $2MnO_4^- + 5HNO_2 + H^+ \rightleftharpoons 2Mn^{2+} + 5NO_3^- + 3H_2O$
  (e) $H_3AsO_4 + 2Ti^{3+} + H_2O \rightleftharpoons H_3AsO_3 + 2TiO^{2+} + 2H^+$
  (f) $VO^{2+} + V^{2+} + 2H^+ \rightleftharpoons 2V^{3+} + H_2O$
  (g) $2MnO_4^- + 3Mn^{2+} + 2H_2O \rightleftharpoons 5MnO_2(s) + 4H^+$

**11–36.** Calculate the equivalence-point potential for the following titrations. Where necessary, use 0.0500 $M$ for the concentrations of analyte and titrant at the outset of the titration, 0.1000 $M$ for [H$^+$] at the equivalence point, and a partial pressure of 1.00 atm for any gas.
  (a) $Tl^{3+} + Sn^{2+} \rightleftharpoons Tl^+ + Sn^{4+}$
  (b) $TiO^{2+} + V^{2+} + 2H^+ \rightleftharpoons Ti^{3+} + V^{3+} + H_2O$
  (c) $2Ce^{4+} + H_3AsO_3 + H_2O \rightleftharpoons 2Ce^{3+} + H_3AsO_4 + 2H^+$ (1 $F$ HNO$_3$)
  (d) $2MnO_4^- + 5U^{4+} + 2H_2O \rightleftharpoons 2Mn^{2+} + 5UO_2^{2+} + 4H^+$
  (e) $2Cr^{2+} + UO_2^{2+} + 4H^+ \rightleftharpoons 2Cr^{3+} + U^{4+} + 2H_2O$
  (f) $V(OH)_4^+ + V^{3+} \rightleftharpoons 2VO^{2+} + 2H_2O$
  (g) $4MnO_4^- + 5N_2H_5^+ + 7H^+ \rightleftharpoons 4Mn^{2+} + 5N_2(g) + 16H_2O$

**11-37.** Generate curves for the following titrations. Calculate the electrode potential of the system after the addition of 5.00, 10.00, 15.00, 19.00, 20.00, 21.00, 25.00, and 30.0 mL of 0.1000 $N$ titrant to 50.00 mL of 0.0400 $N$ analyte. Where necessary, consider that $[H^+]$ has a constant value of 0.200 $M$.

| | Analyte, 0.0400 $N$ | Titrant, 0.1000 $N$ |
|---|---|---|
| *(a) | $Cr^{2+}$ | $Fe^{3+}$ |
| (b) | $Ti^{3+}$ | $Ce^{4+}$ (use 1.44 V for standard potential) |
| *(c) | $U^{4+}$ | $Tl^{3+}$ |
| (d) | $VO^{2+}$ | $V^{2+}$ |
| *(e) | $H_2SO_3$ | $MnO_4^-$ |
| (f) | $H_2O_2$ | $Fe(CN)_6^{4-}$ |

# Chapter 12
# Applications of Oxidation-Reduction Titrations

This chapter is concerned with the preparation of standard solutions of oxidants and reductants and their applications in analytical chemistry. In addition, auxiliary reagents that convert an analyte to a single oxidation state are described. Several excellent oxidizing reagents are available for the preparation of standard solutions. The number of reducing reagents is considerably more limited owing to their susceptibility to air oxidation.

## 12A AUXILIARY OXIDIZING AND REDUCING REAGENTS

The analyte in an oxidation-reduction titration must be in a single oxidation state before the titration can be carried out; otherwise, the results are meaningless. Often, however, the steps that precede the titration (solution of the sample and separation of interferences) convert the analyte to a mixture of oxidation states.[1] An auxiliary reagent is therefore required to transform the analyte quantitatively to a single oxidation state prior to the titration itself. For example, the solution formed when an iron-containing sample is dissolved usually contains a mixture of iron(II) and iron(III). This solution must then be treated with a reducing agent that will convert all of the iron(III) to the +2 state or, alternatively, with an oxidizing agent that forms iron(III) quantitatively.

Ideally, such an auxiliary reagent should not alter the oxidation states of the other components of the solution. Furthermore, some means is needed for getting rid of the excess reagent after its reaction with the analyte is complete. Otherwise, the auxiliary reagent will inevitably interfere by consuming the standard solution. For example, any reagent capable of converting iron(III) quantitatively to the +2 state would certainly react with the standard oxidizing solution used to titrate the iron; a positive determinate error would result.

### 12A–1 Auxiliary Reducing Reagents

Reagents that find general application in the prereduction of analytes are described in this section.

---

[1] For a brief summary of auxiliary reagents, see: J. A. Goldman and V. A. Stenger, in *Treatise on Analytical Chemistry,* I. M. Kolthoff and P. J. Elving, Eds., Part I, Vol. 11, pp. 7204–7206. New York: John Wiley & Sons, 1975.

**Metals**    An examination of standard electrode potential data reveals a number of effective reducing agents among pure metals.[2] Such elements as zinc, cadmium, aluminum, lead, nickel, copper, mercury, and silver have all proved useful reagents for prereduction. Sticks or coils of the metal can be simply lifted from the solution and washed thoroughly after reduction is complete. Filtration is needed to remove granular or powdered forms of the metal. An alternative to filtration is the use of a *reductor* such as that shown in Figure 12 – 1. The finely divided metal is held in a vertical glass tube; the solution to be reduced is then drawn through the column with a moderate vacuum. The metal in a reductor is ordinarily sufficient for hundreds of reductions.

The typical *Jones reductor* has a diameter of about 2 cm and is packed with a 40- to 50-cm column of amalgamated zinc. Amalgamation is accomplished by allowing zinc granules to stand briefly in a solution of mercury(II) chloride:

$$2Zn(s) + Hg^{2+} \rightarrow Zn^{2+} + Zn(Hg)$$

Zinc amalgam is nearly as effective as a reducing agent as the pure metal; it has the important additional virtue of inhibiting the reduction of hydrogen ions by zinc, a parasitic reaction that not only needlessly uses up the reducing agent but also heavily contaminates the sample solution with zinc(II) ion. Solutions that are quite acidic can be passed through a Jones reductor without significant hydrogen formation.

The liquid in a Jones reductor must always cover the packing completely, whether in use or in storage; otherwise, air oxidation will cause formation of basic salts that tend to clog the reductor.

Table 12 – 1 lists the principal applications of the Jones reductor. Also listed in this table are reductions that can be accomplished with the *Walden reductor,* in which granular metallic silver serves as the reductant. The metal is prepared by reducing a solution containing about 30 g of silver nitrate with metallic copper. The resulting suspension of finely divided silver is then poured into a narrow glass column to give about 10 cm of packing. When not in use, the silver is covered with 1 *F* hydrochloric acid.

The reducing strength of silver is enhanced by the presence of chloride ion, and for this reason hydrochloric acid is always used in conjunction with a Walden reductor. The coating of silver chloride produced as the result of use is removed periodically by dipping a zinc rod into the solution that covers the packing.

Table 12 – 1 suggests that the Walden reductor is somewhat more selective in its action than is the Jones reductor.

**Gaseous Reductants**    Both hydrogen sulfide and sulfur dioxide are reasonably effective reducing agents and have been used for prereduction. Excesses of these reagents are readily eliminated from an acidified solution by heating. The

---

[2] For a more extensive discussion of metal reductants, the reader should see: I. M. Kolthoff and R. Belcher, *Volumetric Analysis,* Vol. III, pp. 11 – 23. New York: Interscience Publishers, 1957; W. I. Stephen, *Ind. Chemist,* **1952,** *28,* 13, 55, 197.

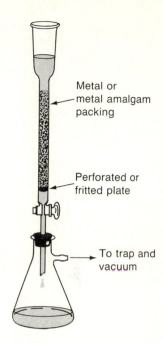

Metal or metal amalgam packing

Perforated or fritted plate

To trap and vacuum

*Figure 12–1* A metal or metal amalgam reductor.

reactions of both gases tend to be slow; 30 min (or more) may be needed to complete the reduction and rid the solution of excess reagent. A further disadvantage is that both gases are noxious and toxic. The use of other reductants is much preferred.

**Tin(II) Chloride**   The reduction of iron(III) to the divalent state is conveniently accomplished with tin(II) chloride. The chemistry of this application is discussed in Chapter 20, Section 20E–3.

**Table 12–1**   *Uses of the Walden Reductor and the Jones Reductor\**

| Walden Reductor<br>$Ag(s) + Cl^- \rightleftarrows AgCl(s) + e$ | Jones Reductor<br>$Zn(s) \rightleftarrows Zn^{2+} + 2e$ |
|---|---|
| $e + Fe^{3+} \rightarrow Fe^{2+}$ | $e + Fe^{3+} \rightarrow Fe^{2+}$ |
| $e + Cu^{2+} \rightarrow Cu^+$ | $Cu^{2+}$ reduced to metallic Cu |
| $e + H_2MoO_4 + 2H^+ \rightarrow MoO_2^+ + 2H_2O$ | $3e + H_2MoO_4 + 6H^+ \rightarrow Mo^{3+} + 4H_2O$ |
| $2e + UO_2^{2+} + 4H^+ \rightarrow U^{4+} + 2H_2O$ | $2e + UO_2^{2+} + 4H^+ \rightarrow U^{4+} + 2H_2O$ |
| | $3e + UO_2^{2+} + 4H^+ \rightarrow U^{3+} + 2H_2O$† |
| $e + V(OH)_4^+ + 2H^+ \rightarrow VO^{2+} + 3H_2O$ | $3e + V(OH)_4^+ + 4H^+ \rightarrow V^{2+} + 4H_2O$ |
| $TiO^{2+}$ not reduced | $e + TiO^{2+} + 2H^+ \rightarrow Ti^{3+} + H_2O$ |
| $Cr^{3+}$ not reduced | $e + Cr^{3+} \rightarrow Cr^{2+}$ |

\* From I. M. Kolthoff and R. Belcher, *Volumetric Analysis,* Vol. III, p. 12. New York: Interscience Publishers, 1957. With permission.

† A mixture of oxidation states is obtained. The Jones reductor may still be used for the analysis of uranium, however, because any $U^{3+}$ produced can be converted to $U^{4+}$ by a brief period of aeration.

## 12A – 2 Auxiliary Oxidizing Reagents

**Sodium Bismuthate**   Sodium bismuthate is an extremely powerful oxidizing agent capable, for example, of converting manganese(II) quantitatively to permanganate. It is a sparingly soluble solid with a formula that is usually written as $NaBiO_3$, although its exact composition is somewhat uncertain. Oxidations are performed by suspending the bismuthate in the analyte solution and boiling for a brief period. The unused reagent is then removed by filtration.

**Ammonium Peroxodisulfate**   Ammonium peroxodisulfate, $(NH_4)_2S_2O_8$, is also a powerful oxidizing agent, which in acid solution will convert chromium(III) to dichromate, cerium(III) to its tetravalent state, and manganese(II) to permanganate. The half-reaction is

$$S_2O_8^{2-} + 2e \rightleftharpoons 2SO_4^{2-} \qquad E^0 = 2.01 \text{ V}$$

The oxidations are catalyzed by traces of silver ion. The excess reagent is readily decomposed by a brief period of boiling:

$$2S_2O_8^{2-} + 2H_2O \rightarrow 4SO_4^{2-} + O_2(g) + 4H^+$$

Potassium peroxodisulfate is an acceptable substitute for the ammonium salt in many applications.

**Sodium Peroxide, Hydrogen Peroxide**   Peroxide is a convenient oxidizing agent either as the solid sodium salt or as a dilute solution of the acid. The half-reaction for hydrogen peroxide in acid solution is

$$H_2O_2 + 2H^+ + 2e \rightleftharpoons 2H_2O \qquad E^0 = 1.78 \text{ V}$$

After oxidation is complete, the solution is freed of excess reagent by boiling:

$$2H_2O_2 \rightarrow 2H_2O + O_2(g)$$

## 12B APPLICATIONS OF STANDARD OXIDANTS

Table 12–2 summarizes the properties of the most widely used volumetric oxidizing reagents. Note that the standard potentials for these reagents vary from 0.5 to 1.6 V. The choice among them depends upon the strength of the analyte as a reducing agent, the rate of reaction between oxidant and analyte, the stability of standard oxidant solutions, and the availability of a satisfactory indicator for end-point detection.

## 12B – 1 Potassium Permanganate

Potassium permanganate, a powerful oxidant, is perhaps the most widely used of all standard oxidizing reagents. The color of permanganate solution is so intense that it can serve as an indicator for most titrations. The reagent is available at modest cost. Disadvantages to the use of permanganate solutions

*Table 12-2  Some Common Oxidants Employed for Standard Solutions*

| Reagent and Formula | Reduction Product | Standard Potential, V | Primary Standard for | Indicator* | Stability† |
|---|---|---|---|---|---|
| Potassium permanganate, $KMnO_4$ | $Mn^{2+}$ | 1.51 | $Na_2C_2O_4$, Fe, $As_2O_3$ | $MnO_4^-$ | (b) |
| Potassium bromate, $KBrO_3$ | $Br^-$ | 1.44 | $KBrO_3$ | (1) | (a) |
| Cerium(IV), $Ce^{4+}$ | $Ce^{3+}$ | 1.44 | $Na_2C_2O_4$, Fe, $As_2O_3$ | (2) | (a) |
| Potassium dichromate, $K_2Cr_2O_7$ | $Cr^{3+}$ | 1.33 | $K_2Cr_2O_7$, Fe | (3) | (a) |
| Periodic acid, $H_5IO_6$ | $IO_3^-$ | 1.60 | $As_2O_3$ | starch | (b) |
| Potassium iodate, $KIO_3$ | $ICl_2^-$ | 1.24 | $KIO_3$ | (4) | (a) |
| Iodine, $I_2$ | $I^-$ | 0.536 | $BaS_2O_3 \cdot H_2O$, $As_2O_3$ | starch | (c) |

* (1) $\alpha$-Naphthoflavone; (2) orthophenanthroline iron(II) complex (ferroin); (3) diphenylamine sulfonic acid; (4) disappearance of $I_2$ from a layer of chloroform.
† (a) Indefinitely stable; (b) moderately stable, requires periodic standardization; (c) somewhat unstable, requires frequent standardization.

include their tendency to oxidize chloride ion, which may preclude their use with hydrochloric acid solutions, and their relatively limited stability. In addition, under some conditions, permanganate ion may yield a mixture of reduction products.

**Reactions of Permanganate Ion**   Most titrations with potassium permanganate ion are carried out in a solution that is 0.1 $N$ (or greater) in mineral acid; under these circumstances, manganese(II) is the product:

$$MnO_4^- + 5e + 8H^+ \rightarrow Mn^{2+} + 4H_2O \qquad E^0 = 1.51 \text{ V}$$

Permanganate oxidations are typically rapid in acidic solution. A notable exception to this statement is the reaction with oxalic acid, which requires elevated temperatures.

When permanganate titrations are performed in solutions that are weakly acidic (pH > 4), neutral, or weakly alkaline, the product is brown manganese dioxide. The half-reaction is

$$MnO_4^- + 3e + 2H_2O \rightarrow MnO_2(s) + 4OH^- \qquad E^0 = 1.695 \text{ V}$$

Titration of certain species can be carried out to advantage under these conditions. For example, cyanide is oxidized to cyanate, manganese(II) is oxidized to manganese dioxide, and hydrazine is oxidized to nitrogen. Finally, sulfide, sulfite, and thiosulfate are converted to sulfate.

Manganese(III) solutions tend to undergo disproportionation

$$2Mn^{3+} + 2H_2O \rightarrow MnO_2(s) + Mn^{2+} + 4H^+$$

and thus lack stability. However, manganese(III) forms several complexes that impart sufficient stability to permit existence of the $+3$ state in aqueous solution. Lingane[3] has made use of this property to titrate manganese(II) with permanganate in highly concentrated solutions of pyrophosphate; the reaction may be expressed as

$$MnO_4^- + 4Mn^{2+} + 15H_2P_2O_7^{2-} + 8H^+ \rightarrow 5Mn(H_2P_2O_7)_3^{3-} + 4H_2O$$

Permanganate ion undergoes a one-electron change to manganate ion, $MnO_4^{2-}$, in solutions that are about $1\ N$ (or greater) in sodium hydroxide. Alkaline solutions of permanganate have proved useful for the analysis of organic compounds.

**End Point**    An obvious property of a potassium permanganate solution is its intense purple color, which is sufficient to serve as an indicator for most titrations. As little as 0.01 to 0.02 mL of a 0.02 $F$ solution will impart a perceptible color to 100 mL of water. If the permanganate solution is very dilute, diphenylamine sulfonic acid or the orthophenanthroline complex of iron(II) (see Table 11–3) will provide a sharper end point.

The permanganate end point is not permanent. Decolorization of acidic solutions results from reaction between the small excess of permanganate ion and the relatively large concentration of manganese(II) that exist at the end point:

$$2MnO_4^- + 3Mn^{2+} + 2H_2O \rightarrow 5MnO_2(s) + 4H^+$$

The equilibrium constant for this reaction is about $10^{47}$. Thus, the concentration of permanganate in equilibrium with manganese(II) is vanishingly small even in highly acidic solutions. Fortunately, the rate at which this equilibrium is approached is so slow that the end point fades only gradually.

**Stability of Permanganate Solutions**    Aqueous solutions of permanganate are not entirely stable because the ion tends to oxidize water. The process may be depicted by the equation

$$4MnO_4^- + 2H_2O \rightarrow 4MnO_2(s) + 3O_2(g) + 4OH^-$$

The constant for this equilibrium indicates that the products are favored in neutral solution. Fortunately, however, the reaction is so slow that a properly

---

[3] J. J. Lingane and R. Karplus, *Ind. Eng. Chem., Anal. Ed.,* **1946,** *18,* 191.

prepared permanganate solution is reasonably stable. The decomposition is catalyzed by light, heat, acids, bases, manganese(II), and manganese dioxide. The stability of a permanganate solution is thus dependent upon minimizing the influence of these effects.

The presence of manganese dioxide greatly accelerates the decomposition of permanganate solutions; since it is also a product of the decomposition, the solid has an autocatalytic effect upon the process. Photochemical catalysis of the decomposition will occur if a permanganate solution is allowed to stand in a buret for an extended period. Evidence for this decomposition is the formation of manganese dioxide as a brown stain on the glass.

Heating of acidic solutions containing an excess of permanganate results in a decomposition error that cannot be compensated for with a blank. On the other hand, it is perfectly permissible to titrate hot, acidic solutions of reductants with standard permanganate solutions because at no time in such a titration is the reagent concentration large enough to cause a measurable uncertainty.

**Preparation and Storage of Permanganate Solutions**  A stable permanganate solution can be produced, provided a number of precautions are observed. Principal among these is the removal of manganese dioxide, an inevitable contaminant in solid potassium permanganate, and a product when permanganate oxidizes organic matter in the water used to prepare the solution. Removal of manganese dioxide by filtration markedly enhances the stability of standard permanganate solutions. Sufficient time must be allowed to permit complete oxidation of contaminants in the water before filtration; the solution may be boiled gently to hasten the process. Paper is an unsatisfactory filtering medium because permanganate ion reacts with it to form more of the undesirable oxide.

Standardized permanganate solutions should be stored in the dark. Filtration and restandardization will be required if any solid is detected in the solution or on the walls of the storage bottle. In any event, restandardization every 1 or 2 weeks is a good precautionary measure.[4]

**Standardization Against Sodium Oxalate**  Sodium oxalate is a widely used primary standard for solutions of permanganate ion. In acidic solutions, potassium permanganate oxidizes oxalic acid to carbon dioxide and water:

$$2MnO_4^- + 5H_2C_2O_4 + 6H^+ \rightarrow 2Mn^{2+} + 10CO_2(g) + 8H_2O$$

This reaction is complex and proceeds slowly even at elevated temperature unless manganese(II) is present as a catalyst. Thus, several seconds are required to decolorize the first additions of permanganate to a hot oxalic acid solution. As soon as the concentration of manganese(II) has become appreciable, however, the reaction proceeds rapidly as a result of autocatalysis.

---

[4] Reports of unusually stable permanganate solutions can be found in the literature. Durham, for example, has reported that the normality of two permanganate solutions changed by less that 0.6% over a period of nine years (B. W. Durham, *Anal. Chem.,* **1979,** *51,* 922A).

The stoichiometry of the reaction between permanganate ion and oxalic acid was investigated in detail by McBride[5] and later by Fowler and Bright.[6] The former devised a procedure in which the oxalic acid is titrated slowly at a temperature between 60 and 90°C until a faint pink color due to the reagent persists. Fowler and Bright subsequently demonstrated that the amount of permanganate used in this titration tends to be from 0.1 to 0.4% less than theoretical, due perhaps to slight air oxidation of the oxalic acid:

$$H_2C_2O_4 + O_2(g) \rightarrow H_2O_2 + 2CO_2(g)$$

The hydrogen peroxide is then believed to decompose spontaneously in the hot solution to oxygen and water.

The Fowler and Bright method calls for the rapid addition of 90 to 95% of the required permanganate to a cool solution of the oxalic acid. After all of this reagent has reacted, as indicated by the disappearance of the color, the solution is heated to about 60°C and titrated as before. Although it minimizes the air oxidation of oxalic acid and gives data that appear to be in exact accord with the theoretical stoichiometry, this method suffers from the disadvantage of requiring knowledge of the approximate normality of the permanganate solution so that a proper initial volume of reagent can be added.

The McBride method will give perfectly adequate standardization data for many purposes (usually 0.2 to 0.3% high). If a more accurate standardization is required, the approximate normality is obtained by this method, following which a pair of titrations can be performed by the Fowler and Bright method. Directions for both procedures are given in Chapter 20, Sections 20E–2(a) and 20E–2(b).

**Other Primary Standards**    Permanganate solutions can also be standardized against potassium iodide, and electrolytic iron wire.[7]

**Applications of Permanganate Titrations**    Table 12–3 demonstrates the versatility of permanganate as a volumetric oxidizing reagent in acidic media. Most of these reactions are sufficiently rapid for direct titration. Directions for the determination of iron and of calcium are provided in Chapter 20, Sections 20E–3 and 20E–4.

## 12B–2 Tetravalent Cerium

Sulfuric acid solutions of cerium(IV) are nearly as effective oxidants as are solutions of potassium permanganate and can be substituted for that reagent in most applications. Advantages of cerium(IV) solutions include indefinite stability, lack of reactivity toward chloride ion that permits its use with hydrochlo-

---

[5] R. S. McBride, *J. Am. Chem. Soc.,* **1912,** *34,* 393.

[6] R. M. Fowler and H. A. Bright, *J. Res. Nat. Bur. Stand.,* **1935,** *15,* 493.

[7] See, for example: I. M. Kolthoff and R. Belcher, *Volumetric Analysis,* Vol. III, pp. 41–59. New York: Interscience Publishers, 1957.

Table 12-3   Some Applications of Potassium Permanganate in Acid Solution

| Substance Sought | Half-Reaction | Condition |
|---|---|---|
| Sn | $Sn^{2+} \rightleftarrows Sn^{4+} + 2e$ | Prereduction with Zn |
| $H_2O_2$ | $H_2O_2 \rightleftarrows O_2(g) + 2H^+ + 2e$ | |
| Fe | $Fe^{2+} \rightleftarrows Fe^{3+} + e$ | Prereduction with $SnCl_2$ or with Jones or Walden reductor |
| $Fe(CN)_6^{4-}$ | $Fe(CN)_6^{4-} \rightleftarrows Fe(CN)_6^{3-} + e$ | |
| V | $VO^{2+} + 3H_2O \rightleftarrows V(OH)_4^+ + 2H^+ + e$ | Prereduction with Bi amalgam or $SO_2$ |
| Mo | $Mo^{3+} + 4H_2O \rightleftarrows MoO_4^{2-} + 8H^+ + 3e$ | Prereduction with Jones reductor |
| W | $W^{3+} + 4H_2O \rightleftarrows WO_4^{2-} + 8H^+ + 3e$ | Prereduction with Zn or Cd |
| U | $U^{4+} + 2H_2O \rightleftarrows UO_2^{2+} + 4H^+ + 2e$ | Prereduction with Jones reductor |
| Ti | $Ti^{3+} + H_2O \rightleftarrows TiO^{2+} + 2H^+ + e$ | Prereduction with Jones reductor |
| $H_2C_2O_4$ | $H_2C_2O_4 \rightleftarrows 2CO_2 + 2H^+ + 2e$ | |
| Mg, Ca, Zn Co, Pb, Ag | $H_2C_2O_4 \rightleftarrows 2CO_2 + 2H^+ + 2e$ | Sparingly soluble metal oxalates filtered, washed, and dissolved in acid; liberated oxalic acid titrated |
| $HNO_2$ | $HNO_2 + H_2O \rightleftarrows NO_3^- + 3H^+ + 2e$ | 15-min reaction time; excess $KMnO_4$ back-titrated |
| K | $K_2NaCo(NO_2)_6 + 6H_2O \rightleftarrows$ $Co^{2+} + 6NO_3^- + 12H^+ + 2K^+ +$ $Na^+ + 11e$ | Precipitated as $K_2NaCo(NO_2)_6$; filtered and dissolved in $KMnO_4$; excess $KMnO_4$ back-titrated |
| Na | $U^{4+} + 2H_2O \rightleftarrows UO_2^{2+} + 4H^+ + 2e$ | Precipitated as $NaZn(UO_2)_3 (OAc)_9$; filtered, washed, dissolved; U determined as above |

ric acid solutions, and simplicity of its half-reaction, with only a single reaction product, cerium(III), being possible. A disadvantage is that, in contrast to permanganate solutions, the color of cerium(IV) solutions is not sufficiently intense to serve as an indicator. Moreover, they cannot be used to titrate neutral or alkaline solutions because of precipitation of basic salts of the cation. Finally, cerium salts are significantly more expensive than most other oxidizing reagents.

Table 12–4   Formal Electrode Potentials for Cerium(IV)

| Acid Concentration, $N$ | Formal Potential vs. Standard Hydrogen Electrode, V | | |
|---|---|---|---|
| | $HClO_4$ Solution | $HNO_3$ Solution | $H_2SO_4$ Solution |
| 1 | +1.70 | +1.61 | +1.44 |
| 2 | 1.71 | 1.62 | 1.44 |
| 4 | 1.75 | 1.61 | 1.43 |
| 8 | 1.87 | 1.56 | 1.42 |

**Properties of Tetravalent Cerium Solutions**   The formal potential for a cerium(IV) solution depends upon the acid that is used in its preparation, and to a lesser extent upon the concentration of that acid. Solutions prepared with sulfuric acid are roughly comparable to permanganate in oxidizing power; those prepared with nitric or perchloric acids are appreciably more potent (Table 12–4).

Solutions of cerium(IV) contain nitrate (or sulfate) complexes as well as various hydrates and dimers. Much remains to be learned concerning the cerium-containing species that exist in such solutions.

**Stability of Cerium(IV) Solutions**   Sulfuric acid solutions of tetravalent cerium are remarkably stable and can be stored for months and heated at $100°C$ for prolonged periods without change in normality. Perchloric and nitric acid solutions of the reagent are by no means as stable. They slowly oxidize water and decrease in normality by 0.3 to 1% over a period of a month. The decomposition is catalyzed by light.

The oxidation of chloride by cerium(IV) is so slow that other reducing agents can be titrated without error in the presence of high concentrations of this ion. Nevertheless, a hydrochloric acid solution of cerium(IV) does not possess sufficient stability for use as a volumetric reagent.

**Indicators for Cerium(IV) Titrations**   Several oxidation-reduction indicators are available for cerium(IV) oxidations. The most widely used of these is the iron(II) complex of 1,10-phenanthroline or one of its substituted derivatives (see Table 11–3). These indicators are frequently referred to as *ferroins*.

**Preparation of Cerium(IV) Solutions**   A number of cerium(IV) salts are available from commercial sources;[8] the most commonly encountered are listed in Table 12–5. Primary-standard grade cerium(IV) ammonium nitrate is avail-

[8] For additional information concerning the preparation, standardization, and use of cerium(IV), see G. Frederick Smith, *Cerate Oxidimetry,* 2nd ed. Columbus, Ohio: G. Frederick Smith Chemical Co., 1964; see also I. M. Kolthoff and R. Belcher, *Volumetric Analysis,* Vol. III, pp. 121–167. New York: Interscience Publishers, 1957.

Table 12–5  *Analytically Useful Cerium(IV) Compounds*

| Name | Formula | Equivalent Weight |
|---|---|---|
| Cerium(IV) ammonium nitrate | $Ce(NO_3)_4 \cdot 2NH_4NO_3$ | 548.2 |
| Cerium(IV) ammonium sulfate | $Ce(SO_4)_2 \cdot 2(NH_4)_2SO_4 \cdot 2H_2O$ | 632.6 |
| Cerium(IV) hydroxide | $Ce(OH)_4$ | 208.1 |
| Cerium(IV) hydrogen sulfate | $Ce(HSO_4)_4$ | 528.4 |

able commercially and can be used to prepare standard solutions of the cation directly by weight. More commonly, less expensive reagent-grade cerium(IV) ammonium sulfate or nitrate is used to give solutions that are subsequently standardized. In either case, the reagent is dissolved in a solution that is at least $0.1\ N$ in sulfuric acid to prevent precipitation of basic salts.

**Standardization of Cerium(IV) Solutions**  Cerium(IV) solutions can be standardized against sodium oxalate, electrolytic iron wire, potassium ferrocyanide, or iron(II) ethylenediamine sulfate (Oesper's salt). With sodium oxalate, the titration is performed at 50°C in hydrochloric acid solution with iodine monochloride as a catalyst.

**Applications of Cerium(IV) Solutions**  In general, the applications of cerium(IV) parallel closely those of potassium permanganate, which are shown in Table 12–3.

## 12B–3 Potassium Dichromate

In its analytical applications dichromate ion is reduced to chromium(III):

$$Cr_2O_7^{2-} + 14H^+ + 6e \rightleftharpoons 2Cr^{3+} + 7H_2O \qquad E^0 = 1.33\ V$$

Potassium dichromate solutions are indefinitely stable, can be boiled without decomposition, and do not react with hydrochloric acid. Moreover, the reagent is available in high purity and at a modest cost. Disadvantages of potassium dichromate compared with cerium(IV) and permanganate ion are its lower electrode potential and the slowness of its reaction with certain reducing agents.

**Preparation, Properties, and Uses of Dichromate Solutions**  For most purposes, commercial reagent-grade potassium dichromate is sufficiently pure to permit the direct preparation of standard solutions; the solid is simply dried at 150 to 200°C before being weighed. A product of even higher quality can be obtained by recrystallizing potassium dichromate two or three times from water. Directions for the preparation of a standard potassium dichromate solution are given in Chapter 20, Section 20F–1.

The orange color of a dichromate solution is not intense enough for use in end-point detection. However, diphenylamine sulfonic acid (see Table 11–3) is an excellent indicator for titrations with this reagent. The oxidized form of the indicator is violet, while its reduced form is essentially colorless; thus, the color change observed in a direct titration is from the green of chromium(III) to violet. An indicator blank cannot be applied because dichromate oxidizes diphenylamine sulfonic acid only slowly in solutions that do not contain other oxidation-reduction systems. Fortunately, the titration error attributable to the indicator is ordinarily negligible. The indicator ceases to be reversible in the presence of large oxidant concentrations and low acidities (above pH 2); yellow or red oxidation products are obtained.

**Determination of Iron**     The principal use of dichromate involves the titration of iron(II):

$$Cr_2O_7^{2-} + 6Fe^{2+} + 14H^+ \rightarrow 2Cr^{3+} + 6Fe^{3+} + 7H_2O$$

The titration is unaffected by moderate amounts of hydrochloric acid. Directions for the analysis of iron are given in Chapter 20, Section 20F–2.

**Other Applications**     A common method for the determination of oxidizing agents calls for treatment of the sample with a measured excess of iron(II), followed by titration of the excess with standard dichromate. The method has been applied to the analysis of nitrate, chlorate, permanganate, dichromate, and organic peroxides, among others.

### 12B–4 Iodimetric Methods

Many volumetric methods are based on the half-reaction

$$I_3^- + 2e \rightleftharpoons 3I^- \qquad E^0 = 0.536 \text{ V}$$

Such methods are grouped into two categories. The first involves use of a standard triiodide solution to titrate analytes that are readily oxidized. These direct, or *iodimetric,* methods have limited applicability because iodine is a relatively weak oxidizing agent. Occasionally, however, the low electrode potential is advantageous because it imparts a degree of selectivity toward strong reducing agents. Another advantage is the availability of a sensitive and reversible indicator for the titrations. Unfortunately, triiodide solutions lack stability and must be restandardized regularly.

Indirect or *iodometric* methods make use of a standard solution of sodium thiosulfate to titrate the iodine liberated by the reaction of an oxidizing analyte with an unmeasured excess of potassium iodide.

**Preparation and Properties of Iodine Solutions**     A saturated aqueous solution of iodine is only about 0.001 $F$ at room temperature. Much more concentrated solutions of the reagent can be prepared, however, by dissolving the iodine in a solution containing an excess of iodide ion, which reacts with the reagent to form triiodide ions:

$$I_2(s) + I^- \rightleftharpoons I_3^- \qquad K = 7.1 \times 10^2$$

It would be better to refer to such solutions as *triiodide solutions,* since $I_3^-$ is the principal iodine-containing solute species. In practice, however, they are called *iodine solutions* because this terminology suffices to account for the stoichiometric behavior of the reagent ($I_2 + 2e \rightleftharpoons 2I^-$).

Iodine dissolves only slowly in solutions of potassium iodide, particularly when the iodide concentration is low. For this reason, iodine is always dissolved in a small volume of concentrated iodide solution and then diluted to the desired volume. All of the element must be dissolved prior to dilution; otherwise, the normality will increase continuously as the remaining solid gradually passes into solution. Directions for the preparation and standardization of iodine solutions are given in Chapter 20, Section 20G.

**Stability**   Iodine solutions lack stability for several reasons, one being the volatility of the solute. Losses of iodine from an open vessel occur in a relatively short time even in the presence of an excess of iodide ion. Iodine slowly attacks most organic materials. Consequently, cork or rubber stoppers are never used to close containers of the reagent, and precautions must be taken to protect standard solutions from contact with organic dusts and fumes.

Air oxidation of iodide ion also causes changes in the normality of an iodine solution:

$$4I^- + O_2(g) + 4H^+ \rightarrow 2I_2 + 2H_2O$$

In contrast to the other effects, oxidation results in an increase in the iodine normality. The oxidation is promoted by acids, heat, and light.

**Completeness of Iodine Oxidations**   Iodine is such a weak oxidizing agent that it cannot be used to titrate many reducing agents unless measures are taken to force their reactions to completion. These measures include pH control and the use of complexing agents.

The oxidation of many substances is accompanied by the production of hydrogen ions. Introduction of a buffer to react with these hydrogen ions as they are formed will result in a more complete overall reaction. An example is the oxidation of arsenic(III) to arsenic(V) for which the reaction is

$$H_3AsO_3 + I_2 + H_2O \rightleftharpoons H_3AsO_4 + 2I^- + 2H^+$$

The equilibrium constant for this reaction is only about 0.17. Nevertheless, iodine cleanly oxidizes arsenic(III) in solutions that are buffered to a pH of 7 or somewhat higher. (It is of interest to note that the reverse reaction occurs quantitatively in a strongly acidic environment.) Although iodine oxidations of this kind are more complete in basic solutions, care must be taken to prevent the formation of hypoiodite ion, which takes place in alkaline solutions:

$$I_2 + OH^- \rightleftharpoons OI^- + I^- + H^+$$

The hypoiodite may be subsequently disproportionate to iodide and iodate:

$$3OI^- \rightleftharpoons IO_3^- + 2I^-$$

The occurrence of these reactions will cause serious errors by disturbing the stoichiometry of the reaction of the standardized reagent with the analyte. Thus, solutions to be titrated with standardized iodine cannot have pH values higher than 9. Occasionally, a pH greater than 7.0 is detrimental.

Complexing reagents can also be used to force some iodine oxidations toward completion. For example, it is clear that the electrode potential of iodine is too low for the quantitative oxidation of iron(II) to the trivalent state. Introduction of a ligand that complexes strongly with iron(III), however, will permit a complete conversion to this state; pyrophosphate and EDTA serve this purpose.

**End Points for Iodine Titrations**   The end point for an iodimetric titration can be detected in any of several ways. For example, the color of triiodide ion can be discerned in a solution that is about $5 \times 10^{-6}$ $F$ in $I_3^-$, which corresponds to an overtitration of less than one drop of a 0.1 $N$ iodine solution. Thus, provided the analyte solution is colorless, the reagent can serve as its own indicator.

A greater sensitivity can be obtained, at the sacrifice of convenience, by adding a few milliliters of chloroform or carbon tetrachloride to the solution. Shaking causes the bulk of any iodine to collect in the immiscible organic layer and impart an intense violet color to it.

The deep blue complex that is formed between starch and iodine is the most widely used indicator for iodimetry. The nature of the species responsible for the color has been the subject of much speculation and controversy. It is now believed that the iodine is held as an adsorption complex within the helical chain of β-amylose, a macromolecular component of most starches. The closely related α-amylose forms a red adduct with iodine that is not readily reversible and is thus undesirable. The so-called *soluble starch,* which is available from commercial sources, consists principally of β-amylose, the α-fraction having been removed; indicator solutions are readily prepared from this product.

Aqueous starch suspensions decompose within a few days, primarily because of bacterial action. The decomposition products tend to interfere with the indicator properties of the preparation and may also be oxidized by iodine. The rate of decomposition can be inhibited by preparing and storing the indicator under sterile conditions and by adding mercury(II) iodide or chloroform as a bacteriostat. Perhaps the simplest alternative is to prepare a fresh suspension of the indicator the same day it is to be used.

Starch tends to decompose irreversibly in solutions containing large concentrations of iodine. Addition of the indicator to solutions containing an excess of iodine should thus be postponed until the titration is nearly complete.

**Standardization of Iodine Solutions**[9]   Iodine solutions may be standardized against barium thiosulfate monohydrate, sodium thiosulfate, or potassium antimony(III) tartrate.

MacNevin and Kriege have demonstrated that barium thiosulfate monohydrate is a satisfactory standard for iodine solutions.[10] Although the salt is soluble to the extent of only about 0.01 $F$, the solid reacts so rapidly with iodine solutions that a direct titration is entirely feasible. It is reported that $BaS_2O_3 \cdot H_2O$ with a purity of 99.85% is readily prepared and is stable at room temperature.[11] Barium thiosulfate monohydrate begins to lose water at temperatures above $50°C$; the anhydrous salt, however, is unsuitable for standardization owing to its low solubility. Therefore, the monohydrate should be used without drying.

A. A. Woolf has reported that sodium thiosulfate pentahydrate can be rendered anhydrous by a period of refluxing in methanol.[12] The product is readily soluble in water and only slowly takes up moisture from the atmosphere.

The reaction between thiosulfate and iodine is discussed in Section 12C–2. Directions for the preparation and standardization of iodine solutions are given in Chapter 20, Sections 20G–1 and 20G–2.

**Applications of Standard Iodine**   Table 12–6 is a summary of methods that make use of iodine as an oxidizing agent. Directions for the determination of antimony in an ore are given in Chapter 20, Section 20G–3.

## 12B–5 Potassium Iodate

A number of useful analytical methods are based upon potassium iodate. The reagent can be obtained in a high state of purity and can be used without the need for further treatment, other than drying, for the direct preparation of standard solutions. Iodate solutions are indefinitely stable.

**Reactions of Iodate**   Iodate ion reacts with iodide to produce iodine:

$$IO_3^- + 5I^- + 6H^+ \rightleftharpoons 3I_2 + 3H_2O$$

This reaction proceeds far to the right in acidic solutions and far to the left in basic media. Potassium iodate is a convenient source for known amounts of iodine (see, for example, p. 322). A measured volume of standard iodate is mixed with an excess of iodide ion in a solution that is 0.1 to 1 $N$ in acid. Exactly six equivalents of iodine are liberated for each mole of iodate; the resulting

---

[9] Arsenic(III) oxide, for many years the preferred primary standard for iodine solutions, has been identified as a carcinogen; its use is now subject to strict regulation.

[10] W. M. MacNevin and O. H. Kriege, *Anal. Chem.*, **1953**, *25*, 767.

[11] Primary-standard grade $BaS_2O_3 \cdot H_2O$, assaying 99.9% or better, is available from G. Frederick Smith Chemical Co., 867 McKinley Ave., Columbus, Ohio 43223.

[12] A. A. Woolf, *Anal. Chem.*, **1982**, *54*, 2134.

Table 12-6    Analyses with Standard Iodine Solutions

| Substance Analyzed | Half-Reaction |
|---|---|
| As | $H_3AsO_3 + H_2O \rightleftharpoons H_3AsO_4 + 2H^+ + 2e$ |
| Sb | $H_3SbO_3 + H_2O \rightleftharpoons H_3SbO_4 + 2H^+ + 2e$ |
| Sn | $Sn^{2+} \rightleftharpoons Sn^{4+} + 2e$ |
| Te | $Te(s) + 3H_2O \rightleftharpoons H_2TeO_3 + 4H^+ + 4e$ |
| $H_2S$ | $H_2S \rightleftharpoons S(s) + 2H^+ + 2e$ |
| $Cd^{2+}, Hg^{2+}, Pb^{2+}$ | $M^{2+} + H_2S \rightleftharpoons MS(s) + 2H^+$ (filter and wash) |
| $Zn^{2+}$, etc. | $MS(s) \rightleftharpoons M^{2+} + S(s) + 2e$ |
| $SO_2, SO_3^{2-}$ | $SO_3^{2-} + H_2O \rightleftharpoons SO_4^{2-} + 2H^+ + 2e$ |
| $S_2O_3^{2-}$ | $2S_2O_3^{2-} \rightleftharpoons S_4O_6^{2-} + 2e$ |
| Phosphorous acid | $H_3PO_3 + H_2O \rightleftharpoons H_3PO_4 + 2H^+ + 2e$ |
| Hydrazine | $N_2H_4 \rightleftharpoons N_2(g) + 4H^+ + 4e$ |
| Ascorbic acid | $C_6H_8O_6 \rightleftharpoons C_6H_6O_6 + 2H^+ + 2e$ |

solution can then be used for the standardization of thiosulfate solutions or for other analytical purposes.

In strongly acidic solutions, iodate will oxidize iodide or iodine to the +1 state, provided some anion such as chloride, bromide, or cyanide is present to stabilize this oxidation state. For example, in solutions that are greater than 3 $F$ in hydrochloric acid, the following reaction proceeds essentially to completion:

$$IO_3^- + 2I_2 + 10Cl^- + 6H^+ \rightleftharpoons 5ICl_2^- + 3H_2O$$

A number of important iodate titrations are performed in strong hydrochloric acid, where the iodate is initially reduced to iodine. As the reducing agent is consumed, however, oxidation of iodine to $ICl_2^-$ takes place. The end point for the process is signaled by the complete disappearance of the iodine. The equivalent weight for the iodate is one-fourth of its formula weight because the final reaction product is iodine in the +1 state. When used in this manner, iodate is a less powerful oxidant than either permanganate or cerium(IV).

**End Points in Iodate Titrations**    The disappearance of iodine from the solution is often sufficient to indicate the end point in an iodate titration. Iodine is nearly always formed in the initial stages of the reaction; only at the equivalence point is it completely oxidized to the +1 state. Starch fails to function as an indicator in the highly acidic medium required for the production of $ICl_2^-$. Instead, a few milliliters of an immiscible organic solvent such as carbon tetrachloride or chloroform are added at the start of the titration. After each addition of iodate the mixture is shaken thoroughly and the organic layer examined after the phases have separated. The bulk of any unreacted iodine remains in the organic layer and imparts an intense violet-red color to it. The titration is judged complete when the minimum amount of iodate needed to discharge this color has been added. This excellent method for detecting iodine is quite comparable in sensitivity to the starch–iodine color but suffers from the disadvantage of being more time consuming. Directions for the iodate titration of iodide and iodine in solutions are given in Chapter 20, Section 20J–3; further applications can be found in reference sources.[13]

### 12B–6  Potassium Bromate Solutions as Sources for Bromine

Primary-standard potassium bromate is available from commercial sources; its solutions are indefinitely stable.

Direct titrations with potassium bromate are relatively few. The principal value of the reagent to chemical analysis is as a convenient and stable source of bromine. In this application, an unmeasured excess of potassium bromide is added to an acidic solution of the analyte. The introduction of standard potassium bromate then releases a known quantity of bromine:

$$\underset{\substack{\text{standard}\\\text{solution}}}{BrO_3^-} + \underset{\text{excess}}{5Br^-} + 6H^+ \rightarrow 3Br_2 + 3H_2O$$

This indirect generation circumvents the problems associated with the use of standard bromine solutions, which lack stability. Note that each bromate is responsible for the formation of three bromine molecules, which in turn require six electrons for reduction to bromide; that is,

$$BrO_3^- \equiv 3Br_2 \equiv 6e$$

The equivalent weight of potassium bromate as a source of bromine is thus one-sixth of its gram-formula weight.

**Indicators for Titrations Involving Bromine**   Several organic indicators, such as methyl orange and methyl red, are readily brominated to yield products with colors that differ from the original species. These brominations are totally irreversible, which eliminates the possibility of back-titration. Moreover, direct titrations are inconvenient because of the need to avoid local excesses of the reagent, which would cause a premature color change.

Quinoline yellow, $\alpha$-naphthoflavone, and $p$-ethoxychrysoidine behave reversibly with respect to bromine. These indicators, which are available commercially, are considerably more convenient to use than methyl orange or methyl red.

**Applications of Standard Potassium Bromate Solutions**[14]   Potassium bromate is a convenient source of bromine for organic analysis. Few organic compounds react sufficiently rapidly with bromine to make direct titration feasible. Instead,

---

[13] For details on the application of iodate titrations, see I. M. Kolthoff and R. Belcher, *Volumetric Analysis,* Vol. III, pp. 449–473. New York: Interscience Publishers, 1957.

[14] For a more complete discussion of bromate titrations, see I. M. Kolthoff and R. Belcher, *Volumetric Analysis,* Vol. III, pp. 501–571. New York: Interscience Publishers, 1957.

a measured excess of the standard bromate is added to the solution that contains the sample and an excess of potassium bromide; the mixture is then acidified and allowed to stand until reaction of the bromine with the analyte is judged complete. The analysis is conveniently completed iodometrically by adding potassium iodide to react with the excess bromine

$$2I^- + Br_2 \rightarrow I_2 + 2Br^-$$

following which the liberated iodine is titrated with standard sodium thiosulfate (p. 320). Bromine is incorporated into an organic molecule either by substitution or by addition.

**Substitution Reactions**   Substitution involves replacement of hydrogen in an aromatic ring by the halogen. For example, the bromination of phenol involves replacement of three hydrogen atoms:

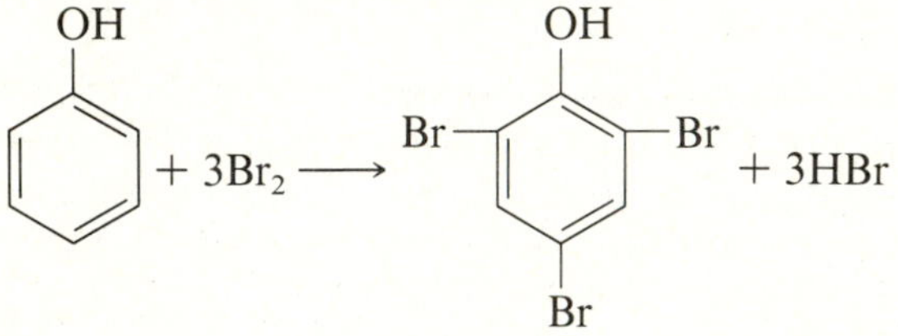

The process requires six atoms of bromine, each of which can be considered to have undergone a one-electron reduction to the $-1$ state; the equivalent weight of phenol is thus one-sixth of its gram-formula weight.

Substitution methods have been successfully applied to the analysis of aromatic compounds that contain strong ortho-para directing groups, particularly amines and phenols.[15] An important application of the method is the titration of 8-hydroxyquinoline:

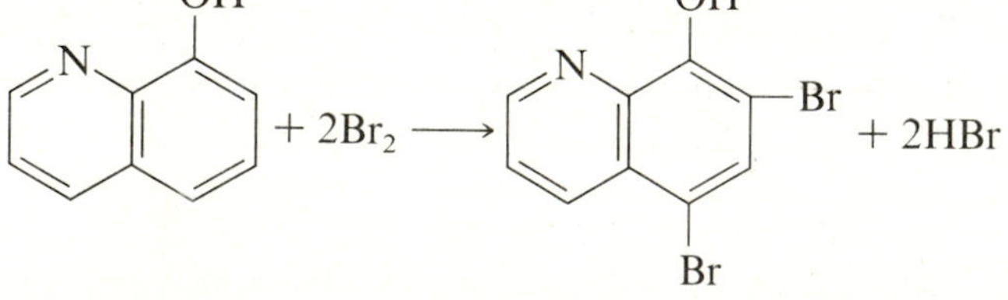

This reaction takes place in hydrochloric acid solution and is sufficiently rapid to permit direct titration; it is of particular interest because 8-hydroxyquinoline is an excellent precipitating reagent for cations (see Table 4–2, p. 83). For

---

[15] See, for example, A. R. Day and W. T. Taggart, *Ind. Eng. Chem., Anal. Ed.,* **1928,** *20,* 545.

example, aluminum can be determined according to the following sequence of reactions:

$$Al^{3+} + 3HOC_9H_6N \xrightarrow{\text{pH } 4-9} Al(OC_9H_6N)_3(s) + 3H^+$$

$$Al(OC_9H_6N)_3(s) \xrightarrow{\text{hot } 4\,F\,HCl} 3HOC_9H_6N + Al^{3+}$$

$$3HOC_9H_6N + 6Br_2 \rightarrow 3HOC_9H_4NBr_2 + 6HBr$$

Because each mole of aluminum is responsible for the reaction of 6 mol (or 12 equivalents) of bromine, the effective equivalent weight for the metal is one-twelfth of its gram-formula weight.

**Addition Reactions**    Addition involves the opening of an olefinic double bond. For example, 1 mol of ethylene reacts with 1 mol (2 equivalents) of bromine:

$$
\begin{array}{ccc}
 & \text{H} \quad \text{H} & \qquad\qquad \text{H} \quad \text{H} \\
 & | \qquad | & \qquad\qquad | \qquad | \\
\text{H}-\text{C}=\text{C}-\text{H} + Br_2 \rightarrow & \text{H}-\text{C}-\text{C}-\text{H} \\
 & & \qquad\qquad | \qquad | \\
 & & \qquad\qquad \text{Br} \quad \text{Br}
\end{array}
$$

The equivalent weight of ethylene is thus one-half of its gram-formula weight.

The literature contains numerous references to the use of bromine for the estimation of olefinic unsaturation in fats, oils, and petroleum products.[16]

## 12C VOLUMETRIC APPLICATIONS OF REDUCTANTS

Standard solutions of reducing agents tend to react with atmospheric oxygen and are, as a consequence, somewhat troublesome to use. It is often necessary to store and use such reagents in an inert atmosphere. Alternatively, an aliquot of the unstandardized reductant is added to the solution containing the analyte, and the excess is quickly back-titrated with a standard solution of a stable oxidizing agent, such as cerium(IV). The concentration of the reductant is then established at the time of the analysis by a similar titration with a blank. Very strong reducing agents, such as chromium(II) and titanium(III), react with oxygen too rapidly for use in this manner; for such reagents, a blanket of an inert gas, such as $N_2$ or $CO_2$, is required.

Iron(II) and sodium thiosulfate are probably the most frequently encountered volumetric reducing reagents.

### 12C-1 Iron(II)

Solutions of iron(II) are readily prepared from iron(II) ammonium sulfate, $Fe(NH_4)_2(SO_4)_2 \cdot 6H_2O$ (Mohr's salt), or from the closely related iron(II) ethyl-

---

[16] See, for example, A. Polgar and J. L. Jungnickel, *Organic Analysis,* Vol. 3. New York: Interscience Publishers, 1956; M. R. F. Ashworth, *Titrimetric Organic Analysis,* Part I, p. 135. New York: Interscience Publishers, 1964.

enediamine sulfate, $FeC_2H_4(NH_3)_2(SO_4)_2 \cdot 4H_2O$ (Oesper's salt). Air oxidation of iron(II) takes place rapidly in neutral solutions but is inhibited in the presence of acids, with the most stable preparations being about 0.5 $F$ in $H_2SO_4$. Iron(II) solutions must be standardized each time they are used.

Numerous oxidizing agents are conveniently determined by reaction with a measured excess of standard iron(II) followed by titration of the excess with a standard solution of potassium dichromate or cerium(IV). This procedure has been applied to the determination of organic peroxides, hydroxylamine, chromium(VI), cerium(IV), molybdenum(VI), and nitrate ion, among others.

### 12C–2 Sodium Thiosulfate: Iodometric Methods

Iodide ion, a moderately effective reducing agent, is used extensively for the analysis of oxidants. In such applications, the iodine liberated by reaction between the analyte and an unmeasured excess of potassium iodide is ordinarily titrated with a standard solution of sodium thiosulfate.

**The Reaction of Iodine with Thiosulfate Ion**    The reaction between iodine and thiosulfate ion is described by the equation

$$I_2 + 2S_2O_3^{2-} \rightarrow 2I^- + S_4O_6^{2-} \tag{12-1}$$

The production of tetrathionate ion involves the loss of two electrons from two thiosulfate ions. Therefore, the equivalent weight of sodium thiosulfate in this reaction is equal to its formula weight.

The quantitative conversion of thiosulfate to tetrathionate ion is unique with iodine; other oxidizing agents tend to carry the oxidation further to sulfate ion or a mixture of tetrathionate and sulfate ions. Thiosulfate titrations of iodine are best performed in neutral or slightly acidic solutions. If strongly acidic solutions must be titrated, air oxidation of the excess iodide must be prevented by blanketing the solution with an inert gas, such as carbon dioxide or nitrogen. One simple way of providing a blanket of carbon dioxide is to introduce a quantity of solid sodium hydrogen carbonate into the solution, which reacts to form a layer of carbon dioxide that excludes oxygen from the titration vessel.

The reaction shown by Equation 12–1 is no longer quantitative in an iodine solution with a pH somewhat greater than 7 because the hypoiodite ion that forms under these conditions (see p. 314) converts the thiosulfate to sulfate ion by the reaction

$$4OI^- + S_2O_3^{2-} + H_2O \rightleftharpoons 2SO_4^{2-} + 4I^- + 2H^+$$

Since one mole of hypoiodite ion is formed from one mole of iodine, we may write

$$1\ I_2 \equiv 1\ OI^- \equiv 1\ S_2O_3^{2-}$$

whereas the stoichiometry shown by Equation 12–1 indicates that

$$1\ I_2 \equiv 2\ S_2O_3^{2-}$$

Thus, when hypoiodite is present, less thiosulfate ion is consumed and a negative determinate error is encountered. It has been shown that an error of about $-4\%$ can be expected when a solution containing 25 mL of 0.1 $N$ iodine and 0.5 g of carbonate is titrated with 0.1 $N$ thiosulfate; this error becomes $-10\%$ or more when 2 g of sodium carbonate are present.[17] Kolthoff recommends maintaining a pH of less than 7.6 when 0.1 $N$ solutions of iodine are being titrated. For 0.01 and 0.001 $N$ solutions the maximum pH should be 6.5 and 5, respectively.

The titration of highly acidic iodine solutions with thiosulfate ion yields quantitative results, provided care is taken to prevent air oxidation of iodide ion. The end point for this titration is readily established by means of a starch solution. It is worth noting again that when titrating solutions of iodine with thiosulfate ion, addition of the starch indicator must be delayed until most of the iodine has been consumed — that is, until the solution has changed from red-brown in color to a pale yellow. This precaution is necessary to avoid decomposition of the starch by the high concentration of iodine.

**Stability of Sodium Thiosulfate Solutions**    Sodium thiosulfate solutions are somewhat unstable and decompose to give sulfur and hydrogen sulfite ion:

$$S_2O_3^{2-} + H^+ \rightleftharpoons HS_2O_3^- \rightarrow HSO_3^- + S(s)$$

Variables that influence the rate of this reaction are pH, the presence of microorganisms, the concentration of the solution, and exposure to oxygen and sunlight. These variables may cause the iodine titer of a thiosulfate solution to change by several percent over a period of a few weeks. On the other hand, proper attention to detail will yield solutions that need only occasional restandardization.

The rate of the decomposition reaction increases markedly as the solution becomes acidic. In fact, when sodium thiosulfate is added to a strongly acidic solution, a cloudiness develops almost immediately as a consequence of precipitation of elemental sulfur. Even in neutral solution, this reaction proceeds at such a rate that standard sodium thiosulfate must be restandardized periodically.

The most important single cause for the instability of thiosulfate solutions can be traced to bacteria that metabolize thiosulfate ion to give sulfite and sulfate ions as well as elemental sulfur. For this reason, standard solutions of the reagent are prepared under reasonably sterile conditions. Bacterial activity appears to be at a minimum at a pH between 9 and 10, which accounts, at least in part, for the reagent's greater stability in slightly basic solutions. The presence of bactericides, such as chloroform, sodium benzoate, or mercury(II) iodide, also slows decomposition.

Many other variables affect the stability of thiosulfate solutions. The decomposition is reported to be catalyzed by copper(II), as well as the decomposition

---

[17] I. M. Kolthoff and R. Belcher, *Volumetric Analysis,* Vol. III, pp. 214–215. New York: Interscience Publishers, 1957.

products themselves. Exposure to sunlight accelerates the process as does atmospheric oxygen. Decomposition effects are most pronounced in dilute solutions.

The decomposition reaction may cause the iodine titer of a thiosulfate solution to increase or decrease. Increases in normality arise because hydrogen sulfite ion formed by the reaction consumes two moles of iodine while the thiosulfate ion from which it was derived consumes but one. On the other hand, decreases in normality sometimes occur when the hydrogen sulfite ion is air oxidized to form the unreactive sulfate ion. As noted earlier, strongly acidic solutions of iodine can be titrated with standard sodium thiosulfate without interference from decomposition of the reagent, provided care is taken to introduce the thiosulfate slowly and with good mixing. Under these circumstances, the reagent is oxidized by the iodine so rapidly that the slower decomposition reaction is precluded.

**Standardization of Thiosulfate Solutions**   Potassium iodate is an excellent primary standard for thiosulfate solutions. As noted previously, iodate rapidly oxidizes iodide ion in acid solution to produce iodine.

$$IO_3^- + 5I^- + 6H^+ \rightleftharpoons 3I_2 + 2H_2O$$

Three moles of iodine are furnished by each formula weight of potassium iodate. The equivalent weight of potassium iodate is one sixth of its formula weight in this application because a six-electron change is associated with the reduction of three iodine molecules, the species actually titrated; that is,

$$IO_3^- \equiv 3I_2 \equiv 6e$$

The sole disadvantage of potassium iodate as a primary standard for thiosulfate is its low equivalent weight (214/6). The amount needed for standardization of a 0.05 $N$ solution is about 0.07 g; the relative error to be expected in weighing this quantity may be prohibitive in some applications. The problem can be circumvented by dissolving a larger quantity of potassium iodate in a known volume of water and taking aliquots of this solution for the standardization. This approach suffers from the disadvantage of providing no duplicate check on the precision of the weighing process.

Other primary standards for sodium thiosulfate include potassium dichromate, potassium bromate, potassium hydrogen iodate, potassium ferricyanide, and metallic copper. Details for their use are to be found in various references.[18] Directions for the standardization of sodium thiosulfate solutions against potassium iodate are found in Chapter 20, Section 20H–2 and against copper in Chapter 20, Section 20H–3.

---

[18] See, for example, I. M. Kolthoff and R. Belcher, *Volumetric Analysis,* Vol. III, pp. 234–243. New York: Interscience Publishers, 1957.

**Table 12–7   Some Applications of the Iodometric Method**

| Substance | Half-Reaction | Conditions |
|---|---|---|
| $IO_4^-$ | $IO_4^- + 8H^+ + 7e \rightarrow \frac{1}{2}I_2 + 4H_2O$ | Acidic solution |
|  | $IO_4^- + 2H^+ + 2e \rightarrow IO_3^- + H_2O$ | Neutral solution |
| $IO_3^-$ | $IO_3^- + 6H^+ + 5e \rightarrow \frac{1}{2}I_2 + H_2O$ | Strong acid |
| $BrO_3^-, ClO_3^-$ | $XO_3^- + 6H^+ + 6e \rightarrow X^- + 3H_2O$ | Strong acid |
| $Br_2, Cl_2$ | $X_2 + 2I^- \rightarrow I_2 + 2X^-$ |  |
| $NO_2^-$ | $HNO_2 + H^+ + e \rightarrow NO(g) + H_2O$ |  |
| $Cu^{2+}$ | $2Cu^{2+} + 4I^- \rightarrow 2CuI(s) + I_2$ |  |
| $O_2$ | $O_2 + 4Mn(OH)_2(s) + H_2O \rightarrow 4Mn(OH)_3(s)$ | Basic solution |
|  | $Mn(OH)_3(s) + 3H^+ + e \rightarrow Mn^{2+} + 3H_2O$ | Acidic solution |
| $O_3$ | $O_3 + 2H^+ + 2e \rightarrow O_2 + H_2O$ |  |
| Organic peroxide | $ROOH + 2H^+ + 2e \rightarrow ROH + H_2O$ |  |

**Errors Associated with Iodometric Analysis**   Three potential sources of error, which have already been mentioned, are decomposition of thiosulfate solutions, alteration in the stoichiometric relationship between thiosulfate and iodine in alkaline solutions, and the premature addition of starch. Additionally, care must be taken to minimize losses of iodine by volatilization. Glass stoppered flasks should be used in situations where iodine solutions must stand, and elevated temperatures must be avoided. The maintenance of a large excess of iodide ion is also helpful in preventing losses.

Air oxidation of iodide ion can also be a serious source of error in iodometric analysis. The process is catalyzed by acid, light, traces of copper(II), and oxides of nitrogen.

**Application of Iodometry**   Numerous substances can be determined iodometrically; typical applications are summarized in Table 12–7. Instructions for the iodometric determination of copper are found in Chapter 20, Sections 20H–4. Section 20H–5 provides details for the iodometric determination of oxygen in natural waters by the Winkler method.

## PROBLEMS

*12–1.  Write balanced net-ionic equations for the following reactions (where necessary, consult tables in Chapter 12 for reaction products):
  (a) the reduction of iron(III) to iron(II) with metallic cadmium.
  (b) the reduction of $H_2MoO_4$ in a Jones reductor.
  (c) the reduction of $V(OH)_4^+$ in a Walden reductor.
  (d) the oxidation of $VOSO_4$ with $K_2S_2O_8$ in acid solution (products: $V(OH)_4^+$, $SO_4^{2-}$).
  (e) the oxidation of U(IV) to $UO_2^{2+}$ with $KMnO_4$ in acid solution.
  (f) the oxidation of $I^-$ with $MnO_4^-$ in strongly basic solution (products: $IO_4^-$, $MnO_4^{2-}$).

(g) the reduction of $VO^{2+}$ by V(II) in acid solution.

(h) the oxidation of $H_2S$ by $I_2$.

(i) the oxidation of $I^-$ by $IO_3^-$ in strong HCl (product: $ICl_2^-$).

(j) the oxidation of selenite ($SeO_3^{2-}$) with hypobromite ($OBr^-$) (products: $SeO_4^{2-}$, $Br^-$).

**12-2.** Use the tables in Chapter 12 to write balanced net-ionic equations for

(a) the reduction of Fe(III) to Fe(II) with metallic Al.

(b) the reduction of $H_2MoO_4$ in a Walden reductor.

(c) the reduction of $V(OH)_4^+$ in a Jones reductor.

(d) the oxidation of Tl(I) to Tl(III) with Ce(IV).

(e) the oxidation of $Mn^{2+}$ with $KIO_4$ in acid solution (products: $MnO_4^-$, $IO_3^-$).

(f) the oxidation of $CN^-$ ion with $MnO_4^-$ in strongly basic solution (products: $CNO^-$, $MnO_4^{2-}$).

(g) the reduction of $V(OH)_4^+$ by $V^{3+}$ in acid solution.

(h) the oxidation of hydrazine ($N_2H_4$) by $I_2$.

(i) the oxidation of $I_2$ by $IO_3^-$ in strong HCl (product: $ICl_2^-$).

(j) the oxidation of dithionate ($S_4O_6^{2-}$) to $SO_4^{2-}$ with $Cr_2O_7^{2-}$.

**\*12-3.** Calculate the normality of a 0.0250 $F$ solution of

(a) $U^{4+}$ with respect to the reaction in Problem 12-1(e).

(b) KI with respect to the reaction in Problem 12-1(f).

(c) $VSO_4$ with respect to the reaction in Problem 12-1(g).

(d) KI with respect to the reaction in Problem 12-1(i).

(e) $H_2SeO_3$ with respect to the reaction in Problem 12-1(j).

**12-4.** Calculate the normality of a 0.0200 $F$ solution of

(a) $Tl^+$ with respect to the reaction in Problem 12-2(d).

(b) $MnO_4^-$ with respect to the reaction in Problem 12-2(f).

(c) $I_2$ with respect to the reaction in Problem 12-2(h).

(d) $KIO_3$ with respect to the reaction in Problem 12-2(i).

(e) $K_2Cr_2O_7$ with respect to the reaction in Problem 12-2(j).

**\*12-5.** Calculate the formal concentration of a solution that is 0.0400 $N$ in

(a) $VOSO_4$ with respect to the reaction in Problem 12-1(d).

(b) $KMnO_4$ with respect to the reaction in Problem 12-1(e).

(c) $I_2$ with respect to the reaction in Problem 12-1(h).

(d) $KIO_3$ with respect to the reaction in Problem 12-1(i).

(e) KOBr with respect to the reaction in Problem 12-1(j).

**12-6.** Calculate the formal concentration of a solution that is 0.0300 $N$ in

(a) $Ce(HSO_4)_4$ with respect to the reaction in Problem 12-2(d).

(b) $KMnO_4$ with respect to the reaction in Problem 12-2(e).

(c) KCN with respect to the reaction in Problem 12-2(f).

(d) hydrazine ($N_2H_4$) with respect to the reaction in Problem 12-2(h).

(e) $Na_2S_4O_6$ with respect to the reaction in Problem 12-2(j).

**\*12-7.** Calculate the normality of the solution produced by dissolving 2.064 g of primary-standard $K_2Cr_2O_7$ in sufficient water to give 500.0 mL.

**12-8.** A standard solution was prepared by dissolving 9.945 g of primary-standard $KIO_3$ in sufficient water to give 1.000 L. Calculate the normality of this solution as an oxidizing agent in strong HCl.

**\*12-9.** Calculate the $Fe(NH_4)_2(SO_4)_2 \cdot 6H_2O$ titer of the $Cr_2O_7^{2-}$ solution in Problem 12-7.

**12–10.** What is the KI titer of the $IO_3^-$ solution in Problem 12–8?

**\*12–11.** What weight of elemental Al will be needed to reduce the Fe(III) in 75.0 mL of 0.0874 $F$ $Fe_2(SO_4)_3$ to the $+2$ state?

**12–12.** The Mo in a solution of molybdic acid ($H_2MoO_4$) was reduced to the $+3$ state by passage through a Jones reductor. Calculate the weight of Zn that was needed to react with 100.0 mL of $8.14 \times 10^{-3}$ $F$ $H_2MoO_4$.

**\*12–13.** Calculate the weight of primary-standard Fe needed for a 40.0-mL titration with 0.0888 $N$ $Ce^{4+}$.

**12–14.** What weight of primary-standard $Na_2C_2O_4$ will be needed to react with 45.0 mL of 0.1025 $N$ $KMnO_4$?

**\*12–15.** What is the normality of a $KMnO_4$ solution if 37.16 mL were needed to titrate 0.2413 g of primary-standard $Na_2C_2O_4$?

**12–16.** Calculate the normality of a Ce(IV) solution if 43.80 mL were needed to titrate the Fe(II) produced from 0.2461 g of electrolytic iron wire.

**\*12–17.** A 0.1165-g sample of primary-standard Cu was dissolved and then treated with an excess of KI. Reaction:

$$2Cu^{2+} + 4I^- \rightarrow Cu_2I_2(s) + I_2$$

Calculate the normality of a $Na_2S_2O_3$ solution if 36.24 mL were needed to titrate the liberated $I_2$.

**12–18.** Calculate the normality of a sodium thiosulfate solution if 47.33 mL were needed to titrate the iodine liberated when an unmeasured excess of KI was added to 20.00 mL of 0.01417 $F$ $KIO_3$.

**\*12–19.** A 0.7120-g specimen of iron ore was brought into solution and passed through a Jones reductor. Titration of the Fe(II) produced required 39.21 mL of 0.1043 $N$ $KMnO_4$. Express the results of this analysis in terms of percent
  (a) Fe.
  (b) $Fe_2O_3$.

**12–20.** The Sn in a 0.4352-g mineral specimen was reduced to the $+2$ state with Pb and titrated with 29.77 mL of 0.1041 $N$ $K_2Cr_2O_7$. Calculate the results of this analysis in terms of percent
  (a) Sn.
  (b) $SnO_2$.

**\*12–21.** The $HNO_2$ in a 25.00-mL aliquot was oxidized to $NO_3^-$ with 50.00 mL of 0.0947 $N$ $KMnO_4$. After the reaction was judged complete, the excess permanganate was titrated with 2.61 mL of 0.1013 $N$ $Fe^{2+}$. Calculate the normality of the $HNO_2$ solution
  (a) as an acid.
  (b) as a reducing agent.

**12–22.** A 25.00-mL aliquot of a solution containing $ClO_3^-$ was acidified and then treated with an equal volume of 0.1147 $N$ $Fe(NH_4)_2(SO_4)_2$. Reaction:

$$6Fe^{2+} + ClO_3^- + 6H^+ \rightarrow 6Fe^{3+} + Cl^- + 3H_2O$$

After the reaction was complete, the excess Fe(II) was titrated with 2.69 mL of 0.0837 $N$ $Ce(SO_4)_2$. Calculate the formal concentration of $NaClO_3$ in the solution.

*12–23. Calculate the percentage of $MnO_2$ in a mineral specimen if the $I_2$ liberated by a 0.1344-g sample in the net reaction

$$MnO_2(s) + 4H^+ + 2I^- \rightarrow Mn^{2+} + I_2 + 2H_2O$$

required 32.30 mL of 0.0722 $N$ $Na_2S_2O_3$.

12–24. Treatment of hydroxylamine ($H_2NOH$) with an excess of Fe(II) results in the formation of $N_2O$ and an equivalent amount of Fe(II):

$$2H_2NOH + 4Fe^{3+} \rightarrow N_2O(g) + 4Fe^{2+} + 4H^+ + H_2O$$

Calculate the formal concentration of an $H_2NOH$ solution if the Fe(II) produced by treatment of a 50.00-mL aliquot required 23.61 mL of 0.1302 $N$ $K_2Cr_2O_7$.

12–25. The Ca in a 0.2741-g limestone sample was precipitated as $CaC_2O_4$. The solid was filtered, washed free of excess $C_2O_4^{2-}$, and redissolved in dilute $H_2SO_4$. Titration of the liberated $H_2C_2O_4$ required 36.98 mL of 0.1201 $N$ $KMnO_4$. Calculate the percentage of $CaCO_3$ in the sample.

12–26. The organic matter in a 0.9280-g sample of burn ointment was eliminated by ashing, following which the solid residue of ZnO was dissolved in acid. Treatment with $(NH_4)_2C_2O_4$ resulted in the formation of the sparingly soluble $ZnC_2O_4$. The solid was filtered, washed, and then redissolved in dilute acid. The liberated $H_2C_2O_4$ required 37.81 mL of 0.07540 $N$ $KMnO_4$. Calculate the percentage of ZnO in the medication.

*12–27. A $KMnO_4$ solution was standardized against $Na_2C_2O_4$ and found to be 0.08147 $N$. This solution was then used in a Volhard Mn determination. Reaction:

$$2MnO_4^- + 3Mn^{2+} + 4OH^- \rightarrow 5MnO_2(s) + 2H_2O$$

Calculate the percentage of Mn in a 0.2140-g sample if a 46.71-mL titration with this $MnO_4^-$ solution was required.

12–28. The Mn in a 0.4137-g sample was converted to the +2 state and titrated with $MnO_4^-$ in the presence of $F^-$. Reaction:

$$MnO_4^- + 4Mn^{2+} + 30\ F^- + 8H^+ \rightarrow 5MnF_6^{3-} + 4H_2O$$

Calculate the percentage of $MnO_2$ in the sample if the titration required 39.56 mL of a $MnO_4^-$ solution that was 0.1274 $N$ with respect to standardization against $Na_2C_2O_4$ in acid solution.

*12–29. A 1.067-g sample of a stibnite-bearing mineral was dissolved in acid, following which the Sb was oxidized from the +3 state to the +5 state with 50.00 mL of 0.1684 $N$ $KMnO_4$. After reaction was complete, the excess $MnO_4^-$ was titrated with 3.65 mL of 0.0411 $N$ Fe(II). Express the results of this analysis as the percentage of $Sb_2S_3$.

12–30. The Cd in a 0.2824-g sample of Ni alloy was precipitated as CdS from an acidic solution with $H_2S$. The solid was filtered, washed, and then redissolved in 50.00 mL of a solution that was 0.1268 $N$ with respect to $KIO_3$ and 9 $N$ with respect to HCl. Reaction:

$$3CdS(s) + 3IO_3^- + 10H^+ + 6Cl^- \rightarrow 3Cd^{2+} + 2S(s) + SO_4^{2-} + 3ICl_2^- + 5H_2O$$

Calculate the percentage of Cd in the alloy if back-titration of the excess $IO_3^-$ used 3.68 mL of 0.0753 $N$ KI. Reaction:

$$IO_3^- + 2I^- + 6H^+ + 6Cl^- \rightarrow 3ICl_2^- + 3H_2O$$

*12–31. The Hg in a 5.00-mL diuretic preparation was freed of organic matter and precipitated as $Hg(IO_3)_2$. The solid was filtered, washed, and redissolved in strong HCl. The liberated $IO_3^-$ required 25.91 mL of 0.0988 $N$ $I_2$. Reaction:

$$IO_3^- + I_2 + 6Cl^- + 6H^+ \rightarrow 3ICl_2^- + 3H_2O$$

Calculate the milligrams of meralluride [$C_{16}H_{22}HgN_6O_7$ (gfw = 611)] contained in each milliliter of the medication.

12–32. The Hf in a 1.109-g mineral specimen was isolated as the sparingly soluble $Hf(SeO_3)_2$ through treatment with $H_2SeO_3$. The solid was collected, washed free of excess precipitant, and redissolved with KI in an acidic medium. Reaction:

$$Hf(SeO_3)_2(s) + 8I^- + 12H^+ \rightarrow Hf^{4+} + 2Se(s) + 4I_2 + 6H_2O$$

The liberated $I_2$ required a 47.78-mL titration with 0.06174 $N$ $Na_2S_2O_3$. Express the results of this analysis in terms of the percentage of hafnia, $HfO_2$.

*12–33. Air drawn from the vicinity of electrical equipment was passed at the rate of 2.75 L/min through a solution of KI, in which $O_3$ was absorbed. Reaction:

$$O_3(g) + 2I^- + H_2O \rightarrow O_2(g) + 2OH^- + I_2$$

The $I_2$ generated during a 14.50-min collection period required a 6.12-mL titration with 0.0156 $N$ $Na_2S_2O_3$.
(a) Calculate the milligrams of $O_3$ per liter of air.
(b) Express the results of this analysis as ppm $O_3$, using 1.20 g/L for the density of the sample.

12–34. The exhaust gases from an internal combustion engine were swept across iodine pentoxide ($I_2O_5$) heated to 150°C. Reaction:

$$I_2O_5(s) + 5CO(g) \rightarrow 5CO_2(g) + I_2(g)$$

The $I_2$ vapors were collected in 25.00 mL of 0.004016 $N$ $Na_2S_2O_3$.
(a) Calculate the milligrams of CO contained in each liter of sample if 1.37 mL of 0.0296 $N$ $I_2$ were needed to titrate the $S_2O_3^{2-}$ remaining after 34.3 L of exhaust gases had been passed.
(b) Using 1.25 g/L as the density of the sample, calculate the concentration of CO in ppm.

*12–35. The tetraethyl lead [$Pb(C_2H_5)_4$] in a 25.00-mL sample of aviation gasoline was shaken with 15.00 mL of 0.04190 $N$ $I_2$. Reaction:

$$Pb(C_2H_5)_4 + I_2 \rightarrow Pb(C_2H_5)_3I + C_2H_5I$$

After the reaction was complete, the unused $I_2$ was titrated with 6.09 mL of 0.03465 $N$ $Na_2S_2O_3$. Calculate the weight (in milligrams) of $Pb(C_2H_5)_4$ (gfw = 323.4) in each liter of the gasoline.

12–36. The nitroglycerin in a 0.2146-g tablet was reduced with 25.00 mL of 0.01574 $N$ $Ti^{3+}$. Reaction:

$$C_3H_5(ONO_2)_3 + 18Ti^{3+} + 18H^+ \rightarrow C_3H_5(ONH_2)_3 + 18Ti^{4+} + 6H_2O$$

The excess Ti(III) was subsequently titrated with 2.55 mL of 0.0611 $N$ $Fe_2(SO_4)_3$. Calculate the milligrams of nitroglycerine (gfw = 227.1) in the tablet.

*12–37. A 25.00-mL aliquot of a solution containing $V(OH)_4^+$ and $VO^{2+}$ was titrated with 15.69 mL of 0.1054 $N$ $KMnO_4$. Titration of a second 25.00-mL aliquot, after passage through a Walden reductor, required 37.37 mL of the permanganate. Calculate the formal concentrations of $V(OH)_4^+$ and of $VO^{2+}$ in the solution.

12–38. The $I_2$ in 25.00 mL of an $I_2$–$I^-$ solution required 24.65 mL of 0.05335 $N$ $Na_2S_2O_3$. Titration of both $I_2$ and KI in 31.64 mL of the sample required 27.21 mL of 0.2392 $N$ $KIO_3$ (product: $ICl_2^-$). Calculate the weight-volume percentages of $I_2$ and of KI in the solution.

*12–39. A 5.00-mL sample of brandy was diluted to 1.000 L in a volumetric flask. The ethanol ($C_2H_5OH$) in a 25.00-mL aliquot of the diluted solution was distilled into 50.00 mL of 0.1104 $N$ $K_2Cr_2O_7$, where oxidation to acetic acid occurred. Reaction:

$$3C_2H_5OH + 2Cr_2O_7^{2-} + 16H^+ \rightarrow 4Cr^{3+} + 3CH_3COOH + 11H_2O$$

Calculate the weight-volume percentage of $C_2H_5OH$ in the brandy if 10.92 mL of 0.1081 $N$ $Fe^{2+}$ were needed to titrate the excess $Cr_2O_7^{2-}$.

*12–40. A 6.044-g sample containing KI, KBr, and other inert substances was diluted to 500.0 mL in a volumetric flask. Treatment of a 50.00-mL aliquot with $Br_2$ resulted in conversion of $I^-$ to $IO_3^-$. The excess $Br_2$ was eliminated by boiling, following which the solution was acidified and treated with an excess of KI. The liberated $I_2$ was titrated with 41.7 mL of 0.0489 $N$ $Na_2S_2O_3$.

The $Br_2$ and $I_2$ produced by oxidation of a 10.00-mL aliquot with strongly acidic $Cr_2O_7^{2-}$ were distilled into a flask containing a concentrated solution of KI. At the conclusion of distillation the $I_2$ in the receiving vessel was titrated with 30.57 mL of the $Na_2S_2O_3$ solution. Calculate the percentages of KI and KBr in the sample.

12–41. A 50.00-mL aliquot of a solution containing $IO_3^-$ and $IO_4^-$ was buffered with $HCO_3^-$ and treated with an excess of KI. Reaction:

$$IO_4^- + 2I^- + 2H^+ \rightarrow I_2 + IO_3^- + 3H_2O$$

Titration of the liberated $I_2$ required 21.83 mL of 0.1006 $N$ $HAsO_3^{2-}$. Reaction:

$$HAsO_3^{2-} + I_2 + 2HCO_3^- \rightarrow HAsO_4^{2-} + 2I^- + 2CO_2(g) + H_2O$$

Acidification of a 10.00-mL aliquot of the sample and the introduction of excess KI resulted in the production of $I_2$ as a result of the reactions

$$IO_3^- + 5I^- + 6H^+ \rightarrow 3I_2 + 3H_2O$$

$$IO_4^- + 7I^- + 8H^+ \rightarrow 4I_2 + 4H_2O$$

The liberated $I_2$ required a 34.24-mL titration with 0.09102 $N$ $Na_2S_2O_3$. Calculate the formal concentrations of $IO_3^-$ and $IO_4^-$ in the original sample.

# Chapter 13
# Potentiometric Methods

In Chapter 11, it was shown how the relative potential of an electrode is determined by the concentration (or strictly, the activity) of one or more species in the solution in which it is immersed. This chapter is concerned with the way in which this concentration sensitivity is employed to provide analytical information.[1]

An electrode employed for the determination of analyte concentration is termed an *indicator electrode;* it is used in conjunction with a second *reference electrode,* whose potential is independent of the concentration of the analyte, or for that matter, any of the other ions present in the solution under study. A potentiometric analysis thus involves determining the voltage of a cell consisting of the two electrodes and the solution under study.

## 13A REFERENCE ELECTRODES

The ideal reference electrode will possess a potential that is known, constant, and completely insensitive to the composition of the analyte solution. In addition, a reference electrode should be rugged and easy to assemble and should maintain a constant in potential even after small currents have been drawn from the cell.

### 13A–1 Calomel Electrodes

A calomel electrode can be represented schematically by

$$\|Hg_2Cl_2(\text{sat'd}), KCl(xF)|Hg$$

where $x$ represents the formal concentration of potassium chloride in the solution. The electrode reaction is given by the equation

$$Hg_2Cl_2(s) + 2e \rightleftharpoons 2Hg(l) + 2Cl^-$$

Table 13–1 lists the composition and the electrode potential for three com-

---

[1] For further reading on potentiometric methods, see: E. P. Serjeant, *Potentiometry and Potentiometric Titrations.* New York: John Wiley & Sons, 1984.

Table 13 – 1    **Potentials for Reference Electrodes in Aqueous Solution**

| Temperature, °C | Potential (vs. SHE) | | | | |
|---|---|---|---|---|---|
| | 0.1 *F* Calomel* | 3.5 *F* Calomel† | Sat'd Calomel* | 3.5 *F* Ag–AgCl† | Sat'd Ag–AgCl† |
| 12 | 0.3362 | | 0.2528 | | |
| 15 | 0.3362 | 0.254 | 0.2511 | 0.212 | 0.209 |
| 20 | 0.3359 | 0.252 | 0.2479 | 0.208 | 0.204 |
| 25 | 0.3356 | 0.250 | 0.2444 | 0.205 | 0.199 |
| 30 | 0.3351 | 0.248 | 0.2411 | 0.201 | 0.194 |
| 35 | 0.3344 | 0.246 | 0.2376 | 0.197 | 0.189 |

\* From R. G. Bates, in *Treatise on Analytical Chemistry,* 2nd ed., I. M. Kolthoff and P. J. Elving, Eds., Part I, Vol. 1, p. 793. New York: John Wiley & Sons, 1978. With permission.
† From D. T. Sawyer and J. L. Roberts, Jr., *Experimental Electrochemistry for Chemists,* p. 42. New York: John Wiley & Sons, 1974. With permission.

mon calomel electrodes. Note that these half-cells differ only in their potassium chloride concentrations; all, in short, are saturated with mercury(I) chloride.

The saturated calomel electrode (SCE) is most commonly used by the analytical chemist because of the ease with which it can be prepared. Its temperature coefficient is somewhat larger than that for electrodes with lower chloride ion concentrations, which is a disadvantage only in those rare circumstances where substantial temperature changes occur during the measurement process. The potential of the saturated calomel electrode is 0.244 V at 25°C.

The saturated calomel electrode shown in Figure 13 – 1 is typical of electrodes that are obtainable from commercial sources. It consists of a 5- to 15-cm tube that is 0.5 to 1.0 cm in diameter. A mercury – mercury(I) chloride paste in

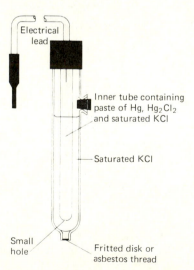

Electrical lead

Inner tube containing paste of Hg, $Hg_2Cl_2$ and saturated KCl

Saturated KCl

Small hole

Fritted disk or asbestos thread

**Figure 13 – 1**    Diagram of a typical saturated calomel electrode.

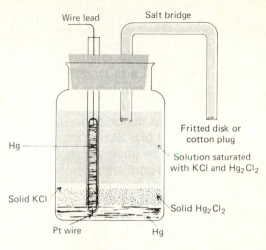

**Figure 13 – 2**  Diagram of a saturated calomel electrode.

Half-reaction:
$$Hg_2Cl_2(s) + 2e \rightleftharpoons 2Hg + 2Cl^-$$

saturated potassium chloride is contained in an inner tube and is connected to the saturated potassium chloride solution in the outer tube through a small opening. Contact with the analyte solution is made through a fritted disk or a porous fiber sealed in the end of the outer tubing.

Figure 13–2 represents a saturated calomel electrode that is simple to construct from materials available in most laboratories. A salt bridge (p. 255) provides electrical contact with the analyte solution. A fritted disk or a wad of cotton at one end of the salt bridge prevents siphoning of liquids into or from the bridge. Alternatively, an agar gel made with potassium chloride is used as the conducting medium in the bridge.[2] This electrode has a lower electrical resistance than the one shown in Figure 13–1.

### 13A – 2 Silver – Silver Chloride Electrodes

A system that is analogous to the calomel electrode consists of a silver electrode immersed in a solution of potassium chloride that has been saturated with silver chloride:

$$\|AgCl(sat'd), KCl(x\ F)|Ag$$

The half-reaction is

$$AgCl(s) + e \rightleftharpoons Ag(s) + Cl^-$$

_____

[2] A conducting gel can be prepared by heating about 5 g of agar in 100 mL of water containing about 35 g of potassium chloride. The resulting liquid is poured into the bridge and allowed to cool.

This electrode is ordinarily prepared with a saturated solution of potassium chloride; its potential is +0.199 V at 25°C (Table 13–1).

A simple and easily constructed silver–silver chloride electrode is shown in Figure 13–3. The electrode consists of a tube fitted with a fritted glass disk. A layer of agar gel saturated with potassium chloride (see Footnote 2) is formed on the disk to prevent loss of solution from the tube. A layer of potassium chloride is introduced, following which the tube is filled with a saturated solution of that salt. A few drops of silver nitrate are added to ensure that the solution is also saturated with silver chloride. Finally, a heavy-gauge (1 to 2 mm in diameter) silver wire is inserted to provide electrical contact.

## 13B INDICATOR ELECTRODES

An ideal indicator electrode would respond rapidly and reproducibly to changes in concentration of a single analyte ion (or a group of ions). Although no electrode has yet been developed that is entirely specific in its response, a few are now available that are remarkably close in their approach to ideal behavior. Two types of indicator electrodes are encountered, namely metallic and membrane.

### 13B–1 Metallic Indicator Electrodes

It is convenient to classify metallic indicator electrodes as being *electrodes of the first kind, electrodes of the second kind,* and *redox electrodes.*

**Electrodes of the First Kind**    An electrode of the first kind is in direct equilibrium with the cation derived from the electrode metal. Here, a single reaction is

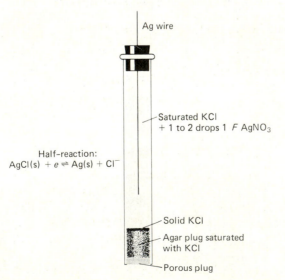

Ag wire

Saturated KCl
+ 1 to 2 drops 1 $F$ AgNO$_3$

Half-reaction:
AgCl(s) + $e$ ⇌ Ag(s) + Cl$^-$

Solid KCl

Agar plug saturated
with KCl

Porous plug

**Figure 13–3**    Diagram of a silver–silver chloride electrode.

involved. For example the equilibrium between a metal M and its ion $M^{n+}$ is given by

$$M^{n+} + ne \rightleftharpoons M(s)$$

for which

$$E_{ind} = E^0 - \frac{0.0591}{n} \log \frac{1}{[M^{n+}]} = E^0 - \frac{0.0591}{n} pM \qquad (13-1)$$

where $[M^{n+}]$ is the concentration of the ion (or more exactly, its activity $a_M$).

Equation 13-1 accurately describes the behavior of a number of common metals that are used as indicator electrodes of the first kind. In contrast, certain harder metals—notably iron, chromium, tungsten, cobalt, and nickel—do not provide reproducible potentials. Moreover, plots of the potential of a metal of this kind as a function of pM often yield slopes that differ significantly from the theoretical $(0.0591/n)$. The nonideal behavior of this type of electrode is attributed to strains and deformations in the crystal structure of the metal or to the presence of oxide films on the surface.

**Electrodes of the Second Kind**  Metals not only serve as indicator electrodes for their own cations, but also respond to the concentration of anions that form sparingly soluble precipitates or highly stable complexes with such cations. The potential of a silver electrode, for example, will respond reproducibly to the concentration of iodide ion in a solution that has been saturated with silver iodide. Here, two equilibria are involved:

$$AgI(s) \rightleftharpoons Ag^+ + I^-$$

and

$$Ag^+ + e \rightleftharpoons Ag(s) \qquad E^0 = 0.799 \text{ V}$$

Combination of these equations (see Chapter 11, Example 11-5) gives

$$AgI(s) + e \rightleftharpoons Ag(s) + I^-$$

The Nernst expression for this process is given by

$$E_{ind} = E^0_{AgI} - 0.0591 \log [I^-] = E^0_{AgI} + 0.0591 \text{ pI} \qquad (13-2)$$

where (see Equation 11-4)

$$E^0_{AgI} = E^0_{Ag} + 0.0591 \log K_{sp} = -0.151 \text{ V}$$

Equation 13-2 shows that the potential of a silver electrode will be proportional to pI, the negative logarithm of the iodide ion concentration. Thus, in a solution that is saturated with silver iodide, a silver electrode can serve as an indicator electrode of the second kind for iodide ion. Note that the sign of the log term for an electrode of this type is opposite that for an electrode of the first kind.

Mercury serves as an indicator electrode of the second kind for measuring the EDTA anion $Y^{4-}$. A small amount of $HgY^{2-}$ is added to the solution containing the analyte. The half-reaction can then be written as

$$HgY^{2-} + 2e \rightleftharpoons Hg(l) + Y^{4-} \qquad E^0 = 0.21V$$

for which

$$E_{ind} = 0.21 - \frac{0.0591}{2} \log \frac{[Y^{4-}]}{[HgY^{2-}]}$$

The formation constant for $HgY^{2-}$ is very large ($K_f = 6.3 \times 10^{21}$) so that the concentration of the complex remains essentially constant over a large range of $Y^{4-}$ concentrations. The Nernst equation for the process can therefore be written as

$$E = K - \frac{0.0591}{2} \log [Y^{4-}] = K + \frac{0.0591}{2} pY \qquad (13-3)$$

where

$$K = 0.21 - \frac{0.0591}{2} \log \frac{1}{[HgY^{2-}]}$$

The mercury electrode is thus a valuable electrode of the second kind for EDTA titrations.

**Indicator Electrodes for Oxidation-Reduction Systems**    Electrodes fashioned from platinum, gold, palladium, or carbon serve as indicator electrodes for oxidation-reduction systems. Of itself, such an electrode is inert; its potential depends solely upon the potential of the system with which it is in contact (p. 281). For example, the potential of a platinum electrode immersed in a solution containing cerium(III) and cerium(IV) is given by

$$E_{ind} = E^0_{Ce(IV)} - 0.0591 \log \frac{[Ce^{3+}]}{[Ce^{4+}]}$$

A platinum electrode is thus a convenient indicator electrode for titrations involving the use of standard cerium(IV) solutions.

### 13B-2 Membrane Electrodes[3]

For many years, the most convenient method for determining pH has involved measurement of the potential developed across a thin glass membrane that separates two solutions with different hydrogen ion concentrations. The phe-

---

[3] Suggested sources for additional information on this topic include *Ion-Selective Electrodes in Analytical Chemistry*, H. Freiser, Ed. New York: Plenum Press, 1978; J. Vesely, D. Weiss, and K. Stulik, *Analysis with Ion-Selective Electrodes*. New York: John Wiley & Sons, 1979; *Ion-Selective Methodology*, A. K. Covington, Ed. Boca Raton, FL: CRC Press, 1979.

nomenon upon which the measurement is based was first reported by Cremer[4] and has been extensively studied by many investigators; as a result, the sensitivity and selectivity of glass membranes toward hydrogen ions are now reasonably well understood. Furthermore, this understanding has led to the development of other types of membranes that respond selectively to more than two dozen other ions.

Membrane electrodes are sometimes called *p-ion electrodes* because the data obtained from such electrodes are usually presented as p-functions, such as pH, pCa, or pNO$_3$. The membranes used in constructing these electrodes are classed as being crystalline or noncrystalline. The latter can be further subdivided into glass, liquid, and immobilized liquid. In this section we will consider all of these types of p-ion membranes. In addition, a gas sensing probe that is also based on a noncrystalline membrane will be described. It, however, is sensitive to molecules rather than ions.

It is important to note at the outset of this discussion that membrane electrodes are *fundamentally different* from metal electrodes both in design and principle. We shall illustrate these differences using the glass electrode for pH measurements as an example.

### 13B-3 The Glass Electrode for pH Measurements

Figure 13-4 shows a typical cell for measuring pH. It consists of a glass indicator electrode and a saturated calomel reference electrode immersed in the solution whose pH is to be determined. The indicator electrode consists of a thin, pH-sensitive glass membrane sealed onto one end of a heavy-walled glass or plastic tube. A small volume of dilute hydrochloric acid that is saturated with silver chloride is contained in the tube (the inner solution in some electrodes is a buffer containing chloride ion). A silver wire in this solution forms a silver-silver chloride reference electrode, which is connected to one of the terminals of a potential-measuring device. A calomel electrode is connected to the other terminal.

A schematic representation of this cell in Figure 13-5 reveals that it contains *two* reference electrodes, one of which is the *external* calomel electrode while the other is the *internal* silver-silver chloride electrode, which, while a part of the glass electrode, is not the pH sensing element. Instead, *it is the thin glass membrane at the tip of the electrode that responds to pH.*

**Composition of Glass Membranes** Much systematic investigation has been devoted to the effects of glass composition on the sensitivity of membranes to protons as well as other cations, and a number of formulations are now used for manufacture of electrodes.[5] Early membranes were fabricated from a glass

[4] M. Cremer, *Z. Biol.,* **1906,** *47,* 562.

[5] For further information on glass compositions, see: J. O. Isard, in *Glass Electrodes for Hydrogen and Other Cations,* G. Eisenman, Ed., Chapter 3. New York: Marcel Dekker, Inc., 1967.

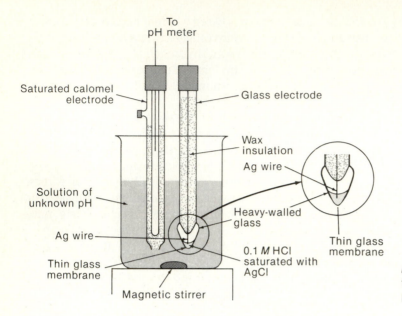

Figure 13–4 Typical electrode system for measuring pH.

consisting of approximately 22% $Na_2O$, 6% CaO, and 72% $SiO_2$. These membranes showed excellent specificity towards hydrogen ions up to a pH of about 9. At higher pH values, however, they became somewhat responsive to sodium, as well as other singly charged cations. Other glass formulations are now used in which sodium and calcium ions are replaced to various degree by barium and lithium ions. These membranes have superior selectivity and lifetimes.

**Structure of Membrane Glasses**    A silicate glass used for membranes consists of an infinite three-dimensional network of $SiO_4^{4-}$ groups in which each silicon is bonded to four oxygens and each oxygen is shared by two silicons. Within the interstices of this structure are sufficient cations to balance the negative charge of the silicate groups. Singly charged cations, such as sodium and lithium, are mobile in the lattice and are responsible for electrical conduction within the membrane. Doubly and triply charged species, such as calcium and aluminum, are strongly held and do not contribute to the conducting properties of the glass.

**Hygroscopicity of Glass Membranes**    It has been shown that the surfaces of a glass membrane must be hydrated before it will function as a pH electrode. The amount of water involved is approximately 50 mg per cubic centimeter of glass. Nonhygroscopic glasses show no pH function. Even hygroscopic glasses lose pH sensitivity after dehydration by storage over a desiccant. The effect is reversible, however, and the response of a glass electrode is restored by soaking in water.

It has also been demonstrated experimentally that hydration of a pH-sensi-

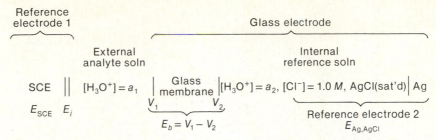

**Figure 13-5** Schematic diagram of a glass–calomel cell for measurement of pH.

tive glass membrane involves an ion-exchange reaction between singly charged cations in the lattice of the glass and protons from the solution. The process involves univalent cations exclusively because di- and trivalent cations are too strongly held within the silicate structure to exchange with ions in the solution. Typically, then, the ion-exchange reaction can be written as

$$\underset{\text{soln}}{H^+} + \underset{\text{glass}}{Na^+Gl^-} \rightleftharpoons \underset{\text{soln}}{Na^+} + \underset{\text{glass}}{H^+Gl^-} \qquad (13-4)$$

where $Gl^-$ represents one of many negatively charged sites in the glass surface. The equilibrium constant for this process is so large that the surface of a hydrated glass membrane will ordinarily consist entirely of silicic acid ($H^+Gl^-$). An exception to this situation exists in highly alkaline media, where the hydrogen ion concentration is vanishingly small and the sodium ion concentration is large; here, a significant fraction of sites are occupied by sodium ions.

Exposure of a membrane to water ultimately causes the formation of a $10^{-5}$ to $10^{-4}$ mm layer of silicic acid gel. All singly charged sites are occupied by hydrogen ions at the outer surface of the gel, and all such sites are occupied by sodium ions in the interior of the glass. Within the gel layer, then, there is a continuous decrease in the number of protons and a corresponding increase in the number of sodium ions. The structure of a hydrated glass membrane is shown schematically in Figure 13-6.

**Electrical Conduction in Membranes**  In order to serve as an indicator for cations, a glass membrane must be capable of conducting electricity. Conduction within the hydrated glass membrane involves movement of sodium and hydrogen ions, the former in the dry interior of the membrane and the latter in the gel layer. Conduction across the solution–gel interfaces occurs by the reactions

$$\underset{\text{soln}_1}{H^+} + \underset{\text{glass}_1}{Gl^-} \rightleftharpoons \underset{\text{glass}_1}{H^+Gl^-} \qquad (13-5)$$

$$\underset{\text{glass}_2}{H^+Gl^-} \rightleftharpoons \underset{\text{soln}_2}{H^+} + \underset{\text{glass}_2}{Gl^-} \qquad (13-6)$$

Subscript 1 refers to the interface between the glass and the analyte solution,

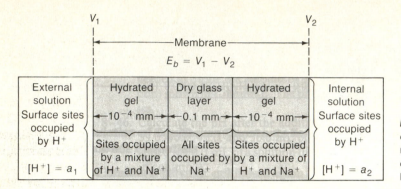

**Figure 13–6** Schematic diagram of a hydrated glass membrane. Note that the dimensions of the three inner layers are not to scale.

while subscript 2 refers to the interface between the internal solution and the glass. The position of these two equilibria will be determined by the hydrogen ion concentrations in the two solutions on either side of the membrane. Where these positions differ, the surface at which the greater dissociation has occurred will be negative with respect to the other surface. A boundary potential thus develops across the membrane, the magnitude of which depends upon the *difference* in the hydrogen ion concentrations between the two solutions. It is this potential difference that serves as the analytical parameter in a potentiometric pH measurement.

The membrane of a typical glass electrode (with a thickness of 0.03 to 0.1 mm) has an electrical resistance of 50 to 500 megohms.

**Membrane Potentials**[6]    The lower part of Figure 13–5 shows the location of four potentials that develop in a cell for the determination of pH with a glass electrode. Two of these are reference electrode potentials, $E_{Ag,AgCl}$ and $E_{SCE}$. A third potential exists across the salt bridge that serves to separate the calomel electrode from the analyte solution. This junction and its associated *junction potential, $E_j$,* are found in all cells that are used for potentiometric measurements of ion concentration. The source and the effect of the junction potential will be discussed later in this chapter. The fourth, and most important potential shown in Figure 13–5, is the *boundary potential, $E_b$, which varies with the pH of the analyte solution.* The two reference electrodes simply provide a means of measuring the magnitude of the boundary potential.

Figure 13–5 reveals that the potential of a glass electrode has two components, the fixed potential of a silver-silver chloride electrode and the pH dependent boundary potential. Not shown is a third potential, called the *asym-*

---

[6] For an extensive discussion of this topic, see: G. Eisenman, in *Glass Electrodes for Hydrogen and Other Cations,* G. Eisenman, Ed., Chapter 4. New York: Marcel Dekker, Inc., 1967.

*metry potential* which is found in most membrane electrodes and which changes slowly with time. The source and characteristics of the boundary potential and the asymmetry potential are discussed in the two sections that follow.

**The Boundary Potential**  As shown in Figure 13–5, the boundary potential consists of two potentials, $V_1$ and $V_2$, each of which is associated with one of the two gel-solution interfaces. The boundary potential is simply the difference between these potentials; that is,

$$E_b = V_1 - V_2 \qquad (13-7)$$

The potential $V_1$ is determined by the ratio of hydrogen ion activity in the analyte solution to the hydrogen ion activity in the surface of the gel, and can be considered to be a measure of the driving force for the reaction shown in Equation 13–5. Similarly, $V_2$ is related to the ratio of hydrogen ion activities between the internal solution and the corresponding gel surface, and is related to the driving force for the reaction shown in Equation 13–6.

Eisenman[7] has shown that the relationship between the two hydrogen ion activities and the potentials $V_1$ and $V_2$ is given by the equations

$$V_1 = j_1 + 0.0591 \log \frac{a_1}{a_1'} \qquad (13-8)$$

$$V_2 = j_2 + 0.0591 \log \frac{a_2}{a_2'} \qquad (13-9)$$

where $a_1$ and $a_2$ are activities of the hydrogen ion *in the two solutions,* and $a_1'$ and $a_2'$ are the corresponding activities *in the surface layers of the two gels.* If the surfaces of the two gels have the same number of sites available to accommodate protons, then the constants $j_1$ and $j_2$ will be identical. In addition, if all of the sodium ions on both surfaces have been replaced by protons, the two activities, $a_1$ and $a_2$, will also be the same. Assuming these equalities, and substituting Equations 13–8 and 13–9 into Equation 13–7 gives

$$E_b = V_1 - V_2 = 0.0591 \log \frac{a_1}{a_2} \qquad (13-10)$$

Thus, *provided that the two gel surfaces are identical,* the boundary potential $E_b$ depends only upon the activities of the hydrogen ion of the solutions on either side of the membrane. For a glass pH electrode, the hydrogen ion activity of the internal solution $a_2$ is held constant so that Equation 13–10 simplifies further to

$$E_b = L' + 0.0591 \log a_1 \qquad (13-11)$$

---

[7] G. Eisenman, *Biophys. J.,* **1962,** *2,* Part 2, 259.

where

$$L' = -0.0591 \log a_2$$

Thus, the potential of the electrode becomes a measure of the hydrogen ion activity of the external solution.

**Asymmetry Potential**    When identical solutions and reference electrodes are placed on either side of a glass membrane, a potential of zero would be expected. In fact, however, a small *asymmetry potential* is frequently encountered when this experiment is performed. Moreover, this potential is found to change gradually with time.

The sources of the asymmetry potential are obscure but undoubtedly include such causes as differences in strains on the two surfaces of the membrane imparted during manufacture, mechanical abrasion on the outer surface during use, and chemical etching of the outer surface. In order to eliminate determinate errors caused by the asymmetry potential, all membrane electrodes must be calibrated against one or more standard analyte solutions. Such calibrations should be carried out at least daily and more often with heavy use.

**Potential of the Glass Electrode**    As noted earlier, the potential of a glass indicator electrode $E_{ind}$ has three components: (1) the boundary potential, as given by Equation 13–11, (2) the potential of the internal Ag–AgCl reference electrode, and (3) a small asymmetry potential; that is,

$$E_{ind} = E_b + E_{Ag-AgCl} + E_{asy}$$

Substitution of Equation 13–11 for $E_b$ gives

$$E_{ind} = L' + 0.0591 \log a_1 + E_{Ag-AgCl} + E_{asy}$$

or

$$E_{ind} = L + 0.0591 \log a_1 = L - 0.0591 \text{ pH} \qquad (13-12)$$

where $L$ is a combination of the three constant terms. Note the similarity between Equations 13–12 and 13–1.

**The Alkaline Error**    Glass electrodes respond to the concentrations of both hydrogen ion and alkali metal ions in basic solution, where the former is necessarily much smaller than the latter. The magnitude of this *alkaline error* for four different glass membranes is shown in Figure 13–7 (curves $C$ to $F$). These curves refer to solutions in which the sodium ion concentration was held constant at 1 $M$ while the pH was varied. Note that the pH error is negative (that is, the measured pH values are lower than the true values), which suggests that the electrode is responding to sodium ions as well as to protons. This observa-

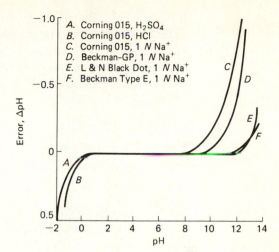

**Figure 13–7** Acid and alkaline
error for selected glass electrodes at
25°C. (From R. G. Bates,
*Determination of pH: Theory and
Practice*, p. 316. New York: John
Wiley & Sons, 1964. With permission.)

tion is confirmed by data obtained for solutions containing different sodium
ion concentrations. Thus, an electrode with a Corning 015 membrane registered a pH of 11.3 for a pH 12 solution in which the sodium ion concentration
was 1 $M$ (see curve $C$ in Figure 13–7), and 11.7 when [Na$^+$] was 0.1 $M$. All
singly charged cations induce alkaline errors; the magnitude of the error depends both upon the cation in question and the composition of the glass
membrane.

The alkaline error can be satisfactorily explained by assuming an exchange
equilibrium between the hydrogen ions on the surface of the glass and the
cations in solution. This process is simply the reverse of that shown in Equation
13–4:

$$H^+Gl^- + B^+ \rightleftharpoons B^+Gl^- + H^+$$

glass    soln    glass    soln

where B$^+$ represents some singly charged cation, such as sodium ion.

The equilibrium constant for this reaction is

$$K_{ex} = \frac{a_1 b_1'}{a_1' b_1} \tag{13–13}$$

where $a_1$ and $b_1$ represent the activities of H$^+$ and B$^+$ in solution, while $a_1'$ and $b_1'$
are the activities of these ions at the surface of the gel. Equation 13–13 can be
rearranged to give the ratio of the activities of B$^+$ to H$^+$ on the surface of the
glass:

$$\frac{b_1'}{a_1'} = \frac{b_1}{a_1} K_{ex}$$

For glasses used for pH electrodes, $K_{ex}$ is so small that the activity ratio $b'_1/a'_1$ is ordinarily minuscule. The situation differs in strongly alkaline media, however. For example, $b'_1/a'_1$ for an electrode immersed in a pH 11 solution that is 1 $M$ in sodium ions (see Figure 13–7) is $10^{11} K_{ex}$. In this case, the activity of sodium ions relative to that of hydrogen ions becomes so large that the electrode responds to both species.

**Selectivity Coefficients**    Eisenman[8] has demonstrated that the effect of an alkali-metal ion on the potential across a membrane can be accounted for by inserting an additional term in Equation 13–12 to give

$$E_b = L + 0.0591 \log (a_1 + k_{H,B}b_1) \qquad (13-14)$$

where $k_{H,B}$ is the *selectivity coefficient* for the electrode. Equation 13–14 applies not only to glass indicator electrodes for hydrogen ion, but to all other types of membrane electrodes as well. Selectivity coefficients range from zero (no interference) to values greater than 1. Thus, if an electrode for ion A responds 20 times more strongly to ion B, $k_{A,B}$ has a value of 20. If the response of the electrode to ion C is 0.001 of its response to A (a much more desirable situation), $k_{A,C}$ is 0.001.[9]

The product $k_{H,B}b_1$ for a glass pH electrode is ordinarily small with respect to $a_1$ provided the pH is less than 9, and Equation 13–14 simplifies to Equation 13–12. At high pH levels and at high concentrations of a singly charged ion, however, the second term in Equation 13–14 assumes a more important role in determining $E$, and an alkaline error is encountered. For electrodes specifically designed for work in highly alkaline media (see curve $E$ in Figure 13–7), the magnitude of $k_{H,B}b_1$ is appreciably smaller than for ordinary glass electrodes.

**The Acid Error**    Curves $A$ and $B$ in Figure 13–7 show that glass electrodes also exhibit an error in strongly acidic solutions (pH < 0.5). This error is opposite in sign to the alkaline error and leads to erroneously high pH readings. The magnitude of the error is not particularly reproducible nor is its source understood.

### 13B–4 Glass Electrodes for the Determination of Other Cations

The existence of the alkaline error in early glass electrodes led to investigations concerning the effect of glass composition upon the magnitude of this error.

---

[8] See R. H. Doremus and G. Eisenman, in *Glass Electrodes for Hydrogen and Other Cations,* G. Eisenman, Ed., Chapters 4, 5. New York: Marcel Dekker, Inc., 1967.

[9] The numerical value of $k$ for a given electrode is affected by the total electrolyte concentration of the solution and by the concentration ratio between the analyte and the interference. Consequently, as a criterion for the response of an electrode to the presence of an interfering ion, the selectivity coefficient is reliable to about one order of magnitude only.

One consequence has been the development of glasses for which the alkaline error is negligible below a pH of about 12. Other studies have been directed toward discovering compositions that would permit the determination of cations other than hydrogen. This application requires that the hydrogen ion activity $a_1$ in Equation 13–14 be negligible with respect to $k_{H,B}b_1$; under such circumstances, the potential is independent of pH and a function of pB instead. A number of investigators have demonstrated that incorporation of $Al_2O_3$ or $B_2O_3$ in the glass has the desired effect.[10] Glass electrodes that permit the direct potentiometric measurement of such singly charged species as $Na^+$, $K^+$, $NH_4^+$, $Rb^+$, $Cs^+$, $Li^+$, and $Ag^+$ have been developed. Some of these glasses are reasonably selective toward particular singly charged cations. Glass electrodes for $Na^+$, $Li^+$, $NH_4^+$, and total univalent cations are now available from commercial sources.

### 13B–5 Liquid-Membrane Electrodes

Liquid-membrane electrodes owe their response to the potential that develops across the interface between the solution containing the analyte and a liquid ion exchanger that selectively bonds with the analyte ion. These electrodes have been developed for the direct potentiometric measurement of numerous polyvalent cations as well as certain anions.

Figure 13–8 is a schematic diagram of a liquid-membrane electrode for calcium. It consists of a conducting membrane that selectively bonds calcium ions, an internal solution containing a fixed concentration of calcium chloride, and a silver electrode that is coated with silver chloride to form an internal reference electrode. The similarity to the glass electrode is obvious. The active membrane ingredient is an ion exchanger, which consists of a calcium dialkyl phosphate that has a limited solubility in water. In the electrode shown in Figure 13–8, the ion exchanger is dissolved in an immiscible organic liquid that is forced by gravity into the pores of a hydrophobic porous disk. This disk then serves as the active membrane that separates the internal solution from that of the analyte. In a more recent design, the ion exchanger is immobilized in a tough polyvinyl chloride gel that is cemented to the end of a tube that holds the internal solution and reference electrode. In either design, a dissociation equilibrium is set up at each membrane interface that is analogous to Equation 13–6. That is,

$$[(RO)_2POO]_2Ca \rightleftharpoons 2(RO)_2POO^- + Ca^{2+}$$
$$\text{organic} \qquad\qquad \text{organic} \qquad \text{aqueous}$$

As with the glass electrode, a potential develops across the membrane when the extent of dissociation of the ion exchanger at the two surfaces differs as a consequence of a difference in the calcium ion activity of the internal and the

---

[10] See, for example, G. Eisenman, in *Advances in Analytical Chemistry and Instrumentation,* C. N. Reilley, Ed., Vol. 4. New York: John Wiley & Sons, 1965.

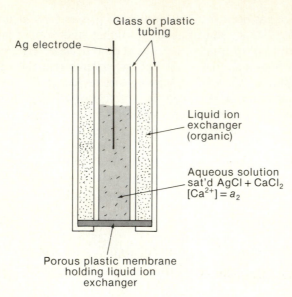

Glass or plastic
tubing

Ag electrode

Liquid ion
exchanger
(organic)

Aqueous solution
sat'd $AgCl + CaCl_2$
$[Ca^{2+}] = a_2$

Porous plastic membrane
holding liquid ion
exchanger

**Figure 13 – 8** Diagram of a liquid-membrane electrode for $Ca^{2+}$.

external solutions. The relationship between this potential and the calcium ion activities is given by an equation that is similar to Equation 13 – 10:

$$E_b = V_1 - V_2 = \frac{0.0591}{2} \log \frac{a_1}{a_2} \qquad (13-15)$$

where $a_1$ and $a_2$ are activities of calcium ion in the external and internal solutions, respectively. Since the calcium ion activity of the internal solution is constant,

$$E_b = C + \frac{0.0591}{2} \log a_1 \qquad (13-16)$$

where $C$ is a constant. Note that because calcium is divalent, a 2 appears in the denominator of the coefficient of the log term.

Figure 13 – 9 compares the structural features of a glass-membrane electrode with a commercially available liquid-membrane electrode for calcium ion. The sensitivity of the latter is reported to be 50 times greater for calcium ion than for magnesium ion, and 1000 times greater than for sodium or potassium ions. Calcium ion activities as low as $5 \times 10^{-7}$ $M$ can be measured. Performance of the electrode is said to be independent of pH in the range between 5.5 and 11. At lower pH levels, hydrogen ions undoubtedly replace some of the calcium ions on the exchanger; the electrode then becomes pH as well as pCa sensitive.

The calcium ion liquid-membrane electrode is a valuable tool for physiological investigations because this ion plays important roles in such processes as nerve conduction, bone formation, muscle contraction, cardiac expansion and contraction, renal tubular function, and perhaps hypertension. At least some of

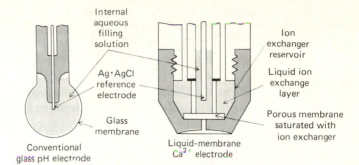

**Figure 13 – 9** Comparison of a glass electrode and a liquid-membrane electrode for calcium ion. (Courtesy of Orion Research Incorporated, Cambridge, MA)

these processes are more influenced by the activity, rather than the concentration, of the calcium ion; activity, of course, is the parameter measured by the membrane electrode.

A liquid-membrane electrode that is specific for potassium is also of great value for physiologists because the transport of neural signals appears to involve movement of this ion across nerve membranes. Investigation of this process requires an electrode that will detect small concentrations of potassium ion in media that contain much larger concentrations of sodium ion. Several liquid-membrane electrodes show promise in meeting this requirement; one is based upon the antibiotic valinomycin, a cyclic ether that has a strong affinity for potassium ion. Of equal importance is the observation that a liquid membrane consisting of valinomycin in diphenyl ether is about $10^4$ times as responsive to potassium ion as it is to sodium ion.[11]

Table 13 – 2 is a list of liquid-membrane electrodes that are available from commercial sources. The anion-sensitive electrodes that are shown make use of a solution containing an anion-exchange resin in an organic solvent. Liquid-membrane electrodes have been developed for $Ca^{2+}$, $K^+$, $NO_3^-$, and $BF_4^-$ in which the exchange liquid is held in a polyvinyl chloride gel. These have the appearance of crystalline electrodes, which are considered in the following section.

## 13B – 6 Crystalline Membrane Electrodes

Considerable work has been devoted to the development of solid membranes that are selective toward anions in the way that some glasses respond toward cations. We have seen that the existence of anionic sites on a glass surface accounts for the selectivity of a membrane for certain cations. By analogy, a membrane with a similar type of cationic sites might be expected to respond

---

[11] M. S. Frant and J. W. Ross, Jr., *Science,* **1970,** *167,* 987.

**Table 13 – 2    Characteristics of Liquid-Membrane Electrodes***

| Analyte Ion | Concentration Range, $M$ | Major Interferences |
|---|---|---|
| $Ca^{2+}$ | $10^0$ to $5 \times 10^{-7}$ | $Pb^{2+}$, $Fe^{2+}$, $Ni^{2+}$, $Hg^{2+}$, $Sr^{2+}$ |
| $Cl^-$ | $10^0$ to $5 \times 10^{-6}$ | $I^-$, $OH^-$, $SO_4^{2-}$ |
| $NO_3^-$ | $10^0$ to $7 \times 10^{-6}$ | $ClO_4^-$, $I^-$, $ClO_3^-$, $CN^-$, $Br^-$ |
| $ClO_4^-$ | $10^0$ to $7 \times 10^{-6}$ | $I^-$, $ClO_3^-$, $CN^-$, $Br^-$ |
| $K^+$ | $10^0$ to $1 \times 10^{-6}$ | $Cs^+$, $NH_4^+$, $Tl^+$ |
| Water hardness $(Ca^{2+} + Mg^{2+})$ | $10^0$ to $6 \times 10^{-6}$ | $Cu^{2+}$, $Zn^{2+}$, $Ni^{2+}$, $Sr^{2+}$, $Fe^{2+}$, $Ba^{2+}$ |

* From *Orion Guide to Ion Analysis.* Cambridge, MA: Orion Research, 1983. With permission.

selectively toward anions. Attempts have been made to prepare membranes containing a sparingly soluble salt of the anion of interest because such salts will bind that anion selectively. It has been difficult, however, to find methods of fabricating membranes from the desired salts that possess adequate physical strength, conductivity, and resistance to abrasion and corrosion.

Membranes prepared from cast pellets of silver halides have been successfully used in electrodes for the selective determination of chloride, bromide, and iodide ions. In addition, an electrode based upon a polycrystalline $Ag_2S$ membrane is offered by one manufacturer for the determination of sulfide ion. In both types of membranes, silver ions are sufficiently mobile to conduct electricity through the solid medium. Mixtures of PbS, CdS, and CuS with $Ag_2S$ provide membranes that are selective for $Pb^{2+}$, $Cd^{2+}$, and $Cu^{2+}$, respectively. Silver ion again acts to transport electricity within these solid membranes. The potential that develops across crystalline solid-state electrodes is described by Equation 13 – 16.

A crystalline electrode for fluoride ion is also available from commercial sources. The membrane consists of a single crystal of lanthanum fluoride that has been doped with europium(II) fluoride to improve its conductivity. The membrane, supported between a reference solution and the solution to be measured, shows a theoretical response to changes in fluoride ion activity in the range from $10^0$ to $10^{-6}$ $M$. The electrode is reported to be selective for fluoride ion over other common anions by several orders of magnitude; only hydroxide ion appears to offer serious interference. Solid-state electrodes available from commercial sources are listed in Table 13 – 3.

### 13B – 7  Gas-Sensing Probes

Figure 13 – 10 illustrates the essential features of a gas-sensing probe, which consists of a tube containing a reference electrode, a specific ion electrode, and an electrolyte solution. A thin, replaceable, gas-permeable membrane, attached to one end of the tube, serves as a barrier between the internal and the analyte solutions. As can be seen from Figure 13 – 10, this device is actually a complete

*Table 13‑3* *Characteristics of Solid-State Crystalline Electrodes**

| Analyte Ion | Concentration Range, $M$ | Major Interferences |
|---|---|---|
| $Br^-$ | $10^0$ to $5 \times 10^{-6}$ | $CN^-, I^-, S^{2-}$ |
| $Cd^{2+}$ | $10^{-1}$ to $1 \times 10^{-7}$ | $Fe^{2+}, Pb^{2+}, Hg^{2+}, Ag^+, Cu^{2+}$ |
| $Cl^-$ | $10^0$ to $5 \times 10^{-5}$ | $CN^-, I^-, Br^-, S^{2-}$ |
| $Cu^{2+}$ | $10^{-1}$ to $1 \times 10^{-8}$ | $Hg^{2+}, Ag^+, Cd^{2+}$ |
| $CN^-$ | $10^{-2}$ to $1 \times 10^{-6}$ | $S^{2-}$ |
| $F^-$ | Sat'd to $1 \times 10^{-6}$ | $OH^-$ |
| $I^-$ | $10^0$ to $5 \times 10^{-8}$ | |
| $Pb^{2+}$ | $10^{-1}$ to $1 \times 10^{-6}$ | $Hg^{2+}, Ag^+, Cu^{2+}$ |
| $Ag^+/S^{2-}$ | $Ag^+$: $10^0$ to $1 \times 10^{-7}$ | $Hg^{2+}$ |
| | $S^{2-}$: $10^0$ to $1 \times 10^{-7}$ | |
| $SCN^-$ | $10^0$ to $5 \times 10^{-6}$ | $I^-, Br^-, CN^-, S^{2-}$ |

* From *Orion Guide to Ion Analysis.* Cambridge, MA: Orion Research, 1983. With permission.

electrochemical cell and is more properly referred to as a probe, rather than an electrode.

**Membrane Composition**  A *microporous membrane* is fabricated from a hydrophobic polymer. As the name implies, the membrane is highly porous (the average pore size is less than 1 $\mu m$) and allows the free passage of gases; on the other hand, the water-repellent polymer prevents water and solute molecules from entering the pores. The thickness of the membrane is about 0.1 mm.

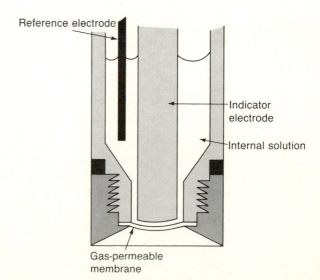

Reference electrode

Indicator electrode

Internal solution

Gas-permeable membrane

*Figure 13‑10* Diagram of a gas-sensing probe.

In contrast, gases dissolve in nonporous *homogeneous membranes* and are extracted from the membrane by the internal solution. Homogeneous membranes are usually fabricated from silicone rubber and range in thickness from 0.01 to 0.03 mm.

**Mechanism of Response**   Using carbon dioxide as an example, we can represent the transfer of gas to the pores of the membrane by the equation

$$\underset{\substack{\text{analyte} \\ \text{solution}}}{CO_2(aq)} \rightleftharpoons \underset{\substack{\text{membrane} \\ \text{pores}}}{CO_2(g)}$$

Because the membrane contains a large number of pores, this equilibrium is rapidly established. The $CO_2$ in the pores is in contact with the internal solution as well, and a second equilibrium is also rapidly established, namely,

$$\underset{\substack{\text{membrane} \\ \text{pores}}}{CO_2(g)} \rightleftharpoons \underset{\substack{\text{internal} \\ \text{solution}}}{CO_2(aq)}$$

These two processes result in equilibration between the analyte solution and the film of internal liquid in contact with the membrane. Here, however, yet another equilibrium

$$\underset{\substack{\text{internal} \\ \text{solution}}}{CO_2(aq) + 2H_2O} \rightleftharpoons \underset{\substack{\text{internal} \\ \text{solution}}}{HCO_3^- + H_3O^+}$$

causes the pH of the internal surface film to change, which is then detected by the internal glass–reference electrode pair. The overall process is obtained by adding the equations for the three individual equilibria to give

$$\underset{\substack{\text{analyte} \\ \text{solution}}}{CO_2(aq) + 2H_2O} \rightleftharpoons \underset{\substack{\text{internal} \\ \text{solution}}}{H_3O^+ + HCO_3^-}$$

The net equilibrium constant is then given by

$$K = \frac{[H_3O^+][HCO_3^-]}{[CO_2(aq)]_{ext}}$$

To make the measured cell potential respond linearly to the logarithm of the carbon dioxide concentration of the external solution, it is necessary that the hydrogen carbonate concentration of the internal solution be sufficiently large so that its concentration is not altered significantly by the carbon dioxide entering from the external solution. Assuming then that $[HCO_3^-]$ is constant, we may rearrange the previous equation to

$$\frac{[H_3O^+]}{[CO_2(aq)]_{ext}} = \frac{K}{[HCO_3^-]} = K_g$$

Letting $a_1$ be the hydrogen ion activity of the internal solution, we may write

$$a_1 = [H_3O^+] = K_g[CO_2(aq)]_{ext} \qquad (13-17)$$

Substitution of Equation 13–17 for $a_1$ in Equation 13–12 yields

$$E_{ind} = L + 0.0591 \log K_g(CO_2(aq)]_{ext}$$

or

$$E_{ind} = L' + 0.0591 \log [CO_2(aq)]_{ext} \qquad (13-18)$$

where

$$L' = L + 0.0591 \log K_g$$

Thus, the potential between the glass electrode and the reference electrode in the internal solution is determined by the $CO_2$ concentration of the external solution. *Note that no electrode comes directly in contact with the analyte solution;* as we have noted, it is better to think of the device as a gas-sensing *cell* or *probe* rather than a gas-sensing electrode. It is also noteworthy that the only species that will interfere are other dissolved gases that permeate the membrane and then affect the pH of the internal solution.

The possibility exists for increasing the selectivity of the gas-sensing probe through use of an indicator electrode that is responsive to some species other than hydrogen ion. For example, a nitrate-sensing electrode could be used to provide a cell that would be sensitive to nitrogen dioxide; the response of this electrode would be determined by the equilibrium

$$2NO_2(aq) + H_2O \rightleftharpoons NO_2^- + NO_3^- + 2H^+$$

external solution          internal solution

Gas-sensing probes for $CO_2$, $NO_2$, $H_2S$, $SO_2$, $HF$, $HCN$, and $NH_3$ are now available from commercial sources.


## 13C INSTRUMENTS FOR MEASUREMENT OF CELL POTENTIALS

An instrument for potentiometric measurements must draw essentially no electricity from the cell being studied. One reason for this requirement is that a current causes changes in reactant concentrations in the cell and thus changes in its potential. Even more important are the effects of *IR* drop and polarization phenomena (Chapter 14) upon the measured potential.

Historically, potential measurements were performed with a potentiometer, a null-point instrument in which the unknown potential was just balanced by a standard reference potential. At null, no electricity is drawn from the cell whose potential is being measured. The potentiometer has now been supplanted by electronic voltmeters, commonly called *pH meters;* with the advent of membrane electrodes for other ions, these instruments could more properly be referred to as *pIon* or *ion meters.*

Numerous direct-reading pH meters are presently on the market. The read-out can either be digital or be indicated by a needle that sweeps a range from 0 to 14 pH units on a scale. Many of the latter are equipped with the capability of scale expansion, which provide full-scale ranges from 0.5 to 2 pH units; precisions on the order of 0.001 to 0.005 pH unit can thus be realized. It should be appreciated, however, that it is seldom if ever possible to measure a pH with a comparable degree of *accuracy.* Indeed, uncertainties of $\pm 0.02$ to $\pm 0.03$ pH unit are typical (p. 357).

## 13D  DIRECT POTENTIOMETRIC MEASUREMENTS

Direct potentiometric measurements can be used to establish the concentration of species for which an indicator electrode is available. The technique is simple, requiring only a comparison of the potential developed by the indicator electrode in the analyte solution with its potential when immersed in one or more solutions of known analyte concentration. Insofar as the response of the electrode is specific for the analyte and independent of matrix effects, no preliminary separation steps are required. Direct potentiometric measurements are also readily adapted to the continuous and automatic recording of analytical data.

### 13D – 1  The Liquid Junction Potential

Notwithstanding these advantages, the user of direct potentiometric measurements must be alert to limitations that are inherent to the method. Principal among these is the existence of a *liquid junction potential* that affects most potentiometric measurements. The junction potential is inconsequential in most electroanalytical methods and can be neglected; its existence, however, places a limitation upon the accuracy that can be attained from a direct potentiometric measurement.

A liquid junction potential is caused by an unequal distribution of cations and anions across the boundary between two dissimilar electrolyte solutions. This inhomogeneity is the result of differences in the rates at which the various charged species diffuse across the boundary between the solutions. Consider, for example, the situation at the interface between 1 $F$ and 0.01 $F$ solutions of hydrochloric acid. This interface can be symbolized as

$$HCl(1\ F)|HCl(0.01\ F)$$

Both hydrogen ions and chloride ions tend to diffuse across this boundary from the more concentrated to the more dilute solution, the driving force for this migration being proportional to the concentration difference. The two species move at different rates, however, under the influence of this force (that is, their *mobilities* are different). In the present example, hydrogen ions are substantially more mobile than chloride ions. Thus, hydrogen ions outstrip chloride ions during the diffusion process and a separation of charge results (see Figure

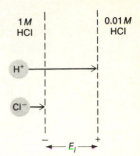

**Figure 13–11** Schematic representation of a liquid junction showing the source of the junction potential $E_j$. The length of the arrows corresponds to the relative mobility of the two ions.

13–11). The more dilute side of the boundary becomes positively charged, owing to the more rapid migration of hydrogen ions, and the concentrated side therefore acquires a negative charge from the excess of slower moving chloride ions. The charge developed tends to counteract the differences in mobilities between the two ions so that a condition of equilibrium is soon attained. The potential difference resulting from this charge separation may amount to several hundredths of a volt.

The magnitude of the potential for a junction as simple as the one under consideration can be calculated from knowledge of the mobilities of the ions involved. However, seldom (if ever) will a cell of analytical importance possess a sufficiently simple composition to permit such a computation. It is an experimental fact that the magnitude of the liquid junction potential can be greatly decreased by interposition of a concentrated electrolyte (a *salt bridge*) between the two solutions. The effectiveness of this device improves as the mobilities of the ions in the bridge approach one another and as their concentrations increase. A saturated solution of potassium chloride is good from both standpoints, its concentration being somewhat greater than 4 $F$ at room temperature and the mobilities of its ions differing by about 4%. A junction potential with such a bridge typically amounts to a few millivolts and is negligible for most electroanalytical measurements; this potential, however, appreciably affects direct potentiometric measurements.

## 13D–2 Equation for Direct Potentiometry

The potential of a cell used for a direct potentiometric measurement can be expressed in terms of the potentials developed by the indicator electrode, the reference electrode, and a junction potential:

$$E_{obs} = E_{ref} - E_{ind} + E_j \qquad (13-19)^{12}$$

The junction potential $E_j$ has two components: the first at the interface

---

[12] Note that the indicator electrode, *by convention,* is treated as the anode and the reference electrode as the cathode.

between the analyte solution and one end of the salt bridge; the second between the reference electrode solution and the other end of the bridge. We have noted that these two potentials tend to cancel one another, but seldom do so completely.

The potential of the indicator electrode is ideally related to the activity $a_1$ of $M^{n+}$, the analyte, by a version of the Nernst equation (see Equations 13–1, 13–2, 13–12, and 13–16). Thus, for the cation $M^{n+}$ at 25°C,

$$E_{ind} = L + \frac{0.0591}{n} \log a_1 = L - \frac{0.0591}{n} \text{pM} \qquad (13-20)$$

where $L$ is a constant. For metallic indicator electrodes, $L$ is ordinarily the standard electrode potential, while for membrane electrodes, $L$ is the summation of several constants, including the time-dependent asymmetry potential (p. 340) of uncertain magnitude.

Combination of Equation 13–19 with Equation 13–20 and rearrangement yields

$$\text{pM} = -\log a_1 = \frac{E_{obs} - (E_{ref} + E_j - L)}{0.0591/n}$$

The constant terms in parentheses can be combined to give a new constant, $K$:

$$\text{pM} = -\log a_1 = \frac{E_{obs} - K}{0.0591/n} \qquad (13-21)$$

Note that this new constant, $K$, includes at least one term $(E_j)$ with a magnitude that cannot be evaluated from theory. Thus, before Equation 13–21 can be used for the determination of pM, $K$ must be evaluated experimentally with a standard solution of the analyte.

For an anion, $A^{n-}$, the signs of Equation 13–21 are reversed; that is,

$$\text{pA} = \frac{K - E_{obs}}{0.0591/n} \qquad (13-22)$$

All direct potentiometric methods for performing an analysis by direct potentiometry are based—directly or indirectly—upon Equation 13–21 or 13–22.

### 13D–3 Electrode Calibration Method

In the electrode calibration method, $K$ in Equation 13–21 is determined by measuring $E_{obs}$ for one or more standard solutions of known pM. The assumption is then made that $K$ is unchanged when the standard is replaced by the analyte solution. The calibration is ordinarily performed at the time pM for the unknown is determined; recalibration may be required if measurements extend over several hours.

The electrode calibration method offers the advantages of simplicity, speed, and applicability to the continuous monitoring of pM. Two important sources of difficulty attend the use of the method, however. One of these is that the results of an analysis are in terms of activities rather than concentrations (to be sure, this factor may be an advantage rather than a disadvantage in some situations.)[13] The other difficulty is the uncertainty introduced by the need to assume that the junction potential $E_j$ remains unchanged when the analyte solution is substituted for the standard; unfortunately, this uncertainty can never be totally eliminated.

**Activity Versus Concentration**   Electrode response is related to analyte activity rather than to analyte concentration. The scientist, however, is ordinarily interested in concentration, and the determination of this quantity from a potentiometric measurement requires activity coefficient data. More often than not, activity coefficients will not be available because the ionic strength of the solution is either unknown or else is so large that the Debye-Hückel equation is not applicable.

The difference between activity and concentration is illustrated by Figure 13–12, in which the response of a calcium ion electrode is plotted against a logarithmic function of the calcium chloride concentration. The nonlinearity in this curve is due to the increase in ionic strength — and the consequent decrease in the activity of calcium ion — with increasing electrolyte concentration. The upper curve is obtained when these concentrations are converted to activities; note that this straight line possesses the theoretical slope of 0.0296 (0.0591/2).

Activity coefficients for singly charged species are less affected by changes in ionic strength than are ions with multiple charges. Thus, the effect shown in Figure 13–12 will be less pronounced for electrodes that respond to $H^+$, $Na^+$, and other univalent ions.

In potentiometric pH measurements, the pH of the standard buffer employed for calibration is generally based on the activity of hydrogen ions. Thus, the results are also on an activity scale. If the unknown sample has a high ionic strength, the hydrogen ion *concentration* will differ appreciably from the activity measured.

**Inherent Error in the Electrode Calibration Method**   A serious disadvantage of the electrode calibration method is the existence of an inherent uncertainty that results from the need to assume that $K$ in Equation 13–21 remains constant after calibration. This assumption can seldom, if ever, be exactly true because the electrolyte composition of the analyte solution will almost inevitably differ from that of the standard that was used for calibration. Consequently, the junction potential that is a component of $K$ must be expected to undergo a slight change, notwithstanding the existence of a salt bridge. The effect of this change,

---

[13] Many chemical reactions of physiological importance depend upon the activity, rather than the concentration, of metal ions. For studies of such reactions, a potentiometric measurement with a reversible indicator electrode is an ideal analytical tool.

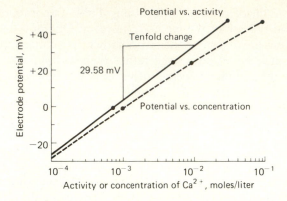

**Figure 13 – 12**    Response of a liquid-membrane electrode to variations in the concentration and the activity of calcium ion. (Courtesy of Orion Research Incorporated, Cambridge, MA)

which can be on the order of 1 mV (or more), upon the measured quantity $a_1$ is given by[14]

$$\frac{\Delta a_1}{a_1} \times 100 = 3.9 \times 10^3 \, n\Delta K$$

The quantity $\Delta a_1 / a_1$ is the relative error in $a_1$ associated with an absolute uncertainty $\Delta K$ in $K$. Note that an uncertainty of $\pm 0.001$ V can be expected to cause a relative error of about $4n\%$ in the value for the analyte concentration. *It is important to appreciate that this uncertainty is characteristic of all measurements involving cells that contain a salt bridge and sets a limit on the accuracy of such measurements.* It does not appear possible to devise a method that will eliminate the uncertainty in $K$ that is the source of this error.

### 13D – 4  Calibration Curves for Direct Potentiometry

An obvious way to convert potentiometric measurements from activity to concentration is to make use of an empirical calibration curve, such as the lower plot in Figure 13 – 12. For this approach to be successful, it is necessary to make the ionic composition of the standards essentially the same as that for the analyte solution. Matching the ionic strength of standards to that of samples is often difficult, particularly for samples that are chemically complex.

Where electrolyte concentrations are not too great, it is often useful to swamp both samples and standards with a measured excess of an inert electrolyte. The added effect of the electrolyte from the sample matrix becomes negligible under these circumstances, and the empirical calibration curve yields results in terms of concentration. This approach has been used, for example, in the potentiometric determination of fluoride ion in public drinking water supplies. Both samples and standards are diluted with a solution that contains sodium chloride, an acetate buffer, and a citrate buffer; the diluent is sufficiently concentrated so that the samples and standards have essentially identi-

---

[14] See, for example, D. A. Skoog and D. M. West, *Fundamentals of Analytical Chemistry,* 4th ed., p. 417. Philadelphia: Saunders College Publishing, 1982.

cal ionic strengths. This method provides a rapid means for measuring fluoride concentrations in the parts-per-million range with an accuracy of about 5% relative.

### 13D–5 The Standard Addition Method

The standard addition method involves determining the potential of the electrode system before and after a measured volume of a standard has been added to a known volume of the analyte solution. Often an excess of an electrolyte is incorporated into the analyte solution at the outset to prevent a major shift in the ionic strength that might accompany the addition of standard. It is also necessary to assume that the junction potential remains constant during the two measurements.

---

**Example 13–1**  A lead ion electrode developed a potential of 0.4706 V (vs. SCE) when immersed in 50.00 mL of a sample. A 5.00-mL addition of standard 0.02000 $M$ lead solution caused the potential to shift to 0.4490 V. Calculate the molar concentration of lead in the sample.

Applying Equation 13–21, and assuming that the activity of $Pb^{2+}$ is approximately equal to its molar concentration $C_x$, we may write

$$pPb = -\log C_x = \frac{E_1 - K}{0.0591/2}$$

where $E_1$ is the measured potential (0.4706 V).

After introduction of the standard addition, the potential is now $E_2$ (0.4490 V) and

$$-\log \frac{50.00 \times C_x + 5.00 \times 0.02000}{50.00 + 5.00} = \frac{E_2 - K}{0.0591/2}$$

which can be written as

$$-\log (0.9091C_x + 1.818 \times 10^{-3}) = \frac{E_2 - K}{0.0591/2}$$

Subtraction of this equation from the first gives

$$-\log \frac{C_x}{0.9091C_x + 1.818 \times 10^{-3}} = \frac{2(E_1 - E_2)}{0.0591}$$

$$= \frac{2(0.4706 - 0.4490)}{0.0591} = 0.7310$$

$$\frac{C_x}{0.9091C_x + 1.818 \times 10^{-3}} = 0.1858$$

$$C_x = 4.06 \times 10^{-4} F$$

## 13D – 6 Potentiometric pH Measurements with a Glass Electrode[15]

The glass electrode is unquestionably the most important indicator electrode for hydrogen ion. It is convenient to use and is subject to few of the interferences that affect other pH-sensing electrodes.

The glass–calomel electrode system is a remarkably versatile tool for the measurement of pH under many conditions. The electrode can be used without interference in solutions containing strong oxidants, reductants, proteins, and gases; the pH of viscous or even semisolid fluids can be determined. Electrodes for special applications are available. Included among these are small electrodes for pH measurements in a drop (or less) of solution, in a cavity of a tooth, or in the sweat on the skin; microelectrodes that permit the measurement of pH inside a living cell; rugged electrodes for insertion in a flowing liquid stream to provide a continuous monitoring of pH; and a small glass electrode that can be swallowed to indicate the acidity of the stomach contents (the calomel electrode is kept in the mouth).

**Summary of Errors that Affect pH Measurements with the Glass Electrode**    The ubiquity of the pH meter and the general applicability of the glass electrode tend to lull the chemist into the attitude that any measurement obtained with such equipment is surely correct. The reader must be alert to the fact that there are distinct limitations to the electrode, some of which have been discussed in earlier sections:

1. *The alkaline error.* The ordinary glass electrode becomes somewhat sensitive to alkali-metal ions and gives low readings at pH values greater than 9. It should be noted that a greater tolerance (to pH 11 to 12) is achieved with membranes in which lithium and barium have replaced sodium and calcium to some extent.
2. *The acid error.* Values registered by the glass electrode tend to be somewhat high when the pH is less than about 0.5.
3. *Dehydration.* Dehydration may cause erratic performance of the electrode.
4. *Errors in unbuffered neutral solutions.* Equilibrium between the surface layer of the electrode and the solution is achieved slowly in poorly buffered, approximately neutral solutions. Time must be allowed for this equilibrium to be established. Before being used to determine the pH of such solutions, the glass electrode should first be thoroughly rinsed with water, following which the electrodes should be immersed in successive portions of the unknown until a constant pH reading is obtained. Good stirring is also helpful, and several minutes should be allowed for the attainment of steady readings.
5. *Variation in junction potential.* A fundamental source of uncertainty for which a correction cannot be applied is the variation in the junction potential resulting from differences in the composition of the standard and the

---

[15] For a detailed discussion of potentiometric pH measurements, see R. G. Bates, *Determination of pH Theory and Practice.* New York: John Wiley & Sons, 1964.

unknown solution. Absolute pH values more reliable than 0.01 unit are generally unobtainable; even reliability to 0.03 unit requires considerable care. To be sure, it is often possible to detect pH *differences* between similar solutions or pH *changes* in a single solution that are as small as 0.001 unit. Many pH meters are designed to permit readings in increments smaller than 0.01 unit for this reason.

6. *Error in the pH of the standard buffer.* Any inaccuracies in the preparation of the buffer used for calibration, or changes in its composition during storage, will propagate an error in a subsequent pH measurement. The action of bacteria on organic components of buffers is a common cause for deterioration.

## 13E POTENTIOMETRIC TITRATIONS

A *potentiometric titration* involves measurement of the potential of a suitable indicator electrode as a function of titrant volume. The information provided by a potentiometric titration is not the same as that obtained from a direct potentiometric measurement. For example, the direct measurement of 0.100 $F$ solutions of hydrochloric and acetic acids would yield substantially different hydrogen ion concentrations because the latter is only partially dissociated. On the other hand, the potentiometric titration of equal volumes of the two acids would require the same amount of standard base, because both solutes have the same number of titratable protons.

Compared with titrations that make use of chemical indicators, the potentiometric end point provides inherently more reliable data. It is particularly useful for the titration of colored or turbid solutions and for detecting the presence of unsuspected species in a solution. Potentiometric titrations suffer from the disadvantage of being more time consuming than those involving indicators; on the other hand, they are readily automated.

Figure 13–13 illustrates a typical apparatus for performing a potentiometric titration. The process ordinarily involves measuring and recording the cell potential (in units of millivolts or pH, as appropriate) after each addition of reagent. The titrant is added in large increments at the outset; these volumes are made smaller as the end point is approached (as indicated by larger changes in response per unit volume).

Sufficient time must be allowed for the attainment of equilibrium after each addition of reagent. Precipitation reactions may require several minutes for equilibration, particularly in the vicinity of the equivalence point. A close approach to equilibrium is indicated by the disappearance of drift in the potential. Effective stirring is often helpful in hastening the achievement of equilibrium.

The first two columns of Table 13–4 show typical potentiometric titration data obtained with the apparatus illustrated in Figure 13–13. The data in the vicinity of the end point are plotted in Figure 13–14a; note that this experimental plot closely resembles titration curves derived from theoretical considerations.

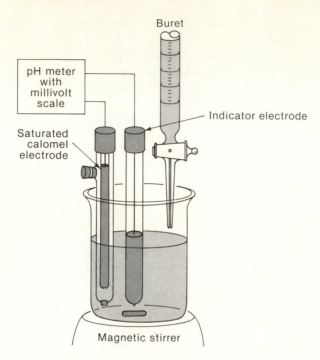

**Figure 13–13** Apparatus for a potentiometric titration.

**Table 13–4** *Potentiometric Titration Data for 2.433 mmol of Chloride with 0.1000 F Silver Nitrate*

| Vol AgNO₃, mL | E vs. SCE, V | $\Delta E/\Delta V$, V/mL | $\Delta^2 E/\Delta V^2$, V/mL² |
|---|---|---|---|
| 5.0 | 0.062 | | |
| | | 0.002 | |
| 15.0 | 0.085 | | |
| | | 0.004 | |
| 20.0 | 0.107 | | |
| | | 0.008 | |
| 22.0 | 0.123 | | |
| | | 0.015 | |
| 23.0 | 0.138 | | |
| | | 0.016 | |
| 23.50 | 0.146 | | |
| | | 0.050 | |
| 23.80 | 0.161 | | |
| | | 0.065 | |
| 24.00 | 0.174 | | |
| | | 0.09 | |
| 24.10 | 0.183 | | |
| | | 0.11 | |
| 24.20 | 0.194 | | 2.8 |
| | | 0.39 | |
| 24.30 | 0.233 | | 4.4 |
| | | 0.83 | |
| 24.40 | 0.316 | | −5.9 |
| | | 0.24 | |
| 24.50 | 0.340 | | −1.3 |
| | | 0.11 | |
| 24.60 | 0.351 | | −0.4 |
| | | 0.07 | |
| 24.70 | 0.358 | | |
| | | 0.050 | |
| 25.00 | 0.373 | | |
| | | 0.024 | |
| 25.5 | 0.385 | | |
| | | 0.022 | |
| 26.0 | 0.396 | | |
| | | 0.015 | |
| 28.0 | 0.426 | | |

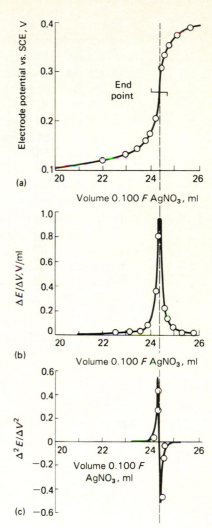

**Figure 13–14** Titration of 2.433 meq of chloride ion with 0.100 *N* silver nitrate. (a) Titration curve. (b) First-derivative curve. (c) Second-derivative curve.

## 13E–1 End-Point Detection

Any of several methods can be used to determine the end point for a potentiometric titration. The most straightforward involves a direct plot of potential with respect to reagent volume, as in Figure 13–14a; the midpoint in the steeply rising portion of the curve is estimated visually and taken as the end point. Various graphical methods have been proposed to aid in the establishment of the midpoint, but it is doubtful that these procedures significantly improve its determination.

The second approach to end-point detection is to calculate the change in potential per unit volume of titrant (that is, $\Delta E/\Delta mL$), as has been done in column 3 of Table 13–4. A plot of these data as a function of the average volume produces a curve with a maximum that corresponds to the point of inflection (Figure 13–14b). Alternatively, this ratio can be evaluated during the titration and recorded in lieu of the potential itself. Inspection of column 3 of Table 13–4 reveals that the maximum is located between 24.30 and 24.40 mL; selection of 24.35 mL would be adequate for most purposes.

Column 4 of Table 13–4 and Figure 13–14c show that the second derivative for the data changes sign at the point of inflection. This change is used as the analytical signal in some automatic titrators.

All of the foregoing methods of end-point evaluation are predicated on the assumption that the titration curve is symmetric about the equivalence point and that the inflection in the curve corresponds to this point. This assumption is perfectly valid, provided the participants in the titration react with one another in an equimolar ratio, and also provided the electrode process is perfectly reversible. The former condition is lacking in many oxidation-reduction titrations; the titration of iron(II) with permanganate (see Chapter 11, Figure 11–4B) is an example. The curve for such titrations is ordinarily so steep that failure to account for asymmetry in the curve results in a vanishingly small titration error.

**Titration to a Fixed Potential**   It is also possible to use the volume needed to reach a predetermined potential as the end point for a titration. The value selected may be the theoretical equivalence-point potential (Chapter 11, Section 11E–1) or a potential determined by the titration of standards. The equivalence-point behavior of the system must be highly reproducible if this approach is to be successful.

## 13E–2 Precipitation Titrations

**Electrode Systems**   The indicator electrode for a precipitation titration is often the metal from which the reacting cation is derived. A membrane electrode that is responsive to this cation or the anion in the titration can also be used.

Silver nitrate is without question the most versatile reagent for precipitation titrations. A silver wire serves as the indicator electrode. For reagent and analyte concentrations that are 0.1 $F$ or greater, a calomel reference electrode can be located directly in the titration vessel without serious error from the slight leakage of chloride ions from the salt bridge. This leakage can, however, be the source of significant error in titrations that involve very dilute solutions or require the highest precision. The difficulty is eliminated by immersing the calomel electrode in a potassium nitrate solution that in turn is connected to the analyte solution by a salt bridge containing potassium nitrate. Reference electrodes with bridges of this type can be purchased from laboratory supply houses.

**Titration Curves**   A theoretical curve for a potentiometric titration is readily derived. For example, the potential of a silver electrode for the argentometric titration of chloride can be described by

$$E_{Ag} = E^0_{AgCl} - 0.0591 \log [Cl^-]$$

where $E^0_{AgCl}$ is the standard potential for the reduction of AgCl to silver. Alternatively, the standard potential for the reduction of silver ion can be used; that is,

$$E_{Ag} = 0.799 - 0.0591 \log \frac{1}{[Ag^+]}$$

The former is convenient for calculating the potential of the silver electrode when an excess of chloride exists, while the latter is preferable for solutions containing an excess of silver ion.

Potentiometric measurements are particularly useful for titrations of mixtures of anions with standard silver nitrate. For example, Figure 7–3A shows a theoretically derived curve for the titration of a chloride–iodide mixture. An experimental curve has the same general appearance, although the ordinate units are different.

### 13E–3 Complex Formation Titrations

Both metallic and membrane electrodes have been used to detect end points for potentiometric titrations involving complex formation. The mercury electrode,[16] illustrated in Figure 13–15, is particularly useful for EDTA titrations of cations forming complexes that are less stable than $HgY^{2-}$ (p. 334). Reilley and co-workers have made a systematic theoretical and experimental study of the mercury electrode as an indicator electrode for the potentiometric titration of 29 di-, tri-, and tetravalent cations with EDTA.[17]

### 13E–4 Neutralization Titrations

Theoretical curves for various neutralization titrations were considered in some detail in Chapter 8. It turns out that the shapes of these curves can be closely approximated experimentally. The experimental curves will ordinarily be somewhat displaced from theoretical along the pH axis because concentrations, rather than activities, were used in their derivation; this displacement is of no consequence insofar as locating the end point is concerned.

Potentiometric neutralization titrations are particularly valuable for the analysis of mixtures of acids or polyprotic acids. The same considerations apply to bases.

**Determination of Dissociation Constants**    An approximate numerical value for the dissociation constant of a weak acid or base can be estimated from

---

[16] These electrodes can be obtained from Kontes Manufacturing Corp., Vineland, NJ.

[17] C. N. Reilley and R. W. Schmid, *Anal. Chem.*, **1958**, *30*, 947; C. N. Reilley, R. W. Schmid, and D. W. Lamson, *ibid.*, 953.

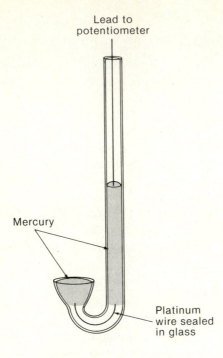

Lead to
potentiometer

Mercury

Platinum
wire sealed
in glass

**Figure 13–15** A typical mercury electrode.

potentiometric titration curves. This quantity can be computed from the pH at any point along the curve, but as a practical matter, the pH at half-neutralization is most convenient. Here,

$$[HA] \cong [A^-]$$

Therefore,

$$K_a = \frac{[H_3O^+][\cancel{A^-}]}{[\cancel{HA}]} = [H_3O^+]$$

or

$$pK_a = pH$$

It is important to note that the use of concentrations instead of activities may cause the value for $K_a$ to differ from its published value by a factor of 2 or more. The more exact form of the dissociation constant for HA is

$$K_a = \frac{a_{H_3O^+} \cdot a_{A^-}}{a_{HA}} = \frac{a_{H_3O^+} \cdot [A^-] \cdot f_{A^-}}{[HA] \cdot f_{HA}}$$

Since the assumption that [HA] and [A⁻] are approximately equal is still valid, we may write

$$K_a = \frac{a_{H_3O^+} \cdot [A^-] \cdot f_{A^-}}{[HA] \cdot f_{HA}}$$

$$= \frac{a_{H_3O^+} \cdot f_{A^-}}{f_{HA}} \qquad (13-23)$$

The glass electrode directly provides a good approximation of $a_{H_3O^+}$; the value of $K_a$ that is measured will differ from the thermodynamic value by the ratio between the two activity coefficients. The activity coefficient in the denominator of Equation 13–23 will not change significantly with ionic strength increases because HA is a neutral species. The activity coefficient for $A^-$, on the other hand, will decrease as the electrolyte concentration increases. Therefore, the observed hydrogen ion activity will be larger than the thermodynamic dissociation constant.

---

**Example 13–2**   In order to determine $K_1$ and $K_2$ for $H_3PO_4$ from titration data, careful pH measurements were made after 0.5 and 1.5 mol of base had been added for each mole of acid. It was then assumed that the hydrogen ion activities computed from these data were identical to the desired dissociation constants. Calculate the relative error incurred by this assumption if the ionic strength was 0.1 at the time of each measurement.

Rearrangement of Equation 13–23 gives

$$K_a(\text{exptl}) = a_{H_3O^+} = K_a \times f_{HA}/f_{A^-}$$

The activity coefficient for $H_3PO_4$ is approximately 1 since it is uncharged. In Table 5–3 (p. 118), we find that the activity coefficient for $H_2PO_4^-$ is 0.78, while that for $HPO_4^{2-}$ is 0.36. Substituting these values into the foregoing equation gives

$$K_1(\text{exptl}) = 7.11 \times 10^{-3} \times 1.0/0.78 = 9.1 \times 10^{-3}$$

$$\text{error} = \frac{9.1 \times 10^{-3} - 7.11 \times 10^{-3}}{9.11 \times 10^{-3}} \times 100 = 22\%$$

$$K_2(\text{exptl}) = 6.34 \times 10^{-8} \times 0.78/0.36 = 1.37 \times 10^{-7}$$

$$\text{error} = \frac{1.37 \times 10^{-7} - 6.34 \times 10^{-8}}{6.34 \times 10^{-8}} \times 100 = 1.2 \times 10^2 \%$$

---

A single titration of a pure acid can provide sufficient data (equivalent weight and dissociation constant) to make identification of the acid possible.

### 13E–5 Oxidation-Reduction Titrations

An inert indicator electrode constructed of platinum is ordinarily used to detect end points in oxidation-reduction titrations. Occasionally, other inert metals are used instead, including silver, palladium, gold, or mercury. Titration curves similar in appearance to those derived in Chapter 11 are ordinarily obtained, although they may be displaced along the ordinate axis as a consequence of the effects of high ionic strengths. End points are determined by the methods described earlier in this chapter.

### 13E–6 Differential Titrations

We have seen (Section 13E–1) that a derivative curve generated from the data of a conventional potentiometric titration curve reaches a maximum in the vicinity of the equivalence point. This derivative curve can be obtained directly through the use of suitable apparatus.

A simple way of performing a differential titration employs two identical *indicator* electrodes, one of which is shielded from the bulk of the solution. Figure 13–16 illustrates a typical arrangement, in which one of the electrodes is housed in a small sidearm test tube. Contact with the bulk of the solution is made through a small (~ 1-mm) hole in the bottom of the tube. Because of this restricted access, the composition of the solution surrounding the shielded electrode will not be immediately affected by an addition of titrant to the bulk of the solution. The consequent difference in composition will result in a difference in potential, $\Delta E$, between the electrodes. The solution is homogenized

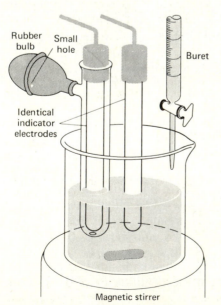

Figure 13–16    Apparatus for a differential potentiometric titration.

after each potential measurement by squeezing the rubber bulb several times, whereupon $\Delta E$ again becomes zero. The error resulting from failure of the final addition of reagent to react with the solution in the tube that shields the one electrode can be shown to be negligible, provided that its volume is small.

Two platinum wires, with one housed in an ordinary medicine dropper, is a convenient electrode system for differential oxidation-reduction titrations. The principal advantage of a differential titration method is the elimination of the need for a reference electrode and a salt bridge.

## PROBLEMS

Unless otherwise noted, the effects of junction potentials are to be neglected.

*13–1.  (a) Calculate $E^0$ for the process

$$AgIO_3(s) + e \rightleftharpoons Ag(s) + IO_3^-$$

(b) Use the shorthand notation (p. 270) to describe a cell with an Ag anode and a saturated calomel cathode that could be used for the measurement of $pIO_3$.

(c) Develop an equation that relates the potential of the cell in (b) to $pIO_3$.

(d) Calculate $pIO_3$ if the cell in (b) has a potential of $-294$ mV.

13–2.  (a) Calculate $E^0$ for the process

$$PbI_2(s) + 2e \rightleftharpoons Pb(s) + 2I^-$$

(b) Use the shorthand notation (p. 270) to describe a cell consisting of a Pb anode and a saturated calomel cathode that could be used for the measurement of pI.

(c) Derive an equation that relates the potential of this cell to pI.

(d) Calculate pI if the cell in (b) develops a potential of 0.248 V.

13–3.  (a) Calculate $E^0$ for the process

$$Hg_2Br_2(s) + 2e \rightleftharpoons 2Hg(l) + 2Br^-$$

(b) Use the shorthand notation to describe a cell consisting of an Hg anode and a saturated calomel cathode that could be used for the measurement of pBr.

(c) Derive an equation that relates the potential of this cell for pBr.

(d) Calculate pBr if the cell in (b) develops a potential of $-125$ mV.

*13–4. Use the shorthand notation to represent a cell consisting of a saturated calomel cathode and

(a) an Ag indicator electrode for the measurement of pSCN.

(b) a Cd indicator electrode for the measurement of $pFe(CN)_6$, given that $K_{sp}$ for $Cd_2Fe(CN)_6$ is $3.2 \times 10^{-17}$.

(c) a Cu indicator electrode for the measurement of $pIO_3$, given that $K_{sp}$ for $Cu(IO_3)_2$ is $7.4 \times 10^{-8}$.

(d) an Ag indicator electrode for the measurement of $pCrO_4$.

*13–5. Calculate

(a) pSCN if the cell in Problem 13–4(a) develops a potential of 74.3 mV.

(b) $pFe(CN)_6$ if the cell in Problem 13–4(b) develops a potential of 0.737 V.

(c) $pIO_3$ if the cell in Problem 13–4(c) develops a potential of $-0.522$ V.

(d) $pCrO_4$ if the cell in Problem 13–4(d) develops a potential of 476 mV.

**13–6.** Use a shorthand notation (unless otherwise noted, specify any needed concentrations as $1.00 \times 10^{-3}$ $M$) to represent a cell consisting of a saturated calomel cathode and

(a) a Pt indicator electrode for the measurement of pFe(II).

(b) a Pt indicator electrode for the measurement of pTl(III).

(c) a Pt indicator electrode for the measurement pTiO.

(d) a Hg indicator electrode for the measurement of pY (where $Y^{4-}$ is the deprotonated anion of EDTA) in a solution buffered to a pH of 10.00.

(e) a Pt indicator electrode for the measurement of $pCr_2O_7$ in a solution buffered to a pH of 2.00.

**13–7.** Calculate

(a) pFe(II) if the cell in Problem 13–6(a) develops a potential of $-603$ mV.

(b) pTl(III) if the cell in Problem 13–6(b) develops a potential of $-955$ mV.

(c) pTiO if the cell in Problem 13–6(c) develops a potential of 533 mV.

(d) pY if the cell in Problem 13–6(d) develops a potential of $-154$ mV.

(e) $pCr_2O_7$ if the cell in Problem 13–6(e) develops a potential of $-830$ mV.

**\*13–8.** Calculate the potential of the cell

$$\text{indicator electrode} \parallel \text{SCE}$$

where the indicator electrode is Hg immersed in a solution that is

(a) $5.00 \times 10^{-4}$ $M$ in $Hg^{2+}$.

(b) $5.00 \times 10^{-4}$ $M$ in $Hg_2^{2+}$.

(c) saturated with $Hg_2Cl_2$ and is $4.00 \times 10^{-3}$ $M$ in $Cl^-$.

(d) $1.00 \times 10^{-3}$ $M$ in $HgI_4^{2-}$ and $2.50 \times 10^{-2}$ $F$ in KI (for $HgI_4^{2-}$, $K_f = 1.0 \times 10^{30}$).

(e) $5.00 \times 10^{-4}$ $F$ in $Hg(NO_3)_2$ and $0.0500$ $F$ in KCl (for $Hg^{2+} + 2Cl^- \rightleftharpoons HgCl_2$(aq), $K_f = 1.6 \times 10^{13}$).

**13–9.** Calculate the potential of the cell

$$\text{indicator electrode} \parallel \text{SCE}$$

where the indicator electrode is Cu and is immersed in a solution that is

(a) $2.84 \times 10^{-3}$ $M$ in $Cu^{2+}$.

(b) $2.84 \times 10^{-3}$ $M$ in $Cu^+$.

(c) saturated with CuI and $3.00 \times 10^{-3}$ $F$ in KI.

(d) $1.00 \times 10^{-3}$ $F$ in $KCu(CN)_2$ and $8.00 \times 10^{-2}$ $F$ in KCN. For the equilibrium

$$Cu^+ + 2CN^- \rightleftharpoons Cu(CN)_2^- \qquad K_f = 5.0 \times 10^{-4}$$

**\*13–10.** The reduction of mercury(II) from its EDTA complex can be written as

$$HgY^{2-} + 2e \rightleftharpoons Hg(l) + Y^{4-}$$

for which the standard electrode potential is 0.21 V. Calculate the potential of the cell

$$Hg|HgY^{2-}(3.00 \times 10^{-5}\ M),\ Y^{4-}(x\ M) \parallel SCE$$

where

(a) $[Y^{4-}]$ is $1.15 \times 10^{-2}$ $M$.

(b) $[Y^{4-}]$ is $1.15 \times 10^{-4}$ $M$.

(c) the formal concentration of uncomplexed Y-containing species, $C_T$, is $4.57 \times 10^{-3}$ and the pH is 5.00.

(d) the formal concentration of uncomplexed Y-containing species, $C_T$, is $4.57 \times 10^{-3}$ and the pH is 10.00.

**13–11.** Calculate $E^0$ for the process

$$Ag(NH_3)_2^+ + e \rightleftharpoons Ag(s) + 2NH_3$$

given that $K_f$ for the complex is $1.0 \times 10^7$.

**13–12.** The formation constant for the 1:1 complex formed between Cu(II) and triethylenetetraamine (trien) is $1.3 \times 10^{20}$. Calculate $E^0$ for the reduction of Cu from this complex.

**\*13–13.** A potential of 0.228 V was registered by a glass electrode–saturated calomel electrode system immersed in a buffer with a pH of 5.75. Solutions of unknown pH were then substituted for the buffer. Calculate the pH of the unknown if the potential was

(a) 0.131 V.          (b) 0.413 V.

(c) 0.509 V.          (d) 0.627 V.

**13–14.** The cell

$$\text{membrane electrode for } Ca^{2+}|Ca^{2+}(a = 2.15 \times 10^{-3} \ M) \| SCE$$

was found to have a potential of 0.354 V. The electrodes were then immersed in solutions containing unknown concentrations of Ca. Calculate $a_{Ca}$ if the potential was

(a) 0.404 V.          (b) 0.537 V.

(c) 0.365 V.          (d) 0.475 V.

**\*13–15.** A 100.0-mL aliquot of $1.00 \times 10^{-3}$ $F$ $Cu^{2+}$ is buffered to a pH of 6.00 and titrated with $2.00 \times 10^{-2}$ $F$ EDTA. Calculate

(a) pCu after the addition of 0.00, 1.00, 2.00, 3.00, 4.00, 4.50, 5.00, 5.50, and 6.00 mL of EDTA.

(b) the potential of the Cu electrode (vs. SHE) after each addition of EDTA.

**\*13–16.** The $Na^+$ concentration of a solution was determined by measurements with a glass-membrane electrode. The electrode system developed a potential of 0.2331 V when immersed in 10.0 mL of the unknown. After addition of 1.00 mL of $2.00 \times 10^{-2}$ $F$ NaCl, the potential decreased to 0.1846 V. Calculate the $Na^+$ concentration of the original solution.

**13–17.** The $F^-$ concentration of a solution was determined by measurements with a liquid-membrane electrode. The electrode system developed a potential of $-0.4965$ V when immersed in 25.00 mL of the sample, and $-0.4117$ V after the addition of 2.00 mL of $5.45 \times 10^{-2}$ $F$ NaF. Calculate pF for the sample.

# Chapter 14
# Other Electroanalytical Methods

The potentiometric methods described in Chapter 13 involved potential measurements under conditions in which currents in the analytical cell were absent. In contrast, the methods to be surveyed in this chapter all require the passage of electricity through a solution of the analyte.[1] When currents are present, the potential of an electrochemical cell is no longer simply the difference between the theoretical electrode potentials because of the effects of *ohmic potential, concentration polarization,* and *kinetic polarization.* It is necessary to consider each of these phenomena before proceeding further.

## 14A INFLUENCE OF CURRENT ON ELECTROCHEMICAL CELL POTENTIALS

### 14A–1 Ohmic Potential: IR Drop

Electrochemical cells, like metallic conductors, offer resistance to the flow of electricity. In both types of conduction, the effect of this resistance is described by Ohm's law. The product of the resistance of a cell in ohms and the current in amperes is called the *ohmic potential* or the *IR* drop of the cell. The effect of *IR* drop is to increase the potential needed to operate an electrolytic cell or to decrease the measured potential of a galvanic cell. Thus, the *IR* drop is always *subtracted* from the theoretical cell potential; that is,[2]

$$E_{cell} = E_{cathode} - E_{anode} - IR \qquad (14-1)$$

**Example 14–1** Consider a cell consisting of a Cu electrode in contact with 1.00 $M$ $Cu^{2+}$, a Cd electrode in contact with 1.00 $M$ $Cd^{2+}$, and a connecting salt bridge. The cell has a resistance of 4.00 $\Omega$.

---

[1] For further information concerning the method in this chapter, see: J. A. Plambeck, *Electroanalytical Chemistry.* New York: John Wiley & Sons, 1982; *Laboratory Techniques in Electroanalytical Chemistry,* P. T. Kissinger and W. R. Heineman, Eds. New York: Marcel Dekker, Inc., 1984.

[2] Strictly, Equation 14–1 should contain a term for the junction potential. For the methods considered in this chapter, however, the magnitude of the junction potential is negligible with respect to other potentials.

(a) Calculate the potential needed to develop a current of 0.200 A in the electrolytic cell

$$Cu|Cu^{2+}(1.00\ M)\|Cd^{2+}(1.00\ M)|Cd$$

Since both cation concentrations are 1.00 $M$, the half-cell potentials and the standard electrode potentials are numerically equal. Substituting into Equation 14–1 gives

$$E_{cell} = E^0_{Cd} - E^0_{Cu} - IR$$
$$= -0.403 - 0.337 - 0.0200 \times 4.00$$
$$= -0.820\ V$$

A potential substantially in excess of theoretical is thus needed to cause the deposition of cadmium at the rate required for a current of 0.0200 A.

(b) Calculate the potential when 0.0200 A is developed in the galvanic cell

$$Cd|Cd^{2+}(1.00\ M)\|Cu^{2+}(1.00\ M)|Cu$$

Here,

$$E_{cell} = E^0_{Cu} - E^0_{Cd} - IR$$
$$= 0.337 - (-0.403) - 0.0200 \times 4.00$$
$$= 0.660\ V$$

Note that the existence of a current causes the potential of this cell to be markedly less than theoretical owing to the effect of $IR$ drop.

## 14A–2 Polarization Effects

Equation 14–1 suggests that a linear relationship should exist between the potential of an electrochemical cell and current. As shown by Figure 14–1, the expected linearity is observed at low currents, but marked departures occur when the currents become larger. When nonlinear behavior is observed, the cell is said to be *polarized*. Note that polarization requires the application of a potential to an electrolytic cell that is greater than theoretical to give a current of the expected magnitude. With a galvanic cell, polarization results in potentials that are smaller than are predicted by Equation 14–1. The current in a cell that is completely polarized is essentially constant and independent of potential.

Polarization is an electrode phenomenon that can affect either or both of the electrodes in a cell. Several factors influence the extent of polarization: (1) the size, shape, and composition of the electrode; (2) the composition of the electrolyte solution, its temperature, and the rate at which it is stirred; (3) the magnitude of the current; and, finally, (4) the physical state of the species involved in the cell reaction. While some of these factors are understood sufficiently to permit their quantitative description, others can be accounted for on an empirical basis only.

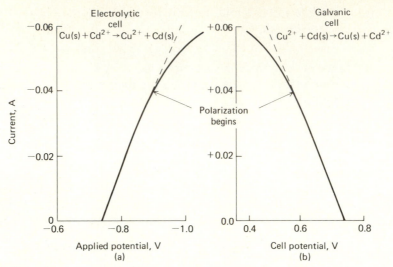

**Figure 14-1**    Current-voltage curves for the cells

(a)    Cu|Cu²⁺(1.00 M), Cd²⁺(1.00 M)|Cd
(b)    Cd|Cd²⁺(1.00 M), Cu²⁺(1.00 M)|Cu

Polarization phenomena are conveniently divided into the two categories of *concentration polarization* and *kinetic polarization.*

**Concentration Polarization**    Concentration polarization occurs when the reactants or products (or both) of an electrode reaction cannot be transported to or away from the electrode surface at a rate demanded by the theoretical current. As an example, consider the reaction at the cadmium cathode in the electrolytic cell described in Example 14-1. The half-reaction involving the deposition of cadmium is rapid and reversible, which means that at any instant the cadmium ion activity in the film of liquid immediately adjacent to the metal surface is that given by the Nernst equation; that is,

$$E_{Cd} = E_{Cd}^0 - \frac{0.0591}{2} \log \frac{1}{[Cd^{2+}]_0}$$

Here, the subscript zero implies that the relationship between the potential of the electrode and the cadmium(II) concentration *applies to the surface film only.* The concentration in the bulk of the solution $[Cd^{2+}]$ will almost certainly be different from $[Cd^{2+}]_0$. Because the rate of the electrode process is so very rapid, any alteration in $E_{Cd}$ results in an immediate adjustment in $[Cd^{2+}]_0$ to its new equilibrium value. In contrast, the concentration of cadmium(II) in the bulk of the solution often changes only slowly with time; minutes, hours, or even days may elapse before this concentration becomes equal to the equilibrium concentration at the electrode surface.

When a potential of appropriate magnitude is applied to the electrolytic cell described in Example 14–1, an instantaneous current develops that immediately adjusts the concentrations of the copper and cadmium ions at the two electrode surfaces to the values demanded by the Nernst equations for the two half-reactions. If this current is to continue at the level required by Equation 14–1, cadmium ions must then be transported from the bulk of the solution to the surface film of the cadmium electrode at the same rate at which they are being deposited. Similarly, copper ions must migrate from the layer of solution immediately adjacent to the anode at a rate that is equal to their formation rate.

Concentration polarization sets in when the rates of either or both of these ion transport processes are insufficient to maintain the theoretical current. The result of this phenomenon is the nonlinear behavior of cells, which is illustrated in Figure 14–1. Note that this phenomenon can affect both galvanic and electrolytic cells.

In order to understand the experimental variables that influence the onset of concentration polarization, we must consider the factors that cause ions or molecules to migrate from one part of a solution to another. These phenomena include (1) diffusion, (2) electrostatic attraction or repulsion, and (3) convection and mechanical mixing. In the discussion that follows, we will focus on mass transport processes as they occur in the cathode cell compartment. The conclusions reached will, however, apply equally well to anodic processes and the transport of ions to and from an anode.

When a concentration gradient develops in a solution, as it does when species are removed or introduced into a solution by electrochemical processes, ions tend to diffuse from the more concentrated region to the more dilute. The rate of diffusion is directly proportional to the concentration difference. For example, when cadmium ions are removed from the surface layer of a cadmium cathode by electrochemical deposition, a concentration gradient is formed, and cadmium ions diffuse from the bulk of the solution to the surface film. The rate of the resulting diffusion is given by the relationship

$$\text{rate of diffusion to cathode surface} = k(C - C_0) \qquad (14-2)$$

where $C$ is the reactant concentration in the bulk of the solution, $C_0$ is its equilibrium concentration at the surface of the cathode, and $k$ is a proportionality constant. *Recall that the value of $C_0$ is fixed by the potential of the electrode and can be calculated through use of the Nernst equation.* As larger potentials are applied to the electrode, $C_0$ becomes smaller, and the rate of diffusion becomes correspondingly larger.

Electrostatic forces also influence the rate at which an ionic reactant migrates to or from an electrode surface. The influence of electrostatic attraction (or repulsion) upon a particular ionic species decreases as the total electrolyte concentration of the solution is increased; it may approach zero if the reactive species is but a small fraction of the total ions with the same charge.

Reactants can also be transferred to (or from) an electrode by mechanical means. Thus, stirring or agitation will tend to decrease concentration polarization. Convection currents resulting from temperature or density differences will also contribute to material transport.

To summarize, then, concentration polarization occurs when the effects of diffusion, electrostatic attraction, and mechanical mixing are insufficient to transport a reactant to or from an electrode surface at a rate that satisfies the theoretical current requirement. Concentration polarization will require applied potentials that are larger than theoretical to maintain a given current in an electrolytic cell (Figure 14–1a); similarly, the phenomenon will cause a galvanic cell potential to be smaller than the value predicted on the basis of the theoretical potential and the *IR* drop (Figure 14–1b).

Concentration polarization is important in several electroanalytical methods. In some applications, its effects are deleterious and steps are taken to eliminate it, while in others, it is essential to the analytical method, and every effort is made to promote its occurrence. Experimental variables that influence the degree of concentration polarization include (1) the reactant concentration, (2) the total electrolyte concentration, (3) mechanical agitation, and (4) the size of the electrode; as the area toward which a reactant can be transported increases, polarization effects become smaller.

**Kinetic Polarization**   In kinetic polarization, the magnitude of the current is limited by the rate of one or both of the electrode reactions. To offset kinetic polarization, an additional potential (the *overvoltage,* or *overpotential*) is required to overcome the energy barrier of the half-reaction. Note that *the current in a kinetically polarized cell is governed by the rate of the electron transfer process* rather than the rate of mass transfer.

Kinetic polarization is most pronounced for electrode processes that yield gaseous products and is often negligible for reactions that involve deposition or solution of a metal. Kinetic effects usually decrease with increasing temperatures and decreasing current densities.[3] These effects also depend upon the composition of the electrode and are most pronounced with softer metals such as lead, zinc, and particularly, mercury. The magnitude of overvoltage effects cannot be predicted from present theory and can only be estimated from empirical information in the literature.[4] In common with *IR* drop, overvoltage effects cause the potential of a galvanic cell to be smaller than theoretical and to require potentials that are greater than theory to operate an electrolytic cell at a desired current.

The overvoltages associated with the formation of hydrogen and oxygen are often as large as several tenths of a volt and are of considerable importance because these molecules are frequently produced in electrochemical reactions. Of particular interest is the high overvoltage of hydrogen on such metals as

---

[3] Current density is defined as amperes per square centimeter ($A/cm^2$) of electrode surface.

[4] Overvoltage data for various gaseous species on different electrode surfaces have been compiled by J. A. Page, in *Handbook of Analytical Chemistry,* L. Meites, Ed., p. 5–184. New York: McGraw-Hill Book Co., 1963.

copper, zinc, and mercury, which permits deposition of these metals and several others without interference from hydrogen evolution. For example, in theory, it should not be possible to deposit zinc from a neutral aqueous solution because hydrogen formation occurs at a potential that is considerably less than that required for zinc deposition. In fact, the metal can be deposited on a copper electrode with no significant hydrogen formation because the rate at which the gas forms on both zinc and copper is negligible.

## 14B ELECTROGRAVIMETRIC METHODS

Gravimetric methods based upon electrodeposition have been used since the middle of the nineteenth century. These methods ordinarily involve determination of the weight of a metal that has deposited upon the surface of a cathode or of a metal oxide on an anode.

### 14B–1 Apparatus

Figure 14–2 shows the components of a typical electrodeposition assembly. The direct-current (dc) power supply may consist of a storage battery, or, more commonly, an alternating-current rectifier with an output of a few volts dc. The magnitude of the potential applied to the cell is controlled by a rheostat and a voltmeter, while an ammeter in series with the cell monitors the current. An entirely adequate electrolysis apparatus can be assembled from equipment that is found in most laboratories. Several instrument manufacturers offer more elaborate equipment.

Electrodes are usually constructed of platinum, although copper, brass, and other metals are used occasionally. Platinum electrodes have the advantage of being inert and can be ignited to remove grease, organic matter, or gases that might otherwise adversely affect the physical properties of the deposit. A limitation of platinum is the need to exclude chloride ion from solutions in which platinum serves as the anode because chloroplatinate complexes are produced and may migrate to the cathode, where they are reduced. Moreover, a few metals — notably bismuth, gallium, and zinc — will cause permanent damage if deposited directly on a platinum cathode; with these metals, a protective coating of copper is applied before undertaking the electrodeposition. A final disadvantage of platinum is its high cost.

A typical *working electrode,* at which the analytical deposition occurs, is a gauze cylinder with a diameter of 2 to 3 cm and a height of about 6 cm. This design minimizes concentration polarization by providing a large surface area about which the solution can freely circulate. The second electrode, called a *counter electrode,* may also be cylindrical with a smaller diameter so that it fits inside the working electrode. Alternatively, the counter electrode may consist of a heavy wire or a paddle that is used to stir the solution.

Efficient stirring will minimize concentration polarization. Tall-form

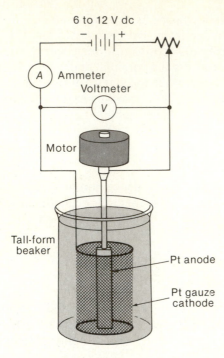

6 to 12 V dc

Ammeter
Voltmeter

Motor

Tall-form
beaker

Pt anode

Pt gauze
cathode

**Figure 14–2** Apparatus for electrodeposition of metals without cathode potential control.

beakers are often employed to avoid losses of solution due to spattering during the electrolysis.

### 14B–2 Physical Properties of Electrolytic Precipitates

Ideal electrolytic deposits are strongly adherent, dense, and smooth so that they can be washed, dried, and weighed without mechanical loss or measurable reaction with the atmosphere. Ideal deposits are fine grained and possess a metallic luster; spongy, powdery, or flaky precipitates are likely to be less pure and less adherent.

The principal factors that influence the physical characteristics of deposits include current density, the presence of complexing reagents, and temperature. Low current densities ($<0.1$ A/cm$^2$) and efficient stirring usually lead to deposits with the best physical properties. Many metals form smoother and more adherent films when deposited from solution in which their ions exist principally as complexes; ammonia and cyanide appear to be particularly effective. The effects of temperature are unpredictable and must be determined experimentally.

Codeposition of hydrogen with a metal at the cathode tends to cause nonadherent deposits and is thus undesirable. The evolution of hydrogen can be controlled through use of a *cathode depolarizer,* which is a substance that is reduced in preference to hydrogen ion and does not interfere with the deposi-

*Table 14-1* *Typical Applications of Electrogravimetric Analysis*

| Analyte | Weighed as | Cathode | Anode | Conditions |
|---------|------------|---------|-------|------------|
| $Ag^+$ | Ag | Pt | Pt | Alkaline $CN^-$ solution |
| $Br^-$ | AgBr (on anode) | Pt | Ag | |
| $Cd^{2+}$ | Cd | Cu on Pt | Pt | Alkaline $CN^-$ solution |
| $Cu^{2+}$ | Cu | Pt | Pt | $H_2SO_4$–$HNO_3$ solution |
| $Mn^{2+}$ | $MnO_2$ (on anode) | Pt | Pt dish | HCOOH–HCOONa solution |
| $Ni^{2+}$ | Ni | Cu on Pt | Pt | Ammoniacal solution |
| $Pb^{2+}$ | $PbO_2$ (on anode) | Pt | Pt | Strong $HNO_3$ solution |
| $Zn^{2+}$ | Zn | Cu on Pt | Pt | Acidic citrate solution |

tion of metals. A small amount of nitrate ion is a useful cathode depolarizer; its reaction is

$$NO_3^- + 10H^+ + 8e \rightarrow NH_4^+ + 3H_2O$$

Hydrazine has been similarly used as an anode depolarizer to prevent the evolution of oxygen or the possible formation of chloro complexes of platinum:

$$H_2NNH_3^+ \rightarrow N_2(g) + 5H^+ + 4e$$

## 14B-3 Applications of Electrogravimetric Methods

Electrogravimetric methods exist for numerous elements; typical applications are given in Table 14-1. Directions are given in Chapter 20, Section 20L-1 for the simultaneous determination of copper and lead in brass.

## 14C COULOMETRY

Coulometric methods are based upon the measurement of the quantity of electricity required to convert an analyte quantitatively to a different oxidation state. Coulometric and gravimetric methods share the common advantage that the proportionality constant between the quantity measured and the weight of analyte can be derived from accurately known physical constants, thus eliminating the need for standards for calibration. In contrast to gravimetric methods, coulometric procedures are usually rapid and do not require that the product of the electrochemical reaction be a weighable solid. Coulometric methods are as accurate as conventional gravimetric and volumetric procedures and in addition are readily automated.[5]

---

[5] For additional information about coulometric methods, see E. Bishop, in *Comprehensive Analytical Chemistry,* C. L. Wilson and D. W. Wilson, Eds., Vol. 11D. New York: Elsevier Scientific Publishing Co., 1975; J. A. Plambeck, *Electroanalytical Chemistry,* Chapter 12, New York: John Wiley & Sons, 1982; G. W. C. Milner and G. Phillips, *Coulometry in Analytical Chemistry,* New York: Pergamon Press, 1967.

### 14C – 1 *Quantity of Electricity*

Units for the quantity of electricity include the coulomb (C) and the faraday (*F*). *The coulomb is the quantity of electricity that is transported by a constant current of one ampere in one second.* The number of coulombs (*Q*) resulting from a constant current of *I* amperes operated for *t* seconds is given by

$$Q = It \qquad (14-3)$$

For a variable current, *i*, the number of coulombs is given by the integral

$$Q = \int_0^t i \, dt \qquad (14-4)$$

*The faraday is the quantity of electricity that will produce one equivalent of chemical change at an electrode.* Since the equivalent in an oxidation-reduction reaction corresponds to the change brought about by one mole of electrons (p. 129), the faraday is equal to $6.20 \times 10^{23}$ electrons. The faraday is also equivalent to 96,487 C. These definitions make it possible to calculate the quantity of chemical change that occurs at an electrode.

---

**Example 14 – 2**  The iodide ion in a 50.0-mL sample was quantitatively precipitated by anodically generated silver ion resulting from a constant current of 44.8 mA that was operated for a total of 386 sec. Calculate the molarity of $I^-$ in the sample solution.

$$44.8 \text{ mA} \times \frac{\text{A}}{1000 \text{ mA}} \times 386 \text{ sec} \times \frac{\text{C}}{\text{A sec}} = 17.29 \text{ C}$$

$$17.29 \text{ C} \times \frac{\text{faraday}}{96,487 \text{ C}} \times \frac{\text{eq}}{\text{faraday}} = 1.792 \times 10^{-4} \text{ eq}$$

Thus,

$$[I^-] = \frac{1.792 \times 10^{-4} \text{ eq } I^-}{50.0 \text{ mL}} \times \frac{1000 \text{ mL}}{\text{L}} \times \frac{\text{mol } I^-}{\text{eq } I^-} = 3.58 \times 10^{-3} \text{ } M$$

---

### 14C – 2 *Types of Coulometric Methods*

Two types of coulometric methods are encountered. The first, called a *coulometric titration,* or *amperostatic coulometry,* makes use of a constant current that is maintained until an indicator signals completion of the reaction. The quantity of electricity is calculated from the magnitude of the current and the time required to reach the end point. The second, known as *controlled potential,* or *potentiostatic coulometry,* involves maintaining the working electrode at a constant potential (relative to a reference electrode) that permits a quantitative oxidation or reduction of the analyte without involvement by a less reactive species from the sample or solvent. Here, the initial current is relatively large but decreases rapidly and approaches zero as the reaction becomes complete. The quantity of electricity, $Q$, needed to convert the analyte quantitatively to a new oxidation state is derived from the area under a curve that relates current with respect to time. This area is determined by graphical, mechanical, or electronic integration. Alternatively, $Q$ can be measured with a chemical cou-lometer, which is simply an electrochemical cell arranged in series with the working cell. The electrochemical reaction selected for the coulometer is one that yields products that are conveniently measured either chemically or physi-cally (see Example 14–3). Of the two methods, coulometric titrations are more widely used.

A fundamental requirement for all coulometric methods is 100% current efficiency; that is, each faraday of electricity must bring about one equivalent of chemical change in the analyte. Note that 100% current efficiency can be achieved without direct participation of the analyte in electron transfer at an electrode. Indeed, it is more common for the species being determined to be involved wholly or in part in a reaction that is secondary to the electrode process. This point is illustrated by the oxidation of iron(II). The sole anode reaction at the outset is

$$Fe^{2+} \rightarrow Fe^{3+} + e$$

As the concentration of iron(II) decreases, concentration polarization causes the anode potential to increase to the point where the decomposition of water becomes a competing process; that is,

$$2H_2O \rightarrow O_2(g) + 4H^+ + 4e$$

The quantity of electricity required to complete the oxidation of iron(II) then exceeds that demanded by theory, and the current efficiency is less than 100%. The lowered current efficiency can be avoided by introducing at the outset an unmeasured quantity of cerium(III), which is oxidized at a lower potential than is water:

$$Ce^{3+} \rightarrow Ce^{4+} + e$$

The cerium(IV) produced diffuses rapidly from the surface of the electrode to the bulk of the solution, where it then oxidizes an equivalent amount of iron(II):

$$Ce^{4+} + Fe^{2+} \rightarrow Ce^{3+} + Fe^{3+}$$

The net effect is an electrochemical oxidation of iron(II) with 100% current efficiency, even though only a fraction of that species is directly oxidized at the electrode surface.

## 14C – 3  Coulometric Titrations[6]

In a coulometric titration, a titrant is generated electrochemically by a constant current of known magnitude. The time required to reach an end point for the reaction between the titrant and the analyte is then measured. The number of equivalents of analyte in the sample is readily computed by dividing the product of current in amperes and time in seconds by the coulombic equivalent of the faraday (96,487 $C/F$). In some analyses, the electrochemical process involves only generation of the reagent; an example is the titration of halide ions by silver ions produced at a silver anode. In other titrations, the analyte may be converted to its oxidation or reduction product, partially by the electrochemically generated titrant and partially by direct reaction at the electrode. An example of the latter is the quantitative oxidation of iron(II) to iron(III) — in part at a platinum anode and in part by electrolytically generated cerium(IV) ions. In either event, the net process must approach 100% current efficiency with respect to a single chemical change in the analyte.

The current in a coulometric titration, the magnitude of which is analogous to the normality of a volumetric reagent, is carefully maintained at a constant and known level by means of a constant current generator called an *amperostat*. An amperostat senses changes in the current and responds by altering the potential applied across the cell in such a way as to maintain the current at the desired level. Thus, if the current begins to decrease as a consequence of concentration polarization, the amperostat immediately increases the applied potential by an amount necessary to offset the current decrease. This behavior and the need for 100% current efficiency requires that an auxiliary reagent *always* be present to assist in the quantitative conversion of the analyte to its oxidation or reduction product.

A consideration of Equation 14 – 1 and the earlier discussion of concentration polarization reveals why the presence of an auxiliary reagent is imperative. Recall that concentration polarization begins when diffusion, electrostatic attraction, and mechanical mixing are no longer able to supply a reactant to the electrode surface at a rate demanded by Equation 14 – 1. With the onset of concentration polarization, then, the current in the cell begins to decrease. The amperostat immediately responds to this decrease, however, by increasing the cell potential. Because $IR$ is being held constant, the increase in $E_{cell}$ must cause $E_{cathode}$ or $E_{anode}$, or both, to change by a corresponding amount (see Equation 14 – 1). Generally, it is the potential of the working electrode that suffers the change because the reaction at the counter electrode usually involves the oxidation or reduction of the solvent water, which is present in large excess. This large excess prevents concentration polarization, which in turn stabilizes the potential of this electrode. The net effect of the increased potential from the ampero-

---

[6] For further details of this technique, see: D. J. Curran, in *Laboratory Techniques in Electroanalytical Chemistry*, P. T. Kissinger and W. R. Heineman, Eds., New York: Marcel Dekker, Inc., 1984.

stat is thus borne entirely by the working electrode. At first, the potential change brings reactant ions to the surface layer at a greater rate due to the increased electrostatic attraction, and the current remains constant. Ultimately, however, as concentration polarization becomes more and more severe, the potential of the working electrode becomes great enough so that in the absence of the auxiliary reagent, oxidation or reduction of the solvent (or some other species in the solution) takes place. At this point 100% current efficiency is lost. Therefore, the desired current efficiency can only be realized by having a large excess of a reagent present that is oxidized or reduced at the onset of concentration polarization to a species that converts the analyte to the desired product.

The analogy between a conventional and a coulometric titration extends well beyond the common requirement of an observable end point. In both, the amount of unknown is determined by measurement of its combining capacity —with a volume of standard reagent in the one and with a quantity of electricity in the other. Similar demands are made of the reactions upon which these titrations are based; that is, they must be rapid, essentially complete, and free of side reactions.

**Electrical Apparatus**   Coulometric titrators are available from several laboratory supply houses; they can also be assembled from apparatus that is available in most laboratories.

Figure 14–3 depicts the principal components of a simple coulometric titrator. Included are a source of constant current and a switch that simultaneously initiates the current and starts an electric timer (position 1). Note that the circuit is so arranged that electricity is also drawn from the source when the switch is moved to position 2. The resistance of $R_1$ is chosen to be about the same as that for the titration cell so that electricity is drawn continuously from the source, a condition that aids in maintaining a constant current.

Heavy-duty B-batteries provide the simplest of all sources for a constant current. More commonly, constant current amperostats are used. As shown in Figure 14–3, the current is determined from the potential drop across the standard resistor, $R_{std}$.

A motor-driven electric clock is inadequate for measurement of the electrolysis time because the rotor of such a device tends to coast when stopped and to lag when started. Modern electronic timers eliminate this problem.

**Cells for Coulometric Titrations**   A typical titration cell is shown in Figure 14–4. It consists of a generator electrode, at which the reagent is produced, and a counter electrode to complete the circuit. The generator electrode— ordinarily a platinum rectangle, a coil of wire, or a gauze cylinder—should have a relatively large surface area.

The products formed at the auxiliary electrode are a potential source of interference. For example, hydrogen is often evolved at the cathode as an oxidizing agent is being generated at an anode. Hydrogen reacts rapidly with most oxidizing agents, however, which leads to a positive determinate error. To

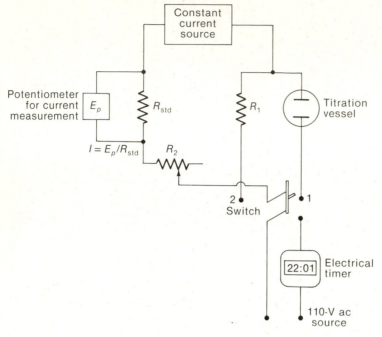

Potentiometer for current measurement

$E_p$

$R_{std}$

$R_1$

Titration vessel

Constant current source

$I = E_p/R_{std}$

$R_2$

2 Switch

1

22:01  Electrical timer

110-V ac source

**Figure 14 – 3** Schematic diagram of a coulometric titrator.

eliminate this difficulty, the auxiliary electrode is generally isolated by a sintered disk or some other porous barrier (see Figure 14 – 4).

**Comparison of Coulometric and Conventional Titrations**   The various components of the titrator in Figure 14 – 3 can be shown to have their counterparts in the reagents and apparatus required for a volumetric titration. The constant current source of known magnitude serves the same function as the standard solution in a volumetric method. The electronic timer and switch correspond to the buret and the stopcock, respectively. Electricity is passed through the cell for relatively long periods at the outset of a coulometric titration, but as chemical equivalence is approached, the time intervals are made smaller and smaller. Note that these steps are analogous to the way a buret is operated in a conventional titration.

A coulometric titration offers several significant advantages over a conventional volumetric procedure. Principal among these is the elimination of problems associated with the preparation, standardization, and storage of standard solutions. This advantage is particularly significant with labile reagents such as chlorine, bromine, or titanium(III) ion. The reactivity of these species seriously complicates their use as volumetric reagents, while their utilization in a coulometric determination is straightforward because they are consumed as soon as they are generated. Coulometric titrations have a clear advantage where small amounts of reagent are involved. Microquantities of reagent can be generated with ease and accuracy through the proper choice of current; in contrast, a

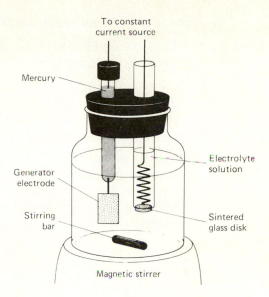

*Figure 14-4* A typical coulometric titration cell.

conventional volumetric titration would require small volumes of very dilute solutions, both of which are inconvenient at best.

A further advantage of the coulometric procedure is that a single constant current source provides reagents for precipitation, neutralization, or oxidation-reduction titrations. Finally, coulometric titrations are more readily automated, since current control is more readily accomplished than is the control of a flow of liquid.

Coulometric titrations are subject to five principal sources of error: (1) variations in current during electrolysis, (2) departure of the process from 100% current efficiency, (3) error in the measurement of current, (4) error in the measurement of time, and (5) titration error due to the difference between the equivalence point and the end point. The last of these difficulties is encountered in conventional volumetric methods as well. Where indicator error is the limiting factor, the two methods possess comparable reliability.

Currents constant to 0.2 to 0.5% relative are easily realized with simple instrumentation such as that shown in Figure 14-4. Control to 0.01% is obtainable with somewhat more sophisticated apparatus. Although generalizations concerning the magnitude of the uncertainty associated with electrode processes are difficult, current efficiencies exceeding 99.5% are reported regularly in the literature. Similarly, modern electronic digital timers permit the measurement of time to within ± 0.1% relative (or better).

To summarize, then, the current–time measurements required for a coulometric titration are inherently as accurate or more accurate than the comparable volume–normality measurements of a conventional volumetric analysis, particularly where small quantities of reagent are involved. When the accuracy of a titration is not limited by these measurements but instead by the sensitivity of the end point, the two titration methods have comparable accuracies.

## 14C–4 Applications of Coulometric Titrations

Coulometric titrations have been developed for all types of volumetric reactions.[7] Selected applications are described in this section.

**Neutralization Titrations**   Hydroxide ion can be generated at the surface of a platinum cathode immersed in a solution containing the analyte acid

$$2H_2O + 2e \rightarrow 2OH^- + H_2(g)$$

The platinum anode must be isolated by some sort of diaphragm to eliminate potential interference from the hydrogen ions simultaneously produced at that electrode. As a convenient alternative, a silver wire can be substituted for the platinum anode, provided chloride or bromide ions are added to the analyte solution. The anode reaction then becomes

$$Ag(s) + Br^- \rightarrow AgBr(s) + e$$

Silver bromide does not interfere with the neutralization reaction.

Both strong and weak acids can be titrated coulometrically with a high degree of accuracy. Either a potentiometric or an indicator end point can be used. The problems associated with the estimation of the equivalence point are identical with those encountered in a conventional neutralization titration. The coulometric method, however, has the advantage that interference from carbonate ion (p. 214) is far less troublesome. The only measure required to avoid a carbonate error is to remove the carbon dioxide from the solvent by boiling or by bubbling an inert gas such as nitrogen through the solution for a brief period (the latter process is called *sparging*).

Hydrogen ions generated at the surface of a platinum anode can be used for the coulometric titration of strong as well as weak bases:

$$2H_2O \rightarrow O_2 + 4H^+ + 4e$$

Here, the cathode must be isolated from the analyte solution to prevent interference from hydroxide ion.

**Precipitation and Complex Formation Reactions**   Titrations based upon the coulometric release of EDTA were first reported by Reilley and Porterfield.[8] The reagent is made available through reduction of the ammine mercury(II) EDTA chelate at a mercury cathode:

$$HgNH_3Y^{2-} + NH_4^+ + 2e \rightarrow Hg(l) + 2NH_3 + HY^{3-}$$

Because the mercury chelate is more stable than the corresponding complexes

[7] For additional applications, see E. Bishop, in *Comprehensive Analytical Chemistry,* C. L. Wilson and D. W. Wilson, Eds., Vol. IID, Chapters XVIII to XXIV. New York: Elsevier, 1975; J. T. Stock, *Anal. Chem.,* **1984,** *56,* 1R, **1982,** *54,* 1R, and **1980,** *52,* 1R.

[8] C. N. Reilley and W. W. Porterfield, *Anal. Chem.,* **1956,** *28*(4), 443.

of such cations as calcium, zinc, lead, or copper, complexation of these ions will occur only after the ligand has been freed by the electrode process.

**Oxidation-Reduction Titrations**   Table 14–2 indicates the variety of reagents that can be generated coulometrically and the analyses to which they have been applied. Of particular interest is bromine, the coulometric generation of which has formed the basis for a large number of methods. Of interest as well are reagents that are not ordinarily encountered in conventional volumetric analysis owing to the instability of their solutions; silver(II), manganese(III), and the chloride complex of copper(I) are examples.

## 14C – 5 Controlled Potential Coulometry

In controlled potential coulometry, 100% current efficiency is achieved by measuring and controlling the potential of the working electrode at a level at which electrolytic oxidation or reduction of the solvent or other species in the solution cannot occur. Control of the potential of the working electrode requires the presence of a third reference electrode, such as a saturated calomel electrode. Figure 14–5 shows how the three electrodes are arranged for a controlled cathode potential electrolysis. As in all coulometric methods, the quantity of electricity passing between the working electrode (in this case a cathode) and the counter electrode provides a measure of the amount of analyte. The control circuit consists of the working electrode, the reference electrode, and the electronic voltmeter ($V$), which monitors the potential of the working electrode (versus the saturated calomel electrode). Because of the very high resistance of the electronic voltmeter, essentially no current is present in the control circuit.

To initiate a controlled potential electrolysis with the apparatus shown in Figure 14–5, the switch is closed, and the sliding contact $C$ is adjusted until the

**Table 14–2**   *Summary of Applications of Coulometric Titrations Involving Oxidation-Reduction Reactions*

| Reagent | Generator Electrode Reaction | Substance Determined |
|---|---|---|
| $Br_2$ | $2Br^- \rightleftharpoons Br_2 + 2e$ | As(III), Sb(III), U(IV), Tl(I), I$^-$, SCN$^-$, NH$_3$, N$_2$H$_4$, NH$_2$OH, phenol, aniline, mustard gas, mercaptans, 8-hydroxyquinoline, olefins |
| $Cl_2$ | $2Cl^- \rightleftharpoons Cl_2 + 2e$ | As(III), I$^-$, styrene, fatty acids |
| $I_2$ | $2I^- \rightleftharpoons I_2 + 2e$ | As(III), Sb(III), S$_2$O$_3^{2-}$, H$_2$S, ascorbic acid |
| $Ce^{4+}$ | $Ce^{3+} \rightleftharpoons Ce^{4+} + e$ | Fe(II), Ti(III), U(IV), As(III), I$^-$, Fe(CN)$_6^{4-}$ |
| $Mn^{3+}$ | $Mn^{2+} \rightleftharpoons Mn^{3+} + e$ | H$_2$C$_2$O$_4$, Fe(II), As(III) |
| $Ag^{2+}$ | $Ag^+ \rightleftharpoons Ag^{2+} + e$ | Ce(III), V(IV), H$_2$C$_2$O$_4$, As(III) |
| $Fe^{2+}$ | $Fe^{3+} + e \rightleftharpoons Fe^{2+}$ | Cr(VI), Mn(VII), V(V), Ce(IV) |
| $Ti^{3+}$ | $TiO^{2+} + 2H^+ + e \rightleftharpoons Ti^{3+} + H_2O$ | Fe(III), V(V), Ce(IV), U(VI) |
| $CuCl_3^{2-}$ | $Cu^{2+} + 3Cl^- + e \rightleftharpoons CuCl_3^{2-}$ | V(V), Cr(VI), IO$_3^-$ |
| $U^{4+}$ | $UO_2^{2+} + 4H^+ + 2e \rightleftharpoons U^{4+} + 2H_2O$ | Cr(VI), Ce(IV) |

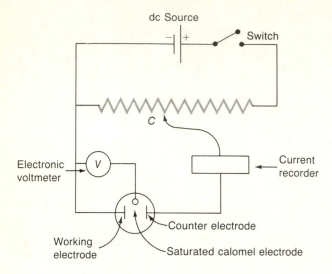

**Figure 14–5** Schematic diagram of a potentiostat.

working electrode potential is at the desired level as indicated by the voltmeter. Note that movement of contact $C$ varies the potential applied across the working and counter electrode system. As the electrolysis progresses, it is necessary to adjust $C$ continually to maintain the working electrode at the desired potential versus the reference electrode.

A controlled potential electrolysis is tedious and time consuming when performed manually. *Potentiostats* are automatic instruments that provide the needed control electronically. Several instrument manufacturers supply potentiostats.

Figure 14–6 illustrates the changes in the cell potential needed to maintain a cathode potential of $-0.36$ V (vs. SCE) throughout the deposition of copper. Also shown is the current in the cell during the deposition process. Initially, the cell potential can be high without causing the cathode potential to become more negative than $-0.36$ V; that is, increasing the cell potential merely increases the current and thus the ohmic potential (large currents are, of course desirable since they shorten the analysis time). With the onset of concentration polarization, however, the applied cell potential must be decreased continuously to maintain the desired cathode potential. A corresponding decrease in current is observed. Ultimately, the current approaches zero as the deposition approaches completion.

In a controlled potential coulometric electrolysis, the current is recorded continuously as a function of time, as in Figure 14–6. The quantity of analyte is then determined from the area under the current-time curve. This area is usually evaluated with an electronic integrator, although as shown in the following example, a chemical coulometer may be employed as well.

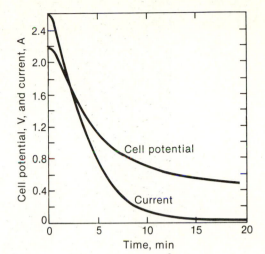

**Figure 14-6** Changes in applied potential and current during a controlled cathode potential deposition of copper. The cathode is maintained at $-0.36$ V vs. SCE throughout the experiment. (Data from J. J. Lingane, *Anal. Chem. Acta*, **1948**, *2*, 590. With permission.)

**Example 14-3** The Fe(III) in a 0.820-g sample was determined by coulometric reduction to Fe(II) at a platinum cathode. Calculate the percentage of $Fe_2(SO_4)_3$ (gfw $= 400$) in the sample if a hydrogen–oxygen coulometer arranged in series with the cell containing the sample evolved 19.37 mL of gas ($H_2 + O_2$) at 23°C and at a pressure of 765 torr (after correction for water vapor). The reactions in the coulometer are

$$4H^+ + 4e \rightarrow 2H_2(g)$$

and

$$2H_2O \rightarrow O_2(g) + 4H^+ + 4e$$

Four moles of electrons thus yield 3 mol of gas in this reaction.

The volume collected is first converted to standard conditions; that is,

$$V = 19.37 \text{ mL} \times \frac{765 \text{ torr}}{760 \text{ torr}} \times \frac{273°\text{K}}{296°\text{K}} = 17.98 \text{ mL (STP)}$$

This volume was produced by

$$17.98 \text{ mL gas} \times \frac{\text{mol gas}}{22,400 \text{ mL}} \times \frac{4 \text{ faradays}}{3 \text{ mol gas}} = 1.070 \times 10^{-3} \text{ faraday}$$

Thus,

$$1.070 \times 10^{-3} \text{ faraday} \times \frac{1 \text{ eq Fe}}{\text{faraday}} \times \frac{\text{fw } Fe_2(SO_4)_3}{2 \text{ eq Fe}} \times \frac{400 \text{ g } Fe_2(SO_4)_3}{\text{fw}} = 0.214 \text{ g}$$

and

$$\frac{0.214}{0.820} \times 100 = 26.1\% \ Fe_2(SO_4)_3$$

## 14C–6 Applications of Controlled Potential Coulometry

Controlled potential coulometric methods have been applied to the determination of some 55 elements in inorganic compounds.[9] Mercury appears to be favored as the cathode, and methods for the deposition of two dozen or more metals at this electrode have been described. The method has found widespread use in the nuclear energy field for the relatively interference-free determination of uranium and plutonium.

The controlled potential coulometric procedure also offers possibilities for the electrolytic determination (and synthesis) of organic compounds. For example, Meites and Meites[10] have demonstrated that trichloroacetic acid and picric acid are quantitatively reduced at a mercury cathode whose potential is suitably controlled:

$$Cl_3CCOO^- + H^+ + 2e \rightarrow Cl_2HCCOO^- + Cl^-$$

Coulometric measurements permit the analysis of these compounds with a relative error of a few tenths of a percent.

Variable current coulometric methods are frequently used to monitor continuously and automatically the concentration of constituents in gas or liquid streams. An important example is the determination of small concentrations of oxygen.[11] A schematic diagram of the apparatus is shown in Figure 14–7. The porous silver cathode serves to break up the incoming gas into small bubbles, with the reduction of oxygen then taking place quantitatively within the pores; that is,

$$O_2(g) + 2H_2O + 4e \rightarrow 4OH^-$$

---

[9] For a summary of the applications, see J. E. Harrar, in *Electroanalytical Chemistry*, A. J. Bard, Ed., Vol. 8. New York: Marcel Dekker, Inc., 1975; and E. Bishop, in *Comprehensive Analytical Chemistry*, C. L. Wilson and D. W. Wilson, Eds., Vol. IID, Chapter XV. New York: Elsevier, 1975.

[10] T. Meites and L. Meites, *Anal. Chem.*, **1955**, *27*, 1531; **1956**, *28*, 103.

[11] F. A. Keidel, *Ind. Eng. Chem.*, **1960**, *52*, 491.

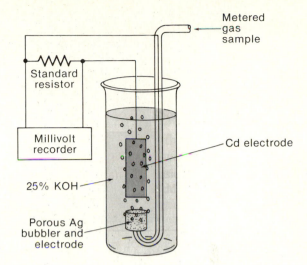

**Figure 14–7** An instrument for continuously monitoring the oxygen content of a gas stream.

The anode is a heavy cadmium sheet, which reacts to form cadmium hydroxide:

$$Cd(s) + 2OH^- \rightarrow Cd(OH)_2(s) + 2e$$

Note that a galvanic cell is formed so that no external power supply is required; nor is a potentiostat necessary because the potential of the working anode can never become great enough to cause oxidation of other species. The electricity produced is passed through a standard resistor and the potential drop is recorded. The oxygen concentration is proportional to this potential, and the chart paper can be made to display the instantaneous oxygen concentration directly. The instrument is reported to provide oxygen concentration data in the range from 1 ppm to 1%.

## 14D VOLTAMMETRY

*Voltammetry* refers to a family of electroanalytical methods in which analytical information is derived from measurements of current as a function of applied potential under conditions that encourage polarization of the indicator or working electrode. Generally, the surface area of indicator electrodes in voltammetry are kept small (a few square millimeters) in order to enhance polarization. Such electrodes are called *microelectrodes.*

Historically, voltammetry evolved from *polarography,* an amperometric method in which the microelectrode consists of small mercury droplets that flow from a fine capillary tubing.[12] This type of electrode is called a *dropping mercury electrode* (DME).

---

[12] Credit for the discovery of polarography belongs to Jaroslav Heyrovsky, a Czechoslovakian chemist. Heyrovsky was awarded the 1959 Nobel prize in chemistry for this discovery and its subsequent development.

## 14E POLAROGRAPHY

Polarography is based upon current–voltage curves that are obtained for a cell consisting of an easily polarized dropping mercury electrode and a reference electrode that is sufficiently large to remain unpolarized throughout the experiment.[13] Ordinarily, the applied voltage is varied over the range of 0 to about $-2.5$ V, while the currents range from perhaps 0.1 to 100 $\mu$A. The resulting current–voltage curves are called *polarograms.*

### 14E–1 Polarograms

Figure 14–8 shows two typical polarograms: one for a solution that is 0.1 $F$ in KCl and $1.0 \times 10^{-3}$ $F$ in Cd(II) (curve *A*), and another for KCl alone (curve *B*). The dropping mercury electrode is connected to the negative terminal of the power supply and thus acts as the cathode in this electrolytic cell. By convention, the applied potential is given a negative sign and currents are supplied with a positive sign when electrons flow from the power supply to the microelectrode. The step-shaped *polarographic wave* in curve *A* is the result of the reaction

$$Cd^{2+} + 2e + Hg(l) \rightarrow Cd(Hg) \qquad (14–5)$$

where Cd(Hg) is an amalgam consisting of elemental cadmium dissolved in mercury. The sharp increase in current at about $-2$ V in both polarograms is caused by the reduction of potassium ions to give a potassium amalgam.

We have noted (Section 14A–2) that ions transport electricity through a solution and that diffusion, electrostatic attraction between the analyte ions and the electrode, and convection contribute to the migration process. For polarographic work, every effort is made to ensure that the currents observed are due solely to diffusion. A high concentration of *supporting electrolyte* (here, the KCl) is maintained to eliminate the effects of electrostatic attraction between the analyte ions and the electrode. Convective forces are minimized by avoiding vibration of the apparatus and temperature inhomogeneities.

Examination of the polarogram for the supporting electrolyte alone reveals the existence of a small *residual current* in the cell even though the solution contains no cadmium ions. This residual current is negative at low applied potentials, passes through zero at about $-0.4$ V, and becomes positive at higher potentials.

A characteristic feature of a polarographic wave is the region in which the current, after rising sharply, becomes essentially independent of the applied potential. This *limiting current* is the result of a restriction in the rate at which the reactant in the electrode process can be brought to the surface of the

---

[13] For a more extensive description of the principles and applications of polarography, see L. Meites, *Polarographic Techniques,* 2nd ed. New York: Interscience Publishers, 1965; P. Zuman, *Topics in Organic Polarography.* New York: Pergamon Press, 1970; A. M. Bond, *Modern Polarographic Methods in Analytical Chemistry.* New York: Marcel Dekker, Inc., 1980.

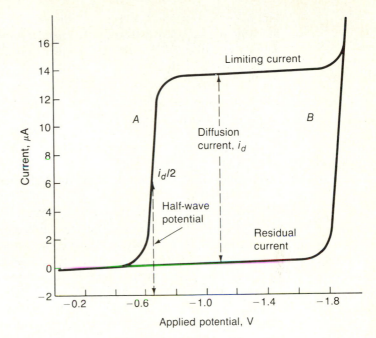

**Figure 14–8** Polarogram for cadmium ion. Curve A: solution is $1.0 \times 10^{-3}$ F with respect to $Cd^{2+}$ and 0.10 F with respect to KCl. Curve B: solution is 0.10 F in KCl only.

microelectrode. Under circumstances where diffusion alone controls the rate of mass transfer, the difference between the limiting current and the residual current is given a special name, the *diffusion current,* and the symbol $i_d$. Ideally, the diffusion current is directly proportional to the reactant concentration and is thus of prime importance from the standpoint of analysis.

The *half-wave potential,* another important quantity, is the potential at which the current is equal to one-half the diffusion current. The half-wave potential is given the symbol $E_{1/2}$; it may permit qualitative identification of the reactant.

## 14E–2 Interpretation of Polarographic Waves

Polarographic waves are readily interpreted provided that the reaction at the dropping mercury electrode is sufficiently rapid that the concentrations of reactants and products at the interface between the solution and mercury drop are determined at any instant by the electrode potential alone. Thus, for the reversible reduction of cadmium(II), the concentrations of $Cd^{2+}$ and of Cd(Hg) at the interface are always those needed to satisfy the equation

$$E_{applied} = E_A^0 - \frac{0.0591}{2} \log \frac{[Cd]_0}{[Cd^{2+}]_0} - E_{ref} \qquad (14–6)$$

Here, $[Cd]_0$ is the activity of metallic cadmium dissolved in the surface film of the mercury, and $[Cd^{2+}]_0$ is the activity of the ion in the aqueous phase. Note that the subscript zero has been employed for the activity terms to emphasize that this relationship *applies to the surface films of the two media only;* the activity of cadmium ion in the bulk of the solution and of elemental cadmium in the interior of the mercury drop *will ordinarily be quite different from the surface activities.* The films we are concerned with are no more than a few atoms or molecules thick.

The term $E_{applied}$ in Equation 14–6 is the potential applied to the cell consisting of the dropping electrode and a reference electrode whose potential is $E_{ref}$; $E_A^0$ is the standard potential for the half-reaction in which a saturated cadmium amalgam is the product. The difference between $E_A^0$ and the standard electrode potential for the reduction to elemental cadmium is about +0.05 V.

Consider what occurs when $E_{applied}$ is sufficiently negative to cause appreciable reduction of cadmium ion. Because the reaction is reversible, the activity of cadmium ion in the film surrounding the electrode decreases, and the activity of cadmium in the outer layer of the mercury drop increases instantaneously to the levels demanded by Equation 14–6; a surge of current results. This current would rapidly decay to zero were it not for the fact that cadmium ions are mobile in the aqueous medium and can migrate to the surface of the mercury. The result is a current, *the magnitude of which depends upon the rate at which the cadmium ions move from the bulk of the solution to the surface where reaction occurs;* that is,

$$i = k' \times v_{Cd^{2+}}$$

where $i$ is the current at an applied potential $E_{applied}$, $v_{Cd^{2+}}$ is the rate of migration of cadmium ions, and $k'$ is a proportionality constant.

In polarography, every effort is made to eliminate electrostatic attraction and thermal and mechanical convection as mechanisms by which ions or molecules migrate through the solution to the electrode. Under these circumstances, diffusion becomes the sole means of transport of cadmium ions to the electrode surface. Because the rate of diffusion is directly proportional to the concentration difference between the two parts of a solution, we may write

$$v_{Cd^{2+}} = k''([Cd^{2+}] - [Cd^{2+}]_0)$$

where $[Cd^{2+}]$ is the concentration *in the bulk of the solution* from which ions are diffusing and $[Cd^{2+}]_0$ is the concentration in the aqueous film surrounding the electrode. As long as diffusion is the only process bringing cadmium ions to the surface, it follows that

$$i = k'v_{Cd^{2+}} = k'k''([Cd^{2+}] - [Cd^{2+}]_0)$$
$$= k([Cd^{2+}] - [Cd^{2+}]_0)$$

Note that $[Cd^{2+}]_0$ becomes smaller as $E_{applied}$ is made more negative (Equation 14–6). Thus, the rate of diffusion as well as the current increases with increases in applied potential. This potential ultimately becomes so negative that essentially every cadmium ion reaching the drop is reduced, and the activity of that ion in the surface film approaches zero; under this circumstance, the rate of diffusion, and thus the current, become constant, and the expression for current becomes

$$i_d = k[Cd^{2+}]$$

where $i_d$ is the potential-independent diffusion current. Note that *the magnitude of the diffusion current is directly proportional to the concentration of the reactant in the bulk of the solution.* Quantitative polarography is based upon this fact.

A state of *complete concentration polarization* is said to exist when the current in a cell is limited by the rate at which a reactant can be brought to the surface of an electrode. The current required to reach this condition with a microelectrode is small—typically, 3 to 10 $\mu$A for a $10^{-3}$ $M$ solution. It is important to note that deposition by such current levels does not significantly alter the reactant concentration in the time required to obtain a polarogram.

**Residual Current** The residual current (Figure 14–8) has two sources. The first is the reduction of trace impurities in the supporting electrolyte or the sample. The second is the so-called *charging* or *condenser current* resulting from a flow of electrons that charge the mercury droplets with respect to the solution. This current can be either positive or negative. At potentials more negative than about $-0.4$ V, an excess of electrons from the dc source provides the surface of each drop with a negative charge that is carried down with the drop as it breaks. Since each new drop is charged as it forms, a small but continuous current results. At potentials less than about $-0.4$ V, the mercury tends to be positive with respect to the solution, and the surface of each drop is slightly deficient in electrons; a positive residual current is thus observed at these potentials. At about $-0.4$ V, the mercury is uncharged, and the residual current is zero.

The ultimate accuracy and sensitivity of the polarographic method depend upon the magnitude of the residual current and the reliability of the correction that is made for its effects.

**Effect of Cell Resistance on Polarographic Waves** The relationship between the potential applied to a polarographic cell and the potential of the dropping mercury electrode, $E_{DME}$, should include a term for $IR$ drop; that is,

$$E_{applied} = E_{DME} - E_{ref} - IR$$

Under many circumstances, the $IR$ drop is negligible with respect to the other two terms in the right-hand side of the equation, and $E_{DME}$ responds linearly to changes in $E_{applied}$. When $R$ is large, however, increases in $I$ require a greater and greater fraction of the applied potential to be used in overcoming the $IR$ drop, and linearity is lost. This effect is illustrated in Figure 14–9. Note that when $R$ is 100 $\Omega$, $IR$ is small enough to have no effect on the slope of the wave. At the

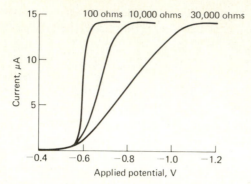

**Figure 14-9**  Effect of cell resistance upon a reversible polarographic wave.

higher resistances, however, the influence of the *IR* drop becomes greater and greater, which leads to drawn-out, less well-defined waves.

The use of potentiostatic control has permitted the extension of polarography to organic solvents with high electrical resistances. Here, three electrodes are used: a dropping mercury electrode, a small counter electrode, and a reference electrode such as a saturated calomel electrode. The arrangement is similar to that described for potentiostatic coulometry in Section 14C-5 and Figure 14-5. The reference electrode, which is located as close as possible to the dropping electrode, serves to control the applied potential such that the change in $E_{DME}$ *is linear with time.* The abscissa for the polarogram now becomes $E_{DME}$ rather than $E_{applied}$. The effect is to produce sharply defined waves similar to the steepest wave in Figure 14-9 even when the cell resistance is orders of magnitude greater.

### 14E-3 Apparatus for Polarographic Measurements

**Cells**    Figure 14-10 shows a cell that is suitable for two-electrode polarography. A mercury pool, which is connected to the electrical source through a sidearm, serves as the anodic counter electrode.

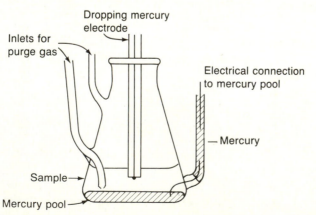

**Figure 14-10**  Heyrovsky cell for two-electrode polarography.

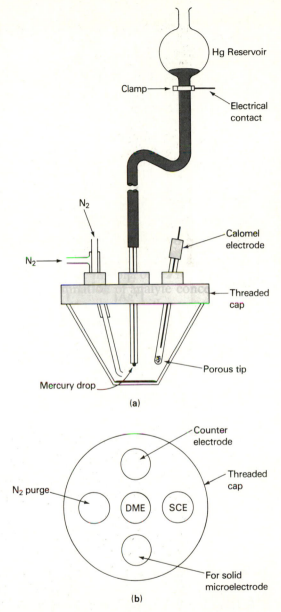

Hg Reservoir

Clamp

Electrical contact

$N_2$

$N_2$

Calomel electrode

Threaded cap

Porous tip

Mercury drop

**(a)**

Counter electrode

Threaded cap

$N_2$ purge

DME

SCE

For solid microelectrode

**(b)**

*Figure 14–11* A dropping mercury electrode and polarographic cell. (a) Cross-sectional view. (b) Top view of cap.

A general-purpose cell that is adaptable to two- or three-electrode polarographic measurements is shown in Figure 14–11. The cell itself is a heavy-walled glass container, which is threaded at the top so that it can be screwed into the polypropylene cap that holds the electrodes. The cell walls are tapered so that as much as 30 mL and as little as 2 mL can be accommodated. The cap contains a purge tube fitted to a three-way stopcock (not shown) that permits nitrogen to be bubbled through the solution to remove oxygen, or over the surface of the liquid to prevent reabsorption of that species.

Modern polarography is generally performed with three electrodes. The reference electrode in this arrangement no longer needs to be as large as in the two-electrode configuration because it no longer serves as the counter electrode. Consequently, the currents in the dropping mercury – reference electrode circuit are negligible. The counter electrode in the three-electrode system is often a mercury pool or a platinum wire.

**The Dropping Mercury Electrode**    A dropping electrode, such as that shown in Figure 14–11, consists of a 5- to 20-cm length of fine capillary tubing (~0.05 mm inside diameter) through which mercury is forced under a head of approximately 50 cm. A continuous series of highly reproducible drops are formed, each with a diameter of 0.1 to 1 mm; typical drop times are 2 to 6 sec. Capillaries are available for purchase from commercial sources.

Figure 14–12a reveals that the current in a cell containing a dropping electrode undergoes periodic fluctuations corresponding in frequency to the drop rate. As a drop breaks, the current falls to zero; it then rises rapidly as the electrode area grows because of the greater surface to which diffusion can occur. The *average current* is the hypothetical constant current that in the drop time $t$ would deliver the same number of coulombs as the fluctuating current does during the same period. In order to determine the average current, it is necessary to reduce the large fluctuations shown in Figure 14–12a to the smaller ones shown in Figure 14–12b. The latter was obtained with a well-damped recording galvanometer. Most modern polarographs achieve damping with an electronic filter, or alternatively sample the current shortly before the drop detaches when its rate of change in size is inconsequential.

Figure 14–12b illustrates how the maximum current is measured for a typical polarogram. Both the average and the maximum currents are used for quantitative analyses. Note the effect of irregular drop times in the limiting current region, probably caused by vibration of the cell.

**Precautions in the Use of Dropping Mercury Electrodes**    With reasonable care, a capillary can be used for several months or even years. Such performance, however, requires the use of scrupulously clean mercury and the maintenance of a mercury head, no matter how slight, at all times. If solution comes in contact with the inner surface of the tip, malfunction of the electrode is inevitable. For this reason, the head of mercury should always be increased to provide a good flow before the tip is immersed in a solution. Cleaning of a malfunctioning electrode is usually not very successful in restoring its performance. When not in use, the electrode is immersed in clean mercury, after which the head can be reduced.

**Advantages and Disadvantages of the Dropping Mercury Electrode**    The dropping mercury electrode offers several advantages over other types of mi-

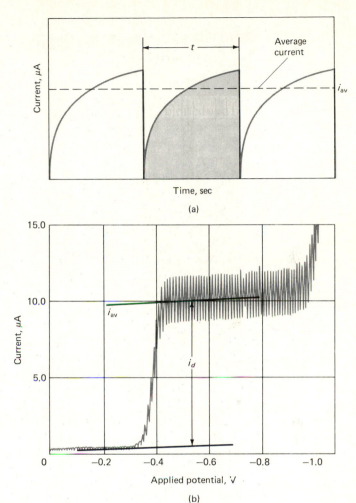

**Figure 14–12** Effect of drop growth on polarographic currents. (a) Current–time relationship during the lifetime, $t$, of drops. The shaded area represents the microcoulombs of electricity associated with each drop. Note that the current increases by only about 20% during the second half of the drop lifetime. (b) A recorded polarogram in which current fluctuations have been damped to permit measurement of the average current, $i_{av}$, or the maximum current, $i_{max}$. (From D. T. Sawyer and J. L. Roberts, Jr., *Experimental Electrochemistry for Chemists.* New York: John Wiley & Sons, 1974. With permission.)

croelectrodes. Principal among these is the high overvoltage for the formation of hydrogen, which permits the reduction of many species from acid solution without interference. Equally important, the behavior of the dropping mercury electrode is independent of its past history because a new metal surface is being continuously generated. Consequently, reproducible current–voltage curves are obtained regardless of how the electrode has previously been used (this behavior is in marked contrast to solid microelectrodes, whose outputs reflect their past history). Finally, the dropping mercury electrode develops reproducible currents essentially instantaneously.

The most serious drawback to the use of the dropping mercury electrode is the ease with which mercury is oxidized, which severely limits the use of mercury as an anode. Mercury(I) is produced at potentials more positive than about

0.4 V (vs. SCE), producing currents that totally mask the waves of other species in the solution. Thus, the dropping mercury electrode is limited to the analysis of reducible or easily oxidizable substances. Additional disadvantages to the dropping mercury electrode include its physical size, which makes its use somewhat cumbersome, and its tendency to malfunction as a result of clogging.

**Electrical Apparatus**   A polarographic measurement requires the means for varying the potential between the electrodes over a range of 0 to $-3$ V. Moreover, it must be possible to measure currents within the range of 0.01 to 100 $\mu A$ with an accuracy of about 0.01 $\mu A$. A manual apparatus that meets these requirements can be assembled from equipment that is available in most laboratories. Recording instruments are more convenient and are available from commercial suppliers.

   A circuit for a simple polarographic instrument is shown in Figure 14–13. Two 1.5-V batteries provide the potential across $R_1$, a 100-$\Omega$ voltage divider, which permits variation in the potential that is applied to the cell. A potentiometer serves not only to measure this potential, but also to monitor the currents through measurement of the potential drop across a precision 10,000-$\Omega$ resistor

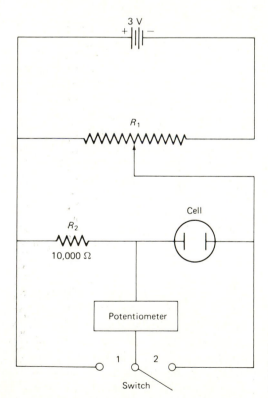

**Figure 14–13**   A simple polarographic circuit. (From J. J. Lingane, *Anal. Chem.*, **1949**, *21*, 47. With permission of the American Chemical Society.)

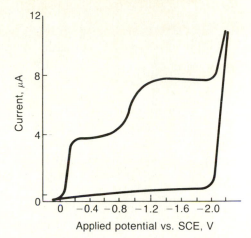

**Figure 14 – 14**  Polarogram for the reduction of oxygen in an air-saturated 0.1 *F* KCl solution. The lower curve is for oxygen-free 0.1 *F* KCl.

(with the switch in position 2). A circuit similar to that shown in Figure 14 – 5 is used for three-electrode polarography.

### 14E – 4 Analytical Details

**Temperature Control**   The rates at which ions diffuse and the diffusion currents that result typically increase by about 2.5%/°C. The temperature of the analyte solution must therefore be controlled to within a few tenths of a degree for accurate polarographic measurements.

**Oxygen Removal**   Dissolved oxygen is reduced to water in two steps, hydrogen peroxide being the intermediate product. Two waves of equal magnitude result, the first with a half-wave potential at about −0.14 V and the second at about −0.9 V (Figure 14 – 14). Because both reactions are slow, the two waves are drawn out over a considerable voltage range. While oxygen waves are useful for the measurement of the dissolved gas, they frequently interfere with the determination of other species. For this reason, oxygen removal is ordinarily the first step in a polarographic analysis. Aeration of the solution with an inert gas for several minutes accomplishes this end. A stream of the same gas, usually nitrogen, is then passed over the surface of the cell during the analysis to prevent the reabsorption of oxygen.

**Current Maxima**   As shown in Figure 14 – 15, polarographic waves are frequently distorted by anomalous *current maxima*. Maxima are troublesome because they interfere with the accurate evaluation of diffusion currents and half-wave potentials. The causes for current maxima are not entirely understood; nevertheless, considerable empirical knowledge exists for their elimination. Small quantities of such high-molecular-weight substances as gelatin, Triton X-100 (a commercial surfactant), methyl red, other dyes, and even

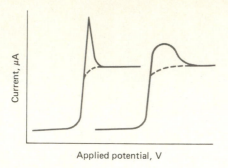

**Figure 14–15**   Typical current maxima.

carpenter's glue, are known to act as maximum suppressors. The first two of these additives are particularly useful.

**Polarogram for Irreversible Waves**   Many polarographic electrode processes, particularly those involving organic analytes, are irreversible and give drawn-out and less well-defined waves as a result (see Figure 14–14). Diffusion currents ordinarily remain linearly related to concentration, however, and can be used for quantitative measurements.

**Evaluation of Diffusion Currents**   Limiting currents must be corrected for the residual current by obtaining a residual current curve before or after the polarogram for the analyte is recorded. The diffusion current is then taken as the difference between the two curves at some potential in the limiting current region (see Figure 14–8). If it is known that the residual current curve increases linearly with the applied potential (as it usually does), the diffusion current can be evaluated by extrapolation of the residual current portion of the curve for the sample, as shown in Figure 14–12b.

Diffusion currents can be derived either from average currents (as in Figure 14-12b) or from maximum currents. The former are often selected because they tend to be less dependent upon the degree of damping. Modern instruments, however, often sample the current just prior to the detachment of the drop. Here, current fluctuations are relatively small, and diffusion currents can be based upon maximum currents.

**Concentration Determination**   Polarographic analyses are generally based on the linear calibration curve produced by the measurement of solutions containing known amounts of analyte. As nearly as possible, these standards should closely resemble the samples to be analyzed in overall composition and encompass a range within which the concentration of the analyte is anticipated. An analysis can, of course, be carried out even if the calibration curve is nonlinear.

The standard addition method (p. 355) is applicable to polarographic analysis. This approach is particularly useful where the diffusion current is sensitive to other components in the sample.

**Analysis of Mixtures** The reactants of a mixture will ordinarily act independently of one another at the dropping mercury electrode; each new wave is simply superimposed on the limiting current of the previous one. A single polarogram may thus provide quantitative information for several components in a mixture. Success depends upon the existence of a sufficient difference between succeeding half-wave potentials to permit evaluation of individual diffusion currents. Approximately 0.2 V is needed if the more reducible species undergoes a two-electron reduction, while a minimum of about 0.3 V is needed if the first reduction is a one-electron process.

**Sensitivity, Accuracy, and Precision of Polarographic Procedures** Classical polarography is most commonly performed on aqueous solutions that are $10^{-2}$ to $10^{-5}$ $M$ in analyte. A polarographic analysis is easily performed on 1 to 2 mL of solution and with a little effort on a volume as small as a single drop.

The accuracy and precision of polarography depends upon the shape of the analyte wave. For a well-defined wave such as that shown in Figure 14–12b, current measurements accurate to 1 to 2% relative can be achieved. Uncertainties from other sources, such as temperature or drop time variations, or instrumental noise, lead to overall deviations of perhaps 3% relative. For nonreversible and ill-defined waves, the uncertainty may lie in the 5 to 20% range.

**Modified Polarographic Methods** By 1950, voltammetry appeared to be a mature and fully developed technique. The decade from 1955 to 1965, however, was marked by the appearance of several major modifications that enhanced the sensitivity of the method by two to three orders of magnitude, increased precision and accuracy by a factor of two to three, and made possible the general application of polarography in organic solvents having high electrical resistance.[14] We have already discussed one of the modifications, namely, the three-electrode cell.

The modifications that have led to enhanced sensitivity and precision have involved avoiding the limitation imposed by the presence of the residual current. This limitation asserts itself whenever the ratio of the diffusion current to the residual current approaches unity or smaller. Under these circumstances, the diffusion current is a small difference between two large experimental numbers, each subject to uncertainty—a situation that always leads to poor precision and reduced sensitivity. The interested reader should consult the references in Footnotes 13 and 14 for further information concerning these modifications.

---

[14] For a brief summary of these modifications, see: J. B. Flato, *Anal. Chem.,* **1972,** *44,* 75A; and D. A. Skoog, *Principles of Instrumental Analysis,* 3rd ed., pp. 681–693. Philadelphia: Saunders College Publishing, 1985.

## 14E-5 Applications of Polarography

**Inorganic Applications**   The polarographic method is widely applicable to the analysis of inorganic species. Most cations, for example, are reduced at the dropping mercury electrode, including ions of the alkali and alkaline earth metals. In order to reach the high reduction potentials that are characteristic of the latter two groups, it is necessary to employ one of the tetraalkyl ammonium halides as supporting electrolyte; these compounds are reduced at even more negative potentials than are the alkali and alkaline earth cations.

The polarographic method is also applicable to the determination of such anions as bromate, iodate, dichromate, vanadate, and nitrite. Generally, these determinations must be carried out from buffered solutions since the half-reactions all consume hydronium ions. If the solution is not buffered, the pH at the electrode surface increases, which causes drawn-out waves or, in some cases, a change in stoichiometry of the reduction process.

**Organic and Biochemical Applications**   Several organic functional groups are reducible at the dropping mercury electrode, which makes polarography applicable to a wide variety of organic and biochemical species.[15] The following functional groups are reactive: (1) the carbonyl groups in aldehydes, ketones, and quinones; (2) the carboxylic acid group when it is conjugated with keto, aldehydo, or another carboxylate group; (3) most peroxides and epoxides; (4) nitro, nitroso, amine oxide, and azo groups; (5) most organic halogen groups; (6) carbon–carbon double bonds, provided they are conjugated with another unsaturated group including aromatic rings; and (7) hydroquinone and mercaptan groups, which give anodic waves.

Generally, organic reductions must be carried out in buffered solutions since these electrode processes ordinarily involve the proton. The absence of a buffer produces the same undesirable side effects that were described in connection with the determination of inorganic anions. A variety of solvent systems have been employed in organic polarography, including aqueous mixtures of glycols, alcohols, dioxane, acetonitrile, cellosolve, and acetic acid. Supporting electrolytes are often lithium salts or tetraalkyl ammonium halides.

## 14F AMPEROMETRIC TITRATIONS

Amperometric titrations are volumetric or coulometric procedures in which end points are established from plots of reagent volume or electrolysis time as a function of current in a cell that is operated at a fixed applied potential. In one type of amperometric titration, a microelectrode and a counter electrode serve as the indicator system. In a second type, twin microelectrodes are used for end-point determination.

---

[15] For a detailed discussion of organic polarographic analysis, see *Polarography of Molecules of Biological Significance,* W. F. Smyth, Ed. New York: Academic Press, 1979; and *Topics in Organic Polarography,* P. Zuman, Ed., New York: Plenum Press, 1970.

Amperometric titrations are inherently more accurate than conventional voltammetric methods and are less dependent upon such experimental variables as temperature, composition and concentration of supporting electrolyte, and electrode characteristics. Moreover, analytes that are not oxidized or reduced at microelectrodes can be determined provided that the reagent or the product of the reaction is reactive.

## 14F–1 Amperometric Titrations with One Microelectrode

**Titration Curves**    Typical amperometric titration curves are displayed in Figure 14–16. The curve in Figure 14–16a is characteristic of a titration in which the analyte is reactive at the microelectrode while the reagent is not; the titration of lead(II) with oxalate or sulfate ion is an example. Here, the applied potential is sufficient (say, −1.0 V) to give a diffusion current for lead, which causes a linear decrease in current as lead ions are removed from the solution by precipitation. Beyond equivalence, the current becomes constant at 0 $\mu A$. The end point is established by extrapolation of the linear portions of the two curves to their intersection, as shown. Incompleteness of the titration reaction is responsible for curvature in the equivalence-point region.

The curve in Figure 14–16b is typical of a reaction in which the reagent is reactive but the analyte is not. An example is the titration of magnesium ion with 8-hydroxyquinoline, a reagent that is reduced at −1.6 V (vs. SCE). Magnesium ion is unreactive at that potential.

The titration of lead with 8-hydroxyquinoline at −1.6 V gives a curve similar to that shown in Figure 14–16c; both reagent and analyte are reactive at this potential, and the end point corresponds to the minimum in the curve.

In order to obtain plots with linear regions before and following the equivalence point, corrections must be made for changes in volume that occur during the titration. By multiplying the measured diffusion currents by $(V + v)/V$, where $V$ is the original volume of the solution containing the analyte and $v$ is the volume of titrant, all measured currents are corrected back to the original volume. An alternative is to make $v$ negligibly small by using a titrant solution that is 20 (or more) times as concentrated as the analyte.

**Figure 14–16**    Typical amperometric titration curves. (a) Analyte is reduced, reagent is not. (b) Reagent is reduced, analyte is not. (c) Both reagent and analyte are reduced.

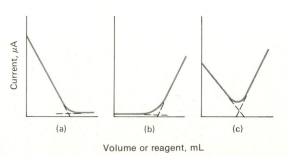

**Apparatus**    A simple manual polarograph is entirely satisfactory for detecting end points in an amperometric titration. The applied potential does not need to be known very accurately ($\pm 0.1$ V) since it is only necessary to select a potential within the diffusion current region of at least one reactant or product in the titration.

The dropping mercury electrode, while useful for many amperometric titrations, cannot be applied to reactions in which one of the reactants is an oxidizing agent because of the ease with which mercury is oxidized. For such analyses, inert microelectrodes manufactured from platinum, graphite, or glossy carbon are used. An example is the *rotating platinum electrode,* which has found widespread use for titrations with such oxidants as bromine, chlorine, and iodine.

Figure 14–17 shows a rotating platinum electrode, which consists of a short length of platinum wire that has been sealed into the side of a glass tube. Mercury within the tube provides contact between the wire and the lead to the polarograph. The tube is held in the hollow chuck of a synchronous motor and is rotated in excess of 600 rpm.

Current–voltage curves with rotating platinum electrodes are similar in appearance to those obtained with the dropping mercury electrode except that current fluctuations due to the growth and detachment of the drop are not observed. Limiting currents tend to be as much as 20 times greater, however, because the reactive species is brought to the surface of the rotating electrode not only by diffusion but by mechanical mixing as well. The rotating electrode also provides steady currents instantaneously, which is in distinct contrast to

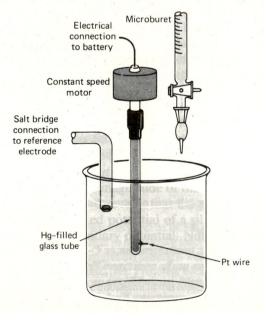

**Figure 14–17** Typical arrangement for amperometric titrations with a rotating platinum electrode.

inert microelectrodes in an unstirred solution that take several minutes to reach a constant current after each reagent addition.

The low overvoltage of hydrogen severely limits the use of the rotating platinum electrode as a cathode in acidic solutions. Moreover, the high currents produced make the electrode particularly sensitive to traces of dissolved oxygen. Diffusion currents obtained with a rotating platinum electrode are influenced in some measure by the previous history of the electrode and are seldom as reproducible as those obtained with the dropping electrode. This lack of reproducibility is not ordinarily a serious problem for most amperometric titrations.

**Applications**   The amperometric end point has been largely confined to titrations in which the product is a sparingly soluble solid or a stable complex and to certain oxidation-reduction titrations, particularly those in which bromine or iodine is the titrant. Selected applications are shown in Table 14–3.

### 14F – 2 Amperometric Titrations with Twin Microelectrodes

An amperometric end point can also be obtained using a pair of identical microelectrodes to which a small potential (0.1 to 0.2 V) is applied. The current is then plotted as a function of titrant volume. The end point is marked by a decrease in the current to zero, by a sudden increase in the current from zero, or as a minimum (at zero) in a V-shaped curve.

The use of two polarizable electrodes for end-point detection was first proposed before 1900. Almost 30 years were to pass, however, before chemists

**Table 14–3**   *Applications of Amperometric Titrations*

| Reagent | Reaction Product | Type Electrode* | Substance Determined |
|---|---|---|---|
| $K_2CrO_4$ | Precipitate | DME | $Pb^{2+}$, $Ba^{2+}$ |
| $Pb(NO_3)_2$ | Precipitate | DME | $SO_4^{2-}$, $MoO_4^{2-}$, $F^-$, $Cl^-$ |
| 8-Hydroxyquinoline | Precipitate | DME | $Mg^{2+}$, $Zn^{2+}$, $Cu^{2+}$, $Cd^{2+}$, $Al^{3+}$, $Bi^{3+}$, $Fe^{3+}$ |
| Cupferron | Precipitate | DME | $Cu^{2+}$, $Fe^{3+}$ |
| Dimethylglyoxime | Precipitate | DME | $Ni^{2+}$ |
| $\alpha$-Nitroso-$\beta$-naphthol | Precipitate | DME | $Co^{2+}$, $Cu^{2+}$, $Pd^{2+}$ |
| $K_4Fe(CN)_6$ | Precipitate | DME | $Zn^{2+}$ |
| $AgNO_3$ | Precipitate | RP | $Cl^-$, $Br^-$, $I^-$, $CN^-$, RSH |
| EDTA | Complex | DME | $Bi^{3+}$, $Cd^{2+}$, $Cu^{2+}$, $Ca^{2+}$, and so on |
| $KBrO_3$, KBr | Substitution, addition, or oxidation | RP | Certain phenols, aromatic amines, olefins; $N_2H_4$, As(III), Sb(III) |

* DME = dropping mercury electrode; RP = rotating platinum electrode.

came to appreciate the potentialities of the method.[16] The name *dead-stop end point* was used to describe the technique, and the term is still occasionally encountered.

**Apparatus** A principal advantage to an amperometric titration with two microelectrodes is the simplicity of the equipment. No reference electrode is required, and the only instrumentation needed, beyond the identical microelectrodes, is a simple voltage divider powered by a dry cell and a galvanometer or microammeter for current detection.

**Typical Applications** Twin silver microelectrodes can be used to detect the end point for many titrations that involve silver nitrate as titrant (see Chapter 7, Table 7 – 1). Consider, for example, what occurs when 0.1 V is applied between two such electrodes immersed in a solution that is being titrated to determine its bromide concentration. No current will be observed short of the equivalence point, owing to the lack of any easily reduced species in the solution; that is, complete cathodic polarization prevents the flow of electricity. Note that the anode is not polarized because the reaction

$$Ag(s) \rightarrow Ag^+ + e$$

could occur if a suitable cathodic reactant existed.

Beyond the equivalence point, depolarization of the cathode takes place because silver ions are now in excess and can be reduced at the electrode:

$$Ag^+ + e \rightarrow Ag(s)$$

A current therefore develops as a result of these two half-reactions. In common with other amperometric methods, the magnitude of the current will be directly proportional to the concentration of excess silver ion, and the titration curve will resemble that shown in Figure 14 – 16b.

A pair of platinum microelectrodes can be used to obtain end points for oxidation-reduction titrations. For example, for the titration of iron(II) with cerium(IV), the reaction is

$$Fe^{2+} + Ce^{4+} \rightarrow Fe^{3+} + Ce^{3+}$$

At the outset, the cathode will be polarized when a small potential is applied to the electrode because the system contains no easily reduced species [the anode is not polarized since oxidation of iron(II) could occur there]. The first addition of cerium(IV) causes the formation of iron(III), which depolarizes the cathode. A current then develops in the cell, the magnitude of which is dependent upon the concentration of iron(III). With further additions of reagent, this current will increase and attain a maximum value when the concentrations of iron(II)

[16] C. W. Foulk and A. T. Bawden, *J. Am. Chem. Soc.,* **1926,** *48,* 2045. For an excellent analysis of this type of end point, see J. J. Lingane, *Electroanalytical Chemistry,* 2nd ed., pp. 280–294. New York: Interscience Publishers, 1958.

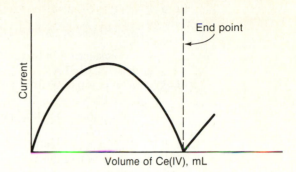

**Figure 14–18** Amperometric titration with twin microelectrodes: curve for titration of Fe(II) with Ce(IV).

and iron(III) are equal. The current then decreases as the supply of iron(II) becomes depleted (that is, as the anode starts to become polarized). The cell becomes completely polarized at the equivalence point. Although the solution contains a reducible species ($Fe^{3+}$) and an oxidizable species ($Ce^{3+}$), a potential of about 0.7 V is required to cause the cell reaction

$$Fe^{3+} + Ce^{3+} \rightarrow Fe^{2+} + Ce^{4+}$$

to take place. Thus, the cell is completely polarized with an applied potential of only 0.1 V, and the current is zero.

With the further addition of cerium(IV), both the cathode and the anode are depolarized as a consequence of the half-reactions

$$Ce^{4+} + e \rightarrow Ce^{3+} \qquad \text{cathode}$$

$$Ce^{3+} \rightarrow Ce^{4+} + e \qquad \text{anode}$$

Now, the current becomes dependent upon the concentration of excess cerium(IV). Figure 14–18 depicts the curve for this titration.

## PROBLEMS

**\*14–1.** Calculate the initial potential needed for a current of 0.078 A in the cell

$$Co|Co^{2+}(6.40 \times 10^{-2}\ M)\|Zn^{2+}(3.75 \times 10^{-3}\ M)|Zn$$

if this cell has a resistance of 5.00 $\Omega$.

**14–2.** The cell

$$Sn|Sn^{2+}(8.22 \times 10^{-4}\ M)\|Cd^{2+}(7.50 \times 10^{-2}\ M)|Cd$$

has a resistance of 3.95 $\Omega$. Calculate the initial potential that will be needed for a current of 0.072 A in this cell.

**\*14–3.** The cell

$$Cu|Cu^{2+}(7.50 \times 10^{-3}\ M)\|Ag^+(1.26 \times 10^{-2}\ M)|Ag$$

has a resistance of 4.25 $\Omega$. Calculate the initial potential if this cell is to be operated with a current of 0.0280 A.

**14–4.** The cell

$$Zn|Zn^{2+}(3.60 \times 10^{-3}\ M)\|Sn^{4+}(1.00 \times 10^{-3}\ M),\ Sn^{2+}(2.85 \times 10^{-2}\ M)|Pt$$

has a resistance of 4.84 $\Omega$. Calculate the initial potential for this cell if the current is 0.0165 A.

**\*14–5.** A solution is 0.0800 $M$ in $Pb^{2+}$ and 0.0600 $M$ in $Cd^{2+}$. Calculate
  (a) the $Pb^{2+}$ concentration at the outset of deposition of Cd.
  (b) the cathode potential needed to lower the $Pb^{2+}$ concentration to $1.00 \times 10^{-5}\ M$.

**14–6.** A solution is originally 0.0963 $F$ in $Cu(NO_3)_2$ and 0.120 $F$ in $AgNO_3$.
  (a) What will be the $Ag^+$ concentration at the onset of deposition by Cu from this solution?
  (b) What cathode potential will be needed to lower the $Ag^+$ concentration to $1.00 \times 10^{-6}\ M$?

**\*14–7.** Silver is to be deposited from a solution that is 0.150 $F$ in $Ag(CN)_2^-$, 0.320 $F$ in KCN, and buffered to a pH of 10.00. Oxygen is evolved at the anode at a partial pressure of 1.00 atm. The cell has a resistance of 2.90 $\Omega$; the temperature is 25°C. Calculate
  (a) the theoretical potential needed to initiate deposition of silver from this solution.
  (b) the $IR$ drop for a current of 0.12 A.
  (c) the initial potential, given that the overvoltage of $O_2$ is 0.80 V.
  (d) the cell potential when $[Ag(CN)_2^-]$ is $1.00 \times 10^{-5}\ M$, assuming no changes in $IR$ and the oxygen overvoltage.

**14–8.** Copper is to be deposited from a solution that is 0.200 $F$ in Cu(II) and is buffered to a pH of 4.00. Oxygen is evolved from the anode at a pressure of 740 torr. The cell has a resistance of 3.60 $\Omega$; $T = 25$°C. Calculate
  (a) the theoretical potential needed to initiate the deposition of Cu.
  (b) the $IR$ drop associated with a current of 0.10 A in this cell.
  (c) the initial potential, given that the overvoltage of $O_2$ is 0.88 V.
  (d) the potential of the cell when $[Cu^{2+}]$ is $5.00 \times 10^{-4}\ M$, all other variables remaining unchanged.

**\*14–9.** A solution is buffered to pH 6.00 and is $8.00 \times 10^{-3}\ F$ with respect to $Pb^{2+}$ and $1.20 \times 10^{-2}\ F$ with respect to $Co^{2+}$. Oxygen is evolved from the anode at a pressure of 765 torr; the overvoltage is 0.70 V. The cell has a resistance of 4.00 $\Omega$; the temperature is 25°C.
  (a) Which ion will be reduced first?
  (b) Estimate the potential needed to initiate deposition in this cell with a current of 0.250 A.
  (c) What will be the concentration of the more reducible cation at the start of deposition by the less reducible species?

**14–10.** A solution is 0.0350 $F$ with respect to Ni(II), 0.0254 $F$ with respect to Fe(II), and has been buffered to pH 4.00. Oxygen is evolved at the anode at a partial pressure of 1.05 atm; the overvoltage is 0.68 V. The resistance of the cell is 3.60 $\Omega$ and the temperature is 25°C.
  (a) Which ion will be reduced first?
  (b) What potential is needed to initiate deposition of the more reducible species with a current of 0.18 A?
  (c) What will be the concentration of the more readily reduced species at the onset of deposition by the other?

**\*14-11.** Calculate the cathode potential (vs. SCE) needed to lower the concentration of Cu(II) to $1.00 \times 10^{-5}$ $F$ in a solution that is
   (a) 0.900 $F$ in $HClO_4$.
   (b) 0.900 $F$ in $Cl^-$, in which $CuCl_3^-$ predominates. The formal potential for the process

$$CuCl_3^- + 2e \rightleftharpoons Cu(s) + 3Cl^-$$

   is 0.178 V under these conditions.

**14-12.** Calculate the cathode potential (vs. SCE) needed to lower the concentration of Hg(II) to $1.00 \times 10^{-6}$ $F$ in
   (a) an aqueous solution.
   (b) a solution that was originally $5.00 \times 10^{-3}$ $F$ in $Hg(SCN)_2$ and $4.00 \times 10^{-2}$ $F$ in $SCN^-$. For the equilibrium

$$Hg^{2+} + 2SCN^- \rightleftharpoons Hg(SCN)_2 \qquad K_f = 1.8 \times 10^7$$

   (c) a solution buffered to a pH of 4.00 in which the formal excess concentration of EDTA-containing species is $1.00 \times 10^{-3}$ $F$.

**\*14-13.** Electrodeposition is to be used to separate the cations in a solution that is $8.00 \times 10^{-3}$ $F$ in $Cr^{3+}$ and $1.20 \times 10^{-2}$ $F$ in $Cd^{2+}$.
   (a) Which cation will be deposited first?
   (b) Using $1.00 \times 10^{-6}$ $F$ as the criterion for quantitative removal, calculate the range of potentials (vs. SCE) within which the cathode potential should be maintained.

**14-14.** Control of the cathode potential is being considered as a way to separate the cations in a solution that is $3.60 \times 10^{-2}$ $F$ in Pb(II) and $9.48 \times 10^{-3}$ $F$ in Tl(I).
   (a) Determine whether this separation is feasible; use $1.00 \times 10^{-6}$ $F$ as the criterion for quantitative removal.
   (b) If this separation is feasible, calculate the range (vs. SCE) within which the cathode potential should be maintained.

**\*14-15.** Electrodeposition is proposed as the means of separating the cations in a solution that is 0.0150 $F$ with respect to Sn(II), 0.180 $F$ with respect to $BiCl_4^-$, and 0.100 $F$ with respect to KCl.
   (a) Using $1.00 \times 10^{-6}$ $M$ as the criterion for quantitative removal, determine whether or not this separation can be achieved through controlled cathode potential analysis.
   (b) If this separation is feasible, establish the range (vs. SCE) within which the cathode potential should be maintained.

**14-16.** Calculate the time needed for a constant current of 1.05 A to deposit 0.251 g of Pb(II) as
   (a) Pb on a cathode.
   (b) $PbO_2$ on an anode.

**\*14-17.** Calculate the time needed for a constant current of 0.88 A to deposit 0.360 g of
   (a) Tl(I) as the element on a cathode.
   (b) Tl(I) as $Tl_2O_3$ on an anode.
   (c) Tl(III) as the element on a cathode.

**14-18.** Calculate the time needed for a constant current of 0.460 A to deposit $Br^-$ (as AgBr on a silver anode) in a 0.1250-g sample of
   (a) NaBr.
   (b) $MgBr_2$.
   (c) $AlBr_3$.

**\*14–19.** Calculate the weight of deposit produced by the electrolysis described in Problem 14–18.

**14–20.** A 0.1684-g sample was dissolved in acid, following which Cu(II) was quantitatively reduced on a Pt cathode. Calculate the percentage of Cu in the sample if 13.77 C were required for quantitative deposition.

**\*14–21.** The arsenic in a 6.16-g sample of an ant-control preparation was converted to the $+3$ state and then oxidized with electrolytically generated $I_2$ in a slightly alkaline medium. Reaction:

$$HAsO_3^{2-} + I_2 + 2HCO_3^- \rightarrow HAsO_4^{2-} + 2I^- + 2CO_2 + H_2O$$

Calculate the percentage of As in the sample if 76.4 C were needed to complete the titration.

**14–22.** A 50.00-mL sample of well water was treated with an unmeasured excess of $HgNH_3Y^{2-}$. Express the hardness of the water (as ppm $CaCO_3$, if the EDTA needed for the titration was generated at a Hg cathode in 10.57 min by a constant current of 40.1 mA. See Section 14C–4 for the reaction.

**\*14–23.** An unmeasured excess of KI was added to 100.0 mL of brackish water. Generation of the $I_2$ needed to react with the $H_2S$ in the sample required a constant current of 61.8 mA for 4.05 min. Reaction:

$$H_2S + I_2 \rightarrow S(s) + 2I^-$$

Express the result in terms of ppm $H_2S$.

**14–24.** A potential of $-1.00$ V (vs. SCE) will cause the reduction of carbon tetrachloride ($CCl_4$) to chloroform ($CHCl_3$) at the surface of a Hg electrode. Reaction:

$$2CCl_4 + 2H^+ + 2e + 2Hg(l) \rightarrow 2CHCl_3 + Hg_2Cl_2(s)$$

A 0.1164-g sample containing $CCl_4$ was dissolved in methanol. A coulometer in series with the working cell indicated that 33.71 C were needed to complete this electrolysis. Calculate the percentage of $CCl_4$ in the sample.

**\*14–25.** Chloroform is reduced to $CH_4$ at a mercury cathode with a potential of $-1.80$ V (vs. SCE). Reaction:

$$2CHCl_3 + 6H^+ + 6e + 6Hg(l) \rightarrow 2CH_4(g) + 3Hg_2Cl_2(s)$$

Calculate the percentage of $CHCl_3$ in a 0.1073-g sample if 186.5 C were needed for this reaction.

**14–26.** Samples consisting of $CCl_4$, $CHCl_3$, and inert materials were electrolyzed at the surface of a Hg cathode, first at $-1.00$ V (vs. SCE) until reaction was complete, and then at $-1.80$ V (see Problems 14–24 and 14–25 for reactions). Calculate the percentages of the two components, based upon the following data:

|  | Weight of Sample, g | Quantity of Electricity, C | |
|---|---|---|---|
|  |  | At $-1.00$ V | At $-1.80$ V |
| (a) | 0.1003 | 11.29 | 108.40 |
| (b) | 0.0923 | 16.84 | 140.87 |
| (c) | 0.0774 | 10.06 | 139.23 |
| (d) | 0.1061 | 31.72 | 242.69 |

**14–27.** Mixtures containing only $CCl_4$ and $CHCl_3$ were dissolved in methanol and subjected to electrolysis at a Hg cathode at $-1.80$ V (vs. SCE); see Problems 14–24 and 14–25 for reactions. Calculate the percent composition of these samples, based on the accompanying data:

| | Weight of Sample, g | Quantity of Electricity, C |
|---|---|---|
| *(a) | 0.0772 | 190.91 |
| (b) | 0.0799 | 195.24 |
| *(c) | 0.0663 | 162.93 |
| (d) | 0.0699 | 173.76 |

**14–28.** Quinone can be reduced with an excess of electrolytically generated Sn(II):

The polarity of the working electrode is then reversed, and the excess Sn(II) is oxidized with $Br^-$ generated in a coulometric titration:

$$Sn^{2+} + Br_2 \rightarrow Sn^{4+} + 2Br^-$$

Appropriate amounts of $SnCl_4$ and KBr were introduced to a 50.00-mL aliquot of sample. Calculate the milligrams of quinone in each milliliter of sample from the accompanying data:

| Working Electrode Functioning as | Generation Time, sec, with a Constant Current of 1.15 mA |
|---|---|
| Cathode | 284.9 |
| Anode | 13.8 |

**\*14–29.** As the terms apply to a polarographic wave, make a distinction between
  (a) the decomposition potential and the half-wave potential.
  (b) the limiting current and the diffusion current.
**14–30.** The accompanying data refer to the polarographic analysis of cadmium(II). Supply the missing numbers.

| [Cd$^{2+}$] | Limiting Current, $\mu$A | Diffusion Current, $\mu$A |
|---|---|---|
| 0.00 | 3.1 | 0.0 |
| 4.00 × 10$^{-4}$ | | 5.3 |
| 1.20 × 10$^{-3}$ | 18.7 | |
| 3.60 × 10$^{-3}$ | | 46.7 |
| 5.00 × 10$^{-3}$ | 68.5 | |
| 6.50 × 10$^{-3}$ | 87.8 | |

*14–31. A linear relationship between concentration and diffusion current requires that

$$i_d = kC$$

Evaluate $i_d/C$ for the data in Problem 14–30. Calculate the standard deviation for this quantity.

14–32. Calculate the concentration of Cd(II) in solutions that were analyzed by the polarographic method in Problem 14–30.

| Solution | Limiting Current, $\mu$A | Solution | Limiting Current, $\mu$A |
|---|---|---|---|
| *(a) | 71.7 | (b) | 39.4 |
| *(c) | 12.6 | (d) | 66.2 |
| *(e) | 48.1 | (f) | 30.0 |

14–33. A standard addition method for the polarographic determination of Pb$^{2+}$ involved addition of 5.00 mL of 8.00 × 10$^{-3}$ Pb$^{2+}$ to one of two 25.00-mL of the sample. Supporting electrolyte and water were then added to each aliquot to give a final volume of 50.00 mL. Calculate the molar Pb$^{2+}$ concentration based upon the following data obtained with the method:

| Sample | Diffusion Current, $i_d$ for Sample | for Sample plus Standard Addition |
|---|---|---|
| *(a) | 78.3 | 88.7 |
| (b) | 13.5 | 23.9 |
| *(c) | 41.0 | 51.4 |
| (d) | 54.6 | 65.0 |

# Chapter 15
# Absorption Spectroscopy

Spectroscopic methods of analysis are based upon measurement of the electromagnetic radiation produced or absorbed by matter. *Emission methods* make use of the radiation given off when an analyte is excited by thermal, electrical, or radiant energy. *Absorption methods,* on the other hand, are based upon the decrease in power (or *attentuation*) of electromagnetic radiation as a consequence of its interaction with and partial absorption by the analyte. Spectroscopic methods rank among the most widespread and powerful tools available to the scientist for the acquisition of qualitative as well as quantitative information.

Spectroscopic methods are conveniently classified according to the region of the electromagnetic spectrum that is involved; principal among these are X-ray, ultraviolet, visible, infrared, microwave, and radio-frequency regions. This chapter is principally concerned with the absorption of visible and ultraviolet radiation by molecules.

## 15A PROPERTIES OF ELECTROMAGNETIC RADIATION

Electromagnetic radiation is a form of energy that is transmitted through space at enormous velocities. Many of the properties of electromagnetic radiation are conveniently described by means of a classical wave model, which employs such parameters as wavelength, frequency, velocity, and amplitude. In contrast to other wave phenomena, such as sound, electromagnetic radiation requires no supporting medium for its transmission; thus, it readily passes through a vacuum.

The wave model fails to account for phenomena associated with the absorption and emission of radiant energy; for these processes, it is necessary to view electromagnetic radiation as a stream of discrete particles or wave packets of energy called *photons* or *quanta.* The energy of a photon is proportional to the frequency of the radiation. These dual views of radiation as particles and as waves are not mutually exclusive but, rather, complementary. Indeed, the duality applies to the behavior of streams of electrons as well as other elementary particles such as protons and is completely rationalized by wave mechanics.

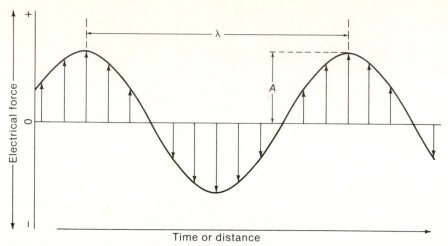

**Figure 15 – 1**   Schematic representation of a beam of monochromatic radiation with wavelength $\lambda$ and maximum amplitude A. The arrows represent the electrical vector of the radiation.

### 15A – 1 Wave Properties of Electromagnetic Radiation

Many of the phenomena associated with electromagnetic radiation, such as transmission, reflection, refraction, scattering, and interference, are best accounted for by viewing a beam of radiation as a sinusoidal wave. In this treatment, radiation is considered to be an electrical force field that oscillates periodically from zero to some maximum amplitude $A$; these oscillations occur at right angles to the direction of propagation of the wave. An electrical field can be treated as a vector quantity, and radiation can be represented by a plot of this vector as a function of time at some fixed point in space or as a function of distance at a fixed time. These coordinates are used in Figure 15 – 1, which is a two-dimensional representation of monochromatic (that is, single-wavelength) radiation. Viewed end-on, the vectors would form a continuous circle having a radius that fluctuates from zero to the maximum amplitude $A$.

**Wave Parameters**   The time required for the passage of successive maxima of a wave through a fixed point is called the *period, p,* of the radiation. The *frequency, v,* is the number of oscillations of the field that occurs per second[1] and is equal to $1/p$. *It is important to realize that the frequency is an invariant quantity that is determined by the source of the radiation.* In contrast, the rate, $v_i$, at which a wave front moves through a medium is dependent upon *both* the density of the medium and the frequency; the subscript $i$ is supplied to indicate this frequency dependence.

The *wavelength, $\lambda_i$,* is the linear distance between successive maxima or

---

[1] The *hertz,* Hz, is the unit of frequency, and is equal to one cycle per second.

minima of a wave. The units commonly used to express wavelength vary with the spectral region in question. The *angstrom unit,* Å ($10^{-10}$ m), is convenient for X-ray and short ultraviolet radiation. The *nanometer,* nm ($10^{-9}$ m), is more convenient for the ultraviolet and visible regions, while the *micrometer, $\mu$m* ($10^{-6}$ m), is useful in describing infrared radiation. (In the older literature, these two terms were called the millimicron, m$\mu$, and the micron, $\mu$, respectively, a usage that is no longer acceptable.)

The product of the frequency (in units of cycles per second) and the wavelength (in centimeters) is the *velocity* of the wave front; that is,

$$v_i = \nu\lambda_i \qquad (15-1)$$

The velocity of electromagnetic radiation becomes independent of frequency in a vacuum and is at its maximum. This velocity, $c$, is equal to $2.99792 \times 10^{10}$ cm/sec, which is usually rounded to $3.00 \times 10^{10}$. Thus, in a vacuum,

$$c = \nu\lambda = 3.00 \times 10^{10} \text{ cm/sec} \qquad (15-2)$$

The rate of propagation in any other medium is less because of interactions between the radiation and bound electrons in the atoms of molecules of the medium. Since the frequency is invariant and fixed by the source, *the wavelength must decrease as radiation passes into matter from a vacuum.* The velocity of electromagnetic radiation in air is only about 0.03% less than $c$, and for most purposes Equation 15–2 is applicable to air as well as to a vacuum.

The *wavenumber, $\sigma$,* is yet another way of describing electromagnetic radiation. It is defined as the number of waves per centimeter and is equal to $1/\lambda$. By definition, $\sigma$ has units of cm$^{-1}$.

The *power, P,* is the energy of a beam that reaches a given area per second; the *intensity* is the power per unit solid angle. Both quantities are related to the amplitude of the radiation (see Figure 15–1). Although it is not strictly correct to do so, power and intensity are frequently used interchangeably.

### 15A–2 Particulate Properties of Electromagnetic Radiation

Some interactions with matter require that electromagnetic radiation be treated as discrete packets of energy, called photons or quanta. The energy of a photon depends upon the frequency of the radiation and is given by

$$E = h\nu \qquad (15-3)$$

where $h$ is the Planck constant ($6.63 \times 10^{-27}$ erg sec). Equation 15–3 can be expressed in terms of wavelength and wavenumber:

$$E = \frac{hc}{\lambda} = hc\sigma \qquad (15-4)$$

Note that the wavenumber, in common with frequency, is directly proportional to energy.

### 15A–3 *The Electromagnetic Spectrum*

The electromagnetic spectrum encompasses an enormous range of wavelengths (or energies). An X-ray photon ($\lambda \sim 10^{-10}$ m), for example, is approximately 10,000 times as energetic as one that is emitted by an incandescent tungsten wire ($\lambda \sim 10^{-6}$ m) and $10^{11}$ times as energetic as a photon in the radio-frequency range ($\lambda \sim 10^{5}$ m).

Figure 15–2 depicts the regions of the electromagnetic spectrum that are used in chemical analysis; also shown are the molecular and atomic transitions responsible for absorption or emission in each region. Note that both the frequency scale and the wavelength scale are logarithmic. Note also that the portion of the spectrum to which the human eye responds (the *visible spectrum*) is remarkably small. Such diverse radiations as gamma rays and radio waves differ from visible light only in the matter of frequency, and hence energy.

## 15B THE GENERATION OF ELECTROMAGNETIC RADIATION

The atoms, ions, or molecules in a sample of matter possess a limited number of discrete and quantized energy levels, the lowest of which is called the *ground state. Excitation* to one or more higher levels is accomplished by supplying

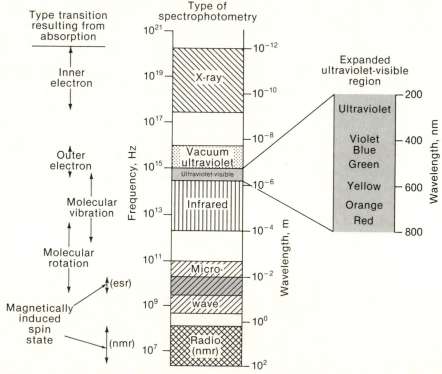

**Figure 15–2**   Analytically useful portions of the electromagnetic spectrum.

energy to the sample. The lifetime of an excited species is brief, with *relaxation* to a lower energy level or to the ground state being accompanied by loss of the excitation energy as heat, fluorescent or phosphorescent radiation, or perhaps both. The energy of an emitted photon is equal to the energy difference between the excited state and the lower level.

An *emission spectrum* is a plot of the radiant energy given off by an excited species as a function of wavelength (or of frequency). The type of spectrum that is observed depends upon the composition and the state of the species responsible for the emission. Excited *atoms* in the gaseous state are sufficiently isolated to behave independently of one another and yield a discontinuous *line spectrum* that consists of relatively few wavelengths. Excited *molecules* exist in a multitude of excited states with energies that differ but slightly from one another and thus emit *band spectra,* which consist of many closely spaced lines. Instruments with very high resolving power are needed to resolve band spectra into their individual lines. Finally, the ions or atoms in an incandescent solid are incapable of independent behavior owing to the small distances that separate them. The result is a *continuous spectrum* that consists of an infinite number of lines that cannot be resolved.

Line and band emission spectra are useful for the identification and determination of emitting species. Continuous spectra are used as sources of electromagnetic radiation for absorption analysis.

## 15C THE ABSORPTION OF ELECTROMAGNETIC RADIATION

*Absorption* refers to the process by which a species in a transparent medium selectively removes certain frequencies of electromagnetic radiation. The absorbed photon converts the species M to an excited state, M*, as shown by the equation

$$M + h\nu \rightarrow M^*$$

After a brief period ($\sim 10^{-8}$ to $10^{-9}$ sec) the excitation energy is lost, ordinarily as heat, as the species relaxes to its former state; that is,

$$M^* \rightarrow M + heat$$

Relaxation can also occur through photochemical decomposition of M* to other products or by emission of fluorescent or phosphorescent radiation. Irrespective of the deactivation route, however, it is important to appreciate that the lifetime of M* is so short that its concentration at any instant is negligible. Moreover, the thermal energy given off during relaxation is ordinarily so small as to be undetectable. Absorption methods thus possess the considerable advantage of creating little or no disturbance to the system under study.

### 15C-1 Quantitative Absorption Measurements: Beer's Law

The principles that govern the absorption of radiation apply to all regions throughout the electromagnetic spectrum from gamma rays to radio frequen-

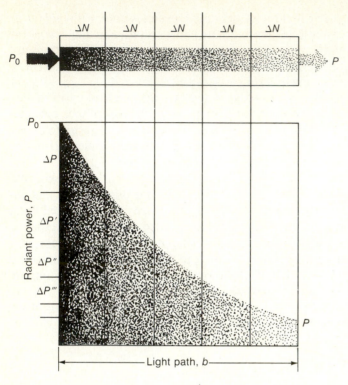

**Figure 15–3** Attenuation of monochromatic radiation as a result of absorption.

cies. Absorption is measured by determining the decrease in power experienced by a beam of radiation as the result of interactions with absorbers in the light path.

The upper portion of Figure 15–3 illustrates how a beam of monochromatic (i.e., a single-wavelength) radiation with incident power $P_0$ is attenuated as it passes through a solution that contains $c$ moles of absorbers per liter. The lower portion of this figure shows that interaction with the same number of absorbers, $\Delta N$, results in an ever-diminishing change, $\Delta P$, in the radiant power, $P$, of the beam; that is,

$$-\Delta P = kP \, \Delta N$$

where $k$ is a constant. A minus sign is supplied to indicate that the change in $P$ represents a diminution in radiant power. From this relationship it is possible to derive *Beer's law*,[2]

$$\log \frac{P_0}{P} = \epsilon bc = A \tag{15–5}$$

The term $\epsilon$ in this equation is a proportionality constant called the *molar*

---

[2] See D. J. Swinehart, *J. Chem. Educ.,* **1972,** *39,* 333.

*absorptivity* and implies that the path length, $b$, is expressed in centimeters and $c$ has the units of moles per liter (the proportionality constant is numerically different and is supplied with a different symbol when $b$ and $c$ have other units). The logarithm (to the base 10) of the ratio between the incident power and the transmitted power is called the *absorbance,* and is given the symbol $A$. The absorbance will clearly increase with increases in either $b$ or $c$.

Beer's law applies equally well to solutions containing more than one absorbing species, provided there is no interaction among such species. Thus, for a multicomponent system, the relationship becomes

$$A_{\text{total}} = A_1 + A_2 + \cdots + A_n$$
$$= \epsilon_1 b c_1 + \epsilon_2 b c_2 + \cdots + \epsilon_n b c_n \tag{15-6}$$

where the subscripts refer to absorbing species 1, 2, $\cdots$, $n$.

**Terminology Associated with Absorption Measurements**   In recent years the attempt has been made to develop a standard nomenclature for the various terms related to the absorption of radiation. Table 15-1 lists the recommendations of the American Society for Testing Materials, along with alternative names and symbols that are encountered in the older literature. An important term in this tabulation is the *transmittance, T,* which is defined as

$$T = \frac{P}{P_0}$$

**Table 15-1**   **Terms and Symbols Used in Absorption Analysis**

| Term and Symbol* | Definition | Alternative Name and Symbol |
|---|---|---|
| Radiant power, $P$, $P_0$ | Energy of radiation reaching a given area of a detector per second | Radiation intensity, $I$, $I_0$ |
| Absorbance, $A$ | $\log \frac{P_0}{P}$ | Optical density, $D$; extinction, $E$ |
| Transmittance, $T$ | $\frac{P}{P_0}$ | Transmission, $T$ |
| Path length of radiation, in cm, $b$ | — | $l, d$ |
| Molar absorptivity,† $\epsilon$ | $\frac{A}{bc}$ | Molar extinction coefficient |
| Absorptivity,‡ $a$ | $\frac{A}{bc}$ | Extinction coefficient, $k$ |

* Terminology recommended by the American Society for Testing Materials and the American Chemical Society (*Anal. Chem.,* **1952**, *24,* 1349; **1985,** *57,* 397-398).
† $c$ is expressed in units of moles per liter.
‡ $c$ may be expressed in units other than moles per liter; $b$ may be expressed in other units of length.

The transmittance, which is the fraction of incident radiation that is transmitted by the solution, is frequently expressed as a percentage. Transmittance is related to absorbance as follows:

$$A = -\log T = \log \frac{1}{T}$$

---

**Example 15–1**    A $7.50 \times 10^{-5}$ $F$ solution of potassium permanganate has an absorbance of 0.439 when measured in a 1.00-cm cell at 525 nm. Calculate the percent transmittance of this solution.

$$0.439 = \log \frac{1}{T}$$

$$\frac{1}{T} = 2.748$$

$$T = \frac{1}{2.748} = 0.364 \text{ or } 36.4\%$$

---

**Measurement of Absorption**    As given by Equation 15–5, Beer's law is not directly applicable to chemical analysis. Neither $P$ nor $P_0$, as defined, can be conveniently measured in the laboratory because the solution to be studied must be held in some sort of container. Interaction between the radiation and the walls is inevitable, with losses in power occurring at each interface as a result of reflection (and possibly absorption as well). Reflection losses are substantial: about 4% of a beam of visible radiation is reflected upon vertical passage across an air-to-glass or glass-to-air interface. In addition to such losses, scattering by large molecules or inhomogeneities in the solvent may cause a diminution in the power of the beam as it traverses the solution.

To compensate for these effects, the power of the beam that has been transmitted through the analyte solution is compared with that which traverses an identical cell containing the solvent for the sample. An experimental absorbance that closely approximates the true absorbance for the solution is thus obtained; that is,

$$A = \log \frac{P_0}{P} \cong \log \frac{P_{\text{solvent}}}{P_{\text{solution}}} \tag{15–7}$$

The terms $P_0$ and $P$, when used henceforth, refer to the power of radiation after it has passed through cells containing the solvent and the analyte, respectively.

### 15C–2 Limitations to the Applicability of Beer's Law

The linear relationship between absorbance and path length through a solution (where $c$ is constant) is a generalization for which there are few exceptions. On the other hand, deviations from the direct proportionality between absorbance and concentration (where $b$ is constant) are frequently observed. Some of these deviations are fundamental and represent real limitations to the law. Others occur as a consequence of the manner in which the absorbance measurements are made (*instrumental deviations*) or as a result of chemical changes associated with concentration changes (*chemical deviations*).

**Real Limitations to Beer's Law**    Beer's law is strictly applicable to solutions in which concentration-dependent interactions among absorbing molecules or ions are at a minimum. Such interactions, which usually begin to appear at concentrations greater than 0.01 $M$, alter molar absorptivities and thus lead to a nonlinear relationship between absorbance and concentration.

Deviations also arise because the molar absorptivity of a substance is dependent upon the refractive index of the solution, which in turn becomes dependent upon the concentration of the analyte at concentrations greater than about 0.01 $M$. These refractive index changes then lead to concentration-dependent variations in the molar absorptivity of the analyte and departures from Beer's law.

**Chemical Deviations**    Apparent deviations from Beer's law appear when the absorbing species undergoes association, dissociation, or reaction with the solvent to give products with absorbing characteristics that differ from those of the analyte. As shown in Example 15–2, the extent of such departures can be predicted from the molar absorptivities of the absorbing species and the equilibrium constants for the reactions that are involved.

**Example 15–2**    Consider the dissociation equilibrium for the acid-base indicator HIn:

$$HIn + H_2O \rightleftharpoons H_3O^+ + In^-$$

for which

$$K_{HIn} = \frac{[H_3O^+][In^-]}{[HIn]} = 1.42 \times 10^{-5}$$

The individual absorbances of HIn and In$^-$ are linear with respect to concentration at 430 and at 570 nm, as demonstrated by measurements in 0.10 $F$ HCl (where, for all practical purposes, the indicator exists solely as HIn) and in 0.10 $F$ NaOH (where In$^-$ is the predominant species). Molar absorptivities are

|  | $\epsilon_{430}$ | $\epsilon_{570}$ |
|---|---|---|
| HIn (HCl soln) | $6.30 \times 10^2$ | $7.12 \times 10^3$ |
| In$^-$ (NaOH soln) | $2.06 \times 10^4$ | $9.60 \times 10^2$ |

Calculate the absorbance of a $2.00 \times 10^{-5}$ $F$ solution of the indicator at 430 and 570 nm in a 1.00-cm cell.

The molar concentrations of HIn and In$^-$ in an unbuffered solution can be calculated through the use of Equation 8–8 (p. 171). Here, however, we are interested in [In$^-$] rather than [H$_3$O$^+$]; that is,

$$[In^-] = [H_3O^+]$$

$$[HIn] = F_{HIn} - [In^-]$$

Substitution of these quantities into the expression for $K_{HIn}$ and rearrangement yields the quadratic equation

$$[In^-]^2 + K_{HIn}[In^-] - K_{HIn}F_{HIn} = 0$$

Thus, when $F_{HIn}$ is $2.00 \times 10^{-5}$ and the measurements are made in 1.00-cm cells,

$$[In^-]^2 + 1.42 \times 10^{-5}[In^-] - 1.42 \times 10^{-5} \times 2.00 \times 10^{-5} = 0$$

$$[In^-] = \frac{-1.42 \times 10^{-5} + \sqrt{(1.42 \times 10^{-5})^2 - 4(-1.42 \times 10^{-5} \times 2.00 \times 10^{-5})}}{2}$$

$$[In^-] = 1.12 \times 10^{-5}$$

$$[HIn] = 2.00 \times 10^{-5} - 1.12 \times 10^{-5} = 8.8 \times 10^{-6}$$

and

$$A_{430} = (2.06 \times 10^4)(1.00)(1.12 \times 10^{-5}) + (6.30 \times 10^2)(1.00)(8.8 \times 10^{-6})$$
$$= 0.236$$

$$A_{570} = (9.60 \times 10^2)(1.00)(1.12 \times 10^{-5}) + (7.12 \times 10^3)(1.00)(8.8 \times 10^{-6})$$
$$= 0.073$$

Additional data, calculated similarly, are plotted in Figure 15–4. Note that both curves are nonlinear and that the direction of curvature for one is opposite that for the other.

**Instrumental Deviations**    Beer's law is a limiting relationship in the sense that true linearity between absorbance and concentration requires the use of monochromatic radiation.[3] As a practical matter, however, significant deviations from Beer's law appear only when the band of radiation used encompasses a spectral region in which large changes in absorbance occur as a function of wavelength. Figure 15–5 illustrates the conditions in which such deviations are likely and unlikely.

---

[3] See, for example, D. A. Skoog, *Principles of Instrumental Analysis,* 3rd ed., pp. 166–168. Philadelphia: Saunders College Publishing, 1985.

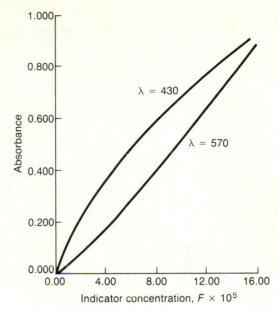

**Figure 15–4** Chemical deviations from Beer's law by unbuffered solutions of the indicator HIn. (See Example 15–2.)

## 15D THE ABSORPTION PROCESS

We have noted that atoms, molecules, and ions have a limited number of discrete, quantized energy levels. For absorption to occur, the energy of the photon must exactly match the energy difference between the ground state of the absorbing species and one of its excited energy levels. These energy differences are unique for each species and result in characteristic spectra that are often useful for qualitative identification and quantitative determination of both organic and inorganic substances.

### 15D–1 Spectral Curves

An *absorption spectrum* is a plot that relates the absorption characteristics of an analyte to the wavelength or frequency of electromagnetic radiation that is being used to investigate it. The abscissa of an absorption spectrum may have

**Figure 15–5** Effect of polychromatic radiation upon the Beer's law relationship. Curve A is linear since ε is nearly constant in the spectral region encompassed by band A. Curve B shows marked deviations from linearity because ε undergoes significant changes within band B.

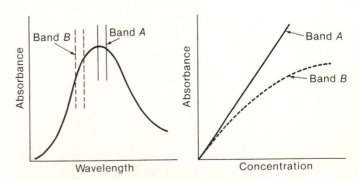

units of wavelength, frequency, or wavenumber, while the ordinate is usually expressed in terms of absorbance, log absorbance, transmittance, or percent transmittance.

The spectra for five different permanganate solutions are shown in Figure 15–6. Note that with absorbance as the ordinate, the greatest difference be-

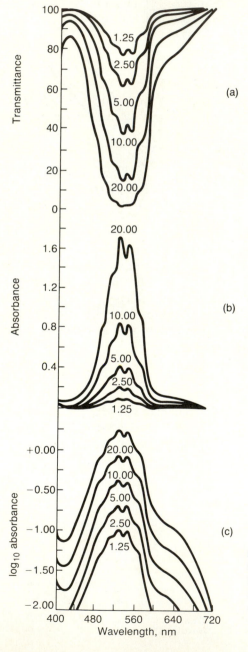

*Figure 15–6* Methods for plotting spectral data. The numbers for the curves indicate the ppm of $KMnO_4$ in the solution; $b = 2.00$ cm. (From M. G. Mellon, *Analytical Absorption Spectroscopy*, pp. 104–106. New York: John Wiley & Sons, 1950. With permission.)

tween curves occurs in the region where the absorbance is high (0.8 to 1.3) and the transmittance is low (<20%). In contrast, transmittance curves show greater differences in the range between 20 and 60%. A plot with log $A$ as ordinate results in a loss of spectral detail but is convenient for comparing solutions that have different concentrations since the curves are displaced equally along the vertical axis.

The typical absorption spectra shown in Figure 15–7 demonstrate the great diversity in the general appearance of such spectra.

## 15D–2 Atomic Absorption Spectra

The atomic spectrum for an element, such as sodium or mercury, is obtained by measuring absorbance as a function of wavelength for a gaseous sample that contains a low concentration of vaporized atoms of the element of interest. For

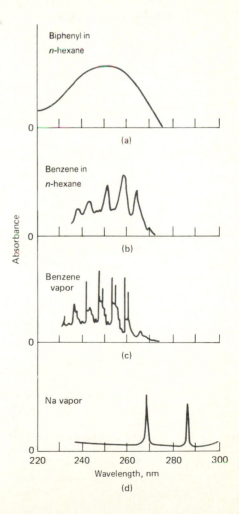

**Figure 15–7** Typical absorption spectra.

most elements, such spectra are relatively simple and consist of a few discrete lines with wavelengths that are characteristic of the element. The simplicity of many atomic spectra is due to the fact that the only possible excitation process involves transitions of electrons from lower to higher energy states, and for many elements the number of such higher excited energy states is limited.[4] For example, sodium vapor exhibits a narrow absorption peak at 589.3 nm, which is due to the transition of the $3s$ ground-state electron to the $3p$ excited state. Several other transitions of this type can also occur and result in a limited number of other absorption lines for sodium. Two of these are shown in Figure 15-7d; the line at about 285 nm results from the transition of the $3s$ electron to the $5p$ state, a process requiring significantly greater energy than the transition responsible for the line at 589.3 nm.

Applications of atomic absorption (and emission) spectra in analytical chemistry are considered in Chapter 16.

## 15D-3 Molecular Absorption

Absorption by polyatomic molecules is a vastly more complex process than is atomic absorption because the number of possible excited energy states is substantially greater. The total energy of a molecule is given by

$$E = E_{\text{electronic}} + E_{\text{vibrational}} + E_{\text{rotational}} \tag{15-8}$$

where $E_{\text{electronic}}$ is the energy associated with the electrons in the various outer orbitals of the molecule, and $E_{\text{vibrational}}$ is the energy of the molecule as a whole due to interatomic vibrations. The last term accounts for the energy associated with rotation of the molecule about its center of gravity.

**Absorption in the Infrared and Microwave Regions**    The three terms on the right in Equation 15-8 are arranged according to decreasing energy, with the average value for each term differing from its neighbor by approximately two orders of magnitude. Microwave radiation, which is less energetic than infrared, will produce pure rotational spectra that are free from electronic and vibrational transitions. Spectroscopic studies of gases in this wavelength region (0.75 to 3.75 nm) provide fundamental information concerning molecular behavior; applications to quantitative analytical problems are limited.

Infrared radiation in the wavelength region between 0.78 and 300 $\mu$m possesses sufficient energy to cause vibrational transitions but is not energetic enough to cause electronic transitions. As shown in Figure 15-8, infrared spectra typically consist of narrow, closely spaced absorption peaks resulting from transitions among the numerous vibrational quantum levels.

---

[4] The spectra for transition metals, while made up of discrete lines, are far from simple because of the large number of possible excited states that result from the unfilled inner orbitals of these elements. Thus, a transition metal spectrum may consist of several thousand individual absorption lines.

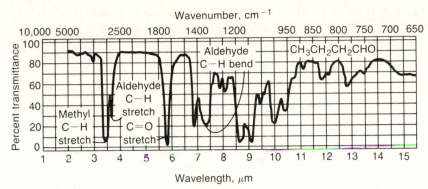

**Figure 15 – 8** Infrared spectrum for *n*-butanal. Note that the ordinate is transmittance rather than absorbance. [From *Catalog of Selected Infrared Spectral Data*, Serial No. 225. Thermodynamics Research Center Data Project. Texas A & M University, College Station, TX 1964 (loose-leaf data sheets extant).]

**Absorption in the Ultraviolet and Visible Regions**   The first term on the right side of Equation 15 – 8 is ordinarily larger than the other two, and energies needed to bring about transitions involving the outer electrons in a molecule correspond to radiation in the ultraviolet and visible regions. In contrast to atomic absorption spectra, molecular spectra are characterized by absorption bands that encompass a substantial wavelength region (see Figure 15 – 7a, b, and c, for example). Numerous rotational and vibrational energy states exist for each electronic state. Thus, for a given value of $E_{electronic}$ in Equation 15 – 8, values for $E$ will exist that differ only slightly due to variations in $E_{vibrational}$ and/or $E_{rotational}$. Often, then, the visible or ultraviolet absorption spectrum for a molecule consists of a band of closely spaced absorption lines such as that shown for benzene vapor in Figure 15 – 7c. Frequently, these lines are so close to one another that they cannot be resolved even with a high-resolution instrument; as a result, the spectra may appear smooth or continuous as in Figure 15 – 7a. Moreover, the individual bands for species in condensed states and in solution tend to broaden to give the type of spectra shown in the upper two curves of Figure 15 – 7.

**Absorption Induced by a Magnetic Field**   When electrons, or the nuclei of certain atoms, are subjected to a strong magnetic field, additional quantized energy levels are produced that owe their origins to magnetic properties possessed by such elementary particles. The difference in energy between the magnetically induced states is small, and transitions associated with them are brought about only by absorption of long wavelengths of radiation. Thus, nuclei absorb radio frequencies in the region from 30 to 500 MHz, while electrons absorb microwaves at about 9500 MHz. Nuclear magnetic resonance (NMR) spectroscopy is concerned with the former region and electron spin resonance (ESR) spectroscopy with the latter.

## 15D – 4 Molecular Absorption of Visible and Ultraviolet Radiation

Absorption measurements in the visible and ultraviolet regions of the spectrum provide qualitative and quantitative information about organic, inorganic, and biochemical molecules.

**Absorption by Organic Compounds**    Absorption of radiation by organic molecules in the wavelength region between 180 and 780 nm results from interactions between photons and those electrons that either participate directly in bond formation (and are thus associated with more than one atom), or are localized about such atoms as oxygen, sulfur, nitrogen, and the halogens.

The wavelength at which an organic molecule absorbs depends upon how tightly its several electrons are bound. The shared electrons in such single bonds as carbon–carbon or carbon–hydrogen are so firmly held that absorption occurs only in a region of the ultraviolet spectrum ($\lambda < 180$ nm) where the components of air also absorb (this region is known as the *vacuum ultraviolet*). Owing to the difficulty of making measurements in this region, such absorption is seldom used for analytical purposes.

Electrons involved in double and triple bonds of organic molecules are not as strongly held and are therefore more easily excited by radiation; thus, species with unsaturated bonds generally exhibit useful absorption peaks. Unsaturated organic functional groups that absorb in the ultraviolet or visible regions are known as *chromophores*. Table 15–2 lists common chromophores and the approximate wavelengths at which they absorb. The data for position and peak intensity can only serve as a rough guide for identification purposes, since both are influenced by solvent effects as well as other structural details of the molecule. Moreover, conjugation between two (or more) chromophores tends to cause shifts in peak maxima to longer wavelengths. Finally, vibrational effects broaden absorption peaks in the ultraviolet and visible regions, which often makes precise determination of a maximum difficult.

The unshared electrons in such elements as sulfur, bromine, and iodine are less strongly held than the shared electrons of a saturated bond. Organic molecules incorporating these elements frequently exhibit useful peaks in the ultraviolet region as a result.

**Absorption by Inorganic Species**    The spectra for most inorganic ions or molecules resemble those of organic compounds, with broad absorption maxima and little fine structure. The ions of the lanthanide and actinide series represent an important exception to this statement. The electrons responsible for absorption by these elements ($4f$ and $5f$, respectively) are shielded from external influences by electrons that occupy orbitals with larger principal quantum numbers. As a result, the bands tend to be narrow and relatively unaffected by the species bonded by the outer electrons.

In general, the ions and complexes of elements in the first two transition series absorb visible radiation in at least one of their oxidation states and are, as

*Table 15 – 2  Absorption Characteristics of Some Common Organic Chromophores*

| Chromophore | Example | Solvent | $\lambda_{max}$, nm | $\epsilon_{max}$ |
|---|---|---|---|---|
| Alkene | $C_6H_{13}CH{=}CH_2$ | n-Heptane | 177 | 13,000 |
| Conjugated alkene | $CH_2{=}CHCH{=}CH_2$ | n-Heptane | 217 | 21,000 |
| Alkyne | $C_5H_{11}C{\equiv}C{-}CH_3$ | n-Heptane | 178 | 10,000 |
| | | | 196 | 2000 |
| | | | 225 | 160 |
| Carbonyl | $CH_3\overset{O}{\overset{\|}{C}}CH_3$ | n-Hexane | 186 | 1000 |
| | | | 280 | 16 |
| | $CH_3\overset{O}{\overset{\|}{C}}H$ | n-Hexane | 180 | Large |
| | | | 293 | 12 |
| Carboxyl | $CH_3\overset{O}{\overset{\|}{C}}OH$ | Ethanol | 204 | 41 |
| Amido | $CH_3\overset{O}{\overset{\|}{C}}NH_2$ | Water | 214 | 60 |
| Azo | $CH_3N{=}NCH_3$ | Ethanol | 339 | 5 |
| Nitro | $CH_3NO_2$ | Isooctane | 280 | 22 |
| Nitroso | $C_4H_9NO$ | Ethyl ether | 300 | 100 |
| | | | 665 | 20 |
| Nitrate | $C_2H_5ONO_2$ | Dioxane | 270 | 12 |
| Aromatic | Benzene | n-Hexane | 204 | 7900 |
| | | | 256 | 200 |

a consequence, colored. Here, absorption involves transitions between filled and unfilled *d*-orbitals with energies that depend upon the ligands bonded to the metal ions. The energy differences between these *d*-orbitals (and thus the position of the corresponding absorption peak) depend upon the position of the element in the periodic table, its oxidation state, and the nature of the ligand bonded to it.

*Charge-transfer absorption* is of particular importance from the analytical standpoint because molar absorptivities of charge-transfer peaks are among the largest observed in spectroscopy ($\epsilon_{max} > 10,000$). These large molar absorptivities impart unusually high sensitivities to methods based on this phenomenon. Many organic and inorganic complexes exhibit this type of absorption and are known as *charge-transfer complexes*. Common examples include the thiocyanate and phenolic complexes of iron(III), the iodide complex of molecular iodine, the iron(II) complex of orthophenanthroline, and the ferro – ferricyanide complex responsible for the color of Prussian blue.

Charge-transfer complexes generally contain an electron-donor and an electron-acceptor group within the complex. Absorption of radiation involves

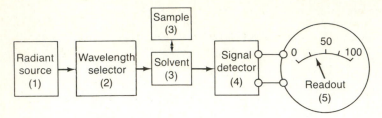

**Figure 15-9** Components of instruments for absorption measurements.

transition of an electron from the donor group to an orbital that is largely associated with the acceptor. The excited state is thus the product of an internal oxidation-reduction process.[5]

## 15E INSTRUMENT COMPONENTS AND INSTRUMENTS FOR THE MEASUREMENT OF ABSORPTION

Regardless of the spectral region involved, instruments that measure the transmittance or absorbance of solutions contain five basic elements: (1) a stable source of radiant energy; (2) a wavelength selector to isolate a restricted wavelength region from the source; (3) transparent containers for the sample and the blank; (4) a radiation detector, or *transducer,* to convert the radiant energy received to a measurable (usually electrical) signal; and (5) a readout device that displays the transduced signal from the detector. Figure 15–9 is a block diagram showing the usual arrangement of these components. The properties and performance characteristics of each of these components are described in Sections 15E through 15G. Section 15H considers briefly how these components are assembled in various types of absorption instruments.

The nature and complexity of the several components in an absorption instrument depend upon the wavelength region involved as well as whether the instrument is to be used principally for qualitative or quantitative measurements. Regardless of the degree of sophistication, however, the function of each component is the same; so also, are the steps involved in performing absorbance measurements. The indicator (Figure 15–9) is first adjusted to read zero transmittance when radiation is blocked from the detector by a shutter (the *0% T adjustment*). The indicator is then caused to read 100% transmittance with the solvent in the light path (the *100% T adjustment*); this adjustment is performed either by varying the intensity of the source or by altering the amplification of the detector signal. Finally, the percent transmittance is obtained directly by replacing the solvent with the solution containing the analyte. Clearly, the substitution of a logarithmic scale will provide a response in terms of absorbance.

---

[5] For further information concerning charge transfer, see C. N. R. Rao, *Ultra-Violet and Visible Spectroscopy, Chemical Applications,* 3rd ed., Chapter 11. London: Butterworth and Co., 1975.

## 15E – 1 Radiation Sources

A source of radiation for molecular absorption measurements must generate a beam with sufficient power in the wavelength region of interest to permit ready detection and measurement. In addition, it should provide continuous radiation that is made up of all wavelengths within the region for which it is to be used. Finally, the source must provide a stable output for the period needed to measure both $P_0$ and $P$. Only under these conditions will absorbance measurements be reproducible. Some instruments are designed to measure $P_0$ and $P$ simultaneously so that all but the most transitory fluctuations are canceled.

**Sources of Ultraviolet Radiation**   Hydrogen and deuterium lamps emit a truly continuous spectrum in the range between 180 and 375 nm. Molecules of the gas are promoted to quantized excited states as a result of electric discharge between a pair of electrodes. The relaxation process involves dissociation of the excited molecule, which produces a photon and two hydrogen atoms in the ground state. The released excitation energy is partitioned between an ultraviolet photon and the kinetic energies of the two hydrogen atoms. Since the kinetic energies can vary continuously, a broad spectrum of photon energies is obtained. Deuterium lamps provide higher intensities of ultraviolet radiation than hydrogen lamps under the same operating conditions.

**Sources of Visible Radiation**   The tungsten filament lamp, without doubt the most common source of visible radiation, provides a continuous spectrum in the wavelength region between 320 and 2500 nm. The output of a tungsten lamp varies approximately as the fourth power of the operating voltage. As a consequence, a constant-voltage transformer or an electronic voltage regulator is needed to realize a stable radiation output. As an alternative, a 6-V storage battery that has been maintained in good condition will provide a remarkably stable voltage for operation of the lamp.

**Sources of Infrared Radiation**   Continuous infrared radiation is produced by the heating of an inert solid. A silicon carbide rod, called a *Globar,* emits radiant energy in the range between 1 and 40 $\mu$m when heated electrically to about 1500°C. A *Nernst glower* is a rod of zirconium and yttrium oxides that is also heated to about 1500°C by the passage of current; radiation in the region between 0.4 and 20 $\mu$m is produced. A heated coil of nichrome wire is yet another useful source of infrared radiation from about 0.7 to 20 $\mu$m.

## 15E – 2 Control of Wavelength

Instruments for absorption measurements generally require a device that restricts the wavelength that is to be used to a narrow band that is absorbed by the analyte. A narrow band of radiation offers three advantages. First, the probability that the absorbing system will adhere to Beer's law is greatly enhanced (p.

420). In addition, a greater selectivity is ensured since substances with absorption peaks in other wavelength regions are less likely to interfere. Finally, use of a narrow band pass causes the greatest change in absorbance per increment of change in concentration; sensitivity is thus increased.

Both *filters* and *monochromators* are used to restrict radiation. *Photometers* make use of filters, which function by absorbing large portions of the spectrum while transmitting relatively limited wavelength regions. *Spectrophotometers* are instruments equipped with monochromators that permit the continuous variation of wavelength.

### 15E – 3 Wavelength Control with Filters

Illumination of a filter results in the transmission of a more or less narrow band of wavelengths. As shown in Figure 15 – 10, a filter can be characterized by the wavelength of its transmission peak, the percent transmittance at its peak, and its *effective bandwidth,* which is the bandwidth in wavelength units at a transmittance that is one-half of the maximum.

Glass *absorption filters* are inexpensive devices that provide bands of radiation throughout the visible spectrum. Absorption filters, which function by removing substantial portions of the spectrum, typically have effective bandwidths that range from 30 to 250 nm (Figure 15–10, curve *A*). Filters that provide the narrowest bandwidths also absorb a significant fraction of the desired radiation and may have transmittances that are 10% or less at their band peaks.

The transmittance of a *cut-off filter* is nearly 100% over a portion of the visible spectrum and decreases rapidly to essentially zero over the remainder (Figure 15–11, curve *B*). A narrow spectral band can often be isolated by coupling a cut-off filter with a second filter (Figure 15–11, curve *C* ).

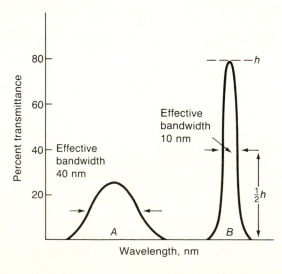

**Figure 15 – 10** Bandwidths for a glass absorption filter (curve A) and an interference filter (curve B).

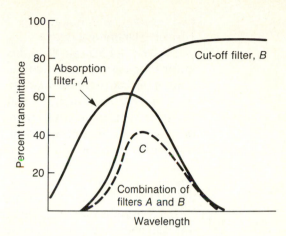

**Figure 15–11** Transmission characteristics of a glass absorption filter (curve *A*), a cut-off filter (curve *B*), and a combination of filters *A* and *B* (curve *C*).

An *interference filter* provides significantly narrower bandwidths (as low as 10 nm) and greater transmission of the desired wavelength than does an absorption filter (Figure 15–10). As its name implies, an interference filter relies upon optical interference to produce relatively narrow bandwidths of radiation. The device consists of a transparent dielectric layer (frequently calcium fluoride or magnesium fluoride) that occupies the space between two semitransparent metallic films that have been coated on the inside surfaces of two glass plates. The thickness of the dielectric layer is carefully controlled and determines the wavelength of the transmitted radiation. When a perpendicular beam of collimated (i.e., parallel) radiation strikes this array, a fraction passes through the first metallic film while the rest is reflected. The portion that is passed undergoes a similar partition upon striking the second metallic film. If the reflected portion from this second interaction is of the proper wavelength, it will be partially reflected from the inner side of the first layer in phase with incoming light of the same wavelength. The result is that this particular wavelength is reinforced, while most others, being out of phase, suffer destructive interference.

Interference filters are available for use in the spectral region from the ultraviolet to about 15 $\mu$m in the infrared. Typically, effective bandwidths are about 1.5% of the wavelength of peak transmittance, although this figure is reduced to 0.15% in some narrow-band filters, which transmit only about 10% of the radiation incident upon them.

## 15E–4 Wavelength Control with Monochromators

A *monochromator* is a device that disperses a polychromatic beam of radiation into its component wavelengths. All monochromators contain an entrance slit, a collimating mirror (or, less commonly, a lens) to produce a parallel beam of radiation, a prism or grating as a dispersing element, and a focusing mirror or lens that projects a series of rectangular images of the entrance slit upon a plane surface (the *focal plane*). In addition, most monochromators have entrance and exit *windows,* which are designed to protect the components from dust and corrosive laboratory fumes.

Figure 15 – 12 shows the optical design of two monochromators, one with a grating for dispersal of radiation and the other a prism. A source of radiation containing but two wavelengths, $\lambda_1$ and $\lambda_2$, is assumed for purposes of illustration. This radiation enters the monochromators via a narrow rectangular opening or *slit,* is collimated, and then strikes the surface of the dispersing element at an angle. For the grating monochromator, angular dispersion results from diffraction, which occurs at the reflective surface; for the prism, refraction at the two faces results in angular dispersion of the radiation, as shown. In both designs, the dispersed radiation is focused on the focal plane $AB$, where it appears as two images of the entrance slit (one for $\lambda_1$ and one for $\lambda_2$).

If a detector is located at the exit slit of the monochromator shown in Figure 15 – 12a and the grating is rotated so that one of the lines shown (say, $\lambda_1$) is scanned across the slit from $\lambda_1 - \Delta\lambda$ to $\lambda_1 + \Delta\lambda$ (where $\Delta\lambda$ is a small wavelength difference), the output of the detector takes a Gaussian shape similar to that shown in curve $B$ of Figure 15 – 10. The effective bandwidth of the monochromator, which is defined in the same way as that for a filter, depends upon the size and quality of the dispersing element, the slit widths, and the focal length of the monochromator. A high-quality monochromator will exhibit an

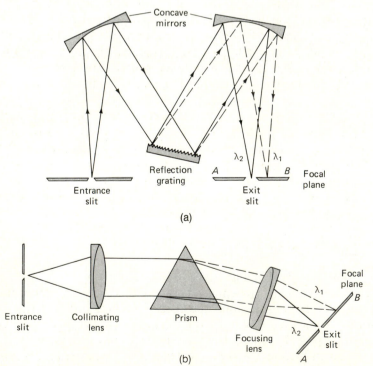

**Figure 15 – 12**
Monochromators. (a) Czerney-Turner grating monochromator. (b) Bunsen prism monochromator. In both diagrams, $\lambda_1 > \lambda_2$.

effective bandwidth of a few tenths of a nanometer or less in the ultraviolet-visible region. The effective bandwidth of a monochromator that is satisfactory for most applications is from about 1 to 5 nm.

**Grating Monochromators** Dispersion of ultraviolet, visible, and infrared radiation can be brought about by directing a polychromatic beam through a *transmission grating* or onto the surface of a *reflection grating;* the latter is by far the more common practice.

*Replica gratings,* which are used in many monochromators, are manufactured from a *master grating.* The master grating consists of a large number of parallel and closely spaced grooves ruled on a hard, polished surface with a suitably shaped diamond tool. For the ultraviolet and visible region, a grating will contain from 300 to 2000 grooves/mm with 1200 to 1400 being most common. For the infrared region, 10 to 200 grooves/mm are encountered. The construction of a good master grating is tedious, time-consuming, and expensive, because the grooves must be identical in size, exactly parallel, and equally spaced over the length of the grating (3 to 10 cm).

Replica gratings are formed from a master grating by evaporating a film of aluminum onto the latter after it has been coated with a parting agent, which permits ready separation of the aluminum from the master. A glass plate is cemented to the aluminum; the plate and film can then be lifted from the master mold, giving the finished replica grating.

One of the products of laser technology has been an optical (rather than mechanical) technique for forming gratings on plane or concave glass surfaces. *Holographic gratings* produced in this way are appearing in ever increasing numbers in modern optical instruments, even in some of the less expensive ones. These gratings, because of their greater perfection with respect to line shape and dimensions, provide spectra that are freer from stray radiation and ghosts (double images).

In the preparation of holographic gratings, the beams from a pair of identical lasers are brought to bear at suitable angles upon a glass surface coated with photoresist. The resulting interference fringes from the two beams sensitize the photoresist, giving sensitized areas that can be dissolved away, leaving a grooved structure that can be coated with aluminum or other reflecting substance to produce a reflection grating. The spacing of the grooves can be changed by changing the angle of the two laser beams with respect to one another. Nearly perfect, large ($\sim 50$ cm) gratings with as many as 6000 lines/mm can be manufactured in this way at a relatively low cost.

**Prism Monochromators** Figure 15–12b is a schematic representation of a monochromator in which the dispersing element is a 60-deg prism. Polychromatic radiation is admitted through an entrance slit, is collimated by a lens, and then strikes the surface of the prism at an angle. Bending (refraction) occurs as the light passes through each interface between the prism and its surroundings. Because the extent of refraction is wavelength dependent, the incoming beam is dispersed into its component wavelengths. A portion of the dispersed radiation then passes through an exit slit. Rotation of the prism permits any desired wavelength to be focused on this slit.

The effective bandwidth of a prism monochromator becomes smaller with increases in the size of the base of the prism. A design that provides a narrow bandwidth while maintaining compactness involves halving the prism in Figure 15–12b along its vertical axis and mirroring the back so that radiation both enters and emerges from the same face. A prism of this type, called a *Littrow prism,* disperses radiation as effectively as a full prism.

**Slit Adjustment**   Many monochromators are equipped with adjustable slits to permit some control over the bandwidth. A narrow slit decreases the effective bandwidth but also diminishes the power of the emergent beam. Thus, the minimum practical bandwidth may be limited by the sensitivity of the detector. For qualitative analysis, narrow slits and minimum effective bandwidths are used in order to obtain the maximum information, particularly when a spectrum consists of narrow peaks. For quantitative work, on the other hand, wider slits permit operation of the detector system at lower amplification, which in turn provides greater reproducibility of response.

**Dispersion by Monochromators**   An important advantage of grating monochromators over their prism counterparts is that the wavelength dispersion along the focal plane of the slit is very nearly *independent of wavelength.* In contrast, the dispersion for a prism monochromator is greater at shorter than at longer wavelengths. This difference is illustrated in Figure 15–13, which shows the positions of various wavelengths along the focal plane *AB* (see Figure 15–12) of a grating monochromator and two prism monochromators. As a conse-

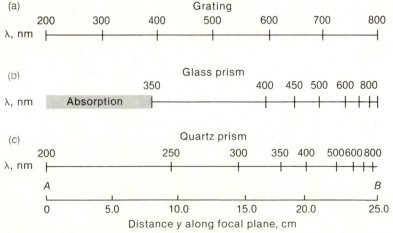

**Figure 15–13**   Dispersion of radiation by monochromators. Dispersing element is (a) a grating, (b) a glass prism, and (c) a quartz prism.

quence of the fixed dispersion of the grating, an entire spectrum can be obtained at a fixed bandwidth without altering the slit width. In contrast, the slit width of a prism monochromator must be changed continually in order to obtain a spectrum at a fixed effective bandwidth.

**Double Monochromators**    Some monochromators are designed with two dispersing elements that consist of two prisms, two gratings, or a prism and a grating. This arrangement markedly decreases the amount of stray radiation[6] and, moreover, provides greater spectral resolution.

## 15F SAMPLE CONTAINERS

In common with the optical elements of monochromators, the *cells* or *cuvettes* that hold the samples must be fabricated from a material that is transparent to radiation in the spectral region of interest. Thus, quartz or fused silica is used for measurements at wavelengths less than 350 nm. Both of these materials are transparent through the visible region and to about 2 $\mu$m in the infrared. Silicate glasses absorb strongly below 350 nm, but are suitable for work between about 350 nm and 2 $\mu$m in the near infrared. Plastic cells have also found application in the visible region. Cells with windows of polished sodium chloride or silver chloride are needed for infrared measurements at wavelengths longer than 2 to 3 $\mu$m.

The best cells have windows that are exactly perpendicular to the light path — an arrangement that minimizes reflection losses. Most instruments are provided with a pair of cells that have been carefully matched with respect to light path and transmission characteristics in order to make possible an accurate comparison of the radiant power that is transmitted through the sample and the solvent. The most common cell path for work in the ultraviolet and visible regions is 1 cm. Other path lengths ranging from 0.1 cm (and shorter) to 10 to 20 cm are employed. Transparent spacers for decreasing the path length of 1-cm cells to 0.1 cm are also available.

For reasons of economy, cylindrical cells are frequently used for measurements in the ultraviolet and visible regions. Care must be taken to duplicate the position of such cells with respect to the light path; otherwise, variations in path length and in reflection losses will introduce appreciable error.

Few solvents are transparent throughout large sections of the infrared region. As a consequence, measurements involving undiluted samples, or perhaps their concentrated solutions, are a necessity. This restriction requires the use of very short light paths (0.1 mm or less) to avoid total absorption of the radiation.

The quality of absorbance data is critically dependent upon the care and maintenance of matched cells. Scratches, fingerprints, grease, or other deposits

---

[6] Stray radiation is radiation that reaches the exit slit as a result of reflections from the surfaces of the various components within the monochromator. Any such radiation with a wavelength different from that which has been isolated by the monochromator is undesirable.

will alter the transmission characteristics of a cell markedly. Thorough cleaning before and after use is imperative. Care must be taken to avoid touching the windows during use. Matched cells should be calibrated against each other regularly with an absorbing solution. They should never be dried by heating in an oven or over a flame because such treatment may cause physical damage or a change in path length.

## 15G RADIATION DETECTORS

Most radiation detectors or transducers convert electromagnetic energy into an electrical signal. Radiation detectors should respond to a broad wavelength range, be sensitive to low levels of radiation, respond rapidly, produce an electrical signal that can be readily amplified, and have a relatively low noise level.[7] Finally, it is essential that the signal produced by the detector be directly proportional to the power of the radiation it receives; that is,

$$G = k'P + k''$$

where $G$ is the response of the detector in units of current, resistance, or electromotive force. The constant $k'$ measures the sensitivity of the detector in terms of electrical response per unit of radiant power. Many detectors exhibit a small constant response, known as a *dark current, $k''$*, even when they receive no radiation. Instruments with detectors that have a dark-current response are ordinarily equipped with a compensating circuit to permit application of a countersignal to decrease $k''$ to zero. Under most circumstances, then,

$$P = \frac{G}{k'} \tag{15-9}$$

and

$$P_0 = \frac{G_0}{k'} \tag{15-10}$$

where $G$ and $G_0$ represent the electrical response of the detector to radiation that has passed through the sample and the solvent, respectively. Substitution of Equations 15–9 and 15–10 into the Beer's law relationship gives

$$\log \frac{P_0}{P} = \log \frac{k'G_0}{k'G} = \log \frac{G_0}{G} = A \tag{15-11}$$

No single detector is capable of providing a useful response over the entire spectral region in which analytical measurements are made. *Photocells,* for

---

[7] *Noise* refers to small, random, and unwanted fluctuations that occur in the signal source, the detector, the amplifier, or the readout device. The term itself has its origins in radio technology, where such emanations manifest themselves as audible static. Common causes for noise include vibration, pickup from 60-Hz lines, temperature variations, and frequency or voltage fluctuations in the power supply.

example, are limited to the visible region. *Phototubes* and *photomultiplier tubes* can be designed to respond to radiation from the ultraviolet to the near infrared. Finally, sensitive heat detectors are needed for measurements in the infrared region.

**Photovoltaic or Barrier-Layer Cells**   The photovoltaic cell consists of a flat copper or iron electrode upon which is deposited a layer of a semiconducting material, such as selenium or copper(I) oxide. The outer surface of the semiconductor is coated with a thin, transparent film of gold, silver, or lead, which serves as a second collector electrode. The entire array is sealed into a plastic envelope for protection from the elements (see Figure 15–14). When radiation is absorbed by the semiconductor, bonding electrons gain enough energy so that they are promoted into a so called *conduction band,* where they are free to move in much the same way as electrons in a metal. Promotion of electrons to the conduction band creates positive *holes,* which can move in the opposite direction. The result is an electrical current whose magnitude is proportional to the number of photons striking the semiconducting surface. Ordinarily these currents are large enough (10 to 100 $\mu A$ are typical) to be measured with a microammeter. If the resistance of the external circuit is kept small, the magnitude of the photocurrent is directly proportional to the radiant power of the radiation that strikes the cell.

As noted, photovoltaic cells are used principally for the detection and measurement of radiation in the visible region. The response of a typical cell is at a maximum at about 550 nm, and decreases continuously to perhaps 10% of the peak value at 250 and 750 nm. The human eye has a similar sensitivity.

The barrier-layer cell is a rugged, low-cost device for the measurement of radiant power. No external source of electrical energy is required. On the other hand, the cell has a low internal resistance, which makes the amplification of its output difficult. The result is that a barrier-layer cell provides a readily measured response at high levels of illumination, but has little sensitivity for low levels. Finally, the photovoltaic cell exhibits *fatigue,* its response decreasing as it is subjected to prolonged illumination. This source of difficulty can be largely minimized or eliminated through proper instrument design and choice of experimental conditions.

Barrier-type cells are widely used in simple, portable instruments where ruggedness and low cost are important. For routine analyses, these instruments often provide reliable analytical data.

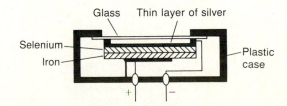

*Figure 15–14*   Schematic diagram of a typical barrier-layer cell.

**Phototubes** The phototube consists of a semicylindrical cathode and a wire anode sealed inside an evacuated transparent envelope. The concave surface of the cathode supports a layer of material that emits electrons upon being irradiated. This coating determines the spectral response of the phototube. Photoemissive materials include the alkali metals or alkali-metal oxides, alone or combined with other oxides. Application of a potential between the electrodes causes the emitted electrons to migrate to the wire anode. For a given radiant intensity, this photocurrent is approximately 25% of that produced by a photovoltaic cell. In contrast, however, amplification is easily accomplished because the photocell has a very high electrical resistance.

Figure 15 – 15 is a schematic diagram of a typical phototube. A potential of about 90 V across the electrodes is sufficient to cause all emitted photoelectrons to be captured and thereby ensure an electrical response that is proportional to the radiant power of the light that strikes the phototube.

**Photomultiplier Tubes** For the measurement of low radiant power, the photomultiplier tube offers a significant advantage over the phototube. Figure 15 – 16 is a schematic diagram of such a device. The composition of the cathode surface is similar to that of a phototube, electrons being emitted upon exposure to radiation. The tube also contains additional electrodes, called *dynodes* (these are labeled 1 to 9 in Figure 15 – 16). Dynode 1 is maintained at a potential 90 V more positive than the cathode, and photoelectrons are accelerated toward it as a consequence. Each photoelectron that strikes the dynode causes the emission of several additional electrons; the emitted electrons are, in turn, accelerated toward dynode 2, which is 90 V more positive than dynode 1. Again, several

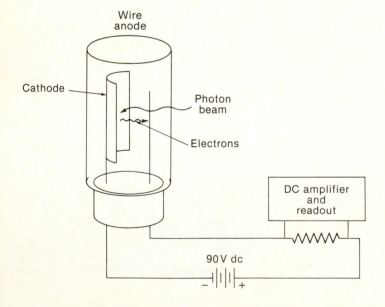

**Figure 15 – 15** Schematic diagram of a phototube and accessory circuit. The photocurrent induced by the radiation causes a potential drop across the resistor; this potential is amplified and measured.

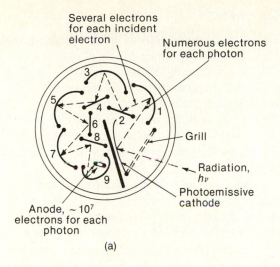

Several electrons for each incident electron

Numerous electrons for each photon

Grill

Radiation, $h\nu$

Photoemissive cathode

Anode, ~ $10^7$ electrons for each photon

(a)

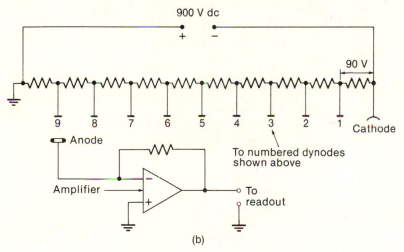

900 V dc

+ −

90 V

9  8  7  6  5  4  3  2  1  Cathode

Anode

Amplifier

To numbered dynodes shown above

To readout

(b)

**Figure 15 – 16**  Schematic diagram of a photomultiplier tube. (a) Cross section of the tube. (b) Electrical circuit.

electrons are emitted for each electron that strikes dynode 2. By the time this process has been repeated nine times, $10^6$ to $10^7$ electrons will have been generated for each photon that originally struck the cathode; this cascade is finally collected at the anode. The resulting current is passed through the resistor $R$ and can be further amplified and measured.

A photomultiplier tube can be used only for the measurement of low radiant power because intense light will cause irreversible changes in its performance. For this reason, the device is housed in a light-tight compartment and care is taken to eliminate the possibility of its being exposed even momentarily to strong light.

## 15H INSTRUMENT DESIGN

The components discussed in the previous sections have been combined in various ways to produce dozens of instruments for absorption measurements. The design of these instruments runs the gamut from very simple to very sophisticated; not surprisingly, substantial differences also exist in their cost. No single instrument is best for all purposes. Selection must be determined by the type of work for which the instrument is intended.

A *colorimeter* is an instrument that relies upon the human eye for radiation detection. The color of the sample is matched against one or more comparison standards.

A *photometer* consists of a source, a slit system, a filter, a photoelectric detector, and a transducer–readout system. Most filter photometers are designed for use in the visible region, although instruments for measurements in the ultraviolet and infrared regions also exist.

A *spectrophotometer* differs from a photometer in that it incorporates a prism or a grating monochromator to permit selection of a narrow band of wavelengths. It also has a slit located in the focal plane of the instrument (see Figure 15–12). We have noted that the spectral purity obtainable with a monochromator is ordinarily superior to that provided by glass filters.

### 15H–1 Single- and Double-Beam Designs

Photometers and spectrophotometers are designed with either one or two light paths. Radiation from the source of a double-beam instrument is split to permit approximately half to traverse the cell containing the sample, while the other half passes through a second cell containing the blank; every effort is made to ensure that the path lengths through the two cells are identical. The power of the two beams is then compared to give the transmittance of the solution. Because this comparison is made simultaneously or nearly simultaneously, a double-beam instrument compensates for all but the most transitory electrical fluctuations as well as for irregular performance of the source, the detector, and the transducer.

Single-beam instruments are particularly suited for quantitative absorption measurements at a single wavelength. Cells containing the sample and the blank are alternately placed in the light path. Here, simplicity of instrumentation and ease of maintenance offer distinct advantages. On the other hand, a double-beam instrument is decidedly more convenient for qualitative analysis, where measurements must be made at numerous wavelengths. Moreover, this design is readily adapted to automatic recording of spectra.

Both single-beam and double-beam instruments for work in the ultraviolet and visible regions are available from several instrument manufacturers. All infrared spectrophotometers incorporate a double beam because this design is more readily adapted to recording of spectra. Infrared spectra are sufficiently complex that recording capability is of prime importance.

## 15H – 2 Colorimeters

Colorimeters make use of the human eye as the detector and the brain as the transducer. The eye and brain, however, can only match colors. As a result, colorimetric methods always require the use of one or more standards at the time an analysis is performed. Other disadvantages are that the human eye responds to a comparatively limited spectral range (400 to 700 nm), and it is incapable of making the required comparison if the analyte solution contains a second colored substance. Finally, the eye is not as sensitive to small differences in absorbance as is a photoelectric detector; normally, concentration differences smaller than about 5% relative cannot be discerned.

Despite these limitations, visual comparison methods are valuable for routine analyses in which the requirements for accuracy are modest. For example, simple colorimetric tests are available to determine the pH and the chlorine content of swimming pool water; kits based on color comparisons are also sold for the analysis of soils. The iron, silicon, fluorine, and chlorine in city water supplies are routinely monitored by means of color comparison tests. A colorimetric reagent is introduced to the sample, and the resulting color is compared with permanent standard solutions or colored glass disks. Errors on the order of 10 to 50% relative are not uncommon and are quite tolerable in these applications.

## 15H – 3 Photometers

A photometer is a simple, relatively inexpensive tool for absorption analysis that offers an ease of maintenance and ruggedness that may not exist in more sophisticated spectrophotometers. Moreover, where high spectral purity is not needed for an analysis (and often it is not), the photometer will provide measurements that are as accurate as those obtained with more complex instrumentation.

A schematic diagram for a single-beam, direct-reading photometer is shown in Figure 15 – 17. The instrument makes use of a tungsten filament lamp, a lens to provide a parallel beam of radiation, a shutter, a filter, a wedge-shaped beam attenuator, and a detector. The attenuator is moved in or out of the beam to vary its intensity when the 100% $T$ adjustment is being made. Ordinarily, an instrument of this type will be equipped with a microammeter with a 5- or 7-in. face scribed with a linear transmittance scale (see Figure 15 – 9).

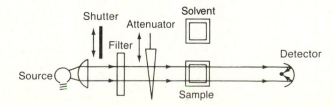

**Figure 15 – 17** Schematic diagram of a filter photometer.

A single-beam instrument must exhibit stable behavior during the time (perhaps 20 to 60 sec) needed to make the 0 and 100% $T$ adjustments and the transmittance measurement with the sample in place. A principal source of instability, as mentioned earlier, is fluctuations in the intensity of the source. For this reason, most single-beam instruments require some type of stabilized power supply.

**Filter Selection for Photometric Analysis**    Photometers are supplied with several filters, each of which transmits a different portion of the spectrum. Selection of the proper filter largely determines the sensitivity of absorbance measurements for any given application. The color of the light that is absorbed is the complement of the color that is transmitted by the solution. Thus, for example, a solution that appears red is transmitting that portion of the spectrum unchanged and is absorbing green radiation. Thus, it is the intensity of the green radiation that changes as a function of concentration, and a green filter should be selected. In general, the most suitable filter for a photometric method will be the color complement of the solution to be analyzed. If several filters with the same general hue are available, the one that causes the sample to exhibit the greatest absorbance (or least transmittance) should be used.

## 15H–4  Spectrophotometers

A spectrophotometer is equipped with a prism or a grating monochromator, which permits selection of any wavelength within the capability of the instrument. Spectrophotometers are available for measurements in the ultraviolet, visible, and infrared regions. While the materials used for optics and cells differ according to the spectral region for which the instrument is intended, the basic design is largely independent of wavelength. A schematic diagram for a single-beam spectrophotometer is similar to that for the filter photometer in Figure 15–17, the principal difference being the substitution of a monochromator for the filter.

The design of one type of double-beam spectrophotometer is shown schematically in Figure 15–18. Light from the monochromator is mechanically split with a rotating *chopper,* which alternately directs the beam through the blank and the sample. The detector thus receives an interrupted signal, which it converts to an ac electrical signal that is readily amplified. A variable *attenuator* is used to match the power of the light passing through the blank with that which passes through the sample. The attenuator, in turn, is calibrated to register the transmittance (or the absorbance) when the power of the two beams are equal and the output of the detector is a direct current. Since $P_0$ and $P$ are measured nearly simultaneously, all but the shortest fluctuations in the output from the source are of no consequence.

All commercial infrared spectrophotometers are double beam because the

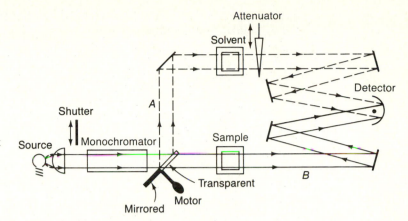

**Figure 15 – 18** Schematic diagram of alternating light paths through a double-beam spectrophotometer. (a) Through the solvent. (b) Through the sample.

primary use of these instruments is for acquiring spectra for qualitative analyses. The complexity of these spectra requires an automatic recording instrument. As was noted earlier, recording capabilities are most easily realized with double-beam instruments. An added reason for employing two beams for infrared measurements is that this design is less demanding in terms of the performance of the source and detector than is the single-beam arrangement. This property is important because of the low energy of infrared radiation, the low stability of sources and detectors, and the need for large signal amplification.

All commercial infrared spectrophotometers incorporate a low-frequency chopper (5 to 13 cycles per minute) to modulate the source output. This feature permits the detector to discriminate between the signal from the source and signals from extraneous radiation, such as infrared emission from various bodies surrounding the detector. Low chopping rates are demanded by the slow response times of most infrared detectors. Infrared instruments are generally of the null type, with the beam being attenuated by a comb or an absorbing wedge.

## 15/ QUANTITATIVE ABSORPTION ANALYSIS

Absorption spectroscopy in the ultraviolet and visible regions is one of the most widely used tools for quantitative analysis.[8] Important characteristics of this method include the following:

1. *Wide applicability.* Numerous inorganic and organic species absorb in the ultraviolet and visible regions and are thus susceptible to quantitative deter-

---

[8] For a wealth of detailed, practical information on spectrophotometric practices, see, *Techniques in Visible and Ultraviolet Spectrometry,* Vol. I, *Standards in Absorption Spectroscopy,* C. Burgess and A. Knowles, Eds. London: Chapman and Hall, 1981; and J. R. Edisbury, *Practical Hints on Absorption Spectrometry.* New York: Plenum Press, 1968.

mination. In addition, many nonabsorbing species can be caused to absorb after suitable chemical treatment.

2. *High sensitivity.* Molar absorptivities ranging between 5000 and 40,000 are often encountered, particularly for charge-transfer complexes of inorganic species. These large absorptivities lead to detection limits in the $10^{-4}$ to $10^{-5}$ $M$ range; with suitable procedural modification, this range can often be extended to $10^{-6}$ or even $10^{-7}$ $M$.

3. *Moderate to high selectivity.* The need for a preliminary separation step is unnecessary if a wavelength region can be found in which the analyte is the only absorbing species in the sample. Moreover, where overlapping bands do occur, corrections based upon measurements at other wavelengths may be possible.

4. *Good accuracy.* The relative error in a typical spectrophotometric (or photometric) measurement is on the order of 1 to 3%. Special techniques often permit reduction of the relative error to a few tenths percent.

5. *Ease and convenience.* Spectrophotometric measurements with modern instruments are rapid and convenient. In addition, the methods frequently lend themselves to automation.

## 15I-1 Scope

The applications of absorption analysis are not only numerous but also touch upon every area in which quantitative information is sought. The reader can obtain a notion of the scope of spectrophotometry by consulting the series of review articles published biennially in *Analytical Chemistry*[9] as well as monographs on the subject.[10]

**Application to Absorbing Species**    Table 15–2 (p. 427) lists many common organic chromophoric groups. Spectrophotometric determination of organic compounds containing one or more of these groups is thus potentially feasible; many such applications can be found in the literature.

A number of inorganic species also absorb. We have noted that many ions of the transition metals possess color in solution and can thus be determined by photometric or spectrophotometric measurement. In addition, a number of other species show characteristic absorption peaks including nitrite, nitrate, and chromate ions, the oxides of nitrogen, the elemental halogens, and ozone.

**Applications to Nonabsorbing Species**    Many nonabsorbing analytes can be

---

[9] L. G. Hargis and J. A. Howell, *Anal. Chem.,* **1980,** *52,* 306R; **1982,** *54,* 171R; **1984,** *56,* 225R.

[10] See, for example, E. B. Sandell and Hiroshi Onishi, *Photometric Determination of Traces of Metals,* 4th ed. New York: John Wiley & Sons, 1978; *Colorimetric Determination of Nonmetals,* 2nd ed., D. F. Boltz, Ed., New York: Interscience Publishers, 1978; F. D. Snell, *Photometric and Fluorometric Methods of Analysis.* New York: John Wiley & Sons, 1978.

determined photometrically by causing them to react with chromophoric reagents to produce products that absorb strongly in the ultraviolet and visible regions. The successful application of these color-forming reagents usually requires that their reaction with the analyte be forced to near completion.

Typical inorganic reagents include the following: thiocyanate ion for iron, cobalt, and molybdenum; the anion of hydrogen peroxide for titanium, vanadium, and chromium; and iodide ion for bismuth, palladium, and tellurium. Of even greater importance are organic chelating reagents that form stable colored complexes with cations. Common examples include 1,10-phenanthroline for the determination of iron, dimethylglyoxime for nickel, diethyldithiocarbamate for copper, and diphenylthiocarbazone for lead; Figure 15–19 shows the color-forming reaction for three of these reagents.

Other reagents are available that react with organic functional groups to produce colors that are useful for quantitative analysis. For example, the red color of the 1:1 complexes that form between low-molecular-weight aliphatic alcohols and cerium(IV) can be used for the quantitative estimation of such alcohols.

**Quantitative Analysis in the Infrared Region**   Quantitative infrared methods are not fundamentally different from the measurements we have just considered. However, several practical problems make it difficult, if not impossible, to attain the same degree of accuracy. The very short path lengths that are ordinarily needed are not easily reproduced. In addition, the complexity of infrared spectra increases the likelihood that absorption peaks for the analyte will overlap those of other components in the sample. Finally, the narrowness of infrared absorption peaks often causes deviations from Beer's law (p. 420).

## 15I–2 Procedural Details

A first step in any photometric or spectrophotometric analysis is the development of conditions that yield a reproducible relationship (preferably linear) between absorbance and analyte concentration.

**Wavelength Selection**   In order to realize maximum sensitivity, spectrophotometric absorbance measurements are ordinarily made at a wavelength corresponding to an absorption peak because the change in absorbance per unit of concentration is greatest at this point. In addition, the absorption curve is often flat at a maximum, which leads to good adherence to Beer's law (see Figure 15–5) and less uncertainty from failure to reproduce precisely the wavelength setting of the instrument.

**Variables that Influence Absorbance**   Common variables that influence the absorption spectrum of a substance include the nature of the solvent, the pH of the solution, the temperature, high electrolyte concentrations, and the presence

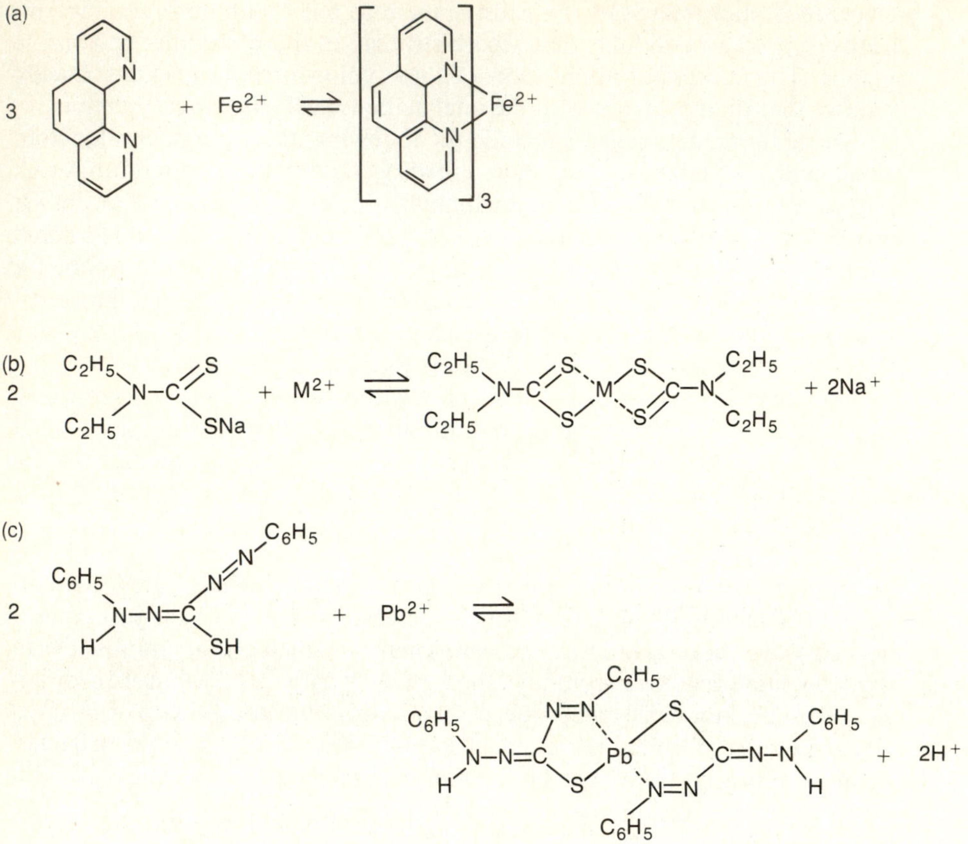

Figure 15-19 Typical chelating reagents for absorption analysis. (a) 1,10-Phenanthroline. (b) Diethyldithiocarbamate. (c) Diphenylcarbazone. (See Section 4C-3 for dimethylglyoxime and its adduct with nickel.)

of interfering substances. The effects of these variables must be known and conditions for the analysis chosen such that the absorbance will not be materially affected by small, uncontrolled variations in their magnitudes.

**Cleaning and Handling of Cells**   Accurate spectrophotometric analysis requires the use of good-quality, matched cells. These should be regularly calibrated against one another to detect differences that can arise from scratches, etching, and wear. Equally important is the use of proper cell cleaning and drying techniques (see Section 20O-1).

**Determination of the Relationship Between Absorbance and Concentration**   The calibration standards for a photometric or a spectrophotometric analysis should approximate as closely as possible the overall composition of the actual samples, and should encompass a reasonable range of analyte concentrations. Seldom, if ever, is it safe to assume adherence to Beer's law and use only a single standard to determine the molar absorptivity; it is even less prudent to base the results of an analysis on a literature value for the molar absorptivity.

The difficulties that attend production of standards with an overall composition closely resembling that of the sample can be formidable, if not insurmountable. Under such circumstances, the standard addition approach may prove useful. Here, a known amount of analyte is introduced to a second aliquot of the sample. Provided Beer's law is obeyed (and this must be confirmed experimentally), the difference in absorbance is used to calculate the analyte concentration of the sample.

---

**Example 15–3**   A 2.00-mL urine specimen was treated with reagents to generate color with phosphate, following which the sample was diluted to 100 mL. Photometric measurement for the phosphate in a 25.0-mL aliquot yielded an absorbance of 0.428. Addition of 1.00 mL of a solution containing 0.0500 mg of phosphate to a second 25.0-mL aliquot resulted in an absorbance of 0.517. Use these data to calculate the milligrams of phosphate in each milliliter of the specimen.

The absorbance of the second measurement must be corrected for dilution. Thus,

$$\text{corrected absorbance} = 0.517 \times \frac{26.0}{25.0} = 0.538$$

$$\text{absorbance caused by 0.0500 mg phosphate} = 0.538 - 0.428 = 0.110$$

$$\text{weight phosphate in } \frac{25.0}{100} \text{ of specimen} = \frac{0.428}{0.110} \times 0.050 = 0.195 \text{ mg}$$

Finally, then,

$$\text{mg phosphate/mL of specimen} = \frac{100}{25.0} \times 0.195 \times \frac{1}{2.00} = 0.390$$

---

**Analysis of Mixtures**   The total absorbance of a solution at any given wavelength is equal to the sum of the absorbances of the individual components in the solution (Equation 15–6). This relationship makes it possible in principle to determine the concentrations of the individual components of a mixture even if total overlap in their spectra exists. For example, Figure 15–20 shows the spectrum of a solution containing a mixture of species A and species B as well as absorption spectra for the individual components. Clearly, no wavelength exists at which the absorbance is due to just one of these components. To analyze the mixture, molar absorptivities for A and B are first determined at wavelengths $\lambda_1$ and $\lambda_2$ with enough standards to be sure that Beer's law is obeyed

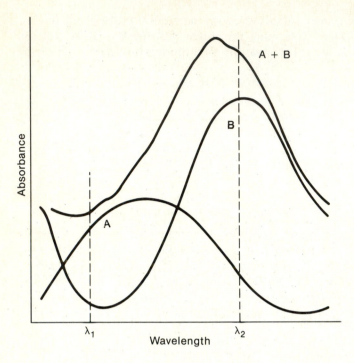

**Figure 15–20** Absorption spectrum for a two-component mixture.

over an absorbance range that encompasses the absorbance of the sample. Note that the wavelengths selected are ones at which the two spectra differ significantly. Thus, at $\lambda_1$, the molar absorptivity of component A is much larger than that for component B. The reverse is true for $\lambda_2$. To complete the analysis, the absorbance of the mixture is determined at the same two wavelengths. Example 15–4 demonstrates how the composition of the mixture is derived from data of this kind.

**Example 15–4** Palladium(II) and gold(III) can be analyzed simultaneously through reaction with methiomeprazine ($C_{19}H_{24}N_2S_2$). The absorption maximum for the Pd complex occurs at 480 nm, while that for the Au complex is at 635 nm. Molar absorptivity data at these wavelengths are

|  | Molar Absorptivity, $\epsilon$ | |
|---|---|---|
|  | 480 nm | 635 nm |
| Pd complex | $3.55 \times 10^3$ | $5.64 \times 10^2$ |
| Au complex | $2.96 \times 10^3$ | $1.45 \times 10^4$ |

A 25.0-mL sample was treated with an excess of methiomeprazine and subsequently diluted to 50.0 mL. Calculate the formal concentrations of Pd(II), $C_{Pd}$, and Au(III), $C_{Au}$, in the sample if the diluted solution had an absorbance of 0.533 at 480 nm and 0.590 at 635 nm when measured in a 1.00-cm cell.

At 480 nm

$$0.533 = (3.55 \times 10^3)(1.00)C_{Pd} + (2.96 \times 10^3)(1.00)C_{Au}$$

or

$$C_{Pd} = \frac{0.533 - 2.96 \times 10^3 C_{Au}}{3.55 \times 10^3}$$

At 635 nm

$$0.590 = (5.64 \times 10^2)(1.00)C_{Pd} + (1.45 \times 10^4)(1.00)C_{Au}$$

Substitution for $C_{Pd}$ in this expression gives

$$0.590 = \frac{5.64 \times 10^2(0.533 - 2.96 \times 10^3 C_{Au})}{3.55 \times 10^3} + 1.45 \times 10^4 C_{Au}$$

$$= 0.0847 - 4.70 \times 10^2 C_{Au} + 1.45 \times 10^4 C_{Au}$$

$$C_{Au} = (0.590 - 0.0847)/(1.403 \times 10^4) = 3.60 \times 10^{-5} \; M$$

and

$$C_{Pd} = \frac{0.533 - (2.96 \times 10^3)(3.60 \times 10^{-5})}{3.55 \times 10^3} = 1.20 \times 10^{-4} \; M$$

Since the analysis involved a two-fold dilution, the concentrations of Pd(II) and Au(III) in the original sample were $7.20 \times 10^{-5}$ and $2.40 \times 10^{-4}$ $M$, respectively.


Mixtures containing more than two absorbing species can be analyzed, in principle at least, if one additional absorbance measurement is made for each added component. The uncertainties in the resulting data become greater, however, as the number of measurements increases. Some of the newer computerized spectrophotometers are capable of minimizing these uncertainties by overdetermining the system; that is, these instruments use many more data points than unknowns and effectively match the entire spectrum of the unknown as closely as possible by deriving synthetic spectra for various concentrations of the components. The derived spectra are then compared with that of the analyte until a close match is found. The spectrum for standard solutions of each component is required, of course.

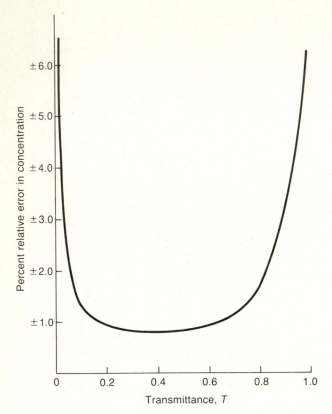

**Figure 15–21**   Effect of a constant error in transmittance upon the relative error in concentration.

## 15J ERRORS IN THE MEASUREMENT OF ABSORBANCE[11]

The accuracy and precision of a spectrophotometric analysis is often limited by the indeterminate uncertainties associated with the instrument. Recall (Section 15E) that a photometric measurement consists of three separate steps, specifically, the 0% $T$ adjustment, the 100% $T$ adjustment with the blank in the light path, and finally the measurement of $T$ with the analyte solution substituted for the blank.

The indeterminate uncertainty associated with each of these steps combines to give an uncertainty in transmittance which is best expressed as the standard deviation for the transmittance measurement, $s_T$. It can be shown[12] that the relationship between the standard deviation in transmittance and the standard

---

[11] For a detailed discussion of the errors associated with spectrophotometric measurements in the visible and ultraviolet regions, see L. D. Rathman, S. R. Crouch, and J. D. Ingle, Jr., *Anal. Chem.,* **1972,** *44,* 1375.

[12] See, for example, D. A. Skoog, *Principles of Instrumental Analysis,* 3rd ed., pp. 168–172. Philadelphia: Saunders College Publishing, 1985.

deviation in concentration (and absorbance, $s_A$) is given by

$$\frac{s_c}{c} = \frac{s_A}{A} = \frac{0.434}{\log T} \times \frac{s_T}{T} \qquad (15-12)$$

Note that the relative standard deviation in concentration is related to both the logarithm of transmittance and the relative standard deviation in the measurement of $T$.

The standard deviation in $T$ tends to be constant and independent of $T$ for many photometers and spectrophotometers. Thus, the resolution for the meter for a direct-reading instrument is ordinarily a few tenths of the full scale and is constant from one end to the other. Similarly, fluctuations in the output of the detector in an infrared spectrophotometer tend to be independent of the magnitude of its signal. Figure 15–21 illustrates how this constant uncertainty in $T$ influences the relative error $s_c/c$ in concentration. It is seen that the optimum transmittance range in instruments subject to this type of uncertainty is between about 0.2 and 0.6 ($A = 0.7$ to 0.2), and that transmittances less than about 0.05 ($A > 1.3$) and greater than about 0.8 ($A < 0.1$) should be avoided wherever possible. It should be noted that the performance of a high-quality spectrophotometer is not ordinarily limited by the resolution of the readout device, but rather by uncertainties in positioning of cells in the light path. With such instruments, the optimum absorbance range is extended to 2.0 or more.

## PROBLEMS

15–1. Complete the following tabulation.

| | $\lambda$, nm | $\lambda$, Å | $\lambda$, cm | $\lambda$, $\mu$m | $\nu$, Hz | $\sigma$, cm$^{-1}$ |
|---|---|---|---|---|---|---|
| *(a) | 350 | | | | | |
| (b) | | 6180 | | | | |
| *(c) | | | $8.69 \times 10^{-5}$ | | | |
| (d) | | | | 0.302 | | |
| *(e) | | | | | $5.01 \times 10^{14}$ | |
| (f) | | | | | | 22779 |
| *(g) | | | | | | 11416 |
| (h) | | | | | $3.02 \times 10^{14}$ | |
| *(i) | | | | 0.717 | | |
| (j) | | | $1.071 \times 10^{-4}$ | | | |
| *(k) | | 2850 | | | | |
| (l) | 679 | | | | | |

15–2. Calculate the frequency (in hertz) of
   *(a) the emission line for $Zn^{2+}$ at 2025.5 Å.
   (b) the line in the X-ray spectrum of Fe at 1.757 Å.
   *(c) an infrared absorption peak at 4.25 $\mu$m.
   (d) microwave radiation with a wavelength of 240 cm.

15–3. Calculate
*(a) the wavenumber of the absorption maximum for benzene at 204 nm.
(b) the frequency of the line in the X-ray spectrum of lead at 0.982 Å.
*(c) the wavelength (in micrometers) of an infrared absorption peak with a wavenumber of 360 cm$^{-1}$.
(d) the frequency of the line in the emission spectrum for Cd at 228.8 nm.

15–4. Convert the accompanying absorbance data into percent transmittance.
*(a) 0.498          (b) 0.667
*(c) 0.144          (d) 0.376
*(e) 0.835          (f) 0.267

15–5. Convert the following percent transmittance data into absorbance.
*(a) 10.4           (b) 22.9
*(c) 69.2           (d) 14.0
*(e) 72.3           (f) 34.8

*15–6. Calculate the percent transmittance of solutions with twice the absorbance of those in Problem 15–4.

*15–7. Calculate the absorbance for solutions with half the percent transmittance of those in Problem 15–5.

15–8. Supply the missing quantities in the accompanying tabulation.

| | Absorbance, $A$ | Molar Absorptivity, $\epsilon$ | Path Length, $b$, cm | Concentration, mol/L |
|---|---|---|---|---|
| *(a) | 0.310 | | 1.00 | $4.89 \times 10^{-5}$ |
| (b) | 0.467 | $1.08 \times 10^4$ | | $1.73 \times 10^{-5}$ |
| *(c) | 0.795 | $5.33 \times 10^3$ | 1.00 | |
| (d) | | $8.54 \times 10^3$ | 1.50 | $4.70 \times 10^{-5}$ |
| *(e) | | $2.61 \times 10^4$ | 1.00 | $3.50 \times 10^{-5}$ |
| (f) | 0.269 | | 1.00 | $4.58 \times 10^{-5}$ |
| *(g) | 0.811 | $4.74 \times 10^3$ | 1.00 | |
| (h) | 0.525 | $7.39 \times 10^3$ | | $2.84 \times 10^{-5}$ |

15–9. Supply the missing quantities in the accompanying tabulation.

| | Absorbance, $A$ (1.00-cm cells) | Molar Absorptivity, $\epsilon$ | Concentration |
|---|---|---|---|
| *(a) | 0.669 | $4.57 \times 10^3$ | g/L (gfw = 737) |
| (b) | 0.420 | | 0.0225 g/L (gfw = 228) |
| *(c) | | $1.09 \times 10^4$ | 0.0313 g/L (gfw = 646) |
| (d) | | $3.64 \times 10^4$ | 0.922 ppm (gfw = 110) |
| *(e) | 0.235 | | 2.65 ppm (gfw = 325) |
| (f) | 0.143 | $7.65 \times 10^3$ | (gfw = 473) |

15–10. What are the units for absorptivity when the path length is given in centimeters and the concentration is expressed in
*(a) parts per million?
(b) micrograms per liter?
*(c) weight-volume percent?
(d) grams per liter?

**\*15–11.** A solution that is $1.04 \times 10^{-4}$ $M$ in $FeSCN^{2+}$ has an absorbance of 0.726 at 580 nm when measured in a 1.00-cm cell. Calculate
  (a) the percent transmittance for this solution.
  (b) the molar absorptivity for the complex at this wavelength.
  (c) the path length needed to match the absorbance of this solution with one that is $4.16 \times 10^{-5}$ $M$ in the complex.
  (d) the absorbance of a solution that is $3.64 \times 10^{-5}$ $M$ in the complex when measured in a 1.50-cm cell.

**15–12.** The complex formed between Cu(I) and 1,10-phenanthroline has a molar absorptivity of 7000 L $cm^{-1}$ $mol^{-1}$ at 435 nm, the wavelength of maximum absorption. Calculate
  (a) the absorbance of an $8.50 \times 10^{-5}$ $M$ solution of the complex when measured in a 1.00-cm cell at 435 nm.
  (b) the percent transmittance of the solution in (a).
  (c) the concentration of a solution that, in a 5.00-cm cell, has the same absorbance as the solution in (a).
  (d) the path length through a $3.40 \times 10^{-5}$ $M$ solution of the complex that is needed for an absorbance that is the same as the solution in (a).

**\*15–13.** The molar absorptivity of the complex that is formed between Bi(III) and thiourea is $9.3 \times 10^3$ L $cm^{-1}$ $mol^{-1}$ at 470 nm. Calculate the range of permissible concentrations for the complex if the absorbance is to be no less than 0.15 nor greater than 0.80 when the measurements are made in a 1.00-cm cell.

**15–14.** The molar absorptivity for aqueous solutions of phenol at 211 nm is $6.17 \times 10^3$ L $cm^{-1}$ $mol^{-1}$. Calculate the range of phenol concentrations that can be used if the transmittance is to be less than 80% and greater than 5% when the measurements are made in 1.00-cm cells.

**\*15–15.** A standard Fe solution was diluted to give the concentrations shown in the accompanying tabulation. The Fe(II)–phenanthroline complex was developed in 10.00-mL aliquots of each solution, water was added to give a total volume of 25.00 mL, and then the absorbance was measured at 510 nm.

| Concentration of Fe, ppm (in 10.00-mL aliquot) | Absorbance, $A$, 510 nm (1.00-cm cells) |
| --- | --- |
| 0.00 | 0.00 |
| 2.00 | 0.160 |
| 4.00 | 0.312 |
| 6.00 | 0.472 |
| 8.00 | 0.627 |
| 10.00 | 0.787 |

  (a) Produce a calibration curve for these data.
  (b) By the method of least squares, derive an equation relating absorbance and concentration of Fe.
  (c) Calculate the standard deviation about regression and the standard deviation of the slope of the calibration curve.

**15–16.** The method developed in Problem 15–15 was used to determine the Fe in samples of waste water. Calculate the ppm of Fe if 10.00-mL aliquots yielded the accompanying absorbance data. Estimate standard deviations for the derived concentrations. Repeat the calculations assuming the absorbance data were the means of three measurements.

| Sample | Absorbance, $A$, 510 nm (1.00-cm cells) |
|---|---|
| *(a) | 0.604 |
| (b) | 0.396 |
| *(c) | 0.171 |
| (d) | 0.248 |
| *(e) | 0.512 |
| (f) | 0.447 |

**15-17.** A standard addition method for the routine analysis of Cu is based upon dilution of the sample, after suitable pretreatment, to 500.0 mL in a volumetric flask. Two aliquots are taken; a known volume of $5.00 \times 10^{-3}$ $F$ $Cu^{2+}$ is added to one. The Cu-1, 10-phenanthroline complex is developed and sufficient water is added to give each solution a volume of 50.00 mL. The absorbance is measured at 508 nm in 1.00-cm cells. Calculate the percentage of Cu in the following samples.

| | | Volumes Taken, mL | | | | |
|---|---|---|---|---|---|---|
| | Wt Sample, g (in 500.0 mL) | Sample | $Cu^{2+}$ 0.00500 $F$ | Ligand | $H_2O$ | Absorbance, $A$, 508 nm (1.00-cm cells) |
| *(a) | 4.137 | 25.00 | 0.00 | 15.00 | 10.00 | 0.374 |
| | | 25.00 | 5.00 | 15.00 | 5.00 | 0.555 |
| (b) | 5.671 | 20.00 | 0.00 | 15.00 | 15.00 | 0.339 |
| | | 20.00 | 4.00 | 15.00 | 11.00 | 0.484 |
| *(c) | 3.798 | 25.00 | 0.00 | 15.00 | 10.00 | 0.176 |
| | | 25.00 | 8.00 | 15.00 | 2.00 | 0.466 |
| (d) | 6.640 | 20.00 | 0.00 | 15.00 | 15.00 | 0.191 |
| | | 20.00 | 6.00 | 15.00 | 9.00 | 0.409 |

**15-18.** Samples of used lubricating oil were freed of organic matter by wet-ashing and subsequently diluted to 100.0 mL in volumetric flasks. Addition of diphenylcarbazide to 25.00-mL aliquots of the diluted samples resulted in formation of a complex with Cr, the absorbance of which was measured at 540 nm. Calculate the milligrams of Cr per liter of oil, based upon the accompanying data.

| | | Volumes Added to 25.00 mL of Diluted Sample | | | |
|---|---|---|---|---|---|
| | Volume Taken, mL | Cr(VI), 2.00 ppm | Ligand | $H_2O$ | Absorbance, $A$, 540 nm (1.00-cm cells) |
| *(a) | 20.00 | 0.00 | 10.00 | 15.00 | 0.525 |
| | | 7.50 | 10.00 | 7.50 | 0.692 |
| (b) | 25.00 | 0.00 | 10.00 | 15.00 | 0.296 |
| | | 5.00 | 10.00 | 10.00 | 0.408 |
| *(c) | 15.00 | 0.00 | 10.00 | 15.00 | 0.317 |
| | | 10.00 | 10.00 | 5.00 | 0.540 |
| (d) | 20.00 | 0.00 | 10.00 | 15.00 | 0.515 |
| | | 12.00 | 10.00 | 3.00 | 0.775 |
| *(e) | 15.00 | 0.00 | 10.00 | 15.00 | 0.345 |
| | | 10.00 | 10.00 | 5.00 | 0.568 |

*15–19. Express the results of the analyses in Problem 15–18 as ppm Cr if the density of the oil samples is 0.800.

*15–20. A. J. Mukhedkar and N. V. Deshpande (*Anal. Chem.*, **1963**, *35*, 47) report on a simultaneous determination for Co and Ni based upon absorption by their respective 8-hydroxyquinoline complexes. Molar absorptivities corresponding to their absorption maxima are as follows:

| | **Molar Absorptivity, $\epsilon$** | |
|---|---|---|
| | 365 nm | 700 nm |
| Co | 3529 | 428.9 |
| Ni | 3228 | 0.00 |

Calculate the ppm of Co and Ni in each of the following solutions, based upon the accompanying data.

| | **Absorbance, $A$ (1.00-cm cells)** | |
|---|---|---|
| Solution | 365 nm | 700 nm |
| *(a) | 0.561 | 0.039 |
| (b) | 0.735 | 0.070 |
| *(c) | 0.640 | 0.026 |
| (d) | 0.990 | 0.108 |
| (e) | 0.684 | 0.057 |

15–21. The acid dissociation constant for the indicator HIn is $6.4 \times 10^{-6}$. The absorption spectra for HIn and In$^-$ overlap throughout the visible region; absorptivity data for each species at wavelengths corresponding to absorption maxima are as follows:

| | **Molar Absorptivity, $\epsilon$** | |
|---|---|---|
| | 440 nm | 620 nm |
| HIn | 2204 | 776 |
| In$^-$ | 408 | 3372 |

Calculate the pH of a solution in which the formal indicator concentration is $3.60 \times 10^{-4}\ F$, given the following absorbance data:

| | **Absorbance, $A$ (1.00-cm cells)** | |
|---|---|---|
| Solution | 440 nm | 620 nm |
| *(a) | 0.416 | 0.824 |
| (b) | 0.286 | 0.617 |
| *(c) | 0.470 | 0.747 |
| (d) | 0.542 | 0.643 |
| *(e) | 0.362 | 0.902 |

**15-22.** Calculate the absorbance at 440 and 620 nm for a solution that is $3.60 \times 10^{-4}$ $F$ with respect to the indicator in Problem 15-21 and has a pH of
   *(a) 4.72
    (b) 5.67
   *(c) 5.00
    (d) 5.25
   *(e) 4.88

**15-23.** A 3.320-g sample of a purified unknown acid was dissolved in sufficient water to give 100.0 mL of solution. Titration of a 25.00-mL aliquot required 26.24 mL of 0.1207 $N$ NaOH. This volume of base was precisely added to a 50.00-mL aliquot of the acid. A sufficient amount of the indicator in Problem 15-21 was added to make the solution $3.60 \times 10^{-4}$ $F$ with respect to the indicator. The absorbance of the resulting solution, measured in a 1.00-cm cell, was 0.506 at 440 nm and 0.695 at 620 nm. Calculate
   (a) the pH of the solution.
   (b) the dissociation constant for the acid.
   (c) the equivalent weight of the acid.

# Chapter 16
# Atomic Spectroscopy

Atomic spectroscopy is based upon the measurement of absorption, emission, or fluorescence of electromagnetic radiation by atoms or elementary ions (such as $K^+$, $Mg^+$, or $Al^+$) in a gaseous medium. Atomic methods are conveniently classified according to their method of *atomization,* a process in which the constituents of a solution are converted into gaseous atoms or elementary ions. Table 16–1 lists the common methods of atomization and the names of the various procedures that are associated with each.

The methods outlined in Table 16–1 offer the advantages of wide applicability, excellent sensitivity, speed, and convenience. They are among the most selective of all analytical methods and are applicable to perhaps 70 of the elements in the periodic table. Atomic methods are generally quite sensitive and can be used to determine analyte concentrations in the parts-per-million to parts-per-billion range.[1]

This chapter is concerned with atomic emission and absorption methods that are based upon the first four atomization methods listed in Table 16–1. In each, a solution of the analyte is evaporated rapidly at an elevated temperature to yield a finely divided solid, which, upon further heating, breaks down into gaseous atoms or elementary ions. The efficiency and reproducibility of the atomization step in large measure determines the method's sensitivity, precision, and accuracy; that is, atomization is by far the most critical step in atomic spectroscopy.

## 16A ATOMIC METHODS BASED ON FLAME ATOMIZATION

As shown in Table 16–1, three types of atomic methods are based on flame atomization, namely, atomic absorption (AAS), atomic emission (AES), and

---

[1] References that deal with the theory and applications of atomic spectroscopy include R. D. Sacks, A. Syty, and J. W. Robinson, in *Treatise on Analytical Chemistry,* P. J. Elving, E. J. Meehan, and I. M. Kolthoff, Eds., Part I, Vol. 7, Chapters 6, 7, and 8. New York: John Wiley & Sons, 1981; K. C. Thompson and R. J. Reynolds, *Atomic Absorption, Fluorescence, and Flame Emission Spectroscopy,* 2nd ed. New York: John Wiley & Sons, 1978; C. Th. J. Alkemade, *et al., Metal Vapors in Flames,* Elmsford, NY: Pergamon Press, 1982; B. Magyar, *Guide-Lines to Planning Atomic Spectrometric Analysis.* New York: Elsevier Publishing Co., 1982.

*Table 16 – 1   Classification of Atomic Spectral Methods*

| Atomization Method | Typical Atomization Temperature, °C | Basis for Method | Common Name and Abbreviation for Method |
|---|---|---|---|
| Flame | 1700 – 3150 | Absorption | Atomic absorption spectroscopy, AAS |
| | | Emission | Atomic emission spectroscopy, AES |
| | | Fluorescence | Atomic fluorescence spectroscopy, AFS |
| Electrothermal | 1200 – 3000 | Absorption | Electrothermal atomic absorption spectroscopy |
| | | Fluorescence | Electrothermal atomic fluorescence spectroscopy |
| Inductively coupled argon plasma | 6000 – 8000 | Emission | Inductively coupled plasma spectroscopy, ICP |
| | | Fluorescence | Inductively coupled plasma fluorescence spectroscopy |
| Direct-current argon plasma | 6000 – 10,000 | Emission | DC plasma spectroscopy, DCP |
| Electric arc | 4000 – 5000 | Emission | Arc-source emission spectroscopy |
| Electric spark | 40,000(?) | Emission | Spark-source emission spectroscopy |

atomic fluorescence (AFS) spectroscopy. The first two of these procedures are treated in this text. Atomic fluorescence spectroscopy will not be considered further.

### 16A – 1 The Flame Atomization Process

In flame atomization, an aqueous solution of the sample is dispersed (or *nebulized*) into small droplets, which are then mixed with a gaseous fuel and oxidant and swept into a laboratory-type burner. Evaporation of solvent occurs in the *base region* of the flame, which is located just above the tip of the burner. The resulting finely divided solid particles are carried to a region in the center of the flame, called the *inner cone.* Here, in this hottest part of the flame, atoms and elementary ions are formed from the solid particles. Finally, the atoms and ions are carried to the outer edge, or *outer cone,* where oxidation may occur before the atomized particles disperse into the atmosphere. Because the velocity of movement of the fuel – oxidant mixture is high, only a fraction of the sample undergoes all of these processes; indeed, a flame is not a very efficient atomizer.

**Emission and Absorption Spectra in Flames**   Both emission and absorption spectra for atoms or elementary ions are obtained from flames. Emission spectra are recorded by locating the inner cone of the flame in front of the entrance slit of a monochromator, such as one of those shown in Figure 15–12, and then monitoring the output signal with a detector as the spectrum is scanned. Figure 16–1 shows a part of a spectrum recorded in this way. The sample was a potassium salt, and the line is one of several for that element. Note that the line is superimposed on a background continuum that results from excitation of molecular species (such as carbon monoxide, cyanogen, or hydroxyl radicals).

Absorption spectra are obtained with spectrophotometers similar to those described in Chapter 15, Section 15H–4. In this application, however, the cell compartment is replaced by the flame and the source is modulated so that the detector circuit can discriminate between radiation from the source and that from the flame (source modulation is described in Section 16B–3).

**Flame Temperatures**   Flame temperatures vary from about 1700 to 3100°C or greater, depending upon the fuels and oxidants used. Natural gas–air flames provide the lowest temperatures, while acetylene–oxygen or acetylene–nitrous oxide flames give the highest. Hydrogen–oxygen flames have intermediate temperatures in the 2500 to 2700°C range.

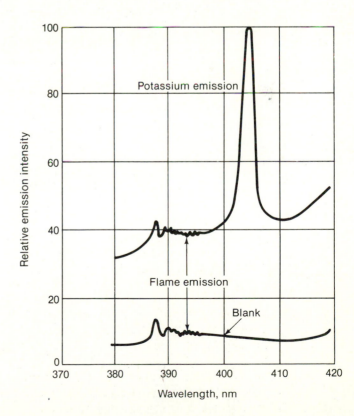

*Figure 16 – 1*   A portion of the emission spectrum for potassium in a hydrogen–oxygen flame. Note that the potassium emission is superimposed upon the background emission from the flame. For clarity, the spectrum for the blank has been displaced downward.

**Effects of Flame Temperature**    As noted in Chapter 15, atomic absorption and emission spectra consist of a series of discrete lines whose wavelengths are unique for each element. Line spectra arise from excitation of the outer electrons of the atom — either by the energy of a radiant source or by the energy of a hot flame. The average number of excited atoms in a flame is ordinarily minuscule with respect to the population of unexcited atoms (typically, the ratio of excited to unexcited atoms at 2000 to 3000°C is $10^{-3}$ to $10^{-5}$). From this standpoint, absorption methods, based as they are on the unexcited atoms, should be significantly more sensitive than emission methods, which depend upon the number of excited atoms. In fact, however, several other variables also influence the sensitivity of absorption and emission procedures and the two methods tend to complement each other in sensitivity. Table 16 – 2 illustrates this point. Here, it can be seen that flame emission methods are more sensitive for 24 of the 62 elements listed, while flame absorption procedures produce higher sensitivities for 21; for 17 elements, the two procedures are similar in sensitivity.

One real advantage of flame absorption methods arises from the fact that the absolute number of unexcited atoms is much less temperature dependent than is the number of excited species. For example, at 2500°K, a 20°K increase in flame temperature causes the number of sodium atoms in the excited $3p$ state to increase by approximately 8%, while the corresponding decrease in the much larger number of ground-state sodium atoms is only about 0.02%. Therefore, emission methods, based as they are on atoms in the excited state, require closer control of flame temperatures. It should be noted, however, that secondary temperature effects do influence atomic absorption measurements as well. One temperature effect, which is important in both emission and absorption measurements, is the increase in atom population in the flame that accompanies higher flame temperatures. As we will see later, higher temperatures also increase line broadening, which causes decreases in peak heights.

Table 16 – 2    Comparison of Flame Emission and Atomic Absorption Spectrophotometry*

| Flame Emission More Sensitive | | | Approximately Equal Sensitivity | | Atomic Absorption More Sensitive | | |
|---|---|---|---|---|---|---|---|
| Al | Li | Sr | Cr | Pd | Ag | Hg | Sn |
| Ba | Lu | Tb | Cu | Rh | As | Ir | Te |
| Ca | Na | Tl | Dy | Sc | Au | Mg | Zn |
| Eu | Nd | Tm | Er | Ta | B | Ni | |
| Ga | Pr | W | Gd | Ti | Be | Pb | |
| Ho | Rb | Yb | Ge | V | Bi | Pt | |
| In | Re | | Mn | Y | Cd | Sb | |
| K | Ru | | Mo | Zr | Co | Se | |
| La | Sm | | Nb | | Fe | Si | |

* From E. E. Pickett and S. R. Koirtyohann, *Anal. Chem.,* **1969,** *41*(14), 28A. With permission of the American Chemical Society.

**Flame Atomizers**   Two types of burners are used in flame spectroscopy. Figure 16–2a is a schematic diagram of a commercially available *turbulent flow* or *total consumption* burner. Here, the nebulizer and burner are combined into a single unit. The sample is drawn up the capillary and nebulized by Venturi action caused by the flow of gases around the capillary tip. Typical sample flow rates are 1 to 3 mL/min.

Turbulent flow burners offer the advantage of introducing a relatively large

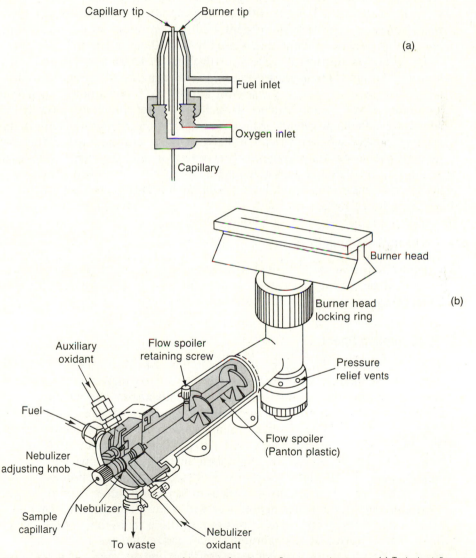

**Figure 16–2**   Schematic diagrams of burners for atomic flame spectroscopy. (a) Turbulent-flow or total consumption burner. (Courtesy of Beckman Instruments, Fullerton, CA.) (b) Laminar premix burner. (Courtesy of Perkin-Elmer Corporation, Norwalk, CT.)

and representative sample into the flame. Disadvantages include a relatively short path length through the flame and problems with clogging of the tip. In addition, these burners are noisy both from the electronic and auditory standpoint. Although sometimes used for emission and fluorescence, turbulent flow burners find little use in present-day absorption instruments.

Figure 16–2b is a diagram of a typical commercial *laminar flow,* or *premix,* burner. The sample is nebulized by the flow of oxidant past a capillary tip. The resulting aerosol is then mixed with fuel and flows past a series of baffles that remove all but the finest droplets. As a result of the baffles, the majority of the sample collects in the bottom of the mixing chamber, where it is drained to a waste container. The aerosol, oxidant, and fuel are then burned in a slotted burner that provides a flame that is usually 5 or 10 cm in length.

Laminar flow burners provide a relatively quiet flame and a significantly longer path length. These properties tend to enhance sensitivity and reproducibility. Furthermore, clogging is seldom a problem. Disadvantages include a lower rate of sample introduction (which may offset the longer path length advantage) and the possibility of selective evaporation of mixed solvents in the mixing chamber, which may lead to analytical uncertainties. Furthermore, the mixing chamber contains a potentially explosive mixture that can be ignited by a flashback. Note that the burner in Figure 16–2b is equipped with pressure-relief vents for this reason. In addition, the burner head is sometimes held in place by stainless steel cables.

## 16B  ATOMIC ABSORPTION SPECTROSCOPY

Atomic absorption spectroscopy based on flames is currently the most widely used of all the atomic methods listed in Table 16–1 because of its simplicity, effectiveness, and relatively low cost.

### 16B–1  Atomic Absorption Spectra

**Source of Spectra**    The energy of radiation absorbed by a vaporized atom is identical with that needed to bring about excitation to a higher electronic state. Because the majority of atoms in a flame are unexcited, the most probable transitions will involve excitation of an electron from the ground state to a higher energy level. For example, when a sodium atom absorbs radiation of 589 nm, an outer $3s$ electron is promoted to the $3p$ state. Transitions from one excited state to another, while not forbidden, are far less likely, owing to the very small population of excited atoms. As a consequence, absorption associated with such transitions is usually so weak as to be undetectable. Typically, then, an atomic absorption spectrum produced with a flame consists predominantly of *resonance* lines which are the result of transitions from the ground state to higher energy levels. It is important to emphasize that *the wavelength of a resonance line is identical to that of the emission line associated with the same electronic transition.*

**Spectral Line Widths**   The natural width of an atomic absorption or atomic emission line is on the order of $10^{-5}$ nm. However, two effects tend to cause observed widths to be larger by a factor of 100 (or more). *Doppler broadening* results from the rapid motion of species in the flame. Atoms moving toward the detector emit wavelengths that are slightly shorter, while the reverse is true for atoms moving in the opposite direction. The net effect is an increase in the width of the emission line. Doppler broadening is also responsible for the widening of absorption lines and for precisely the same reasons. *Pressure broadening* is due to collisions among atoms that result in slight variations in their ground-state energies. Increases in temperature cause both effects to become more pronounced; broader absorption and emission peaks are thus associated with elevated temperatures.

## 16B–2 An Instrumental Problem Created by Narrow Atomic Absorption Lines

Since transition energies for atomic absorption lines are unique for each element, analytical methods based upon atomic absorption have the potential of being highly specific. On the other hand, the narrow line widths create a problem in quantitative analysis that is not encountered in molecular absorption. No ordinary monochromator is capable of yielding a band of radiation as narrow as the peak width of an atomic absorption line (0.002 to 0.005 nm). Consequently, the use of radiation that has been isolated from a continuous source by a monochromator will inevitably cause instrumental departures from Beer's law (see the discussion of instrumental deviations from Beer's law in Chapter 15, Section 15C–2). In addition, since the fraction of radiation absorbed from such a beam is small, the detector receives a signal that is only slightly attenuated (i.e., $P \rightarrow P_0$) and sensitivity of the measurement is lost.

**Solution to the Problem**   The problem of narrow line widths has been overcome by making use of radiation from a source that not only emits a *line of the same wavelength* as the one selected for the absorption analysis but also one that is *narrower*. For example, the output of a mercury-vapor lamp can be used for the determination of mercury. Gaseous mercury atoms, electrically excited in such lamps, return to the ground state by emitting radiation with wavelengths that are characteristic for that element. One of these emission lines, such as the one at 546.1 nm, can be isolated by filters or a monochromator and used in the determination of mercury by atomic absorption. Since the lamp is operated at a temperature lower than that of the flame, Doppler and pressure broadening of the line from the source is less than that for the corresponding absorption peak in the flame. As a consequence, the effective bandwidth of the source is significantly less than the corresponding bandwidth of the absorption peak. Figure 16–3 illustrates the principle of the procedure. Figure 16–3a depicts four narrow *emission* lines from a typical atomic lamp source. Figure 16–3b shows the flame *absorption spectrum* for the analyte between the wavelengths $\lambda_1$ and $\lambda_2$; note that the width of the absorption peak is significantly greater than that

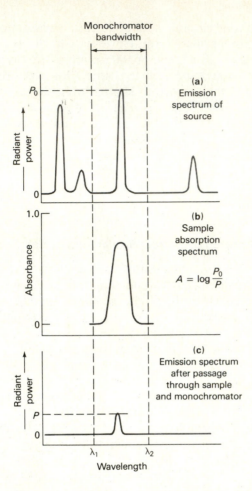

**Figure 16-3** Atomic absorption of a resonance line.

for the emission peak. As shown in Figure 16-3c, the intensity of the incident beam $P_0$ has been decreased to $P$ by passage through the sample. Since the bandwidth of the source is now significantly less than the bandwidth of the peak, $\log P_0/P$ is more likely to be linearly related to concentration.

**Hollow-Cathode Lamps**    The most useful source for atomic absorption spectroscopy is the *hollow-cathode lamp,* shown schematically in Figure 16-4. It consists of a tungsten anode and a cylindrical cathode sealed in a glass tube containing an inert gas, such as argon, at a pressure of 1 to 5 torr. The cathode is fabricated from the metal whose spectrum is sought or else serves to support a coating of that metal.

Application of a potential of about 300 V across the electrodes causes ionization of the argon and generation of a current of 5 to 10 mA as the ions and electrons migrate to the two electrodes. If the potential is sufficiently large, the cations of the gas strike the cathode with sufficient energy to dislodge some of

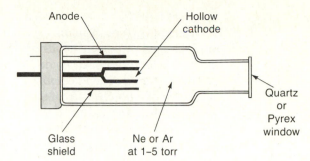

**Figure 16–4** Schematic diagram of a hollow-cathode lamp.

the metal atoms and thereby produce an atomic cloud; this process is called *sputtering*. A fraction of the sputtered metal atoms are in excited states and emit their characteristic wavelengths as they return to the ground state. The vaporized metal atoms eventually diffuse back to the cathode surface (or to the walls of the lamp) and are deposited.

Hollow-cathode lamps for about 40 elements are available from commercial sources. Some are fitted with cathodes containing more than one element; such lamps provide spectral lines for the determination of several species.

**Electrodeless Discharge Lamps** Electrodeless discharge lamps are useful sources of atomic line spectra and provide radiant intensities that are usually one to two orders of magnitude greater than their hollow-cathode counterparts. A typical lamp is constructed from a sealed quartz tube containing a few torr of an inert gas such as argon and a small quantity of the metal (or its salt) whose spectrum is of interest. The lamp contains no electrode, but instead is energized by an intense field of radio-frequency or microwave radiation. Ionization of the argon occurs, giving ions that are accelerated by the high-frequency component of the field until they gain sufficient energy to excite by collision the atoms of the metal whose spectrum is sought.

Electrodeless discharge lamps are available commercially for several elements. Their performance does not appear to be as reliable as that of the hollow-cathode lamp.

## 16B–3 Source Modulation

In an atomic absorption measurement it is necessary to eliminate the effects of radiation from the flame. Most of such radiation is removed by the monochromator, which is located between the flame and the detector. This arrangement does not, however, remove radiation with the same wavelength as that selected for the analysis. The flame will emit such radiation as the result of thermal excitation of a fraction of the analyte atoms. The effect of analyte emission is overcome by *modulating* the output from the hollow-cathode lamp so that its intensity fluctuates at a constant frequency. The detector thus receives an alternating signal from the hollow-cathode lamp and a continuous signal from the flame. These signals are converted into the corresponding types of electrical current. A relatively simple electronic system then eliminates the unmodulated dc current produced by the flame and passes the ac signal from the source for amplification.

Modulation is often accomplished by interposing a mechanical chopper, consisting of a notched circular disk, in the light path between the source and the flame. Rotation of this disk at a constant speed causes modulation of the output from the hollow-cathode lamp. As an alternative, the power supply for the source can be designed for intermittent or ac operation.

### 16B-4 Instruments

An atomic absorption instrument contains the same basic components as an instrument designed for absorption analysis of solutions. These components include a source, a sample container (here, a flame or a hot surface), a wavelength selector, and a detector–readout system. Instruments for atomic absorption work are offered by numerous manufacturers; both single- and double-beam designs are available. The range of sophistication and cost (upward from a few thousand dollars) is substantial.

In general, the instrument must be capable of providing a sufficiently narrow bandwidth to isolate the line chosen for the measurement from other lines that may interfere with or diminish the sensitivity of the analysis. A glass filter is adequate for some of the alkali metals, which have only a few widely spaced resonance lines in the visible region. An instrument equipped with readily interchangeable interference filters is available commercially. A separate filter (and light source) is used for each element. Satisfactory results for the analysis of 22 metals are claimed. Most instruments, however, incorporate a good-quality ultraviolet-visible monochromator.

**Spectrophotometers**    Figure 16–5 is a schematic diagram for a typical double-beam atomic absorption spectrophotometer. Radiation from the hollow-cathode lamp is mechanically split into two beams, one of which passes through the flame and the other around the flame. A half-silvered mirror returns both beams to a single path, by which they pass alternately through the monochromator and thence to the detector. Note that the monochromator is located between the sample and the detector, rather than between the source and the sample, as in ultraviolet-visible spectrophotometers for solution measurement. This arrangement eliminates most of the radiation emanating from the flame. It is of interest to note that a similar arrangement is not used in conventional spectrophotometers for liquids because such geometry exposes the sample to the full output from the source, which may cause photochemical decomposition. Infrared spectrophotometers also locate the sample between the source and the monochromator because infrared radiation is not sufficiently energetic to cause photodecomposition. This geometry is particularly advantageous, however, because it minimizes the effects of scattered radiation, which is often serious in the infrared region.

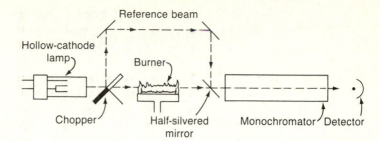

**Figure 16 – 5** Optical paths in a double-beam atomic absorption spectrophotometer.

**Atomizers**   Two types of atomizers are employed in atomic absorption spectroscopy, namely *flame* and *electrothermal* or *nonflame*. The former type was described earlier. Electrothermal atomizers are considered in the next section.

## 16B – 5 Electrothermal Atomization

Flame atomization is not very efficient for two reasons. First, a large portion of the sample flows down the drain (laminar burner) or is not completely atomized (turbulent burner). Second, the residence time of individual atoms in the optical path of the flame is brief ($\sim 10^{-4}$ sec). Electrothermal atomizers, which first appeared on the market in the early 1970s, generally provide enhanced sensitivity because the entire sample is atomized in a short period, and the average residence time of the atoms in the optical path is 1 sec or more.[2]

**Instrumentation**   In an electrothermal atomizer, a few microliters of sample are first evaporated at a low temperature and then ashed at a somewhat higher temperature on an electrically heated surface of carbon, tantalum, or other conducting material. The conductor is formed as a hollow tube, a strip or rod, a cup, or a trough. After ashing, the current is rapidly increased to several hundred amperes, which causes the temperature to soar to perhaps 2000 to 3000°C; atomization of the sample occurs in a period of a few milliseconds to seconds. The absorbance of the atomized particles is then determined in the region immediately above the heated conductor.

Figure 16 – 6a illustrates a commercially available electrothermal atomizer, which fits in front of the entrance slit of a monochromator. In this device, the sample is contained in a graphite cup (A) supported between two electrodes (B) of the same material. The cup, which is surrounded by a sheath of inert gas, has holes on either side to permit passage of radiation from the hollow-cathode lamp over the surface of the sample. The metal mounting (C) is water cooled.

**Detector Output**   With the monochromator set at a wavelength at which absorption occurs, the detector output during electrothermal atomization rises to

[2] See R. E. Sturgeon, *Anal. Chem.,* **1977,** *49,* 1255A; S. R. Koirtyohann and M. L. Kaiser, *Anal. Chem.,* **1982,** *54,* 1515A; and C. W. Fuller, *Electrothermal Atomization for Atomic Absorption Spectroscopy.* London: The Chemical Society, 1978.

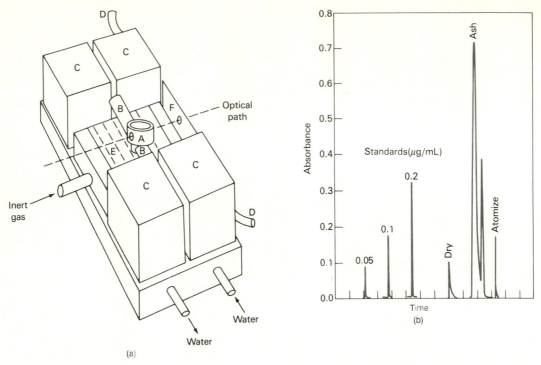

**Figure 16–6**   (a) Schematic diagram of a graphite cup atomizer. (b) Typical output from a spectrophotometer equipped with an electrothermal atomizer. The time for drying and ashing are 20 and 60 sec, respectively. (Courtesy of Varian Instrument Division, Palo Alto, CA.)

a maximum after a few seconds of ignition and then rapidly decays back to zero as the atomization products escape into the surroundings. The change is rapid enough (often < 1 sec) to require a high-speed recorder. Quantitative analyses are usually based on peak height although peak area has also been used.

Figure 16–6b shows the output signals (as a function of time) from an atomic absorption spectrophotometer equipped with a graphite cup atomizer. The four peaks on the right were recorded at the wavelength of a lead peak as a 2-$\mu$L sample of canned orange juice was dried, ashed, and then atomized. The peaks produced during drying and ashing were caused by particulate ignition products. Comparison of the peak for lead on the far right with peaks recorded for standards indicates a lead concentration of about 0.1 $\mu$g/mL of juice.

Electrothermal atomizers offer the advantage of unusually high sensitivity for small volumes of sample. Typically, sample volumes are between 0.5 and 10 $\mu$L; under these circumstances, absolute detection limits typically lie in the range of $10^{-10}$ to $10^{-13}$ g of analyte.

The relative precision of nonflame methods is generally in the range of 5 to 10%, compared with the 1 to 2% that can be expected for flame atomization. Furthermore, interference problems, which are discussed in the next section, tend to be more severe with electrothermal than with flame atomizers.

## 16B–6 Interferences

Interferences of two types are encountered in atomic absorption methods that employ flame and electrothermal atomization. *Spectral interferences* arise when particulate matter from atomization scatters the incident radiation from the source, or when the absorption or emission of an interfering species either overlaps or lies so close to the analyte absorption that resolution by the monochromator becomes impossible. *Chemical interferences* result from various chemical processes occurring during atomization that alter the absorption characteristics of the analyte.

**Spectral Interferences**   Interference due to overlapping lines is rare because the emission lines of hollow cathode sources are so very narrow. Nevertheless, such an interference can occur if the separation between two lines is on the order of 0.1 Å. For example, a vanadium line at 3082.11 Å interferes in an analysis based upon the aluminum absorption line at 3082.15 Å. The interference is readily avoided, however, by making use of the aluminum line at 3092.7 Å instead.

Spectral interferences also result from the presence of *molecular* combustion products that exhibit broad-band absorption or of particulate products that scatter radiation. Both diminish the power of the transmitted beam and lead to positive analytical errors. Where the source of these products is the fuel and oxidant mixture alone, corrections are readily obtained from absorbance measurements while a blank is aspirated into the flame.

A much more troublesome problem is encountered when the source of absorption or scattering originates in the sample matrix. In this type of interference, the power of the transmitted beam, $P$, is reduced by the matrix components, but the incident beam power, $P_0$, is not; a positive error in absorbance and thus concentration results. An example of a potential matrix interference due to absorption occurs in the determination of barium in alkaline-earth mixtures. The wavelength of the barium line used for atomic absorption analysis appears in the center of a broad absorption *band* for molecular CaOH; interference of calcium in a barium analysis results. The effect is readily eliminated by substituting nitrous oxide for air as the oxidant; the higher temperature decomposes the CaOH and eliminates the absorption band.

Spectral interference due to scattering by products of atomization often occurs when concentrated solutions containing elements such as Ti, Zr, and W—which form stable oxides—are aspirated into the flame. Metal oxide particles with diameters greater than the wavelength of light appear to be formed, which cause scattering of the incident beam.

Fortunately, with flame atomization, spectral interferences by matrix products are not widely encountered and usually can be avoided by variations in such analytical parameters as temperature and fuel-to-oxidant ratio. Alternatively, if the source of interference is known, an excess of the interfering substance can be added to both sample and standards; provided the excess is large with respect to the concentration from the sample matrix, the contribution of the latter will become insignificant. The added substance is sometimes called a *radiation buffer.*

The matrix interference problem is greatly exacerbated with electrothermal atomization and is one of the major causes of the poor accuracy associated with nonflame methods.[3]

**Chemical Interferences**    Chemical interferences can frequently be minimized by a suitable choice of operating conditions. Perhaps the most common type of chemical interference is by anions that form compounds of low volatility with the analyte and thus decrease the rate at which it is atomized. Low results are the consequence. An example is the decrease in calcium absorbance that is observed with increasing concentrations of sulfate or phosphate ions, which form nonvolatile compounds with calcium ion.

Interferences due to formation of species of low volatility can often be eliminated or moderated by use of higher temperatures. Alternatively, *releasing agents,* which are cations that react preferentially with the interference and prevent its interaction with the analyte, can be introduced. For example, addition of an excess of strontium or lanthanum ion minimizes the interference of phosphate in the determination of calcium. Here, the strontium or lanthanum replaces the analyte in the nonvolatile compound formed with the interfering species.

*Protective agents* prevent interference by preferentially forming stable but volatile species with the analyte. Three common reagents for this purpose are EDTA, 8-hydroxyquinoline, and APDC (the ammonium salt of 1-pyrrolidine-carbodithioic acid). For example, the presence of EDTA has been shown to eliminate the interference of silicon, phosphate, and sulfate in the determination of calcium.

Ionization of atoms and molecules is usually inconsequential in combustion mixtures that involve air as the oxidant. In high-temperature flames where oxygen or nitrous oxide serves as the oxidant, however, ionization becomes important, and a significant concentration of free electrons exists as a consequence of the equilibrium

$$M \rightleftharpoons M^+ + e^-$$

where M represents a neutral atom or molecule and $M^+$ is its ion. Ordinarily, the spectrum of $M^+$ is quite different from that of M, so that ionization of the analyte ions leads to low results. It is important to appreciate that treatment of the ionization process as an equilibrium — with free electrons as one of the products — immediately implies that the degree of ionization of an analyte atom will be strongly influenced by the presence of other ionizable metals in the

---

[3] For a discussion of methods for overcoming matrix interference problems, see D. A. Skoog, *Principles of Instrumental Analysis,* 3rd ed., pp. 270–273. Philadelphia: Saunders College Publishing, 1985.

flame. Thus, if the medium contains not only species M, but species B as well, and if B ionizes according to the equation

$$B \rightleftharpoons B^+ + e^-$$

then the degree of ionization of M will be decreased by the mass-action effect of the electrons formed from B. The errors caused by ionization of the analyte can frequently be eliminated by addition of an *ionization suppressor,* which provides a relatively high concentration of electrons to the flame; suppression of ionization of the analyte results.

### 16B–7 Applications of Atomic Absorption Spectroscopy

Atomic absorption spectroscopy provides a sensitive means for the determination of more than 60 elements. The method is well suited for routine measurements by relatively unskilled operators.

**Quantitative Techniques**   Quantitative atomic absorption analyses are often based on calibration curves, which, in principle, are linear. Departures from linearity occur, however, and analyses should *never* be based on the measurement of a single standard with the assumption that Beer's law is being followed. In addition, production of an atomic vapor involves sufficient uncontrolled variables to warrant measuring the absorbance of at least one standard solution each time an analysis is performed. Any deviation of the standard from its original calibration value can then be applied as a correction to the analytical results.

The standard addition method is extensively used in atomic absorption spectroscopy. Here, two or more aliquots of the sample are transferred to volumetric flasks. One is diluted to volume directly, while a known amount of analyte is introduced to the other before dilution to the same volume. The absorbance of each is measured (several different standard additions are recommended if the method is unfamiliar). If a linear relationship exists between absorbance and concentration (and this must be verified experimentally) the following relationships apply:

$$A_x = \frac{kV_xC_x}{V_T}$$

$$A_T = \frac{k(V_xC_x + V_sC_s)}{V_T}$$

where $V_x$ and $C_x$ are the volume and concentration of the analyte solution, $V_s$ and $C_s$ are the same parameters for the standard, and $V_T$ is the total volume; $A_x$ and $A_T$ are the absorbances of the sample alone and the sample plus standard, respectively. These two equations are readily combined to give

$$C_x = \frac{A_x}{(A_T - A_x)} \times \frac{C_sV_s}{V_x}$$

If several standard additions have been made, $A_T$ can be plotted against $C_s$. The resulting straight line can then be extrapolated to $A_T = 0$, at which point,

$$C_x = \frac{-C_s V_s}{V_x}$$

Use of the standard addition method tends to compensate for variations caused by physical and chemical interferences in the analyte solution.

**Detection Limits**   Columns two and three of Table 16–3 provide information on detection limits for a number of common elements by flame and electro-thermal atomic absorption. For comparison purposes, limits for some of the other atomic procedures are also included. Small differences among the quoted values are not significant. Thus, whereas an order of magnitude is probably meaningful, a factor of 2 or 3 certainly is not.

**Accuracy**   Under usual conditions, the relative error associated with a flame absorption analysis is of the order of 1 to 2%. With special precautions, this

**Table 16–3   Sensitivity (nm/mL)* for Selected Elements†**

| Element | AAS,‡ Flame | AAS,§ Electrothermal | AES,‡ Flame | AES,‡ ICP |
|---------|-------------|----------------------|-------------|-----------|
| Al | 30 | 0.005 | 5 | 2 |
| As | 100 | 0.02 | 0.0005 | 40 |
| Ca | 1 | 0.02 | 0.1 | 0.02 |
| Cd | 1 | 0.0001 | 800 | 2 |
| Cr | 3 | 0.01 | 4 | 0.3 |
| Cu | 2 | 0.002 | 10 | 0.1 |
| Fe | 5 | 0.005 | 30 | 0.3 |
| Hg | 500 | 0.1 | 0.0004 | 1 |
| Mg | 0.1 | 0.00002 | 5 | 0.05 |
| Mn | 2 | 0.0002 | 5 | 0.06 |
| Mo | 30 | 0.005 | 100 | 0.2 |
| Na | 2 | 0.0002 | 0.1 | 0.2 |
| Ni | 5 | 0.02 | 20 | 0.4 |
| Pb | 10 | 0.002 | 100 | 2 |
| Sn | 20 | 0.1 | 300 | 30 |
| V | 20 | 0.1 | 10 | 0.2 |
| Zn | 2 | 0.00005 | 0.0005 | 2 |

* Nanogram/milliliter $= 10^{-3}\ \mu g/mL = 10^{-3}$ ppm.
† AAS = atomic absorption spectroscopy; AES = atomic emission spectroscopy; ICP = inductively coupled plasma.
‡ From V. A. Fassel and R. N. Kniseley, *Anal. Chem.*, **1974**, *46*, 1111A. With permission of the American Chemical Society.
§ From C. W. Fuller, *Electrothermal Atomization for Atomic Absorption Spectroscopy*, pp. 65–83. London: The Chemical Society, 1977. With permission of The Royal Society of Chemistry.

figure can be lowered to a few tenths of one percent. Errors encountered with nonflame atomization usually exceed those for flame atomization by a factor of 5 to 10.

## 16C FLAME EMISSION SPECTROSCOPY

Atomic emission spectroscopy employing flames (also called flame emission spectroscopy or flame photometry) has found widespread application to elemental analysis. Its most important uses have been in the determination of sodium, potassium, lithium, and calcium, particularly in biological fluids and tissues. For reasons of convenience, speed, and relative freedom from interferences, flame emission spectroscopy has become the method of choice for these otherwise difficult to determine elements. The method has also been applied, with varying degrees of success, to the determination of perhaps half the elements in the periodic table.

### 16C – 1 Instrumentation

Instruments for flame emission work are similar in design to flame absorption instruments except that the flame now acts as the radiation source; the hollow-cathode lamp and chopper are therefore unnecessary. Many modern instruments are adaptable to either emission or absorption measurements. Much of the early work in atomic emission analyses was accomplished with turbulent flow burners. Laminar flow burners, however, are becoming more and more widely used.

**Spectrophotometers**   For nonroutine analysis, a recording, ultraviolet-visible spectrophotometer with a resolution of perhaps 0.5 Å is desirable. The recording feature provides a simple means for making background corrections (Figure 16 – 7).

**Photometers**   Simple filter photometers often suffice for routine determinations of the alkali and alkaline-earth metals. A low-temperature flame is employed to eliminate excitation of most other metals. As a consequence, the spectra are simple, and interference filters can be used to isolate the desired emission line.

   Several instrument manufacturers supply flame photometers designed specifically for the analysis of sodium, potassium, and lithium in blood serum and other biological samples. In these instruments, the radiation from the flame is split into three beams of approximately equal power. Each then passes into a separate photometric system consisting of an interference filter (which transmits an emission line of one of the elements while absorbing those of the other two), a phototube, and an amplifier. The outputs can be measured separately if desired. Ordinarily, however, lithium serves as an *internal standard* for the analysis. For this purpose, a fixed amount of lithium is introduced into each

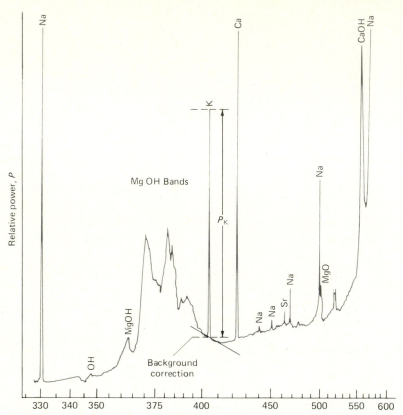

**Figure 16 – 7**   Flame emission spectrum for a natural brine showing the method used for correcting for background radiation. (From: R. Herrmann and C. T. J. Alkemade, *Chemical Analysis by Flame Photometry*, 2nd ed., p. 484. New York: Interscience, 1963. Reprinted by permission of John Wiley & Sons, Inc.)

standard and sample. The ratios of outputs of the sodium and lithium transducer and the potassium and lithium transducer then serve as analytical parameters. This system provides improved accuracy because the intensities of the three lines are affected in the same way by most analytical variables, such as flame temperature, fuel flow rates, and background radiation. Clearly, lithium must be absent from the sample.

**Automated Flame Photometers**   Fully automated photometers now exist for the determination of sodium and potassium in clinical samples. In one of these, the samples are withdrawn sequentially from a sample turntable, dialyzed to remove protein and particulates, diluted with a lithium internal standard, and aspirated into a flame. Sample and reagent transport is accomplished with a roller-type pump. Air bubbles serve to separate samples. Results are printed out on a paper tape. Calibration is performed automatically after every nine samples.

### 16C – 2 Interferences

The interferences encountered in flame emission spectroscopy have the same sources as those in atomic absorption methods (see Section 16B – 6); the severity of any given interference will often differ for the two procedures, however.

### 16C – 3 Self-Absorption

The center of a flame is hotter than the exterior, so that atoms emitting in the center are surrounded by a cooler region which contains a greater fraction of unexcited atoms; *self-absorption* of the resonance wavelengths by the atoms in the cooler layer results. Doppler broadening affects the emission line more than it does the resonance absorption line, however, because the particles are moving more rapidly in the hotter emission zone. Thus, self-absorption tends to alter the center of a line more than its edges. In the extreme, the center may become less intense than the edges, or it may even disappear.

As shown by Figure 16 – 8, self-absorption and ionization sometimes produce an S-shaped calibration curve that has three distinct segments. At intermediate concentrations, a linear relationship between intensity and concentration is observed. The curvature at low concentrations is due to the increased degree of ionization that occurs in the more dilute solution. Self-absorption, on the other hand, causes negative departures from a straight line at higher concentrations.

### 16C – 4 Analytical Techniques

The analytical techniques for flame emission spectroscopy are similar to those described earlier for atomic absorption spectroscopy. Both calibration curves and the standard addition method are employed. In addition, internal standards may be used to compensate for flame variables.

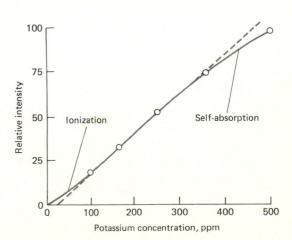

**Figure 16 – 8** Influence of ionization and self-absorption on a calibration curve for potassium.

## 16D ATOMIC EMISSION METHODS BASED ON PLASMA SOURCES

Plasma sources, which became available commercially in the mid-1970s, offer several advantages over both flame emission and atomic absorption procedures.[4] By definition, a plasma is a conducting gaseous mixture containing a significant concentration of cations and electrons. In the argon plasma employed for emission analyses, argon ions and electrons are the principal conducting species, although cations from the sample will also contribute. Argon ions, once formed in a plasma, are capable of absorbing sufficient power from an external source to maintain the temperature at a level at which further ionization sustains the plasma indefinitely; temperatures as great as $10,000°K$ are encountered. Three power sources have been employed in argon plasma spectroscopy. One is a dc electrical source capable of maintaining a current of several amperes between electrodes immersed in the argon plasma. The second and third are powerful radio-frequency and microwave frequency generators through which the argon flows. Of the three, the radio-frequency, or *inductively coupled plasma* (ICP), source appears to offer the greatest advantage in terms of sensitivity and freedom from interference. On the other hand, the *dc plasma source* (DCP) has the virtue of simplicity and lower cost. Both will be described here.

### 16D-1 The Inductively Coupled Plasma Source[5]

Figure 16–9a is a schematic drawing of an inductively coupled plasma source. It consists of three concentric quartz tubes through which streams of argon flow at a total rate between 11 and 17 L/min. The diameter of the largest tube is about 2.5 cm. Surrounding the top of this tube is a water-cooled induction coil that is powered by a radio-frequency generator, which is capable of producing 2 kW of energy at about 27 MHz. Ionization of the flowing argon is initiated by a spark from a Tesla coil. The resulting ions, and their associated electrons, then interact with the fluctuating magnetic field (labeled $H$ in Figure 16–9) produced by the induction coil. This interaction causes the ions and electrons within the coil to flow in the closed annular paths depicted in the figure; ohmic heating is the consequence of their resistance to this movement.

The temperature of this plasma is high enough (several thousand degrees Kelvin) to require thermal isolation from the outer quartz cylinder. Isolation is achieved by flowing argon tangentially around the walls of the tube, as indicated by the arrows in Figure 16–9a. The tangential flow cools the inside walls of the central tube and centers the plasma radially.

---

[4] For a detailed discussion of the various plasma sources, see P. Tschopel, in *Comprehensive Analytical Chemistry*, G. Svehla, Ed., Vol. IX, Chapter 3. New York: Elsevier, 1979; R. D. Sacks in *Treatise on Analytical Chemistry*, 2nd ed., P. J. Elving, E. J. Meehan, and I. M. Kolthoff, Eds., Part I, Vol. 7, pp. 516–526. New York: John Wiley & Sons, 1981.

[5] For a more complete discussion, see V. A. Fassel, *Science,* **1978,** *202,* 183; V. A. Fassel, *Anal. Chem.,* **1979,** *51,* 1290A; M. Thompson and J. N. Walsh, *Inductively Coupled Plasma Spectrometry.* London: Blackie, 1983; R. M. Barnes, *CRC Crit. Rev. Anal. Chem.,* **1978,** *7,* 203.

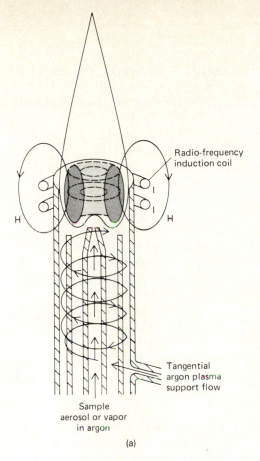

Radio-frequency
induction coil

H                    H

Tangential
argon plasma
support flow

Sample
aerosol or vapor
in argon

(a)

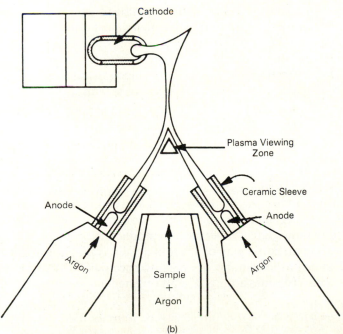

Cathode

Plasma Viewing
Zone

Ceramic Sleeve

Anode                                    Anode

Argon                                    Argon

Sample
+
Argon

(b)

**Figure 16–9** Plasma sources.
(a) Inductively coupled plasma
source. (From: V. A. Fassel,
*Science,* **1978,** *202,* 185. With
permission. Copyright 1978 by
the American Association for the
Advancement of Science.) (b) A
three-electrode dc plasma jet.
(Courtesy of SpectraMetrics Inc.,
Haverhill, MA.)

**Sample Injection**     The sample is carried into the hot plasma at the head of the tubes by argon flowing at about 1 L/min through the central quartz tube. The sample may be an aerosol, a thermally generated vapor, or a fine powder.

**Plasma Appearance and Spectra**     The typical plasma has a very intense, brilliant white, nontransparent core topped by a flamelike tail. The core, which extends a few millimeters above the tube, is made up of a continuum upon which is superimposed the atomic spectrum for argon. The continuum apparently results from recombination of argon and other ions with electrons. In the region 10 to 30 mm above the core, the continuum fades, and the plasma is optically transparent. Spectral observations are generally made at a height of 15 to 20 mm above the induction coil. Here, the background radiation is remarkably free of argon lines and is well suited for analysis. Many of the most sensitive analyte lines in this region of the plasma are from ions such as $Ca^+$, $Ca^{2+}$, $Cd^+$, $Cr^{2+}$, and $Mn^{2+}$.

**Analyte Atomization and Ionization**     By the time the sample atoms have reached the observation point in the plasma, they will have had a residence time of about 2 msec at temperatures ranging from 6000 to 8000°K. These times and temperatures are two to three times as great as those attainable in the hotter combustion flames (acetylene–nitrous oxide). As a consequence, atomization is more complete, and fewer chemical interferences are encountered. Surprisingly, ionization interference effects are small or nonexistent, perhaps because the large concentration of electrons from ionization of the argon represses ionization of the sample components.

Several other advantages are associated with the plasma source. First, atomization occurs in a chemically inert environment, which should also enhance the lifetime of the analyte. In addition, and in contrast to flame sources, the temperature cross section of the plasma is relatively uniform; therefore, self-absorption and self-reversal effects (Section 16C–3) are not encountered. Calibration curves thus tend to remain linear over several orders of magnitude of concentration.

### 16D–2 The Direct-Current Argon Plasma Source

Direct-current plasma jets were first described in the 1920s and have been systematically investigated as sources for emission spectroscopy for more than two decades. It was not until recently, however, that a source based on this principle has been designed that can successfully compete with flame and inductively coupled plasma sources in terms of reproducible behavior.[6]

Figure 16–9b is a schematic diagram of a commercially available dc plasma source that is well suited for excitation of emission spectra. This plasma jet source consists of three electrodes arranged in an inverted Y configuration. A

---

[6] For additional details, see G. W. Johnson, H. E. Taylor, and R. K. Skogerboe, *Anal. Chem.,* **1979,** *51,* 2403; *Spectrochim. Acta, Part B,* **1979,** *34,* 197; J. Reednick, *Amer. Lab.,* **1979,** *11*(3), 53.

graphite anode is located in each arm of the Y and a tungsten cathode at the inverted base. Argon flows from the two anode blocks toward the cathode. The plasma jet is formed by bringing the cathode momentarily in contact with the anodes. Ionization of the argon occurs and a current develops ($\sim 14$ A) that generates additional ions to sustain the current indefinitely. The temperature at the arc core is perhaps $10,000\,°K$, and at the viewing region $5000\,°K$. The sample is aspirated into the area between the two arms of the Y, where it is atomized, excited, and viewed.

Spectra produced from the plasma jet tend to have fewer lines than those produced by the inductively coupled plasma, and the lines formed are largely from atoms rather than ions. Sensitivities achieved with the dc jet plasma appear to range from an order of magnitude lower to about the same as those found with the inductively coupled plasma. Reproducibilities of the two systems are similar. Significantly less argon is required for the dc plasma, and the auxiliary power supply is simpler and less expensive. On the other hand, the graphite electrodes must be replaced every few hours, while the inductively coupled plasma source requires little or no maintenance.

### 16D-3 Instruments for Plasma Spectroscopy

Several instrument makers offer instruments for plasma emission spectroscopy. In general, these consist of a high-quality grating spectrophotometer for the ultraviolet and visible regions with a photomultiplier detector. Many are automated, so that an entire spectrum can be scanned sequentially. Others have several photomultiplier tubes located in the focal plane, so that the lines for several elements (two dozen or more) can be monitored simultaneously. Such instruments are very expensive.

### 16D-4 Quantitative Applications of Plasma Sources

Unquestionably, the inductively coupled and the dc plasma sources yield significantly better quantitative analytical data than other emission sources. The excellence of these results stems from the high stability, low noise, low background, and freedom from interferences of the sources when operated under appropriate experimental conditions. The performance of the inductively coupled plasma source is somewhat better than that of the dc plasma source in terms of detection limits. The latter, however, is less expensive to purchase and operate and is entirely adequate for many applications.

In general, the detection limits with the inductively coupled plasma source appear comparable to or better than other atomic spectral procedures. Table 16-3 compares the sensitivity of several of these methods.

## PROBLEMS

**16-1.** Define the following terms:
*(a)  atomization                              (b)  pressure broadening
*(c)  Doppler broadening                  (d)  source modulation

*(e)  flame photometry                  (f)  hollow-cathode lamp
*(g)  sputtering                        (h)  spectral interference
*(i)  chemical interference             (j)  radiation buffer
*(k)  releasing agent                   (l)  protective agent
*(m)  ionization suppressor             (n)  plasma

*16–2. Why is atomic emission more sensitive to flame instability than atomic absorption?

16–3. Why is a nonflame atomizer more sensitive than a flame atomizer?

*16–4. Why is source modulation employed in atomic absorption spectroscopy?

16–5. In a hydrogen–oxygen flame, large concentrations of $SO_4^{2-}$ caused an atomic absorption peak for Fe to decrease.
  (a) Suggest an explanation for this observation.
  (b) Suggest possible methods for overcoming the potential interference of $SO_4^{2-}$ in a quantitative determination of Fe.

*16–6. In the concentration range of 500 to 2000 ppm of U, a linear relationship exists between absorbance at 351.5 nm and concentration. At lower concentrations the relationship becomes nonlinear unless about 2000 ppm of an alkali-metal salt are introduced. Explain.

*16–7. A 5.00-mL sample of blood was treated with trichloroacetic acid to precipitate proteins. After centrifugation, the resulting solution was brought to a pH of 3 and extracted with two 5-mL portions of methyl isobutyl ketone containing the organic Pb complexing agent APCD. The extract was aspirated directly into an air–acetylene flame, yielding an absorbance of 283.3 nm of 0.502. Five-milliliter aliquots of standard solutions containing 0.400 and 0.600 ppm Pb were treated in the same way and yielded absorbances of 0.396 and 0.599. Calculate the ppm Pb in the sample, assuming that Beer's law is followed.

16–8. The Na in a series of cement samples was determined by flame emission spectroscopy. The flame photometer was calibrated with a series of standards containing 0.0, 20.0, 40.0, 60.0, and 80.0 $\mu$g $Na_2O$ per mL. The instrument readings for these solutions were 18.4, 37.8, 54.0, 74.2, and 92.1.
  (a) Plot the data.
  (b) Derive a least-squares line for the data.
  (c) Calculate standard deviations for the slope and about regression for the line in (b).
  (d) The following data were obtained for replicate 1.000-g samples of cement that were dissolved in HCl and diluted to 100.0 mL after neutralization.

**Emission Reading**

|             | Sample A | Sample B | Sample C |
|-------------|----------|----------|----------|
| Replicate 1 | 28.6     | 40.7     | 73.1     |
| Replicate 2 | 28.2     | 41.2     | 72.1     |
| Replicate 3 | 28.9     | 40.2     | Spilled  |

Calculate the percentage of $Na_2O$ in each sample. What is the absolute and relative standard deviation for the average of each determination?

# Chapter 17
# Analytical Separations

An interference in a chemical analysis arises whenever a species in the sample matrix either produces a signal that is indistinguishable from that of the analyte or, alternatively, attenuates the analyte signal. Few if any analytical signals are so specific as to be free of interference. As a consequence, most analytical methods require one or more preliminary steps to eliminate the effects of interferences.

Two general methods are available for dealing with interferences. The first makes use of a *masking agent* to immobilize or chemically bind the interfering species in a form in which it no longer contributes to or attenuates the signal from the analyte. Clearly, a masking agent must not affect the behavior of the analyte significantly. In earlier chapters, we have encountered several masking agents. An example is the use of fluoride ion to prevent iron(III) from interfering in the iodometric determination of copper(II). Here, masking results from the strong tendency of fluoride ions to complex iron(III) but not copper(II). The consequence is a decrease in the electrode potential of the iron(III) system to the point where only copper(II) ions from the sample oxidize iodide to iodine.

The second approach involves physically separating the analyte and the interference in separate phases. Applications of this technique are considered in this chapter and the next.

## 17A PRECIPITATION METHODS

Precipitation methods of separation depend upon differences in solubility between the analyte and potential interferences. Solubility product calculations permit the determination of the feasibility of such separations. It has been pointed out, however (Chapter 4), that such other factors as rate of precipitate formation and coprecipitation problems are crucial in determining the success or failure of a precipitation separation. The effects of these variables can only be established through experiment. The literature of analytical chemistry abounds with descriptions of separations that make use of precipitating reagents. The discussion that follows will be limited to those with very general applicability.

### 17A – 1 Separations Based on Control of pH

Enormous differences exist among the solubilities of the hydroxides, hydrous oxides, and acids of elements. Moreover, the hydrogen or hydroxide ion concentration of aqueous solutions can be varied over a range of $10^{15}$ $M$ or more. As a practical matter, separations based on the control of pH can be divided into three categories: (1) precipitations that are quantitative in relatively concentrated solutions of strong acids; (2) those that are made in intermediate pH regions through the use of buffers; and (3) those that require concentrated solutions of sodium or potassium hydroxide. Typical examples of separations based upon the control of acidity are given in Table 17 – 1.

### 17A – 2 Sulfide Separations

For many years, chemists have used the large differences that exist in the solubilities of sulfides to bring about separations of inorganic species. Sulfide separations are particularly useful because of the relative ease by which the sulfide ion concentration can be regulated through control of pH. In addition, sulfide ion can be generated homogeneously through the hydrolysis of thioacetamide (see Table 4 – 2, p. 83).

The formation of sulfides was treated from an equilibrium standpoint in Chapter 4. Such calculations are likely to provide overly optimistic conclusions regarding the feasibility of separations, because they fail to account for coprecipitation problems and the slow rate at which many sulfides form. Again, it is only through experiment that the feasibility of a separation based on differences in the solubilities of sulfides can be confirmed. Typical sulfide separations are listed in Table 17 – 2.

Table 17 – 1    Separations Based upon Control of pH

| Reagent | Species That Form Precipitates | Species That Are Not Precipitated |
|---|---|---|
| Hot, conc $HNO_3$ | Oxides of W(VI), Ta(V), Nb(V), Si(IV), Sn(IV), Sb(V) | Most other cations |
| $NH_3 – NH_4Cl$ buffer | Fe(III), Cr(III), Al(III) | Ions of the alkali and alkaline-earth metals, Mn(II), Zn(II), Cu(II), Ni(II), Co(II) |
| $HOAc – NH_4OAc$ buffer | Fe(III), Cr(III), Al(III) | Common divalent cations |
| $NaOH – Na_2O_2$ | Fe(III), most divalent cations, ions of the rare-earth elements | Al(III), Cr(VI), Zn(II), U(VI), V(V) |

*Table 17 – 2    Precipitation of Sulfides*

| Elements | Conditions for Precipitation* | Conditions for No Precipitation* |
|---|---|---|
| Hg(II), Cu(II), Ag(I) | 1, 2, 3, 4 | |
| As(V), As(III), Sb(V), Sb(III) | 1, 2, 3 | 4 |
| Bi(III), Cd(II), Pb(II), Sn(II) | 2, 3, 4 | 1 |
| Sn(IV) | 2, 3 | 1, 4 |
| Zn(II), Co(II), Ni(II) | 3, 4 | 1, 2 |
| Fe(II), Mn(II) | 4 | 1, 2, 3 |

* Conditions:
  1. $3 F$ HCl.
  2. $0.3 F$ HCl.
  3. Buffered to pH 6 with acetate.
  4. Buffered to pH 9 with $NH_3$–$(NH_4)_2S$.

## 17A – 3 Other Inorganic Precipitants

No other inorganic ion is as generally useful for separations as the ones considered in the previous sections. Phosphate, carbonate, and oxalate ions find some use as precipitating agents for cations, but their behavior is nonselective. Chloride and sulfate ions, on the other hand, are relatively selective as precipitants. Silver can be isolated from most other cations with the former; lead, barium, and strontium can be selectively precipitated with the latter.

## 17A – 4 Organic Precipitants

Selected organic reagents that are useful for the isolation of cations were considered in Chapter 4. Some organic precipitants, such as dimethylglyoxime, are remarkably specific. Others, such as 8-hydroxyquinoline, react with numerous cations to form products with solubilities that encompass a substantial range. Separations with such reagents are achieved through control of the reagent concentration by pH adjustment with buffers.

## 17A – 5 Electrolytic Separations

Electrolytic precipitation is a highly useful method for bringing about separations. Here, the more easily reduced species, which may either be the analyte or an unwanted component, is isolated as a second phase. A mercury cathode is especially useful for the electrolytic removal of many metal ions prior to the analysis of the residual solution. In general, metals more easily reduced than zinc are conveniently deposited in the mercury; ions of the alkaline earths, the alkali metals, aluminum, and beryllium are unaffected. The potential needed to lower the concentration of a metal ion to a particular level can often be determined polarographically.

## 17B SEPARATION BY EXTRACTION

The extent to which solutes, both inorganic and organic, distribute themselves between two immiscible liquids differs enormously, and these differences have been used for decades to accomplish separations of chemical species. This section considers applications of the distribution phenomenon to analytical separations.

### 17B-1 Theory

The partition of a solute between two immiscible phases is an equilibrium phenomenon that is governed by the *distribution law*. If the solute species A is allowed to distribute itself between water and an organic phase, the resulting equilibrium may be written as

$$A_{aq} \rightleftharpoons A_{org}$$

where the subscripts refer to the aqueous and the organic phases, respectively. Ideally, the ratio of activities for A in the two phases will be constant and independent of the total quantity of A; that is, at any given temperature,

$$K = \frac{[A_{org}]}{[A_{aq}]} \qquad (17-1)$$

The equilibrium constant $K$ is known as the *partition, or distribution, coefficient*. As with many other equilibria, molar concentrations can often be substituted for activities without serious error. Generally, the numerical value for $K$ approximates the ratio of the solubility of A in each solvent.

If the solute exists in different states of aggregation in the two solvents, the equilibrium becomes

$$x(A_y)_{aq} \rightleftharpoons y(A_x)_{org}$$

and the partition coefficient is now

$$K = \frac{[(A_x)_{org}]^y}{[(A_y)_{aq}]^x}$$

Partition coefficients are useful because they provide guidance as to the most efficient way to perform an extractive separation. Consider, for example, a simple system that is described by Equation 17-1.[1] Suppose, further, that $a_0$ mmol of the solute A in $V_{aq}$ mL of aqueous solution are extracted with $V_{org}$ mL of an immiscible organic solvent. At equilibrium, $a_1$ mmol of A will remain in the aqueous layer, and $(a_0 - a_1)$ mmol will have been transferred to the organic

---

[1] This treatment can be modified to take account of other equilibria; see H. A. Laitinen and W. E. Harris, *Chemical Analysis,* 2nd ed., pp. 443–453. New York: McGraw-Hill Book Co., 1975.

layer. The concentrations of A in the two layers will then be

$$[A_{aq}] = \frac{a_1}{V_{aq}}$$

and

$$[A_{org}] = \frac{(a_0 - a_1)}{V_{org}}$$

Substitution of these quantities into Equation 17-1 and rearrangement gives

$$a_1 = \left(\frac{V_{aq}}{V_{org}K + V_{aq}}\right)a_0 \qquad (17-2)$$

Similarly, the number of millimoles, $a_2$, remaining after a second extraction with the same volume of solvent will be

$$a_2 = \left(\frac{V_{aq}}{V_{org}K + V_{aq}}\right)a_1$$

Substitution of Equation 17-2 for $a_1$ in this expression gives

$$a_2 = \left(\frac{V_{aq}}{V_{org}K + V_{aq}}\right)^2 a_0$$

By the same argument, the number of millimoles, $a_n$, that remain after $n$ extractions will be given by the expression

$$a_n = \left(\frac{V_{aq}}{V_{org}K + V_{aq}}\right)^n a_0 \qquad (17-3)$$

Finally, Equation 17-3 can be written in terms of the initial and final concentrations of $a$ in the aqueous layer by substituting the relationships

$$a_n = [A_{aq}]_n V_{aq} \qquad \text{and} \qquad a_0 = [A_{aq}]_0 V_{aq}$$

Thus,

$$[A_{aq}]_n = \left(\frac{V_{aq}}{V_{org}K + V_{aq}}\right)^n [A_{aq}]_0 \qquad (17-4)$$

Example 17-1 demonstrates that several small volumes provide a more efficient extraction than does a single large one.

**Example 17-1**   The distribution coefficient for iodine between $CCl_4$ and $H_2O$ is 85. Calculate the concentration of $I_2$ remaining in the aqueous layer after extraction of 50.0 mL of $1.00 \times 10^{-3}$ $F\,I_2$ with the following quantities of $CCl_4$:

(a) 50.0 mL.
(b) two 25.0-mL portions.
(c) five 10.0-mL portions.

Substitution into Equation 17–4 gives

(a) $[I_{2_{aq}}] = \left(\dfrac{50.0}{50.0 \times 85 + 50.0}\right)^1 \times 1.00 \times 10^{-3} = 1.16 \times 10^{-5}$

(b) $[I_{2_{aq}}] = \left(\dfrac{50.0}{25.0 \times 85 + 50.0}\right)^2 \times 1.00 \times 10^{-3} = 5.28 \times 10^{-7}$

(c) $[I_{2_{aq}}] = \left(\dfrac{50.0}{10.0 \times 85 + 50.0}\right)^5 \times 1.00 \times 10^{-3} = 5.29 \times 10^{-10}$

Figure 17–1 shows that the improved efficiency of multiple extractions falls off rapidly as a total fixed volume is subdivided into smaller and smaller portions. Clearly, little is to be gained by dividing the extracting solvent into more than five or six portions.

## 17B–2 Applications

An extraction is frequently more attractive than a precipitation method for separating inorganic species. The processes of equilibration and separation of phases in a separatory funnel are less tedious and time consuming than conventional precipitation, filtration, and washing. Moreover, difficulties associated with coprecipitation are avoided. Finally, and in contrast to the precipitation process, extraction procedures are ideally suited for the isolation of trace quantities of analytes.

A number of inorganic species can be separated by extraction with suitable solvents. For example, a single ether extraction of a 6 $F$ hydrochloric acid solution will cause better than 50% of several ions to be transferred to the organic medium; included among these are iron(III), antimony(V), titanium(III), gold(III), molybdenum(VI), and tin(IV). Other ions, such as aluminum(III) and the divalent cations of cobalt, lead, manganese, and nickel are not extracted.

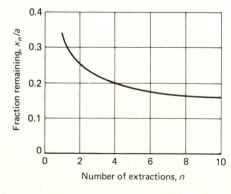

**Figure 17–1**  Plot of Equation 17–4, with $K = 2$ and $V_{aq} = 100$ mL. The total volume of organic solvent is also 100 mL; thus, $V_{org} = 100/n$.

Uranium(VI) can be separated from such elements as lead and thorium by ether extraction of a solution that is 1.5 $F$ in nitric acid and saturated with ammonium nitrate. Bismuth and iron(III) are also extracted to some extent from this medium.

Numerous organic reagents form chelates with metal ions; these products are frequently soluble in such solvents as chloroform, carbon tetrachloride, benzene, and ether. The quantitative transfer of a metal ion to an organic phase can thus be accomplished through the use of chelation.[2]

Organic compounds are frequently separated by extraction. For example, carboxylic acids are readily separated from phenolic compounds by extracting a nonaqueous solution of the mixture with dilute aqueous sodium hydrogen carbonate. The phenolic compounds remain in the nonaqueous phase, while the carboxylic acids are almost completely transferred, as their sodium salts, to the aqueous phase.

## 17C SEPARATION BY ION EXCHANGE

Ion exchange is a process by which ions held on an essentially insoluble solid are exchanged for ions in a solution that is brought in contact with the solid. The ion-exchange properties of clays and zeolites have been recognized and studied for more than a century. Synthetic ion-exchange resins were first produced in 1935 and have since found widespread application in water softening, water deionization, solution purification, and ion separation.

### 17C – 1 Ion-Exchange Resins

Synthetic ion-exchange resins are high-molecular-weight polymers that contain large numbers of an ionic functional group per molecule. Cation-exchange resins contain acidic groups, while anion-exchange resins have basic groups. Strong-acid type exchangers have sulfonic acid groups ($-SO_3^-H^+$) attached to the polymeric matrix and have wider application than weak-acid type exchangers, which owe their action to carboxylic acid ($-COOH$) groups. Similarly, strong-base anion exchangers contain quaternary amine [$-N(CH_3)_3^+OH^-$] groups, while weak-base types contain secondary or tertiary amines.

Cation exchange is illustrated by the equilibrium

$$x\text{RSO}_3^-\text{H}^+ + \text{M}^{x+} \rightleftharpoons (\text{RSO}_3^-)_x\text{M}^{x+} + x\text{H}^+$$
$$\quad\text{solid}\qquad\quad\text{soln}\qquad\quad\text{solid}\qquad\quad\text{soln}$$

where $\text{M}^{x+}$ represents a cation and R represents *that part* of a resin molecule which contains one sulfonic acid group. The analogous equilibrium involving a

---

[2] See, for example, A. K. De, S. M. Khopar, and R. A. Chalmers, *Solvent Extraction of Metals.* New York: D. Van Nostrand Co., 1970; E. B. Sandell and H. Onishi, *Photometric Determination of Traces of Metals,* 4th ed., Chapter 9. New York: John Wiley and Sons, 1978.

strong-base anion exchanger and an anion $A^{x-}$ is

$$xRN(CH_3)_3^+OH^- + A^{x-} \rightleftharpoons [RN(CH_3)_3^+]_xA^{x-} + xOH^-$$

### 17C-2 Ion-Exchange Equilibria

Ion-exchange equilibria can be treated by the law of mass action. For example, when a dilute solution containing calcium ions is passed through a column packed with a sulfonic acid resin, the following equilibrium is established:

$$Ca_{aq}^{2+} + 2H_{res}^+ \rightleftharpoons Ca_{res}^{2+} + 2H_{aq}^+$$

for which

$$K = \frac{[Ca_{res}^{2+}][H_{aq}^+]^2}{[Ca_{aq}^{2+}][H_{res}^+]^2} \qquad (17-5)$$

As usual, the bracketed terms are molar concentrations (strictly, activities) of the species in the two phases. Note that $[Ca_{res}^{2+}]$ and $[H_{res}^+]$ are molar concentrations of the two ions *in the solid phase*. In contrast to most solids, however, these concentrations can vary from zero to some maximum value when all of the negative sites on the resin are occupied by one species only.

Ion-exchange separations are ordinarily performed under conditions in which one ion predominates in *both* phases. Thus, in the removal of calcium ions from a dilute and somewhat acidic solution, the calcium ion concentration will be much smaller than that of hydrogen ion in both the aqueous and resin phases; that is,

$$[Ca_{res}^{2+}] \ll [H_{res}^+]$$

and

$$[Ca_{aq}^{2+}] \ll [H_{aq}^+]$$

As a consequence, the hydrogen ion concentration is essentially constant in both phases, and Equation 17-5 can be rearranged to

$$K = \frac{[Ca_{res}^{2+}]}{[Ca_{aq}^{2+}]} = K \frac{[H_{res}^+]^2}{[H_{aq}^+]^2} = K_D \qquad (17-6)$$

where $K_D$ is a distribution constant analogous to the constant that governs an extraction equilibrium (Equation 17-1). Note that $K_D$ in Equation 17-6 represents the affinity of the resin for calcium ion relative to another ion (here, $H^+$). In general, where $K_D$ for an ion is large, a strong tendency for the stationary phase to retain that ion exists; where $K_D$ is small, the opposite obtains. Selection of a common reference ion (such as $H^+$) permits a comparison of distribution ratios for various ions on a given type of resin. Such experiments reveal that polyvalent ions are much more strongly retained than singly charged species. Within a given charge group, differences that exist among values for $K_D$ appear to be related to the size of the hydrated ion, as well as other properties. Thus, for a typical sulfonated cation-exchange resin, values of $K_D$ for univalent ions decrease in the order $Ag^+ > Cs^+ > Rb^+ > K^+ > NH_4^+ > Na^+ > H^+ > Li^+$. For divalent cations the order is $Ba^{2+} > Pb^{2+} > Sr^{2+} > Ca^{2+} > Ni^{2+} > Cd^{2+} > Cu^{2+} > Co^{2+} > Zn^{2+} > Mg^{2+} > UO_2^{2+}$.

## 17C–3 Applications

Ion-exchange resins are used to eliminate ions that would otherwise interfere with an analysis. For example, iron(III), aluminum(III), as well as many other cations, tend to coprecipitate with barium sulfate during the determination of sulfate ion. Passage of a solution containing sulfate through a cation-exchange resin results in the retention of these ions and the release of an equivalent number of hydrogen ions. Sulfate ions pass freely through the column and can be precipitated as barium sulfate from the effluent.

Another valuable application of ion-exchange resins involves the concentration of ions from a very dilute solution. Thus, traces of metallic elements in large volumes of natural waters can be collected on a cation-exchange column and subsequently liberated from the resin by treatment with acid; the result is a considerably more concentrated solution for analysis.

The total salt content of a sample can be determined by titrating the hydrogen ion released as an aliquot of sample passes through a cation exchanger in the acidic form. Similarly, a standard hydrochloric acid solution can be prepared by diluting to known volume the effluent resulting from treatment of a cation-exchange resin with a known weight of sodium chloride. Substitution of an anion-exchange resin in its hydroxide form will permit the preparation of a standard base solution.

## 17D SEPARATIONS BY DISTILLATION

Distillation permits the resolution of a mixture with partition coefficients that differ significantly between the solution and vapor phases. The process is quite simple in situations where the partition coefficient for a single species is large compared with other components; the Kjeldahl analysis (Chapter 9) is an important example.

## PROBLEMS

17–1. Aluminum can be separated from $Cd^{2+}$, $Co^{2+}$, $Zn^{2+}$, and several other cations by precipitation with 8-hydroxyquinoline (see Chapter 4, Section 4C–3) provided reagents such as $CN^-$ or $NH_3$ are added to complex divalent cations and prevent precipitation of the 8-hydroxyquinolates of these ions. What are reagents such as $CN^-$ and $NH_3$ called?

17–2. Suggest a method for the separation of
*(a) $Fe^{3+}$ from $Al^{3+}$.
 (b) $Zn^{2+}$ from $Hg^{2+}$.

*(c) Si(IV) from divalent cations.

(d) $Cr^{3+}$ from $Zn^{2+}$.

*(e) U(VI) from Pb(II).

(f) $NH_4^+$ from $Na^+$ and $K^+$.

*(g) $SO_4^{2-}$ from $Al^{3+}$.

(h) $Ba^{2+}$ from $Cu^{2+}$.

**17–3.** How do strong- and weak-acid synthetic ion-exchange resins differ in structure?

**\*17–4.** A standard HCl solution was prepared by passing 0.4642 g of primary-standard NaCl in 200 mL of water through a column packed with a strong-acid ion-exchange resin. The resulting solution and washings were collected in a 500.0-mL volumetric flask and diluted to volume. What was the normality of the acid?

**17–5.** A standard solution of KOH was prepared by dissolving 1.397 g of purified KCl in water and passing the solution through a strong-base ion-exchange resin. The resulting solution and washings were diluted to 2.000 L. What was the normality of the base?

**\*17–6.** A diluted solution containing $Ca^{2+}$, $Ba^{2+}$, and $Sr^{2+}$ ions as chlorides was passed through a strong-acid ion-exchange column. The resulting solution required 12.15 mL of 0.02021 $N$ NaOH to titrate the HCl formed. Calculate the number of millimoles of divalent cations present in the sample.

**17–7.** A 0.2166-g sample containing only $Al(NO_3)_2$ and $Mg(NO_3)_2$ was dissolved in water and passed through a column containing a strong-acid ion-exchange resin. The resulting solution and washings were titrated with 36.89 mL of 0.08081 $N$ NaOH. Calculate the percent of the two nitrates in the original sample.

**\*17–8.** The distribution coefficient for a metal halide between water and ether was determined to be 12.3. Calculate the concentration of the cation $M^{2+}$ remaining after 50.0 mL of 0.125 $M$ $M^{2+}$ were extracted with the following quantities of ether:

(a) one 40.0-mL portion.  (b) two 20.0-mL portions.

(c) four 10.0-mL portions.  (d) eight 5.00-mL portions.

**17–9.** The distribution coefficient for the 8-hydroxyquinoline chelate of the cation $M^{3+}$ between water and methyl isobutyl ketone is 7.76. Calculate the percentage of $M^{3+}$ remaining in 50.0 mL of the aqueous layer, which was originally 0.715 $M$ in $M^{3+}$, after extraction with the following quantities of the organic solvent:

(a) one 25.0-mL portion.  (b) two 12.5-mL portions.

(c) five 5.00-mL portions.  (d) ten 2.5-mL portions.

**\*17–10.** What volume of ether would be required to decrease the concentration of $M^{2+}$ in Problem 17–8 to $1.00 \times 10^{-4}$ $M$ if 25.0 mL of 0.0500 $M$ $M^{2+}$ were extracted with the following quantities of ether:

(a) 25.0-mL portions?

(b) 10.0-mL portions?

(c) 3.00-mL portions?

**17–11.** What volume of methyl isobutyl ketone would be required to decrease the concentration of $M^{3+}$ in Problem 17–9 to $1.00 \times 10^{-5}$ $M$ if 40.0 mL of 0.0200 $M$ $M^{3+}$ were treated with an excess of 8-hydroxyquinoline and extracted with

(a) 50.0-mL portions of the solvent?

      (b) 25.0-mL portions of the solvent?

      (c) 10.0-mL portions of the solvent?

**\*17–12.** A 0.2000 $M$ aqueous solution of the weak organic acid HA was prepared from the pure compound, and three 50.0-mL aliquots were transferred to 100-mL volumetric flasks. Solution 1 was diluted to 100 mL with 2.0 $N$ HClO$_4$, while solution 2 was diluted to the mark with 2.0 $N$ NaOH; solution 3 was diluted to the mark with water. Aliquots of 25.0 mL each were extracted with 25.0 mL of $n$-hexane. The extract from the basic solution contained no detectable trace of A-containing species, which indicated that A$^-$ is not soluble in the organic solvent. The extract from solution 1 contained no ClO$_4^-$ or HClO$_4$ but was found to be 0.0846 $M$ in HA (by extraction with standard NaOH and back-titration with standard HCl). The extract of solution 3 was found to be 0.0372 $M$ in HA. Assume HA does not associate or dissociate in the organic solvent, and calculate

      (a) the distribution coefficient for HA between the two solvents.

      (b) the molar concentrations of the *species* HA and A$^-$ in the aqueous solution 3 after extraction.

      (c) the dissociation constant for HA in water.

**17–13.** To determine the equilibrium constant for the reaction

$$I_2 + X^- \rightleftharpoons I_2X^-$$

a 0.00825 $M$ solution of aqueous I$_2$ was prepared, and 50.0 mL of this solution were extracted with 20.0 mL of $n$-hexane. After extraction, spectrophotometric measurements showed that the I$_2$ concentration of the *aqueous phase* was $1.04 \times 10^{-4}$ $M$. An aqueous solution was then prepared that was 0.00793 $M$ in I$_2$ and 0.0750 $M$ in NaX. After extraction of 50.0 mL of this solution with 20.0 mL of hexane, the concentration of I$_2$ *in the n-hexane* was found by spectrophotometric measurement to be $9.47 \times 10^{-4}$ $M$.

      (a) What is the partition coefficient for I$_2$ between $n$-hexane and water?

      (b) What is the formation constant for I$_2$X$^-$?

**\*17–14.** What pH range should be used to separate Pb$^{2+}$ from Mn$^{2+}$ by a sulfide precipitation procedure? Assume each ion is present at a concentration of 0.050 $M$ and that quantitative removal of the less soluble species requires its concentration to be decreased to $5.0 \times 10^{-6}$ $M$.

**17–15.** What range of pH values could be used to separate Mg$^{2+}$ from Pb$^{2+}$ as the hydroxide? Assume that the solutions are initially 0.10 $M$ in each ion and that a quantitative separation involves lowering the concentration of the less soluble species to $1.0 \times 10^{-5}$ $M$.

# Chapter 18
# Chromatographic Methods

Chromatography encompasses a diverse and important group of separation methods that permit the scientist to separate, identify, and determine related components of complex mixtures that cannot be resolved by other means. The term "chromatography" is difficult to define rigorously, owing to the variety of systems and techniques to which it has been applied. All of these methods, however, make use of a *stationary phase* and a *mobile phase.* Components of a mixture are carried through the stationary phase by the flow of the mobile one; separations are based on differences in migration rates among the sample components.

## 18A CLASSIFICATION OF CHROMATOGRAPHIC METHODS

Chromatographic methods can be categorized in two ways. The first is based upon the physical means by which the stationary and mobile phases are brought into contact. In *column chromatography,* the stationary phase is held in a narrow tube through which the mobile phase is forced under pressure or by gravity. In *planar chromatography,* the stationary phase is supported on a flat plate or in the interstices of a paper. In this case, the mobile phase moves through the stationary phase by capillary action or under the influence of gravity. This discussion focuses on column chromatography only. It is important to note, however, that the equilibria upon which the two types of chromatography are based are identical and that the theory developed for column chromatography is readily adapted to planar as well.

A more fundamental classification of chromatographic methods is one based upon whether the mobile phase is a liquid *(liquid chromatography)* or a gas *(gas chromatography)*. Table 18 – 1 lists several subdivisions of each method according to the type of equilibrium by which solutes distribute themselves between the mobile and stationary phases. It should be noted that the stationary phase can be a liquid that has been immobilized in various ways. For example, a thin film of the liquid may be adsorbed upon or chemically bonded to the surface of a finely divided, inert solid or retained in the pores or interstices of such a solid. Alternatively, the liquid may be immobilized by adsorption or bonding on the inner walls of a capillary tubing. Ideally, the solid plays no direct

Table 18–1  Classification of Column Chromatographic Methods

| General Classification | Specific Type | Stationary Phase | Type of Equilibrium |
|---|---|---|---|
| Liquid chromatography (LC) (liquid mobile phase) | Liquid–liquid (partition) | Liquid adsorbed on a solid | Partition between immiscible liquids |
| | Liquid–solid | Solid | Adsorption |
| | Liquid–bonded phase (partition) | Organic species bonded to a solid surface | Partition/adsorption |
| | Ion exchange | Ion-exchange resin | Ion exchange |
| | Gel permeation | Liquid in interstices of a polymeric solid | Partition/adsorption |
| Gas chromatography (GC) (gas mobile phase) | Gas–liquid (GLC) | Liquid adsorbed on a solid | Partition between gas and liquid |
| | Gas–solid | Solid | Adsorption |
| | Gas–bonded phase | Organic species bonded to a solid surface | Partition/adsorption |

part in the separation, serving only as a support for the liquid. Often, however, the nature of the solid does have an effect on the separation. It is also noteworthy that liquid chromatography can be performed in columns and on plane surfaces, while gas chromatography is restricted to column procedures.

## 18B  COLUMN CHROMATOGRAPHY

Figure 18–1 shows schematically how two components, A and B, are resolved on a column by *elution chromatography*. Elution involves washing a solid through a column by additions of fresh solvent. A single portion of the sample, dissolved in the mobile phase, is introduced at the head of a column (time $t_0$ in Figure 18–1), whereupon components A and B distribute themselves between the two phases. Introduction of additional mobile phase (the *eluent*) forces the dissolved portion of the sample down the column, where further partition between the mobile phase and fresh portions of the stationary phase occurs (time $t_1$). Partitioning between the fresh solvent and the stationary phase takes place simultaneously at the site of the original sample. Further additions of

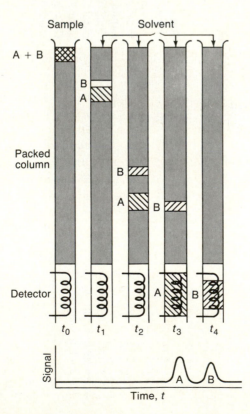

**Figure 18–1**  Column elution chromatography. Schematic diagram of the separation of a mixture of components A and B. The lower figure shows the output of a signal detector at the indicated stages of elution in the upper part of the figure.

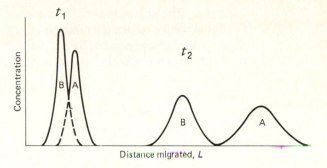

**Figure 18-2**  Concentration profiles for solutes A and B on the chromatographic column shown in Figure 18-1. The profile on the left shows the distribution of the solutes along the length of the column at time $t_1$. Note that separation is quite incomplete. The profile on the right shows the distribution at a later time $(t_2)$, when the two species have just been resolved.

solvent carry solute molecules down the column in a continuous series of transfers between the two phases. Because solute movement can only occur in the mobile phase, the average *rate* at which a solute migrates *depends upon the fraction of time it spends in that phase.* This fraction is small for solutes that are strongly retained by the stationary phase (component B in Figure 18-1, for example) and is large where retention in the mobile phase is more likely (component A). Ideally, the resulting differences in rates cause the components in a mixture to separate into *bands* or *zones* situated along the length of the column (Figure 18-2). Isolation of the separated species is then accomplished by passing a sufficient quantity of mobile phase through the column to cause the individual bands to pass out the end, where they can be collected (times $t_3$ and $t_4$ in Figure 18-1).

If a detector that responds to solute concentration is placed at the end of the column and its signal is plotted as a function of time (or of volume of the added mobile phase), a series of symmetric peaks is obtained, as shown in the lower part of Figure 18-1. Such a plot, called a *chromatogram,* is useful for both qualitative and quantitative analysis. The positions of peaks on the time axis may serve to identify the components of the sample, while the areas under the peaks provide a quantitative measure of the amount of each species.

## 18C THEORIES OF ELUTION CHROMATOGRAPHY

Figure 18-2 shows concentration profiles for solutes A and B at an early and a late stage of elution from the chromatographic column shown in Figure 18-1.[1] Species B is more strongly retained than A and thus lags during the migration process. It is apparent that movement down the column increases the distance

---

[1] Note that the relative positions of bands for A and B in the concentration profile shown in Figure 18-2 appear to be reversed from their positions in the chromatogram shown in the lower part of Figure 18-1. The difference is that the abscissa in Figure 18-2 is distance along the column, while in Figure 18-1 it is time. Thus, in the chromatogram in Figure 18-1, the *front* of a peak lies to the left and the *tail* to the right; for the concentration profile in Figure 18-2 the reverse obtains.

between the two bands. At the same time, however, broadening of both bands occurs, which lowers the efficiency of the column as a separating device. While band broadening is inevitable, conditions can often be found where broadening occurs more slowly than band separation. Thus, as shown in Figure 18–2, a clean resolution of two compounds is possible provided the column is sufficiently long. Numerous chemical and physical variables influence the rates of band separation and band broadening. As a consequence, improved separations can be realized by control of those variables that increase the former and/or decrease the latter.

The earliest theory of chromatographic separations was based upon the concept of the *theoretical plate,* which had been applied with considerable success in the early part of this century to interpreting the fractionation that occurs in a distillation column. In this theory, a chromatographic column is envisioned as being made up of a series of narrow contiguous horizontal layers called theoretical plates. During elution, equilibria develop in each plate that involve transfer of molecules between the mobile and stationary phases. The migration of a species during elution then takes place via a series of stepwise transfers of molecules from the mobile phase to the stationary phase, back to the mobile phase, and so forth. Because separation occurs as a result of these transfers, column efficiency depends upon the number of equilibrations that occur during elution, which in turn depends upon the number of theoretical plates $N$ that are contained in the column. The narrower the theoretical plate, the greater will be their number in a given length of column. Thus, efficiency is inversely related to the thickness of the theoretical plate $H$, which is termed the *plate height* or the *height equivalent of the theoretical plate (HETP)*. The product of the plate height and the number of plates is equal to the column length $L$; that is, $L = NH$, which rearranges to

$$N = \frac{L}{H} \tag{18–1}$$

The plate theory, while providing a quantitative description of the variables that influence the rate at which solutes migrate down a column, fails to reveal the causes of zone broadening. As a consequence, the plate theory has been entirely supplanted by the *rate* or *kinetic theory* of chromatography, which accounts for not only those variables that affect the rate of migration of a species but also the rate at which bands broaden as they move down the column.

### 18C – 1 The Rate Theory of Chromatography[2]

The fundamental difference between the rate theory and the plate theory is that in the latter, it is postulated that throughout a column, a state of true equilib-

---

[2] For a detailed presentation of the rate theory, see J. C. Giddings, in *Chromatography,* 3rd ed., E. Heftmann, Ed., Chapter 3. New York: Van Nostrand-Reinhold, 1975. For a shorter presentation, see J. C. Giddings, *J. Chem. Educ.,* **1958,** *35,* 588; **1967,** *44,* 704.

rium exists for the distribution of molecules between the mobile and the stationary phases and that the position of these equilibria determine the appearance of chromatograms. In contrast, the rate theory assumes that the equilibrium state can never be reached in a chromatographic column because of the dynamic nature of the elution process. Furthermore, the breadths of chromatographic peaks are determined by the rates at which various mass transfer processes occur in and between the mobile and stationary phases.

It is important to note that while the kinetic theory is by now fully accepted, the efficiency parameters $N$ and $H$ from the plate theory (as well as Equation 18–1, which relates the two) still appear in the kinetic theory. It should be understood, however, that retention of these terms does not imply the existence of plates as real physical entities in a column; instead, the terms are retained for historical reasons only.

**Movement of Solutes Through a Column**   Examination of the peaks in a typical chromatogram (Figure 18–1) or the bands on a column (Figure 18–2) reveals a similarity to normal error or Gaussian curves (Figure 3–3, p. 39), which are obtained when replicate values of a measurement are plotted as a function of the frequency of their occurrence. As was shown in Chapter 3, Section 3D–1, normal error curves can be rationalized by assuming that the uncertainty associated with any single measurement is the summation of a much larger number of small, individually undetectable and random uncertainties, each of which has an equal probability of being positive or negative. The most common occurrence is for these uncertainties to cancel one another, thus leading to the mean value. With less likelihood, the summation may cause results that are greater or smaller than the mean. The consequence is a symmetric distribution of data around the mean value shown in Figure 3–3. In a similar way, the typical Gaussian shape of a chromatographic band can be attributed to the additive combination of the random motions of the myriad solute particles in the chromatographic band or zone.

It is instructive to consider the behavior of an individual solute molecule, which undergoes many thousands of transfers between the stationary and the mobile phase during migration. The residence time in either phase after a transfer is highly irregular and depends upon the molecule accidentally gaining sufficient thermal energy from its environment to accomplish a reverse transfer. Thus, in some instances, the residence time in a given phase may be transitory; in others, the period may be relatively long. Recall that the particle is eluted *only while it is in the mobile phase.* As a consequence, certain individual particles travel rapidly by virtue of their accidental inclusion in the mobile phase for a majority of the time. Others, in contrast, may lag because they happen to have been incorporated in the stationary phase for a greater-than-average time. The result of these random individual processes is a symmetric spread of velocities around the mean value, which represents the behavior of the average particle.

The breadth of a band increases as it moves down the column because more time is allowed for spreading to occur. Thus, the zone breadth is directly related to residence time in the column and inversely related to the velocity at which the mobile phase flows.

**Variance and Standard Deviation as Measures of Column Efficiency**   As shown in Chapter 3, Section 3D–2, the standard deviation $\sigma$ and the variance $\sigma^2$ define the breadth of a Gaussian curve. Because chromatographic bands are also Gaussian, and because the efficiency of a column is reflected in the breadth of chromatographic peaks, it is convenient to define column efficiency in terms of variance per unit length of column. By definition, then,

$$H = \frac{\sigma^2}{L} \qquad (18-2)$$

where $H$, the plate height in centimeters, is the measure of efficiency and $L$ is the column length, also in centimeters. Note that $\sigma^2$ then carries units of centimeters squared.

**Experimental Evaluation of $N$ and $H$**   Figure 18–3 shows a typical chromatogram with time as the abscissa. Here, $t_R$, the *retention time,* is defined as the time required after sample injection for the solute peak to appear at the end of the column. The variance of this peak, which can be obtained by a simple graphical procedure, will, however, have units of seconds squared and is usually designated as $\tau^2$ to distinguish it from $\sigma^2$, which has units of centimeters squared. The two standard deviations $\tau$ and $\sigma$ are related by

$$\tau = \frac{\sigma}{L/t_R} \qquad (18-3)$$

where $L/t_R$ is the average linear velocity of the solute in centimeters per second.

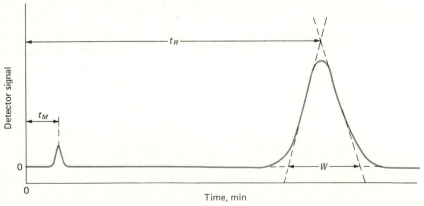

**Figure 18–3**   Determination of standard deviation $\tau$ from a chromatographic peak. Here, $W = 4\tau$; $t_R$ is the retention time for a solute that is retained by the column packing, and $t_M$ is the time for one that is not. Thus, $t_M$ is approximately equal to the time required for a molecule of the mobile phase to pass through the column.

Figure 18–3 illustrates a simple means for approximating $\tau$ and $\sigma$ from an experimental chromatogram. Tangents at the inflection points on the two sides of the chromatographic peak are extended to form a triangle with the abscissa. The area of this triangle can be shown to be approximately 96% of the total area under the peak. In Chapter 3, Section 3D–2, it was shown that about 96% of the area under a Gaussian peak is included within plus or minus two standard deviations ($\pm 2\sigma$) of its maximum. Thus, the intercepts shown in the figure occur at approximately $\pm 2\tau$ from the maximum, and

$$W = 4\tau$$

where $W$ is the magnitude of the base of the triangle. Substituting this relationship into Equation 18–3 and rearranging yields

$$\sigma = \frac{LW}{4t_R}$$

Substitution of this expression for $\sigma$ into Equation 18–2 gives

$$H = \frac{LW^2}{16t_R^2}$$

To obtain $N$, we substitute into Equation 18–1 and rearrange, giving

$$N = 16\left(\frac{t_R}{W}\right)^2 \qquad (18–4)$$

Thus, $N$ can be calculated from two time measurements $t_R$ and $W$; to obtain $H$, the length of the column packing $L$ must also be known.

**Band Broadening**   Band broadening is the consequence of the finite rate at which several mass-transfer processes occur during migration of a solute down a column. Some of the variables that affect these rates are controllable and can be exploited to improve separations. It has been found that narrower bands can be realized by decreasing the particle size of column packings, by employing thinner layers of the immobilized film (where the stationary phase is an adsorbed liquid), and by lowering the mobile-phase viscosity. Increases in temperature also reduce band broadening under most circumstances. Figure 18–4 illustrates how particle size and mobile-phase flow rate affect plate heights and thus band broadening. Quite generally, plate heights and broadening pass through a minimum as the velocity of the mobile phase is increased. Over the past 30 years, an enormous effort, both experimental and theoretical, has been devoted to developing quantitative relationships that account for the effects of these several variables on band broadening; much, however, remains to be learned.[3] Thus, we will restrict our discussion to some qualitative explanations

---

[3] See J. Hawkes, *J. Chem. Educ.,* **1983,** *60,* 393.

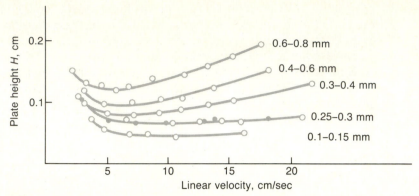

**Figure 18 – 4**    Effect of particle size on plate height for a packed gas – liquid chromatographic column. Particle diameters are shown on the right. (From J. Boheman and J. H. Purnell, in *Gas Chromatography*, D. H. Desty, Ed. New York: Academic Press, 1958. With permission.)

for the effects of important variables on band broadening and thus plate heights.

One cause of band broadening is *longitudinal diffusion,* which results from the tendency of solutes to diffuse from the concentrated center of a band to the more dilute regions on either side, that is, both in and opposed to the direction of flow. This type of diffusion occurs both in the mobile and stationary phases but its effects are only significant in gaseous mobile phases. The contribution of longitudinal diffusion to plate height is inversely proportional to the linear velocity of the mobile phase because higher velocities provide less time in which diffusion can occur. This effect accounts for the initial decrease in plate height with increasing flow rates shown in Figure 18 – 4.

The finite rates at which solutes transfer into and out of the stationary phase is another kinetic source of zone broadening. Broadening occurs when either of these rates is slow. As the rate of movement of the mobile phase increases, less and less time is available for mass transfer to occur; consequently, broadening is enhanced at higher flow rates. The rate of mass transfer out of liquid stationary phases is dependent upon the thickness of the liquid film on the support particle. With thick films, the average distance that a molecule must travel to reach the interface with the mobile phase is larger. Consequently, the average rate of mass transfer is greater, and band broadening results.

Zone broadening in the mobile phase is also caused in part by *eddy diffusion,* which results from the multitude of pathways by which a molecule can find its way down a packed column. As shown in Figure 18 – 5, the length of these paths may differ significantly; thus, the residence time of molecules of the same species is variable. Eddy diffusion would be independent of flow rate if it were not partially offset by ordinary diffusion, which results in molecules being transferred from a stream following one pathway to that following another (for example, from path 1 to path 2 in Figure 18 – 5). A large number of these transfers occurs when the flow rate is low, and each molecule samples several

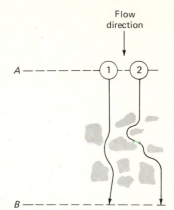

**Figure 18–5** Typical pathways for two solute molecules during elution. Note that the distance traveled by molecule 2 is greater than that traveled by molecule 1. Thus, molecule 2 will arrive at B later than molecule 1.

flow paths spending a brief time in each. As a consequence, the solute migration rates tend toward an average value and molecules are not significantly dispersed as a result of the many possible pathways through the packing. At moderate or high flow rates, however, sufficient time is not available for diffusion averaging to occur, and band broadening is observed. At sufficiently high flow rates, the effect of eddy diffusion becomes independent of flow velocity.

Superimposed upon eddy diffusion is an effect caused by the presence of *stagnant pools* of mobile phase that are retained in pores in the stationary phase. Solute molecules must diffuse through these stagnant pools before transfer to the stationary phase can take place. The presence of such pools of mobile phase slows the exchange process and results in zone broadening that is directly proportional to the flow rate and inversely proportional to the diffusion rate of the solute in the mobile phase. Increases in the diameter of the particles of the packing result in significantly wider bands due to the enhanced internal stagnant volume associated with larger particle size.

### 18C–2 Migration Rates of Solutes

In addition to band broadening, the relative rates at which solutes migrate determine the effectiveness of a column for separating two solutes. These rates are determined principally by the chemical nature and relative amounts of the stationary and mobile phases.

**The Partition Coefficient**   All chromatographic separations are based upon differences in the extent to which solutes are partitioned between the mobile and the stationary phase. The equilibrium involved can be described quantitatively by means of a *partition coefficient, K,* which for chromatography is defined as

$$K = \frac{C_S}{C_M} \qquad (18–5)$$

Here, $C_S$ is the molar concentration of a solute in the stationary phase and $C_M$ is its concentration in the mobile phase. Ideally, the partition ratio is constant over a wide range of solute concentrations; that is, $C_S$ is directly proportional to $C_M$. At high solute concentrations, however, marked departures from linearity are often encountered. Fortunately, most chromatography is performed under conditions in which Equation 18–5 does apply, which greatly simplifies the derivation of expressions to describe the separation process. Chromatography carried out under conditions in which $K$ is more or less constant is termed *linear chromatography*. The discussions that follow will deal exclusively with separations of this type.

**Retention Times**    For the chromatogram shown in Figure 18–3, zero on the time axis corresponds to the instant the sample is injected onto the column and elution is started. The peak at $t_M$ is for a species that is *not* retained by the column; its rate of migration will be the same as the average rate of motion of the molecules of the mobile phase. The *retention time, $t_R$,* for the solute responsible for the second peak is the time for that peak to reach the detector at the end of the column.

The average linear rate of migration, $\bar{v}$, of the solute is given by

$$\bar{v} = \frac{L}{t_R} \qquad (18-6)$$

Similarly, the average linear rate of movement, $u$, of molecules of the mobile phase will be

$$u = \frac{L}{t_M} \qquad (18-7)$$

**Relationship Between Retention Time and Partition Coefficient**    In order to relate the retention time of a solute to its partition coefficient, we express the migration rate as a fraction of the velocity of the mobile phase; that is,

$\bar{v} = u \times$ (fraction of the time the solute spends in the mobile phase)

This fraction, however, equals the average number moles of solute in the mobile phase at any instant compared with the total number of moles in the column:

$$\bar{v} = u \times \frac{\text{no. moles solute in mobile phase}}{\text{total no. moles of solute}}$$

or

$$\bar{v} = u \frac{C_M V_M}{C_M V_M + C_S V_S} = u \left( \frac{1}{1 + C_S V_S / C_M V_M} \right)$$

where $C_M$ and $C_S$ are the molar concentrations of the solute in the mobile and stationary phases, respectively; similarly, $V_M$ and $V_S$ are the volumes of the two phases in the column. Substitution of Equation 18–5 into this expression gives

$$\bar{v} = u \left( \frac{1}{1 + KV_S/V_M} \right) \tag{18–8}$$

Finally, substituting Equations 18–6 and 18–7 into Equation 18–8 and rearranging yields

$$t_R = t_M(1 + KV_S/V_M) = t_M(1 + k') \tag{18–9}$$

where $k'$, which is equal to $KV_S/V_M$, is the *capacity factor,* a constant that is related to the rate at which an analyte is eluted from a column. When $k'$ is much less than unity, the analyte is eluted so rapidly that a satisfactory separation cannot be realized. When the capacity factor is larger than perhaps 20, elution times become inordinately long. Ideally, optimum separations are achieved when $k'$ lies in the range of 1 to 5.

## 18D  QUALITATIVE AND QUANTITATIVE CHROMATOGRAPHIC ANALYSES

Chromatography has grown to be the premiere method for separating closely related chemical species. In addition, it can be employed for qualitative identification and quantitative determination of separated species.

### 18D–1  Qualitative Analysis

Chromatography is widely used for recognizing the presence or absence of components in mixtures that contain a limited number of possible species whose identities are known. For example, 30 or more amino acids in a protein hydrolysate can be detected with a reasonable degree of certainty by means of a chromatogram. On the other hand, because a chromatogram provides but a single piece of information about each species in a mixture (the retention time), its application to the qualitative analysis of complex samples of unknown composition is limited. Nevertheless, chromatography often serves as a first step in a qualitative analysis by various spectroscopic techniques.

It is important to note that while it may not lead to positive identification of species present in a sample, a chromatogram often provides sure evidence of the *absence* of such compounds. Thus, failure of a sample to produce a peak at the same retention time as a standard obtained under identical conditions is good evidence the compound in question is absent (or present at a concentration level below the detection limit of the procedure).

### 18D – 2 Quantitative Analysis

Chromatography owes its enormous growth in part to its speed, simplicity, relatively low cost, and wide applicability as a separating tool. It is doubtful, however, if its use would have become as widespread had it not been for the fact that it can also provide quantitative information about separated species.

Quantitative column chromatography is based upon a comparison of either peak heights or peak areas with those of several standards. The height of a chromatographic peak is readily obtained by connecting the baselines on either side of the peak by a straight line and measuring the perpendicular distance from this line to the peak. This measurement can ordinarily be made with reasonably high precision and will yield accurate results, provided that variations in column conditions do not alter the peak width during the period required to obtain chromatograms for sample and standards. Column temperature, eluent flow rate, and rate of sample injection must be closely controlled. In addition, care must be taken to avoid overloading the column. Uncertainties in sample injection are perhaps the most common source of error. Syringe injection causes relative uncertainties of 5 to 15%. Precisions of 1 to 3% are often realized with rotary sampling valves (see p. 517).

Peak areas are independent of broadening due to the variables just mentioned. From this standpoint, then, peak areas are more satisfactory than peak heights for quantitative work. On the other hand, peak heights are more easily measured and, for narrow peaks, more accurately determined. Most modern chromatographic instruments are equipped with a mechanical or electronic integrator that provides accurate estimates of peak areas. Lacking such equipment, manual estimates can be made in several ways. A simple method, which works well with symmetric peaks of reasonable widths, is to multiply the height of the peak by its width at one-half the peak height. Another approach involves cutting out the peak and determining its weight relative to the weight of a known area of recorder paper. Finally, areas can be determined with a planimeter.

## 18E  GAS – LIQUID CHROMATOGRAPHY

In *gas – liquid chromatography* (GLC), the components of a vaporized sample are fractionated as a consequence of being partitioned between a mobile *gaseous* phase and a liquid stationary phase held in a column.[4]

### 18E – 1 Apparatus

The basic components of a typical instrument for performing gas chromatography are shown in Figure 18 – 6 and are described briefly in this section.

---

[4] For detailed treatment of GLC, see J. A. Perry, *Introduction to Analytical Gas Chromatography,* New York: Marcel Dekker, 1981; *Modern Practice of Gas Chromatography,* 2nd ed., R. L. Grob, Ed. New York: John Wiley & Sons, 1985; J. Q. Walker, M. T. Jackson, Jr., and J. B. Maynard, *Chromatographic Systems,* 2nd ed., Part II. New York: Academic Press, 1977.

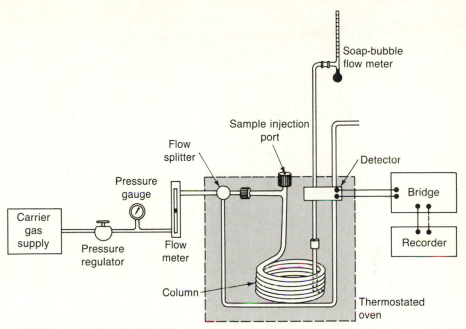

**Figure 18-6** Schematic diagram of a gas chromatograph.

**Carrier Gas Supply** The gaseous mobile phase must be chemically inert. Helium is the commonest mobile phase, although argon, nitrogen, and hydrogen are also used. These gases are available in pressurized tanks. Associated with the gas supply are pressure regulators, gauges, and flow meters.

Pressures at the column inlet usually range from 10 to 50 psi (lb/in² above room pressure) and provide flow rates of 25 to 50 mL/min. Flow rates are generally measured by a simple soap-bubble meter, such as that shown at the end of the column in Figure 18-6. A soap film is formed in the path of the gas when the rubber bulb containing a solution of soap is squeezed; the time required for this film to move between two graduations on the buret is measured and converted to a flow rate.

**Sample Injection System** Column efficiency requires that the sample be of a suitable size and be introduced as a "plug" of vapor; slow injection or oversized samples cause band spreading and poor resolution. A microsyringe is used to inject liquid samples through a rubber or silicone diaphragm or septum into a heated sample port located at the head of the column (the sample port is ordinarily about 50°C above the boiling point of the least volatile component of the sample). For ordinary packed analytical columns, sample sizes range from a few tenths of a microliter to 20 μL. Capillary columns require samples that are smaller by a factor of 100 or more. Here, a sample splitter is often needed to deliver only a small known fraction (1:100 to 1:500) of the injected sample, with the remainder going to waste.

**Packed Columns**    Two types of columns are encountered in gas–liquid chromatography, *packed,* and *open tubular* or *capillary.* The former, which are discussed in this section, can accommodate larger samples and are generally more convenient to use. The latter, which are described in the next section, are of considerable importance because of their unparalleled resolution.

Present-day packed columns are fabricated from glass or metal tubing; they are typically 2 to 3 m long and have inside diameters of 2 to 4 mm. The tubes are ordinarily formed as coils with diameters of roughly 15 cm to permit convenient thermostating in an oven.

The packing for a column holds the liquid stationary phase in place, so that the surface area exposed to the mobile phase is as large as possible. The ideal solid packing consists of small, uniform, spherical particles with good mechanical strength and with a specific surface of at least 1 $m^2/g$. In addition, the material should be inert at elevated temperatures and be uniformly wetted by the liquid phase. No substance that meets all of these criteria perfectly is yet available.

The first, and still the most widely used, supports for gas chromatography were prepared from naturally occurring diatomaceous earth, which consists of the skeletons of thousands of species of single-celled plants that inhabited ancient lakes and seas. Two types of supports are derived from diatomaceous earth. The first, which is generally known by the trade name of Chromosorb P, is prepared by grinding and screening diatomaceous earth that has been briquetted at 900°C. The second, called Chromosorb W or G, is prepared from diatomaceous earth that has been heated to 900°C with a sodium carbonate flux. This product is more rugged than the first and also has less tendency to adsorb solutes. On the other hand, its specific surface area is only about 1 $m^2/g$, compared with 4 $m^2/g$ for Chromosorb P. Both types of support materials are often treated chemically with dimethylchlorosilane, which gives a surface layer of methyl groups. This treatment reduces the tendency of the support to adsorb polar molecules.

The particle size of packings for gas chromatography typically fall in the range of 60 to 80 mesh (250 to 170 $\mu$m) or 80 to 100 mesh (170 to 149 $\mu$m). The use of smaller particles is not practical because the pressure drop within the column becomes prohibitively high.

**Open Tubular Columns**    Open tubular or capillary columns were first described in the 1950s, when it became apparent from theoretical considerations that such columns should provide separations that were unprecedented in terms of speed and number of theoretical plates.[5] At that time, several investigations demonstrated that columns with as many as 300,000 plates or more were practical. Despite such spectacular results, the use of capillary columns

---

[5] For a detailed description of open tubular columns, see M. L. Lee, F. J. Yang, and K. D. Bartie, *Open Tubular Column Gas Chromatography.* New York: John Wiley & Sons, 1984.

was delayed until recently because of a number of problems associated with their use. These problems have now become manageable and a number of instrument vendors offer open tubular equipment for routine use. Capillary columns, which are constructed of stainless steel, glass, or fused silica, typically have inside diameters of 0.25 to 0.50 mm and lengths of 25 to 50 m. Their inner surfaces are coated with a thin layer of the stationary phase, which may be any of the liquids described in Section 18E–2.

The most recent columns are constructed of fused silica. Their manufacture is based on techniques developed for the production of optical fibers. Silica capillaries, which have much thinner walls than their glass or metal counterparts, have outside diameters of about 0.3 mm. The tubes are given added strength by an outside polyimide coating. The resulting columns are quite flexible and strong, and can be bent into coils with diameters of a few inches. An important advantage of fused silica columns is their minimal tendency to adsorb analyte molecules.

**Column Thermostating**    Reproducible retention times require control of the column temperature to within a few tenths of a degree. For this reason, the coiled column is ordinarily housed in a thermostated oven. The optimum temperature depends upon the boiling points of the sample components. A temperature that is roughly equal to or slightly above the average boiling point of a sample results in a reasonable elution period (2 to 30 min). For samples with a broad boiling range, it may be necessary to employ temperature programming, whereby the column temperature is increased either continuously or in steps as the separation proceeds.

In general, optimum resolution is associated with minimal temperature. Lower temperatures, however, result in longer elution times and hence slower analyses. Figure 18–7 illustrates this principle.

**Detectors**    Detection devices for gas–liquid chromatography must respond rapidly to minute concentrations of solutes as they exit the column. The solute concentration in the carrier gas at any instant is no more than a few parts per thousand and often is smaller by one or two orders of magnitude. Moreover, the time during which a peak passes the detector is typically 1 sec (or less), which requires that the device be capable of exhibiting its full response during this brief period.

Other desirable properties for a detector include linear response, stability, and uniform response for a wide variety of chemical species or, alternatively, a predictable and selective response toward one or more classes of solutes. No single detector fulfills all of these requirements. Two of the most widely used detectors are discussed in the sections that follow.

**The Thermal Conductivity Detector**    The *thermal conductivity detector,* or *katharometer,* which was one of the earliest detectors for gas chromatography,

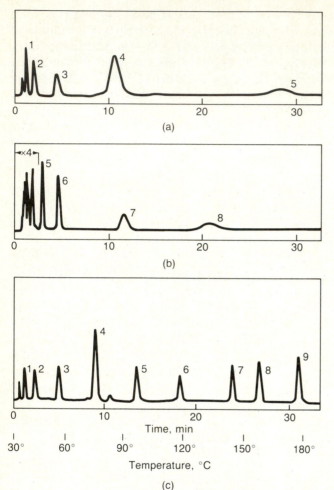

**Figure 18–7** Effect of temperature on gas chromatograms. (a) Isothermal conditions at 45°C. Note that a good separation of compounds 1 through 4 is realized but that the higher boiling compounds are not eluted. (b) Isothermal conditions at 145°C. Here, compounds 4 through 8 are separated but compounds 1 through 3 are not; compound 9 is not eluted. (c) Linear programmed temperature variations from 30° to 180°C. Note that all nine compounds are eluted with good separations at both ends of the temperature range. (From W. E. Harris and H. W. Habgood, *Programmed Temperature Gas Chromatography*, p. 10. New York: John Wiley & Sons, 1966. With permission.)

still finds wide application. This device consists of an electrically heated source whose temperature at constant electric power depends upon the thermal conductivity of the surrounding gas. The heated element may be a fine platinum, gold, or tungsten wire or, alternatively, a small thermistor. The electrical resistance of this element depends on the thermal conductivity of the gas. Twin detectors are ordinarily used, one being located ahead of the sample injection chamber and the other immediately beyond the column; alternatively, the gas stream is split as in Figure 18–6. The detectors are incorporated in two arms of a simple bridge circuit such that the thermal conductivity of the carrier is canceled. In addition, the effects of variations in temperature, pressure, and electric power are minimized. The thermal conductivities of helium and hydrogen are roughly six to ten times greater than those of most organic compounds. Thus, even small amounts of organic species cause relatively large decreases in the thermal conductivity of the column effluent, which results in a marked rise in the temperature of the detector. Detection by thermal conductivity is less satisfactory with carrier gases whose conductivities more closely resemble those of most samples.

The advantages of the thermal conductivity detector are its simplicity, its large linear dynamic range (about five orders of magnitude), its general response to both organic and inorganic species, and its nondestructive character, which permits collection of solutes after detection. The chief limitation of the katharometer is its relatively low sensitivity ($\sim 10^{-8}$ g/mL carrier). Other detectors exceed this sensitivity by factors of $10^4$ to $10^7$.

**The Flame Ionization Detector**   One of the most sensitive and widely used detectors for gas chromatography is based upon the fact that most organic compounds, when pyrolyzed in a hot flame, produce ionic intermediates that can conduct electricity through the flame. Hydrogen is used as the carrier gas with this detector, and the eluent is mixed with oxygen and combusted in a burner equipped with a pair of electrodes. Detection involves monitoring the conductivity of the combustion products. The ionization detector exhibits a high sensitivity ($\sim 10^{-13}$ g/mL), a large linear response, and low noise. It is also rugged and easy to use. A disadvantage of the ionization detector is that it destroys the sample.

### 18E–2  Liquid Phases for Gas–Liquid Chromatography

Several hundred liquids have been used as the stationary phase in gas–liquid chromatography. The successful separation of closely related compounds is often critically dependent upon the proper choice among these. The retention time for a solute depends upon its partition coefficient (Equation 18–9), which in turn is related to the nature of the stationary phase. Clearly, to be useful in gas–liquid chromatography, the immobilized liquid must generate different partition coefficients among solutes; additionally, however, these coefficients must be neither extremely large nor extremely small. For species with large coefficients, the time required for elution becomes prohibitive, while compounds with low coefficients pass so rapidly through the column that little or no separation occurs.

To have a reasonable residence time in the column, a species must show at least some degree of compatibility (solubility) with the stationary phase. Thus, the polarities of the two substances should be at least somewhat alike. For example, a stationary phase such as squalene (a high-molecular-weight, nonpolar, saturated hydrocarbon) might be chosen for separation of members of a nonpolar homologous series such as hydrocarbons or ethers. On the other hand, a more polar stationary phase, such as polyethylene glycol, would be more effective for separating alcohols or amines. For aromatic hydrocarbons, benzyldiphenyl might prove more appropriate.

Among analytes of similar polarity, the elution order usually follows the order of boiling points; where these differ sufficiently, clean separations are feasible. Solutes with nearly identical boiling points but different polarities frequently require a stationary phase that will selectively retain one or more of the components by dipole interaction or by adduct formation. Another important interaction, which often enhances selectivity, is hydrogen bonding. For this effect to operate, the solute must be a proton donor and the stationary phase must contain a proton acceptor group (such as oxygen, fluorine, or nitrogen), or the converse. Table 18–2 lists a few of the most widely used stationary phases.

## 18E–3 Applications of Gas–Liquid Chromatography

Gas–liquid chromatography is applicable to species that are appreciably volatile and thermally stable at temperatures up to a few hundred degrees Celsius. An enormous number of compounds of interest to man possess these qualities. Consequently, gas chromatography has been widely applied to the separation and determination of the components in a variety of sample types. Figure 18–8 shows chromatograms for a few such applications. Data on the conditions employed for obtaining these chromatograms are found in Table 18–3. Note that the chromatogram in Figure 18–8a was obtained with an open tubular column and illustrates the power of this technique. Here, nine heptane isomers were separated in just under 1 min.

## 18F HIGH-PERFORMANCE LIQUID CHROMATOGRAPHY

Early liquid chromatographic columns were glass tubes with diameters of perhaps 10 to 50 mm that held 50- to 500-cm lengths of solid particles of the stationary phase. To ensure reasonable flow rates, the particle size of the solid was kept larger than 150 to 200 $\mu$m; even then, flow rates were, at best, a few tenths of a milliliter per minute. Attempts to speed up this classic procedure by application of vacuum or pressure were not effective because increases in flow rates were accompanied by increases in plate heights and accompanying decreases in column efficiency.

Early in the development of liquid chromatography, it was realized that large decreases in plate heights could be expected to accompany decreases in the particle size of packings. It was not until the late 1960s, however, that the technology for producing and using packings with particle diameters as small as 5 to 10 $\mu$m was developed. This technology required sophisticated instruments that contrasted markedly with the simple devices that preceded them. The

Table 18–2  Some Common Stationary Phases for Gas–Liquid Chromatography

| Trade Name | Chemical Composition | Maximum Temperature, °C | Polarity* | Type of Separation |
|---|---|---|---|---|
| Squalene | $C_{30}H_{62}$ | 150 | NP | Hydrocarbons |
| OV-1 | Polymethyl siloxane | 350 | NP | General purpose nonpolar |
| DC 710 | Polymethylphenyl siloxane | 300 | NP | Aromatics |
| QF-1 | Polytrifluoropropylmethyl siloxane | 250 | P | Amino acids, steroids, nitrogen compounds |
| XE-30 | Polycyanomethyl siloxane | 275 | P | Alkaloids, halogenated compounds |
| Carbowax 20M | Polyethylene glycol | 250 | P | Alcohols, esters, essential oils |
| DEG adipate | Diethylene glycol adipate | 200 | SP | Fatty acids, esters |
| | Dinonyl phthalate | 150 | SP | Ketones, ethers, sulfur compounds |

* NP = nonpolar; SP = semipolar; P = polar.

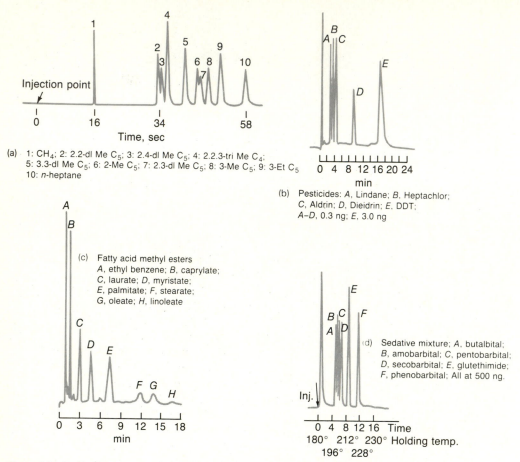

(a)  1: $CH_4$; 2: 2,2-dl Me $C_5$; 3: 2,4-dl Me $C_5$; 4: 2,2,3-tri Me $C_4$;
     5: 3,3-dl Me $C_5$; 6: 2-Me $C_5$; 7: 2,3-dl Me $C_5$; 8: 3-Me $C_5$; 9: 3-Et $C_5$
     10: n-heptane

(b)  Pesticides: A, Lindane; B, Heptachlor;
     C, Aldrin; D, Dieidrin; E, DDT;
     A–D, 0.3 ng; E, 3.0 ng

(c)  Fatty acid methyl esters
     A, ethyl benzene; B, caprylate;
     C, laurate; D, myristate;
     E, palmitate; F, stearate;
     G, oleate; H, linoleate

(d)  Sedative mixture; A, butalbital;
     B, amobarbital; C, pentobarbital;
     D, secobarbital; E, glutethimide;
     F, phenobarbital; All at 500 ng.

**Figure 18–8**  Typical gas–liquid chromatograms. See Table 18–3 for experimental conditions.

name *high-performance liquid chromatography* (HPLC) is often employed to distinguish these newer procedures from their predecessors, which still find considerable use for preparative purposes.[6]

Figure 18–9 illustrates the improved column efficiency that can be obtained with smaller diameter particles. It is of interest to note that in none of these plots is the minimum shown in Figure 18–4 reached. The reason for this difference is that longitudinal diffusion in liquids is much slower than in gases; consequently, its effect on plate heights is observed only at extremely low flow rates.

---

[6] References for HPLC include L. R. Snyder and J. J. Kirkland, *Introduction to High-Performance Liquid Chromatography,* 2nd ed. New York: Chapman and Hall, 1982; *High-Performance Liquid Chromatography,* C. Horvath, Ed., Vols. 1 and 2. New York: Academic Press, 1980.

*Table 18 – 3  Conditions for Chromatograms Shown in Figure 18 – 8*

| Chromatogram | Column* | Stationary Phase | Temperature, °C | Carrier | Flow Rate |
|---|---|---|---|---|---|
| a | 50′ × 0.005″ open tubular | Squalene | 20 | $H_2$ | 95 cm/sec |
| b | 6′ × ¼″ G | 1.5% OV-17† on Chromosorb G | 22 | — | — |
| c | 6′ × ⅛″ S | Chromosorb W | 190 | He | 30 mL/min |
| d | 6′ × ¼″ G | 1.5% OV-17 on HP Chromosorb G | 180–230 | — | — |

\* S = stainless steel; G = glass.

† OV-17 is a registered trade name for a mixture of methyl and phenyl silicones.

513

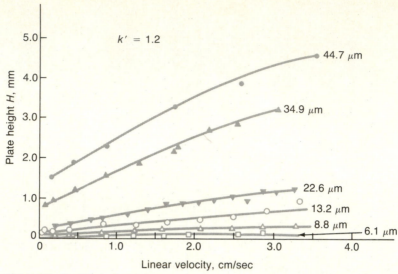

**Figure 18–9** Effect of particle size of packing and flow rate upon plate height in high-performance liquid chromatography. Column dimensions: 30 cm × 2.4 mm. Solute: N,N-diethyl-n-aminoazobeneze. Mobile phase: mixture of hexane, methylene chloride, and isopropyl alcohol. (From R. E. Majors, *J. Chromatogr. Sci.*, **1973**, *11*, 92. With permission.)

## 18F–1 Apparatus

Pumping pressures of several hundred atmospheres are required to achieve reasonable flow rates with packings in the 3 to 10 $\mu$m size range, which are common in modern liquid chromatography. As a consequence of these high pressures, the equipment for high-performance liquid chromatography tends to be considerably more elaborate and expensive than that encountered in other types of chromatography. Figure 18–10 is a schematic diagram showing the important components of a typical high-performance chromatograph.

**Mobile Phase Reservoirs and Solvent Treatment Systems**   A modern HPLC apparatus is equipped with one or more glass or stainless steel reservoirs, each of which contains 500 mL or more of a solvent (Figure 18–10). Provisions are often included to remove dissolved gases and dust from the liquids. The former produce bubbles in the column and thereby cause band spreading; in addition, both bubbles and dust interfere with the performance of detectors. Degassers may consist of a vacuum pumping system, a distillation system, a device for heating and stirring, or, as shown in Figure 18–10, a system for *sparging,* in which the dissolved gases are swept out of solution by fine bubbles of an inert gas that is not soluble in the mobile phase.

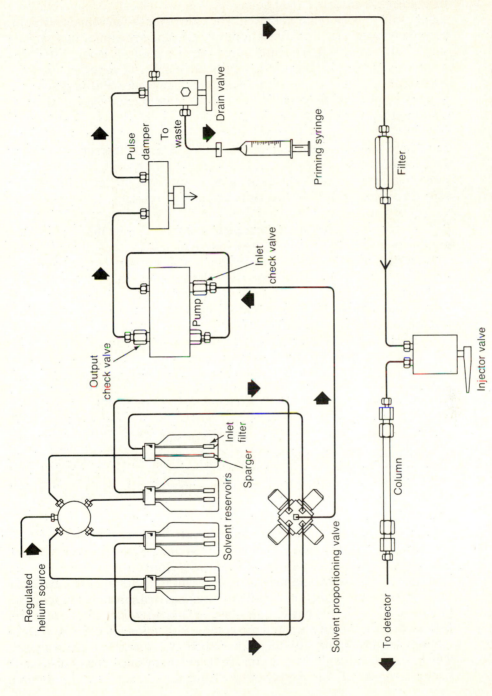

**Figure 18-10** Schematic diagram of an apparatus for high-performance liquid chromatography. (Courtesy of Perkin-Elmer Corporation, Norwalk, CT.)

Regulated helium source

Solvent reservoirs

Inlet filter

Sparger

Solvent proportioning valve

Output check valve

Pump

Inlet check valve

Pulse damper

To waste

Drain valve

Priming syringe

Filter

Column

To detector

Injector valve

An elution with a single solvent of constant composition is termed *isocratic*. In *gradient elution*, two (and sometimes more) solvent systems that differ significantly in polarity are employed. The ratio of the two solvents is varied in a preprogrammed way, sometimes continuously and sometimes in a series of steps. Gradient elution frequently improves separation efficiency, just as temperature programming helps in gas chromatography. Modern high-performance liquid chromatography instruments are often equipped with proportionating valves that introduce liquids from two or more reservoirs at rates that vary continuously (Figure 18–10).

**Pumping Systems**   The requirements for liquid chromatographic pumps are severe and include (1) the generation of pressures of up to 6000 psi ($lb/in^2$), (2) pulse-free output, (3) flow rates ranging from 0.1 to 10 mL/min, (4) flow reproducibilities of 0.5% relative or better, and (5) resistance to corrosion by a variety of solvents.

It should be noted that the high pressures generated by liquid chromatographic pumps do not constitute an explosion hazard because liquids are not very compressible. Thus, rupture of a component results only in solvent leakage. To be sure, such leakage may constitute a fire hazard.

**Sample Injection Systems**   Although syringe injection through an elastomeric septum is often used in liquid chromatography, this procedure is not very reproducible and is limited to pressures less than about 1500 psi. In *stop-flow* injection, the solvent flow is stopped momentarily, a fitting at the column head is removed, and the sample is injected directly onto the head of the packing by means of a syringe.

Although syringe ejection finds considerable use owing to its simplicity, the most widely used method of sample introduction in liquid chromatography is based upon sampling loops such as that shown in Figure 18–11. These devices are often an integral part of modern liquid chromatography equipment and have interchangeable loops that provide a choice of sample sizes ranging from 5 to 500 $\mu$L. The reproducibility of injections with a typical sampling loop is a few tenths of a percent relative.

**Columns for High-Performance Liquid Chromatography**   Liquid chromatographic columns are usually constructed from stainless steel tubing, although heavy-walled glass tubing is sometimes employed for lower pressure applications (< 600 psi). Most columns range in length from 10 to 30 cm and have inside diameters of 4 to 10 mm. Column packings typically have particle sizes of 5 or 10 $\mu$m. Columns of this type often contain 40,000 to 60,000 plates/m. Recently, high-performance microcolumns with inside diameters of 1 to 4.6 mm and lengths of 3 to 7.5 cm have become available. These columns, which are packed with 3- or 5-$\mu$m particles, contain as many as 100,000 plates/m and have the advantage of speed and minimal solvent consumption.

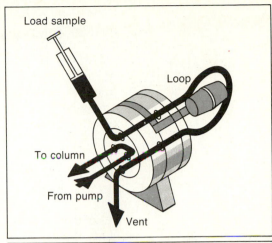

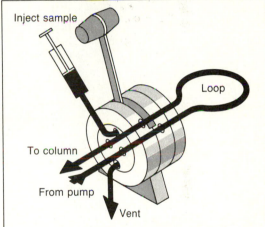

**Figure 18–11** A sampling loop for a liquid chromatograph. (Courtesy of Beckman Instruments, Fullerton, CA.)

The most common packing for liquid chromatography is prepared from silica particles, which are synthesized by agglomerating submicron silica particles under conditions that lead to larger particles with highly uniform diameters. The resulting particles are often coated with thin organic films, which are chemically or physically bonded to the surface. Other packing materials include alumina particles, porous polymer particles, and ion-exchange resins.

**Detectors**    No highly sensitive, universal detector system, such as those for gas chromatography, is available for high-performance liquid chromatography. Thus, the system used will depend upon the nature of the sample. Table 18–4 lists some of the common detectors and their properties.

*Table 18-4*  *Characteristics of Detectors for High-Performance Liquid Chromatography*\*

| Basis for Detection | Type† | Maximum Sensitivity‡ | Flow-Rate Sensitive? | Temperature Sensitivity | Applicable to Gradient Elution? | Available Commercially? |
|---|---|---|---|---|---|---|
| UV absorption | S | $2 \times 10^{-10}$ | No | Low | Yes | Yes |
| IR absorption | S | $10^{-6}$ | No | Low | Yes | Yes |
| Fluorometry | S | $10^{-11}$ | No | Low | Yes | Yes |
| Refractive index | G | $1 \times 10^{-7}$ | No | $\pm 10^{-4}/°C$ | No | Yes |
| Electrical conductivity | S | $10^{-8}$ | Yes | $2\%/°C$ | No | Yes |
| Mass spectrometry | G | $10^{-10}$ | No | None | Yes | Yes |
| Electrochemical | S | $10^{-12}$ | Yes | $1.5\%/°C$ | No | Yes |
| Radiochemical | S | — | No | None | Yes | No |

\* Most of these data were taken from L. R. Snyder and J. J. Kirkland, *Introduction to Modern Liquid Chromatography*, 2nd ed., p. 162. New York: Wiley-Interscience, 1979. With permission.

† G = general; S = selective.

‡ Sensitivity for a favorable sample in grams per milliliter.

The most widely used detectors for liquid chromatography are based upon absorption of ultraviolet or visible radiation. Both photometers and spectrophotometers, specifically designed for use with chromatographic columns, are available from commercial sources. The former often make use of the 254- and 280-nm lines from a mercury source because many organic functional groups absorb in this region (see Chapter 15, Section 15D – 3). Deuterium or tungsten filament sources with interference filters also provide a simple means of detecting absorbing species. Some modern instruments are equipped with wheels that contain several filters, which can be rapidly switched into place. Spectrophotometric detectors are considerably more versatile than photometers and are also widely used in high-performance instruments.

Another detector, which has found considerable application, is based upon the changes in the refractive index of the solvent that is caused by analyte molecules. In contrast to most of the other detectors listed in Table 18 – 4, the refractive index indicator is general rather than selective and responds to the presence of all solutes.

### 18F – 2 High-Performance Partition Chromatography

Partition chromatography has become the most widely used of all liquid chromatographic procedures. This technique can be subdivided into *liquid – liquid* and *bonded-phase* chromatography. The difference between the two lies in the method by which the stationary phase is held on the support particles of the packing. With liquid – liquid, retention is by physical adsorption, while with bonded-phase, covalent bonds are involved. Early partition chromatography was exclusively liquid – liquid; now, however, bonded-phase methods predominate, with liquid – liquid separations being relegated to certain special applications.

**Bonded-Phase Packings**   Most bonded-phase packings are prepared by reaction of an organochlorosilane with the —OH groups formed on the surface of silica particles by hydrolysis in hot, dilute hydrochloric acid. The product is an organosiloxane. The reaction for one such SiOH site on the surface of a particle can be written as

$$-\overset{|}{\underset{|}{Si}}OH \quad + \quad Cl-\overset{\overset{\displaystyle CH_3}{|}}{\underset{\underset{\displaystyle CH_3}{|}}{Si}}-R \quad \longrightarrow \quad -\overset{|}{\underset{|}{Si}}-O-\overset{\overset{\displaystyle CH_3}{|}}{\underset{\underset{\displaystyle CH_3}{|}}{Si}}-R \quad + \quad HCl$$

where R is often a straight chain octyl- or octyldecyl-group. Other organic functional groups that have been bonded to silica surfaces include aliphatic amines, ethers, and nitriles, as well as aromatic hydrocarbons. Thus, a variety of polarities for the bonded stationary phase are available.

Bonded-phase packings have the advantage of markedly greater stability compared with physically held stationary phases. With the latter, periodic recoating of the solid surfaces is required because the stationary phase is gradually dissolved away in the mobile phase. Furthermore, gradient elution is not practical with liquid–liquid packings, again because of losses by solubility to the mobile phase. The main disadvantage of bonded-phase packings is their somewhat limited sample capacity.

**Normal- and Reversed-Phase Packings**   Two types of partition chromatography are distinguishable based upon the relative polarities of the mobile and stationary phases. Early work in liquid chromatography was based upon highly polar stationary phases such as triethylene glycol or water; a relatively nonpolar solvent such as hexane or *i*-propyl ether then served as the mobile phase. For historic reasons, this type of chromatography is now called *normal-phase chromatography.* In *reversed-phase chromatography,* the stationary phase is nonpolar, often a hydrocarbon, and the mobile phase is a relatively polar solvent (such as water, methanol, or acetonitrile).[7] In normal-phase chromatography, the *least* polar component is eluted first; *increasing* the polarity of the mobile phase then *decreases* the elution time. In contrast, in the reversed-phase method, the *most* polar component elutes first, and *increasing* the mobile phase polarity *increases* the elution time.

It has been estimated that more than three-quarters of all HPLC separations are currently performed with reversed-phase, bonded, octyl- or octyldecyl siloxane packings. With such preparations, the long-chain hydrocarbon groups are aligned parallel to one another and perpendicular to the surface of the particle, giving a brushlike, nonpolar, hydrocarbon surface. The mobile phase used with these packings is often an aqueous solution containing various concentrations of such solvents as methanol, acetonitrile, or tetrahydrofuran.

**Choice of Mobile and Stationary Phases**   Successful partition chromatography requires a proper balance of intermolecular forces among the three participants in the separation process—the solute, the mobile phase, and the stationary phase. These intermolecular forces are described qualitatively in terms of the relative polarity possessed by each of the three reactants. In general, the polarities of common organic functional groups in increasing order are as follows: aliphatic hydrocarbons < olefins < aromatic hydrocarbons < halides < sulfides < ethers < nitro compounds < esters ≈ aldehydes ≈ ketones < alcohols ≈ amines < sulfones < sulfoxides < amides < carboxylic acids < water.

As a rule, most chromatographic separations are achieved by matching the polarity of the analyte to that of the stationary phase; a mobile phase of considerably different polarity is then used. This procedure is generally more successful than one in which the polarities of the analyte and the mobile phase are

---

[7] For a detailed discussion of reversed-phase HPLC, see A. M. Krstulovic and P. R. Brown, *Reversed-Phase High-Performance Liquid Chromatography.* New York: John Wiley & Sons, 1982.

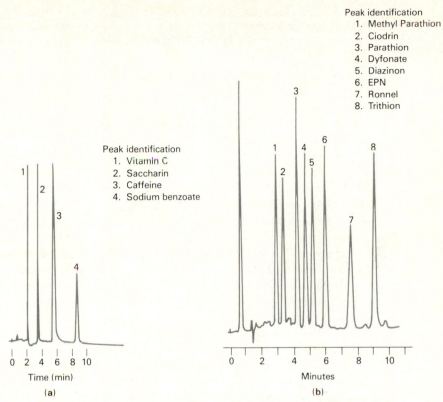

Peak identification
1. Methyl Parathion
2. Ciodrin
3. Parathion
4. Dyfonate
5. Diazinon
6. EPN
7. Ronnel
8. Trithion

Peak identification
1. Vitamin C
2. Saccharin
3. Caffeine
4. Sodium benzoate

Time (min)

(a)

Minutes

(b)

**Figure 18–12** Typical applications of bonded-phase chromatography. (a) Soft-drink additives. Column: 4.6 × 250 mm packed with polar (nitrile) bonded-phase packing. Isocratic elution: 6% HOAC–94% $H_2O$. Flow rate: 1.0 mL/min. (Courtesy of Du Pont Instrument Systems, Wilmington, DE.) (b) Organophosphate insecticides. Column: 4.5 × 250 mm, packed with 5-$\mu$m $C_8$ bonded-phase particles. Gradient elution: 67% $CH_3OH$–33% $H_2O$ to 80% $CH_3OH$–20% $H_2O$. Flow rate: 2 mL/min. (Courtesy of IBM Instruments, Danbury, CT.)

matched but are different from that of the stationary phase. Here, the stationary phase often cannot compete successfully for the sample components; retention times then become too short for practical application. At the other extreme is the situation where the polarities of the analyte and stationary phase are too much alike; here, retention times become inordinately long.

**Applications**   Figure 18–12 shows typical applications of bonded-phase partition chromatography. Table 18–5 further illustrates the variety of samples to which the technique is applicable.

### 18F–3 High-Performance Adsorption Chromatography

All of the pioneering work in chromatography was based upon liquid–solid adsorption, in which the stationary phase is the surface of a finely divided polar

*Table 18–5    Typical Applications of High-Performance Partition Chromatography*

| Field | Typical Mixtures |
| --- | --- |
| Pharmaceuticals | Antibiotics, sedatives, steroids, analgesics |
| Biochemical | Amino acids, proteins, carbohydrates, lipids |
| Food products | Artifical sweetners, antioxidants, aflatoxins, additives |
| Industrial chemicals | Condensed aromatics, surfactants, propellants, dyes |
| Pollutants | Pesticides, herbicides, phenols, PCBs |
| Forensic chemistry | Drugs, poisons, blood alcohol, narcotics |
| Clinical medicine | Bile acids, drug metabolites, urine extracts, estrogens |

solid. With such a packing, the analyte competes with the mobile phase for sites on the surface of the packing; retention is the result of adsorption forces.

**Stationary and Mobile Phases**    Finely divided silica and alumina are the only stationary phases that find extensive use for adsorption chromatography. Silica is preferred for most (but not all) applications because of its higher sample capacity and its wider range of useful forms. The adsorption characteristics of the two substances parallel one another. For both, retention times become longer as the polarity of the analyte increases.

In adsorption chromatography, the only variable that affects the partition coefficient of analytes is the composition of the mobile phase (in contrast to partition chromatography, where the polarity of the stationary phase can also be varied). Fortunately, enormous variations in retention and thus resolution accompany variations in the solvent system, and only rarely is a suitable mobile phase not available.

**Applications of Adsorption Chromatography**    Currently, liquid–solid HPLC is used extensively for the separations of relatively nonpolar, water-insoluble organic compounds with molecular weights that are less than about 5000. A particular strength of adsorption chromatography, which is not shared by other methods, is its ability to resolve isomeric mixtures such as meta and para substituted benzene derivatives.

### 18F–4 High-Performance Ion-Exchange Chromatography

We have already considered some of the applications of ion-exchange resins to analytical separations. In addition, these materials are useful as stationary phases for liquid chromatography, where they are used to separate charged species.[8]

---

[8] For a brief review of ion chromatography, see: H. Small, *Anal. Chem.* **1983,** *55,* 235A. For a detailed description of the method, see: F. C. Smith, Jr. and R. C. Chang, *The Practice of Ion Chromatography.* New York: John Wiley & Sons, 1983.

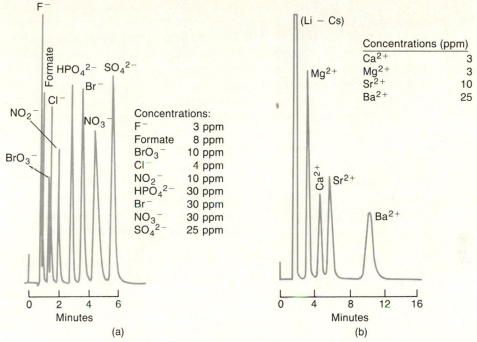

**Figure 18–13** Typical applications of ion-exchange chromotography. (a) Separation of anions on an anion-exchange column. Eluent: 0.0028 F NaHCO₃–0.0023 F Na₂CO₃. Sample size: 50 μL. (b) Separation of alkaline-earth ions on a cation-exchange column. Eluent: 0.0025 F phenylenedi-amine dihydrochloride–0.0025 F HCl. Sample size: 100 μL. (Courtesy of Dionex, Sunnyvale, CA.)

The techniques for fractionating ions with $K_D$ values that are relatively close to one another are analogous to those described for partition and adsorption chromatography. For example, Figure 18–13b shows a separation of several cations on a sulfonic acid resin. Figure 18–13a illustrates an anion-exchange separation of some common anions. In both cases, detection was based on conductometric measurements.

### 18F–5 High-Performance Size-Exclusion Chromatography

Size-exclusion chromatography is the newest of the liquid chromatographic procedures. It is a powerful technique that is particularly applicable to high-molecular-weight species.[9]

**Packings** Packings for size-exclusion chromatography consist of small (~10 μm) silica or polymer particles containing a network of uniform pores

---

[9] See W. W. Yao, J. J. Kirkland, and D. D. Bly, *Modern Size Exclusion Liquid Chromatography.* New York: John Wiley & Sons, 1979.

into which solute and solvent molecules can diffuse. While in the pores, molecules are effectively trapped and removed from the flow of the mobile phase. The average residence time of analyte molecules depends upon their effective size. Molecules that are significantly larger than the average pore size of the packing are excluded and thus suffer no retention; that is, they travel through the column at the rate of the mobile phase. Molecules that are appreciably smaller than the pores can penetrate throughout the pore maze and are thus entrapped for the greatest time; they are last to be eluted. Between these two extremes are intermediate-size molecules whose average penetration into the pores of the packing depends upon their diameters. Fractionation occurs within this group that is directly related to molecular size and, to some extent, molecular shape. Note that size-exclusion separations differ from the other chromatographic procedures in the respect that no chemical or physical interactions between analytes and the stationary phase are involved. Indeed, every effort is made to avoid such interactions because they lead to impaired column efficiencies.

Numerous size-exclusion packings are on the market. Some are hydrophilic for use with aqueous mobile phases; others are hydrophobic and are used with nonpolar organic solvents. Chromatography based on the former is sometimes called *gel filtration,* while techniques based on the latter are termed *gel permeation.* With both types of packings, a wide variety of pore diameters are available. Ordinarily, a given packing will accommodate a 2- to 2.5-decade range of

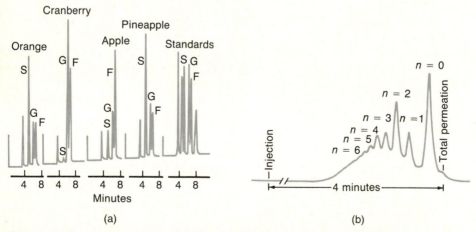

**Figure 18 – 14**   Typical applications of size-exclusion chromatography. (a) Determination of glucose (G), fructose (F), and sucrose (S) in canned juices. A 25-cm column was packed with hydrophilic sulfonated polymer particles with an exclusion limit of 1000. (b) Analysis of a commercial epoxy resin ($n$ = number of monomeric units in the polymer). A porous silica column, 6.2 × 250 mm, was used with a mobile phase of tetrahydrofuran. (Courtesy of Du Pont Instrument Systems, Wilmington, DE.)

*Table 18-6  Comparison of High-Performance Liquid Chromatography and Gas-Liquid Chromatography*

**Characteristics of both methods**

Efficient, highly selective, and widely applicable
Only small sample required
May be nondestructive of sample
Readily adapted to quantitative analysis

**Advantages of HPLC**

Can accommodate nonvolatile and thermally unstable samples
Generally applicable to inorganic ions

**Advantages of GLC**

Simple and inexpensive equipment
Rapid
Unparalleled resolution (with capillary columns)
Easily interfaced with mass spectroscopy

molecular weight. The average molecular weight suitable for a given packing may be as small as a few hundred or as large as several million.

**Applications**   Figure 18-14a illustrates the application of gel filtration to the determination of three sugars in an aqueous medium. The hydrophilic packing used excluded molecular weights greater than 1000. Figure 18-14b is a chromatogram obtained with a hydrophobic packing in which the eluent was tetrahydrofuran. The sample was a commercial epoxy resin in which each monomer unit had a molecular weight of 280.

Another important application of size-exclusion chromatography is to the rapid determination of the molecular weight or the molecular weight distribution of large polymers or natural products. Here, the elution volumes of the sample are compared with elution volumes for a series of standard compounds that have the same chemical characteristics.

## 18G COMPARISON OF HIGH-PERFORMANCE LIQUID CHROMATOGRAPHY WITH GAS-LIQUID CHROMATOGRAPHY

Table 18-6 provides a comparison between high-performance liquid chromatography and gas-liquid chromatography. When either is applicable, gas-liquid chromatography offers the advantage of speed and simplicity of equipment. On the other hand, high-performance liquid chromatography is applicable to nonvolatile substances (including inorganic ions) and thermally unstable materials, whereas gas-liquid chromatography is not. Often, the two methods are complementary.

## PROBLEMS

*18-1. Describe the differences between gas–liquid chromatography and high-performance liquid chromatography.

18-2. How do planar and column chromatography differ?

*18-3. What is meant by elution?

18-4. What is a chromatogram?

*18-5. What is the fundamental difference between $N$ and $H$ as measures of separation efficiencies of columns?

18-6. What variables must be controlled if satisfactory quantitative data are to be obtained from chromatograms?

*18-7. Describe the physical differences between open tubular and packed columns. What are the advantages and disadvantages of each?

18-8. Define the following terms used in HPLC:
  *(a) sparging                    (b) gradient elution
  *(c) isocratic elution           (d) reversed-phase packing
  *(e) stop-flow injection         (f) sampling loops
  *(g) bonded-phase packings

18-9. Indicate the order in which the following compounds would be eluted from an HPLC column containing a reversed-phase packing:
  *(a) benzene, diethyl ether, *n*-hexane
   (b) acetone, dichloroethane, acetamide

18-10. Indicate the order of elution of the following compounds from a normal-phase packed HPLC column:
  *(a) ethyl acetate, acetic acid, dimethylamine
   (b) propylene, hexane, benzene, dichlorobenzene

*18-11. Describe the fundamental difference between adsorption and partition chromatography.

18-12. Describe the fundamental difference between ion-exchange and size-exclusion chromatography.

*18-13. Describe the difference between gel-filtration and gel-permeation chromatography.

18-14. What types of species can be separated by HPLC but not by GLC?

*18-15. One method for quantitative determination of the concentration of constituents in a sample analyzed by gas chromatography is the area normalization method. Here, complete elution of all of the sample constituents is necessary. The area of each peak is then measured and corrected for differences in detector response to the different eluates. This correction involves multiplication of the area by an empirically determined correction factor. The concentration of the analyte is found from the ratio of its corrected area to the total corrected area of all peaks.

  The relative peak areas obtained from a gas chromatogram of a mixture of methyl acetate, methyl propionate, and methyl *n*-butyrate were 18.4, 30.6, and 49.3, respectively. Calculate the percentage of each compound if the respective relative detection responses were 0.65, 0.83, and 0.92.

18-16. The peak areas and relative detector responses for a sample are to be used to determine the concentration of the five species it contains. The area normalization method described in Problem 18-15 is to be used. The relative areas for the five gas chromatographic peaks are given below. Also shown are the relative responses of the detector to the five compounds. Calculate the percentage of each component in the mixture.

| Compound | Peak Area, Relative | Detector Response, Relative |
|---|---|---|
| A | 27.6 | 0.70 |
| B | 32.4 | 0.72 |
| C | 47.1 | 0.75 |
| D | 40.6 | 0.73 |
| E | 27.3 | 0.78 |

**\*18-17.** The following data apply to a column for liquid chromatography:

| | |
|---|---|
| Length of packing | 24.7 cm |
| Flow rate | 0.313 mL/min |
| $V_M$ | 1.37 mL |
| $V_S$ | 0.164 mL |

A chromatogram of a mixture of species A, B, C, and D provided the following data:

| | Retention Time, min | Width of Peak Base ($W$), min |
|---|---|---|
| Nonretained | 3.1 | — |
| A | 5.4 | 0.41 |
| B | 13.3 | 1.07 |
| C | 14.1 | 1.16 |
| D | 21.6 | 1.72 |

Calculate
(a) the number of plates from each peak.
(b) the mean and the standard deviation for $N$.
(c) the plate height for the column.

**\*18-18.** From the data in Problem 18-17, calculate for A, B, C, and D
(a) the capacity factor.
(b) the partition coefficient.

**18-19.** The following data were obtained by GLC on a 40-cm packed column:

| Compound | $t_R$, min | $W_{1/2}$, min |
|---|---|---|
| Air | 1.9 | — |
| Methylcyclohexane | 10.0 | 0.76 |
| Methylcyclohexene | 10.9 | 0.82 |
| Toluene | 13.4 | 1.06 |

Calculate
(a) the average number of plates from the data.
(b) the standard deviation for the average in (a).
(c) an average plate height for the column.

**18-20.** List variables that lead to
(a) band broadening.
(b) band separation.

# Chapter 19
# Chemicals, Apparatus, and Unit Operations for Analytical Chemistry

This chapter is concerned with practical aspects of the unit operations that are used in an analytical laboratory as well as a description of the apparatus and the chemicals required for these operations.

## 19A SELECTION AND HANDLING OF CHEMICALS AND REAGENTS

The purity of the reagents has an important bearing upon the accuracy that can be attained in an analysis. Care must therefore be taken to be sure that the quality of a reagent is consistent with the use for which it is intended.

### 19A–1 Classification of Commercial Chemicals

**Technical Grade**  Chemicals labeled technical grade are of indeterminate quality and should be used only where purity is not of paramount importance. Thus, a cleaning solution could be prepared from technical-grade potassium dichromate and sulfuric acid. In general, however, technical-grade chemicals are not used in the analytical laboratory.

**USP Grade**  USP chemicals have been found to conform to the tolerances set forth in the United States Pharmacopoeia.[1] The specifications are designed to limit the contaminants that are dangerous to health; thus, chemicals that pass USP criteria may still contain significant amounts of impurities that are not considered to be physiological hazards.

**Reagent Grade**  Reagent-grade chemicals conform to the minimum specifications of the Reagent Chemical Committee of the American Chemical Society;[2]

---

[1] U.S. Pharmacopoeial Convention, *Pharmacopoeia of the United States of America,* 20th rev. ed. Easton, PA: Mack Publishing Co., 1980.

[2] Committee on Analytical Reagents, *Reagent Chemicals,* 6th ed. Washington, D.C.: American Chemical Society, 1982.

these are used wherever possible in analytical work. Some suppliers label their products with the maximum limits of impurity allowed by these specifications; others print the actual results of analyses for the various impurities.

**Primary-Standard Grade**   The qualities required of a *primary standard*—in addition to extraordinary purity—were described in Chapter 6, Section 6B–1. Primary-standard reagents, which are available from commercial sources, have been carefully analyzed, and the assay is printed on the label of their containers. An excellent source for primary-standard chemicals is the National Bureau of Standards. This agency also supplies *reference standards,* which are complex mixtures that have been exhaustively analyzed.[3]

**Special-Purpose Reagents**   Reagent chemicals that have been prepared for a specific application are also available. Included among these are solvents for spectrophotometry and for HPLC, as well as reagents for nonaqueous atomic spectroscopy and for electron microscopy. Accompanying these reagents is information that is pertinent to the specific use for which they are intended; data provided with a spectrophotometric solvent, for example, will specify its absorbance at various wavelengths, its ultraviolet cut-off wavelength, and its assay.

### 19A–2 Rules for the Handling of Reagents and Solutions

The availability of reagents and solutions with established purity is crucial to the performance of successful analytical work. A freshly opened bottle of a reagent-grade chemical can be used with confidence in most applications; whether the same confidence is justified when this bottle is half-full depends entirely on the way it has been handled after being opened. Only conscientious adherence to the accompanying rules will prevent the accidental contamination of reagents and solutions.

1. Select the best available grade of chemical for analytical work. If there is a choice, pick the smallest bottle that will supply the desired quantity.
2. Replace the top of every container *immediately* after the removal of reagent; do not rely on someone else to do this.
3. Hold stoppers of reagent bottles between the fingers; stoppers should never be set on the desk top.
4. Unless specifically directed to the contrary, *never return any excess reagent or solution to a bottle.* The minor saving represented by the return of an excess is overshadowed by the risk of contaminating the entire bottle.
5. Again, unless specifically directed otherwise, do not insert spoons, spatulas, or knives into a bottle containing a solid chemical. Instead, shake the capped

---

[3] United States Department of Commerce, *NBS Standard Reference Materials Catalog 1984–85, NBS Special Publ 260.* Washington, D.C.: Government Printing Office, 1984.

bottle vigorously or tap it sharply on a wooden table to break up any encrustations; then pour out the desired quantity. These measures are occasionally insufficient and a clean porcelain spoon must be used.
6. Keep the reagent shelf and the laboratory balance neat. Clean up any spillage immediately, even though someone else is waiting to use the same chemical.

## 19B THE CLEANING AND MARKING OF LABORATORY WARE

A chemical analysis is ordinarily performed in duplicate or triplicate. All equipment used in performance of such an analysis must be marked so that each sample is kept separate from the others. Flasks, beakers, and some filtering crucibles have small etched areas on which semipermanent markings can be made with a pencil. Special marking inks are available for porcelain surfaces; the marking is baked permanently into the glaze by heating at high temperature. A saturated solution of iron(III) chloride, while not as satisfactory as the commercial preparation, can also be used for marking.

Any beaker, flask, or crucible that will contain the sample must be thoroughly cleaned before being used. The apparatus should first be washed with a hot detergent solution, followed by rinsing, first with copious amounts of tap water, and finally, with several small portions of distilled water. A properly cleaned object will be coated with a uniform and unbroken film of water. *It is seldom necessary to dry glassware before use;* ordinarily, drying is a waste of time at best, and a potential source of contamination at worst.

If a grease film persists after thorough cleaning with detergent, a cleaning solution consisting of potassium dichromate in concentrated sulfuric acid may be used. After use of this preparation, extensive rinsing is required to remove the last traces of dichromate ions, which adhere strongly to glass and porcelain surfaces. Cleaning solution is most effective when warmed to about 70°C; at this temperature it rapidly attacks plant and animal matter and is thus a *potentially dangerous preparation.* Any spillages should be promptly diluted with copious volumes of water.

---

**Preparation of Cleaning Solution**   Mix 10 to 15 g of potassium dichromate with about 15 mL of water in a 500-mL conical flask. Add concentrated sulfuric acid *slowly* and in small increments; swirl the flask thoroughly between additions. The contents will become a semisolid mass; add just enough sulfuric acid to dissolve this mass. Allow the solution to cool before attempting to transfer it to a glass-stoppered storage bottle. The solution may be reused until it acquires the green color of chromium(III) ion, at which time it should be discarded. *CAUTION: Cleaning solution is highly corrosive and must be used with extreme care.*

---

Alternatives to dichromate cleaning solution are commercially available.

## 19C THE EVAPORATION OF LIQUIDS

In the course of an analysis the chemist often finds it necessary to decrease the volume of a solution without loss of a nonvolatile solute. An arrangement such as that shown in Figure 19–1 is satisfactory for most evaporations. The ribbed watch glass permits vapors to escape and protects the solution from accidental contamination. The use of glass hooks to provide space between the lip of the container and a conventional cover glass is less satisfactory.

The evaporation process is occasionally difficult to control, owing to the tendency of some solutions to overheat locally. The bumping that results can be sufficiently vigorous to cause partial loss of the solution. The danger of such loss is minimized through careful and gentle heating; the introduction of glass beads, where permissible, is also helpful.

Unwanted constituents in a solution can frequently be removed by evaporation. For example, chloride and nitrate ions can be eliminated by adding sulfuric acid and evaporating until copious white fumes of sulfur trioxide are observed. Nitrate ion and nitrogen oxides can be eliminated from acidic solutions by adding urea, evaporating to dryness, and gently igniting the residue. The removal of ammonium chloride is best accomplished by adding concentrated nitric acid and evaporating the solution to a small volume. Rapid oxidation of the ammonium ion occurs upon heating; the solution is then evaporated to dryness.

Organic constituents can frequently be eliminated by adding sulfuric acid and heating the solution until copious fumes of sulfur trioxide are observed; this process is known as *wet-ashing*. Nitric acid may be added toward the end of heating to hasten oxidation of the last traces of organic matter.

## 19D THE MEASUREMENT OF MASS

Highly reliable weighing data are ordinarily required at one stage or another in a chemical analysis; an *analytical balance* is used to acquire such data. Approximate weights are perfectly satisfactory for other purposes and a less precise but more rugged *laboratory balance* suffices.

**Figure 19 – 1**    Arrangement for the evaporation of liquids.

### 19D – 1  The Distinction between Mass and Weight

The reader should appreciate the difference between mass and weight. *Mass,* which is the more fundamental, is an invariant measure of the quantity of matter in an object. *Weight,* on the other hand, is the force of attraction exerted between the object and its surroundings, principally the earth. Because gravitational attraction is subject to geographical variation with altitude as well as latitude, the weight of an object is a somewhat variable quantity. For example, the weight of a crucible would be less in Denver than in Atlantic City because the attractive force between it and the earth is less at the higher altitude. Similarly, it would weigh more in Seattle than in Panama because the earth is somewhat flattened at the poles, and the force of attraction increases measurably with latitude. The mass of the crucible, on the other hand, remains constant regardless of the location in which it is measured.

Weight and mass are related by the familiar expression

$$W = Mg$$

where the weight of an object, $W$, is simply the product of its mass, $M$, and the acceleration due to gravity, $g$.

Chemical analyses are always based upon mass in order to free the results from a dependence on locality. In the laboratory, mass is determined with a *balance,* a device in which the weight of an object is compared with the weight of a set of standard masses. Because $g$ affects both the known and unknown to the same extent, an equality of weight indicates an equality of mass.

The terminological distinction between mass and weight is seldom observed in the sense that the operation of comparing masses is generally called *weighing,* and the objects of known mass as well as the results of weighing are called *weights.* It should always be borne in mind, however, that analytical data are based on mass rather than weight.

### 19D – 2  The Analytical Balance

By definition, an analytical balance is a weighing instrument with a capacity that ranges from 1 g to a few kilograms and with a precision of at least 1 part in $10^5$ at maximum capacity. Many modern analytical balances have precisions that are better than one part in $10^6$ at full capacity and have comparable accuracies as well.

The most commonly encountered analytical balance has a maximum load that ranges from 160 to 200 g; measurements can be made with a standard deviation of $\pm 10^{-4}$ g (or $\pm 0.1$ mg). *Semimicro balances* have maximum loadings of 10 to 30 g with precisions of $\pm 0.01$ mg. Typical *micro balances* have capacities of 1 to 3 g and precisions of $\pm 0.001$ mg.

Early analytical balances consisted of a light-weight beam that pivoted around a sharp central knife-edge made of a hard material. The object to be weighed was placed on a pan located at one end of the beam, which created a moment of force that tended to rotate the beam about its central pivot. This motion was offset by a gravitational counterforce in the form of standard weights added to a pan at the other end of the beam. Weighing with such an *equal-arm* balance was tedious and time-consuming.

The first single-pan balance appeared on the market in 1946. In terms of speed and convenience, the single-pan balance was vastly superior to its equal-arm counterpart and rapidly replaced it in most laboratories. The single-pan balance is described in the next section.

Another revolution in balance design began in the early 1970s with the appearance of what is now called the *hybrid electronic balance*. This balance retained the beam and knife-edge but employed a solenoid in place of weights to supply the counterforce to bring the beam back to its original position. The current in a solenoid is proportional to the electromagnetic force it creates and thus to the weight of the object on the opposite end of the beam.

The era of the hybrid electronic balances was brief and terminated with the appearance of the true electronic balance, which contains neither a beam nor a knife-edge. It seems likely that both mechanical and hybrid balances will ultimately be replaced completely by electronic balances.

In Section 19D–3, the construction and use of mechanical balances will be considered because these devices are still widely used, particularly in teaching laboratories. Following this discussion, consideration will be given (Section 19D–4) to the modern electronic balance, which is so simple to use that no instructions as to its operation are really required.

### 19D–3  The Single-Pan Mechanical Analytical Balance

**Components**    Although differing considerably in appearance and performance characteristics, all mechanical analytical balances—equal arm or single pan—contain several common components. Figure 19–2 is a schematic diagram of a typical single-pan balance. Fundamental to this instrument is a light-weight beam, which is a lever that pivots about a prism-shaped knife-edge. When an object is placed on the pan, a force moment is created that tends to rotate the beam about its pivot. This moment is offset by lifting one or more of the built-in weights from the beam by means of a mechanical device that is controlled from the exterior of the balance housing.

The knife-edge and the plane surface upon which it rests form a bearing with minimal friction. The performance of a mechanical balance is ultimately limited by the mechanical perfection of this bearing. Both knife-edge and bearing are fabricated from extraordinarily hard materials (agate or synthetic sapphire). Great care must be taken to avoid damage to the knife-edges and bearing surfaces of any mechanical balance.

A single-pan balance has a second bearing about which motion of the pan can occur. Here the flat bearing surface is part of the *stirrup* which rests upon an outer knife-edge. The stirrup couples the pan with the beam.

In weighing an object on a single-pan balance, the equilibrium position of

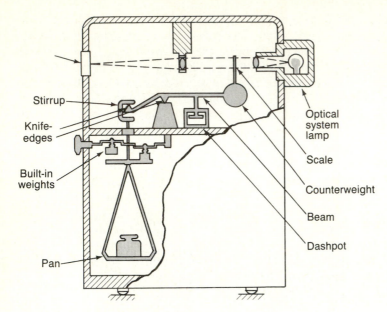

Stirrup

Knife-
edges

Built-in
weights

Pan

Optical
system
lamp

Scale

Counterweight

Beam

Dashpot

**Figure 19 – 2**   A modern
single-pan mechanical analytical
balance. (From R. M. Schoonover, *Anal. Chem.*, **1982**, *54*,
973A. Published 1982
American Chemical Society.)

the beam is approached by removing weights systematically until the imbalance is less than 100 mg. The additional weight removal needed to bring the beam into balance is then determined from its angle of deflection relative to its equilibrium position. This determination is made possible by the optical system shown in the upper right-hand corner of Figure 19 – 2. This system consists of a glass plate or reticle that is mounted on the beam and is scribed with a vertical scale that reads from 1 to 100 mg. A beam of radiation passes through the scale to an enlarging lens, which focuses a small part of the enlarged scale onto a frosted glass plate located in the front of the balance. By means of a vernier, it is possible to read the scale to the nearest 0.1 mg.

An air damper (sometimes called a *dashpot*) is mounted on the end of the beam opposite the pan. This device consists of a piston that moves within a concentric cylinder that is attached to the balance case. When the beam is set in motion, the enclosed air undergoes slight expansions and contractions because of the close spacing between the piston and the cylinder; the beam rapidly comes to rest as a result of this opposition to its motion.

Most analytical balances are equipped with *beam arrests* and *pan arrests*. The beam arrest is a mechanical device that raises the beam so that the central knife edge no longer touches its bearing surface; simultaneously, the stirrup is freed from contact with the outer knife-edge. The purpose of the arrest mechanism is to prevent damage to the bearings of the balance as objects are placed upon or removed from the pan. When engaged, the pan arrest supports most of the weight of the pan and its contents and thus prevents oscillation. Both arrests are controlled by a lever mounted on the outside of the balance case and should be engaged when the balance is not in use.

Protection from air currents is needed to permit discrimination between small differences in weight (< 1 mg). An analytical balance is thus enclosed in a case equipped with doors to allow the introduction or removal of objects.

**Stability of a Balance**  An important property of a balance is *stability*. Application of a small imbalance causes the beam to rotate about the central knife-edge. For a balance to be stable, the center of gravity of the beam (including the stirrup and the pan) must be below the central knife-edge so that the weight of the beam acts as a restoring force to offset the displacement.

---

**Weighing with a Single-Pan Balance**  An object is easily weighed with a single-pan balance.

1.  To zero the empty balance, rotate the pan arrest control to the full release position. Make any necessary adjustment.
2.  Arrest the balance again, and place the object to be weighed on the pan. Turn the arrest knob to its partially released position. Rotate the dial controlling the heaviest likely weight for the object until the illuminated scale changes or the instruction "remove weight" appears; then turn the knob back one stop. Repeat this procedure with *all* of the other dials, working successively through the lighter weights.
3.  Finally, turn the arrest control to its full release position, and allow the balance to come to equilibrium. The weight of the object is found by taking the sum of the weights indicated on the dials and that which is displayed on the illuminated scale. A vernier will permit the reading of this scale to the nearest tenth of a milligram.

**NOTE**

Directions for operation vary with the make and model of a single-pan balance. Consult the instructor for any modification to this procedure that may be required.

---

## 19D–4  The Electronic Analytical Balance[4]

Figure 19–3 is a schematic diagram that illustrates the operating principle of an electronic balance. The pan of the balance rides above a hollow metal cylinder that is surrounded by a coil. The cylinder fits over the inner pole of a cylindrical permanent magnet. A current in the coil creates a magnetic field that supports or levitates the cylinder and the pan as well as its accouterments. With the pan empty, the current is adjusted so that the level of the indicator arm is in the null position. When an object is added, a downward motion of the pan and indicator arm occurs, which increases the amount of radiation striking the photocell of the null detector. The increased current is amplified and fed into the coil,

---

[4] See R. M. Schoonover, *Anal. Chem.,* **1982,** *54,* 973A; K. M. Lang, *Amer. Lab.,* **1983,** *15*(3), 72.

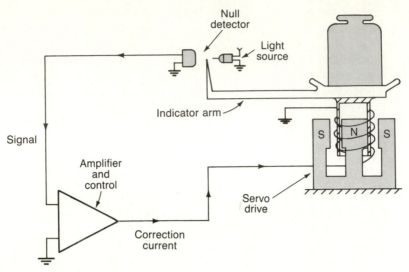

**Figure 19–3** Schematic diagram showing the principles of an electronic balance. (From R. M. Schoonover, *Anal. Chem.*, **1982**, *54*, 974A. Published 1982 American Chemical Society.)

creating a larger magnetic field, which returns the pan to its original null position. A device such as this, in which a small current exerts control over a mechanical system in a way that maintains it at a null position, is termed a *servo system* or *servomotor.* The current required to keep the pan and object in the null position is directly proportional to the weight of the object and is readily measured, digitized, and displayed as the weight of the object in grams. Calibration of an electronic balance involves weighing a standard weight of known mass and adjusting the current so that the weight of the standard is exhibited on the display.

Figure 19–4 shows two configurations for electronic analytical balances. In each the pan is tethered to a system of constraints or guides formed in the shape of a parallelogram called a *cell.* The cell contains several *flexures* that permit restricted movement of the pan and prevent torsional forces (caused by off-center loading) from disturbing the alignment of the balance mechanism. It should be noted that the null point is chosen such that the flexure pivots are in a relaxed position when the beam is parallel to the gravitational horizon.

Figure 19–4a shows one configuration of an electronic balance in which the pan is located in a more or less conventional position below the cell. Higher precision is achieved with this arrangement than with the top-loading configuration shown in Figure 19–4b. Nevertheless, balances of the latter design have precisions that are equal to or better than the best mechanical balances, and in addition provide unencumbered access to the pan.

Electronic balances generally have an automatic *taring control,* which forces the display to read zero with a container such as a boat or a weighing bottle on the pan. Most balances permit taring to 100% of capacity. Some electronic

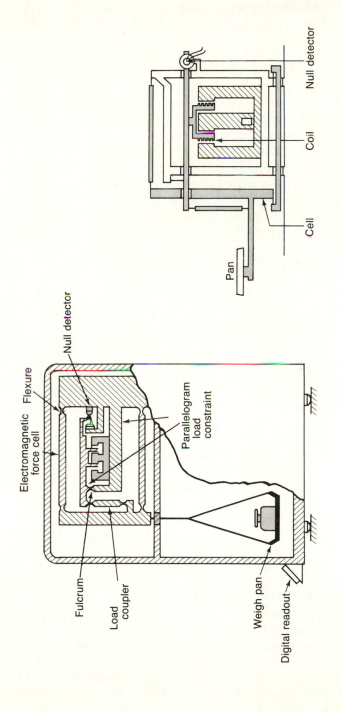

**Figure 19-4** (a) An electronic balance of a classical design. (From R. M. Schoonover, *Anal. Chem.*, **1982**, *54*, 976A. Published 1982 American Chemical Society.) (b) A top-loading electronic balance. Note that the mechanism shown is enclosed in a windowed case. (Reprinted with permission from K. M. Lang, *Amer. Lab.*, **1983**, *15(3)*, 72. Copyright 1983 by International Scientific Communications, Inc.)

balances have dual capacities and dual precisions. This feature permits the capacity to be decreased to that of a semimicro balance (30 g) with a gain of precision to 0.01 mg. Thus, the chemist has two types of balances available in one.

The great advantage of the modern electronic balance is its ease and speed of use. For example, one instrument is controlled by touching a control bar at various positions along its length. One position turns the instrument on or off, another automatically calibrates the instrument against a standard weight, and a third zeros the display with or without an object on the pan. Essentially no instructions are required to teach a novice how to obtain highly precise weights with such balances.

A few disadvantages attend the use of an electronic balance. Problems are sometimes encountered in weighing ferromagnetic materials. Strong electromagnetic radiation may also interfere with the performance of the balance. Finally, the precision and accuracy of electronic balances appear to be somewhat more susceptible to the effects of dust.

### 19D–5 Summary of Rules for the Use of an Analytical Balance

*To avoid damage or to minimize wear on the balance:*
1. Be certain that the arresting mechanisms(s) for the beam are engaged whenever the loading of the balance is being changed and when the balance is not in use.
2. Center the load on the pan insofar as possible.
3. Protect the balance from corrosion. Only vitreous and nonreactive metal or plastic objects should be placed directly on the pans.
4. Use special precautions (p. 544) when weighing volatile materials.
5. Do not attempt to adjust the balance without the prior consent of the instructor.
6. Keep the balance and its case scrupulously clean. A camel's hair brush is useful for the removal of spilled material or dust.

   *To obtain reliable weighing data:*
7. Do not attempt to weigh an object that has been heated until it has returned to room temperature.
8. Do not touch a dried object with bare hands: use tongs or finger pads to prevent the uptake of moisture.

### 19D–6 Sources of Error in the Weighing Operation

**Correction for Buoyancy**[5]    When the density of the object weighed differs considerably from that of the weights, a *buoyancy error* is introduced. This error has its origins in the difference in the buoyant force of the medium (air) that is

---

[5] For a recent discussion of buoyancy correction and calibration of balances, see R. Battino and A. G. Williamson, *J. Chem. Educ.*, **1984**, *61*, 51.

acting on the object and the weights. It is seldom necessary to apply a buoyancy correction to the weight of a typical solid because its density approaches that of the weights. The same cannot be said for liquids, gases, or low-density solids; for these, the effects of buoyancy are significant. The corrected weight is calculated by means of the equation

$$W_1 = W_2 + W_2 \left( \frac{d_{air}}{d_1} - \frac{d_{air}}{d_2} \right) \tag{19-1}$$

where $W_1$ is the corrected weight of the object and $W_2$ is the mass of the weights. The terms $d_1$ and $d_2$ represent the densities of the object and the weights, respectively, and $d_{air}$ is the density of the air displaced by them; its value is 0.0012 g/mL. The density of weights used in single-pan balances varies from manufacturer to manufacturer. So also does the density of weights used for the calibration of electronic balances.[6] Since 1973, Mettler weights (and weights for standardizing its electronic balances) are manufactured from a stainless steel that has been adjusted to have a density of 8.0 g/cm³. Ainsworth uses nickel-plated brass weights with a density of 8.4 g/cm³ in some balances and stainless steel with a density of 7.8 g/cm³ in others. The weights for Sartorius balances are reported to have a density of 7.88 g/cm³.

---

**Example 19-1**    A bottle weighed 8.6500 g empty, and 9.8600 g after addition of an organic liquid with a density of 0.92 g/mL. The balance was equipped with stainless steel weights with a density of 7.8 g/cm³. Correct the weight of the sample for the effects of buoyancy.

The apparent weight ($W_2$) of the liquid is (9.8600 − 8.6500) 1.2100 g. The same buoyant force acted on the container during both weighings; thus, we need to consider only the force that acted upon the 1.2100 g of liquid. Substituting 0.0012 g/mL for $d_{air}$, 7.8 g/cm³ for the density of the weights ($d_2$) into Equation 18-1, we obtain

$$W_1 = 1.2100 + 1.2100 \left( \frac{0.0012}{0.92} - \frac{0.0012}{7.8} \right)$$

$$= 1.2100 + 1.2100(0.0013 - 0.00015)$$

$$= 1.2114 \text{ g}$$

---

**Temperature Effects**    Attempts to weigh an object whose temperature is different from that of its surroundings will result in a significant error. Failure to permit sufficient time for a heated object to return to room temperature is the commonest cause for this problem. The error has two sources. First, convection

---

[6] See Footnote 5.

currents within the balance case exert a bouyant effect on the pan and object. Second, warm air entrapped within a closed container weighs less than the same volume at a lower temperature. Both effects cause the apparent weight of the object to be low. For a typical weighing bottle or a glass crucible, this error may amount to as much as 10 or 15 mg. Heated objects must always be cooled to room temperature before they are weighed.

**Other Sources of Error**    A porcelain or glass object will occasionally acquire a static charge sufficient to cause the balance to perform erratically; this problem is particularly serious where the humidity is very low. Spontaneous discharge frequently occurs after a short period. The use of a faintly damp chamois to wipe the object is a recommended preventative.

The optical scale of a single-pan balance should be checked regularly for accuracy, particularly under loading conditions that require essentially the full range of the scale; a standard 100-mg weight is used to make this check.

### 19D-7 Auxiliary Balances

Balances with lower sensitivities than analytical balances are found in the laboratory. These offer the advantages of speed, ruggedness, large capacity, and convenience; such balances should be used for all weighings where maximum sensitivity is not required.

Auxiliary balances designed for top loading are particularly convenient. A sensitive top-loading balance will accommodate loads as large as 150 to 200 g with a precision of about 1 mg—an order of magnitude less than that of a macro analytical balance. Some balances of this type can be used for loads as large as 25,000 g ($\pm 0.05$ g). Most are equipped with taring devices, which will bring the balance reading to zero with an empty sample container on the pan. Some are fully automatic, require no manual dialing or weight handling, and provide a digital readout of the weight.

A triple-beam balance with a sensitivity that is less than most top-loading balances is also useful. This is a single-pan balance equipped with three decades of weights that can be introduced by movement of sliding objects along three calibrated scales. The precision of such balances is one or two orders of magnitude less than that of a top-loading instrument, but is ample for many weighing operations; the balances themselves have the advantages of simplicity, durability, and low cost.

### 19E EQUIPMENT AND MANIPULATIONS ASSOCIATED WITH WEIGHING

Most solids absorb moisture and, as a consequence, change in composition. This effect is appreciable when a large surface area is exposed, as with a sample or a reagent chemical that has been ground to a fine powder. As a first step in an analysis, then, the sample is usually dried to free the results from dependence upon the humidity of the surrounding atmosphere.

***Figure 19–5*** Typical weighing bottles.

The drying (or ignition) of a sample, a precipitate, or a container to *constant weight* is a process in which the object whose weight is to be determined is first heated at an appropriate temperature (ordinarily for an hour or more), following which it is cooled and weighed. The heating and cooling cycle is repeated as many times as may be needed to achieve successive weighings that agree within 0.2 or 0.3 mg of one another. The establishment of constant weight provides some assurance that the chemical or physical processes that occur during heating (or ignition) are complete.

## 19E–1 Weighing Bottles

Solids are conveniently dried and stored in weighing bottles, two common varieties of which are shown in Figure 19–5. The ground-glass portion of the cap-style bottle shown on the left is on the outside and does not come in contact with the contents; the possibility that some of the sample might become entrained on and subsequently lost from a ground-glass surface is thus eliminated.

Plastic weighing bottles are available; ruggedness is the principal advantage possessed by these bottles over their glass counterparts.

## 19E–2 Desiccators, Desiccants

Oven drying is the commonest way of removing absorbed moisture from solids. To be sure, this approach is not appropriate for substances that undergo decomposition or for those in which the water is not removed at the ambient temperature of the oven.

While they cool, dried materials are stored in *desiccators,* which help prevent the uptake of moisture. Figure 19–6 depicts the components of a typical desiccator. The base section contains a quantity of a chemical drying agent, such as anhydrous calcium chloride, calcium sulfate (Drierite[7]), anhydrous magnesium perchlorate (Anhydrone[8] or Dehydrite[9]), or phosphorus pentoxide. Whether it is being removed or replaced, the lid of a desiccator is properly moved by a sliding, rather that a lifting motion. An airtight seal is achieved by slight rotation and downward pressure upon the positioned lid.

---

[7] ® W. A. Hammond Drierite Co.
[8] ® J. T. Baker Chemical Co.
[9] ® Arthur H. Thomas Co.

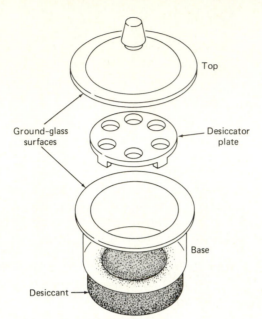

Top

Ground-glass
surfaces

Desiccator
plate

Base

Desiccant

**Figure 19–6**  Components of a typical
desiccator.

When a heated object is placed in a desiccator, the increase in pressure of the
enclosed air may be great enough to break the seal between lid and base, and
cause the lid to slide off and break. As objects cool in a desiccator, on the other
hand, the interior becomes a partial vacuum. Both of these conditions can cause
the contents of the desiccator to be physically lost or contaminated. Although it
defeats the purpose of the desiccator somewhat, it is advisable to allow some
cooling to occur before finally sealing the lid. It also helps to break the seal once
or twice during cooling to relieve any excessive vacuum that may be developing.
Finally, it is prudent to lock the lid in place with one's thumbs while moving the
desiccator to prevent loss through breakage.

Very hygroscopic materials should be stored in containers equipped with
snugly fitting covers, such as weighing bottles; the covers remain in place during
storage in the desiccator. Most substances can be safely stored with container
covers removed.

### 19E–3  Manipulation of Weighing Bottles

Heating at 105 to 110°C for about 1 hr is sufficient to remove the moisture held
on the surface of most solids. Figure 19–7 depicts the arrangement recommended for drying a sample in a weighing bottle. The bottle is contained in a
labeled beaker, which in turn is covered by a ribbed watch glass; the sample is
thus protected from accidental contamination while free access of air is maintained. This arrangement is also satisfactory for crucibles which contain precip-

*Figure 19 – 7*    Arrangement for the drying of samples.

itates that can be freed of moisture by simple drying. It is important to remember the need for identifying markings on the beaker.

Moisture transferred from one's fingers to an object can significantly alter the weight of that object. For this reason, weighing bottles should be manipulated with tongs, chamois finger cots, clean cotton gloves, or strips of paper. The last of these procedures is illustrated in Figure 19 – 8.

**Weighing by Difference**    Weighing by difference is a simple method for obtaining the weights of a series of samples. The weight of the bottle and its contents is first determined. An individual sample is then transferred from the bottle to a container; gentle tapping of the weighing bottle with its top, and slight rotation of the bottle itself provides control over the amount of sample that is removed. The utmost care is taken to avoid losses during the transfer. The bottle and its residual contents are again weighed; the weight of the sample taken is simply the difference between these two weighings.

**The Weighing of Hygroscopic Solids**    Special precautions must be taken in weighing hygroscopic substances because of their rapid rate of equilibration with moisture in the atmosphere. Here, individual samples are placed in separate weighing bottles that have been previously dried and weighed. The bottles

*Figure 19 – 8*    Method for transferring a sample.

and samples are then dried, capped, and cooled in a desiccator. After the cap is loosened momentarily to equalize pressure, the bottles and contents are weighed.

### 19E – 4 The Weighing of Liquids

The weight of a liquid is always obtained by difference. Samples that are non-corrosive and relatively nonvolatile can be weighed into containers with snugly fitting covers, such as weighing bottles; the mass of the container is subtracted from the total weight. A volatile or a corrosive sample should be sealed in a weighed glass ampoule. The bulb is first heated, following which the neck is placed in the sample; as cooling occurs, the liquid is drawn into the bulb. The neck is then sealed off with a small flame. After cooling, the bulb and its contents, as well as any glass removed during the sealing operation, are weighed. The ampoule is then transferred to a suitable vessel and broken. A small volume correction for the glass from the ampoule may be needed if the transfer has been made to a volumetric flask.

## 19F EQUIPMENT AND MANIPULATIONS FOR FILTRATION AND IGNITION

### 19F – 1 Apparatus

**Simple Crucibles**    Simple crucibles serve only as containers. The more common types maintain constant weight, within the limits of experimental error, during use. These crucibles are made of porcelain, aluminum oxide, silica, or platinum and are used to convert precipitates into suitable weighing forms. The solid is first isolated on filter paper; the paper is then eliminated by ignition.

Simple crucibles of nickel, iron, silver, or gold are used as containers for the high-temperature fusion of samples that are not soluble in aqueous reagents. Both the atmosphere and the contents may attack these crucibles, with the result that they suffer weight changes. In addition, such attack will contaminate the sample with the species derived from the crucible. The analyst selects a crucible whose components will offer the least interference in subsequent steps of the analysis.

**Filtering Crucibles**    Filtering crucibles differ from simple crucibles in that they not only act as containers but serve as the filtering medium as well. Filtrations with a filtering crucible are frequently more rapid than those with paper because a vacuum can be employed to hasten the passage of liquid through the filtering medium. A tight seal between the crucible and the filter flask is accomplished with any of several types of rubber adapters (Figure 19 – 9). (A diagram for a complete filtration train is shown in Figure 19 – 14.)

*Sintered* or *fritted-glass* crucibles are manufactured in three porosities (and are marked *f*, *m*, or *c* for fine, medium, or coarse). The upper temperature limit for a sintered-glass crucible is ordinarily about 200°C. Fritted crucibles made

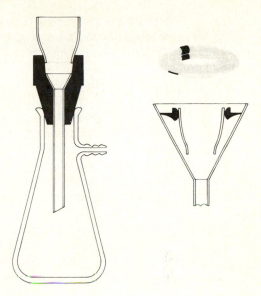

**Figure 19-9**  Adaptors for filtering crucibles.

entirely of quartz may be taken to higher temperatures and cooled without damage. Filtering crucibles with unglazed porcelain or alumina disks can also tolerate high temperatures and are not as costly as fused quartz.

A *Gooch crucible* has a perforated bottom which supports a fibrous mat. Asbestos was at one time the filtering medium for a Gooch crucible; however, current regulations regarding this material have virtually eliminated its use in the laboratory. Small circles of glass matting now substitute for asbestos; they are used in pairs to protect against disintegration during the filtration operation. Glass mats can tolerate temperatures in excess of 500°C and are substantially less hygroscopic than asbestos.

**Filter Paper**    Paper is an important filtering medium. Ashless paper is manufactured from cellulose fibers that have been treated with hydrochloric and hydrofluoric acids to remove metallic impurities as well as silica; neutralization with ammonia completes the process. The residual ammonium salts in many papers are sufficient to affect the analysis for amine nitrogen (Chapter 9).

All papers tend to pick up moisture from the atmosphere, and ashless paper is no exception. For this reason, a filter paper is destroyed by ignition if its contents are to be weighed. Typically, 9- or 11-cm circles of paper will leave an ash that weighs less than 0.1 mg, an amount that is ordinarily negligible. Ashless paper is manufactured in various grades of porosity (see Appendix 8).

Gelatinous precipitates, such as hydrous iron(III) oxide, present special problems because they tend to clog the pores of the paper upon which they are being collected. This problem can be minimized by mixing a dispersion of ashless filter paper with the precipitate prior to filtration. The pulp may be prepared by treating a piece of ashless paper with concentrated hydrochloric acid and washing the disintegrated mass free of acid. Tablets of pulp are also available from chemical suppliers. Table 19–1 summarizes the characteristics of common filtering media.

**Heating Equipment**   Many precipitates can be weighed directly after the low-temperature removal of moisture by electrically heated drying ovens, which maintain a constant temperature within a degree or less. The maximum attainable temperature will range from 140 to 260°C, depending upon make and model; for many precipitates, 110°C is a satisfactory drying temperature. The efficiency of a drying oven is greatly increased by the forced circulation of air. Further improvements are achieved by predrying the air to be circulated, and by using vacuum ovens through which a small flow of predried air is maintained.

Ordinary heat lamps will provide temperatures capable of charring filter paper. A convenient method of treating precipitates that have been collected on paper is to use heat lamps for the initial drying and charring, followed by ignition at an elevated temperature in a muffle furnace.

Burners are convenient sources of intense heat. The maximum temperature attainable depends upon the design of the burner and the combustion properties of the gas that is available. Of the three common laboratory burners, the Meker provides the highest temperatures, followed by the Tirril and Bunsen types, in that order.

Heavy-duty electric furnaces are capable of maintaining temperatures at 1100°C or higher with greater control than is obtainable with a burner. Special long-handled tongs and heat-resistant gloves are required for protection while transferring objects to or from such furnaces.

## 19F – 2 Manipulations Associated with Filtration and Ignition

**Preparation of Crucibles**   A crucible that is used to convert a precipitate into a form that is suitable for weighing must maintain a substantially constant weight throughout the drying or ignition process. To demonstrate this property, the crucible is first cleaned thoroughly (filtering crucibles are conveniently cleaned by backwashing in a filtration train) and then brought to constant weight (p. 541) under the same conditions of heating and cooling as will be required for the precipitate. This process is repeated until consecutive weighings differ by 0.2 mg or less.

**Filtration and Washing of Precipitates**   The three steps in a filtration are *decantation, washing,* and *transfer.* In decantation, as much supernatant liquid as possible is carefully poured through the filter while the precipitated solid is left essentially undisturbed. Because the pores of any filtering medium become clogged with precipitate, the longer the transfer of the precipitate can be delayed, the more rapid will be the overall filtering operation. A stirring rod is employed to direct the flow of the decantate (see Figure 19–10a and b). When

Table 19-1  Comparison of Filtering Media for Gravimetric Analysis

| Characteristic | Paper | Glass Fiber (Gooch Crucible) | Glass Filtering Crucible | Porcelain Filtering Crucible | Aluminum Oxide Filtering Crucible |
|---|---|---|---|---|---|
| Speed of filtration | Slow | Rapid | Rapid | Rapid | Rapid |
| Convenience, ease of preparation | Somewhat troublesome | Somewhat troublesome | Convenient | Convenient | Convenient |
| Maximum allowable temperature | None | 540°C | 200–500°C | 1100°C | 1450°C |
| Chemical reactivity | Hot carbon has reducing properties | Inert | Inert | Inert | Inert |
| Control of porosity | Many porosities available | Many porosities available | Several porosities available | Several porosities available | Several porosities available |
| Suitability with gelatinous precipitates | Satisfactory | Unsuitable: pores tend to clog | Unsuitable: pores tend to clog | Unsuitable: pores tend to clog | Unsuitable: pores tend to clog |
| Cost | Low | Low | High | High | High |

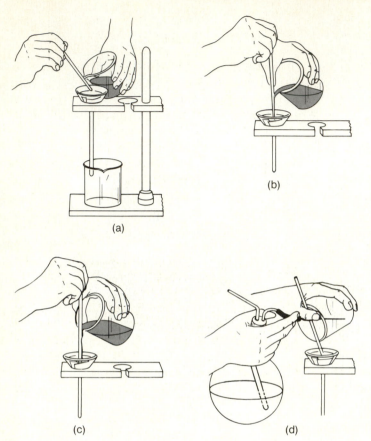

(a)

(b)

(c)

(d)

*Figure 19 – 10*  The filtering operation: techniques for decantation and transfer of precipitates.

flow ceases, the drop of liquid at the tip of the pouring spout is collected on the stirring rod and returned to the beaker. Wash liquid is next added to the beaker and thoroughly mixed with the precipitate; after the solid has again settled, this liquid is also decanted through the filter. It may be necessary to repeat this washing step several times, depending upon the precipitate. It is clear that the principal washing takes place *before* the solid is transferred; a more thoroughly washed precipitate and a more rapid filtration are the result.

The transfer process is illustrated in Figures 19 – 10c and 19 – 10d. The bulk of the precipitate is moved from beaker to filter by suitably directed streams of liquid. As before, a stirring rod is used to provide direction for the flow of liquid to the filtering medium.

The last traces of precipitate that cling to the walls of the beaker are dislodged with a *rubber policeman,* a small section of rubber tubing that has been crimped shut at one end; this device is fitted on the end of a stirring rod. A rubber policeman should be wetted with wash liquid before use. Any solid collected is combined with the main portion on the filter. Small pieces of ashless paper can be used to wipe the last traces of hydrous oxide precipitates from the wall of a beaker; these papers are ignited along with the cone containing the bulk of the precipitate.

Many precipitates have the exasperating property of *creeping* or spreading over wetted surfaces against the force of gravity. Filters are never filled to more than three-quarters of their capacity, owing to the possibility of losses of solid as a result of creeping. Sometimes creeping can be prevented by addition of a small amount of a nonionic detergent to the solution before filtration.

A gelatinous precipitate should not be allowed to dry until the washing cycle is complete because the dried mass shrinks and develops cracks. Any wash liquid that is subsequently added merely passes through these channels and accomplishes little or no washing.

### 19F – 3 Directions for Filtration and Ignition with Ashless Filter Paper

**Preparation of a Filter Paper**     Figure 19 – 11 shows the sequence followed in folding a filter paper and seating it in a 58-deg funnel (or 60-deg fluted funnel). The paper is first folded exactly in half (a), firmly creased, and then folded loosely again (b). A triangular portion is torn from one of the corners parallel to the second fold (c). The paper is then opened to form a cone (d); the second fold is creased after the cone has been fitted in the funnel (e). Seating is completed by dampening the cone with water from a wash bottle and *gentle* patting with a finger. When the cone is properly seated, there will be no leakage of air between the paper and funnel, and the stem of the funnel will be filled with an unbroken column of liquid.

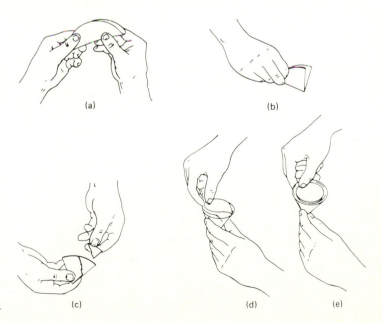

(a)                    (b)

(c)                    (d)          (e)

**Figure 19 – 11**  Method for folding and seating a filter paper.

**Transfer of Paper and Precipitate to a Crucible**    After filtration and washing have been completed, the filter cone and its contents must be transferred from the funnel to a weighed crucible. Ashless paper has very low wet strength and must be handled with considerable care during transfer. The danger of tearing is lessened considerably if the paper is allowed to dry partially before it is removed from the funnel.

Figure 19–12 illustrates the transfer process: the triple-thick portion is first drawn across the funnel to flatten the cone along its upper edge (a); the corners are next folded inward (b). The top edge is then folded over (c). Finally, the paper and its contents are eased into the crucible (d) so that the bulk of the precipitate is near the bottom.

**Ashing of a Filter Paper**    If a heat lamp is to be used, the crucible is placed on a clean, nonreactive surface such as a wire screen covered with aluminum foil. The lamp is then positioned about 1 cm above the rim of the crucible and turned on. Charring of the paper will take place without further intervention. The process is considerably accelerated if the paper can be moistened with no more than one drop of concentrated ammonium nitrate solution. Removal of the residual carbon is accomplished with a burner, as described in the next paragraph.

Considerably more attention is required when a burner is used to ash a filter paper. Because the burner produces much higher temperatures, there is the possibility that moisture will be expelled so rapidly in the initial stages of heating that mechanical loss of the precipitate occurs. The same problem exists if the paper is allowed to flame. Finally, partial reduction of the precipitate can occur through reaction with the hot carbon of the charring paper; reduction is a serious problem if reoxidation following ashing of the paper is not convenient. These difficulties can be minimized by arranging the crucible as illustrated in Figure 19–13; the tilted position allows for the ready access of air. A clean

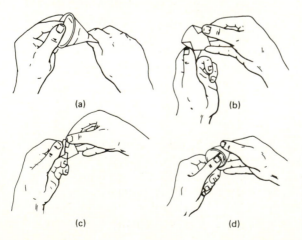

(a)

(b)

(c)

(d)

**Figure 19–12**  Method for transferring a filter paper and precipitate from funnel to crucible.

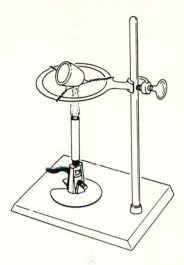

**Figure 19-13** Ignition of the precipitate: arrangement of the crucible for preliminary charring of the paper.

crucible cover should be nearby for use, if needed. Heating is commenced with a small burner flame. The temperature is gradually increased as moisture is evolved and the paper begins to char. The intensity of heating that can safely be tolerated can be gauged by the smoke that is given off. Thin wisps are normal. A significant increase in volume indicates that the paper is about to flash; heating should then be temporarily discontinued. Any flame that does develop should be immediately snuffed out with the crucible cover. (The cover may become discolored, owing to condensation of carbonaceous products; these products must ultimately be removed by ignition to confirm the absence of entrained particles of precipitate.) Finally, when no further smoking can be detected, heating is increased to remove the residual carbon. Strong heating, as necessary, can then be undertaken. Care must be taken to avoid exposing the crucible to the reducing portion of the flame.

This sequence will ordinarily precede the final ignition of a precipitate in a muffle furnace, in which a reducing atmosphere is equally undesirable.

**Use of Filtering Crucibles**   A suction filtration train (Figure 19-14) is used where a filtering crucible can be used instead of paper. The trap isolates the filter flask from the source of vacuum.

### 19F-4 Rules for the Manipulation of Heated Objects

1. Hot objects should *never* be placed directly on the desk top but instead should be allowed to cool on a heat-resistant ceramic plate or a wire gauze.
2. A crucible that has been subjected to the full flame of a burner or to a muffle furnace should be allowed to cool momentarily (on a heat-resistant ceramic plate or a wire gauze) before it is transferred to a desiccator.
3. The tongs and forceps that are used to handle heated objects should be kept

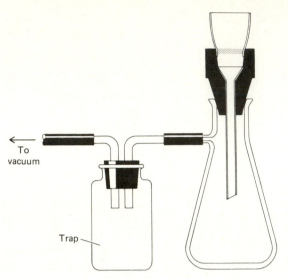

**Figure 19-14**   Train for vacuum filtration.

scrupulously clean. In particular, the tips should not be allowed to come in contact with the desk top.

4. Manipulations should be practiced before they are put to use.

## 19G   THE MEASUREMENT OF VOLUME

The precise measurement of volume is as important to many analytical methods as is the precise measurement of mass.

### 19G-1   Units of Volume

The fundamental unit of volume is the *liter* (L), defined as one cubic decimeter. The *milliliter* (mL) is one one-thousandth of a liter and is used where the liter represents an inconveniently large volume unit.

### 19G-2   Effect of Temperature upon Volume Measurements

The volume occupied by a given mass of liquid varies with temperature; to a lesser extent, so also does the volume of the container that holds the liquid. The accurate measurement of volume may require that both of these effects be taken into account.

Most volumetric measuring devices are made of glass, a material that fortunately has a small temperature coefficient. The volume of a soft-glass vessel will change by about 0.003%/°C; a heat-resistant glass will change about one-third this amount. Clearly, variations in volume of a container resulting from changes in temperature need be considered only for the most exacting work.

The coefficient of expansion for dilute aqueous solutions (approximately 0.025%/°C) is such that a 5°C change will measurably affect the reliability of ordinary volumetric measurements.

---

**Example 19-2**   A 40.00-mL sample is taken from an aqueous solution at 5°C; what volume will this sample occupy at 20°C?

$$V_{20°} = V_{5°} + 0.00025\,(20 - 5)(40.00) \qquad (19-2)$$
$$= 40.00 + 0.15$$
$$= 40.15 \text{ mL}$$

---

Volumetric measurements must be referred to some standard temperature; this reference point is ordinarily 20°C. The ambient temperature of most laboratories is sufficiently close to 20° to eliminate the need for temperature corrections in the measurement of aqueous solutions. In contrast, the coefficient of cubic expansion for many organic liquids may require corrections for temperature variations of 1°C or less.

---

**Example 19-3**   The coefficient of expansion for an alcoholic KOH solution is 0.11%/°C. Calculate the relative error in the volume of an aliquot delivered at 25.2°C by a 50.00-mL volumetric pipet if no temperature correction is applied.

$$V_{20°} = 50.00 - 0.0011\,(25.2 - 20.0)(50.00) = 49.71 \text{ mL}$$

$$E_{rel} = \frac{49.71 - 50.00}{50.00} \times 1000 = -5.7 \text{ parts per thousand}$$

---

### 19G-3 *Apparatus for the Precise Measurement of Volume*

The reliable measurement of volume is performed with the *pipet,* the *buret,* and the *volumetric flask.*

Volumetric equipment is marked by the manufacturer to indicate not only the manner of calibration (usually with a TD for "to deliver" or a TC for "to contain"), but also the temperature at which the calibration strictly applies. Pipets and burets are ordinarily designed and calibrated to deliver specified volumes, whereas volumetric flasks are calibrated on a "to contain" basis.

**Pipets**   Pipets are devices that permit the transfer of accurately known volumes from one container to another. Common types are shown in Figure 19-15; Table 19-2 provides information regarding their use. A *volumetric* or *transfer* pipet (Figure 19-15a) will deliver a single, fixed volume between 0.5 and 200 mL; many such pipets are color coded by volume for convenience in

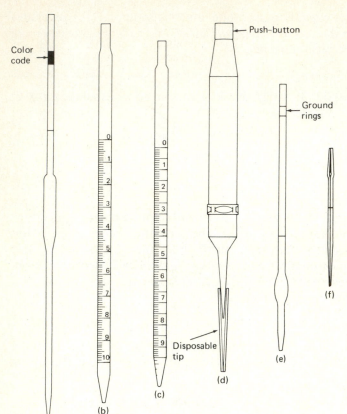

**Figure 19–15** Typical pipets. (a) Volumetric. (b) Mohr. (c) Serological. (d) Syringe. (e) Ostwald-Folin. (f) Lambda.

identification and sorting. *Measuring* pipets (Figure 19–15b and c) are calibrated in convenient units to permit delivery of any volume up to the maximum capacity; sizes range from 0.1 to 25 mL.

Volumetric and measuring pipets are filled to an initial calibration mark at the outset; the manner in which the transfer is completed depends upon the particular type. Because an attraction exists between most liquids and glass, a small amount of liquid tends to remain in the tip of a drained transfer pipet. This drop is blown from some pipets but not from others (Table 19–2).

Numerous *automatic* pipets are available for situations that call for the repeated delivery of a particular volume. Hand-held syringe pipets (Figure 19–15d) deliver volumes in the range between 1 and 1000 $\mu$L (1 mL). A useful feature of this type of pipet is that the liquid is contained in a disposable plastic tip. A volume of liquid is drawn into the tip by means of a spring-operated piston that is activated by a push button at the top of the syringe. The liquid is then delivered by reversing the action of the push button on the piston. Remarkable precision is claimed for these devices ($\pm 0.02$ $\mu$L for a 1 $\mu$L measurement, $\pm 0.3$ $\mu$L at 1000 $\mu$L).

*Table 19-2   Characteristics of Pipets*

| Name | Type of Calibration* | Function | Available Capacities (mL) | Type of Drainage |
|------|----------------------|----------|----------------------------|------------------|
| Volumetric | TD | Delivery of a fixed volume | 1-200 | Free drainage |
| Mohr | TD | Delivery of a variable volume | 1-25 | Drain to lower calibration line |
| Serological | TD | Delivery of a variable volume | 0.1-10 | Blow out last drop† |
| Serological | TD | Delivery of a variable volume | 0.1-10 | Drain to lower calibration line |
| Ostwald-Folin | TD | Delivery of a fixed volume | 0.5-10 | Blow out last drop† |
| Lambda | TC | To contain a fixed volume | 0.001-2 | Wash out with suitable solvent |
| Lambda | TD | Delivery of a fixed volume | 0.001-2 | Blow out last drop† |
| Syringe | TD | Delivery of a variable or a fixed volume | 0.001-1 | Tip emptied by syringe |

* TD = to deliver; TC = to contain.
† A frosted ring near the top of recently manufactured pipets indicates that the last drop is to be blown out.

**Burets**   Burets, like measuring pipets, enable the analyst to delivery any volume up to their maximum capacity. The precision attainable with a buret is substantially better than that with a pipet.

A buret is a calibrated tube that holds the titrant and a valve arrangement by which flow from the tip can be controlled. This valve is the principal source of difference among burets. The simplest consists of a closely fitting glass bead within a short length of rubber tubing; only when the tubing is deformed can liquid flow past the bead. Burets equipped with glass stopcocks rely upon a lubricant between the ground-glass surfaces of stopcock and barrel for a liquid-tight seal. Some solutions, notably bases, will cause a glass stopcock to freeze upon long contact; thorough cleaning is necessary after each use. Valves made of Teflon are commonly encountered; these are unaffected by most common reagents and require no lubricant.

**Volumetric Flasks**   Volumetric flasks are manufactured with capacities ranging from 5 mL to 5 L and are usually calibrated to contain a specified volume when filled to a line etched on the neck. They are used in the preparation of standard solutions and the dilution of samples to a fixed volume prior to taking aliquots with a pipet. Some are also calibrated on a "to deliver" basis; these are readily distinguished by two reference lines on the neck. If delivery of the stated volume is desired, the flask is filled to the upper of the two lines.

## 19G – 4 General Considerations Concerning the Use of Volumetric Equipment

Volume calibrations were blazed by the manufacturer upon clean volumetric equipment. An equal degree of cleanliness is needed if these markings are to have their stated meanings. Only clean glass surfaces will support a uniform film of liquid; the presence of dirt or oil will tend to cause breaks in this film. The existence of breaks is a certain indication of an unclean surface.

**Cleaning**   A brief soaking in warm detergent solution is ordinarily sufficient to remove the grease and dirt responsible for water breaks. Prolonged soaking should be avoided, because a rough area or ring is likely to develop at the detergent – air interface. This ring is impossible to remove and causes a film break that destroys the usefulness of the equipment. If detergent solution is ineffective, soaking for a few minutes in warm cleaning solution will usually help.

After being cleaned, the apparatus must be thoroughly rinsed with tap water and then with three or four portions of distilled water. It is seldom necessary to dry volumetric ware. As a general rule, calibrated glass equipment should not be heated.

**Avoidance of Parallax**   The surface of a liquid that is confined in a narrow tube (such as a pipet or a buret) will exhibit a marked curvature, or *meniscus*. It is common practice to use the bottom of the meniscus as the point of reference in calibrating and using volumetric equipment; this minimum can often be established more exactly by holding an opaque card or piece of paper behind the graduations (Figure 19 – 16).

In judging volumes, the eye must be at the same level as the surface of the liquid to avoid an error due to *parallax*. Parallax will cause the volume to appear smaller than is actually the case if the meniscus is observed from above, and larger if viewed from below (Figure 19 – 16).

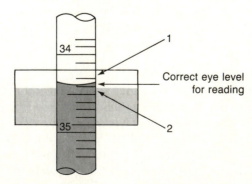

Correct eye level for reading

**Figure 19 – 16**   Method for reading a buret. The eye should be level with the meniscus. The black-white line on the card should be slightly below the meniscus. The reading shown is 34.39 mL. If viewed from position 1, the reading will appear smaller than 34.39 mL; from 2 it will appear larger.

### 19G – 5  Directions for Use of a Pipet

The accompanying directions pertain specifically to transfer pipets, but can be modified for the use of other types as well. Liquids are drawn into pipets through application of a slight vacuum. The mouth should *never* be used for suction because of the danger of accidentally ingesting liquids. Instead, a rubber suction bulb or a rubber tube connected to a vacuum source should be used (Figure 19 – 17a).

**Cleaning**   Use a rubber bulb to draw detergent (or cleaning) solution to a level that is an inch or two above the calibration mark of the pipet. Rinse with several portions of tap water. Inspect for film breaks, and repeat this portion of the cleaning cycle if necessary. Finally, fill the pipet to perhaps one-third of its capacity with distilled water, and carefully rotate it so that the entire interior surface is wetted. Repeat this rinsing step *at least* twice more.

**Measurement of an Aliquot**   As in rinsing, use a pipet bulb to draw in a small quantity of the liquid to be sampled, and thoroughly rinse the interior surfaces. Repeat with *at least* two more portions. Then carefully fill the pipet to a level somewhat above the graduation mark (Figure 19 – 17a). Quickly replace the bulb with a *forefinger* to arrest the outflow of liquid (Figure 19 – 17b). Make certain that there are no bubbles in the bulk of the liquid or foam at the surface. Tilt the pipet slightly from the vertical, and wipe the exterior free of adhering liquid (Figure 19 – 17c). Touch the tip of the pipet to the wall of a glass vessel *(not the container into which the aliquot is to be transferred),* and slowly allow the liquid level to drop by partially releasing the forefinger (see Note 1). Halt further flow as the bottom of the meniscus coincides exactly with the graduation mark. Then place the tip within the receiving vessel, and allow the aliquot to drain. When free flow ceases, rest the tip against the inner wall for a full 10 sec (Figure 19 – 17d). Finally, withdraw the pipet with a rotating motion to remove any liquid that adheres to the tip. *The small volume remaining inside the tip should not be blown or rinsed into the receiving vessel.*

**NOTES**
1. The liquid can best be held at a constant level if one's forefinger is *faintly* moist; too much moisture makes control difficult.
2. Do not hold the pipet by its bulb; the warmth of one's hand can alter the temperature of the contents.
3. Rinse the pipet thoroughly after use.

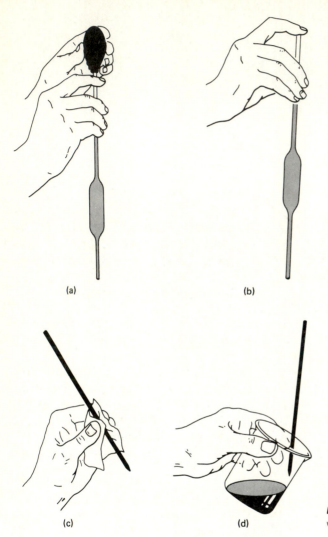

(a)

(b)

(c)

(d)

**Figure 19–17**  Technique for use of a volumetric pipet.

### 19G–6  *Directions for Use of a Buret*

Before it is placed in service, a buret must be scrupulously clean; in addition, its valve must be liquid tight.

**Cleaning**    Thoroughly clean the tube of the buret with detergent and a long brush. If water breaks persist after rinsing, clamp the buret in an inverted position, with the end in a beaker of warm cleaning solution. Connect a hose from the buret tip to a source of vacuum. Gently pull the cleaning solution into the buret, stopping well short of the stopcock (see Note 1). Allow the cleaning solution to stand for about 15 min and then drain. Rinse the buret thoroughly with tap water and then with distilled water. Inspect again for water breaks. Repeat the treatment if necessary.

**Lubrication of a Stopcock**   Carefully remove all old grease from the stopcock and barrel with a paper towel, and dry both parts completely. Lightly grease the stopcock, taking care to avoid the area near the hole. Insert the stopcock into the barrel and rotate it vigorously with a slight inward pressure. A proper amount of lubricant has been used when the area of contact between stopcock and barrel appears nearly transparent, the seal is liquid tight, and no grease has worked its way into the tip.

**NOTES**

1. Cleaning solution often disperses more stopcock lubricant than it destroys and thus leaves a buret with a heavier grease film than before treatment. For this reason, cleaning solution should *never* be allowed to come in contact with lubricated stopcock assemblies.
2. Grease films that are unaffected by cleaning solution may yield to treatment with such organic solvents as benzene or acetone. Thorough washing with detergent should follow such treatment. The use of silicone lubricants is not recommended; contamination by such preparations is difficult, if not impossible, to remove.
3. So long as the flow of liquid is not impeded, fouling of a buret tip with stopcock grease is not a serious matter. Removal is best accomplished with organic solvents. A stoppage in the middle of a titration can be freed by *gently* warming the tip of the buret with a lighted match.
4. Before a buret is returned to service after reassembly, it is advisable to test for leakage. Simply fill the buret with water and establish that the volume reading does not change with time.

**Filling**   Make certain the stopcock is closed. Add 5 to 10 mL of the solution, and carefully rotate the buret to wet the interior completely. Allow the liquid to drain through the tip. Repeat this procedure *at least two more times.* Then fill the buret well above the zero marking. Free the tip of air bubbles by rapidly rotating the stopcock and allowing small quantities of solution to pass. Finally, lower the level of the liquid to, or somewhat below, the zero marking. Allow about 1 min for drainage; then record the initial volume reading.

**Titration**   Figure 19–18 illustrates the preferred method for the manipulation of a stopcock; any tendency for lateral movement by the stopcock will be in the direction of firmer seatings. Be sure the tip of the buret is well within the titration vessel. Introduce the titrant in increments of 1 mL or so. Swirl (or stir) the sample constantly to ensure thorough mixing. Decrease the volume of titrant added as the titration progresses; the solution should be added dropwise in the immediate vicinity of the end point (Note 3). When it is judged that only a few more drops are needed, rinse the walls of the container (see Note 2). Allow at least 30 sec for drainage before reading the final volume of the buret after the titration is complete.

**NOTES**

1. When unfamiliar with a particular titration, many analysts prepare an extra sample. No care is lavished on its titration since its functions are to reveal the nature of the end point and to provide a rough estimate of titrant requirements. This deliberate sacrifice of one sample frequently results in an overall saving of time.
2. Instead of being rinsed at the end of the titration, a flask can be carefully tipped and rotated so that the bulk of the liquid picks up any droplets that adhere to the inner surface.
3. Volume increments smaller than an ordinary drop can be taken by allowing a small volume of titrant to form on the tip of the buret and then touching the tip to the wall of the flask. This droplet is then combined with the bulk of the solution as in Note 2.

### 19G – 7 Directions for the Use of a Volumetric Flask

Before they are used, volumetric flasks should be thoroughly washed with detergent (and cleaning solution if necessary) and then thoroughly rinsed. Only rarely do they need to be dried. If required, however, drying is best accomplished by clamping the flasks in an inverted position. The insertion of a glass tube connected to a vacuum line will hasten the process.

**Figure 19–18** Recommended method for manipulation of a buret stopcock.

**Direct Weighing into a Volumetric Flask**    The direct preparation of a standard solution requires that a known weight of solute be introduced into a volumetric flask. To minimize the possibility of loss during transfer, insert a powder funnel into the neck of the flask. Be certain to wash the funnel free of solid.

The foregoing procedure is inappropriate where heating is needed to dissolve the solute. Instead, weigh the solid into a beaker or flask, dissolve it, and allow the solution to cool to room temperature. Transfer this solution quantitatively to the volumetric flask, as described in the next section.

**Quantitative Transfer of a Liquid to a Volumetric Flask**    Insert a funnel into the neck of the flask; use a stirring rod to direct the flow of the solution. Tip off the last drop of liquid in the spout of the beaker with the stirring rod. Rinse both the stirring rod and the interior of the beaker with distilled water, and transfer the washings to the volumetric flask, as before. Repeat this rinsing process *at least* two more times.

**Dilution to the Mark**    After the solute has been transferred, fill the flask about half-full and swirl the contents to achieve solution. Add more solvent and again mix well. Bring the level almost to the mark, and allow time for drainage; then use a medicine dropper to make such final additions of solvent as are necessary (see Note). Firmly stopper the flask, and invert it repeatedly to ensure thorough mixing. Finally, transfer the contents to a storage bottle that is either initially dry or has been thoroughly rinsed with several small portions of the solution from the flask.

**NOTE**

If, as sometimes happens, the liquid level accidentally exceeds the calibration mark, the solution can be saved by correcting for the excess volume. Use a gummed label to mark the actual location of the meniscus. After the flask has been emptied, carefully refill to the etched graduation mark with water. Then, with a buret, determine the additional volume needed to duplicate the actual volume of the solution. This volume, of course, must be added to the nominal value for the flask when calculating the concentration of the solution.

## 19H THE CALIBRATION OF VOLUMETRIC WARE

The reliability of a volumetric analysis depends upon agreement between the volumes actually and purportedly contained (or delivered) by the apparatus. Calibration simply verifies this agreement or provides a correction if agreement is lacking. The latter involves the assignment of corrections to the existing volume markings or the striking of new markings that more closely agree with the nominal value.

A calibration consists of determining the mass of a liquid of known density contained (or delivered) by volumetric ware. Although calibration appears to be a straightforward process, a number of important variables must be controlled. Principal among these is temperature, which influences a calibration in two ways. First, and more important, the volume occupied by a given mass of liquid varies with temperature. Second, the volume of the apparatus itself is variable, owing to the tendency of the glass to contract or expand with changes in temperature.

We have noted (p. 588) that the effect of buoyancy upon weighing data is most pronounced when the density of the object is significantly less than that of the weights. As a general rule, a buoyancy correction must be applied to data where water has been used as the calibration fluid.

Finally, the liquid selected for calibration requires consideration. Water is the liquid of choice for most work. Mercury is also useful, particularly for small volumes. Because mercury does not wet glass surfaces, the volume contained by the apparatus will be identical with that which is delivered. In addition, a small correction must be applied to account for the convex meniscus of mercury; the magnitude of this correction depends upon the diameter of the apparatus at the graduation mark.

The calculations associated with calibrations, while not difficult, are somewhat involved. The raw weighing data are first corrected for buoyancy with Equation 19–1 (p. 539). Next, the volume of the apparatus at the temperature *(t)* of calibration is obtained by dividing the density of the liquid at that temperature into the corrected weight. Finally, this volume is corrected to the standard temperature of 20°C with Equation 19–2 (p. 553).

Table 19–3 is provided to ease the computational burden of calibration. Corrections for buoyancy with respect to stainless steel or brass weights (the density difference between the two is small enough to be neglected) and for the volume change of the water, as well as the glass container, have been incorporated into these data. Multiplication by the appropriate factor from Table 19–3 converts the mass of water measured at temperature $t$ to the volume it would occupy at 20°C.

---

**Example 19–4**  A 25-mL pipet was found to deliver 24.976 g of water when calibrated against stainless steel weights at 25°C. Use the data in Table 19–3 to calculate the volume delivered by the pipet at this temperature and at 20°C.

At 25°C:  $V = 24.976 \text{ g} \times 1.0040 \text{ mL/g} = 25.08 \text{ mL}$

At 20°C:  $V = 24.976 \text{ g} \times 1.0037 \text{ mL/g} = 25.07 \text{ mL}$

---

**General Directions for Calibration Work**  All volumetric ware should be painstakingly freed of water breaks before being tested. Burets and pipets need not be dried. Volumetric flasks should be thoroughly drained and dried (room temperature). The water used for calibration should be in thermal equilibrium

**Table 19–3** *Volume Occupied by 1.000 g of Water Weighed in Air Against Stainless Steel Weights\**

| Temperature, t, °C | Volume, mL | |
|---|---|---|
| | At Temperature t | Corrected to 20°C |
| 10 | 1.0013 | 1.0016 |
| 11 | 1.0014 | 1.0016 |
| 12 | 1.0015 | 1.0017 |
| 13 | 1.0016 | 1.0018 |
| 14 | 1.0018 | 1.0019 |
| 15 | 1.0019 | 1.0020 |
| 16 | 1.0021 | 1.0022 |
| 17 | 1.0022 | 1.0023 |
| 18 | 1.0024 | 1.0025 |
| 19 | 1.0026 | 1.0026 |
| 20 | 1.0028 | 1.0028 |
| 21 | 1.0030 | 1.0030 |
| 22 | 1.0033 | 1.0032 |
| 23 | 1.0035 | 1.0034 |
| 24 | 1.0037 | 1.0036 |
| 25 | 1.0040 | 1.0037 |
| 26 | 1.0043 | 1.0041 |
| 27 | 1.0045 | 1.0043 |
| 28 | 1.0048 | 1.0046 |
| 29 | 1.0051 | 1.0048 |
| 30 | 1.0054 | 1.0052 |

\* Corrections for buoyancy (stainless steel weights) and the change in volume of the container have been applied.

with its surroundings. This condition is best established by drawing the water well in advance, noting its temperature at frequent intervals, and waiting until no further changes occur.

An analytical balance can be used for calibrations involving 50 mL or less. Weighings to the nearest milligram are perfectly satisfactory; this order of reliability is well within the capabilities of many modern top-loading, single-pan laboratory balances. Weighing bottles or small, well-stoppered conical flasks are convenient receivers for small volumes.

**Calibration of a Volumetric Pipet**   Determine the empty weight of the stoppered receiver (to the nearest milligram). Transfer an aliquot of equilibrated water to the receiver with the pipet (Section 19G–5), weigh the receiver and its contents (again, to the nearest milligram), and calculate the weight of water in the aliquot from the differences in these weights. Calculate the volume delivered with the aid of Table 19–3. Repeat the calibration several times.

**Calibration of a Buret**   Fill the buret with equilibrated water and make certain that no air bubbles are entrapped in the tip. Allow about 1 min for drainage; then lower the liquid level to bring the bottom of the meniscus to the 0.00-mL graduation, and touch the tip of the buret to the wall of a beaker to remove any adhering drop. Wait 10 min and recheck the volume; if the stopcock is tight, there should be no perceptible change. During this interval, weigh (to the nearest milligram) a 125-mL conical flask fitted with a rubber stopper.

Once tightness of the stopcock has been established, slowly transfer (at about 10 mL/min) approximately 10 mL to the flask. Touch the tip to the wall of the flask. Wait 1 min, record the volume that was apparently delivered, and refill the buret. Weigh the flask and its contents (to the nearest milligram); the difference between this and the initial weight gives the mass of the water that was actually delivered. Use Table 19–3 to convert this mass into the true volume. Subtract the apparent volume from the true volume. This difference is the correction that must be applied to the apparent volume to give the true volume. Repeat the calibration until agreement within ±0.02 mL is achieved.

Starting again from the zero mark, repeat the calibration, this time delivering about 20 mL to the receiver. Test the buret at 10-mL intervals over its entire volume. Prepare a plot of the corrections to be applied as a function of volume. Corrections associated with any interval can be determined from this plot.

**Calibration of a Volumetric Flask with a Single-Pan Balance**   Weigh the clean, dry flask (to the nearest milligram). Then add equilibrated water until the meniscus coincides with the graduation mark (see Note) and reweigh. Calculate the volume contained with the aid of Table 19–3.

**NOTE**
A glass tube with one end drawn to a tip is useful in making final adjustments to the liquid level.

**Calibration of a Volumetric Flask Relative to a Pipet**   The calibration of a volumetric flask relative to a pipet provides an excellent method for partitioning a sample into aliquots. The directions given here pertain to a 50-mL pipet and a 500-mL volumetric flask; other combinations are equally convenient. Carefully transfer ten 50-mL aliquots from the pipet to a 500-mL volumetric flask. Mark the location of the meniscus with a gummed label. Coat the label with paraffin to ensure permanence. Dilution to this mark will permit the same pipet to deliver a one-tenth aliquot of the solution in the flask.

## 19I THE LABORATORY NOTEBOOK

A laboratory notebook is needed to record measurements and observations concerning an analysis. The book itself should be permanently bound with consecutively numbered pages (numbering, if needed, should be done before any entries are made in the notebook). Most notebooks have more than ample room; the crowding of entries is unnecessary.

The first few pages should be reserved for a table of contents, which should be conscientiously kept up to date.

## 19I–1 Rules for the Use of a Laboratory Notebook

1. *All data should be entered directly into the notebook and in ink.* The virtues of neatness cannot be denied. However, it is far preferable to have a permanent record of observations that may be somewhat lacking in this quality than to have an experiment ruined because a piece of paper upon which some crucial data were recorded was misplaced or thrown away.
2. Entries should be liberally provided with labels. A series of weights that refer to a set of empty crucibles, for example, should be identified by the heading "Empty Crucible Weights" or something similar (needless to say, each crucible should also be identified). The significance of an entry is obvious at the time it is recorded but may become unclear with the passage of time.
3. Each notebook page should be dated as it is used.
4. An erroneous entry should *never* be erased or obliterated. Instead, it should be crossed out with a single horizontal line, and the corrected entry should be located as nearby as possible. Numbers should never be written over; with time, it may be impossible to distinguish between the number that is correct from the one that is in error.
5. Pages should never be removed from the notebook. It is sufficient to draw a single line diagonally across a page that is to be disregarded. A brief notation of the reason for striking out the page is useful.

## 19I–2 Format

The instructor should be consulted with regard to the details of keeping the laboratory notebook. A suggested format is the consecutive use of all pages for the recording of data. The analysis is then summarized on the next pair of facing pages. One of these pages should contain the following entries:

1. The title of the experiment (e.g., "The Gravimetric Determination of Chloride").
2. A brief statement of the principles upon which the analysis is based.
3. A summary of the weighing and/or volume data needed to calculate the result of the analysis for each sample in the set.
4. A report of the best value for the set and a statement of the precision that was attained in the analysis.

A sample page is shown in Figure 19–19.
The second page should contain the following items:

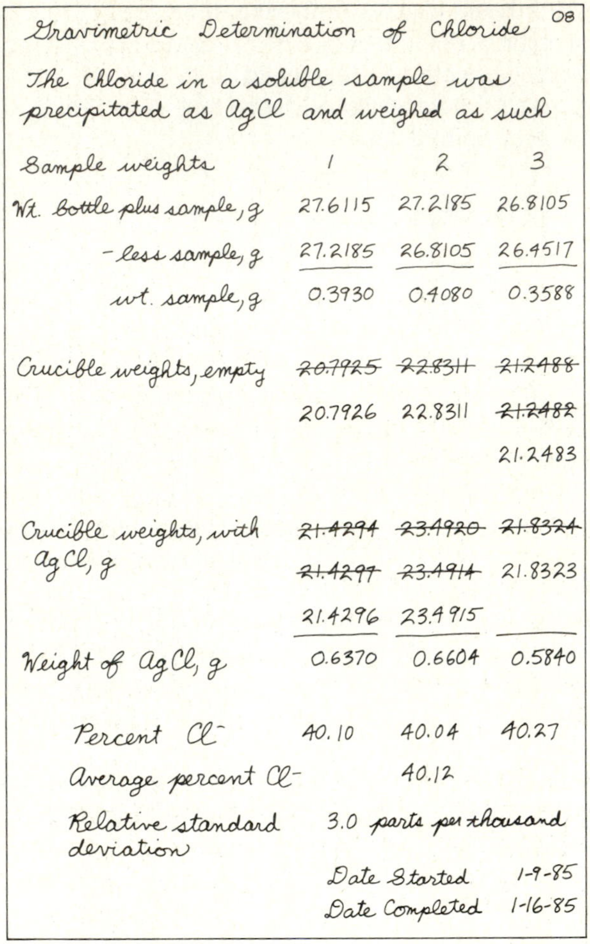

Gravimetric Determination of Chloride

The Chloride in a soluble sample was precipitated as AgCl and weighed as such

| Sample weights | 1 | 2 | 3 |
|---|---|---|---|
| Wt. Bottle plus sample, g | 27.6115 | 27.2185 | 26.8105 |
| − less sample, g | 27.2185 | 26.8105 | 26.4517 |
| wt. sample, g | 0.3930 | 0.4080 | 0.3588 |
| Crucible weights, empty | ~~20.7925~~ | ~~22.8311~~ | ~~21.2488~~ |
| | 20.7926 | 22.8311 | ~~21.2482~~ |
| | | | 21.2483 |
| Crucible weights, with AgCl, g | ~~21.4294~~ | ~~23.4920~~ | ~~21.8324~~ |
| | ~~21.4297~~ | ~~23.4914~~ | 21.8323 |
| | 21.4296 | 23.4915 | |
| Weight of AgCl, g | 0.6370 | 0.6604 | 0.5840 |
| Percent Cl⁻ | 40.10 | 40.04 | 40.27 |
| Average percent Cl⁻ | | 40.12 | |
| Relative standard deviation | 3.0 parts per thousand | | |
| Date Started | 1-9-85 | | |
| Date Completed | 1-16-85 | | |

**Figure 19–19** Laboratory notebook data page.

1. Equations for the principal reactions in the analysis.
2. An equation that shows how the results of the analysis were calculated.
3. A *summary* of observations that appear to bear upon the validity of a particular result or the analysis as a whole. *Any such observations must have been originally recorded in the notebook at the time the observation was made.*

## 19J LABORATORY SAFETY

As in all laboratories, hazards exist in the analytical laboratory, particularly for the careless and unthinking. The sources of these hazards are (1) corrosive and poisonous chemicals, (2) broken glass, (3) explosion, (4) fire, and (5) electrical shock. Although the last three are always of concern in any laboratory environment, they are generally of less hazard in the beginning analytical laboratory than the first two.

## 19J–1 Safety Rules

1. At the outset, learn the locations of the nearest eye fountain, fire blanket, shower, and fire extinguisher. Learn the proper use of each, and do not hesitate to use this equipment if the need arises.
2. *EYE PROTECTION MUST BE WORN AT ALL TIMES.* The potential for serious and often permanent injury to the eyes makes it imperative that adequate eye protection be worn *at all times* by everyone (students, instructors, and visitors) in the laboratory. Eye protection should be donned before entering the laboratory and not be removed until after leaving; serious eye injuries have occurred to people performing such innocuous operations as computing or writing in a laboratory notebook. Usually, such incidents are the result of someone else's experiment getting out of control.

   Regular prescription glasses or contact lenses are not an adequate substitute for eye protection approved by the federal Office of Safety and Health Administration (OSHA).
3. Most of the chemicals in a laboratory are toxic; some are very toxic and some, such as concentrated solutions of strong acids and bases, are corrosive to the skin. In handling all chemicals, avoid contact with the skin. In the event of such contact, *immediately* flood the affected area with copious amounts of water; with corrosive solutions speed is particularly important. If corrosive solutions are spilled on clothing, the garment should be removed immediately; modesty should be of no concern in this circumstance.
4. Unauthorized experiments should *NEVER* be performed. Such activity is grounds for disqualification at many institutions.
5. Never work alone in the laboratory.
6. Never bring food or beverages into the laboratory. Do not drink from laboratory glassware. No smoking in the laboratory.
7. Always use a bulb to draw liquids into a pipet; *NEVER* use the mouth to provide suction.
8. Wear adequate foot covering (no open-toed sandals). Use a net to contain long hair. Avoid flammable clothing. A laboratory coat or apron will provide a measure of protection and may be required.
9. Be extremely tentative in touching objects that have been heated.
10. Always fire polish the ends of freshly cut glass tubing. *NEVER* attempt to force glass tubing through a hole in a stopper. Instead, make sure that both tubing and hole are thoroughly wet with soapy water and protect hands

with several layers of towel or with heavy gloves while inserting the glass into the stopper.

11. Use fume hoods whenever toxic or noxious gases are likely to be evolved. Use care in testing for odors; use the hand to waft vapors above containers toward the nose.

12. In the event of an injury or burn, notify the instructor.

13. Dispose of solutions and chemicals as instructed. In many areas, it is illegal to flush heavy metal solutions or organic liquids down the drain; alternative arrangements are required for disposal of such liquids.

# Chapter 20
# Selected Methods of Analysis

This chapter contains detailed directions for performing selected methods of chemical analysis. The procedures have been chosen with the idea of introducing the student to a variety of analytical laboratory techniques that are widely used by scientists for gaining information about the composition of matter.

Before undertaking any one of these experiments, it is wise to identify which measurements must be made with the highest precision, and thus with the greatest care, and which measurements can be carried out rapidly and with less concern for accuracy. Ordinarily, the greatest care should be given only to those measurements that yield data that appear in the equation used for computing the results of an analysis. The remaining measurements can and *should* be made less carefully because it is a waste of time and effort to measure, let us say, a volume to $\pm 0.02$ mL when an uncertainty of $\pm 0.5$ mL or even $\pm 5$ mL will have no detectable effect on the outcome of the analysis.

In a similar vein, it is worthwhile at the outset of an analysis to study the time requirements associated with the various unit operations that make up the procedure. Such a study often reveals that certain operations require periods of waiting—while, for example a weighing bottle cools, a solution evaporates, or a sample dries in an oven. The experienced experimentalist uses these periods to carry out other operations in the analysis at hand or in another analysis. Time planning is also important in revealing places where an experiment can be conveniently interrupted for the day as well as those operations that must be performed without a break.

Preliminary planning of the types just described can reduce significantly the amount of time required to complete a laboratory course in analytical chemistry (or for that matter, any laboratory course). For the greatest efficiency, such planning should be carried out before stepping into the laboratory so that maximum use is made of the time allotted for laboratory work.

## 20A GRAVIMETRIC METHODS OF ANALYSIS

The calculations of gravimetric analysis, as well as typical applications, are discussed in Chapter 4.

## 20A – 1  Determination of Water in Barium Chloride Dihydrate

**Discussion**   The water in a crystalline hydrate is readily determined by measuring the weight that is lost as a result of heating. A known weight of sample is heated at a suitable temperature; its water content is taken as the difference in its weight before and after heating. Less frequently, the evolved water is collected and weighed.

**Procedure**   Perform all weighings in this determination to the nearest 0.1 mg. See page 543 for instructions on the handling and weighing of samples.

Carefully clean two or three weighing bottles. Dry them in a covered beaker (p. 542) at 105 to 110°C for at least 1 hr. Allow the bottles to cool to room temperature in a desiccator and then determine the weight of each. Repeat this cycle of heating, cooling, and weighing until successive measurements agree within 0.2 mg of one another (this process is called bringing an object to constant weight). Next, introduce an appropriate amount (see Note 1) of unknown to each bottle and reweigh. Heat the samples for about 2 hr at 105 to 110°C; then cool and weigh as before. Repeat the heating cycle until the bottles and their contents have been brought to constant weight. Report the percent water in the sample.

**NOTES**
1. Samples of pure $BaCl_2 \cdot 2H_2O$ should weigh approximately 1 to 1.5 g. Consult with the instructor for a suitable weight if the sample consists of the dihydrate and some anhydrous diluent.
2. Barium chloride can be heated to elevated temperatures without decomposition. As an alternative to the foregoing procedure, the analysis can be performed in crucibles with a Bunsen flame as the source of heat.
3. Magnesium chloride heptahydrate can be used instead of barium chloride; heating to 140°C is needed to eliminate the last traces of moisture.

## 20A – 2  Gravimetric Determination of Chloride in a Soluble Sample

**Discussion**   The chloride content of a soluble salt can be determined by precipitation as silver chloride:

$$Ag^+ + Cl^- \rightarrow AgCl(s)$$

The precipitate is collected in a weighed filtering crucible, washed, and brought to constant weight at 105 to 110°C. The solution is kept somewhat acidic during the precipitation step to eliminate potential interference from anions of weak acids (such as $CO_3^{2-}$), which form sparingly soluble silver salts in neutral solution. A moderate excess of silver ion is needed to diminish the solubility of silver chloride; a large excess will result in serious coprecipitation and should be avoided.

Silver chloride forms first as a colloid, which is then coagulated with heat. Nitric acid and the small excess of silver nitrate aid in this process by providing a relatively high electrolyte concentration. Nitric acid is added to the wash liquid to maintain a high electrolyte concentration and thus eliminate the possibility of peptization during the washing step; the acid is volatilized during the subsequent heat treatment. Additional information concerning the properties of colloidal precipitates can be found on pages 76 to 80.

In common with other silver halides, finely divided silver chloride will undergo photodecomposition:

$$AgCl(s) \rightarrow Ag(s) + \tfrac{1}{2} Cl_2(g)$$

This reaction causes the results to be low. The silver that is produced in this reaction is responsible for the violet color that develops in the precipitate. If photochemical decomposition occurs in the presence of excess silver ion, the additional reaction

$$3Cl_2(aq) + 3H_2O + 5Ag^+ \rightarrow 5AgCl(s) + ClO_3^- + 6H^+$$

may cause the results of the analysis to be high. Dry silver chloride is virtually unaffected by exposure to light.

Some photodecomposition of silver chloride is inevitable as the analysis is ordinarily performed. The effect is negligible, provided exposure to direct sunlight is avoided; storage in a dark space is also worthwhile.

Iodide, bromide, and thiocyanate, if present, will be precipitated along with silver chloride and cause high results. Additional interference can be expected from tin and antimony, which are likely to precipitate as oxychlorides under the conditions of the analysis.

Silver nitrate is very expensive. Any unused reagent should be collected in a storage container; similarly, silver chloride should be removed from the crucibles at the completion of the analysis and retained.

---

**Procedure**   Clean three sintered-glass crucibles (or porcelain filtering crucibles) by introducing about 5 mL of concentrated $HNO_3$ to each and letting them stand for a few minutes. Use a vacuum (see Figure 19–14, p. 552) to draw the acid through the crucible. Draw three portions of tap water through each crucible, and then discontinue suction. Next, add about 5 mL of 6 $F$ $NH_3$, and again wait for a few minutes before drawing it through the filter. Finally, rinse the crucible six to eight times with distilled (or deionized) water. Provide each crucible with an identifying mark (p. 530); then bring all to constant weight by heating at 110°C for at least 1 hr during periods of waiting in the analysis.

Transfer the unknown to a weighing bottle and dry it for 1 to 2 hr at 105 to 110°C (see Figure 19–7, p. 543); store the bottle in a desiccator while cooling. Weigh (to the nearest 0.1 mg) individual 0.15- to 0.20-g samples, by difference (p. 543), into 400-mL beakers. Dissolve each in about 100 mL of distilled water to which 2 to 3 mL of 6 $F$ $HNO_3$ have been added.

Slowly, and with good stirring, add 0.2 $F$ $AgNO_3$ to the cold solution until the AgCl is observed to coagulate (see Notes 1 and 2); then introduce an additional 3 to 5 mL. Heat almost to boiling, and digest the solid (p. 80) for about 10 min. Check for completeness of precipitation by adding a few drops of $AgNO_3$ to the supernatant liquid. If further precipitation occurs, add about 3 mL of $AgNO_3$, digest, and again test for completeness of precipitation. Pour any unused $AgNO_3$ into the waste container (NOT into the original reagent bottle). Cover each beaker and store for at least 2 hr, preferably until the next laboratory period.

Read the instructions for filtration and washing on pages 546 to 549. Decant the supernatant liquid through a weighed filtering crucible. Wash the precipitate several times (while it is still in the beaker) with a cold solution consisting of 2 to 5 mL of 6 $F$ $HNO_3$ per liter of distilled water; decant these washings through the filter as well. Finally, use a stream of wash liquid to transfer the bulk of the solid to the crucible; use a rubber policeman to dislodge any particles that adhere to the walls of the beaker. Continue washing until the filtrate is essentially free of silver ion (see Note 3).

Dry the precipitates at 105 to 110°C for at least 1 hr. Store crucibles in a desiccator while they cool. Determine the weight of the crucibles and their contents. Repeat the cycle of heating, cooling, and weighing until consecutive weighings agree within 0.2 mg. Report the percentage of chloride in the sample.

After the experiment has been completed, remove the precipitates by tapping the crucibles over a piece of glazed paper. Transfer the collected AgCl to a container for silver wastes. Clean the crucibles as instructed at the beginning of the procedure.

**NOTES**

1. Determine the approximate amount of $AgNO_3$ needed by calculating the volume that would be required if the sample were pure NaCl.
2. Use a separate stirring rod for each sample and leave it in the beaker throughout the determination.
3. Washings are readily tested for the presence of $Ag^+$ by collecting a small volume in a test tube and adding a few drops of HCl. Washing is judged complete when little or no turbidity is observed.

### 20A–3 Determination of Nickel in Steel

**Discussion**   The nickel in a steel sample can be precipitated from a slightly alkaline medium with an alcoholic solution of dimethylglyoxime (p. 86). Interference from iron(III) is eliminated by masking with tartaric acid. The product is freed of moisture by drying at 110°C.

The bulky character of nickel dimethylglyoxime sets an upper limit on the amount of nickel that can be conveniently handled; the weight of sample taken for analysis is governed by this consideration. The excess of reagent must also be controlled, not only because its solubility in water is relatively low, but also because nickel dimethylglyoxime becomes appreciably more soluble as the alcohol content of the solution is increased.

**Procedure**   Clean and mark three sintered-glass crucibles (see Note 1); bring them to constant weight by drying at 110°C for at least 1 hr. Weigh (to the nearest 0.1 mg) samples containing between 30 and 35 mg of nickel into individual 400-mL beakers, and dissolve by warming with about 50 mL of 6 $F$ HCl. Carefully add about 15 mL of 6 $F$ HNO$_3$ and boil gently to expel any oxides of nitrogen that may have been produced. Dilute the resulting solution to about 200 mL and heat nearly to boiling. Introduce about 30 mL of 25% tartaric acid (see Note 2), and sufficient concentrated NH$_3$ until a faint odor can be detected in the vapors above the solution (see Note 3); add 1 to 2 mL in excess. If the solution is not clear at this stage, proceed as directed in Note 4. Make the solution slightly acidic with HCl (no odor of NH$_3$), heat to 60 to 80°C, and add about 20 mL of a 1% (w/v) alcoholic solution of dimethylglyoxime. Then, with good stirring, introduce sufficient 6 $F$ NH$_3$ until a slight excess exists (as indicated by the odor) plus an additional 1 to 2 mL. Digest for 30 to 60 min, cool for at least 1 hr, and filter.

Wash the solid with water until the washings are free of Cl$^-$ (see Note 6). Finally, bring the crucibles and contents to constant weight by drying at 110°C. Report the percentage of nickel in the sample; the dried precipitate has the composition C$_8$H$_{14}$N$_4$NiO$_4$ (gfw = 288.9).

**NOTES**

1. Porcelain filtering crucibles or Gooch crucibles with glass pads can be substituted for sintered-glass crucibles in this determination.
2. The tartaric acid solution should be clear; filter it, if necessary.
3. The presence (or absence) of excess NH$_3$ is readily established by odor; waft the vapors over the beaker toward your nose with a waving motion.
4. If Fe$_2$O$_3 \cdot x$H$_2$O forms upon the addition of ammonia, acidify the solution, introduce additional tartaric acid, and neutralize again. Alternatively, remove the solid by filtration. Thorough washing of the entire filter paper with hot, dilute NH$_3$–NH$_4$Cl solution is required; the washings should be combined with the filtrate.
5. Use a separate stirring rod for each sample and leave it in the beaker throughout the determination.
6. Washings can be tested for Cl$^-$ by collecting a small volume in a test tube, acidifying with HNO$_3$, and adding one or two drops of 0.1 $F$ AgNO$_3$. Washing is complete when little or no turbidity develops.

## 20B PRECIPITATION TITRATIONS

### 20B – 1 Determination of Chloride: The Fajans Method

**Discussion**    The Fajans method is based upon the titration of chloride with a standard solution of silver nitrate. The anion of the indicator becomes bound into the counter-ion layer of silver chloride with the first excess of titrant and imparts a color to the solid (p. 157). Since this is a surface phenomenon, the particles of silver chloride should have colloidal dimensions; dextrin is added to prevent coagulation.

The silver nitrate solution is prepared by diluting a known weight of the primary-standard salt to a known volume. *The reagent is very expensive,* and every effort must be made to avoid waste. Unused solutions should be collected rather than being discarded.

---

### Preparation of Reagents

*(a) Silver nitrate.*    Use a top-loading balance to transfer the approximate weight (see Note 1) of $AgNO_3$ to a clean, dry weighing bottle. Dry at 110°C for 1 hr, but not much longer (see Note 2), and cool to room temperature in a desiccator. Weigh (to the nearest 0.1 mg) the bottle and its contents. Quickly transfer the silver nitrate to a volumetric flask; use of a powder funnel is recommended. Reweigh the bottle and any residual solid. Dissolve the $AgNO_3$, dilute to the mark, and mix well (see Note 3).

**NOTES**

1. Consult with the instructor concerning the volume and concentration of $AgNO_3$ to be used.

| Concentration Required, $N$ | Grams of $AgNO_3$ to Be Taken for | | |
|---|---|---|---|
| | 1000 mL | 500 mL | 250 mL |
| 0.10 | 16.9 | 8.5 | 4.2 |
| 0.05 | 8.5 | 4.2 | 2.1 |
| 0.01 | 1.7 | 0.9 | 0.5 |

2. Silver nitrate is perceptibly decomposed by prolonged heating. Some discoloration may occur, even after 1 hr at 110°C; the effect on the purity of the reagent is ordinarily negligible.
3. Silver nitrate solutions should be stored in a dark place when not actually in use.

*(b) 2',7'-Dichlorofluorescein.*    Dissolve 0.1 g of dichlorofluorescein in a solution that contains 75 mL of ethanol and 25 mL of water.

**Procedure**   Dry the unknown for 1 hr at 110°C. Weigh individual samples (to the nearest 0.1 mg) into individual conical flasks, and dissolve them in an appropriate volume of distilled water (see Note 1). To each, introduce about 0.1 g of dextrin and 5 drops of indicator. Titrate (see Note 2) with $AgNO_3$ to the first permanent appearance of the pink color of the indicator. Report the percent chloride in the sample.

**NOTES**

1. Use 0.25-g samples for 0.1 $N$ $AgNO_3$, and about half that amount for 0.05 $N$ reagent. Dissolve the former in about 200 mL of water and the latter in about 100 mL. For 0.01 $N$ $AgNO_3$, it is recommended that a 0.4-g sample be weighed into a 1-L volumetric flask and that 50.0-mL aliquots be taken for titration.
2. Silver chloride is particularly sensitive to photodecomposition in the presence of the indicator; the titration will fail if attempted in direct sunlight. If photodecomposition appears to be a problem, establish the approximate end point with a rough, preliminary titration, and use this information to calculate the volumes of $AgNO_3$ needed for the other samples. Add the indicator and dextrin only after most of the $AgNO_3$ has been added to each subsequent sample, and then complete the titration without delay.

## 20C NEUTRALIZATION TITRATIONS

Neutralization titrations are performed with standard solutions of strong acids or bases. While in principle a single solution (of either acid or base) is sufficient for a given kind of analyte, in practice it is convenient to have standard solutions of both available to locate end points more exactly. The concentration of one solution is established by titrations against a primary standard; the normality of the other is then found by determining the acid-base ratio (that is, the volume of acid needed to neutralize 1.000 mL of the base).

When atmospheric carbon dioxide is in equilibrium with aqueous carbonic acid, the concentration of the latter is about $1.5 \times 10^{-5}$ $F$ at ordinary temperatures. The concentrations of the acids and bases used in the determinations that follow are such that this small concentration of carbonic acid causes no significant error. On the other hand, distilled water is sometimes supersaturated with the gas and thus contains sufficient carbonic acid to affect the results of an analysis.[1] The instructions that follow are based on the assumption that dissolved $CO_2$ is negligible.

---

[1] Water to be used for neutralization titrations can be tested by adding 5 drops of phenolphthalein to a 500-mL sample. Less than 0.2 to 0.3 mL of 0.1 $N$ base should suffice to cause the appearance of the first faint pink color. If a larger volume is required, the water should be boiled and cooled before it is used to prepare standard solutions or to dissolve samples.

### 20C – 1 Preparation of Indicators for Neutralization Titrations

**Discussion**    The theory of acid-base indicator behavior is discussed in Chapter 8, Section 8C – 4. An indicator exists for virtually any pH range between 1 and 13.[2] Directions follow for the preparation of indicator solutions that will permit the performance of most common analyses.

---

**Procedure**    Stock solutions ordinarily contain 0.5 to 1.0 g of the indicator per liter.

   *(a) Methyl orange, methyl red.*    Dissolve the sodium salt directly in distilled water.

   *(b) Phenolphthalein, thymolphthalein.*    Dissolve the solid indicator in a solution containing 800 mL of ethanol and 200 mL of water.

   *(c) Sulfonphthaleins.*    Dissolve the sodium salts directly in water. Alternatively, grind 100 mg of indicator with the following volumes of 0.100 $N$ NaOH and then dilute to 100 mL with distilled water: *bromocresol green,* 1.45 mL; *bromothymol blue,* 1.6 mL; *bromophenol blue,* 1.5 mL; *thymol blue,* 2.15 mL; *cresol red,* 2.65 mL; *phenol red,* 2.85 mL.

---

### 20C – 2 Preparation of 0.1 N Hydrochloric Acid

**Discussion**    See Chapter 9, Sections 9A – 1 and 9A – 2 for additional information concerning the preparation and standardization of acids.

---

**Procedure**    Add about 8 mL of 12 $F$ HCl reagent to about 1 L of distilled water. Mix thoroughly, and store in a glass-stoppered bottle.

**NOTE**

For the preparation of very dilute ($<0.05\ N$) HCl solutions, it is advisable to eliminate $CO_2$ from the water by a preliminary boiling.

---

### 20C – 3 Preparation of Carbonate-Free 0.1 N Sodium Hydroxide

**Discussion**    See Chapter 9, Sections 9A – 3 and 9A – 4 for further information concerning the preparation and standardization of bases.

---

[2] See, for example, J. Beukenkamp and W. Rieman III, in *Treatise on Analytical Chemistry,* I. M. Kolthoff and P. J. Elving, Eds., Part I, Vol. 11, pp. 6987–7001. New York: John Wiley & Sons, 1974.

**Procedure**   If directed by the instructor, prepare a bottle for protected storage (see Figure 9–1, p. 215, and Note 1). Boil about 1 L of distilled water for a few minutes, allow it to cool, and then transfer it to the storage bottle. Decant 4 to 5 mL of 50% NaOH (see Note 2), add to the water and *mix thoroughly* (*CAUTION!* 50% NaOH is highly corrosive to the skin. If the reagent comes in contact with skin, *immediately* flush the area with *copious* amounts of water.) Protect the solution from unnecessary contact with the atmosphere.

**NOTES**

1. A solution of base that will be used over a period of one or two weeks only can be prepared and stored in a tightly capped polyethylene bottle. The bottle should be squeezed as the cap is tightened after each removal of base to minimize the air space above the solution.
2. Be certain that any solid sodium carbonate in the 50% NaOH has settled to the bottom of the container so that none is transferred during the decantation step. If necessary, filter the base through a glass mat in a Gooch crucible; collect the clear filtrate in a test tube in the filtering flask.

### 20C–4 Determination of the Acid-Base Ratio

**Discussion**   If both acid and base solutions have been prepared, it is convenient to establish the volume ratio between the two before one of them is standardized.

**Procedure**   Thoroughly rinse the inside of each buret with three or four portions of the solution it is to contain. Rotate the buret to be sure that the entire inner surface is wetted after each addition; drain each portion through the tip.

Fill each buret well above the zero volume line. Remove any air bubbles in the tip by opening the stopcock briefly. Place a test tube or small beaker over the top of the buret that contains the sodium hydroxide solution. After allowing at least 30 sec for drainage, record the initial buret volumes; estimate the volume to the nearest 0.01 mL.

Deliver a 35- to 40-mL portion of the acid into a 250-mL conical flask. Touch the tip of the buret to the inside wall of the flask, and rinse down with a little distilled water. Add two drops of phenolphthalein (see Note 1), and then introduce sufficient base until the solution is definitely pink. Introduce acid dropwise to discharge the color, and rinse down the walls of the flask. Carefully add base until the solution acquires a faint pink hue that persists for 30 sec (see Notes 2 and 3). Record the final buret volumes (again, to the nearest 0.01 mL). Repeat the titration. Calculate the volume ratio that exists between the acid and the base (see Note 4). Duplicate titrations should agree within 1 to 2 parts per thousand. Perform additional titrations, if necessary, to achieve this order of precision.

NOTES

1. The volume ratio can also be obtained with an indicator that has an acidic transition range, such as bromocresol green (see Chapter 8, Section 8B – 1, p. 165). If the base solution is contaminated with carbonate, the ratio with this indicator will differ significantly from the value obtained with phenolphthalein.
2. Fractional drops can be delivered by forming them on the buret tip, touching the tip to the wall of the flask, and then rinsing down with a small amount of water.
3. The phenolphthalein end point will fade slowly as $CO_2$ is absorbed from the atmosphere.
4. The normality of one solution can be evaluated with the acid-base ratio and the normality of the other.

## 20C – 5 Standardization of Hydrochloric Acid Against Sodium Carbonate

**Discussion** See Chapter 9, Section 9A – 2.

**Procedure** Dry a quantity of primary-standard $Na_2CO_3$ for 2 hr at 110°C (see Figure 19 – 7, p. 543) and cool in a desiccator. Weigh individual 0.2- to 0.25-g samples (to the nearest 0.1 mg) into 250-mL conical flasks, and dissolve them in about 50 mL of distilled water. Introduce 3 drops of bromocresol green, and titrate with HCl until the solution just changes from blue to green. Boil the solution for 2 to 3 min, cool to room temperature (see Note 1), and complete the titration (see Note 2).

Determine an indicator correction by titrating approximately 100 mL of 0.05 $F$ NaCl and 3 drops of indicator. Subtract any volume of acid needed for the blank from the titration data. Calculate the normality of the HCl solution.

NOTES

1. The indicator should change from green to blue as $CO_2$ is removed during the heating step. If no color change occurs, an excess of acid was added originally. This excess can be back-titrated with base, provided that its combining ratio with the acid is known; otherwise, the sample must be discarded.
2. It is permissible to back-titrate with standard base to establish the end point with greater certainty.

## 20C – 6 Standardization of Sodium Hydroxide Against Potassium Hydrogen Phthalate

**Discussion** See Chapter 9, Section 9A – 4.

---

**Procedure** Dry a quantity of primary-standard potassium hydrogen phthalate for 2 hr at 110°C (see Figure 19 – 7, p. 543), and cool in a desiccator. Weigh individual 0.7- to 0.9-g samples (to the nearest 0.1 mg) into 250-mL conical flasks, and dissolve in 50 to 75 mL of distilled water. Introduce 2 drops of phenolphthalein; titrate with base until the pink color of the indicator persists for 30 sec (see Note). Calculate the normality of the sodium hydroxide solution.

**NOTE**
If a dilute HCl solution has been prepared, it is permissible to back-titrate to establish the end point more precisely. The volume of acid must, of course, be recorded; a correction to the volume of base used in the titration can then be calculated from the acid-base ratio.

---

## 20C – 7 Determination of Potassium Hydrogen Phthalate in an Impure Sample

**Discussion** The sample is a mixture of potassium hydrogen phthalate and a neutral salt.

---

**Procedure** Consult with the instructor concerning an appropriate sample size. Then, follow the directions in Section 20C – 6. Report the percent $KHC_8H_4O_4$ in the sample.

---

## 20C – 8 Determination of the Equivalent Weight of a Weak Acid

**Discussion** The equivalent weight of an acid, which is an aid in establishing its identity, is readily determined by titrating a weighed quantity of the purified acid with standard sodium hydroxide.

---

**Procedure** Consult the instructor for an appropriate sample size. Weigh (to the nearest 0.1 mg) individual samples into 250-mL conical flasks, and dissolve in 50 to 75 mL of distilled water (see Note). Add 2 drops of phenolphthalein, and titrate with standard base to the first persistent (~ 30 sec) color of the indicator. Calculate the equivalent weight of the acid.

**NOTE**

Acids with limited solubility in water may dissolve more readily in ethanol or an ethanol–water mixture. The alcohol may be measurably acidic and should be rendered faintly alkaline to phenolphthalein before it is used as a solvent. As an alternative, a sparingly soluble weak acid can be dissolved in a measured excess of standard base, the excess being determined by back-titration with standard acid.

### 20C–9 Determination of Acetic Acid in Vinegar

**Discussion** The total acid content of vinegar is conveniently determined by titration with standard base. Even though vinegar contains other acids, the results of the analysis are customarily reported in terms of acetic acid, its principal acidic constituent. Vinegars are approximately 5% acid (w/v), expressed as acetic acid.

**Procedure** Pipet a 25.00-mL aliquot of the vinegar into a 250-mL volumetric flask and dilute to the mark with distilled water; mix thoroughly. Transfer 50.00-mL aliquots of the diluted sample to 250-mL conical flasks. Add about 50 mL of water and 2 drops of phenolphthalein; then titrate with standard base until the pink color of the indicator persists for at least 30 sec. Calculate the acidity of the vinegar as grams of $CH_3COOH$ per 100 mL of sample.

### 20C–10 Determination of Sodium Carbonate in an Impure Sample

**Discussion** See Chapter 9, Section 9A–2.

**Procedure** Consult with the instructor concerning an appropriate sample size. Then follow the directions in Section 20C–5. Report the percent $Na_2CO_3$ in the sample.

### 20C–11 Determination of Sodium Carbonate and Sodium Hydrogen Carbonate in a Mixture

**Discussion** The sample consists of $Na_2CO_3$, $NaHCO_3$, and inert materials. The principles of the determination are discussed in Chapter 9, Section 9B–2.

**Procedure** If the sample is a solid, weigh dried portions into 250-mL volumetric flasks, dilute to the mark, and mix well. If the sample is a liquid, pipet suitable aliquots into 250-mL volumetric flasks, dilute to the mark, and mix well.

*(a) Determination of total basicity.* Transfer 25.00-mL aliquots to 250-mL conical flasks. Titrate the carbonate and hydrogen carbonate with standard HCl, using the directions in Section 20C – 5.

*(b) Determination of hydrogen carbonate only.* Pipet additional 25.00-mL aliquots to 250-mL conical flasks; *treat each sample individually from this point.* Use a pipet to transfer 50.00 mL of standard base to the sample. Immediately add 10 mL of 10% (w/v) $BaCl_2$ and 2 drops of phenolphthalein. Titrate the excess NaOH without delay; the color change will be from pink to colorless.

Titrate a blank consisting of 25 mL $H_2O$, 10 mL $BaCl_2$, and *exactly* the same volume of NaOH as used with the sample. The difference between the volumes of HCl needed for the blank and the sample is chemically equivalent to the amount of $NaHCO_3$ in the aliquot. Calculate the percent $Na_2CO_3$ and $NaHCO_3$ in the sample.

## 20D COMPLEX FORMATION TITRATIONS

The uses of complex formation reactions in volumetric analysis are discussed in Chapter 10. The directions that follow make use of ethylenediaminetetraacetic acid as a chelating agent.

### 20D – 1 Preparation of an Eriochrome Black T Indicator Solution

**Procedure** Dissolve 100 mg of solid Eriochrome Black T in a solution consisting of 15 mL of triethanolamine and 5 mL of ethanol. Solutions of the indicator should be freshly prepared every two weeks; refrigeration will slow deterioration somewhat.

### 20D – 2 Preparation of Standard 0.01 F EDTA

**Discussion** Superficial moisture can be eliminated from $Na_2H_2Y \cdot 2H_2O$ by drying at 80°C; a standard solution can then be prepared by dissolving a weighed quantity in sufficient water to give a known volume. This direct preparation requires the total exclusion of polyvalent cations. Distilled water should be passed through a cation-exchange resin if any doubt exists regarding its quality.

Alternatively, an EDTA solution of approximately the desired concentration can be prepared and subsequently titrated against a standard magnesium solution.

**Procedure**   Dry purified $Na_2H_2Y \cdot 2H_2O$ for 1 to 2 hr in an oven that has been set to 80°C (see Notes 1 and 2). Cool in a desiccator. Then weigh about 3.8 g (to the nearest milligram) into a 1-L volumetric flask; use a powder funnel to ensure a quantitative transfer of the solid. Introduce 600 to 800 mL of water and swirl periodically until solution is complete; this process may take 15 min (or more). Dilute to the mark and mix well.

**NOTES**

1. W. J. Blaedel and H. T. Knight (*Anal. Chem.*, **1954,** *26,* 741) give specific directions for the purification of commercial preparations of the disodium salt.
2. If desired, the anhydrous salt can be used instead of the dihydrate. The weight should be adjusted accordingly. Anhydrous $Na_2H_2Y$ is appreciably hygroscopic under ordinary atmospheric conditions and is thus less satisfactory than the dihydrate.

### 20D – 3 Preparation of a pH 10 Buffer

**Procedure**   Dilute 570 mL of concentrated $NH_3$ and 70 g of $NH_4Cl$ to about 1 L.

### 20D – 4 Titration of Magnesium with EDTA

**Discussion**   This is a typical example of a direct titration with EDTA. As performed here, the method is straightforward. It should be appreciated that many other polyvalent cations will interfere by reacting with the titrant under the conditions of this titration. Pretreatment with ammonium carbonate is reported to eliminate interference from alkaline-earth metals. Other cations should be precipitated as their hydroxides in a preliminary step.

**Procedure**   The sample will be issued as an aqueous solution; dilute it to 500 mL in a volumetric flask and mix thoroughly. Transfer 50.00-mL aliquots to individual conical flasks, add 1 to 2 mL of pH 10 buffer and 4 to 5 drops of Eriochrome Black T indicator to each. Titrate with 0.01 $F$ $Na_2H_2Y$ to a color change from red to a pure blue (see Note). Express the concentration of the diluted sample in ppm of Mg.

**NOTE**

The color change of the indicator is slow in the vicinity of the end point, and care must be taken to avoid overtitration.

## 20D-5 Determination of Total Hardness in Water

**Discussion**   See Chapter 10, Section 10B-8.

**Procedure**   Acidify 100-mL aliquots of the sample with a few drops of HCl, and boil gently for a few minutes to eliminate $CO_2$. Cool, add 3 to 4 drops of methyl red, and neutralize the sample with NaOH. Introduce 2 mL of pH 10 buffer and 4 to 5 drops of Eriochrome Black T indicator; titrate with standard 0.01 $F$ EDTA to a color change from red to blue (see Note). Express the results of this analysis in terms of ppm of $CaCO_3$.

**NOTE**

A sluggish color change suggests that the sample contains little or no magnesium ion. In this event, add 1 to 2 mL of $MgY^{2-}$ solution. This solution is prepared by adding 2.645 g of $MgSO_4 \cdot 7H_2O$ to 3.722 g of $Na_2H_2Y \cdot 2H_2O$ in 50 mL of distilled water. Introduce a few drops of phenolphthalein, followed by sufficient sodium hydroxide to turn the solution faintly pink. Dilute to about 100 mL. Introduction of a few drops of Eriochrome Black T to a small portion of this solution that has been buffered with the pH 10 buffer solution should produce a dull violet color. Moreover, a single drop of 0.01 $F$ $Na_2H_2Y$ should cause a color change to blue, while an equal volume of 0.01 $F$ $Mg^{2+}$ should cause a change to red. If necessary, the composition of the solution should be adjusted with 0.01 $F$ $Na_2H_2Y$ or 0.01 $F$ $Mg^{2+}$ until these criteria are met. This procedure produces a solution in which the mole ratio of $Mg^{2+}$ to EDTA is 1:1.

## 20E TITRATIONS WITH POTASSIUM PERMANGANATE

The theory of oxidation-reduction reactions is discussed in Chapter 11, and typical volumetric oxidation-reduction methods are described in Chapter 12. This section contains directions for the preparation and standardization of potassium permanganate solutions, as well as for the use of this reagent in the determination of iron in an ore and calcium in a limestone.

### 20E-1 Preparation of Approximately 0.1 N Potassium Permanganate

**Discussion**   The precautions that must be observed in the preparation and storage of permanganate solutions are discussed in Chapter 12, Section 12B-1.

**Procedure**   Dissolve about 3.2 g of $KMnO_4$ in about 1 L of distilled water. Bring the solution to a gentle boil, and maintain at this temperature for about 1 hr. Then cover, and let stand overnight. Remove $MnO_2$ by vacuum filtration (see Notes 1 and 2). Transfer the solution to a clean, glass-stoppered bottle, and store in a dark place when not in use.

**NOTES**

1. The $MnO_2$ that collects in a filtering crucible can be removed with dilute $H_2SO_4$ containing a few milliliters of 3% $H_2O_2$, or with dilute HCl (Hood). Washing with copious amounts of water should follow either treatment.
2. The heating and filtering steps can be omitted if the solution can be standardized and used on the same day.

### 20E – 2 Standardization of Potassium Permanganate Against Sodium Oxalate

**Discussion**   Directions are provided for standardization by both the method of McBride and that proposed by Fowler and Bright. These methods as well as those for other primary standards for permanganate solutions are discussed on pages 307 to 308.

**Procedure**   Dry primary-standard $Na_2C_2O_4$ for at least 1 hr at 110 to 120°C. Cool to room temperature in a desiccator and weigh (to the nearest 0.1 mg) individual 0.2- to 0.3-g samples of the dried $Na_2C_2O_4$ into 400-mL beakers. Dissolve in approximately 250 mL of 0.9 $F$ $H_2SO_4$.

*(a) Method of McBride.*   Heat the solution to 80 to 90°C, and titrate with $KMnO_4$, stirring vigorously with a thermometer. Allow sufficient time for the pink color of the reagent to disappear before making further additions (see Notes 1 and 2); reheat if the temperature drops below 60°C. Take the first persistent pink color as the end point (see Notes 3 and 4). Determine an end-point blank by titrating about 250 mL of 0.9 $F$ sulfuric acid. Correct (if necessary) for the blank, and calculate the normality of the permanganate solution.

*(b) Method of Fowler and Bright.*   Introduce from a buret sufficient permanganate to react with 90 to 95% of the oxalate (about 40 mL of 0.1 $N$ reagent for a 0.3-g sample; a preliminary titration by the McBride method will provide an appropriate volume). Allow the solution to stand until the permanganate color has disappeared. Warm to 55 to 60°C, and complete the titration, taking the first faint pink color that persists for 30 sec as the end point. Determine an end-point blank by titrating about 250 mL of 0.9 $F$ sulfuric acid. Correct (if necessary) for the blank, and calculate the normality of the permanganate solution.

**NOTES**

1. Any $KMnO_4$ that adheres to the side of the beaker should be immediately washed into the bulk of the solution with a stream of water.
2. Manganese dioxide (as well as $Mn^{2+}$) will be produced if $KMnO_4$ is added too rapidly in the McBride titration. Evidence for $MnO_2$ formation is a faint brown discoloration of the solution. The existence of $MnO_2$ is not a serious problem so long as sufficient oxalate remains to reduce it to $Mn^{2+}$; the titration is temporarily discontinued until the solution clears. The solution must be free of $MnO_2$ at the equivalence point.
3. The surface of the $KMnO_4$ solution can be used for reading volumes in the buret. Alternatively, sufficient backlighting with a flashlight will permit use of the bottom of the meniscus as the point of reference.
4. Permanganate solutions should not be allowed to stand in burets any longer than necessary because decomposition to $MnO_2$ may occur. Freshly formed $MnO_2$ can be removed with 3% $H_2O_2$ in dilute $H_2SO_4$ or with HCl.

### 20E–3 Determination of Iron in an Ore by Titration with Potassium Permanganate

**Discussion**   Common iron ores include hematite ($Fe_2O_3$), magnetite ($Fe_3O_4$), and limonite ($3Fe_2O_3 \cdot 3H_2O$). Volumetric methods for the analysis of iron in ore samples consist of three steps: (1) solution of the sample, (2) reduction of the iron to the divalent state, and (3) titration with a standard oxidizing reagent.

Many iron ores are completely decomposed by concentrated hydrochloric acid. A small amount of tin(II) chloride accelerates the rate of attack, probably by reducing iron(III) oxides on the surface of the particles to more soluble iron(II) species. Because iron(III) tends to form stable complexes with chloride ion, hydrochloric acid is a more efficient solvent than either nitric or sulfuric acid.

Iron ores that contain silicates may be entirely decomposed by treatment with hydrochloric acid. If decomposition is complete, a white residue of hydrated silica remains, which will not interfere with subsequent steps of the analysis. A dark residue is indicative of incomplete decomposition.

Part or all of the iron will exist in the trivalent state in the dissolved sample, and prereduction must therefore precede the titration step. A Jones or a Walden reductor (see Chapter 12, Section 12A–1) can be used for this purpose. Perhaps the most satisfactory of all prereductants for iron, and the one that is used in this experiment, is tin(II) chloride. The only other species that are reduced by this reagent include ions of vanadium, copper, molybdenum, arsenic, and tungsten. The excess tin(II) is eliminated by the addition of mercury(II) chloride:

$$Sn^{2+} + HgCl_2 \rightarrow Hg_2Cl_2(s) + Sn^{4+} + 2Cl^-$$

The sparingly soluble mercury(I) chloride that is produced will not reduce permanganate, nor will the excess mercury(II) reoxidize the iron(II). Care must be taken, however, to avoid the formation of elemental mercury by the alternative reaction:

$$Sn^{2+} + HgCl_2 \rightarrow Hg(l) + Sn^{4+} + 2Cl^-$$

Metallic mercury reacts with permanganate and causes the results of the analysis to be high. The formation of mercury is favored by an appreciable excess of tin(II) and is prevented by the rapid addition of excess mercury(II) chloride. Introduction of this reagent should result in the formation of a small amount of white mercury(I) chloride. A gray precipitate indicates the presence of mercury while the total absence of precipitate indicates that insufficient tin(II) chloride was added. In either event, the sample must be discarded.

The reaction between iron(II) and permanganate proceeds smoothly and rapidly to completion. Hydrochloric acid, if present, is oxidized to chlorine by permanganate ion, which leads to high results. This parasitic reaction, which ordinarily does not proceed rapidly enough to cause serious errors, is *induced* by the presence of divalent iron; an intermediate manganese(III) ion that is formed during the reduction by iron(II) appears to be responsible for the oxidation of chloride ion. This problem can be eliminated through the preliminary removal of chloride by evaporation with sulfuric acid or by the use of the *Zimmermann-Reinhardt reagent.* The latter is a solution of manganese(II) in moderately concentrated sulfuric and phosphoric acids. Both the manganese(II) and the phosphoric acid in the Zimmermann-Reinhardt reagent act to decrease the potential of the manganese(III)–manganese(II) couple and thereby inhibit the formation of chlorine, the former through its action as a common ion, and the latter through the formation of a stable phosphate complex with manganese(III). Phosphoric acid also forms nearly colorless complexes with the iron(III) formed in the titration and prevents the intense yellow color of iron(III)–chloride complexes from interfering with the end point.[3]

**Preparation of Reagents**   (sufficient for 100 titrations)

(a)  *0.25 F SnCl₂.*   Dissolve 60 g of iron-free $SnCl_2 \cdot 2H_2O$ in 100 mL of concentrated HCl; warm, if necessary. After solution is complete, dilute to about 1 L and store in a well-stoppered bottle. A few pieces of metallic tin in the bottle will prevent air oxidation of Sn(II).

(b)  *5% (w/v) HgCl₂.*   Dissolve 50 g of $HgCl_2$ in about 1 L of water.

(c)  *Zimmermann-Reinhardt reagent.*   Dissolve 300 g of $MnSO_4 \cdot 4H_2O$ in about 1 L of water. Cautiously add 400 mL of concentrated $H_2SO_4$ and 400 mL of 85% $H_3PO_4$. Dilute to about 3 L.

(d)  *0.2 F KMnO₄.*   Dissolve about 3 g of $KMnO_4$ in about 100 mL of water.

---

[3] The mechanism by which the Zimmermann-Reinhardt reagent acts has been the subject of much study; see, for example, H. A. Laitinen and W. E. Harris, *Chemical Analysis,* 2nd ed., p. 331. New York: McGraw-Hill Book Co. 1975.

**Procedure**

*(a) Sample preparation.* Dry the ore for at least 3 hr at 105 to 110°C and allow to cool in a desiccator. Consult the instructor for a sample size that will require 25 to 40 mL of 0.1 $N$ KMnO$_4$. Weigh individual samples (to the nearest 0.1 mg) into 500-mL conical flasks. Introduce 10 mL of concentrated HCl and 3 mL of 0.25 $F$ SnCl$_2$ solution (see Note 1). Cover the flask with a small watch glass, and heat just below boiling until the sample is decomposed (as indicated by the disappearance of all dark particles; see Note 2); a pure white residue may remain. A blank containing 10 mL of concentrated HCl and 3 mL of the SnCl$_2$ solution should be carried through the heating step. Add more SnCl$_2$ to any flask in which the solution becomes perceptibly yellow as it is heated. After decomposition is complete, remove any excess SnCl$_2$ by the dropwise addition of 0.2 $F$ KMnO$_4$ until the solutions turn pale yellow. Dilute to about 15 mL. Add KMnO$_4$ to the blank until the solution just turns pink, and then just decolorize with 1 drop of SnCl$_2$.

*Carry samples individually through the remaining steps of the analysis to minimize air oxidation of iron(II).*

*(b) Reduction of iron.* Heat the solution containing the sample nearly to boiling. Then introduce 0.25 $F$ SnCl$_2$ dropwise until the yellow color just disappears; add 2 drops in excess (see Note 3). Cool to room temperature and *rapidly* introduce 10 mL of 5% HgCl$_2$ solution. A small quantity of silky white Hg$_2$Cl$_2$ should be produced. If no precipitate forms, or if it is gray, the sample must be discarded. The blank should also be treated with 10 mL of the HgCl$_2$ solution.

*(c) Titration.* Approximately 2 to 3 min after adding the HgCl$_2$, introduce 25 mL of Zimmermann-Reinhardt reagent and 300 mL of distilled water. *Titrate immediately* with standard KMnO$_4$ to the first faint pink that persists for 15 to 20 sec. Do not titrate rapidly at any time. Subtract the volume of KMnO$_4$ used by the blank from the volume of reagent needed for the titration. Calculate the percentage of Fe$_2$O$_3$ in the sample.

**NOTES**

1. The sample dissolves more quickly if SnCl$_2$ is present to reduce iron(III) oxides to iron(II). Insufficient SnCl$_2$ is indicated by the appearance of yellow iron(III) complexes in the solution.
2. If dark particles persist after treatment with acid for several hours, filter the solution through ashless paper, wash the residue with 5 to 10 mL of 6 $F$ HCl, and retain the filtrate and washings. Place the paper in a small platinum crucible and ignite. Mix the ignited residue with about 0.5 g of finely ground anhydrous Na$_2$CO$_3$ and heat until a clear liquid melt is obtained. Cool, add 5 mL of water, and then cautiously add an equal volume of 6 $F$ HCl. Warm the crucible until the cake dissolves, and combine the contents with the original filtrate. Evaporate the solution to about 15 mL, and proceed with the reduction.

3. The solution may not become entirely colorless but may instead acquire a pale yellow-green hue. Further additions of $SnCl_2$ will not alter this color. If too much $SnCl_2$ is added, again add 0.2 $F$ $KMnO_4$ until the yellow color returns, and repeat the reduction.
4. The absence of a precipitate means that insufficient $SnCl_2$ was added; the reduction of iron(III) was thus incomplete.

## 20E–4 Determination of Calcium in a Limestone by Titration with Potassium Permanganate

**Discussion**   In common with a number of other cations, calcium is conveniently precipitated with oxalate ion. The solid is filtered, washed free of excess precipitating reagent, and dissolved in dilute sulfuric acid; the liberated oxalic acid is then titrated with a standard solution of potassium permanganate (or perhaps some other oxidizing reagent). This method is applicable to samples that contain magnesium and the alkali metals. Most other cations must be absent, however, since they either precipitate or coprecipitate as oxalates and cause positive errors in the analysis.

The success of this analysis is critically dependent upon the existence of a 1 : 1 molar ratio between calcium and oxalate in the precipitate. Several precautions are needed to ensure this condition. For example, calcium oxalate formed in a neutral or ammoniacal solution is likely to be contaminated with calcium hydroxide or a basic calcium oxalate, either of which will cause low results. This problem is eliminated by adding the oxalate to an acidic solution of the sample and slowly forming the precipitate by the dropwise addition of ammonia. The coarsely crystalline solid that is produced is readily filtered.

Losses due to the solubility of calcium oxalate are negligible at pH 4, provided washing is restricted to freeing the precipitate of excess oxalic acid.

Coprecipitation by sodium oxalate is a potential source of positive error in the analysis. This becomes a problem when the sodium ion concentration exceeds that of calcium ion; it is eliminated by double precipitation of the calcium oxalate.

Magnesium, if present in high concentrations, may also be a source of contamination. This interference is minimized if the excess of oxalate is sufficient to allow the formation of a soluble magnesium complex and if filtration is performed promptly after completion of the precipitation. A double precipitation will be required if the magnesium content of the sample exceeds that of calcium.

Limestones are composed principally of calcium carbonate; dolomitic limestones contain large amounts of magnesium carbonate in addition. Silicates of calcium, magnesium, iron, aluminum, manganese, titanium, and the alkali metals also exist in smaller amounts. Hydrochloric acid will decompose most limestones completely, leaving only silica, which does not interfere with the analysis. Some limestones are more readily dissolved if first ignited; a few will yield only to a carbonate fusion.

The accompanying instructions are remarkably effective for the analysis of calcium in most limestones, as well as impure calcium carbonate. Iron and aluminum, in amounts that are chemically equivalent to calcium, do not interfere; small amounts of titanium and manganese can also be tolerated.

### Preparation of Reagents

*(a)* *6% Ammonium oxalate* (100 mL needed for each sample).   Dissolve an appropriate amount of $(NH_4)_2C_2O_4 \cdot 2H_2O$ in distilled water and filter if the solution is not clear.

*(b)* *Saturated bromine water.*

### Procedure

*(a)* *Sample preparation.*   Dry the unknown for 1 to 2 hr at 110°C and cool in a desiccator. Weigh (to the nearest 0.1 mg) individual 0.25- to 0.30-g samples (see Note 1) into 250-mL beakers and cover each with a watch glass. Add 10 mL of distilled water, and then 10 mL of concentrated HCl; add the acid slowly to prevent losses by spattering.

Add 5 drops of saturated bromine water to oxidize any iron present, and then boil gently for 5 min to eliminate the excess $Br_2$. Dilute to about 50 mL, and add 100 mL of hot 6% $(NH_4)_2C_2O_4$ solution (see Note 2). Add 3 to 4 drops of methyl red; then introduce 6 $F$ $NH_3$ at the rate of 1 drop every 3 to 4 sec until the intermediate orange-red color of the indicator is observed (pH 4.5 to 5.5). Allow the solutions to stand for 30 min but not much longer (see Note 3), and collect the solid by vacuum filtration using a medium-porosity filtering crucible. Wash the precipitate with several 10-mL portions of cold water. Rinse the outside of the crucible, and return it to the beaker in which the $CaC_2O_4$ was originally formed. Dissolve the solid in about 200 mL of 0.8 $F$ $H_2SO_4$.

*(b)* *Titration.*   Heat the solution to 80 to 90°C and titrate with 0.1 $N$ $KMnO_4$; keep the temperature above 60°C throughout the titration. Report the percent calcium oxide in the sample.

### NOTES

1. If the unknown is not readily decomposed with acid, weigh individual samples into porcelain crucibles. Ignite by raising the temperature slowly to 800 to 900°C, and maintain this temperature for about 30 min. Allow crucibles and contents to cool, place each in a 250-mL beaker, add 5 mL of distilled water, and cover with a watch glass. Carefully add 10 mL of concentrated HCl to each beaker and heat to boiling. Remove the crucibles with a stirring rod, rinsing thoroughly with water.

2. If $CaC_2O_4$ is produced upon addition of $(NH_4)_2C_2O_4$, the sample was not

sufficiently acidic. Add enough HCl to redissolve the solid before proceeding.
3. The standing period is not at all critical if it is known that the sample contains no magnesium.

## 20F  TITRATION WITH POTASSIUM DICHROMATE

The properties and applications of potassium dichromate are described in Chapter 12, Section 12B–3.

### 20F–1  Preparation of a 0.1 N Potassium Dichromate Solution

**Discussion**  Solutions are conveniently prepared by diluting a known weight of primary-standard $K_2Cr_2O_7$ to a known volume.

**Procedure**  Dry primary-standard $K_2Cr_2O_7$ for 2 hr at 150 to 200°C, and then cool in a desiccator. Weigh (to the nearest milligram) 4.9 g of the solid into a 1-L volumetric flask; use a powder funnel to ensure a quantitative transfer. Dissolve the solid, dilute to the mark, and mix well.

### 20F–2  Determination of Iron in an Ore by Titration with Potassium Dichromate

**Discussion**  The titration of iron(II) with dichromate is straightforward. Diphenylamine sulfonic acid is an excellent indicator for the titration, the color change being from the green of chromium(III) to the oxidized form of the indicator, which is purple.

The problems associated with preparation of the sample and prereduction of the iron are no different from the permanganate method; the discussion in Section 20E–3 should be consulted. Note that the Zimmermann-Reinhardt reagent is not needed for the titration with potassium dichromate.

**Preparation of Reagents**

*(a) Sodium diphenylamine sulfonate* (sufficient for 150 to 200 titrations).   Dissolve 0.2 g of sodium diphenylamine sulfonate in 100 mL of distilled water.
*(b) Remaining reagents.*   See Section 20E–3 for preparation.

**Procedure**  Follow the directions labeled *Sample preparation* and *Reduction of iron* in the Procedure for Section 20E–3. About 2 to 3 min after adding $HgCl_2$, introduce 60 mL of 3 $F$ $H_2SO_4$, 15 mL of 85% $H_3PO_4$, and 100 mL of water. Cool, add 8 drops of diphenylamine sulfonate indicator, and titrate with dichromate to a violet-blue end point. Report the percent $Fe_2O_3$ in the sample.

## 20G IODIMETRIC METHODS OF ANALYSIS

The oxidizing properties of iodine, the composition and stability of triiodide solutions, and the applications of this reagent to volumetric analysis are considered in Chapter 12, Section 12B–4.

### 20G–1 Preparation of Approximately 0.1 N Triiodide Solution

**Procedure**   Weigh about 40 g of KI into a 100-mL beaker. Add 13 g of $I_2$, 10 mL of water, and stir for several minutes (see Note 1). Introduce an additional 20 mL of water and again stir for several minutes. Carefully decant the bulk of the liquid into a storage bottle containing about 1 L of water. Avoid transferring any iodine that may remain undissolved in the last milliliter or two of the concentrated solution (see Note 2).

**NOTES**
1. Iodine dissolves relatively slowly in the concentrated KI solution. Thus, the $KI$–$I_2$ mixture should be thoroughly stirred to hasten the solution process.
2. Any solid $I_2$ that is inadvertently transferred to the storage bottle will cause a gradual increase in the normality. Filtration through a sintered-glass filter will eliminate this source of difficulty.

### 20G–2 Standardization of Triiodide Solutions

**Discussion**   As noted in Chapter 12, Section 12B–4, arsenic(III) oxide, long the favored primary standard for triiodide solutions, is now seldom used because of the elaborate precautions that must be taken to protect against this carcinogen. Barium thiosulfate monohydrate and anhydrous sodium thiosulfate have been proposed as alternative standards. Perhaps the most convenient method for standardizing triiodide solutions involves determining the concentration of a sodium thiosulfate solution against primary-standard potassium iodate and using this solution for the standardization of triiodide.

**Preparation of Solutions**

*(a) Standard sodium thiosulfate, 0.1 N.*   Follow the directions in Sections 20H–1 and 20H–2 for the preparation and standardization of this solution.
*(b) Starch indicator.*   Make a paste by rubbing about 2 g of soluble starch and 10 mg of $HgI_2$ in about 30 mL of water. Pour this suspension into 1 L of boiling water and heat until clarification occurs. Cool and transfer to a storage bottle. Between 3 and 5 mL of this preparation will suffice for most titrations.

**Procedure**   Transfer 25.00-mL aliquots of the triiodide solution to 250-mL conical flasks and dilute to about 50 mL. Introduce approximately 1 mL of 3 *F*

$H_2SO_4$ and titrate immediately with standard sodium thiosulfate until the solution becomes a faint straw yellow. Then add 5 mL of starch indicator; complete the titration, taking as the end point the volume at which the solution changes from blue to colorless. Calculate the normality of the iodine solution.

### 20G–3 Determination of Antimony in Stibnite

**Discussion**   The analysis of stibnite, a common antimony ore, is a typical application of iodimetry. The stoichiometry of the reaction can be represented as

$$SbO_3^{3-} + I_2 + H_2O \rightarrow SbO_4^{3-} + 2I^- + 2H^+$$

In practice, sodium hydrogen carbonate is added to react with the acid produced and force the reaction to the right.

The ore is primarily antimony(III) sulfide, although silica and other contaminants may be present as well. The sample is decomposed in hot, concentrated hydrochloric acid to eliminate the sulfide as $H_2S$. Some care is required to prevent losses of volatile antimony(III) chloride during this step. The presence of potassium chloride minimizes volatilization losses by forming such non-volatile complexes as $SbCl_4^-$ and $SbCl_6^{3-}$. Basic antimony salts, such as the sparingly soluble SbOCl, often form in hydrochloric acid solutions and react incompletely with iodine, causing low results. This source of difficulty is eliminated by adding tartaric acid, which forms a tartrate complex, $(SbOC_4H_4O_6^-)$, that is rapidly and completely oxidized by iodine.

**Procedure**   Dry the unknown for 1 hr at 110°C and cool in a desiccator. Weigh (see Note 1) individual samples into 500-mL conical flasks. Introduce about 0.3 g of KCl and about 10 mL of concentrated HCl to each flask. Heat the mixture (Hood) just below boiling until only a white or slightly gray residue of $SiO_2$ remains. Add more HCl, if needed, to keep the mixture from going dry.

Add 3 g of solid tartaric acid to the solution and heat for an additional 10 to 15 min. Then, with good swirling, slowly add water (see Note 2) from a pipet until the volume is about 100 mL. If reddish $Sb_2S_3$ forms, stop the dilution with water and heat further to remove the $H_2S$, adding more acid if necessary.

Add 3 drops of phenolphthalein and 6 $F$ NaOH until the first faint pink is observed. Introduce 6 $F$ HCl dropwise until the color is discharged, and then add 1 mL in excess. Add 4 to 5 g of $NaHCO_3$, taking care to avoid losses of solution by spattering during the addition. Add 5 mL of starch indicator, rinse down the walls of the flask, and titrate with standard 0.1 $N$ $I_2$ to the first blue color that persists for about 30 sec. Report the percent $Sb_2S_3$ in the sample.

**NOTES**

1. Samples should contain between 3 and 4 meq of antimony; consult the instructor for an appropriate size. Weighings to the nearest milligram are adequate for samples over 1 g.
2. The slow addition of water, with efficient stirring, is essential to prevent the formation of unwanted SbOCl.

## 20H IODOMETRIC METHODS OF ANALYSIS

As discussed in Chapter 12, Section 12C–2, iodometric methods of analysis are based upon the reducing properties of iodide ion:

$$2I^- \rightleftharpoons I_2 + 2e$$

The iodine produced is ordinarily titrated with a standard thiosulfate solution.

### 20H–1 Preparation of a 0.1 N Sodium Thiosulfate Solution

**Discussion**    See Chapter 12, Section 12C–2.

**Procedure**    Boil about 1 L of distilled water for 5 to 10 min. After it has cooled, add about 25 g of $Na_2S_2O_3 \cdot 5H_2O$ and 0.1 g of $Na_2CO_3$. Stir until the solid is dissolved. Transfer the solution to a clean glass or plastic bottle, and store in a dark place.

### 20H–2 Standardization of Sodium Thiosulfate Against Potassium Iodate

**Discussion**    The standardization is based upon titration of the iodine liberated when an excess of potassium iodide is added to an acidic solution that contains a known amount of potassium iodate (see Chapter 12, Section 12B–5).

**Procedure**    Dry primary-standard potassium iodate for at least 1 hr at 110°C, and cool in a desiccator. Weigh (to the nearest 0.1 mg) individual 0.15-g samples into 250-mL conical flasks (see Note). Dissolve each in about 75 mL of distilled water. *Treat each sample individually from this point to minimize errors resulting from the air oxidation of iodide.* Add about 2 g of iodate-free KI and swirl the flask to hasten solution. Then add about 2 mL of 6 *F* HCl, and titrate the liberated iodine immediately with thiosulfate until the color of the solution becomes pale yellow. Introduce 5 mL of starch indicator, and titrate to the disappearance of the blue color.

**NOTE**

Weighing uncertainties can be minimized by weighing (to the nearest 0.1 mg) 0.6 g of $KIO_3$ into a 250-mL volumetric flask, dissolving in distilled water, and diluting to the mark. The standardization is then performed with 50.00-mL aliquots of this solution.

## 20H-3 Standardization of Sodium Thiosulfate Against Copper

**Discussion**    Copper wire or foil is used advantageously for the standardization of thiosulfate solutions when the reagent is to be employed subsequently for copper determinations because any determinate error in the analysis tends to be canceled.

Copper(II) is quantitatively reduced to copper(I) by iodide ion:

$$2Cu^{2+} + 4I^- \rightarrow 2CuI(s) + I_2$$

Extensive studies of this reaction have revealed that quantitative reduction of the copper occurs only if the solution contains at least 4% excess iodide ion to force the reaction toward completion. Furthermore, the pH must be less than 4 to prevent the formation of basic copper complexes that react only slowly and incompletely with iodide ion. On the other hand, the hydrogen ion concentration cannot be greater than about 0.3 $M$ because copper-catalyzed air oxidation of iodide occurs at an appreciable rate in more acidic solutions.

Oxides of nitrogen also catalyze the air oxidation of iodide ion. A common source of these oxides is nitric acid, which is often used as a solvent for metallic copper and other copper-containing samples. A convenient way of removing nitrogen oxides is by the addition of urea to the solution:

$$(NH_2)_2CO + 2HNO_2 \rightarrow 2N_2 + CO_2 + 3H_2O$$

It has been found experimentally that the titration of iodine by thiosulfate in the presence of copper(I) iodide tends to yield slightly low results because small but appreciable quantities of iodine are physically adsorbed upon the solid. The adsorbed iodine is released only slowly, even in the presence of thiosulfate ion; transient and premature end points result. This difficulty is largely overcome by the addition of thiocyanate ion, which forms a sparingly soluble copper(I) salt. Part of the copper(I) iodide is converted to the corresponding thiocyanate at the surface of the solid:

$$CuI(s) + SCN^- \rightleftharpoons CuSCN(s) + I^-$$

Accompanying this reaction is the release of the adsorbed iodine, thus making it available for titration. Early addition of thiocyanate must be avoided, however, because that ion tends to reduce iodine slowly.

### Preparation of Solutions

*(a) 5% Urea (w/v).*   About 10 mL are needed for each titration.

*(b) Starch indicator solution.*   See page 591.

**Procedure**   Use scissors to cut copper wire or foil into 0.20- to 0.25-g portions. Wipe the metal free of dust and grease with a piece of filter paper; avoid directly touching the metal henceforth. Use a tared watch glass (or weighing bottle) to obtain the weight (to the nearest 0.1 mg) of each sample. Transfer each to a 250-mL conical flask, add 5 mL of 6 $F$ $HNO_3$, and cover with a small watch glass. Warm the mixture (Hood) until solution is complete. Add 25 mL of distilled water and 10 mL of 5% (w/v) urea, and boil briefly to eliminate oxides of nitrogen. Cool.

Make dropwise additions of concentrated $NH_3$ with thorough mixing to produce the intensely blue tetraammine–copper(II) complex; the solution will smell faintly of $NH_3$. Introduce 3 $F$ $H_2SO_4$ dropwise until the color of the complex just disappears, then add 2.0 mL of 85% $H_3PO_4$. Cool to room temperature.

*Treat each sample individually from this point on to minimize the air oxidation of iodide ion.* Add 4.0 g of KI to the sample and titrate immediately with $Na_2S_2O_3$ until the solution becomes pale yellow. Add 5 mL of starch indicator and titrate until the blue begins to fade. Add 2 g of KSCN; swirl vigorously for a few seconds. Complete the titration, using the disappearance of blue starch–$I_2$ color as the end point. Calculate the normality of the $Na_2S_2O_3$ solution.

## 20H–4 Determination of Copper in Brass

**Discussion**   The standardization procedure described in Section 20H–3 is readily adapted to the determination of copper in brass, an alloy that also contains appreciable amounts of tin, zinc, and lead (and perhaps minor amounts of iron and nickel). The method is relatively simple and applicable to brasses containing less than 2% iron. The sample is dissolved with nitric acid, which causes tin to precipitate as the hydrated oxide (known also as metastannic acid). Evaporation and fuming with sulfuric acid eliminates the excess nitrate, redissolves the metastannic acid, and often causes part of the lead to precipitate as the sulfate. The pH is then adjusted through the addition of ammonia, followed by acidification with a measured amount of phosphoric acid. An excess of potassium iodide is added and the liberated iodine is titrated with thiosulfate. See Section 20H–3 for additional discussion.

**Procedure**    If so directed, free the metal of oils by treatment with an organic solvent; briefly heat in an oven to drive off the solvent. Weigh (to the nearest 0.1 mg) 0.3-g samples into 250-mL conical flasks, and introduce 5 mL of 6 $F$ $HNO_3$. Warm (Hood) until solution is complete. Then add 10 mL of concentrated $H_2SO_4$ and evaporate (Hood) until copious white fumes of $SO_3$ are observed. Allow the mixture to cool. Carefully add 30 mL of distilled water, boil for 1 to 2 min, and cool.

Add concentrated $NH_3$ dropwise until the first dark blue color of the tetraammine–copper(II) complex appears; the solution should smell faintly of $NH_3$. Add 3 $F$ $H_2SO_4$ dropwise until the color of the complex just disappears. Add 2.0 mL of 85% $H_3PO_4$ and cool to room temperature.

*Treat each sample individually from this point.* Add 4.0 g of KI, and titrate immediately with standard 0.1 $N$ $Na_2S_2O_3$ until the iodine color is no longer distinct. Add 5 mL of starch solution, and titrate until the blue color starts to fade. Add 2 g of KSCN, swirl vigorously for several seconds, and complete the titration. Report the percent Cu in the brass sample.

### 20H–5 Determination of Dissolved Oxygen by the Winkler Method

**Discussion**    The *Winkler method* for the determination of dissolved oxygen involves treating the sample with an excess of manganese(II), potassium iodide, and sodium hydroxide. The white manganese(II) hydroxide produced reacts rapidly with oxygen to form brown manganese(III) hydroxide; that is,

$$4Mn(OH)_2(s) + O_2 + 2H_2O \rightarrow 4Mn(OH)_3(s)$$

Upon being acidified, manganese(III) oxidizes iodide to iodine:

$$2Mn(OH)_3(s) + 2I^- + 6H^+ \rightarrow I_2 + 3H_2O + 2Mn^{2+}$$

The success of the method is critically dependent upon the manner in which the sample is manipulated; every effort must be made at all stages to ensure that oxygen is neither introduced to nor lost from the sample. Biological Oxygen Demand (BOD) bottles are designed to minimize the entrapment of air, but ordinary glass-stoppered 250-mL bottles can also be used.

The sample must be free of any solutes that will either oxidize iodide or reduce iodine. Numerous modifications have been devised to permit application of the Winkler method to such samples.

#### Preparation of Solutions

*(a) Manganese(II) sulfate reagent.*    Dissolve 48 g of $MnSO_4 \cdot 4H_2O$ in sufficient water to give 100 mL of solution.

*(b) Potassium iodide–sodium hydroxide.*    Dissolve 15 g of KI in about

25 mL of distilled water, add 66 mL of 50% NaOH, and dilute to 100 mL. (*CAUTION!* Concentrated NaOH is highly corrosive to the skin. Wash exposed areas *immediately* with copious amounts of water.)

*(c) Sodium thiosulfate, 0.025 N.* Follow the directions in Section 20H–1, using about 6.2 g of $Na_2S_2O_3 \cdot 5H_2O$ to prepare 1 L of solution.

*(d) Starch indicator solution.* See page 591.

**Procedure** Fill the BOD bottle with the sample, taking care to minimize exposure to air. For water from a pressurized source, use a tube to introduce the water near the bottom of the bottle. When the water overflows, slowly remove the tube, and immediately add 1 mL of the $MnSO_4$ reagent with a dropper; discharge the reagent well below the surface (some overflow will occur). Similarly, introduce 1 mL of the KI–NaOH solution (see Note 1). Stopper the bottle, taking care that no air is entrapped. Invert the bottle to distribute the precipitate uniformly.

After the precipitate has settled at least 3 cm below the stopper, introduce 1 mL of 18 $F$ $H_2SO_4$, again, well below the surface (see Note 1). Replace the stopper and mix until the precipitate dissolves (see Note 2). Use a graduated cylinder to transfer exactly 200 mL of the acidified sample to a 500-mL conical flask (see Note 3). Titrate promptly with 0.025 $N$ $Na_2S_2O_3$ until the iodine color becomes faint. Then add 5 mL of starch indicator and complete the titration. Report the milliliters of oxygen (STP) in each liter of sample (see Note 4).

**NOTES**
1. Care should be taken to avoid exposure to the overflow, as the solution is quite alkaline.
2. A magnetic stirrer can be used to hasten solution of the precipitate.
3. The accuracy demands of the analysis are ordinarily such that careful measurement of the sample volume with a graduated cylinder is sufficient. Moreover, correction for the volume of the added reagents is unnecessary.
4. Suggested samples include water taken from different depths in a lake (particularly in summer), or from above and below rapids in a stream. Samples drawn sequentially from a little-used faucet should also show interesting differences in oxygen content.

## 20I TITRATIONS WITH POTASSIUM BROMATE

The properties of potassium bromate as an analytical reagent are described in Chapter 12, Section 12B–6. Directions follow for the determination of ascorbic acid in vitamin C tablets.

### 20I–1 Preparation of a Standard 0.05 N Potassium Bromate Solution

**Procedure**  Transfer about 0.8 g of reagent-grade potassium bromate to a weighing bottle, dry at 110 to 120°C for at least 1 hr, and cool in a desiccator. Weigh (to the nearest 0.1 mg) 0.7 g into a 500-mL volumetric flask, using a powder funnel to ensure quantitative transfer. Rinse down the funnel, dissolve the solid, dilute to the mark, and mix well.

### 20I–2 Standardization of 0.05 N Sodium Thiosulfate Against Potassium Bromate

**Discussion**  Iodine is generated by the reaction between a known volume of potassium bromate and an unmeasured excess of potassium iodide:

$$BrO_3^- + 6I^- + 6H^+ \rightarrow Br^- + 3I_2 + 3H_2O$$

The iodine produced is then titrated with the sodium thiosulfate solution.

**Preparation of Solutions**

(a) *Sodium thiosulfate, 0.05 N.*  Follow the directions in Section 20H–1, using about 12.5 g of $Na_2S_2O_3 \cdot 5H_2O$ per liter.
(b) *Starch indicator.*  See page 591.

**Procedure**  Pipet 25.00-mL aliquots of the $KBrO_3$ solution into conical flasks and rinse down the walls. *Treat samples individually from this point.* Introduce 2 to 3 g of KI and 5 mL of 3 $F$ $H_2SO_4$. Immediately titrate with the $Na_2S_2O_3$ solution until the solution is a pale yellow. Then add 5 mL of starch indicator and titrate to the disappearance of the blue color. Calculate the normality of the thiosulfate solution.

### 20I–3 Determination of Ascorbic Acid in Vitamin C Tablets by Titration with Potassium Bromate

**Discussion**  Ascorbic acid, $C_6H_8O_6$, is cleanly oxidized to dehydroascorbic acid by bromine:

An unmeasured excess of potassium bromide is added to an acidified solution of the sample. The solution is titrated with standard potassium bromate to the first appearance of bromine; the excess is then determined iodometrically with standard sodium thiosulfate. The analysis must be performed without delay to prevent air oxidation of the ascorbic acid.

---

**Preparation of Reagents**

*(a) Starch indicator.* See page 591.

**Procedure**  Weigh (to the nearest milligram) 3 to 5 vitamin C tablets (see Note 1). Pulverize thoroughly in a mortar and transfer the powder to a dry weighing bottle. Weigh (to the nearest 0.1 mg) 0.20- to 0.25-g samples into dry 250-mL conical flasks. *Treat samples individually from this point.* Dissolve (see Note 2) the sample in a solution consisting of 10 mL of 3 $F$ $H_2SO_4$, 25 mL of distilled water, and 5 g of KBr. Titrate immediately with standard $KBrO_3$ to the first faint yellow due to excess $Br_2$. Record the volume of oxidant, add 5 g of KI, 5 mL of starch indicator, and back-titrate (see Note 3) with standard 0.05 $N$ $Na_2S_2O_3$. Calculate the average weight (in milligrams) of ascorbic acid (gfw = 176.1) in each tablet.

**NOTES**
1. This method is not applicable to chewable vitamin C tablets.
2. The binder in many vitamin C tablets will remain in suspension throughout the titration. If the binder is starch, the characteristic color with iodine will be observed upon addition of KI.
3. The volume needed for back-titration will typically be no more than a few milliliters.

---

## 20J TITRATIONS WITH POTASSIUM IODATE

The oxidizing properties of potassium iodate are discussed in Chapter 12, Section 12B–5. Directions follow for the preparation of a standard iodate solution, its use as a primary standard for sodium thiosulfate solutions, and its application to the titration of an iodine–iodide mixture.

### 20J–1 Preparation of a Standard 0.020 F Potassium Iodate Solution

---

**Procedure**  Dry reagent-grade $KIO_3$ at 100 to 110°C for at least 1 hr; cool in a desiccator. Use a powder funnel to weigh (to the nearest milligram) about 2.1 g into a 500-mL volumetric flask. Dissolve the solid, dilute to the mark, and mix thoroughly.

## 20J–2 Standardization of Sodium Thiosulfate Against 0.020 F Potassium Iodate

**Discussion**  Standardization of thiosulfate is accomplished by titration of the iodine produced when an unmeasured excess of potassium iodide is added to a known volume of the potassium iodate solution. Reaction:

$$IO_3^- + 5I^- + 6H^+ \rightarrow 3I_2 + 3H_2O$$

Note that each iodate results in the formation of three molecules of iodine. In this application, then, the equivalent weight of potassium iodate is one-sixth of its gram-formula weight; that is,

$$IO_3^- \equiv 3I_2 \equiv 6S_2O_3^{2-} \equiv 6e$$

and the normality of the iodate solution is six times its formal concentration.

---

**Procedure**  Pipet 25.00-mL aliquots of the standard iodate solution into 250-mL conical flasks; to each, add about 50 mL of water and 2 g of iodate-free KI. *Treat each sample individually from this point.* Introduce about 2 mL of 6 F HCl and titrate with $Na_2S_2O_3$ until the color becomes pale yellow. Add about 5 mL of starch indicator, and titrate to the disappearance of the blue color. Calculate the normality of the thiosulfate solution.

---

## 20J–3 Determination of Iodine and Iodide in an Aqueous Mixture

**Discussion**  The iodine in an aliquot of the mixture is determined by direct titration with a standard sodium thiosulfate solution. The concentration of iodine plus iodide is then determined with standard potassium iodate. This titration is performed in a solution that contains a sufficiently high concentration of chloride ion so that the reaction product is exclusively $ICl_2^-$:

$$IO_3^- + 2I_2 + 10Cl^- + 6H^+ \rightarrow 5ICl_2^- + 3H_2O$$

$$IO_3^- + 2I^- + 6Cl^- + 6H^+ \rightarrow 3ICl_2^- + 3H_2O$$

Note that the oxidation state of iodine in $ICl_2^-$ is $+1$. Therefore, the equivalent weight of $KIO_3$ is one-fourth its formula weight in this reaction, while the equivalent weights of $I_2$ and KI are one-half of their formula weights. Thus, the equivalent weight of iodine is the same when it reacts with thiosulfate ion and with iodate ion in spite of the fact that it is reduced in one reaction and oxidized in the other.

Starch ceases to function as an indicator in the high concentration of hydrochloric acid that is needed for this titration. Instead, the solution is shaken with a small portion of chloroform. Iodine is extracted from the aqueous solution and imparts an intense violet color in the nonaqueous layer. The end point is taken as the volume of iodate needed to cause the color to just disappear. There is a tendency to overtitrate with iodate, owing to the tenacity with which iodine is held by the chloroform. For this reason, it is prudent to have the sample in one buret, the iodate in another, and to titrate back and forth until one drop of the sample causes formation of the color and one drop of iodate discharges it.

### Preparation of Solutions

*(a) Sodium thiosulfate, 0.05* N.   Follow the directions in Sections 20H–1 and 20H–2; use about 12.5 g of $Na_2S_2O_3 \cdot 5H_2O$ per liter of solution.

*(b) Potassium iodate, 0.05* F.   See Section 20K–1. Weigh about 5.4 g (to the nearest milligram) of $KIO_3$ into a 500-mL volumetric flask. Dilute an aliquot of this solution tenfold for standardization of the thiosulfate.

*(c) Starch indicator.*   See page 591.

**Procedure**   Obtain the unknown in a clean, 500-mL volumetric flask. Dilute to the mark and mix thoroughly.

*(a) Determination of iodine.*   Transfer 50.00-mL aliquots of the unknown to conical flasks. Add no more than 1 mL of 3 $F\,H_2SO_4$. Titrate promptly with standard $Na_2S_2O_3$; swirl the solution constantly during the titration. When the iodine color becomes faint, add 5 mL of starch indicator and complete the titration.

*(b) Determination of iodine plus iodide.*   Fill one buret with the unknown solution; fill another with standard 0.05 $F\,KIO_3$. Run about 25 mL of unknown into a 250-mL conical flask and cool in ice water. Add about 40 mL of concentrated HCl and titrate to the disappearance of the brown iodine color (see Note 2). Introduce about 10 mL of chloroform and add sufficient additional unknown to cause the nonaqueous layer to become purple. Titrate, as before, with $KIO_3$. Repeat, as necessary, to establish the end point (see Note 3). Report the weight-volume percentages of $I_2$ and KI in the sample.

### NOTES

1. A stock $I_2$–KI solution can be prepared by dissolving 110.0 g of $I_2$ and 360.0 g of KI in a minimum volume of water and subsequently diluting to 1.000 L. Individual samples will require 20 to 27 mL of this solution.
2. The residual color of $ICl_2^-$ is yellow.
3. There is a tendency to overtitrate with $KIO_3$, owing to the relative difficulty with which $I_2$ is removed from the chloroform layer. Vigorous swirling is essential. At the end point, introduction of 1 drop of unknown should cause the nonaqueous layer to become faintly purple, and 1 drop of $KIO_3$ should discharge the color.

## 20K POTENTIOMETRIC METHODS

The uses of potential measurements in analytical chemistry are discussed in Chapter 13. A general approach for the performance of a potentiometric titration is given here, along with directions for typical applications. A procedure is also given for the direct potentiometric determination of fluoride ion with a specific-ion electrode.

### 20K – 1 General Instructions for Performing Potentiometric Titrations

1. Dissolve the sample in 50 to 250 mL of water. Rinse the electrodes with distilled water, and immerse them in the sample solution. Provide magnetic (or mechanical) stirring. Position the buret so that reagent can be delivered without splashing.
2. Connect the electrodes to the meter and commence stirring. Record the initial potential as well as the initial buret volume.
3. Measure and record the potential and the volume after each addition of reagent. Introduce fairly large volumes (say, 5 mL) at the outset. Judge the volume of reagent to be added by calculating an approximate value of $\Delta E/\Delta V$ after each addition. Allow time for the potential to become constant within 1 to 2 mV (or 0.05 pH unit). A stirring motor will occasionally cause erratic potential readings; it may be advisable to turn off the motor during the actual measuring process. As the equivalence point is approached, make additions in exact 0.1-mL increments. Continue the titration 2 to 3 mL beyond the equivalence point; increase the volumes added as $\Delta E/\Delta V$ once again acquires small values.
4. Locate the end point by any of the methods described in Chapter 13, Section 13E – 1.

### 20K – 2 Potentiometric Determination of Chloride and Iodide in a Mixture

**Discussion**    The titration of halide mixtures is discussed in Chapter 7, Section 7A – 3. The indicator electrode can be a polished silver wire or a commercial billet type. A calomel electrode can serve as reference, although diffusion of chloride ion from the salt bridge may cause results that are slightly high. This problem can be eliminated by locating the calomel electrode in a saturated solution of potassium nitrate and using a $KNO_3$ bridge to make contact with the sample solution. Commercial electrodes with nitrate bridges are also available. Alternatively, the solution can be made acidic with several drops of nitric acid; a glass electrode can then be used as the reference because the pH of the solution will remain essentially constant throughout the titration.

As shown by Figure 7 – 3 (p. 154), the potentiometric end point lends itself to the determination of the individual components in certain halide mixtures such as chloride and iodide ions. The potential measured in a potentiometric titration is directly proportional to the negative logarithm of the silver ion concentration; thus, the experimental curve relating potential to reagent volume has the same shape as Figure 7 – 3, although the ordinate units will be different.

### Preparation of Solutions

*(a) Silver nitrate.*   Prepare 250 mL of $AgNO_3$ according to the instructions in Section 20B – 1.

**Procedure**   The sample will be issued as a solution. Dilute to known volume, transfer aliquots to 250-mL beakers, and dilute with $100 \pm 10$ mL of water. Acidify with nitric acid and titrate (see Note 1) with standard silver nitrate as described in the General Instructions (Section 20K – 1). Use small increments of titrant in the end-point regions. Plot the data and establish the end point for each ion. Plot a theoretical titration curve, assuming that the measured concentrations of the two constituents are correct. Report the number of milligrams of $I^-$ and $Cl^-$ in the sample.

#### NOTES

1. If 0.01 *F* silver nitrate is to be used, a potassium nitrate bridge will be needed to isolate the calomel electrode from the sample. Bend an 8-mm glass tube into a U-shape with arms long enough to reach near the bottom of the beakers. Mix 1.8 g of agar with 50 mL of boiling water and stir to obtain a uniform suspension. Dissolve 12 g of $KNO_3$ in the hot suspension. Clamp the tube in an upright position; use a medicine dropper to fill it with the warm suspension. Cool until a gel is formed. When not in use, the bridge should be stored with each end immersed in 2.5 *F* $KNO_3$.
2. If so directed, transfer the precipitate and unused silver nitrate to a container designated for such wastes.

## 20K – 3  Potentiometric Titration of a Weak Acid

**Discussion**   A glass – calomel electrode system is convenient for locating end points in acid-base titrations, as well as establishing approximate values for the dissociation constants of weak acids and bases (see Section 13E – 4). For the latter application, it is necessary to calibrate the electrode system with a buffer of known pH (see Section 13D – 3).

### Preparation of Solutions

*(a) Standardized 0.1 N carbonate-free NaOH.*   See Sections 20C – 3 and 20C – 6.

*(b) Standard buffer solutions.*   Buffer solutions for calibrating glass-calomel electrode systems are available from chemical supply houses.

*(c) Phenolphthalein indicator solution.*   See Section 20C – 1.

**Procedure**   Dissolve between 2 and 4 meq of the acid in about 100 mL of distilled water (see Note). Rinse the electrodes with distilled water and immerse them in a standard buffer solution; follow the instrument instructions for adjusting the meter to read the pH of the buffer.

Again rinse the electrodes thoroughly and immerse them in the solution containing the sample. Add 1 or 2 drops of phenolphthalein. Titrate according to the General Instructions (Section 20K – 1). Some samples may contain more than one replaceable hydrogen; be alert for more than one break in the titration curve. Note the volume at which the indicator changes color.

Plot the titration data, and determine the end point (or end points); if so instructed, use the derivative method to establish the end point. Compare the volumes needed for the potentiometric and the indicator end points. Calculate the number of milliequivalents of replaceable hydrogen in the sample. Evaluate approximate dissociation constant(s) for the species titrated.

**NOTE**

Acids that have limited solubility can be dissolved in ethanol or an ethanol – water mixture. As received, the alcohol may be measurably acidic. Titration of a blank containing the same volumes of water and ethanol as were used to dissolve the sample is recommended.

## 20K – 4 Potentiometric Titration of the Species in a Phosphate Mixture

**Discussion**   The sample will be issued as a liquid and will contain HCl, $H_3PO_4$, $NaH_2PO_4$, $Na_2HPO_4$, $Na_3PO_4$, and NaOH, alone or in a compatible combination. The object is to establish which species are present and their amount(s).

The analysis of some mixtures can be completed by a titration with either standard acid or standard base. Others will require titration of separate aliquots, one with acid and the other with base. The initial pH and inspection of the titration curve for phosphoric acid (see Figure 8 – 9, p. 200) will establish whether a single titrant will suffice. A review of Chapter 8, Section 8G may assist in the decision.

### Preparation of Solutions

*Standard 0.1 N HCl and/or 0.1 N NaOH.*   See Sections 20C – 2 and 20C – 5, and/or Sections 20C – 3 and 20C – 6, as appropriate.

**Procedure**   Dilute the sample to 250 mL in a volumetric flask and mix well. Transfer a 50.00-mL aliquot of the sample to a beaker and determine its pH. Titrate with standard HCl or standard NaOH, as appropriate, until one or two end points have been passed; follow the General Instructions (Section 20K – 1). If necessary, titrate a second 50.00-mL aliquot with the reagent that was not used in the first titration. Repeat the titration(s) to obtain duplicate data.

Based upon the titration curve(s), identify the principal solute species in the sample that was issued, and report the formal concentration of each. In addition, calculate and report any dissociation constant(s) for $H_3PO_4$ that can be extracted from the titration data.

## 20K–5 Direct Potentiometric Determination of Fluoride Ion

**Discussion** The solid-state fluoride electrode (p. 346) has been widely used for the determination of fluoride ion in a variety of materials. In this section, methods are given for the determination of this ion in drinking water and toothpaste. For both, a total ionic strength adjustment buffer is added to the samples and standards to bring them to approximately the same ionic strength; in this way, the concentration rather than the activity of fluoride ion is determined. The pH of the buffer is about 5, a level at which fluoride ion is the predominant fluorine-containing species. The buffer also contains an amino-carboxylic acid complexing agent, cyclohexylenedinitrilotetraacetic acid, which forms stable complexes with heavy metal ions such as iron(III) and aluminum(III), thus freeing the fluoride ion from its complexes with these cations. Before undertaking this experiment, it is suggested that Chapter 13, Section 13D, on direct potentiometry be studied.

**Apparatus** The apparatus for this experiment will consist of a solid-state fluoride ion electrode, a saturated calomel electrode, and a pH meter. For the toothpaste analysis, a sleeve-type calomel electrode is required because the measurement is made upon a suspension that tends to clog the liquid junction. After each series of samples has been analyzed, the interface must be renewed by loosening the sleeve momentarily.

### Preparation of Reagents

*(a) Total ionic strength adjustment buffering solution (TISAB).* This solution can be purchased from a commercial source under the label TISAB.[4] It can be prepared by dissolving with stirring approximately 57 mL of glacial acetic acid, 58 g of NaCl, and 4 g of cyclohexylenedinitrilotetraacetic acid in 500 mL of distilled water in a 1-L beaker. Cool the beaker in a water or ice bath and carefully add 5 $M$ NaOH until the solution reaches a pH of 5.0 to 5.5 (glass electrode). Dilute to about 1 L, mix, and store in a stoppered plastic bottle.

*(b) 100-ppm fluoride standard solution.* Prepare a 100-ppm standard solution of $F^-$ by weighing into a 1-L volumetric flask 0.22 g (to the nearest milligram) of NaF (*CAUTION!* NaF is highly toxic) that has been dried for 2 hr at 110°C and cooled in a desiccator. Dissolve, dilute to the mark, mix well, and store in a plastic bottle. Calculate its exact concentration in parts per million of $F^-$. This solution is available commercially.

---

[4] Orion Research, Cambridge, MA.

**Procedure**

*(a) Determination of fluoride in drinking water.* Transfer 50.0-mL samples of the water to 100-mL volumetric flasks and dilute to volume with the TISAB solution. Prepare a 5-ppm F⁻ solution by diluting 25.0 mL of the 100-ppm solution to 500 mL in a volumetric flask. Transfer 5.00-, 10.0-, 25.0-, and 50.0-mL aliquots of the 5-ppm solution to 100-mL volumetric flasks, add 50 mL of the TISAB solution, and dilute to 100 mL (these solutions correspond to 0.5, 1.0, 2.5, and 5.0 ppm F⁻ in the sample). After thorough rinsing and drying with tissue, immerse the electrodes in the 0.5-ppm standard solution and stir mechanically for 3 min. Measure and record the potential. Repeat for each of the remaining standards and the samples. Construct a plot of the potential versus the log of the concentration of the standards. Calculate and report the slope of this plot. Determine the parts per million of F⁻ in the unknown samples from the measured potentials.

*(b) Determination of fluoride in a sample of toothpaste.*[5] Weigh (to the nearest 1 mg) approximately 0.2-g samples of the toothpaste into 250-mL beakers. Add 50 mL of the TISAB solution, and boil for 2 min with good mixing. After cooling, transfer the suspension quantitatively to a 100-mL volumetric flask, and dilute to the mark. Follow the directions beginning with the second sentence in the preceding paragraph. Report the percent F⁻ in the sample.

## 20L  ELECTROGRAVIMETRIC ANALYSIS

The general aspects of electrogravimetric analysis are discussed in Chapter 14. Directions follow for the simultaneous determination of copper and lead in a sample of brass.

### 20L–1  Determination of Copper and Lead in a Sample of Brass

**Discussion**    This procedure is based upon the deposition of Cu at a cathode and $PbO_2$ at an anode. Before deposition, it is necessary to separate by filtration the hydrous oxide $SnO_2 \cdot xH_2O$ that forms when the sample is dissolved in $HNO_3$. Lead oxide forms quantitatively at an anode in the presence of high concentrations of nitrate ions. In order to achieve quantitative deposition of the Cu, it is necessary to remove the excess nitrate ion after deposition of the $PbO_2$ is complete. Addition of urea lowers the nitrate ion concentration by the reaction

$$6NO_3^- + 6H^+ + 5(NH_2)_2CO \rightleftharpoons 8N_2 + 5CO_2 + 13H_2O$$

---

[5] This procedure was taken from T. S. Light and C. C. Cappuccino, *J. Chem. Educ.*, **1975**, *52*, 247.

**Procedure** It is not necessary to dry the unknown. A rinse with acetone is recommended if there is evidence of oil on the surface of the metal. Weigh (to the nearest 0.1 mg) approximately 1-g samples into 250-mL beakers. Cover the beakers with watch glasses and introduce about 20 mL of 8 $F$ HNO$_3$. Digest for at least 30 min; add more HNO$_3$ if necessary. Evaporate to about 5 mL, but never to dryness.

Add about 5 mL of 3 $F$ HNO$_3$, 25 mL of water, and about one-quarter tablet of filter paper pulp; digest without boiling for about 45 min. Use fine-porosity filter paper to remove the SnO$_2 \cdot x$H$_2$O (see Note 1) and collect the filtrate in tall-form electrolysis beakers. Use many small washes with hot 0.3 $F$ HNO$_3$ to remove the last traces of copper; test for completeness of removal with a few drops of NH$_3$. The final volume should be between 100 and 125 mL. If necessary, evaporate to attain this volume.

Clean the electrodes, if necessary, with warm HNO$_3$; rinse in distilled water followed with acetone (see Note 2). Dry by heating at 110°C for a few minutes, cool in air, and weigh. With the current switch off, attach the larger electrode to the negative terminal and the stirrer electrode to the positive terminal. Briefly turn on the stirring motor to be certain that the electrodes do not touch. Cover the beakers with split watch glasses, and electrolyze with a current of 1.3 A for 35 min.

Rinse off the cover glasses, add 10 mL of 3 $F$ H$_2$SO$_4$ and 0.5 g of urea to each beaker. Maintain a current of 2.0 A until the solutions are colorless. Test a few drops of the solution for completeness of deposition with NH$_3$. If necessary, return the test solution to the beaker and continue the electrolysis for an additional 10 min. Repeat the test until no blue Cu(NH$_3$)$_4^{2+}$ is produced.

When electrolysis is complete, discontinue stirring, but leave the current on (see Note 3). Rinse the electrodes with distilled water as they are removed from the solution. After rinsing is complete, turn off the current, remove the electrodes, and dip them in clean acetone. Dry the cathodes for about 3 min, and the anodes for about 15 min at 110°C. Cool the electrodes and weigh.

Report the percentages of lead (see Note 4) and copper in the brass.

**NOTES**
1. If desired, the tin content can be determined by ignition of the SnO$_2 \cdot x$H$_2$O to SnO$_2$ at 900°C.
2. Grease and organic material, if present, can be removed by bringing the electrode to red heat in a flame. The electrode surfaces should not be touched with the fingers because grease and oil cause nonadherent deposits.
3. It is important to maintain the potential until the electrodes are clear of the solution and washed free of acid; otherwise, some Cu may redissolve.
4. Experience has shown that a small amount of moisture is retained by the PbO$_2$, and that better results are obtained if 0.8643 is used instead of 0.8660, the stoichiometric factor.

## 20M COULOMETRIC TITRATIONS

### 20M – 1 Coulometric Determination of Cyclohexene

**Discussion**[6]   In a largely nonaqueous environment and in the presence of mercury(II) as a catalyst, most olefins react rapidly enough with bromine to make their direct titration feasible. A convenient way of performing this titration is to introduce an excess of bromide ions and generate the bromine at an anode connected to a constant current source. The electrode processes are

$$2Br^- \rightleftharpoons 2Br_2 + 2e$$

$$2H^+ + 2e \rightleftharpoons H_2$$

The hydrogen does not react with bromine rapidly enough to constitute an interference. The bromine reacts with an olefin such as cyclohexene to give the addition product; that is,

The amperometric method with twin-polarized electrodes (see Chapter 14, Section 14F – 2) provides a convenient way to determine the end point for this titration. Here, two smaller platinum electrodes are maintained at a potential difference of 0.2 to 0.3 V. This potential is not sufficient to cause the generation of bromine and hydrogen. Thus, at the outset the indicator electrodes are polarized and no current is observed. The first excess of bromine depolarizes the system, giving a current that is proportional to the bromine concentration. The indicator electrode reactions are then

$$2Br^- \rightleftharpoons Br_2 + 2e$$

$$Br_2 + 2e \rightleftharpoons 2Br^-$$

The current is readily measured with a microammeter.

A convenient way to perform replicate analyses is to generate sufficient bromine in a blank to give a readily measured current, say, 20.0 $\mu$A. Upon introduction of the sample, bromine is consumed and the current immediately decreases and approaches zero. Generation is again commenced, and the time needed to regain a current of 20.0 $\mu$A is measured. A second portion of the sample is then added to the same solution and the process is repeated. Several samples can thus be analyzed without changing the solvent system.

The procedure that follows has been designed for the determination of the cyclohexene content of a methanol solution. It can be applied to numerous other olefins.

---

[6] This procedure was described by D. H. Evans, *J. Chem. Educ.,* **1968,** *45,* 88.

### Preparation of Reagents

*(a) Solvent.* Mix 300 mL of glacial acetic acid, 130 mL of methanol, and 65 mL of water; dissolve 9 g of KBr and 0.5 g of mercury(II) acetate in this mixture. (*CAUTION!* Mercury compounds are highly toxic and the solvent is a skin irritant. If inadvertent contact with the solvent occurs, wash the affected area with copious amounts of water.)

*(b) Methanol.*

**Procedure** Obtain the sample in a 100-mL volumetric flask and dilute to a few milliliters below the mark with methanol. Before diluting and taking aliquots of this sample, be sure the temperature of the mixture is between 18 and 22°C. (The temperature coefficient of expansion is 0.11%/°C; thus, significant volumetric errors can result if temperature is not controlled.) Dilute to the mark and mix well.

Add sufficient acetic acid–methanol solvent to the electrolysis vessel to cover the indicator and generator electrodes. Apply about 0.2 V to the indicator electrodes. Activate the generator electrode system and generate bromine until a current of about 20 $\mu$A is indicated on the microammeter. Stop the generator, record the indicator electrode current to the nearest 0.1 $\mu$A, and set the timer to zero. Introduce a 10-mL aliquot of the sample with a transfer pipet; the indicator current should decrease and approach zero. Titrate the sample by turning on the generator. As the indicator current rises and approaches the previously recorded current, add the reagent in smaller and smaller increments by turning the generator on for shorter and shorter periods. When the original indicator current is reached, read and record the time. Then, reset the timer, add another 10-mL aliquot (or larger or smaller aliquots if the time for the first titration was too short or too long for convenient and accurate measurement). Titrate additional aliquots if desired. Report the number of milligrams of cyclohexene in the original sample.

## 20N VOLTAMMETRY

Various aspects of polarography and amperometric titrations were considered in Chapter 14, Section 14D. Typical applications of the procedures are found in this section. Details concerning operation of the instruments for these determinations are not given because the wide variety of equipment available makes it impossible to present meaningful general instructions.

### 20N–1 Polarographic Determination of Copper and Zinc in Brass

**Discussion** The polarographic procedure makes possible the rapid determination of Cu and Zn in brass. The accuracy, however, is substantially less than that attainable by classic methods of analysis for these elements.

The sample is dissolved in a minimum amount of nitric acid, which converts at least part of the tin in the sample to a precipitate of hydrous stannic oxide ($SnO_2 \cdot xH_2O$); this precipitate is not removed. The solution is then made basic with an ammonia–ammonium chloride buffer, which results in precipitation of lead as the basic oxide. The polarogram of the supernatant liquid consists of two waves for copper at $-0.2$ and $-0.5$ V versus the saturated calomel electrode (SCE); one corresponds to the reduction of $Cu^{2+}$ to $Cu^+$, and the second to the formation of Cu. The analysis is based upon the total diffusion current for the two waves. The zinc concentration is obtained from its wave at $-1.3$ V. For instruments equipped with a current compensation device, the Cu wave is first measured at the highest feasible sensitivity. Its wave is then suppressed by the compensator and the Zn wave is obtained at the highest possible sensitivity setting.

### Preparation of Reagents

*(a) Standard $2.5 \times 10^{-2}$ M $Cu^{2+}$.*  Cut pure copper wire into short pieces with scissors and weigh (to the nearest milligram) 0.4 g into a weighing bottle. Transfer to a small flask and dissolve in 5 mL of concentrated $HNO_3$. Boil to remove oxides of nitrogen, cool, dilute with water, and transfer quantitatively into a 250-mL volumetric flask. Dilute to the mark.

*(b) Standard $2.5 \times 10^{-2}$ M $Zn^{2+}$.*  Dry reagent-grade ZnO at 110°C, cool in a desiccator, and weigh (to the nearest milligram) about 0.5 g into a small beaker. Dissolve in 25 mL of water containing 5 mL of boiled, concentrated $HNO_3$. Transfer to a 250-mL volumetric flask and dilute to the mark.

*(c) 0.1% Gelatin.*  Dissolve about 0.1 g of gelatin in 100 mL of boiling water.

*(d) $NH_3$–$NH_4Cl$ buffer.*  Prepare 500 mL of a solution that is approximately 1 $M$ in $NH_3$ and 2 $M$ in $NH_4Cl$.

*(e) Concentrated $HNO_3$.*

**Procedure**  With a buret, transfer 0-, 1-, 8-, and 15-mL aliquots of the standard $Cu^{2+}$ solution to 50-mL volumetric flasks; add 5 mL of the gelatin solution and 30 mL of the buffer to each. Dilute to the mark. For each standard, rinse the polarographic cell with three small portions of the solution and then fill. Bubble nitrogen through the solution for 10 to 15 min to remove oxygen. Apply a potential of about $-1.6$ V (see Note), and adjust the sensitivity of the instrument so that the detector is at nearly full scale. Then obtain a polarogram by varying the potential between 0 and $-1.5$ V (versus the SCE). Evaluate the diffusion current by measuring the limiting current at a potential just beyond the second wave; measure the current for the blank at the same potential and calculate $i_d/C$.

Repeat the foregoing measurements with the standard Zn solution.

Analyze the brass sample by dissolving 0.10- to 0.15-g samples (weighed to the nearest 0.5 mg) in 2 mL of concentrated $HNO_3$ and boiling briefly to remove oxides of nitrogen. Cool, add 10 mL of water, and transfer quantitatively to a 50-mL volumetric flask. Dilute to the mark and mix well. Transfer a 10.0-mL aliquot to a 50.0-mL volumetric flask, add 5 mL of gelatin, 30 mL of buffer, and dilute to the mark. Obtain the polarogram in the same way as for the standards. Determine diffusion currents from the Cu and Zn waves and calculate the percent of each element in the sample.

**NOTE**
The polarogram for a blank, consisting of 0.6 $F$ $HNO_3$, should be obtained at the sensitivity setting used for the standard of lowest concentration.

### 20N – 2 Determination of Lead by an Amperometric Titration

**Discussion**   Amperometric titrations are discussed in Chapter 14, Section 14F – 1. In the procedure that follows, the lead concentration of an aqueous solution is determined by titration with a standard potassium dichromate solution. The reaction is

$$2Pb^{2+} + Cr_2O_7^{2-} + H_2O \rightarrow 2PbCrO_4(s) + 2H^+$$

The titration can be performed with a dropping electrode that is maintained at either 0 or $-1.0$ V (versus SCE). At 0 V, the current remains near zero until equivalence; then it rises rapidly as a consequence of reduction of the excess dichromate ion. At $-1.0$ V both dichromate and lead ions are reduced. Thus, the current decreases to a minimum and then rises as the equivalence point is passed. In principle, the titration error should be less with the V-shaped curve. The advantage of the titration at 0 V is that oxygen does not have to be removed.

**Preparation of Reagents**

   *(a) Buffered supporting electrolyte.*   Dissolve 10 g of $KNO_3$ and 8.2 g of sodium acetate in about 500 mL of water. Add glacial acetic acid until a pH of 4.2 is reached (pH meter). About 10 mL of the acid will be required.
   *(b) 0.1% (w/v) gelatin.*   Dissolve 0.1 g gelatin in 100 mL of boiling water.
   *(c) Standard 0.01 M $K_2Cr_2O_7$.*   Weigh (to the nearest milligram) 1.47 g of primary-standard grade $K_2Cr_2O_7$ into a 500-mL volumetric flask. Dissolve, dilute to the mark, and mix well.

**Procedure**   The titration can be carried out in a 100-mL beaker. A saturated $KNO_3$ salt bridge similar to the one described in Section 20K – 2, Note 1, should be used to provide contact between the saturated calomel electrode and the analyte solution.

Obtain the sample in a 100-mL volumetric flask, dilute to the mark, and mix well (the resulting solution will be 0.01 to 0.02 $M$ in $Pb^{2+}$). Transfer a 10.00-mL aliquot to the titration vessel, add 25 mL of the buffer and 5 mL of the gelatin. Determine the current at zero applied potential. Then add $K_2Cr_2O_7$ in 1-mL increments, measuring the current after each addition. Continue the additions to 5 mL beyond the equivalence point. Correct the currents for the volume change, and plot the data. Determine the end point and calculate the number of milligrams of Pb in the sample.

Repeat the titrations at $-1.0$ V. Here, however, it will be necessary to bubble nitrogen through the solution for 10 to 15 min before the titration and while additions of reagents are made. The flow of nitrogen must, of course, be stopped during the current measurement. Again, correct the currents for dilution, plot the data, establish the end point, and report the number of milligrams of Pb in the sample.

## 20D METHODS BASED ON MOLECULAR ABSORPTION SPECTROSCOPY

Molecular absorption methods are considered in Chapter 15. Laboratory directions follow for (1) cleaning and handling of spectrophotometric cells, (2) a determination of iron that requires the generation of a calibration curve, (3) a determination of manganese in steel that makes use of a standard addition, and (4) an experiment involving the spectrophotometric determination of pH with an acid-base indicator.

### 20D–1 Cleaning and Handling of Cells

It is apparent that accurate spectrophotometric analysis requires the use of good quality, matched cells. These should be regularly calibrated against one another to detect differences that can arise from scratches, etching, and wear. Equally important is proper cleaning of the outside windows of the cells just prior to their insertion into the photometer or spectrophotometer. Erickson and Surles[7] recommend cleaning the outer surfaces with a lens paper soaked in methanol. After wiping, the methanol is allowed to evaporate, leaving the cell surfaces free of contaminants. The authors showed that this method was far superior to the usual procedure of wiping the cell surfaces with a dry lens paper, which apparently leaves lint and a film on the surface.

### 20D–2 Determination of Iron in Water

**Discussion**  A sensitive method for determining the iron content of household or commercial water supplies is based upon the formation of the red-orange iron(II)–orthophenanthroline complex. Because orthophenanthroline is a weak base, it exists in an acid solution as the phenanthrolium ion $PhH^+$.

---

[7] J. O. Erickson and T. Surles, *Amer. Lab.,* **1976,** *8*(6), 50.

Complex formation is thus best described by the equation

$$Fe^{2+} + 3PhH^+ \rightleftharpoons Fe(Ph)_3^{2+} + 3H^+$$

The equilibrium constant for this reaction is $2.5 \times 10^6$ at $25°C$. Quantitative formation of the complex occurs in the pH range of 3 to 9. A pH of about 3.5 is ordinarily recommended to prevent the precipitation of various iron salts, such as phosphates.

An excess of a reducing agent, such as hydroxylamine or hydroquinone, is needed to maintain the iron in the $+2$ state. Once formed, the complex is very stable. The experiment can be performed with a spectrophotometer set at 508 nm or with a photometer equipped with a green filter. Consult the manu-facturer's instructions for operating details.

---

### Preparation of Reagents

*(a) Standard iron solution, 0.01 mg/mL.* Weigh (to the nearest milligram) 0.0702 mg of reagent-grade $Fe(NH_4)_2(SO_4)_2 \cdot 6H_2O$ into a 1-L volumetric flask. Dissolve in 50 mL of water that contains 1 to 2 mL of concentrated $H_2SO_4$. Dilute to the mark and mix well.

*(b) Hydroxylamine hydrochloride solution.* Dissolve 10 g of $H_2NOH$ in about 100 mL of distilled water.

*(c) Orthophenanthroline solution.* Dissolve 0.1 g of orthophenanthroline monohydrate in 100 mL of distilled water. Warm gently, if necessary, and stir well. Store in a dark place. The solution must be discarded if it darkens. (Each milliliter of this solution is sufficient for no more than about 0.09 mg of Fe.)

*(d) Sodium acetate, 1.2 F.* Dissolve about 10 g of NaOAc in about 100 mL of distilled water.

### Procedure

*(a) Preparation of a calibration curve.* Transfer a 25.00-mL aliquot of the standard iron solution to one 100-mL volumetric flask and about 25 mL of distilled water to a second. Add 1 mL of hydroxylamine solution, 10 mL of sodium acetate, and 10 mL of orthophenanthroline to each flask. Allow the mixtures to stand for about 5 min; then dilute to the mark.

Clean a pair of matched cells for the instrument. Rinse one with at least three small portions of the blank, and the other with the standard. Determine the absorbance of the standard with respect to the blank.

Repeat the above with three other volumes of iron, such that an absorbance range between about 0.1 and 1.0 is covered. Plot a calibration curve.

*(b) Analysis for iron.* Transfer a 10-mL aliquot of the sample to a 100-mL volumetric flask; treat in exactly the same way as the standards, measuring the absorbance with reference to the blank. Obtain replicate measurements after adjusting the sample volume so that the absorbance is between 0.1 and 1.0. Report the ppm of iron in the sample.

## 200–3 Determination of Manganese in Steel

**Discussion** Small quantities of manganese are readily determined colorimetrically by oxidation of Mn(II) to the intensely colored permanganate ion. Potassium periodate is an effective oxidizing reagent for this purpose:

$$5IO_4^- + 2Mn^{2+} + 3H_2O \rightarrow 5IO_3^- + 2MnO_4^- + 6H^+$$

Permanganate solutions that contain an excess of periodate are quite stable.

Interferences to this method are few. The existence of colored ions can be compensated for with a blank that contains an aliquot of the unoxidized sample. This method fails to correct for the effects of cerium(III) and chromium(III); both are oxidized to some extent by periodate, and their reaction products absorb in the region that is ordinarily used for the measurement of permanganate.

The accompanying method is applicable to steels that do not contain large amounts of chromium. The sample is dissolved in nitric acid. Any carbon that may be present is eliminated with peroxodisulfate. Phosphoric acid is added to complex iron(III) and prevent the color of this species from interfering with the analysis. The standard addition method (p. 447) is used to establish the relationship between absorbance and concentration.

A spectrophotometer set at 525 nm or a photometer with a green filter may be used for absorbance measurements. The operating manual for the instrument should be consulted for specific instructions.

### Preparation of Reagents

*(a) Standard manganese(II) solution.* Dissolve 0.1 g (to the nearest 0.1 mg) of manganese in about 10 mL of 6 $F$ $HNO_3$. Boil the solution gently to eliminate oxides of nitrogen. Cool; then transfer the solution quantitatively to a 1-L volumetric flask, and dilute to the mark. Each milliliter of this solution will cause an increase of about 0.09 in the absorbance of a total volume of 50.0 mL.

**Procedure** The steel sample does not require drying. If there is evidence of oil, a rinse with acetone, followed by brief drying, is recommended. Weigh (to the nearest 0.1 mg) duplicate (see Note 1) samples into 150-mL beakers. Introduce about 50 mL of 6 $F$ $HNO_3$ and boil gently; heating for about 5 min should suffice. Cautiously introduce about 1 g of ammonium peroxodisulfate, and boil gently for 10 to 15 min. If the solution is pink or contains brown $MnO_2$, add approximately 1 mL of $NH_4HSO_3$ (or about 0.1 g of $NaHSO_3$) and heat for an additional 5 min. Cool, and dilute each (see Note 2) to exactly 250 mL in a volumetric flask.

Pipet three 20.0-mL aliquots of each sample into small beakers. Treat as follows (see Note 3):

| | 85% H$_3$PO$_4$, mL | Std Mn Soln, mL | KIO$_4$, g |
|---|---|---|---|
| Aliqout 1 | 5 | 0.00 | 0.4 |
| Aliquot 2 | 5 | 5.00 | 0.4 |
| Aliquot 3 | 5 | 0.00 | 0.0 |

Boil each solution gently for 5 min, cool, and dilute to 50.0 mL in volumetric flasks. Measure the absorbance of aliquots 1 and 2, with aliquot 3 serving as the reference for setting zero absorbance. Report the percentage of manganese in the sample.

**NOTES**
1. The sample size will depend upon the manganese content; consult the instructor.
2. If turbidity exists, the solutions should be filtered as they are transferred to volumetric flasks.
3. The volume of the standard addition will be dictated by the absorbance of the sample. It may be advisable to get a rough estimate by generating permanganate in about 20 mL of the sample, diluting to about 50 mL, and measuring the absorbance.

## 200–4 Spectrophotometric Determination of the pH of a Buffer Mixture

**Discussion**   The absorptivities of the acid and conjugate base forms of an indicator are measured at two different wavelengths. The concentration of each form in an unknown buffer is determined at the two wavelengths (p. 419); the hydronium ion concentration can be calculated from these data. The accompanying directions make use of bromocresol green as the indicator; others are equally satisfactory.

The relationship between the two forms of bromocresol green in aqueous solution can be described by the equilibrium

$$\text{HIn} + \text{H}_2\text{O} \rightleftharpoons \text{In}^- + \text{H}_3\text{O}^+$$
$$\text{yellow} \qquad\qquad \text{blue}$$

for which

$$K_a = 1.6 \times 10^{-5} = \frac{[\text{H}_3\text{O}^+][\text{In}^-]}{[\text{HIn}]}$$

The spectrophotometric measurement of [In$^-$] and [HIn] permits the calculation of [H$_3$O$^+$].

**Preparation of Solution**

*(a) Bromocresol green.*    See Section 20C–1.

*(b) 0.4 F HCl.*    Dilute about 7 mL of concentrated HCl to approximately 200 mL.

*(c) 0.4 F NaOH.*    Dissolve about 3 g of NaOH in 200 mL of distilled or deionized water.

**Procedure**

*(a) Determination of individual absorption spectra.*    Transfer 25.0-mL aliquots of stock bromocresol green indicator to each of two 100-mL volumetric flasks. To one add 25.0 mL of 0.4 *F* HCl; to the other add 25.0 mL of 0.4 *F* NaOH. Dilute each to the mark; mix thoroughly. Obtain the absorption spectra for the acid and conjugate-base forms of the indicator between 400 and 600 nm, using water as a blank; record absorbance values at 10-nm intervals routinely and at smaller intervals as necessary to define maxima or minima. Evaluate the absorptivity for each species at wavelengths that correspond to their absorption maxima.

*(b) Determination of the pH of an unknown buffer.*    Transfer a 25.0-mL aliquot of the stock bromocresol green indicator to a 100-mL volumetric flask. Add 50.0 mL of the unknown buffer, dilute to the mark, and measure the absorbance of the diluted solution at the wavelengths for which absorptivity data were calculated. Report the pH of the buffer.

## 20P METHODS BASED ON ATOMIC ABSORPTION SPECTROSCOPY

Atomic absorption and emission spectroscopy are discussed in Chapter 16. This section contains two methods that illustrate the application of atomic absorption spectroscopy to the determination of lead.

### 20P–1 Determination of Lead in Brass

**Discussion**    Brass and other copper-based alloys contain from 0 to 10% lead as well as zinc, tin, and copper. Atomic absorption spectroscopy permits the rapid quantitative estimation of these elements. The weighed sample is dissolved in nitric acid that contains enough hydrochloric acid to prevent precipitation of the tin as metastannic acid. After suitable dilution, the sample is aspirated into the flame, and the various elements are determined using an appropriate hollow-cathode lamp for each constituent.

**Preparation of Reagents**

*Standard solution of Pb²⁺, 100 μg/mL.* Dry a quantity of reagent-grade $Pb(NO_3)_2$ for 1 hr at 110°C. Cool; then weigh 0.17 g (to the nearest 0.1 mg) into a 1-L volumetric flask. Dissolve in water containing 1 to 3 mL of concentrated $HNO_3$; dilute to the mark with distilled or deionized water.

**Procedure** Weigh duplicate samples of brass (see Note 1) into 100-mL beakers. Dissolve in a mixture consisting of about 4 mL of concentrated $HNO_3$, 4 mL of water, and 4 mL of concentrated HCl (see Note 2). Boil gently to remove oxides of nitrogen. Cool, transfer the solutions to 250-mL volumetric flasks, and dilute to the mark.

Use a buret to measure 0-, 5-, 10-, 15-, and 20-mL portions of the standard Pb solution into individual 50-mL volumetric flasks. Add 4 mL of concentrated $HNO_3$ and 4 mL of concentrated HCl to each, and dilute to the mark with distilled or deionized water. Transfer a 10-mL aliquot of each sample into a 50-mL volumetric flask and dilute to the mark.

Set the monochromator at 283.3 nm and measure absorbances for each standard and each sample at that wavelength. Take at least three readings for each sample and standard. Plot the data and determine the μg Pb²⁺/mL for each sample solution. Report the percentage of Pb in the sample.

**NOTES**
1. The amount of sample that will be convenient will depend upon the lead content of the brass as well as the sensitivity of the instrument to be used; 6 to 10 mg of Pb per sample is a reasonable amount. Consult the instructor.
2. Brasses that contain large amounts of tin will require additional HCl to prevent the formation of metastannic acid (p. 595). Upon prolonged standing the diluted samples may develop some turbidity; small amounts have no effect on the analysis for lead.

## 20P–2 Determination of Lead in a Pottery Glaze

**Discussion** Glazes used in the decoration of pottery may contain metals— notably lead, cadmium, and barium—that are harmful to humans. Atomic absorption provides an excellent means of determining whether such species can be leached from a glaze and thus become a health hazard. The procedure recommended by the Association of Official Analytical Chemists[8] calls for analysis of a dilute acetic acid solution that has been in contact with the glaze for 24 hr at room temperature.

[8] Association of Official Analytical Chemists, *Official Methods of Analysis,* 14th ed., W. Williams, Ed., p. 448. Arlington, VA: Association of Official Analytical Chemists, 1984.

**Preparation of Reagents**

*(a) Acetic acid, 4%.*  Dilute 80 mL of glacial acetic acid to 2 L with distilled water.

*(b) Standard lead solution.*  Dry a quantity of reagent-grade $Pb(NO_3)_2$ for 1 hr at 110°C. Cool; then weigh 0.17 g (to the nearest 0.1 mg) into a 1-L volumetric flask. Dissolve the solid and dilute to the mark with 4% acetic acid.

*(c) Working standards.*  Dilute 0, 5.00, 10.00, 15.00, and 20.00 mL of the standard lead solution to 100 mL with 4% acetic acid.

**Procedure**  Use a graduated cylinder to fill the specimen to be tested with 4% acetic acid to within $\frac{1}{4}$ in. of overflowing (measure this distance along the surface of the utensil, not the vertical distance). Record the volume of the acid. Cover the utensil to retard evaporation. Let stand for 24 hr at room temperature.

Set the monochromator to 283.3 nm and determine the absorbances for the sample and the standards. Calculate the micrograms of Pb extracted from the specimen.

## 20Q  SEPARATION OF CATIONS BY ION-EXCHANGE

In this experiment, nickel and zinc are separated on an anion-exchange column. The separated ions are then determined by titration with standard EDTA.

### 20Q–1  Separation of Nickel and Zinc

**Discussion**  The separation of the two ions is made possible by the fact that in 2 *F* HCl zinc is present as the stable anionic complexes $ZnCl_3^-$ and $ZnCl_4^{2-}$ that are retained on an anion-exchange column. Nickel, which does not form stable negatively charged complexes under these conditions, is not retained by the packing and can be washed through the column with the 2 *F* acid. The zinc is then eluted from the column with water, which decomposes the anionic zinc complexes. The positively charged zinc ions are then free to pass through the column.

**Preparation of Ion-Exchange Columns**  It is recommended that two columns be prepared so that duplicate samples can be treated simultaneously. Ion-exchange columns are typically 25 to 40 cm in length and 1 to 1.5 cm in diameter. A 50-mL buret may be used. Insert a plug of glass wool to retain the resin particles and fill the column to a height of about 10 cm with 50- to 100-mesh strong base anion exchange resin. Wash the column with about 50 mL of 6 *F* $NH_3$ followed by 100 mL of water and then 100 mL of the 2 *F* HCl. When washing is complete, leave about 1 cm of the HCl solution standing above the top of the resin. *At no time should the liquid level be allowed to drop below the top of the resin.*

**Procedure** Transfer the sample containing 2 to 4 mfw of Zn and of Ni quantitatively to a 100-mL volumetric flask, add 16 mL of concentrated HCl, and dilute to the mark with water. The resulting solution is approximately 2 $F$ in acid. With a volumetric pipet transfer a 10-mL aliquot onto the column. Place a 250-mL conical flask under the end of the column, and let the liquid drain until it barely covers the surface of the resin. Then rinse down the walls with 2 to 3 mL of the acid, and again lower the liquid level to the resin surface. Repeat this operation several times. Then elute the nickel with about 50 mL of 2 $F$ HCl using a flow rate of 2 to 3 mL/min. After elution is complete, place the flask containing the nickel chloride onto a hot plate or a steam bath for evaporation.

Elute the zinc ions into a 500 mL conical flask by passing about 100 mL of water through the column.

## 20Q – 2 Titration of Nickel and Zinc with EDTA

**Discussion** Both nickel and zinc are determined by titration with standard EDTA. The zinc titration is carried out in a solution buffered to pH 10; Eriochrome Black T serves as the indicator. The nickel containing eluate is evaporated to dryness to remove the excess HCl. The residue of nickel chloride is then dissolved in water and diluted with a pH 10 buffer. The resulting solution is titrated with EDTA with bromopyrogallol or murexide as the indicator.

**Preparation of Solutions**

(a) *Eriochrome Black T indicator.* See Preparation 20D – 1.
(b) *Bromopyrogallol red indicator.* Dissolve 0.05 g of the solid indicator in 100 mL of 50% (v/v) alcohol.
(c) *Murexide indicator.* The indicator is a solid that has been mixed with NaCl so that the indicator concentration is 0.2% by weight. Approximately 0.2 g of this mixture is used for each titration. The solid indicator is used because its solutions are quite unstable.
(d) *Standard 0.01 F EDTA.* See Preparation 20D – 2.
(e) *pH 10 buffer.* See Preparation 20D – 3.

**Titration of Nickel** Evaporate the solution containing the nickel to dryness on a hot plate, taking care to avoid heating the residual nickel chloride to a high enough temperature to cause decomposition to nickel oxide. Dissolve the nickel chloride residue in 100 mL of water and add 10 to 20 mL of the pH 10 buffer. Add 15 drops of the bromopyrogallol indicator (or 0.2 g of the murexide indicator). Titrate to the color change, which for the bromopyrogallol is blue to purple. For the murexide indicator the color changes from yellow of the nickel murexide complex to purple. Report the milligrams of nickel in the sample.

**Titration of Zinc**   Dilute the eluate to about 100 mL and add 10 to 20 mL of the pH 10 buffer. Add 1 to 2 drops of Eriochrome Black T indicator and titrate with the standard EDTA solution to a color change from red to blue. Report the milligrams of Zn in the sample.

## 20R GAS-CHROMATOGRAPHIC DETERMINATION OF ETHANOL IN BEVERAGES[9]

**Discussion**   This procedure provides a simple and rapid method for the quantitative determination of ethanol in aqueous solutions; the method is readily extended to the determination of the *proof* of alcoholic beverages. By definition, the proof of a beverage is two times its volume percent ethyl alcohol at 60°F.

The column used in the development of this procedure was a $\frac{1}{4}$ in. o.d. $\times \frac{1}{2}$ meter Poropak®T column containing an 80 to 100 mesh packing. Detection was by means of a thermal conductivity detector. Other columns would presumably provide equally satisfactory data. A flame ionization detector cannot be used, however, because of its insensitivity to water.

The analysis is based on a calibration curve in which the ratio of the area of the ethanol peak to the sum of the areas of the ethanol peak and the water peak is plotted as a function of the volume percent ethanol; that is,

$$\text{vol \% EtOH} = \frac{\text{vol EtOH}}{\text{vol soln}} \times 100$$

The resulting curve is not linear but instead is concave upward. At least two reasons can be found for the curvature. First, the thermal conductivity detector responds linearly to mass ratios and not volume ratios, and the two are not proportional to one another. Second, at the high concentrations involved, the volumes are not strictly additive as would be required for linearity. That is,

$$\text{volume alcohol} + \text{volume alcohol} \neq \text{volume mixture}$$

### 20R–1 Preparation of Standards

With a buret, measure 10.00, 20.00, 30.00, and 40.00 mL of absolute alcohol into 50-mL volumetric flasks (see Note). Dilute to volume with distilled water.

**NOTE**
Because the coefficient of thermal expansion of ethyl alcohol is roughly five times that of water, the temperature of solutions used in this experiment should be kept constant to $\pm 1$°C when volume measurements are being made.

---

[9] Adapted from J. J. Leary, *J. Chem. Educ.* **1983,** *60,* 675.

### 20R – 2 Operating Procedure

The following set of operating conditions have yielded satisfactory chromatograms for this analysis:

| | |
|---|---|
| Column temperature | 100°C |
| Detector temperature | 130°C |
| Injection port temperature | 120°C |
| Bridge current | 100 mA |
| Flow rate | 60 mL/min |

Inject a 1 $\mu$L sample of the 20% by volume standard and record the chromatogram. Obtain additional chromatograms, adjusting the recorder speed until the water peak is about 20 mm wide at half height. Then vary the volume of the sample injected and the attenuation until peaks with heights of at least 40 mm are obtained. Obtain chromatograms for the remainder of the standards (including pure water and pure ethanol) and the samples in the same way. Determine the area under each of the peaks and plot $\text{area}_{alc}/(\text{area}_{alc} + \text{area}_{H_2O})$ as a function of volume percent ethanol (include the data for 0 and 100% ethanol). Report the volume percent ethanol and proof for the unknowns.

Obtain several chromatograms for the 60% and 20% standards and determine the standard deviations for the areas as well as for the area ratios.

# Appendix 1
## Generation of Net-Ionic Equations

A net-ionic equation contains terms for only those species that are actually involved in a chemical reaction. The chemical formula used in a net-ionic equation is usually that of the species that is present in greatest amount. A sparingly soluble participant is symbolized by its chemical formula followed by (s). Similarly, a (g) is used to identify a gaseous species and (l) is used to indicate a liquid second phase. The symbol (aq) can be employed to indicate that the species is a solute in aqueous solution.

As with other chemical equations, a net-ionic equation must be balanced with respect to both quantity of material and charge.

**Examples**

1. Mixing solutions of silver nitrate and sodium carbonate results in the precipitation of silver carbonate. The net-ionic equation is

$$2Ag^+ + CO_3^{2-} \rightleftharpoons Ag_2CO_3(s)$$

   Terms for $Na^+$ and $NO_3^-$ do not appear because they are not involved in the process. Note also that $CO_3^{2-}$ is used in the equation as the principal carbonate-containing species, even though the solution contains small amounts of $HCO_3^-$ and $H_2CO_3$ as well.

2. Silver carbonate also is produced by mixing solutions containing silver nitrate and sodium hydrogen carbonate. Here, $HCO_3^-$ is the principal carbonate-containing species and the net-ionic equation is

$$2Ag^+ + HCO_3^- \rightleftharpoons Ag_2CO_3(s) + H^+$$

   Substitution of $H_3O^+$ for $H^+$ as a product of this reaction would require inclusion of $H_2O$ on the left to provide balance; the net-ionic equation would then be

$$2Ag^+ + HCO_3^- + H_2O \rightleftharpoons Ag_2CO_3(s) + H_3O^+$$

3. Silver carbonate is produced by bubbling carbon dioxide through a silver nitrate solution:

$$2Ag^+ + CO_2(g) + H_2O \rightleftharpoons Ag_2CO_3(s) + 2H^+$$

As in Example 2, three $H_2O$ would be needed on the left if $H_3O^+$ were to appear on the right.

4. When solutions containing iron(II) ammonium sulfate and potassium dichromate are mixed the products are $Fe^{3+}$ and $Cr^{3+}$. The net reaction involves only $Fe^{2+}$ and $Cr_2O_7^{2-}$ and occurs in acid solution:

$$6Fe^{2+} + Cr_2O_7^{2-} + 14H^+ \rightleftharpoons 6Fe^{3+} + 2Cr^{3+} + 7H_2O$$

Note that it was necessary to introduce $H^+$ and $H_2O$ to achieve balance.

# Appendix 2
## Use of Exponential Numbers

Scientists frequently find it necessary (or convenient) to make use of exponential notation to express numerical data. A brief review of this notation follows.

### EXPONENTIAL NOTATION

An exponent is used to describe the process of repeated multiplication or division. For example, $3^5$ means

$$3 \times 3 \times 3 \times 3 \times 3 = 3^5 = 243$$

The 5 is the exponent of the number (or base) 3; thus, 3 raised to the fifth power is equal to 243.

A negative exponent represents repeated division. For example, $3^{-5}$ means

$$\frac{1}{3} \times \frac{1}{3} \times \frac{1}{3} \times \frac{1}{3} \times \frac{1}{3} = \frac{1}{3^5} = 3^{-5} = 0.00412$$

Note that changing the sign of the exponent yields the *reciprocal* of the number; that is,

$$3^{-5} = \frac{1}{3^5} = \frac{1}{243} = 0.00412$$

It is important to note that a number raised to the first power is the number itself, and any number raised to the zero power has a value of 1. For example,

$$4^1 = 4$$

$$4^0 = 1$$

$$67^0 = 1$$

### Fractional Exponents

A fractional exponent symbolizes the process of extracting or taking the root of a number. The fifth root of 243 is 3; this process is expressed exponentially as

$$(243)^{1/5} = 3$$

Other examples are

$$25^{1/2} = 5$$

$$25^{-1/2} = \frac{1}{25^{1/2}} = \frac{1}{5}$$

## Combination of Exponential Numbers in Multiplication and Division

Multiplication and division of exponential numbers having the same base is accomplished by adding and subtracting the exponents. For example,

$$3^3 \times 3^2 = (3 \times 3 \times 3)(3 \times 3) = 3^{(3+2)} = 3^5 = 243$$

$$3^4 \times 3^{-2} \times 3^0 = (3 \times 3 \times 3 \times 3)(\tfrac{1}{3} \times \tfrac{1}{3}) \times 1 = 3^{(4-2+0)} = 3^2 = 9$$

$$\frac{5^4}{5^2} = \frac{5 \times 5 \times 5 \times 5}{5 \times 5} = 5^{(4-2)} = 5^2 = 25$$

$$\frac{2^3}{2^{-1}} = \frac{(2 \times 2 \times 2)}{1/2} = 2^4 = 16$$

Note that in the last example the exponent is given by the relationship

$$3 - (-1) = 3 + 1 = 4$$

## Extraction of the Root of an Exponential Number

To obtain the root of an exponential number, the exponent is divided by the desired root. Thus,

$$(5^4)^{1/2} = (5 \times 5 \times 5 \times 5)^{1/2} = 5^{(4/2)} = 5^2 = 25$$

$$(10^{-8})^{1/4} = 10^{(-8/4)} = 10^{-4}$$

$$(10^9)^{1/2} = 10^{(9/2)} = 10^{4.5}$$

The procedure that can be followed when the exponent is not evenly divisible by the root is shown in the next section.

## USE OF EXPONENTS IN SCIENTIFIC NOTATION

Scientists and engineers are frequently called upon to use very large or very small numbers for which ordinary decimal notation is either awkward or impossible. For example, to express Avogadro's number in decimal notation would require 21 zeros following the number 602. In scientific notation the number is written as a multiple of two numbers, the one number in decimal notation, and the other expressed as a power of 10. Thus, Avogadro's number is written as $6.02 \times 10^{23}$. Other examples are

$$4.32 \times 10^3 = 4.32 \times 10 \times 10 \times 10 = 4320$$

$$4.32 \times 10^{-3} = 4.32 \times \frac{1}{10} \times \frac{1}{10} \times \frac{1}{10} = 0.00432$$

$$0.002002 = 2.002 \times \frac{1}{10} \times \frac{1}{10} \times \frac{1}{10} = 2.002 \times 10^{-3}$$

$$375 = 3.75 \times 10 \times 10 = 3.75 \times 10^2$$

It should be noted that the scientific notation for a number can be expressed in any of several equivalent forms. Thus,

$$4.32 \times 10^3 = 43.2 \times 10^2 = 432 \times 10^1 = 0.432 \times 10^4 = 0.0432 \times 10^5$$

The number in the exponent is equal to the number of places the decimal must be shifted to convert a number from scientific to purely decimal notation. The shift is to the right if the exponent is positive and to the left if it is negative. The process is reversed when decimal numbers are converted to scientific notation.

## ARITHMETIC OPERATIONS WITH SCIENTIFIC NOTATION

The use of scientific notation is helpful in preventing decimal errors in arithmetic calculations. Some examples follow.

### Multiplication

Here, the decimal parts of the numbers are multiplied and the exponents are added; thus,

$$420,000 \times 0.0300 = (4.20 \times 10^5)(3.00 \times 10^{-2})$$
$$= 12.60 \times 10^3 = 1.26 \times 10^4$$

$$0.0060 \times 0.000020 = 6.0 \times 10^{-3} \times 2.0 \times 10^{-5}$$
$$= 12 \times 10^{-8} = 1.2 \times 10^{-7}$$

### Division

Here, the decimal parts of the numbers are divided; the exponent in the denominator is subtracted from that in the numerator. For example,

$$\frac{0.015}{5000} = \frac{15 \times 10^{-3}}{5 \times 10^3} = 3.0 \times 10^{-6}$$

## Addition and Subtraction

Addition or subtraction in scientific notation requires that all numbers be expressed to a common power of 10. The decimal parts are then added or subtracted, as appropriate. Thus,

$$2.00 \times 10^{-11} + 4.00 \times 10^{-12} - 3.00 \times 10^{-10} =$$
$$2.00 \times 10^{-11} + 0.400 \times 10^{-11} - 30.0 \times 10^{-11} = -27.6 \times 10^{-11}$$
$$= -2.76 \times 10^{-10}$$

## Raising a Number Written in Exponential Notation to a Power

Here, each part of the number is raised to the power separately. For example,

$$(2 \times 10^{-3})^4 = (2.0)^4 \times (10^{-3})^4 = 16 \times 10^{-(3 \times 4)}$$
$$= 16 \times 10^{-12} = 1.6 \times 10^{-11}$$

## Extraction of the Root of a Number Written in Exponential Notation

In this case, the number is written in such a way that the exponent of 10 is evenly divisible by the root. Thus,

$$(4.0 \times 10^{-5})^{1/3} = \sqrt[3]{40 \times 10^{-6}} = \sqrt[3]{40} \times \sqrt[3]{10^{-6}}$$
$$= 3.4 \times 10^{-2}$$

# Appendix 3
# Logarithms

In this discussion, we will assume that the reader has available an electronic calculator for obtaining logarithms and antilogarithms of numbers. It is desirable, however, to understand what a logarithm is as well as some of its properties. The discussion that follows provides this information.

A logarithm (or log) of a number is the power to which some base number (usually 10) must be raised in order to give the desired number. Thus, a logarithm is an exponent of the base 10. From the discussion in Appendix 2 with respect to exponents, we can draw the following conclusions with respect to logs:

1. The logarithm of a product is the sum of the logarithms of the individual numbers in the product.
2. The logarithm of a quotient is the difference between the logarithms of the individual numbers.
3. The logarithm of a number raised to some power is the logarithm of the number multiplied by that power.
4. The logarithm of a root of a number is the logarithm of that number divided by the root.

The following examples illustrate these statements:

$$\log (100 \times 1000) = \log 10^2 + \log 10^3 = 2 + 3 = 5$$

$$\log (100/1000) = \log 10^2 - \log 10^3 = 2 - 3 = -1$$

$$\log (1000)^2 = 2 \times \log 10^3 = 2 \times 3 = 6$$

$$\log (0.01)^6 = 6 \times \log 10^{-2} = 6 \times (-2) = -12$$

$$\log (1000)^{1/3} = \frac{1}{3} \times \log 10^3 = \frac{1}{3} \times 3 = 1$$

$$\log 40 \times 10^{20} = \log 4.0 \times 10^{21} = \log 4.0 + \log 10^{21}$$
$$= 0.60 + 21 = 21.60$$

$$\log 2.0 \times 10^{-6} = \log 2.0 + \log 10^{-6} = 0.30 + (-6)$$
$$= -5.70$$

For some purposes it is helpful to dispense with the subtraction step shown in the last example and report the log as a *negative* integer and a *positive* decimal number; that is,

$$\log 2.0 \times 10^{-6} = \log 2.0 + \log 10^{-6} = \overline{6}.30$$

The last two examples demonstrate that the logarithm of a number is the sum of two parts, a *characteristic* located to the left of the decimal point and a *mantissa* that lies to the right. The characteristic is the logarithm of 10 raised to a power and serves to indicate the location of the decimal point in the original number when that number is expressed in decimal notation. The mantissa is the logarithm of a number in the range between 0.00 and 9.99 . . . Note that the mantissa is *always positive.* As a consequence, the characteristic in the last example is $-6$ and the mantissa is $+0.30$.

# Appendix 4
## The Quadratic Equation

Problems involving chemical equilibrium frequently require solution of a quadratic equation. The expression in question is first written in the form $ax^2 + bx + c = 0$, where $a$, $b$, and $c$ are numerical constants. Numerical values for the constants are then substituted into the equation

$$x = \frac{-b \pm \sqrt{b^2 - 4ac}}{2a} \tag{1}$$

**Example** The hydronium ion concentration of a solution is given by the equation

$$2.0 \times 10^{-2} = \frac{4.0 \times 10^{-3}}{2[H_3O^+]} - [H_3O^+]$$

This expression is rearranged to

$$2[H_3O^+]^2 + 4.0 \times 10^{-2}[H_3O^+] - 4.0 \times 10^{-3} = 0$$

Substitution into Equation 1 gives

$$[H_3O^+] = \frac{-4.0 \times 10^{-2} \pm \sqrt{(4.0 \times 10^{-2})^2 - 4(2)(-4.0 \times 10^{-3})}}{2 \times 2}$$

$$= \frac{-4.0 \times 10^{-2} \pm \sqrt{0.16 \times 10^{-2} + 3.2 \times 10^{-2}}}{4}$$

$$= \frac{-4.0 \times 10^{-2} \pm 1.83 \times 10^{-1}}{4}$$

Of the two solutions to this equation, only the positive value has significance in the present context. We therefore conclude that

$$[H_3O^+] = \frac{-4.0 \times 10^{-2} + 18.3 \times 10^{-2}}{4}$$

$$= 3.6 \times 10^{-2}$$

# Appendix 5
## Solution of Higher Order Equations

Although methods exist for the rigorous solution of equations that contain terms raised to powers greater than 2, it is usually simpler to obtain a solution by means of systematic approximations. This approach entails the following steps: arrange the relationship such that all terms are collected on one side of the equal sign; substitute values for the unknown quantity; and take as the answer that value which brings the equation closest to zero.

**Example** The solubility, $s$, for a precipitate is given by the expression

$$s^2(1 \times 10^{-3} + s) = 2 \times 10^{-6}$$

As a first step, the equation is rearranged to

$$s^3 + 1 \times 10^{-3}s^2 - 2 \times 10^{-6} = 0$$

Because $s$ represents a solubility, we need to be concerned with positive roots only. If a value of 0.0 is assumed for $s$, the equation becomes

$$-2 \times 10^{-6} = 0$$

and if 1.0 is used

$$1 + 1 \times 10^{-3} - 2 \times 10^{-6} > 0$$

Clearly, $s$ will have a value between 0 and 1. Substitution of 0.1 for $s$ gives

$$1 \times 10^{-3} + 1 \times 10^{-5} - 2 \times 10^{-6} = 0.001008 > 0$$

With a trial value of 0.01,

$$1 \times 10^{-6} + 1 \times 10^{-7} - 2 \times 10^{-6} = -0.9 \times 10^{-6}$$

We can now assert that $s$ is smaller than 0.1 and larger than 0.01. Moreover, it appears that the latter represents the closer estimate.

Taking $s = 0.012$, we obtain

$$1.73 \times 10^{-6} + 1.44 \times 10^{-7} - 2 \times 10^{-6} = 1.87 \times 10^{-6} - 2 \times 10^{-6}$$

which is still slightly negative.

With $s = 0.013$,

$$2.20 \times 10^{-6} = 1.69 \times 10^{-7} - 2 \times 10^{-6} = 2.37 \times 10^{-6} - 2 \times 10^{-6} > 0$$

Thus, $s$ is somewhat larger than 0.012 and smaller than 0.013. Trial values between these bounds would refine the estimate further, if desired.

This process can be performed quite rapidly with a simple electronic calculator.

# Appendix 6

# Simplification of Equations Through Neglect of Terms

A term that consists of a sum or difference can frequently be simplified without serious error, provided the quantities in the term have sufficiently different magnitudes. For example, two possibilities exist for simplification of the term $(0.001 + x)$:

$$(0.001 + x) \cong x \qquad \text{provided } x \gg 0.001$$

or

$$(0.001 + x) \cong 0.001 \qquad \text{provided } x \ll 0.001$$

A knowledge of the physical significance of $x$ will frequently provide the basis for choice between these assumptions. Needless to say, the validity of either assumption will ultimately depend upon the numerical value of $x$ and the accuracy with which $x$ must be known.

*It is always worthwhile to make all simplifying assumptions that may seem reasonable. The answer obtained, however, must be considered provisional until it is confirmed that the assumption was indeed justified.*

**Example** Consider again the example used in Appendix 5:

$$s^2(1 \times 10^{-3} + s) = 2 \times 10^{-6}$$

If the assumption is made that $s$ is negligible with respect to 0.001, the equation simplifies to

$$1 \times 10^{-3} \, s^2 = 2 \times 10^{-6}$$

and

$$s = 4.5 \times 10^{-2}$$

Here, the value calculated for $s$ is larger than $1 \times 10^{-3}$; the original assumption was clearly unjustified.

The other possibility to consider is that 0.001 is small with respect to $s$. The equation now becomes

$$s^3 = 2 \times 10^{-6}$$

and

$$s = 1.26 \times 10^{-2}$$

This assumption provides a value for $s$ that is about 13 times greater than $1 \times 10^{-3}$. Whether the approximation is justified or not will depend upon the magnitude of the error that can be tolerated in the answer. If an uncertainty on the order of one part in ten is acceptable, this value for $s$ is satisfactory. If, on the other hand, a greater accuracy is required, it will be necessary to solve the unsimplified equation. It should be noted that the approximate value for $s$ obtained here ($1.26 \times 10^{-2}$) is an excellent starting point for a more rigorous solution by the method of successive approximations (Appendix 5).

# Appendix 7
# Designations and Porosities for Filtering Crucibles

| Type | Designation and Pore Size* | | | | |
|---|---|---|---|---|---|
| Glass, Pyrex®† | C (60) | M (15) | F (5.5) | | |
| Glass, Kimax®‡ | EC (170-220) | C (40-60) | M (10-15) | F (4-4.5) | VF (2-2.5) |
| Porcelain, Coors U.S.A.®§ | Medium (40) | Fine (5) | VF (1.2) | | |
| Porcelain, Selas‖ | XF (100) | XFF (40) | #10 (8.8) | #01 (6) | |
| Aluminum oxide, ALUNDUM®¶ | Extra Coarse (30) | Coarse (20) | Medium (5) | Fine (0.1) | |

\* Nominal maximum pore diameter in micrometers is given in parentheses.
† Corning Glass Works, Corning, N.Y.
‡ Owens-Illinois, Toledo, Ohio.
§ Coors Porcelain Company, Golden, Colo.
‖ Selas Corporation of America, Dresher, Pa.
¶ Norton Company, Worcester, Ma.

# Appendix 8
## Designations Carried by Ashless Filter Papers*

| Manufacturer | Fine Crystals | Moderately Fine Crystals | Coarse Crystals | | Gelatinous Precipitates | |
|---|---|---|---|---|---|---|
| Schleicher and Schuell† | 507, 590 589 blue ribbon | 589 white ribbon | 589 green ribbon | 589 black ribbon | 589 black ribbon | 589-1H |
| Munktell‡ | OOH | OK  OO | OOR | | OOR | |
| Whatman§ | 42 | 44, 40 | 41 | | 41 | 41H |
| Eaton-Dikeman‖ | 94 | 74 | | | 54 | |

* Manufacturers' literature should be consulted for more complete specifications. Tabulated are manufacturer designations of papers suitable for filtration of the indicated type of precipitate.
† Schleicher and Schuell, Inc., Keene, N.H.
‡ E. H. Sargent and Company, Chicago, Ill., agents.
§ Whatman, Inc., Clifton, N.J.
‖ Filtration Sciences Corporation, Eaton-Dikeman Division, Mount Holly Springs, Pa.

# Appendix 9
## Solubility Product Constants*

| Substance | Formula | $K_{sp}$ |
|---|---|---|
| Aluminum hydroxide | $Al(OH)_3$ | $2 \times 10^{-32}$ |
| Barium carbonate | $BaCO_3$ | $5.1 \times 10^{-9}$ |
| Barium chromate | $BaCrO_4$ | $1.2 \times 10^{-10}$ |
| Barium iodate | $Ba(IO_3)_2$ | $1.57 \times 10^{-9}$ |
| Barium manganate | $BaMnO_4$ | $2.5 \times 10^{-10}$ |
| Barium oxalate | $BaC_2O_4$ | $2.3 \times 10^{-8}$ |
| Barium sulfate | $BaSO_4$ | $1.3 \times 10^{-10}$ |
| Bismuth oxide chloride | $BiOCl$ | $7 \times 10^{-9}$ |
| Bismuth oxide hydroxide | $BiOOH$ | $4 \times 10^{-10}$ |
| Cadmium carbonate | $CdCO_3$ | $2.5 \times 10^{-14}$ |
| Cadmium hydroxide | $Cd(OH)_2$ | $5.9 \times 10^{-15}$ |
| Cadmium oxalate | $CdC_2O_4$ | $9 \times 10^{-8}$ |
| Cadmium sulfide | $CdS$ | $2 \times 10^{-28}$ |
| Calcium carbonate | $CaCO_3$ | $4.8 \times 10^{-9}$ |
| Calcium fluoride | $CaF_2$ | $4.9 \times 10^{-11}$ |
| Calcium oxalate | $CaC_2O_4$ | $2.3 \times 10^{-9}$ |
| Calcium sulfate | $CaSO_4$ | $2.6 \times 10^{-5}$ |
| Copper(I) bromide | $CuBr$ | $5.2 \times 10^{-9}$ |
| Copper(I) chloride | $CuCl$ | $1.2 \times 10^{-6}$ |
| Copper(I) iodide | $CuI$ | $1.1 \times 10^{-12}$ |
| Copper(I) thiocyanate | $CuSCN$ | $4.8 \times 10^{-15}$ |
| Copper(II) hydroxide | $Cu(OH)_2$ | $1.6 \times 10^{-19}$ |
| Copper(II) sulfide | $CuS$ | $6 \times 10^{-36}$ |
| Iron(II) hydroxide | $Fe(OH)_2$ | $8 \times 10^{-16}$ |
| Iron(II) sulfide | $FeS$ | $6 \times 10^{-18}$ |
| Iron(III) hydroxide | $Fe(OH)_3$ | $4 \times 10^{-38}$ |
| Lanthanum iodate | $La(IO_3)_3$ | $6.2 \times 10^{-12}$ |
| Lead carbonate | $PbCO_3$ | $3.3 \times 10^{-14}$ |
| Lead chloride | $PbCl_2$ | $1.6 \times 10^{-5}$ |
| Lead chromate | $PbCrO_4$ | $1.8 \times 10^{-14}$ |
| Lead hydroxide | $Pb(OH)_2$ | $2.5 \times 10^{-16}$ |
| Lead iodide | $PbI_2$ | $7.1 \times 10^{-9}$ |
| Lead oxalate | $PbC_2O_4$ | $4.8 \times 10^{-10}$ |
| Lead sulfate | $PbSO_4$ | $1.6 \times 10^{-8}$ |

| Substance | Formula | $K_{sp}$ |
|---|---|---|
| Lead sulfide | PbS | $7 \times 10^{-28}$ |
| Magnesium ammonium phosphate | $MgNH_4PO_4$ | $3 \times 10^{-13}$ |
| Magnesium carbonate | $MgCO_3$ | $1 \times 10^{-5}$ |
| Magnesium hydroxide | $Mg(OH)_2$ | $1.8 \times 10^{-11}$ |
| Magnesium oxalate | $MgC_2O_4$ | $8.6 \times 10^{-5}$ |
| Manganese(II) hydroxide | $Mn(OH)_2$ | $1.9 \times 10^{-13}$ |
| Manganese(II) sulfide | MnS | $3 \times 10^{-13}$ |
| Mercury(I) bromide | $Hg_2Br_2$ | $5.8 \times 10^{-23}$ |
| Mercury(I) chloride | $Hg_2Cl_2$ | $1.3 \times 10^{-18}$ |
| Mercury(I) iodide | $Hg_2I_2$ | $4.5 \times 10^{-29}$ |
| Silver arsenate | $Ag_3AsO_4$ | $1 \times 10^{-22}$ |
| Silver bromide | AgBr | $5.2 \times 10^{-13}$ |
| Silver carbonate | $Ag_2CO_3$ | $8.1 \times 10^{-12}$ |
| Silver chloride | AgCl | $1.82 \times 10^{-10}$ |
| Silver chromate | $Ag_2CrO_4$ | $1.1 \times 10^{-12}$ |
| Silver cyanide | AgCN | $7.2 \times 10^{-11}$ |
| Silver iodate | $AgIO_3$ | $3.0 \times 10^{-8}$ |
| Silver iodide | AgI | $8.3 \times 10^{-17}$ |
| Silver oxalate | $Ag_2C_2O_4$ | $3.5 \times 10^{-11}$ |
| Silver sulfide | $Ag_2S$ | $6 \times 10^{-50}$ |
| Silver thiocyanate | AgSCN | $1.1 \times 10^{-12}$ |
| Strontium oxalate | $SrC_2O_4$ | $5.6 \times 10^{-8}$ |
| Strontium sulfate | $SrSO_4$ | $3.2 \times 10^{-7}$ |
| Thallium(I) chloride | TlCl | $1.7 \times 10^{-4}$ |
| Thallium(I) sulfide | $Tl_2S$ | $1 \times 10^{-22}$ |
| Zinc hydroxide | $Zn(OH)_2$ | $1.2 \times 10^{-17}$ |
| Zinc oxalate | $ZnC_2O_4$ | $7.5 \times 10^{-9}$ |
| Zinc sulfide | ZnS | $4.5 \times 10^{-24}$ |

* Data from R. P. Frankenthal, in *Handbook of Analytical Chemistry,* L. Meites, Ed., pp. **1**-13–**1**-19. New York: McGraw-Hill Book Co., 1963. With permission.

# Appendix 10
## Dissociation Constants for Acids*

| Name | Formula | Dissociation Constant, 25°C | | |
|------|---------|-----------------------------|---|---|
| | | $K_1$ | $K_2$ | $K_3$ |
| Acetic | $CH_3COOH$ | $1.75 \times 10^{-5}$ | | |
| Arsenic | $H_3AsO_4$ | $6.0 \times 10^{-3}$ | $1.05 \times 10^{-7}$ | $3.0 \times 10^{-12}$ |
| Arsenious | $H_3AsO_3$ | $6.0 \times 10^{-10}$ | $3.0 \times 10^{-14}$ | |
| Benzoic | $C_6H_5COOH$ | $6.14 \times 10^{-5}$ | | |
| Boric | $H_3BO_3$ | $5.83 \times 10^{-10}$ | | |
| 1-Butanoic | $CH_3CH_2CH_2COOH$ | $1.51 \times 10^{-5}$ | | |
| Carbonic | $H_2CO_3$ | $4.45 \times 10^{-7}$ | $4.7 \times 10^{-11}$ | |
| Chloroacetic | $ClCH_2COOH$ | $1.36 \times 10^{-3}$ | | |
| Citric | $HOOC(OH)C(CH_2COOH)_2$ | $7.45 \times 10^{-4}$ | $1.73 \times 10^{-5}$ | $4.02 \times 10^{-7}$ |
| Ethylene-diamine-tetraacetic | $H_4Y$ | $1.0 \times 10^{-2}$ | $2.1 \times 10^{-3}$ | $6.9 \times 10^{-7}$ |
| | | | $K_4 = 5.5 \times 10^{-11}$ | |
| Formic | $HCOOH$ | $1.77 \times 10^{-4}$ | | |
| Fumaric | $trans$-$HOOCCH:CHCOOH$ | $9.6 \times 10^{-4}$ | $4.1 \times 10^{-5}$ | |
| Glycolic | $HOCH_2COOH$ | $1.48 \times 10^{-4}$ | | |
| Hydrazoic | $HN_3$ | $1.9 \times 10^{-5}$ | | |
| Hydrogen cyanide | $HCN$ | $2.1 \times 10^{-9}$ | | |
| Hydrogen fluoride | $H_2F_2$ | $7.2 \times 10^{-4}$ | | |
| Hydrogen peroxide | $H_2O_2$ | $2.7 \times 10^{-12}$ | | |
| Hydrogen sulfide | $H_2S$ | $5.7 \times 10^{-8}$ | $1.2 \times 10^{-15}$ | |
| Hypochlorous | $HOCl$ | $3.0 \times 10^{-8}$ | | |
| Iodic | $HIO_3$ | $1.7 \times 10^{-1}$ | | |
| Lactic | $CH_3CHOHCOOH$ | $1.37 \times 10^{-4}$ | | |
| Maleic | $cis$-$HOOCCH:CHCOOH$ | $1.20 \times 10^{-2}$ | $5.96 \times 10^{-7}$ | |
| Malic | $HOOCCHOHCH_2COOH$ | $4.0 \times 10^{-4}$ | $8.9 \times 10^{-6}$ | |
| Malonic | $HOOCCH_2COOH$ | $1.40 \times 10^{-3}$ | $2.01 \times 10^{-6}$ | |
| Mandelic | $C_6H_5CHOHCOOH$ | $3.88 \times 10^{-4}$ | | |
| Nitrous | $HNO_2$ | $5.1 \times 10^{-4}$ | | |
| Oxalic | $HOOCCOOH$ | $5.36 \times 10^{-2}$ | $5.42 \times 10^{-5}$ | |

| Name | Formula | Dissociation Constant, 25°C | | |
|------|---------|------|------|------|
| | | $K_1$ | $K_2$ | $K_3$ |
| Periodic | $H_5IO_6$ | $2.4 \times 10^{-2}$ | $5.0 \times 10^{-9}$ | |
| Phenol | $C_6H_5OH$ | $1.00 \times 10^{-10}$ | | |
| Phosphoric | $H_3PO_4$ | $7.11 \times 10^{-3}$ | $6.34 \times 10^{-8}$ | $4.2 \times 10^{-13}$ |
| Phosphorous | $H_3PO_3$ | $1.00 \times 10^{-2}$ | $2.6 \times 10^{-7}$ | |
| o-Phthalic | $C_6H_4(COOH)_2$ | $1.12 \times 10^{-3}$ | $3.91 \times 10^{-6}$ | |
| Picric | $(NO_2)_3C_6H_2OH$ | $5.1 \times 10^{-1}$ | | |
| Propanoic | $CH_3CH_2COOH$ | $1.34 \times 10^{-5}$ | | |
| Pyruvic | $CH_3COCOOH$ | $3.24 \times 10^{-3}$ | | |
| Salicylic | $C_6H_4(OH)COOH$ | $1.05 \times 10^{-3}$ | | |
| Succinic | $HOOCCH_2CH_2COOH$ | $6.21 \times 10^{-5}$ | $2.32 \times 10^{-6}$ | |
| Sulfamic | $H_2NSO_3H$ | $1.03 \times 10^{-1}$ | | |
| Sulfuric | $H_2SO_4$ | Strong | $1.20 \times 10^{-2}$ | |
| Sulfurous | $H_2SO_3$ | $1.72 \times 10^{-2}$ | $6.43 \times 10^{-8}$ | |
| Tartaric | $HOOC(CHOH)_2COOH$ | $9.20 \times 10^{-4}$ | $4.31 \times 10^{-5}$ | |
| Trichloroacetic | $Cl_3CCOOH$ | $1.29 \times 10^{-1}$ | | |

* Data from V. E. Bowers and R. G. Bates, in *Handbook of Analytical Chemistry,* L. Meites, Ed., pp. **1**-21 – **1**-27. New York: McGraw-Hill Book Co., 1963. With permission.

# Appendix 11
## Dissociation Constants for Bases*

| Name | Formula | Dissociation Constant, $K$, 25°C |
|------|---------|----------------------------------|
| Ammonia | $NH_3$ | $1.76 \times 10^{-5}$ |
| Aniline | $C_6H_5NH_2$ | $3.94 \times 10^{-10}$ |
| 1-Butylamine | $CH_3(CH_2)_2CH_2NH_2$ | $4.0 \times 10^{-4}$ |
| Dimethylamine | $(CH_3)_2NH$ | $5.9 \times 10^{-4}$ |
| Ethanolamine | $HOC_2H_4NH_2$ | $3.18 \times 10^{-5}$ |
| Ethylamine | $CH_3CH_2NH_2$ | $4.28 \times 10^{-4}$ |
| Ethylenediamine | $NH_2C_2H_4NH_2$ | $K_1 = 8.5 \times 10^{-5}$ |
| | | $K_2 = 7.1 \times 10^{-8}$ |
| Hydrazine | $H_2NNH_2$ | $1.3 \times 10^{-6}$ |
| Hydroxylamine | $HONH_2$ | $1.07 \times 10^{-8}$ |
| Methylamine | $CH_3NH_2$ | $4.8 \times 10^{-4}$ |
| Piperidine | $C_5H_{11}N$ | $1.3 \times 10^{-3}$ |
| Pyridine | $C_5H_5N$ | $1.7 \times 10^{-9}$ |
| Trimethylamine | $(CH_3)_3N$ | $6.25 \times 10^{-5}$ |

* Data from V. E. Bowers and R. G. Bates, in *Handbook of Analytical Chemistry,* L. Meites, Ed., pp. 1-21 – 1-27. New York: McGraw-Hill Book Co., 1963. With permission.

# Appendix 12
## Selected Standard Electrode Potentials and Formal Potentials*

| Half-Reaction | $E^0$, V | $E^f$, V |
|---|---|---|
| $F_2(g) + 2H^+ + 2e \rightleftharpoons 2HF(aq)$ | 3.06 | |
| $O_3(g) + 2H^+ + 2e \rightleftharpoons O_2(g) + H_2O$ | 2.07 | |
| $S_2O_8^{2-} + 2e \rightleftharpoons 2SO_4^{2-}$ | 2.01 | |
| $Co^{3+} + e \rightleftharpoons Co^{2+}$ | 1.808 | |
| $H_2O_2 + 2H^+ + 2e \rightleftharpoons 2H_2O$ | 1.776 | |
| $MnO_4^- + 4H^+ + 3e \rightleftharpoons MnO_2(s) + 2H_2O$ | 1.695 | |
| $Ce^{4+} + e \rightleftharpoons Ce^{3+}$ | | 1.70, 1 $F$ HClO$_4$;<br>1.61, 1 $F$ HNO$_3$;<br>1.44, 1 $F$ H$_2$SO$_4$;<br>1.28, 1 $F$ HCl |
| $HClO + H^+ + e \rightleftharpoons \frac{1}{2}Cl_2(g) + H_2O$ | 1.63 | |
| $H_5IO_6 + H^+ + 2e \rightleftharpoons IO_3^- + 3H_2O$ | 1.601 | |
| $BrO_3^- + 6H^+ + 5e \rightleftharpoons \frac{1}{2}Br_2(1) + 3H_2O$ | 1.52 | |
| $MnO_4^- + 8H^+ + 5e \rightleftharpoons Mn^{2+} + 4H_2O$ | 1.51 | |
| $Mn^{3+} + e \rightleftharpoons Mn^{2+}$ | | 1.51, 7.5 $F$ H$_2$SO$_4$ |
| $ClO_3^- + 6H^+ + 5e \rightleftharpoons \frac{1}{2}Cl_2(g) + 3H_2O$ | 1.47 | |
| $PbO_2(s) + 4H^+ + 2e \rightleftharpoons Pb^{2+} + 2H_2O$ | 1.455 | |
| $BrO_3^- + 6H^+ + 6e \rightleftharpoons Br^- + 3H_2O$ | 1.44 | |
| $Cl_2(g) + 2e \rightleftharpoons 2Cl^-$ | 1.359 | |
| $Cr_2O_7^{2-} + 14H^+ + 6e \rightleftharpoons 2Cr^{3+} + 7H_2O$ | 1.33 | |
| $Tl^{3+} + 2e \rightleftharpoons Tl^+$ | 1.25 | 0.77, 1 $F$ HCl |
| $IO_3^- + 2Cl^- + 6H^+ + 4e \rightleftharpoons ICl_2^- + 3H_2O$ | 1.24 | |
| $MnO_2(s) + 4H^+ + 2e \rightleftharpoons Mn^{2+} + 2H_2O$ | 1.23 | 1.24, 1 $F$ HClO$_4$ |
| $O_2(g) + 4H^+ + 4e \rightleftharpoons 2H_2O$ | 1.229 | |
| $IO_3^- + 6H^+ + 5e \rightleftharpoons \frac{1}{2}I_2(s) + 3H_2O$ | 1.196 | |
| $IO_3^- + 6H^+ + 5e \rightleftharpoons \frac{1}{2}I_2(aq) + 3H_2O$ | 1.178† | |
| $SeO_4^{2-} + 4H^+ + 2e \rightleftharpoons H_2SeO_3 + H_2O$ | 1.15 | |
| $Br_2(aq) + 2e \rightleftharpoons 2Br^-$ | 1.087† | |
| $Br_2(l) + 2e \rightleftharpoons 2Br^-$ | 1.065 | 1.05, 4 $F$ HCl |
| $ICl_2^- + e \rightleftharpoons \frac{1}{2}I_2(s) + 2Cl^-$ | 1.056 | |
| $V(OH)_4^+ + 2H^+ + e \rightleftharpoons VO^{2+} + 3H_2O$ | 1.00 | 1.02, 1 $F$ HCl, HClO$_4$ |
| $HNO_2 + H^+ + e \rightleftharpoons NO(g) + H_2O$ | 1.00 | |
| $Pd^{2+} + 2e \rightleftharpoons Pd(s)$ | 0.987 | |

| Half-Reaction | $E^0$, V | $E^f$, V |
|---|---|---|
| $NO_3^- + 3H^+ + 2e \rightleftharpoons HNO_2 + H_2O$ | 0.94 | 0.92, 1 $F$ HNO$_3$ |
| $2Hg^{2+} + 2e \rightleftharpoons Hg_2^{2+}$ | 0.920 | 0.907, 1 $F$ HClO$_4$ |
| $HO_2^- + H_2O + 2e \rightleftharpoons 3OH^-$ | 0.88 | |
| $Cu^{2+} + I^- + e \rightleftharpoons CuI(s)$ | 0.86 | |
| $Hg^{2+} + 2e \rightleftharpoons Hg(l)$ | 0.854 | |
| $Ag^+ + e \rightleftharpoons Ag(s)$ | 0.799 | 0.228, 1 $F$ HCl; 0.792, 1 $F$ HClO$_4$; 0.77, 1 $F$ H$_2$SO$_4$ |
| $Hg_2^{2+} + 2e \rightleftharpoons 2Hg(l)$ | 0.788 | 0.274, 1 $F$ HCl; 0.776, 1 $F$ HClO$_4$, 0.674, 1 $F$ H$_2$SO$_4$ |
| $Fe^{3+} + e \rightleftharpoons Fe^{2+}$ | 0.771 | 0.700, 1 $F$ HCl; 0.732, 1 $F$ HClO$_4$; 0.68, 1 $F$ H$_2$SO$_4$ |
| $H_2SeO_3 + 4H^+ + 4e \rightleftharpoons Se(s) + 3H_2O$ | 0.740 | |
| $PtCl_4^{2-} + 2e \rightleftharpoons Pt(s) + 4Cl^-$ | 0.73 | |
| $O_2(g) + 2H^+ + 2e \rightleftharpoons H_2O_2$ | 0.682 | |
| $PtCl_6^{2-} + 2e \rightleftharpoons PtCl_4^{2-} + 2Cl^-$ | 0.68 | |
| $Hg_2SO_4(s) + 2e \rightleftharpoons 2Hg(l) + SO_4^{2-}$ | 0.615 | |
| $I_2(aq) + 2e \rightleftharpoons 2I^-$ | 0.615† | |
| $Sb_2O_5(s) + 6H^+ + 4e \rightleftharpoons 2SbO^+ + 3H_2O$ | 0.581 | |
| $MnO_4^- + e \rightleftharpoons MnO_4^{2-}$ | 0.564 | |
| $H_3AsO_4 + 2H^+ + 2e \rightleftharpoons H_3AsO_3 + H_2O$ | 0.559 | 0.577, 1 $F$ HCl, HClO$_4$ |
| $I_3^- + 2e \rightleftharpoons 3I^-$ | 0.536 | |
| $I_2(s) + 2e \rightleftharpoons 2I^-$ | 0.5355 | |
| $Cu^+ + e \rightleftharpoons Cu(s)$ | 0.521 | |
| $H_2SO_3 + 4H^+ + 4e \rightleftharpoons S(s) + 3H_2O$ | 0.450 | |
| $Ag_2CrO_4(s) + 2e \rightleftharpoons 2Ag(s) + CrO_4^{2-}$ | 0.446 | |
| $Fe(CN)_6^{3-} + e \rightleftharpoons Fe(CN)_6^{4-}$ | 0.36 | 0.71, 1 $F$ HCl; 0.72, 1 $F$ HClO$_4$, H$_2$SO$_4$ |
| $VO^{2+} + 2H^+ + e \rightleftharpoons V^{3+} + H_2O$ | 0.359 | |
| $Cu^{2+} + 2e \rightleftharpoons Cu(s)$ | 0.337 | |
| $UO_2^{2+} + 4H^+ + 2e \rightleftharpoons U^{4+} + 2H_2O$ | 0.334 | |
| $BiO^+ + 2H^+ + 3e \rightleftharpoons Bi(s) + H_2O$ | 0.320 | |
| $Hg_2Cl_2(s) + 2e \rightleftharpoons 2Hg(l) + 2Cl^-$ | 0.268 | 0.244, sat'd KCl; 0.282, 1 $F$ KCl; 0.334, 0.1 $F$ KCl |
| $AgCl(s) + e \rightleftharpoons Ag(s) + Cl^-$ | 0.222 | 0.228, 1 $F$ KCl |
| $SO_4^{2-} + 4H^+ + 2e \rightleftharpoons H_2SO_3 + H_2O$ | 0.172 | |
| $BiCl_4^- + 3e \rightleftharpoons Bi(s) + 4Cl^-$ | 0.16 | |
| $Sn^{4+} + 2e \rightleftharpoons Sn^{2+}$ | 0.154 | 0.14, 1 $F$ HCl |
| $Cu^{2+} + e \rightleftharpoons Cu^+$ | 0.153 | |
| $S(s) + 2H^+ + 2e \rightleftharpoons H_2S(s)$ | 0.141 | |
| $TiO^{2+} + 2H^+ + e \rightleftharpoons Ti^{3+} + H_2O$ | 0.099 | 0.04, 1 $F$ H$_2$SO$_4$ |
| $S_4O_6^{2-} + 2e \rightleftharpoons 2S_2O_3^{2-}$ | 0.08 | |
| $AgBr(s) + e \rightleftharpoons Ag(s) + Br^-$ | 0.073 | |

| Half-Reaction | $E^0$, V | $E^f$, V |
|---|---|---|
| $Ag(S_2O_3)_2^{3-} + e \rightleftharpoons Ag(s) + 2S_2O_3^{2-}$ | 0.017 | |
| $2H^+ + 2e \rightleftharpoons H_2(g)$ | 0.000 | $-0.005$, 1 $F$ HCl, HClO$_4$ |
| $Pb^{2+} + 2e \rightleftharpoons Pb(s)$ | $-0.126$ | $-0.14$, 1 $F$ HClO$_4$; $-0.29$, 1 $F$ H$_2$SO$_4$ |
| $Sn^{2+} + 2e \rightleftharpoons Sn(s)$ | $-0.136$ | $-0.16$, 1 $F$ HClO$_4$ |
| $AgI(s) + e \rightleftharpoons Ag(s) + I^-$ | $-0.151$ | |
| $CuI(s) + e \rightleftharpoons Cu(s) + I^-$ | $-0.185$ | |
| $N_2(g) + 5H^+ + 4e \rightleftharpoons N_2H_5^+$ | $-0.23$ | |
| $Ni^{2+} + 2e \rightleftharpoons Ni(s)$ | $-0.250$ | |
| $V^{3+} + e \rightleftharpoons V^{2+}$ | $-0.256$ | $-0.21$, 1 $F$ HClO$_4$ |
| $Co^{2+} + 2e \rightleftharpoons Co(s)$ | $-0.277$ | |
| $Ag(CN)_2^- + e \rightleftharpoons Ag(s) + 2CN^-$ | $-0.31$ | |
| $Tl^+ + e \rightleftharpoons Tl(s)$ | $-0.336$ | $-0.551$, 1 $F$ HCl; $-0.33$, 1 $F$ HClO$_4$, H$_2$SO$_4$ |
| $PbSO_4(s) + 2e \rightleftharpoons Pb(s) + SO_4^{2-}$ | $-0.350$ | |
| $Ti^{3+} + e \rightleftharpoons Ti^{2+}$ | $-0.369$ | |
| $Cd^{2+} + 2e \rightleftharpoons Cd(s)$ | $-0.403$ | |
| $Cr^{3+} + e \rightleftharpoons Cr^{2+}$ | $-0.408$ | |
| $Fe^{2+} + 2e \rightleftharpoons Fe(s)$ | $-0.440$ | |
| $2CO_2(g) + 2H^+ + 2e \rightleftharpoons H_2C_2O_4$ | $-0.49$ | |
| $Cr^{3+} + 3e \rightleftharpoons Cr(s)$ | $-0.744$ | |
| $Zn^{2+} + 2e \rightleftharpoons Zn(s)$ | $-0.763$ | |
| $Mn^{2+} + 2e \rightleftharpoons Mn(s)$ | $-1.180$ | |
| $Al^{3+} + 3e \rightleftharpoons Al(s)$ | $-1.662$ | |
| $Mg^{2+} + 2e \rightleftharpoons Mg(s)$ | $-2.363$ | |
| $Na^+ + e \rightleftharpoons Na(s)$ | $-2.714$ | |
| $Ca^{2+} + 2e \rightleftharpoons Ca(s)$ | $-2.866$ | |
| $Ba^{2+} + 2e \rightleftharpoons Ba(s)$ | $-2.906$ | |
| $K^+ + e \rightleftharpoons K(s)$ | $-2.925$ | |
| $Li^+ + e \rightleftharpoons Li(s)$ | $-3.045$ | |

Sources for $E^0$ values:

* G. Milazzo, S. Caroli, and V. K. Sharma, *Tables of Standard Electrode Potentials.* New York: John Wiley & Sons, 1978. Source for formal potentials $E^f$: E. H. Swift and E. A. Butler, *Quantitative Measurements and Chemical Equilibria.* New York: W. H. Freeman and Co., 1972.

† These potentials are hypothetical because they correspond to solutions that are 1.00 $M$ in Br$_2$ or I$_2$. The solubilities of these species at 25°C are 0.18 $M$ and 0.0020 $M$, respectively. In saturated solutions containing an excess of Br$_2$(l) or I$_2$(s), the standard potentials for the half-reactions Br$_2$(l) + 2$e$ $\rightleftharpoons$ 2Br$^-$ and I$_2$(s) + 2$e$ $\rightleftharpoons$ 2I$^-$ should be used. On the other hand, the hypothetical potentials should be used when solutions are undersaturated with Br$_2$ or I$_2$.

# Answers to Problems

**2-1.** (a) base; $NH_4^+$    (b) acid; $H_2PO_4^-$    (c) acid; $CN^-$
      (d) base; $H_2PO_4^-$, or acid; $PO_4^{3-}$    (e) base; $HOCl$    (f) base; $CH_3CH_2NH_3^+$

**2-4.** $6.90 \times 10^{22}$ sodium ions

**2-6.** (a) 71.2 mfw    (b) 0.821 mfw    (c) 52.9 mfw    (d) 0.686 mfw

**2-8.** (a) 9.54 mfw    (b) 13.8 mfw    (c) 0.0122 mfw    (d) $1.63 \times 10^3$ mfw

**2-10.** (a) 25.5 mg    (b) $1.310 \times 10^4$ mg    (c) $1.28 \times 10^6$ mg
      (d) $2.63 \times 10^3$ mg

**2-12.** (a) $1.23 \times 10^3$ mg    (b) $2.06 \times 10^3$ mg    (c) 3.89 mg    (d) 25.6 mg

**2-14.** (a) pNa = 0.658; pCl = 1.086; pOH = 0.860
      (c) pH = $-0.176$; pCl = $-0.241$; pZn = 0.921
      (e) pK = 4.247; pOH = 4.498; $pFe(CN)_6$ = 5.206

**2-15.** (a) $2.3 \times 10^{-11}$    (c) $1.8 \times 10^{-1}$    (e) $3.2 \times 10^{-7}$    (g) 6.5

**2-16.** (a) $[Na^+] = 4.79 \times 10^{-2}$; $[SO_4^{2-}] = 2.87 \times 10^{-3}$
      (b) pNa = 1.320; $pSO_4$ = 2.542

**2-18.** (a) $1.73 \times 10^{-3}\ F$    (b) $1.73 \times 10^{-3}\ M$    (c) $5.20 \times 10^{-3}\ M$
      (d) 0.0482 % (w/v)    (e) 0.130 mmol    (f) 67.6 ppm    (g) 2.761
      (h) 2.284

**2-20.** (a) $0.351\ F$    (b) $1.05\ M$    (c) 85.0 g/L

**2-22.** (a) Dilute 60.0 g $C_2H_5OH$ to exactly 500 mL with $H_2O$
      (b) Dilute 60.0 g $C_2H_5OH$ with 440 g $H_2O$
      (c) Dilute 60.0 mL $C_2H_5OH$ to 500 mL with $H_2O$

**2-24.** Dilute 102 mL of the reagent to 250 mL with $H_2O$

**2-26.** (a) Dissolve 6.79 g $AgNO_3$ in $H_2O$ and dilute to 500 mL
      (b) Dilute 33.3 mL of the $6.00\ F$ reagent to 1.00 L
      (c) Dissolve 4.14 g $K_4Fe(CN)_6$ in $H_2O$ and dilute to 600 mL
      (d) Dilute 270 mL of the $0.400\ F$ solution to 750 mL
      (e) Dilute 25.1 mL of the reagent to 2.00 L
      (f) Dissolve 0.926 g $Na_2SO_4$ and dilute to 5.00 L

**2-28.** (a) 2.42 g KI    (b) 48.7 mL    (c) 35.9 mL    (d) 1.50 g $PbI_2$

**CHAPTER 3**

**3–1.**

| | | A | C | E |
|---|---|---|---|---|
| **(a)** | Mean | $611.6 = 612$ | $0.07173 = 0.0717$ | $38.803 = 38.80$ |
| **(b)** | Median | 613 | 0.0721 | 38.74 |
| **(c)** | Range | 28 | 0.0021 | 0.35 |
| **(d)** | Abs dev from mean | 8.1 | 0.00071 | 0.14 |
| **(e)** | Rel dev from mean | 13 ppt | 9.9 ppt | 3.6 ppt |
| **(f)** | Abs std dev | 10.7 | 0.00097 | 0.18 |
| **(g)** | Rel std dev, % | 1.8 | 1.4 | 0.47 |

**3–3.**

| | | A | C | E |
|---|---|---|---|---|
| **(a)** | Abs error | $-2.4$ | 0.00013 | $-0.047$ |
| **(b)** | Rel error | $-0.39\%$ | 0.18% | $-0.12\%$ |

**3–4.** **(a)** 0.05%;  **(c)** 0.4%

**3–6.**

| | | A | B |
|---|---|---|---|
| **(a)** | $s$ | 0.31 | 0.030 |
| | $(s)_r$ | 2.6% | 6.4% |
| **(b)** | Abs error | 0.075 | $-0.073$ |
| | Rel error | 0.63% | $-13\%$ |

**3–8.** **(a)**

| Sample No. | $s$ | Sample No. | $s$ |
|---|---|---|---|
| 1 | 0.099 | 4 | 0.099 |
| 2 | 0.12 | 5 | 0.10 |
| 3 | 0.091 | | |

**(b)** pooled $s = 0.10$

**3–10.** $s = 4.7$ ppb Zn

**3–12.** **(a)** $3.64 \pm 1.96 \times 0.08/\sqrt{3} = 3.64 \pm 0.09$
**(b)** $3.64 \pm 1.64 \times 0.08/\sqrt{3} = 3.64 \pm 0.08$
**(c)** $3.64 \pm 1.29 \times 0.08/\sqrt{3} = 3.64 \pm 0.06$

**3–15.** $5.71 \pm 2.13 \times 0.6/\sqrt{5} = 5.7 \pm 0.6$

**3–16.** **(a)** $612 \pm 1.53 \times 10.7/\sqrt{5} = 612 \pm 7$
**(c)** $0.0717 \pm 3.18 \times 0.00097/\sqrt{4} = 0.0717 \pm 0.0015 = 0.072 \pm 0.002$
**(e)** $38.80 \pm 2.92 \times 0.18/\sqrt{3} = 38.8 \pm 0.3$

**3–17.** **(a)** $Q_{exp} = 0.39$; retain    **(c)** $Q_{exp} = 0.76$; retain    **(e)** $Q_{exp} = 0.77$; retain

**3–19.** **(a)** $R = 0.0067 + 0.0622$ mg Mn, where $R =$ instrument response
**(b)** $s_b = 0.0011$    $s_r = 0.0073$

**3–20.** **(a)** (i) $= 0.721\%$ Mn; (iii) $= 1.07\%$ Mn
**(b)** (i) $= 0.014\%$ Mn; (iii) $= 0.017\%$ Mn
**(c)** (i) $= 0.010\%$ Mn; (iii) $= 0.014\%$ Mn

**3–22.** **(a)** mg $K^+/100$ mL $= 3.43 \pm 0.13$    **(c)** mg $K^+/100$ mL $= 3.76 \pm 0.12$

**3–23.** **(a)** 4    **(c)** 6    **(e)** 4    **(g)** 5    **(i)** 6

**3–24. (a)** $s = 0.050$; $s_r = 0.09\%$; $53.86 \pm 0.05$
**(c)** $s = 0.054$; $s_r = 0.15\%$; $36.46 \pm 0.05$
**(e)** $s = 0.16$; $s_r = 0.04\%$; $396.3 \pm 0.2$

**3–25. (a)** $s = 0.46$; $s_r = 1.9\%$; $24.5 \pm 0.5$
**(c)** $s = 4.7$; $s_r = 0.86\%$; $543 \pm 5$

**3–26. (a)** $s = 0.035$; $s_r = 0.29\%$; $1.89 \pm 0.0055$
**(c)** $s = 1.2 \times 10^{-6}$; $s_r = 1.2\%$; $9.8 \pm (0.1) \times 10^{-5}$
**(e)** $s = 0.00034$; $s_r = 0.26\%$; $0.1311 \pm 0.0003$

**3–27. (a)** $-2.514$    **(c)** $23.780$    **(e)** $2.6 \times 10^4$    **(g)** $2.4 \times 10^{-15}$

## CHAPTER 4

**4–1. (a)** $\dfrac{\text{gfw ZnCl}_2}{2 \times \text{gfw AgCl}}$    **(c)** $\dfrac{\text{gfw (NH}_4)_2\text{SO}_4}{\text{gfw BaSO}_4}$    **(e)** $\dfrac{\text{gfw Mn}_3\text{O}_4}{3 \times \text{gfw MnO}_2}$

**(g)** $\dfrac{\text{gfw Fe}_6\text{Si}_{17}}{3 \times \text{gfw Fe}_2\text{O}_3}$    **(i)** $\dfrac{\text{gfw Cu}_2\text{HgI}_4}{\text{gfw HgO}}$

**4–2.** Letting $x = \text{gfw (NH}_4)_3\text{Fe(C}_2\text{O}_4)_3 \cdot 3\text{H}_2\text{O}$,
**(a)** $\dfrac{2x}{\text{gfw Fe}_2\text{O}_3}$    **(c)** $\dfrac{x}{3 \times \text{gfw (C}_6\text{H}_5)_4\text{BNH}_4}$    **(e)** $\dfrac{x}{6\text{ gfw CO}_2}$

**4–3.** Letting $x = \text{gfw MgSnCl}_6 \cdot 6\text{H}_2\text{O}$
**(a)** $\dfrac{x}{\text{gfw MgO}}$    **(c)** $\dfrac{x}{6 \times \text{gfw AgCl}}$    **(e)** $\dfrac{2x}{\text{gfw Mg}_2\text{P}_2\text{O}_7}$

**4–6. (a)** $\dfrac{\text{gfw CaC}_2\text{O}_4 \cdot 2\text{H}_2\text{O}}{\text{gfw CaC}_2\text{O}_4}$    **(b)** $\dfrac{\text{gfw CaC}_2\text{O}_4 \cdot 2\text{H}_2\text{O}}{2 \times \text{gfw H}_2\text{O}}$    **(e)** $\dfrac{\text{gfw CaC}_2\text{O}_4 \cdot 2\text{H}_2\text{O}}{\text{gfw CaO}}$

**4–8. (a)** 0.994 g;    **(c)** 1.81 g;    **(e)** 2.10 g

**4–9.** 5.77 g

**4–11.** 0.320 g

**4–13.** 0.266 g

**4–14. (a)** 36.17%;    **(c)** 16.31%;    **(e)** 24.21%

**4–17.** 0.154%

**4–20.** 0.08672 $F$

**4–22.** 0.925%

**4–24.** 65.14%

**4–26.** 19.95% $NH_4Cl$ and 28.72% KI

**4–28.** 51.3% $NH_4NO_3$ and 31.9% $(NH_4)_2SO_4$

**4–30. (a)** 80.00% Ag and 20.00% Cu    **(c)** 90.00% Ag and 10.00% Cu
**(e)** 50.00% Ag and 50.00% Cu

**CHAPTER 5**

| **5–1** | **5–2** |
|---|---|
| **(a)** $AgSCN \rightleftarrows Ag^+ + SCN^-$ | $K_{sp} = [Ag^+][SCN^-]$ |
| **(c)** $Ag_2S \rightleftarrows 2Ag^+ + S^{2-}$ | $K_{sp} = [Ag^+]^2[S^{2-}]$ |
| **(e)** $Ag_3AsO_4 \rightleftarrows 3Ag^+ + AsO_4^{3-}$ | $K_{sp} = [Ag^+]^3[AsO_4^{3-}]$ |
| **(g)** $PbClF \rightleftarrows Pb^{2+} + Cl^- + F^-$ | $K_{sp} = [Pb^{2+}][Cl^-][F^-]$ |
| **(i)** $Ca_3(AsO_4)_2 \rightleftarrows 3Ca^{2+} + 2AsO_4^{3-}$ | $K_{sp} = [Ca^{2+}]^3[AsO_4^{3-}]^2$ |

**5–3.**

| | | | Molar Concentration | | |
|---|---|---|---|---|---|
| | | Solubility, mg/L | of Cations | of Anions | $K_{sp}$ |
| **(a)** | $AgI$ | $2.14 \times 10^{-3}$ | | $9.1 \times 10^{-9}$ | $8.3 \times 10^{-17}$ |
| **(c)** | $Pb(IO_3)_2$ | 24 | $4.3 \times 10^{-5}$ | $8.6 \times 10^{-5}$ | |
| **(e)** | $BiI_3$ | 7.8 | $1.32 \times 10^{-5}$ | $3.95 \times 10^{-5}$ | |
| **(i)** | $In_4[Fe(CN)_6]_3$ | 0.18 | $6.4 \times 10^{-7}$ | | $1.9 \times 10^{-44}$ |

**5–5.** **(a)** $1.7 \times 10^{-4} M$;   **(b)** $1.2 \times 10^{-5} M$;   **(c)** $2.0 \times 10^{-8} M$

**5–6.**

| | **(i)** | **(ii)** | **(iii)** |
|---|---|---|---|
| **(a)** | $7.2 \times 10^{-5} M$ | $6.5 \times 10^{-8} M$ | $6.5 \times 10^{-8} M$ |
| **(c)** | $1.2 \times 10^{-3}$ | $1.5 \times 10^{-4}$ | $1.1 \times 10^{-6}$ |
| **(e)** | $2.2 \times 10^{-3}$ | $3.7 \times 10^{-4}$ | $6.7 \times 10^{-6}$ |
| **(g)** | $1.2 \times 10^{-6}$ | $1.2 \times 10^{-19}$ | $3.1 \times 10^{-8}$ |

**5–7.** **(a)** $6.0 \times 10^{-7} M$   **(c)** $6.4 \times 10^{-4} M$

**5–8.** **(a)** $7.5 \times 10^{-7} M$   **(c)** $2.0 \times 10^{-7} M$

**5–9.** **(a)** no precipitate;   **(c)** precipitate forms;   **(e)** precipitate forms

**5–11.** **(a)** $[Ag^+] = 5.9 \times 10^{-15} M$; $[K^+] = 0.024 M$
**(b)** $[Hg_2^{2+}] = 2.8 \times 10^{-24} M$; $[K^+] = 0.0240 M$
**(c)** $[Bi^{3+}] = 0.0020 M$; $[K^+] = 0.0240 M$

**5–13.** **(a)** $2.0 \times 10^{-5} M$;   **(c)** $1 \times 10^{-24} M$;   **(e)** $1 \times 10^{-15} M$;
**(g)** $1 \times 10^{-22} M$

**5–14.** **(a)** $[K^+] = [Cl^-] = 0.050$
$[K^+] + [H_3O^+] = [Cl^-] + [OH^-]$
**(c)** $[HClO] + [OCl^-] = 0.075$
$[H_3O^+] = [OCl^-] + [OH^-]$
**(e)** $[Ba^{2+}] = 0.020 + 0.100 = 0.120$
$[Cl^-] = 2 \times 0.100 = 0.200$
$2[Ba^+] + [H_3O^+] = [Cl^-] + [OH^-]$
**(g)** $[Na^+] = 0.10$
$[Mg^{2+}] = 0.040$
$[Cl^-] = 0.080$
$[SO_4^{2-}] + [HSO_4^-] = 0.050$
$[Na^+] + 2[Mg^+] + [H_3O^+] = [Cl^-] + [HSO_4^-] + 2[SO_4^{2-}] + [OH^-]$
**(i)** $[Na^+] = 0.060 + 2 \times 0.050 = 0.16$
$[SO_3^{2-}] + [HSO_3^-] + [H_2SO_3] = 0.060 + 0.050 = 0.11$
$[Na^+] + [H_3O^+] = 2[SO_3^{2-}] + [HSO_3^-] + [OH^-]$

**5–16. (b)** $s = [Pb^{2+}] = \frac{1}{2}[I^-]$
$[H_3O^+] + 2[Pb^{2+}] = [I^-] + [OH^-]$
**(d)** $s = [Cl^-] = [Ag^+] + [Ag(CN)_2^-]$
$[Na^+] = 0.100; [Cl^-] = 0.0200$
$[CN^-] = 0.100 - 2[Ag(CN)_2^-] - [HCN]$
$[Ag^+] + [Na^+] + [H_3O^+] = [CN^-] + [Ag(CN)_2^-] + [OH^-] + [Cl^-]$

**5–17. (a)** $5.9 \times 10^{-4} F$     **(b)** $5.9 \times 10^{-5} F$

**5–18. (a)** $1.9 \times 10^{-2} F$     **(b)** $1.1 \times 10^{-3} F$

**5–19. (a)** $5.2 \times 10^{-3} F$

**5–20. (a)** $1.0 \times 10^{-2} F$

**5–22. (a)** AgSCN precipitates first. Separation not feasible.
**(b)** $Hg_2I_2$ precipitates first.
Separation feasible if $[I^-] = 6.7 \times 10^{-12} M$ and $< 2.2 \times 10^{-6}$
**(e)** $In(OH)_3$ precipitates first.
Separation feasible if $[OH^-] = 8 \times 10^{-10} M$ and $< 2.2 \times 10^{-8}$

**5–23. (a)** $Cd^{2+}$ precipitates first.
Separation feasible if $[H^+] = 0.18$ to $2.4 \times 10^{-4} M$
**(c)** $Tl^+$ precipitates first. Separation not feasible.
**(e)** $Zn^{2+}$ precipitates first. Separation not feasible.

**5–24. (a)** $5.01 \times 10^{-3}$     **(c)** $1.00 \times 10^{-2}$     **(e)** $0.100$     **(g)** $0.100$

**5–25. (a)** $2.45 \times 10^{-9}$     **(c)** $1.5 \times 10^{-15}$

**5–26.** $0.0500$

## CHAPTER 6

**6–1. (a)** acid/base: gfw $HClO_4$; gfw $CO_2/2$; gfw $Fe_2(CO_3)_3/6$
**(c)** oxidation/reduction: gfw $MnSO_4 \cdot 7H_2O/5$; gfw $SnF_2/2$;
gfw $SnO \cdot Sn(NO_3)_2/4$; gfw $Mn_2O_3/10$
**(e)** oxidation/reduction: gfw $Na_3VO_4/1$; gfw $H_2V_4O_{11}/4$; gfw $Fe_2O_3/2$
**(g)** oxidation/reduction: gfw $I_2/2$; gfw $Na_2S_2O_3 \cdot 5H_2O/1$;
gfw $Na_3Ag(S_2O_3)_2/2$

**6–2. (a)** gfw $C_2H_5NS/2$     **(c)** gfw $C_6H_{10}S_2/4$     **(e)** gfw $C_2H_4S_5/10$
**(g)** gfw $C_4H_{12}N_2S_2/4$

**6–4. (a)** gfw $Ca/2$; gfw $CaCl_2/2$; gfw $Ca_3Al_2O_6/6$; gfw $Ca_2Fe(CN_6)/4$

**6–5. (a)** $F = 0.04991$     **(b)** $N = 0.04991$     **(c)** $N = 0.3993$
**(d)** $16.39$ mg $H_2SO_3$/mL

**6–7. (a)** $F = 0.0783$     **(b)** $N = 0.1566$     **(c)** $N = 0.626$
**(d)** $4.240$ mg HgO/mL

**6–8. (a)** meq $= 25.0 \times 0.0187 \times 4 = 1.87$
**(c)** meq $= 25.0 \times 0.0244 = 0.610$
**(e)** meq $= 25.0 \times 0.0316 = 0.790$

6–9.  (a) 0.0506 g    (c) 0.138 g    (e) 0.201 g

6–10. (a) 3.880 meq    (c) 5.838 meq    (e) 2.299 meq

6–11. (a) 82.8 mg/mL    (c) 8.70 mg/mL    (e) 0.0452 $N$    (g) 0.0669 $N$

6–12. (a) 0.0250 $N$    (c) 0.0250 $N$

6–13. (a) 29.06 mL    (c) 45.47 mL    (e) 19.94 mL

6–14. (a) Dilute 6.3 mL to 1.00 L
      (c) Dissolve 13.4 g in $H_2O$ and dilute to 1.00 L
      (e) Dissolve 2.44 g in $H_2O$ and dilute to 1.00 L

6–15. (a) 0.1255 $N$    (c) 0.0927 $N$    (e) 0.0652 $N$    (g) 0.1061 $N$

6–16. 0.1203 $N$

6–18. 0.1176 $N$

6–20. 0.08411 $N$

6–22. 93.7%

6–24. 80.1%

6–26. 10.5%

CHAPTER 7

7–1.  (a) 0.0978 $F$    (c) 0.102 $F$

7–2.  (a) 0.01355 $F$    (c) 0.02050 $F$    (e) 0.1320 $F$

7–3.  (a) 8.23 mg/mL    (c) 5.44 mg/mL    (e) 0.845 mg/mL

7–4.  73.3%

7–6.  18.0 ppm

7–10. 90.6%

7–12. 71.61%

7–14. 0.174%

7–16. % KCl = 29.4; % $KClO_4$ = 58.4

7–18.

|              | (a) | (c) |
|--------------|-----|-----|
| mL $AgNO_3$  | $[Ag^+]$ | $[Ag^+]$ |
| 5.00  | $1.30 \times 10^{-15}$ | $2.86 \times 10^{-9}$ |
| 20.0  | $2.90 \times 10^{-15}$ | $6.37 \times 10^{-9}$ |
| 30.0  | $6.64 \times 10^{-15}$ | $1.46 \times 10^{-8}$ |
| 35.0  | $1.41 \times 10^{-14}$ | $3.09 \times 10^{-8}$ |
| 39.0  | $7.39 \times 10^{-14}$ | $1.62 \times 10^{-7}$ |
| 40.0  | $9.11 \times 10^{-9}$ | $1.35 \times 10^{-5}$ |
| 41.0  | $1.10 \times 10^{-3}$ | $1.10 \times 10^{-3}$ |
| 45.0  | $5.26 \times 10^{-3}$ | $5.26 \times 10^{-3}$ |
| 50.0  | $1.00 \times 10^{-2}$ | $1.00 \times 10^{-2}$ |

**7-19.** $6.0 \times 10^{-3} M$

**7-20.** **(a)** 0.022%

**7-21.** $1.65 \times 10^2$

CHAPTER 8

Many of the answers for this chapter are followed by the label **A** or **Q**. The former have been derived with the approximation that $[H_3O^+]$ and/or $[OH^-]$ are small with respect to the formal concentration of the acid and/or the base. The label **Q** indicates that the more exact quadratic solution has been used. All calculated values for pH or pOH have been rounded to two places to the right of the decimal point.

**8-1.** **(a)** $[H_3O^+] = 6.03$; $[OH^-] = 1.66 \times 10^{-15}$
**(b)** $[H_3O^+] = 2.40$; $[OH^-] = 4.18 \times 10^{-15}$
**(c)** 9.0% KOH; $[H_3O^+] = 5.78 \times 10^{-15}$
**(d)** 4.5% $HNO_3$; $[OH^-] = 1.37 \times 10^{-14}$
**(e)** $d = 1.64$ g/mL; $[H_3O^+] = 5.29 \times 10^{-15}$

**8-2.** **(a)** 1.44    **(c)** 12.38    **(e)** 1.44

**8-3.** **(b)** 11.86    **(d)** 2.44    **(f)** 12.40

**8-4.** **(a)** 12.30    **(c)** 12.60    **(e)** 12.30

**8-5.**

|  | A | Q |
|---|---|---|
| **(a)** | 4.84 | 4.84 |
| **(c)** | 2.88 | 2.88 |
| **(e)** | 1.93 | 1.96 |

**8-6.**

|  | A | Q |
|---|---|---|
| **(a)** | 5.34 | 5.34 |
| **(c)** | 3.38 | 3.39 |
| **(e)** | 2.43 | 2.51 |

**8-7.**

|  | A | Q |
|---|---|---|
| **(a)** | 6.34 | 6.34 |
| **(b)** | 4.38 | 4.47 |
| **(c)** | 3.43 | 4.03 |

**8-8.**

|  | A | Q |
|---|---|---|
| **(a)** | 12.06 | 12.03 |
| **(c)** | 10.84 | 10.84 |
| **(e)** | 10.56 | 10.56 |

**8-9.**

|  | A | Q |
|---|---|---|
| **(a)** | 11.56 | 11.48 |
| **(c)** | 10.34 | 10.34 |
| **(e)** | 10.06 | 10.05 |

**8–10.**

| | A | Q |
|---|---|---|
| (a) | 10.56 | 9.97 |
| (c) | 9.34 | 9.29 |
| (e) | 9.06 | 9.03 |

**8–11.**

| | A | Q |
|---|---|---|
| (a) | 2.34 | 2.54 |
| (c) | 8.04 | 8.04 |
| (e) | 3.29 | 3.30 |

**8–12.**

| | A | Q |
|---|---|---|
| (a) | 9.72 | 9.72 |
| (c) | 5.26 | 5.26 |
| (e) | 9.45 | 9.45 |

**8–13.**

| | A | Q |
|---|---|---|
| (a) | 6.92 | 6.92 |
| (c) | 3.15 | 3.19 |

**8–14.**

| | A | Q |
|---|---|---|
| (a) | 3.68 | 3.70 |
| (c) | 4.72 | 4.72 |

**8–15.**

| | A | Q |
|---|---|---|
| (a) | 3.50 | 3.50 |
| (b) | 3.14 | 3.15 |
| (c) | 3.34 | 3.35 |
| (d) | 3.12 | 3.13 |

**8–17.**

| | A | Q |
|---|---|---|
| (a) | 2.99 | 3.00 |
| (c) | 4.08 | 4.08 |
| (d) | 3.98 | 3.98 |

**8–18.** (a) 0.029  (c) $1.3 \times 10^2$  (e) 4.3

**8–19.** 31.2 g

**8–21.** (a) $\Delta pH = 3.25 - 2.36 = 0.89$ (Q)
(b) $\Delta pH = 7.79 - 8.64 = -0.85$ (Q)
(c) $\Delta pH = 4.133 - 4.116 = 0.017$ (Q)
(d) $\Delta pH = 4.253 - 4.119 = 0.134$ (Q)

**8–23.** (a) $\Delta pH = 2.02 - 2.36 = -0.34$ (Q)
(b) $\Delta pH = 5.26 - 8.64 = -3.38$ (Q)
(c) $\Delta pH = 4.070 - 4.116 = -0.046$ (Q)
(d) $\Delta pH = 3.687 - 4.119 = -0.432$ (Q)

**8–25.** 9.34 g

**8–27.** 83.5 mL

**8–29.** (a) 1.66 (Q)   (b) 3.83 (A)   (c) 1.75 (Q)

**8–31.** (a) 4.54   (b) 8.34   (c) 9.70

**8–33.** (a) 9.95 (A)   (b) 11.51 (A)   (c) 9.20 (A)

**8–35.** (a) 7.07 (A)   (c) 2.04 (Q)   (d) 2.04 (Q)   (f) 7.46 (A)

**8–36.** (a) $[H_2A]/[HA^-] = 0.25$   (b) $[HA^-]/[A^{2-}] = 1.8$   (c) $[HA^-]/[A^{2-}] = 2.3$

**8–38.** (a) 7.20 (A)   (c) 4.69   (f) 2.27 (Q)

**8–39.** (a) 12.52 (Q)   (c) 2.77 (Q)   (e) 9.75

**8–40.** (a) Dissolve 44.5 g $Na_2CO_3$ and 75.0 g $NaHCO_3$ in $H_2O$ and dilute to 1.00 L
(c) Dissolve 95.6 g $NaHCO_3$ in 750 mL 0.450 $F$ $Na_2CO_3$ and dilute to 1.00 L
(e) Dissolve 7.62 g in 600 mL of 0.0100 $F$ $H_3PO_4$ and dilute to 1.00 L

**8–41.** (a)

| pH | $D$ | $\alpha_0$ | $\alpha_1$ | $\alpha_2$ |
|---|---|---|---|---|
| 2.00 | $1.09 \times 10^{-4}$ | $9.15 \times 10^{-1}$ | $8.42 \times 10^{-2}$ | $3.63 \times 10^{-4}$ |
| 4.00 | $1.42 \times 10^{-7}$ | $7.06 \times 10^{-2}$ | $6.50 \times 10^{-1}$ | $2.80 \times 10^{-1}$ |
| 6.00 | $4.06 \times 10^{-8}$ | $2.46 \times 10^{-5}$ | $2.27 \times 10^{-2}$ | $9.77 \times 10^{-1}$ |

(c)

| pH | $D$ | $\alpha_0$ | $\alpha_1$ | $\alpha_2$ |
|---|---|---|---|---|
| 6.00 | $1.44 \times 10^{-12}$ | $6.92 \times 10^{-1}$ | $3.08 \times 10^{-1}$ | $1.45 \times 10^{-5}$ |
| 8.00 | $4.57 \times 10^{-15}$ | $2.19 \times 10^{-2}$ | $9.74 \times 10^{-1}$ | $4.58 \times 10^{-3}$ |
| 10.00 | $6.54 \times 10^{-17}$ | $1.53 \times 10^{-4}$ | $6.80 \times 10^{-1}$ | $3.20 \times 10^{-1}$ |

(e)

| pH | $D$ | $\alpha_0$ | $\alpha_1$ | $\alpha_2$ |
|---|---|---|---|---|
| 6.00 | $6.88 \times 10^{-12}$ | $1.4 \times 10^{-5}$ | $1.2 \times 10^{-1}$ | $8.8 \times 10^{-1}$ |
| 8.00 | $9.20 \times 10^{-11}$ | $1.1 \times 10^{-2}$ | $9.2 \times 10^{-1}$ | $6.6 \times 10^{-2}$ |
| 10.00 | $1.85 \times 10^{-8}$ | $5.4 \times 10^{-1}$ | $4.6 \times 10^{-1}$ | $3.3 \times 10^{-4}$ |

**8–42.** (a) $[H_2CO_3] = \alpha_0 \times 0.0500 = 1.53 \times 10^{-4} \times 0.0500 = 7.65 \times 10^{-6}\ M$
(c) $[HT^-] = 0.650 \times 0.0400 = 0.0260\ M$

**8–43.** (a) $D = 1.83 \times 10^{-15}$; $\alpha_0 = 2.9 \times 10^{-1}$; $\alpha_1 = 7.1 \times 10^{-1}$; $\alpha_2 = 3.7 \times 10^{-8}$

(c)

| pH | $D$ | $\alpha_0$ | $\alpha_1$ | $\alpha_2$ | $\alpha_3$ |
|---|---|---|---|---|---|
| 10.70 | $9.2 \times 10^{-21}$ | $8.6 \times 10^{-13}$ | $3.1 \times 10^{-4}$ | $9.8 \times 10^{-1}$ | $2.1 \times 10^{-2}$ |
| 2.40 | $1.8 \times 10^{-7}$ | $3.6 \times 10^{-1}$ | $6.4 \times 10^{-1}$ | $1.0 \times 10^{-5}$ | $1.1 \times 10^{-15}$ |

**8–44.**

| Vol HCl | pH | Vol HCl | pH |
|---|---|---|---|
| 0.0 | 13.00 | 49.0 | 11.00 |
| 10.0 | 12.82 | 50.0 | 7.00 |
| 25.0 | 12.52 | 51.0 | 3.00 |
| 40.0 | 12.05 | 55.0 | 2.32 |
| 45.0 | 11.72 | 60.0 | 2.04 |

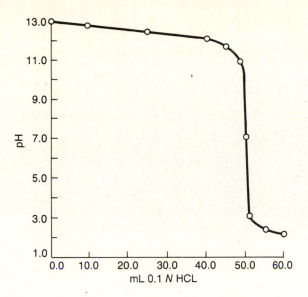

**8–45.**

| Vol Reagent | (a) pH | (c) pH |
|---|---|---|
| 0.00 | 2.21 (Q) | 11.12 (A) |
| 10.0 | 2.76 (Q) | 9.85 (A) |
| 25.0 | 3.31 (Q) | 9.25 (A) |
| 40.0 | 3.89 (A) | 8.64 (A) |
| 49.0 | 4.98 (A) | 7.56 |
| 50.0 | 7.95 (A) | 5.27 |
| 51.0 | 10.90 | 3.00 |
| 55.0 | 11.58 | 2.32 |
| 60.0 | 11.86 | 2.04 |

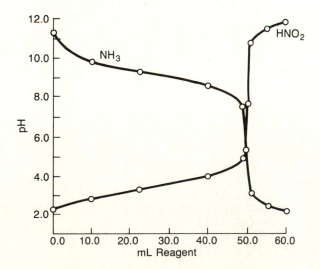

**8-46.**

| Vol Reagent | (a) pH | (c) pH |
|---|---|---|
| 0.00 | 1.53 (Q) | 1.62 (Q) |
| 5.00 | 1.75 (Q) | 1.88 (Q) |
| 10.0 | 2.02 (Q) | 2.18 (Q) |
| 15.0 | 2.40 (Q) | 2.58 (Q) |
| 19.0 | 3.17 (Q) | 3.35 (Q) |
| 20.0 | 4.54 | 4.33 |
| 21.0 | 5.91 (A) | 5.31 (A) |
| 30.0 | 7.19 (A) | 6.58 |
| 39.0 | 8.48 (A) | 7.86 |
| 40.0 | 9.92 (A) | 9.62 |
| 41.0 | 11.34 | 11.34 |
| 50.0 | 12.30 | 12.30 |

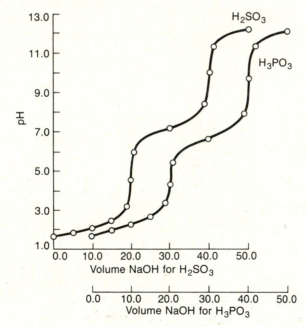

CHAPTER 9

**9-1.** (a) gfw $H_2SO_4/2$    (c) gfw $H_5IO_6/1$    (e) gfw $H_2SO_4/2$    (g) gfw HCl/1

**9-2.** (a) gfw $Na_2CO_3/2$    (c) gfw KOH/1    (e) gfw $HONH_2/1$
      (g) gfw $Na_3PO_4/2$

**9-3.** (a) 2.68 meq    (c) 2.83 meq    (e) 3.03 meq    (g) 9.00 meq

**9-4.** (a) 22.1 mL    (c) 19.5 mL    (e) 28.3 mL    (g) 45.8 mL

**9-5.** (a) Dilute about 25 mL to 2.0 L
      (b) Dilute 28.84 g to 2.000 L
      (c) Dilute 41.5 mL to 500 mL
      (d) Dissolve 22.2 g in $H_2O$ and dilute to 1.000 L

**9–7.** (a) 0.0876 $N$ (b) 0.1581 $N$ (c) 0.1368 $N$ (d) 0.1148 $N$

**9–9.** (a) $x_i$ = 0.10839, 0.10883, 0.10766, 0.10847
$\bar{x}$ = 0.1083
(b) $s = 4.9 \times 10^{-4} = 0.0005$
(c) $Q_{exp} = 0.64$; $Q_{crit} = 0.76$; ∴ retain

**9–11.** (a) 7.00 mg $H_3PO_7$/mL (b) 14.0 mg $H_3PO_4$/mL

**9–13.** (a) 0.86 to 1.1 g (c) 0.55 to 0.70 g (e) 0.23 to 0.30 g

**9–15.**

|  | **Equiv Point, pH** | **Indicator** |
|---|---|---|
| **(a)** | 7.0 | Bromothymol blue |
| **(b)** | 6.0 | Methyl red or Bromocresol purple |
| **(c)** | 5.7 | Methyl red or Bromocresol purple |
| **(d)** | 9.7 | Phenolphthalein or Thymolphthalein |
| **(e)** | 3.2 | Methyl orange or Bromophenol blue |
| **(f)** | 8.6 | Phenolphthalein or Cresol purple |

**9–17.** 102.1 g/equiv

**9–19.** 4.02 g/100 mL

**9–22.** 77.99% = 78.0%

**9–24.** 0.103%

**9–25.** (a) 10.09% (c) 47.61%

**9–26.** (a) 4.000% N (b) 25.0% protein

**9–28.** 26.16% $NH_4Cl$ and 31.71% $NaNO_3$

**9–30.** 33.1 ppm; in compliance

**9–32.** 27.26%

**9–34.** (a) 10.98 mL (c) 10.98 mL (e) 14.69 mL

**9–35.** (a) 13.02 mL (c) 11.77 mL (e) 46.38 mL (g) 45.13 mL

**9–36.** 0.04960 $F$ $H_3PO_4$ and 0.0836 $NaH_2PO_4$

**9–38.** (a) 12.46 mg $Na_2CO_3$/mL
(b) 9.08 mg $NaHCO_3$/mL
(c) 6.545 mg NaOH/mL
(d) 7.813 mg $Na_2CO_3$/mL and 7.765 mg $NaHCO_3$/mL
(e) 5.423 mg NaOH/mL and 6.263 mg $Na_2CO_3$/mL

**CHAPTER 10**

**10–1.** Dissolve 18.67 g of the salt in water and dilute to exactly 1 L

**10–2.** (a) 2.80 mg CaO/mL (c) 7.62 mg $Zn_2P_2O_7$/mL

**10–3.** (a) 39.1 mL (c) 41.6 mL (e) 31.2 mL

**10–5.** 7.869%

**10-7.** (a) 39.0 mg/mL    (b) 123 mg/mL

**10-9.** (a) 31.62 mL    (b) 16.80 mL    (c) 31.62 mL

**10-11.** 0.998 mg Cr/cm$^2$

**10-13.** 0.0978%

**10-15.** 98.7 mg Ca$^{2+}$ and 24.1 mg Mg$^{2+}$

**10-17.** 58.04%

**10-20.** (a) $4 \times 10^{16}$    (b) $1.4 \times 10^{14}$

**10-21.** (a) 5.0    (c) 3.0

**10-22.** $\alpha_4 = 2.2 \times 10^{-5}$ and $K'_{CuY} = 1.39 \times 10^{14}$

| mL Reagent | pCu | mL Reagent | pCu |
|---|---|---|---|
| 0.00 | 2.00 | 25.00 | 8.16 |
| 10.00 | 2.30 | 25.10 | 11.74 |
| 24.00 | 3.57 | 26.00 | 12.74 |
| 24.90 | 4.57 | 30.00 | 13.44 |

## CHAPTER 11

**11-1.** (a) $2Fe^{3+} + Sn^{2+} \rightleftarrows 2Fe^{2+} + Sn^{4+}$
(b) $Cr(s) + 3Ag^+ \rightleftarrows Cr^{3+} + 3Ag(s)$
(c) $2NO_3^- + Cu(s) + 4H^+ \rightleftarrows 2NO_2(g) + Cu^{2+} + 2H_2O$
(d) $2MnO_4^- + 5H_2SO_3 \rightleftarrows 2Mn^{2+} + 5SO_4^{2-} + 3H_2O + 4H^+$
(e) $Fe(CN)_6^{3-} + Ti^{3+} + H_2O \rightleftarrows TiO^{2+} + Fe(CN)_6^{4-} + 2H^+$
(f) $2Ce^{4+} + H_2O_2 \rightleftarrows 2Ce^{3+} + O_2 + 2H^+$
(g) $2Ag(s) + 2I^- + Sn^{4+} \rightleftarrows 2AgI(s) + Sn^{2+}$
(h) $UO_2^{2+} + Zn(s) + 4H^+ \rightleftarrows U^{4+} + Zn^{2+} + 2H_2O$
(i) $2MnO_4^- + 5HNO_2 + H^+ \rightleftarrows 2Mn^{2+} + 5NO_3^- + 3H_2O$
(j) $H_2NNH_2 + IO_3^- + 2Cl^- + 2H^+ \rightleftarrows ICl_2^- + N_2 + 3H_2O$

**11-3.** (a) $Fe^{3+} + e \rightleftarrows Fe^{2+}$
(b) $Ag^+ + e \rightleftarrows Ag(s)$
(c) $NO_3^- + 2H^+ + e \rightleftarrows NO_2 + H_2O$
(d) $MnO_4^- + 8H^+ + 5e \rightleftarrows Mn^{2+} + 4H_2O$
(e) $Fe(CN)_6^{3-} + e \rightleftarrows Fe(CN)_6^{4-}$
(f) $Ce^{4+} + e \rightleftarrows Ce^{3+}$
(g) $Sn^{4+} + 2e \rightleftarrows Sn^{2+}$
(h) $UO_2^{2+} + 4H^+ + 2e \rightleftarrows U^{4+} + 2H_2O$
(i) $MnO_4^- + 8H^+ + 5e \rightleftarrows Mn^{2+} + 4H_2O$
(j) $IO_3^- + 2Cl^- + 6H^+ + 4e \rightleftarrows ICl_2^- + 3H_2O$

**11-4.** (a) $Sn^{2+} \rightleftarrows Sn^{4+} + 2e$    (b) $Cr(s) \rightleftarrows Cr^{3+} + 3e$
(c) $Cu(s) \rightleftarrows Cu^{2+} + 2e$    (d) $H_2SO_3 + H_2O \rightleftarrows SO_4^{2-} + 4H^+ + 2e$
(e) $Ti^{3+} + H_2O \rightleftarrows TiO^{2+} + 2H^+ + e$    (f) $H_2O_2 \rightleftarrows O_2 + 2H^+ + 2e$
(g) $Ag(s) + I^- \rightleftarrows AgI(s) + e$    (h) $Zn(s) \rightleftarrows Zn^{2+} + 2e$
(i) $HNO_2 + H_2O \rightleftarrows NO_3^- + 3H^+ + 2e$    (j) $H_2NNH_2 \rightleftarrows N_2 + 4H^+ + 4e$

**11-7.** (a) gfw $Fe^{3+}/1$ and gfw $Sn^{2+}/2$      (b) gfw $Ag^+/1$ and gfw $Cr/3$
(c) gfw $NO_3^-/1$ and gfw $Cu/2$      (d) gfw $MnO_4^-/5$ and gfw $H_2SO_3/2$
(e) gfw $Fe(CN)_6^{3-}/1$ and gfw $Ti^{3+}/1$      (f) gfw $Ce^{4+}/1$ and gfw $H_2O_2/2$
(g) gfw $Sn^{4+}/2$ and gfw $Ag/1$      (h) gfw $UO_2^{2+}/2$ and gfw $Zn/2$
(i) gfw $MnO_4^-/5$ and gfw $HNO_2/2$      (j) gfw $IO_3^-/4$ and gfw $H_2NNH_2/4$

**11-9.** (a) $Sn(s) \rightleftarrows Sn^{2+} + 2e$    **(b),(c)**                   $E^0$

| | $E^0$ |
|---|---|
| $2H^+ + 2e \rightleftarrows H_2(g)$ | |
| $Ag^+ + e \rightleftarrows Ag(s)$ | $Ag^+ + e \rightleftarrows Ag$   0.799 |
| $Fe^{2+} \rightleftarrows Fe^{3+} + e$ | $Fe^{3+} + e \rightleftarrows Fe^{2+}$   0.771 |
| $Sn^{4+} + 2e \rightleftarrows Sn^{2+}$ | $Sn^{4+} + 2e \rightleftarrows Sn^{2+}$   0.154 |
| $H_2(g) \rightleftarrows 2H^+ + 2e$ | $2H^+ + 2e \rightleftarrows H_2(g)$   0.00 |
| $Fe^{3+} + e \rightleftarrows Fe^{2+}$ | $Sn^{2+} + 2e \rightleftarrows Sn(s)$   $-0.14$ |
| $Sn^{2+} \rightleftarrows Sn^{4+} + 2e$ | $Co^{2+} + 2e \rightleftarrows Co(s)$   $-0.277$ |
| $Sn^{2+} + 2e \rightleftarrows Sn(s)$ | |
| $Co \rightleftarrows Co^{2+} + 2e$ | |

**11-11.** (a) 0.305 V    (b) 0.248 V    (c) $[Cu^{2+}] = 2.56 \times 10^{-16}$; $E = -0.124$ V
(d) $[Cu^{2+}] = 8.82 \times 10^{-19}$; $E = -0.197$ V
(e) $[Cu^+] = 3.67 \times 10^{-8}$; $E = 0.082$ V

**11-13.** (a) 0.204 V    (b) 0.42 V    (c) 0.81 V    (d) 0.198 V    (e) 0.247 V
(f) $-0.514$ V    (g) 1.183 V    (h) 1.143 V    (i) 0.145 V    (j) 0.781 V

**11-15.** (a) anode    (b) cathode    (c) cathode    (d) anode    (e) cathode
(f) anode    (g) cathode

**11-17.** (a) 0.589 V galvanic      (b) $-0.491$ V electrolytic
(c) 0.121 V galvanic      (d) 1.299 V galvanic
(e) $-0.131$ V electrolytic      (f) 0.064 V galvanic
(g) 0.029 V galvanic      (h) 0.883 V galvanic
(i) 0.747 V galvanic      (j) 0.249 V galvanic

**11-19.** (a) $[Hg^{2+}]/[Pd^{2+}] = 3.2 \, (\pm 0.2) \times 10^4$
(c) $[Cd^{2+}]/[Sn^{2+}] = 1.09 \, (\pm 0.08) \times 10^9$
(e) $[Cr^{3+}]^2/([Sn^{2+}][Cr^{2+}]^2) = 1.6 \, (\pm 0.1) \times 10^9$
(h) $[V^{3+}]^2/([V^{2+}][VO^{2+}][H^+]^2) = 2.5 \, (\pm 0.1) \times 10^{10}$
(j) $[Mn^{2+}]^2[Tl^{3+}]^5/([MnO_4^-]^2[Tl^+]^5[H^+]^{16}) = 10^{41}$ to $10^{46}$
(l) $[VO^{2+}]^2[UO_2^{2+}]/[V(OH)_4^+]^2[U^{4+}] = 3 \, (\pm 3) \times 10^{22}$

**11-21.** $-1.041$ V

**11-23.** 0.140 V

**11-25.** $8.9 \, (\pm 0.7) \times 10^{-11}$

**11-27.** $1.21 \, (\pm 0.09) \times 10^{12}$

**11-29.** (a) $2.4 \, (\pm 0.2) \times 10^9$    (b) 1.273 V

**11-31.** $1.94 \times 10^{-5}$

**11-33.** $4.2 \times 10^{-11}$

**11-35.** (a) 0.182 V    (b) 0.62 V    (c) 0.573 V    (d) 1.25 V    (e) 0.327 V
(f) $-0.008$ V    (g) 1.41 V

**11–37.**

| | (a) | (c) | (e) |
|---|---|---|---|
| **Vol Reagent** | $E$ | $E$ | $E$ |
| 5.00 mL | $-0.436$ V | 0.237 V | 0.075 V |
| 10.00 | $-0.408$ | 0.251 | 0.089 |
| 15.00 | $-0.380$ | 0.265 | 0.103 |
| 19.00 | $-0.332$ | 0.289 | 0.127 |
| 20.00 | $+0.182$ | 0.751 | 1.06 |
| 21.00 | 0.694 | 1.21 | 1.43 |
| 25.00 | 0.735 | 1.23 | 1.44 |
| 30.00 | 0.753 | 1.24 | 1.44 |

## CHAPTER 12

**12–1.** (a) $2Fe^{3+} + Cd(s) \rightleftarrows 2Fe^{2+} + Cd^{2+}$
(b) $2H_2MoO_4 + 3Zn(s) + 12H^+ \rightleftarrows 2Mo^{3+} + 3Zn^{2+} + 8H_2O$
(c) $V(OH)_4^+ + Ag(s) + Cl^- + 2H^+ \rightleftarrows VO^{2+} + AgCl(s) + 3H_2O$
(d) $2VO^{2+} + S_2O_8^{2-} + 6H_2O \rightleftarrows 2V(OH)_4^+ + 2SO_4^{2-} + 4H^+$
(e) $2MnO_4^- + 5U^{4+} + 2H_2O \rightleftarrows 2Mn^{2+} + 5UO_2^{2+} + 4H^+$
(f) $8MnO_4^- + I^- + 8OH^- \rightleftarrows 8MnO_4^{2-} + IO_4^- + 4H_2O$
(g) $VO^{2+} + V^{2+} + 2H^+ \rightleftarrows 2V^{3+} + H_2O$
(h) $H_2S + I_2 \rightleftarrows S(s) + 2I^- + 2H^+$
(i) $IO_3^- + 2I^- + 6Cl^- + 6H^+ \rightleftarrows 3ICl_2^- + 3H_2O$
(j) $SeO_3^{2-} + OBr^- \rightleftarrows SeO_4^{2-} + Br^-$

**12–3.** (a) $0.0500\ N$ (b) $0.200\ N$ (c) $0.0250\ N$ (d) $0.0500\ N$
(e) $0.0500\ N$

**12–5.** (a) $0.0400\ F$ (b) $0.00800\ F$ (c) $0.0200\ F$ (d) $0.0100\ F$
(e) $0.0200\ F$

**12–7.** $0.08419\ N$

**12–9.** $33.01$ mg/mL

**12–11.** $0.1179$ g

**12–13.** $0.1984$ g

**12–15.** $0.09692\ N$

**12–17.** $0.05059\ N$

**12–19.** (a) $32.08\%$ Fe (b) $45.86\%$ $Fe_2O_3$

**12–21.** (a) $0.0894$ (b) $0.1788$

**12–23.** $75.4\%$

**12–27.** $29.31\%$

**12–29.** $65.82\%$

**12–31.** $39.10$ mg/mL

**12–33.** (a) $0.0575$ mg/L (b) $47.9$ ppm

**12–35.** 2700 mg/L

**12–37.** 0.06615 $F$ $VO^{2+}$ and 0.09140 $F$ $V(OH)_4^+$

**12–39.** 40.0%

**12–40.** 9.34% KI and 52.2% KBr

CHAPTER 13

**13–1.** (a) 0.354V

  (b) Ag|AgIO$_3$(sat'd), $IO_3^-(xM)$‖SCE

  (c) $pIO_3 = (E_{SCE} - E_{AgIO_3} - E_{obs})/0.0591$
  $= -(0.110 + E_{obs})/0.0591$

  (d) 3.11

**13–4.** (a) Ag|AgSCN(sat'd), $SCN^-(xM)$‖SCE

  (b) Cd|Cd$_2$Fe(CN)$_6$(sat'd), $Fe(CN)_6^{4-}(xM)$‖SCE

  (c) Cu|Cu(IO$_3$)$_2$(sat'd), $IO_3^-(xM)$‖SCE

  (d) Ag|Ag$_2$CrO$_4$(sat'd), $CrO_4^{2-}(xM)$‖SCE

**13–5.** (a) $E^0_{AgSCN} = 0.092$ V and pSCN = 1.31

  (b) $E^0_{Cd_2Fe(CN)_6} = -0.647$ V and pFe(CN)$_6$ = 10.42

  (c) $E^0_{Cu(IO_3)_2} = 0.126$ V and pIO$_3$ = 10.83

  (d) $E^0_{Ag_2CrO_4} = 0.446$ V and pCrO$_4$ = 9.27

**13–8.** (a) $-0.512$ V  (b) $-0.446$ V  (c) $-0.166$ V  (d) 0.176 V
  (e) $-0.200$ V

**13–10.** (a) 0.11 V  (b) 0.05 V  (c) $-0.09$ V  (d) 0.08 V

**13–13.** (a) 3.25  (b) 5.10  (c) 6.72  (d) 13.27

**13–15.**

| mL EDTA | $C_T$ | $[Cu^{2+}]$ | pCu | $E$, V |
|---|---|---|---|---|
| 0.00 | 0.00 | $1.00 \times 10^{-3}$ | 3.000 | 0.248 |
| 1.00 | 0.00 | $7.92 \times 10^{-4}$ | 3.101 | 0.245 |
| 2.00 | 0.00 | $5.88 \times 10^{-4}$ | 3.230 | 0.242 |
| 3.00 | 0.00 | $3.88 \times 10^{-4}$ | 3.411 | 0.236 |
| 4.00 | 0.00 | $1.92 \times 10^{-4}$ | 3.716 | 0.227 |
| 4.50 | 0.00 | $9.57 \times 10^{-5}$ | 4.019 | 0.218 |
| 5.00 | $2.62 \times 10^{-9}$ | $2.6 \times 10^{-9}$ | 8.58 | 0.083 |
| 5.50 | $9.48 \times 10^{-5}$ | $7.2 \times 10^{-14}$ | 13.14 | $-0.051$ |
| 6.00 | $1.89 \times 10^{-4}$ | $3.6 \times 10^{-14}$ | 13.44 | $-0.060$ |

**13–16.** $1.59 \times 10^{-4}$ $M$

CHAPTER 14

**14–1.** $-0.913$ V

**14–3.** 0.294

**14–5.** (a) $3 \times 10^{-11}$ $M$  (b) $-0.274$ V

**14–7.** (a) $-0.94$ V  (b) 0.348 V  (c) $-2.09$ V  (d) $-2.38$ V

**14–9.** (a) $Pb^{2+}$ reduced first
  (b) $E = -0.188 - 0.874 - 1.000 - 0.70 = -2.76$ V
  (c) $9.1 \times 10^{-8}$

**14–11.** (a) $-0.055$ V   (b) $-0.210$ V

**14–13.** (a) $Cd^{2+}$   (b) $-0.824$ to $-1.029$ V vs. SCE

**14–15.** (a) Separation feasible   (b) $-0.123$ to $-0.434$ V vs. SCE

**14–17.** (a) 3.22 min   (b) 6.44 min   (c) 9.66 min

**14–19.** (a) 0.2281 g   (b) 0.2550 g   (c) 0.2640 g

**14–21.** 0.482%

**14–23.** 26.5 ppm

**14–25.** 71.7%

**14–27.** (a) 57% $CCl_4$ and 43% $CHCl_3$   (c) 39% $CCl_4$ and 61% $CHCl_3$

**14–29.** (a) The decomposition potential is the potential at which an analyte begins to react at the dropping electrode at a significant rate. It is indicated by the departure of the analyte polarogram from the residual current curve.

  (b) A limiting current is a current that is essentially independent of applied potential; it occurs immediately after the rapid rise in current that results from reduction or oxidation of the analyte.
    A diffusion current is a limiting current that is observed when mass transfer of the analyte to the electrode surface is brought about exclusively by diffusion.

**14–31.** $i_d/C = 13.25, 13.00, 12.97, 13.08, 13.03\ \mu A/(mmol/L)$
  mean $= 13.066 = 13.1$ and $s = 0.11$

**14–32.** (a) 5.49 mmol/L   (c) 0.96 mmol/L   (e) 3.68 mmol/L

**14–33.** (a) $1.20 \times 10^{-2}\ M$   (c) $6.31 \times 10^{-3}\ M$

## CHAPTER 15

**15–1.**

| | $\lambda$, nm | $\lambda$, Å | $\lambda$, cm | $\lambda$, $\mu$m | $\nu$, Hz | $\sigma$, cm$^{-1}$ |
|---|---|---|---|---|---|---|
| (a) | 350 | $3.50 \times 10^3$ | $3.50 \times 10^{-5}$ | 0.350 | $8.57 \times 10^{14}$ | $2.86 \times 10^4$ |
| (c) | 869 | $8.69 \times 10^3$ | $8.69 \times 10^{-5}$ | 0.869 | $3.45 \times 10^{14}$ | $1.15 \times 10^4$ |
| (e) | 598 | $5.98 \times 10^3$ | $5.98 \times 10^{-5}$ | 0.598 | $5.01 \times 10^{14}$ | $1.67 \times 10^4$ |
| (g) | 877 | $8.77 \times 10^3$ | $8.77 \times 10^{-5}$ | 0.877 | $3.42 \times 10^{14}$ | $1.14 \times 10^4$ |
| (i) | 717 | $7.17 \times 10^3$ | $7.17 \times 10^{-5}$ | 0.717 | $4.18 \times 10^{14}$ | $1.39 \times 10^4$ |
| (k) | 285 | $2.85 \times 10^3$ | $2.85 \times 10^{-5}$ | 0.285 | $1.05 \times 10^{15}$ | $3.51 \times 10^4$ |

**15–2.** (a) $1.4801 \times 10^{15}$ Hz   (c) $7.06 \times 10^{13}$ Hz

**15–3.** (a) $4.90 \times 10^4$ cm$^{-1}$   (c) 27.8 $\mu$m

**15–4.** (a) 31.8%   (c) 71.8%   (e) 14.6%

**15–5.** (a) 0.983   (c) 0.160   (e) 0.141

**15–6.** (a) 10.1%   (c) 51.5%   (e) 2.14%

**15-7.** **(a)** 1.28    **(c)** 0.461    **(e)** 0.442

**15-8.** **(a)** $\epsilon = 6.34 \times 10^3$ L mol$^{-1}$ cm$^{-1}$    **(c)** $c = 1.49 \times 10^{-4}$ $M$
   **(e)** $A = 0.914$                **(g)** $c = 1.71 \times 10^{-4}$ $M$

**15-9.** **(a)** 0.108 g/L    **(c)** 0.528    **(e)** $2.88 \times 10^4$ L/cm mol

**15-10.** **(a)** cm$^{-1}$ ppm$^{-1}$ or L/mg cm    **(c)** mL/g cm

**15-11.** **(a)** 18.8    **(b)** $6.98 \times 10^3$ L/mol cm    **(c)** 0.400 cm    **(d)** 0.381

**15-13.** $1.6 \times 10^{-5}$ to $8.6 \times 10^{-5}$

**15-15.** **(a)**

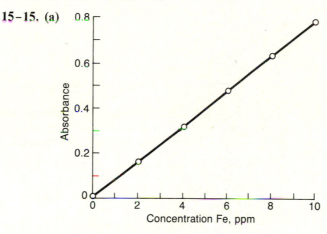

   **(b)** $A = 0.00043 + 0.0785\,C_{\text{Fe}}$
   **(c)** $s_r = 0.0021$ and $s_b = 0.00025$

**15-16.**

| | ppm Fe | s, ppm Fe | s, ppm Fe (mean of 3 measurements) |
|---|---|---|---|
| **(a)** | 7.69 | 0.03 | 0.02 |
| **(c)** | 2.17 | 0.03 | 0.02 |
| **(e)** | 6.52 | 0.03 | 0.02 |

**15-17.** **(a)** 1.59%    **(c)** 0.812%

**15-18.** **(a)** 9.43 mg Cr/L oil    **(c)** 7.58 mg/L

**15-19.** **(a)** 11.79 ppm    **(c)** 9.48 ppm

**15-20.** **(a)** 5.4 ppm Co and 4.37 ppm Ni
   **(c)** 3.6 ppm Co and 7.75 ppm Ni

**15-21.** **(a)** 5.34    **(c)** 5.19    **(e)** 5.49

**15-22.** **(a)** $A_{440} = 0.631$    $A_{620} = 0.514$
   **(c)** $A_{440} = 0.541$    $A_{620} = 0.644$
   **(e)** $A_{440} = 0.582$    $A_{620} = 0.584$

**CHAPTER 16**

**16-1.** **(a)** Atomization is the process in which a sample is vaporized and
   decomposed to its atoms or elementary ions, usually by heat.

(c) Doppler broadening arises because atoms moving toward or away from the monochromator give rise to absorption or emission lines at frequencies that differ slightly from the natural frequencies for the element.

(e) Flame photometry is an emission method in which the sample is excited in a flame and the intensity of the emitted radiation measured with a filter photometer.

(g) Sputtering is the process in which gaseous cations bombard a cathode and eject atoms from the cathode into the gas phase. Some of the ejected atoms are in an excited state.

(i) Chemical interference is the result of any chemical reaction which decreases or increases the absorption or emission characteristics of the analyte.

(k) A releasing agent is a cation which preferentially reacts with a species that would otherwise react with the analyte to cause a chemical interference.

(m) An ionization suppressor is an atomic species that readily ionizes in a flame to give cations and electrons. It is added to a solution of an analyte to repress the ionization of the atoms of the analyte.

16–2. The population of excited atoms, from which emission arises, is small and very sensitive to the flame temperature. The population of the much larger number of ground state atoms, from which absorption originates, is not very sensitive to temperature.

16–4. Source modulation is employed to distinguish between the component of radiation arising from the source and the component of radiation arising from flame background.

16–6. The nonlinearity apparently arises from the tendency of uranium to ionize. The alkali metal salt provides an excess of electrons, which suppress the ionization.

16–7. 0.504 ppm Pb

**CHAPTER 17**

17–2. (a) Extract the iron from a 6 $F$ HCl solution with ether or precipitate iron from a strong NaOH solution. Aluminum remains in solution as $Al(OH)_4^-$.

(c) Precipitate the silicon from hot concentrated $HNO_3$.

(e) Extract the uranium from 1.5 $F$ $HNO_3$ with ether.

(g) Pass the mixture through an ion-exchange column packed with a strong acid resin. The $Al^{3+}$ will be retained while $SO_4^{2-}$ will pass unhindered through the column.

17–4. 0.01589 $N$

17–6. 0.1228 mmol

17–8. (a) 0.0115 $M$;    (b) 0.00357 $M$;    (c) $8.72 \times 10^{-4}\ M$;
(d) $2.04 \times 10^{-4}\ M$

17–10. (a) 75 mL;    (b) 40 mL;    (c) 21 mL

17–12. (a) 5.49    (b) $[HA]_{aq} = 6.78 \times 10^{-3}\ M$; $[A^-]_{aq} = 0.0560\ M$
(c) $4.62 \times 10^{-1}$

17–14. pH = 0.7 to 6

**CHAPTER 18**

**18-1.** The primary difference between gas-liquid chromatography and high-performance chromatography is that in the former the mobile phase is a gas, while in the latter it is a liquid. As a consequence, gas-liquid chromatography is limited to samples that are volatile and thermally stable.

**18-3.** Elution is the process whereby a solute is washed through a column by fresh portions of a solvent.

**18-5.** The efficiency of a column increases as the number of plates $N$ increases or as the plate height $H$ decreases.

**18-7.** In an open tubular column, the stationary phase is physically or chemically held on the inner walls of a capillary tubing. In a packed column, the stationary phase is held on the surface of finely divided particles that are supported on tubing of somewhat larger diameter. The advantages of open tubular columns include their enormous efficiencies and high rate of throughput. The advantages of packed columns are their greater sample capacities and their greater ruggedness and ease of use.

**18-9.** (a) Diethyl ether, benzene, $n$-hexane

**18-10.** (a) Ethyl acetate, dimethylamine, acetic acid

**18-11.** Adsorption chromatography is based upon the distribution of solutes between a mobile solution phase and a solid stationary phase such as silica or alumina. The solute is retained on the stationary phase by adsorption. Partition chromatography is based upon the distribution of solutes between two immiscible solvents, one of which is mobile and one of which is stationary.

**18-13.** Gel filtration chromatography is a type of size-exclusion chromatography that is based upon the use of a hydrophilic packing and aqueous solvents. Gel permeation chromatography, on the other hand, employs a hydrophobic packing and nonpolar organic solvents.

**18-15.** 14% methyl acetate; 31% methyl propionate; 55% methyl $n$-butyrate

**18-17.** (a) $N_A = 2775$; $N_B = 2472$; $N_C = 2364$; $N_D = 2523$
(b) $\overline{N} = 2.5 \times 10^3$; $s = 0.2 \times 10^3$
(c) $9.8 \times 10^{-3}$ cm

**18-18.** (a) $k'_A = 0.74$; $k'_B = 3.3$; $k'_C = 3.5$; $k'_D = 6.0$
(b) $K_A = 6.2$; $K_B = 27$; $K_C = 30$; $K_D = 50$

# Index

Entries in boldface refer to specific laboratory directions; *t* refers to a table.

# FOUR-PLACE LOGARITHMS  OF NUMBERS

| n | 0 | 1 | 2 | 3 | 4 | 5 | 6 | 7 | 8 | 9 |
|---|---|---|---|---|---|---|---|---|---|---|
| **10** | 0000 | 0043 | 0086 | 0128 | 0170 | 0212 | 0253 | 0294 | 0334 | 0374 |
| 11 | 0414 | 0453 | 0492 | 0531 | 0569 | 0607 | 0645 | 0682 | 0719 | 0755 |
| 12 | 0792 | 0828 | 0864 | 0899 | 0934 | 0969 | 1004 | 1038 | 1072 | 1106 |
| 13 | 1139 | 1173 | 1206 | 1239 | 1271 | 1303 | 1335 | 1367 | 1399 | 1430 |
| 14 | 1461 | 1492 | 1523 | 1553 | 1584 | 1614 | 1644 | 1673 | 1703 | 1732 |
| **15** | 1761 | 1790 | 1818 | 1847 | 1875 | 1903 | 1931 | 1959 | 1987 | 2014 |
| 16 | 2041 | 2068 | 2095 | 2122 | 2148 | 2175 | 2201 | 2227 | 2253 | 2279 |
| 17 | 2304 | 2330 | 2355 | 2380 | 2405 | 2430 | 2455 | 2480 | 2504 | 2529 |
| 18 | 2553 | 2577 | 2601 | 2625 | 2648 | 2672 | 2695 | 2718 | 2742 | 2765 |
| 19 | 2788 | 2810 | 2833 | 2856 | 2878 | 2900 | 2923 | 2945 | 2967 | 2989 |
| **20** | 3010 | 3032 | 3054 | 3075 | 3096 | 3118 | 3139 | 3160 | 3181 | 3201 |
| 21 | 3222 | 3243 | 3263 | 3284 | 3304 | 3324 | 3345 | 3365 | 3385 | 3404 |
| 22 | 3424 | 3444 | 3464 | 3483 | 3502 | 3522 | 3541 | 3560 | 3579 | 3598 |
| 23 | 3617 | 3636 | 3655 | 3674 | 3692 | 3711 | 3729 | 3747 | 3766 | 3784 |
| 24 | 3802 | 3820 | 3838 | 3856 | 3874 | 3892 | 3909 | 3927 | 3945 | 3962 |
| **25** | 3979 | 3997 | 4014 | 4031 | 4048 | 4065 | 4082 | 4099 | 4116 | 4133 |
| 26 | 4150 | 4166 | 4183 | 4200 | 4216 | 4232 | 4249 | 4265 | 4281 | 4298 |
| 27 | 4314 | 4330 | 4346 | 4362 | 4378 | 4393 | 4409 | 4425 | 4440 | 4456 |
| 28 | 4472 | 4487 | 4502 | 4518 | 4533 | 4548 | 4564 | 4579 | 4594 | 4609 |
| 29 | 4624 | 4639 | 4654 | 4669 | 4683 | 4698 | 4713 | 4728 | 4742 | 4757 |
| **30** | 4771 | 4786 | 4800 | 4814 | 4829 | 4843 | 4857 | 4871 | 4886 | 4900 |
| 31 | 4914 | 4928 | 4942 | 4955 | 4969 | 4983 | 4997 | 5011 | 5024 | 5038 |
| 32 | 5051 | 5065 | 5079 | 5092 | 5105 | 5119 | 5132 | 5145 | 5159 | 5172 |
| 33 | 5185 | 5198 | 5211 | 5224 | 5237 | 5250 | 5263 | 5276 | 5289 | 5302 |
| 34 | 5315 | 5328 | 5340 | 5353 | 5366 | 5378 | 5391 | 5403 | 5416 | 5428 |
| **35** | 5441 | 5453 | 5465 | 5478 | 5490 | 5502 | 5514 | 5527 | 5539 | 5551 |
| 36 | 5563 | 5575 | 5587 | 5599 | 5611 | 5623 | 5635 | 5647 | 5658 | 5670 |
| 37 | 5682 | 5694 | 5705 | 5717 | 5729 | 5740 | 5752 | 5763 | 5775 | 5786 |
| 38 | 5798 | 5809 | 5821 | 5832 | 5843 | 5855 | 5866 | 5877 | 5888 | 5899 |
| 39 | 5911 | 5922 | 5933 | 5944 | 5955 | 5966 | 5977 | 5988 | 5999 | 6010 |
| **40** | 6021 | 6031 | 6042 | 6053 | 6064 | 6075 | 6085 | 6096 | 6107 | 6117 |
| 41 | 6128 | 6138 | 6149 | 6160 | 6170 | 6180 | 6191 | 6201 | 6212 | 6222 |
| 42 | 6232 | 6243 | 6253 | 6263 | 6274 | 6284 | 6294 | 6304 | 6314 | 6325 |
| 43 | 6335 | 6345 | 6355 | 6365 | 6375 | 6385 | 6395 | 6405 | 6415 | 6425 |
| 44 | 6435 | 6444 | 6454 | 6464 | 6474 | 6484 | 6493 | 6503 | 6513 | 6522 |
| **45** | 6532 | 6542 | 6551 | 6561 | 6571 | 6580 | 6590 | 6599 | 6609 | 6618 |
| 46 | 6628 | 6637 | 6646 | 6656 | 6665 | 6675 | 6684 | 6693 | 6702 | 6712 |
| 47 | 6721 | 6730 | 6739 | 6749 | 6758 | 6767 | 6776 | 6785 | 6794 | 6803 |
| 48 | 6812 | 6821 | 6830 | 6839 | 6848 | 6857 | 6866 | 6875 | 6884 | 6893 |
| 49 | 6902 | 6911 | 6920 | 6928 | 6937 | 6946 | 6955 | 6964 | 6972 | 6981 |
| **50** | 6990 | 6998 | 7007 | 7016 | 7024 | 7033 | 7042 | 7050 | 7059 | 7067 |
| 51 | 7076 | 7084 | 7093 | 7101 | 7110 | 7118 | 7126 | 7135 | 7143 | 7152 |
| 52 | 7160 | 7168 | 7177 | 7185 | 7193 | 7202 | 7210 | 7218 | 7226 | 7235 |
| 53 | 7243 | 7251 | 7259 | 7267 | 7275 | 7284 | 7292 | 7300 | 7308 | 7316 |
| 54 | 7324 | 7332 | 7340 | 7348 | 7356 | 7364 | 7372 | 7380 | 7388 | 7396 |